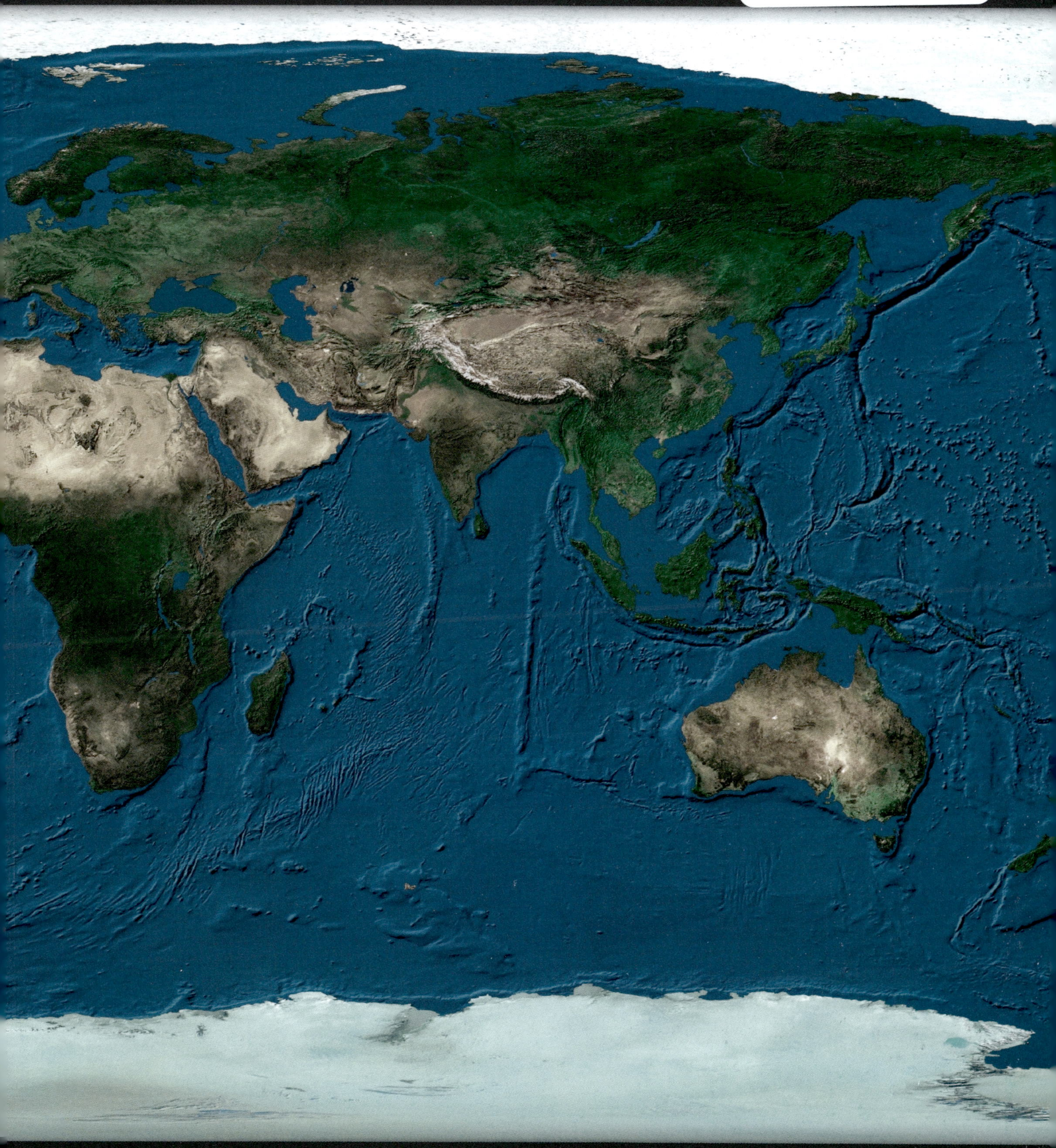

INTERNATIONAL PETROLEUM ENCYCLOPEDIA 2010

International Petroleum Encyclopedia is published annually by:
PennWell Corporation
1421 South Sheridan Road
Tulsa, Oklahoma 74112, USA
Phone: 918-835-3161
Fax: 918-831-9555
www.pennwellbooks.com

800.752.9764
+1.918.831.9421
sales@pennwell.com
www.pennwellbooks.com
www.pennwell.com

ISBN 9781593702199

Printed in the United States of America

1 2 3 4 5 14 13 12 11 10

PennWell

Contents

Key Stats & Tables

Information Sources

Guest Essay

The Challenges Facing America's 21st Century Oil and Gas Workforce

Lane Sloan

Anybody associated with the energy business for any length of time knows that there are booms and bust periods that present significant challenges for this long-cycle industry. And while this is a hugely capital intensive business, it's the people who drive the business with their talent, knowledge, and skills that are crucial to its success. Unfortunately, the volatility of the business can wreak havoc with the "people pipeline"—the entry and subsequent career development of highly skilled workers, managers, and executives.

A history of price volatility

The past is rarely an accurate predictor of the future, but lessons can always be learned. The last major upturn in the energy industry emanated from the restrained supply environment caused by the Arab oil embargo of 1973. Bullish forecasts that foresaw crude oil prices rising to $100/bbl brought on a ballooning of the workforce in the second half of the 1970s and early 1980s. People in the upstream part of the business were in high demand, with bonuses, cars, and even country club memberships being handed out to lure top performers away from the majors.

The up-cycle came to a grinding halt when crude prices dropped to less than $10/bbl by mid-1986 as the Saudis abandoned their role as the swing producer. The prolonged down-cycle through the 1990s didn't really recover until 2002. It was a devastating period for talent management and skill-pool development as ongoing workforce reductions and restructurings created lopsided demographics. The U.S. oil and gas workforce of over 850,000 in 1982 fell by some 500,000 by 2000. This devastating reduction was traumatic both for those who stayed and for those who left. The industry veteran who recommended that a son or daughter pursue an oil and gas career was the exception rather than the historic rule from prior generations.

Demographics a cause for concern

With signs of a cycle upturn, a number of us began to sound the manpower warning alarm, such as my 2002 article in *World Energy* magazine, "Energy Careers in Crisis? A Time for Renaissance."[1] At the time, the demographic issues were clearly alarming; the average age of the energy worker was around 50, and it was possible that half would retire within a decade.

Recruiting the best and brightest to rebalance the workforce was difficult in view of the industry's negative environmental image and concerns about cyclic hiring and firing. On top of that, the number of universities offering petroleum-engineering courses fell almost in half since the upbeat early 1980s. Additionally, America's youth were avoiding science, technology, engineering, and math courses—essential skill competencies for this highly technical industry.

With such a prolonged down period, oil and gas management developed a conservative mindset that hampered proactively rebuilding the workforce pipeline. Over the next several years, oil and gas workforce conferences transitioned from (a) debating whether the industry had a workforce dilemma, to (b) addressing what to do about the impending brain drain, and then to (c) evaluating who had the best staff recruiting, development, and retention practices.

Prices are everything in the fortunes of oil and gas companies, and they rose to surpass the long-foretold $100/bbl mark, reaching almost $150/bbl in mid-2008. With the price rise came unprecedented starting salaries for new petroleum engineers, some approaching $100,000 per year, coupled with lucrative starting bonuses.

The mind-boggling collapse of the global financial system in the second half of 2008 clearly took a toll on crude oil prices, which fell to levels approaching $30/bbl by January 2009. However, prices rather quickly rebounded to the $70–$80/bbl range by the third quarter. A decade ago, if one had forecast today's price range, it would have been considered optimistic, but now there are underlying worries about volatility and direction. On the workforce side, there is a clear divide between those companies hunkering down and trimming staff and those cautiously staying the course with their staffing strategies.

Clearly, the past several decades have been fraught with workforce challenges in the oil and gas industry. Many of the people issues identified in the late 1990s are still with us in terms of lopsided demographics and an impending crew change over the coming decade. Moreover, the propensity of the current U.S. Administration to promote "green jobs" over "brown jobs" is certainly troubling for the oil and gas industry.

Leadership during a "phase change"

As we try to envision America's oil and gas workforce challenges of the future, it appears that a pretty wild ride is likely.

First and foremost, we are entering an energy "phase change" as described in the book, *Terra Incognita: A Navigation Aid for Energy Leaders*, which I co-authored with Chris Ross.[2] Energy supply was the driver as in the last upturn, but this time, we are in a period of increasing energy demand from developing countries. China and India together have almost 2.5 billion people, twice the number of the OECD nations. And they are going through the industrialization process that the developed countries pursued in the last 150 years.

Michael Economides points out in his book *Energy: China's Choke Point* that "in the United States there are more than 110 cars per one hundred people; in China it is still fewer than ten cars per hundred people."[3] Not surprisingly, energy demand is forecast to double over the next 50 years.

How will we meet this burgeoning demand? Many of us believe it will take a portfolio of energy sources, as was noted in the National Petroleum Council report of 2007.[4] This is a phase change. Higher prices will certainly facilitate development and use of alternative energy choices. Oil and gas undoubtedly will play a major role for decades, but hydrocarbon market dynamics will change.

There will continue to be a shakeout of business models for the oil and gas industry as well as the overall energy industry. Even today, the super majors hold less than 10% of world crude oil reserves, which will be under continuing pressure in a higher-priced environment as oil-rich countries seek to capture more of the margin. The national oil companies in developing countries will continue to compete fiercely with their American and European counterparts to fuel their countries' economic growth. And the role of service companies will become even more blurred competitively with traditional E&P operators in seeking business from oil-rich nations.

What does all this mean for the 21st century oil and gas workforce? Quite bluntly, these business-model dynamics will put a premium on business and leadership skills.

Meeting future energy requirements will involve more than just recognizing the need for a portfolio of energy sources. There will be many changing dynamics in the interplay of energy sources in a complex network of intersubstitutability. We also need to recognize the complexity of meeting U.S.

Lane Sloan

Lane Sloan is co-chairman of the Greater Houston Partnership's Energy Collaborative. The mission of the collaborative is to perpetuate Greater Houston as the energy capital of the world by leveraging the strengths of Texas.

He is also president of Sloan Consulting Services, Inc. (Houston), and co-author with Chris Ross of the book, *Terra Incognita: A Navigation Aid for Energy Leaders*, published in April 2007.

In a career of more than 29 years with various units of Shell Oil Company, Mr. Sloan served most recently as president of Shell Chemical Company, from 1997–1999.

His prior executive positions with Shell included: vice president, corporate planning; chief financial officer; and regional coordinator for the Far East, then director of the East Zone in oil products (both with Royal Dutch Shell).

After retiring from Shell in 1999, Mr. Sloan joined Science Applications International Corp. (SAIC) as executive vice president, responsible for their energy sector activities. He later became chief executive officer of GrandBasin, a joint venture of SAIC and Halliburton.

From 2002–2007, Mr. Sloan was an executive professor at Bauer College of Business at the University of Houston, teaching corporate strategy, strategic leadership and energy strategy/leadership courses. He served as executive director of the university's Global Energy Management Institute.

In 2007 and 2008, Mr. Sloan was the special assistant to the provost and director of the University of Houston's Strategic Energy Alliance, whose mission is to establish UH as a recognized leader locally and globally in energy education, research, and outreach.

Mr. Sloan also has been involved with the energy programs of the Houston Technology Center and serves on the HTC board and executive committee, as well as its energy screening committee (which evaluates energy start-up companies).

Mr. Sloan earned a B.S. degree in business and an M.S. degree in management science from the University of Colorado. He also holds an M.S. degree in accountancy and an MBA degree from the University of Houston.

needs for affordability, national security, and environmental protection. For these reasons, energy-industry leaders must also possess deftness in energy policy as well as the ability to persuasively articulate for the public the role of oil and gas in America's energy portfolio.

Technology, geopolitics, and environment

A first vital aspect of this energy-demand-driven world is that technology will be a critical success factor. There is a competitive race under way across the globe for breakthroughs, particularly in "clean" technologies. At the same time, a rejuvenation of R&D in the oil and gas sector will continue the search for next-generation technologies for finding, recovering, and transforming hydrocarbons. New skills sets from the molecular revolution in nanotechnology and biotechnology should be part of the energy technology renaissance. This technology thrust may well reshape workforce requirements, because the demand for specialized researchers and engineers is likely to grow. PhDs may again be as popular as they were in the 1970s.

Second, the "clash of nations" will continue, putting a premium on energy security. Availability and affordability of crude oil in particular will be a continuing subject of news and commentary. As the world economy rebounds and scarcity economics comes back into play, the volatility of oil prices will increase as rather inelastic supply and demand curves rub up and down. Likely disturbances throughout the world, but particularly in the Middle East and Africa, will exacerbate price movements, causing difficulties in planning. In this volatile price world, well-thought-out staffing strategies may separate the winners from the losers. Hiring and then firing in this up-and-down world is not likely to be a winning formula.

Third, environmental concerns about greenhouse gases likely will not go away, despite the December 2009 "Climategate" scandal about the e-mails of global climate-change researchers. Green will still be "in," imposing a new twist on the oil and gas industry's reputation problem in the competitive hunt for new talent in the United States. But as the late Nobel Laureate Richard Smalley often highlighted in his talks on the crucial challenges facing the world in the 21st century, energy is the biggest because it is integral to the solution of every other challenge.[5] The energy industry must develop smarter messaging. It should stop differentiating "green" jobs from "brown" jobs and talk instead about "energy" jobs that can contribute to meeting this global challenge. At the end of the day, it will be energy engineers—with specializations in different aspects of the energy equation—who find the solutions needed.

Education gap

Unfortunately, North America may find it difficult to grow its own talent. The percentage of U.S. undergraduate degrees awarded in science and engineering is well under 20%. This compares to about 50% in China and to about 66% in the global trendsetter nation of Singapore. Most of us are familiar with the statistics from 2006 indicating that China graduated 650,000 engineers and India 500,000, compared to 70,000 in the United States. The oil and gas industry may well need to bring in technical talent from abroad, as well as to outsource some activities. This will likely lead to a commoditization of general engineering skills. People with very specialized capabilities and with project management and leadership talents will be the most sought-after with attractive compensation packages.

Evolution in power, transportation markets

In transportation markets, high oil prices coupled with energy security concerns will continue to spur development of alternative fuel sources. These will include compressed natural gas, biofuels, and GTL products. In addition, we can expect continuing evolution of hybrids and electric vehicles. Nevertheless, getting the right talent into the oil industry, which supplies 97% of U.S. transportation fuels, will be essential to help avoid gasoline prices of $5–6/gal. Expect petroleum engineers, geologists, and geophysicists to be in demand over the short- and mid-term.

The U.S. power generation market will be a rollercoaster ride as old coal-fired and nuclear plants are replaced with alternatives: next-generation coal and nuclear facilities; renewable wind, solar, and perhaps ocean-wave energy; and natural gas combined-cycle plants. Balancing costs, time-to-build, and climate-change concerns will bring on chaotic change. At times of such major transformation, leadership is even more critical. This is a gigantic talent development challenge for the energy industry overall. The interplay of natural gas in both power and transportation markets in the United States will require significant business and leadership skills in the oil and gas sector as highlighted earlier.

Training the next generation(s)

One key question: Is the Gen Y generation ready to meet this challenge? They certainly think so! This confident group has high expectations and strong demands for personal flexibility, and they perform well when goals and pathways are made clear. The sophisticated staff development processes discarded by many companies in the 1990s will need to be recreated. But Gen Y also will need the wisdom that can be imparted by baby-boomer mentors who worked their way up the hierarchy and weathered both up and down business cycles.

Smart human resource strategists will also figure out other innovative ways to capitalize on the knowledge and experience of retiring alumni. Expect to see new mechanisms created that enable retirees to continue to share their experience and provide tactical help without taking on the full burden of their former jobs.

Developing the oil and gas workforce pipeline also requires a long-term focus, addressing the next generation of students, currently in school. No longer can the oil and gas industry concentrate just on universities with recruiting programs, scholarships, and internships; that will be too late in the educational cycle.

More effort is needed to develop the so-called "STEM" skills (science, technology, engineering, and mathematics) in grades K–12. This will lay the foundation for later university training of the geologists, engineers, and scientists that are the core human resource in this technically savvy industry.

Significant efforts are already under way to try to shape the future workforce. For example:

- The Energy Collaborative of the Greater Houston Partnership has a very progressive Workforce Development Committee, focused particularly on K–12 initiatives.
- The Independent Petroleum Association of America has sponsored three impressive Petroleum Academy programs at the high school level in Houston and is now branching out to Fort Worth.
- The American Petroleum Institute has supported the Rockies Energy Workforce Collaborative task force that links common interests and goals of the energy industry, education, and state workforce systems in Colorado, Montana, New Mexico, North Dakota, Utah, and Wyoming.

As these examples make clear, the industry must work together to solve the upcoming talent development challenge.

A growing number of oil and gas companies, both operators and service suppliers, have recognized the need for thoughtful staffing and development strategies that put responsibilities with the senior leadership team. Several conferences and workshops in 2009 showcased some of the best practices in this regard. These strategies emphasize adaptability and accountability and also recognize that this is a short-, mid-, and long-term challenge not unlike capital investment. They look beyond their own company to recognize that the industry must pursue collaborative initiatives.

Beyond the classroom

A massive advertising campaign has been undertaken over the past several years by the oil and gas industry in the United States. It seeks to protect the industry's interests from punitive legislation but also aims to build company and industry reputations with workers of today and potential workers of tomorrow.

But also there needs to be a more concerted focus on educating youth.

- Can a television program be developed (perhaps modeled after crime-scene investigation dramas) to show the excitement of solving global energy challenges?
- Can we make high school students aware of the incredible sophistication of technology used for hydrocarbon development—that their joystick skills from computer games have application in advanced drilling?
- Can the oil and gas industry use evolving social networking channels to inform young people about energy careers?

At the same time, my own experience tells me that the best medium for effective communication is often face-to-face. Talking directly with middle and high school students can clear up misperceptions about the industry through personalized, open, two-way dialogue.

Extraordinary demands on leaders

The challenges facing America's 21st century oil and gas workforce will be daunting in this rapidly evolving energy world. The current dominant role of oil and gas likely will give way to the next energy phase, in which we develop multiple networked energy sources in response to high energy prices, energy security concerns, and global issues of environmental impact and economic sustainability.

As highlighted in *Terra Incognita*, "change of a scope and magnitude that are difficult to grasp will make extraordinary demands on the next generation of energy leaders. The leaders of the future will be called upon to originate, capture, and implement a series of massive growth projects. They will face great uncertainties about the demands of society, the course of geopolitics, and the progress of technology."[2]

References

1: Sloan, Lane. 2002. Energy Careers in Crisis? A Time for Renaissance. *World Energy*, Vol. 5, No. 2, pp. 151–154.

2: Ross, Christopher E.H., and Lane E. Sloan. 2007. *Terra Incognita: A Navigation Aid for Energy Leaders*, PennWell, p. 485.

3: Economides, Michael J. and Xina Xie; 2009, *Energy: China's Choke Point*, ET Publishing, p. 16.

4: National Petroleum Council (Washington, DC). *Hard Truths: Facing the Hard Truths about Energy: A Comprehensive View to 2030 of Global Oil and Natural Gas*, p. 5.

5: Smalley, Richard. 2005. Future Global Energy Prosperity: The Terawatt Challenge. *MRS Bulletin*, Vol. 30, p. 412.

Chronology

2009 In Brief

January

Exploration/Development

- The 2010–2013 strategic plan of Petrobras projects oil production of 2.68 MMb/d in 2013 vs. 2.05 MMb/d in 2009, including pre-salt discoveries.
- U.S. MMS releases a draft of the proposed 2010–2015 U.S. OCS oil and gas leasing program.
- Egypt announces that proved gas reserves in the country total 76 tcf, up from the 2007 estimate of 72.3 tcf.

Production

- North Caspian Operating Co. BV replaces Agip KCO consortium as operator of Kazakhstan's Kashagan oil field development project.
- Pemex reports that Mexico's oil production dropped 9.2% in 2008 to just under 2.8 MMb/d.
- Oil production begins from the Grouse field in the UK Central North Sea.

Natural Gas/LNG

- FERC approves with conditions a proposed LNG import terminal at Sparrows Point, near Baltimore, MD.
- Woodside Petroleum suspends its proposed OceanWay LNG import terminal offshore Los Angeles due to market conditions.

Refining/Petrochemicals

- Pemex announces a delay until yearend 2009 for start of construction on a 300,000-b/d refinery.
- LyondellBasell Industries filed for Chapter 11 bankruptcy protection for its U.S. operations and Basell Germany Holdings.
- BP agrees to supply as much as 70,000 b/d of oil to Vietnam's 130,000-b/d Dung Quat refinery

Pipelines/Transportation

- Gas flow to Europe from Russia via Ukraine stops amid a dispute between Gazprom and Ukraine over unpaid bills and prices for deliveries in 2009.

Government

- Barack H. Obama takes office as President of the United States.
- U.S. DOE arranges for first direct oil purchases for the Strategic Petroleum Reserve since 1994.

Markets

- IEA reduces its projection of 2009 oil demand to 85.3 MMb/d, due to expected global economic weakness.
- The Gas Exporters Countries Forum—a group of the world's largest natural gas suppliers—establishes a charter and chooses Doha, Qatar as home for its secretariat.

February

Exploration/Development

- Iraq reverses earlier plans, says that it will let IOCs hold stakes of up to 75% in oil drilling projects.
- Noble Energy raises gross mean resource potential at the Tamar gas discovery offshore Israel to 5 tcf.

- Heritage Oil PLC reports sufficient recoverable oil found in Uganda's Albert basin to be commercially viable.
- Gazprom secures rights to explore the onshore El Assel concession in Algeria, the company's first project in that country.

Production

- TNK-BP reports the start of commercial production at Urna and Ust-Tegus fields in the Tyumen region of Siberia.
- OPEC delays 35 of 150 planned oil drilling projects by at least four years due to deteriorating economic conditions in member companies.

Natural Gas/LNG

- Chevron reports that Wheatstone and Iago fields off Western Australia hold enough gas to support a two-train export LNG development onshore.
- Operations begin at the 9-mtpy Sakhalin II LNG liquefaction plant at Prigorodnoye.

Pipelines/Transportation

- Alaska Gas Pipeline LLC lets a contract for design of a treatment plant to process gas delivered from the North Slope through its proposed 4-bcfd pipeline.

Government

- U.S. Dept. of the Interior delays a five-year OCS leasing plan that was launched in mid-2008, saying a more orderly process is needed.
- U.S. SEC updates reserves reporting requirements for oil and natural gas companies.

Markets

- Benchmark U.S. light sweet crude price jumps $4.86 to $39.48/bbl on February 19, biggest one-day gain since 2008.

March

Production

- Gas production begins from the Alve field in the Norwegian Sea and the Rita field in the UK North Sea.
- Shell Oil completes installation of the drilling and production platform atop a 555-ft cylindrical spar at its Perdido development in the Gulf of Mexico.
- Production starts from Nigeria's Apko deepwater gas condensate field.
- Alberta offers royalty incentives to oil and gas producers in an effort to revive drilling.

Natural Gas/LNG

- Dominion Cove Point LNG starts up the 1.8-MMcfd expansion of its terminal on Chesapeake Bay.
- Algeria delays again, until 2013, start-up of the Skikda LNG plant.

Refining/Petrochemicals

- U.S. ethanol supporters petition the EPA to raise the limit on the amount of ethanol blended into gasoline to 15% from 10%.
- China approves construction of a 10-mtpy refinery at Jieyang in southern Guangdong Province.

Pipelines/Transportation

- Alaska officials review a proposal for a 1,287-km (800-mi) intrastate gas pipeline from the North Slope to the Cook Inlet area.

Government

- Iraq's oil ministry signs its first JV agreement with a foreign firm (Mesopotamia Petroleum Co. Ltd.) to focus on increasing oil and gas production.
- Russia declines an invitation join OPEC, proposing instead to send a permanent envoy.
- Venezuela says it will let Panama to join the PetroCaribe fuel assistance program.

April

Exploration/Development

- USGS estimates that Colorado's Piceance basin holds 1.525 trillion bbl of in-place oil shale.
- India invites bids for 70 blocks under its eighth New Exploration Licensing Policy round.
- Afghanistan launches its first-ever bid round, offering one oil and two gas blocks in the north.
- U.S. DOE's initial estimate: U.S. OCS holds 66–115 billion bbl of oil and 362–565 tcf of natural gas (technically recoverable).

> "Long-term energy demand will double between now and 2060 because of population growth."
>
> *—Royal Dutch Shell PLC Chief Executive Officer Jeroen van de Veer*

Production

- API reports that 1Q-2009 U.S. oil and gas drilling fell to levels not seen since 2004.
- Production begins from the Al Gasant gas-condensate discovered on the West Manzala concession in the Nile Delta.
- TNK-BP says it plans to invest $1 billion/yr through 2011 to sustain production from the Samotlor supergiant oil field in Western Siberia.
- A London energy analyst estimates that Iraq will need to invest $28–43 billion to raise its oil production capacity to 6 MMb/d.

Natural Gas/LNG

- State officials approve the environmental impact statement for a facility that would be the world's first for processing coal-seam methane into LNG, near Gladstone, Australia.
- Qatargas inaugurates its much-delayed 15.6-mtpy Qatargas 2 facility in Ras Laffan City.
- Operators of Russia's Sakhalin II project send their first LNG cargo to Tokyo.

Refining/Petrochemicals

- Kuwait cancels plans to build a 630,000-b/d fourth refinery at Al Jour.

Pipelines/Transportation

- Construction is completed on the first phase of the Eastern Siberia-Pacific Ocean (ESPO) pipeline, running 2,000 km (1,243 mi) from Taishet to Skovorodino.

Government

- A swine flu outbreak in Mexico prompts a warning from the World Health Organization of a potential global pandemic.
- Both countries agree to start increased Venezuelan oil exports to China in 2010 rather than 2013.
- Japan and Venezuela agree to invest $33.5 billion to develop oil and gas projects in Venezuela for Japanese markets.

> "No matter how much you try to accelerate the transition to a new energy economy, gas will play a major role for a significant period. But it will need a bigger infrastructure."
>
> —*INGAA President Donald F. Santa*

Markets

- *Oil & Gas Journal* surveys indicate oil and gas industry capital spending in the United States will fall 26% in 2009 to $242 billion.
- IEA projects average 2009 oil demand of 83.4 million b/d, down 2.4 million b/d from its estimate for 2008.
- Suncor Energy Inc. and Petro-Canada agree to merge in a $19.3-billion (Can.) deal and to retain the Suncor name.
- U.S. automaker Chrysler LLC files for Chapter 11 bankruptcy protection.

May

Exploration/Development

- Uruguay offers 11 blocks for exploration in the Punta del Este, Pelotas, and Oriental del Plata basins.
- ExxonMobil begins drilling at the Point Thomson (Alaska) condensate project.
- ConocoPhillips and Anadarko discover oil at two wells in the National Petroleum Reserve-Alaska (NPR-A) on the North Slope.
- Heritage Oil PLC reports a giant oil field in Iraqi Kurdistan, with up to 4.2 billion bbl of oil in place.
- Peru signs 13 exploration agreements with IOCs.

Production

- Chevron starts oil production starts from Tahiti, deepest field in the Gulf of Mexico.
- Colorado creates new laws to reduce environmental impacts of oil and gas development.
- Extended production testing begins at the giant pre-salt Tupi discovery off Brazil.
- EIA says unconventional sources could provide nearly half the growth in global liquid fuel supply, 2006–2030.
- Indonesian officials say oil production from Cepu block will not reach 20,000 b/d until late 2010.

Natural Gas/LNG

- CNOOC receives the first LNG commercial cargo at its Fujian terminal.

Refining/Petrochemicals

- Iran and Uganda agree to build a refinery in Uganda that could be as large as 100,000 b/d.

- Kuwait and China agree to establish a $9-billion, 300,000-b/d refinery at Zhanjiang in Guangdong Province.
- China's industry blueprint calls for raising ethylene production capacity by 51.2% to 15.5 mtpy by 2011.

Pipelines/Transportation

- Pakistan and Iran agree to build a 2,100-km (1,305-mi) pipeline to bring 750 MMcfd of Iranian gas to Pakistan for the next 25 years.
- TransCanada wins a contract to build, own and operate a 500-MMcfd gas pipeline in Mexico, from an LNG plant near Manzanillo to Guadalajara.
- Partners suspend the Skanled pipeline project that was to deliver Norwegian gas to Sweden and Denmark, due to global economic uncertainty.
- EOG Resources says it plans to ship crude from the Bakken formation in North Dakota to Cushing, OK, by rail car, starting early 2010.

Government

- Petrobras completes negotiations with the China Development Bank for a $10-billion, 10-year bilateral credit line to support exploration and development.
- Venezuela's Pres. Hugo Chavez nationalizes nearly 40 domestic and international oil field services companies operating in that country.
- Brazil says it will increase emphasis on renewable resources (chiefly biofuels).
- Argentina claims 1.7 million sq km (656,400 sq mi) of seabed, including the Falkland Islands, South Georgia, and South Sandwich Islands, plus part of Antarctica.

Markets

- July contract for benchmark U.S. light sweet crudes reaches $62.04/bbl on May 20, its first close above $60/bbl since early November 2008.

June

Production

- First oil is pumped from Tawke field in northern Iraq—the first newly developed oil field to come on stream in that country in more than 30 years.

Natural Gas/LNG

- Canada's Kitimat LNG Inc. signs an MOU to supply to Korea Gas Corp. up to 40% of Kitimat production.
- Commissioning begins at Canaport LNG in St. John, NB; Cameron LNG near Lake Charles; and Quintero LNG north of Valparaiso, Chile.

Pipelines/Transportation

- TransCanada agrees to buy ConocoPhillips's remaining interest in the Keystone pipeline system.
- TransCanada and ExxonMobil agree to work together on TransCanada's Alaska Gas Pipeline Project.
- Trinidad proceeds with plans to build a natural gas pipeline to sister island Tobago.
- Kurdistan announces plans to export crude by pipeline from Taq Taq field through Turkey to a terminal at Ceyhan.

Government

- U.S. House of Representatives passes the Waxman-Markey Clean Energy Bill (HR-2454) June 26, including provisions related to mitigating climate change.
- French government authorizes Total to start injecting CO_2 into a depleted reservoir at Lacq field in the country's first carbon capture and storage (CCS) pilot.
- Iraq's oil ministry approves oil exports from Kurdistan but says it considers contracts between Kurdistan and IOCs illegal.
- U.S. EPA proposes new regulations for renewable fuels for 2010 and beyond, requiring annual production of 36 billion gal of such fuels.

Markets

- General Motors Corp. files for Chapter 11 bankruptcy protection.
- July NYMEX contract for benchmark U.S. light sweet crude hits an intraday high of $73.23/bbl on June 11.

July

Exploration/Development

- Occidental Petroleum Corp. says a giant gas and oil field in California's San Joachin basin could hold 150–250 MMboe of recoverable reserves.
- Ghana approves a first phase of development of the giant Jubilee offshore oil field.
- An ExxonMobil affiliate spuds the first deepwater exploration well in Libya's offshore Sirte basin.

> "We must avoid inadvertently creating unattainable public expectations. An energy transition will not occur overnight, at little cost and with no inconvenience."
>
> — *ConocoPhillips CEO James J. Mulva*

- Total and Russia's Novatek set up JV to develop onshore Termokarstovoye gas field in the Russian Arctic.
- USGS releases its 2008 assessment that the Barents Sea shelf may hold more than 76 boe of technically recoverable resources.
- Algeria launches a licensing round for 25 blocks said to hold unconventional gas resources.
- Marathon Oil Corp. subsidiary agrees to sell its 20% interest in Block 32 off Angola to China's CNOOC and Sinopec for $1.3 billion.

Production

- Supergiant Khurais oil field comes on stream in Saudi Arabia.
- Cabinda Gulf Oil Co. Ltd. brings Mafumeira Norte oil project off Angola on stream.
- Production of oil and associated gas begins from Thunder Hawk field in the Gulf of Mexico and from the Tyrihans oil and gas field in the Norwegian Sea.
- Various forecasters reduce expectations for higher bitumen production from Alberta's oil sands.

Natural Gas/LNG

- The 4.4-mtpy Dragon LNG terminal at Milford Haven, South Wales, begins commercial operation.
- Quintero LNG regasification terminal in Chile receives its first cargo.
- Taiwan's CPC Corp. inaugurates the country's second LNG terminal, at Taichung.
- Shell and Vopak announce plans to develop an LNG terminal at Fos-sur-Mer, France.
- The Indonesian Tangguh LNG project in Papua Barat sends out its first cargo.
- Gazprom signs agreement to import 500 million cu m/yr of natural gas from Azerbaijan and transport it to Europe, starting January 2010.

Refining/Petrochemicals

- Fluor Corp. completes a five-year expansion of the petrochemicals complex at Shaiba, Kuwait, doubling its capacity.
- Kurdistan opens its first oil refinery, at Erbil.
- Irving Oil Corp. suspends plans to build a new 300,000-b/d refinery at St. John, NB.

Pipelines/Transportation

- Gazprom and Korea Gas Corp. sign an MOU to study extending the Sakhalin-Khabarovsk-Vladivostok pipeline to deliver Russian gas to South Korea.

Markets

- IEA forecasts global oil demand of 89 MMb/d in 2014, up from 85.8 MMb/d in 2008.
- BP analysts report oil demand in the developed world fell by 1.6% in 2008, the greatest decline since 1982.

Government

- Loans up to $1.7 billion to help Ukraine buy, store and move gas are approved by the European Bank for Reconstruction and Development, the World Bank, and the European Investment Bank.
- Egyptian authorities arrest 26 on charges of plotting attacks on oil pipelines and on ships moving through the Suez Canal.
- Ethnic clashes between Muslim Uighers and Chinese Han in Xinjiang Province kill 150 people but do not disrupt oil field and pipeline operation.
- Russia redesigns its fiscal policy to boost oil exploration and development.

August

Exploration/Development

- Western Gulf of Mexico Lease Sale 210 draws 189 bids totaling just over $145 million.
- Nicaragua signs an agreement allowing hydrocarbon exploration by a Russian consortium.

Production

- Oil production begins from both the Affleck and Shelley fields in the central UK North Sea.
- Murphy Oil Corp. starts production from Azurite field off Congo (Brazzaville) through what it says is the industry's

first floating drilling, production, storage and offloading (FDPSO) facility.

- Total declares the giant Itau gas-condensate field in Bolivia commercial.

Natural Gas/LNG

- Australia Pacific LNG JV chooses Curtis Island near Gladstone, Australia as site of its proposed LNG plant. Coalseam methane will feed the plant.
- Petrobras gets an operating license for its 14-MMcmd LNG regasification terminal at Guanabara Bay, Rio de Janeiro State.
- Train 6 at Ras Laffan LNG Co. Ltd. III in Qatar starts production. Capacity is 7.8 mtpy.
- The 8-bcm/yr Adriatic LNG regasification terminal offshore Italy gets its first LNG cargo.

Refining/Petrochemicals

- Pemex chooses Tula in Mexico's Hidalgo State as the site for a $9-billion refinery.

Pipelines/Transportation

- U.S. State Dept. authorizes construction of the Alberta Clipper pipeline to carry crude from Alberta to Superior, WI.
- The 805-km (500-mi) Midcontinent Express pipeline begins full operation, carrying gas from the Barnett shales and Bossier sands to eastern U.S. markets.
- TAQA Bratani assumes operatorship from Shell UK E&P for the Brent System pipeline in the UK North Sea, the first change in management in almost 30 years.

Markets

- On August 18, near-term U.S. futures prices for natural gas fall to $3.10/MMBtu—the lowest since August 2002—as supply expands and industrial use drops.
- Suncor Energy Inc. and Petro-Canada merge in a $19.3 billion (Can.) deal. The new company will retain the Suncor name.

Government

- A federal appeals court rules that MMS did satisfy its environmental-review duties when it conducted a 2007 sale of leases in the Beaufort Sea off Alaska.
- Ecuador agrees to sell 3 MMbbl/month of oil to PetroChina International over a two-year period.

September

Exploration/Development

- Petrobras reports that the Guara field in Brazil's Santos basin could hold 1.1–2.0 billion boe.
- BP announces a major oil discovery with its ultradeep Tiber well in the Gulf of Mexico.
- Libya announces plans to invest $9.92 billion to develop 24 wells, including work at the Nafoora-Oujlaa-Knleej and Waha-Jalou oil fields.
- Consulting engineers estimate a prospective 4.28 tcf of natural gas in place in the Ordovician Utica shale in Quebec's St. Lawrence Lowlands.

Production

- Oil production begins from the $3.8-billion Tombua-Landana project off Angola, the Volund field in the Norwegian North Sea, and Mangala field in India.
- Mexico's energy minister predicts oil production by Pemex will average 2.5 million b/d in 2010, 4% lower than in the first half of 2009.
- Athabasca Oil Sands Corp. enters JV agreements with PetroChina for financing in situ oil sands development in Canada.

Natural Gas/LNG

- JV partners Chevron, Exxon Mobil, and Shell give final investment approval to development of the $43-billion (Aus.) Gorgon-Jansz-Io LNG and domestic gas project off Western Australia.
- The first Persian Gulf' LNG regasification terminal, Mina Al-Ahmadi GasPort offshore Kuwait, receives its first LNG cargo.
- Train 5 at the Qatargas 2 JV complex starts up, with a capacity of 7.8 mtpy.

Refining/Petrochemicals

- Irving Oil Ltd. announces plans for a major upgrade of its 250,000-b/d refinery at St. John, NB.
- Valero Energy says it will close its 210,000-b/d refinery in Delaware City, DE.

Pipelines/Transportation

- Gazprom completes the second segment of the Minsk-Vilnius-Kaunas-Kalingrad gas pipeline.

> "This economic crisis, in depressing demand, has given us a window of opportunity to bring about change."
>
> — *Didier Houssin, Director of the Directorate of Energy Markets and Security for the IEA*

- Tesoro Corp. ships first oil through the reversed Petroterminal de Panama Trans-Panamanian Pipeline, from the Atlantic to the Pacific side of the isthmus.

Markets

- October contract for natural gas jumps 43 cents to $3.26/MMBtu on September 10—the largest one-day gain in almost five years on the NYMEX.

Government

- U.S. EPA proposes new rules requiring firms emitting more than 25,000 tpy of greenhouse gases to obtain construction and operating permits.
- California imposes the first statewide carbon tax in the United States.
- U.S. Senate Energy and Natural Resources Committee begins hearings on climate change legislation. The House passed its version of such a bill in June.
- Brazil's president proposes new legislation concerning development of its large reserves of oil in the offshore pre-salt layer.

October

Exploration/Development

- Eni confirms a "world class" natural gas discovery with its exploration well in Perla field in the Gulf of Venezuela.
- MMS approves, with stringent environmental conditions, Shell's exploration plan for two leases in the Beaufort Sea off Alaska.

Production

- Gas production begins from Longtom field in Australia's Bass Strait and from Oyong field offshore Indonesia.
- Argentina approves the first contract (with Apache Corp.) under an incentive program to encourage development of unconventional natural gas reservoirs.
- EIA expects U.S. natural gas storage inventories to reach a record high 3.85 tcf October 31, the end of the storage injection season.

Refining/Petrochemicals

- Petrobras begins refining the first load of oil from the offshore presalt Tupi field.
- Southern Sudan approves plans to build a $2-billion, 50,000-b/d refinery in Akon, Warap State.

Natural Gas/LNG

- China begins commissioning its third LNG terminal at Shanghai.
- Production begins from the first of two planned trains at the 6.7-mtpy Yemen LNG plant.

Markets

- October 22, the price of benchmark U.S. crudes for December hits a 12-month high, settling at $81.19/bbl.
- Enterprise Product Partners LP and TEPPCO Partners LP merge, creating a pipeline, storage, and gas processing firm valued at more than $26 billion.

Government

- Iraq says it will allow up to 100 British Royal Navy personnel into the country to help protect its southern oil export terminals.
- The U.S. government conducts a "Cash for Clunkers" rebate program to boost car sales and get older, polluting vehicles off the road.
- France and Kazakhstan sign agreements on oil and gas, including construction of an oil pipeline from Kashagan field and Baku, and part ownership of Khvalynskoye gas field.

November

Exploration/Development

- Seven operators in Greenland form a trade association to more actively pursue oil and gas development in that nation.
- Gabon sets an oil licensing round for May 2010 for more than 40 deepwater blocks.

Production

- Oil production begins in Ecuador's Sacha field via the Rio Napo JV.

- Gas production starts from Longhorn field on Mississippi Canyon Blocks 502/546 and from Savonette field off Trinidad and Tobago.
- Encore Acquisition Co. and Denbury Resources Inc. announce plans to merge and apply their skills in EOR using carbon dioxide as well as in CO_2 sequestration.

Refining/Petrochemicals

- PDVSA buys a 49% stake in the state refiner of the Dominican Republic.
- Petrobras and PDVSA complete negotiations for joint construction and operation of Abreu e Lima refinery in Brazil. It will process 230,000 b/d of heavy oil.
- OSHA proposes to fine BP Products North America $87.4 million for a deadly explosion in 2005 at its Texas City, TX, refinery.
- Sinopec and Saudi Basic Industries Corp. complete construction of a $3-billion petrochemicals complex in Tianjin, China.

Natural Gas/LNG

- Construction starts on Saudi Aramco's Karan project. It will process 1.8 bscfd of Karan Khuff gas.
- U.S. Maritime Administration officials okay construction of the offshore Gulf of Mexico LNG terminal project planned by Port Dolphin Energy LLC.
- Construction starts on China's fifth LNG terminal at Ningbo.

Pipelines/Transportation

- The 2,692-km (1,679-mi) Rockies Express pipeline becomes operational, from northwestern Colorado to eastern Ohio.

Markets

- Saudi Aramco says that, starting January 2010, it will switch from WTI to an index of Gulf Coast sour crudes as the benchmark for pricing its oil for sale in the United States.
- EnCana Corp. shareholders split the firm into two companies, one focused on natural gas, the other on the Alberta oil sands.

> "In just the past 25 years, America's oil production has fallen by nearly 4 million barrels per day . . . and U.S. oil demand grew by nearly 4 million barrels per day."
>
> *—Incoming Chevron CEO John W. Watson*

December

Production

- Iraq auctions rights to several IOCs to develop selected oil and gas fields.
- Mexico's KMZ field produces a record-high 802,000 b/d on December 29.
- Oil production starts from Oyo field, 75 km (47 mi) offshore Nigeria.

Natural Gas/LNG

- InterOil Corp. gets approval from Papua New Guinea for a $7-billion LNG project. The government earlier gave the go-ahead for a $15-billion LNG project by ExxonMobil and its partners.

Pipelines/Transportation

- An 1,800-km (1,119-mi) pipeline is commissioned in Central Asia, to carry 30 bcm of gas from Turkmenistan through Uzbekistan and Kazakhstan to China.
- The 661-km (411-mi) Urucu-Corari-Manaus pipeline begins operation, carrying 4.1 million cu m/d (14.5 MMcfd) of gas from Brazil's Amazonian Solimoes basin.

Refining/Petrochemicals

- Shell withdraws from talks with Sinopec and Kuwait Petroleum Corp. regarding construction of a $9-billion, 300,000-b/d refinery in China's Guangdong Province.

Government

- Iraq charges that Iranian forces seized one of its oil wells along their disputed southern border.
- U.S. EPA issues a formal finding that greenhouse gas emissions endanger public health and therefore can be regulated under the federal Clean Air Act.
- Climate change legislation awaits action in the U.S. Congress. A House bill passed in June, but no Senate bill has been approved at yearend.

Markets

- J.D. Power and Associates projects that more than 12.7 million cars and trucks will be sold in China in 2009, surpassing the 10.3 million forecast for the United States.
- ExxonMobil agrees to buy XTO Energy, a major U.S.-based natural gas producer, for $41 billion.
- NYMEX January 2010 natural gas price surges to $5.82/MMBtu on December 18, highest level in almost a year, as storage levels drop.

World Energy Market: The Big Picture

Turbulence continues

The wild oil price swing and global economic collapse of 2008 created huge questions at the start of 2009. Where is the bottom of the economic slump? When will oil and gas prices rebound? Where and when, if ever, will prices stabilize?

The economy

The Dow Jones Industrials Average fell to a 12-year low of 6,547 on March 9, as financial markets and the global economy struggled to recover from the collapse of 2008.

The United States, the UK, Europe, and Japan all began the year in recession, for the first time since World War II.

Deutsch Bank predicted that 2009 would see the worst economic performance among industrialized countries since the Great Depression. They forecast a rise of just 1.2% in global growth—the lowest rate since the early 1980s—as economic problems spread to emerging markets.

Barclays Capital Resources said it expected global spending for exploration and production to drop 12% to $400 billion in 2009—"a reversal after six years of global growth."

But some optimists talked of a possible rebound in the second half of 2009 as energy supplies fall below projected demand levels.

In fact, from April forward, indicators of financial health showed uneven but gradual improvement, with the DJIA recovering to close 2009 at 10,428. The price of gold topped $1,100/ounce at the end of the year.

In June 2009, the World Bank said that it expected the global economy to shrink by nearly 3% in 2009, far deeper than the 1.7% contraction it predicted about two months earlier.

In December, the International Monetary Fund said it expects the global economy to contract by 1.1% for 2009 over 2008, and then to grow by 3.1% in 2010.

Oil markets

Price volatility continued to disrupt oil markets in 2009.

Having reached $147/bbl in July 2008, the NYMEX front-month contract for West Texas Intermediate (WTI) crudes (closing price) started 2009 at $46.30/bbl.

It fell to a yearly low of $32.70/bbl in late January, rebounded to a yearly high of $81.40/bbl on October 21, and closed the year at $79.40/bbl. For the year, said the Energy Information

Administration of the U.S. Department of Energy, the WTI price averaged $62/bbl.

Overall, world oil prices soared about 70% during 2009 on signs of global economic recovery. WTI crude futures closed at a 15-month high of $81.77/bbl on January 5, 2010.

In a January 2010 review, EIA noted that global oil demand dropped in 2009 for a second consecutive year—the first time since 1983 that this had occurred.

Other notable events impacting energy markets and oil prices in 2009 included:

- Warnings in April that Mexico's swine-flu outbreak could become a global pandemic
- Declarations of bankruptcy by both General Motors and Chrysler Corp., reflecting dismal U.S. automobile sales
- A quiet hurricane season in the Gulf of Mexico
- Discovery of two very large oil fields in the Gulf of Mexico: Tiber and Guara
- The start of bidding for permits to develop Iraq's oil resources
- The UN climate change summit in Copenhagen in December
- A Saudi decision to use the Argus Sour Crude Index instead of the WTI benchmark for U.S. sales

A Look Back at 2008

BP: Statistical Review of World Energy

In June 2009, BP PLC released its annual review of global energy markets.

According to the firm's *Statistical Review of World Energy: June 2009*, global oil production climbed by 0.4% in 2008. But worldwide demand fell by 1.6%—its fastest decline since 1982.

BP expected global oil production would continue to fall because of dwindling demand and improvements in energy efficiency.

Oil demand dropped by 1.5 million b/d in the developed world, spurred first by record oil prices and then by the collapsing global economy. Non-OECD countries also registered slower growth in demand, at just 1.1 million b/d.

In 2008, for the first time, the developing world led by China consumed more energy than OECD countries. China represented nearly 75% of global growth, and its energy usage was 17.7%.

BP expected that oil prices would rise to $60–90/bbl in the future and said that producers need at least $60/bbl to support continued investment.

In 2008, gas consumption crept along, below the decade average at 2.5%. China experienced the fastest rise in gas consumption, at a rate of 15.8%. U.S. consumption rose by 0.6%, the UK by 3%.

BP noted that the United States recorded its highest-ever annual increase in gas production in 2008 as strong development of unconventional gas resources raised output by 7.5%.

OGJ 100 firms: 2008 performance

Oil & Gas Journal released in September 2009 its review of the 2008 performance of the "OGJ 100"—the 100 leading oil and gas producers based outside the United States.

Most posted improved financial results from 2007, but higher operating costs offset strong oil and gas prices for some. The full survey is accessible through the *OGJ* Web site at www.ogj.com.

OGJ100: OIL PRODUCTION AND RESERVE LEADERS

Rank	Company	Production, million bbl
1	Saudi Arabian Oil Co.	3,248.5
2	National Iranian Oil Co.	1,423.5
3	Petroleos de Venezuela SA	1,180.8
4	Petroleos Mexicanos	1,019.1
5	Abu Dhabi National Oil Co.	894.3
6	BP PLC	876.4
7	PetroChina Co. Ltd.	870.7
8	Iraq National Oil Co.	863.2
9	Kuwait Petroleum Corp.	846.8
10	OAO Rosneft	776.3
11	Petroleo Brasileiro SA	722.7
12	Nigerian National Petroleum Corp.	708.1
13	OAO Lukoil	702.9
14	Sonangol	668.0
15	National Oil Corp. (Libya)	627.8
16	Royal Dutch Shell	617.9
17	Total SA	531.4
18	Sonatrach	481.8
19	OJSC Surgutneftegas	453.7
20	StatoilHydro	380.0
Total		**17,893.8**

Rank	Company	Reserves, million bbl
1	Saudi Arabian Oil Co.	259,900.0
2	National Iranian Oil Co.	136,150.0
3	Iraq National Oil Co.	115,000.0
4	Kuwait Petroleum Corp.	101,500.0
5	Petroleos de Venezuela SA	99,377.0
6	Abu Dhabi National Oil Co.	92,200.0
7	National Oil Corp. (Libya)	43,660.0
8	Nigerian National Petroleum Corp.	36,220.0
9	OAO Rosneft	17,694.0
10	Qatar Petroleum Corp.	15,210.0
11	OAO Lukoil	14,458.0
12	Sonatrach	12,200.0
13	PetroChina Co. Ltd.	11,221.3
14	Petroleos Mexicanos	10,501.3
15	Petroleo Brasileiro SA	9,109.0
16	Sonangol	9,040.0
17	Total SA	5,695.0
18	BP PLC	5,665.0
19	Petroleum Development Oman LLC	5,500.0
20	Petronas	5,460.0
Total		**1,005,760.6**

Global View: 2009

As noted earlier, ongoing volatility characterized oil and gas markets in 2009 as the widespread financial crisis that began in third quarter 2008 continued to rock nations around the world.

Predictions of future energy-market performance by various agencies varied as the year unfolded. For that reason, no attempt is made in this book to report in detail the projections of oil and gas supply, demand and prices made in the first half of 2009.

EIA: Short-term outlook

In its January 2009 *Short-Term Energy Outlook (STEO)*, the Energy Information Administration (EIA) of the U.S. Department of Energy projected that world oil consumption would fall by 800,000 b/d in 2009.

EIA also saw a generally balanced supply and demand situation near-term, with a modest rebound of global oil consumption in 2010.

Later in the year, in its November 2009 STEO, EIA said WTI prices should average $77/bbl from October 2009 to March 2010, and could reach $81/bbl by December 2010 if U.S. and world economic conditions keep improving.

The forecast assumes U.S. real domestic product will grow by 1.9% in 2010 while world oil consumption-weighted GDP increases by 2.6%.

EIA projected growth of 1.26 million b/d in world oil consumption during 2010, with non-OEDC countries representing the largest share.

EIA reported that oil production outside OPEC averaged 50.1 million b/d for the first three quarters of 2009, and that Russia produced more than 10 million b/d during August, a new post-Soviet era record.

EIA expects non-OPEC production to grow by about 250,000 b/d in 2010, as higher output from the FSU and the United States offsets falling production in Mexico and the North Sea.

It also expects OPEC to raise output gradually in 2010 to an average of 29.4 million b/d.

IEA: Early-2009 view

In its March 2009 *Oil Market Report*, the International Energy Agency (IEA) projected that oil from non-OPEC sources in 2009 would total 50.6 million b/d, unchanged from 2008. This view was at odds with OPEC, which said it expected an increase in oil supply from non-OPEC sources in 2009.

The agency also projected that global oil demand would fall by 1.3 million b/d in 2009 to 84.4 million b/d compared to 2008.

In May, IEA said it expected global oil consumption to contract in 2009 by 2.6 million b/d, the steepest decline since 1981.

That same month, at the G8 energy summit, IEA estimated that global upstream oil and gas investment budgets for 2009 could be cut by about 21% compared to 2008, due to the global economic crisis.

OGJ: Mid-2009 view

Oil & Gas Journal evaluated 2009 global and U.S. energy supply and demand at mid-year.

(Details about its perspective on the U.S. energy market are in the United States section of the North America chapter of this book.)

Global oil demand was expected to decline from 2008, led by a decrease in oil consumption by the world's advanced economies.

Volumes of crude and products in storage were abundant, and *OGJ* expected them to remain that way even though it forecast a drop in worldwide oil production in 2009 of more than 2 million b/d on average.

World crude prices plunged from mid-2008 to the start of 2009, and then rebounded a bit by mid-year. *OGJ* pointed to EIA figures showing that all OPEC crudes averaged $121.52/bbl on June 1, 2008, dropped to $35.48/bbl on January 1, 2009, and rose to $65.88/bbl on June 1, 2009.

World Oil Consumption

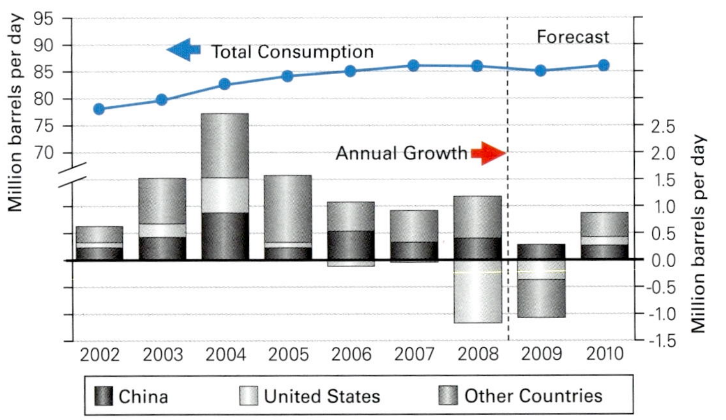

Short-Term Energy Outlook, January 2009

Worldwide Oil Supply and Demand

	2008					2009				
	1st Qtr.	2nd Qtr.	3rd Qtr.	4th Qtr.	Year	1st Qtr.	2nd Qtr.	3rd Qtr.	4th Qtr.	Year
	Million b/d									
DEMAND										
OECD										
North America	24.8	24.4	23.6	23.9	24.2	23.5	22.6	22.7	22.9	22.9
Europe	15.3	15.0	15.4	15.3	15.3	14.9	14.2	14.8	14.8	14.7
Asia-Pacific	8.9	7.9	7.5	8.0	8.1	8.1	7.2	7.0	7.5	7.5
Total OECD	**49.0**	**47.3**	**46.6**	**47.2**	**47.5**	**46.6**	**44.0**	**44.6**	**45.2**	**45.1**
Non-OECD										
FSU	4.3	4.1	4.3	4.1	4.2	4.0	3.9	4.1	4.1	4.0
Europe	0.8	0.7	0.7	0.7	0.7	0.7	0.7	0.7	0.7	0.7
China	7.9	8.0	8.1	7.6	7.9	7.7	8.4	8.0	7.9	8.0
Other Asia	9.9	9.9	9.4	9.5	9.7	9.9	9.7	9.4	9.6	9.6
Latin America	5.7	6.0	6.0	5.9	5.9	5.7	5.9	6.0	6.0	5.9
Middle East	6.7	7.1	7.6	6.9	7.1	6.7	7.3	7.8	7.2	7.2
Africa	3.2	3.2	3.1	3.2	3.2	3.2	3.2	3.1	3.3	3.2
Total non-OECD	**38.5**	**39.0**	**39.3**	**38.1**	**38.7**	**37.9**	**39.1**	**39.0**	**38.7**	**38.7**
TOTAL DEMAND	**87.5**	**86.3**	**85.8**	**85.3**	**86.2**	**84.4**	**83.1**	**83.6**	**83.9**	**83.8**
SUPPLY										
OECD										
North America	14.2	14.0	13.6	13.8	13.9	14.2	13.8	13.8	14.0	14.0
Europe	4.9	4.8	4.6	4.8	4.8	4.9	4.4	4.1	4.4	4.4
Asia	0.6	0.7	0.7	0.7	0.6	0.7	0.7	0.7	0.7	0.7
Total OECD	**19.7**	**19.5**	**18.8**	**19.3**	**19.3**	**19.7**	**18.9**	**18.6**	**19.0**	**19.1**
Non-OECD										
FSU	12.8	12.9	12.7	12.7	12.8	12.8	13.1	12.9	12.9	13.0
Europe	0.1	0.1	0.1	0.1	0.1	0.1	0.1	0.1	0.1	0.1
China	3.8	3.8	3.8	3.8	3.8	3.7	3.8	3.9	3.9	3.8
Other Asia	3.7	3.6	3.7	3.7	3.7	3.7	3.6	3.7	3.7	3.7
Latin America	4.1	4.1	4.2	4.2	4.1	4.3	4.4	4.4	4.4	4.4
Middle East	1.7	1.6	1.7	1.6	1.6	1.6	1.6	1.6	1.6	1.6
Africa	2.5	2.5	2.5	2.5	2.5	2.5	2.5	2.5	2.5	2.5
Total non-OECD	**28.7**	**28.7**	**28.6**	**28.7**	**28.7**	**28.8**	**29.2**	**29.2**	**29.2**	**29.1**
Processing gain	2.2	2.2	2.3	2.3	2.2	2.3	2.3	2.3	2.3	2.3
Other biofuels	0.3	0.4	0.4	0.4	0.4	0.3	0.3	0.4	0.4	0.3
Total non-OPEC	**51.0**	**50.8**	**50.1**	**50.7**	**50.6**	**51.2**	**50.7**	**50.5**	**50.9**	**50.8**
OPEC										
Crude	31.5	31.4	31.5	30.5	31.2	28.5	28.5	28.4	28.0	28.4
NGL	4.6	4.6	4.7	4.8	4.7	4.9	5.1	5.4	5.6	5.2
Total OPEC	**36.1**	**36.0**	**36.2**	**35.3**	**35.9**	**33.4**	**33.6**	**33.8**	**33.6**	**33.6**
Total supply	**87.1**	**86.8**	**86.3**	**86.0**	**86.5**	**84.6**	**84.3**	**84.3**	**84.5**	**84.4**
Stock change	(0.4)	0.5	0.5	0.7	0.3	0.1	1.2	0.7	0.6	0.5

Totals may not add due to rounding.

Source: International Energy Agency; *OGJ* estimates for OPEC 3rd and 4th qtr 2009 crude supply.

World Crude Prices[1]

Country	Type of crude and API gravity°	June 1, 2009, $/bbl	% change June 2009 Jan. 2009	In effect Jan. 1, 2009, $/bbl	% change June 2009 June 2008	In effect June 1, 2008, $/bbl	In effect June 1, 2007, $/bbl	In effect June 1, 2006, $/bbl	In effect June 1, 2005, $/bbl	In effect June 1, 2004, $/bbl
OPEC										
Saudi Arabia	Arabian Light 34	64.17	82.2	35.21	–46.9	120.79	63.68	45.96	33.77	23.52
Abu Dhabi	Murban 39	68.37	75.4	38.97	–46.0	126.63	69.77	52.67	37.53	26.69
Algeria	Saharan 44	67.42	80.1	37.44	–46.6	126.32	70.07	50.07	37.73	27.17
Nigeria	Bonny Light 37	68.58	72.1	39.85	–47.2	129.82	71.48	51.61	37.87	27.44
Libya	Es Sider 37	67.03	82.8	36.66	–45.0	121.80	67.55	49.88	37.34	26.57
Venezuela	Tia Juana 31	66.81	88.6	35.42	–45.8	122.26	64.71	47.69	36.92	27.81
Total OPEC		**65.88**	**85.7**	**35.48**	**–45.2**	**121.52**	**65.11**	**47.29**	**35.09**	**25.04**
OTHER										
Indonesia	Minas 34	70.18	91.6	36.63	–45.2	127.97	69.85	50.81	38.06	28.24
UK	Brent Blend 38	66.55	93.9	34.33	–47.0	125.47	69.43	51.21	37.78	27.58
Norway	Ekofisk 42	67.67	81.8	37.22	–47.1	127.92	69.02	50.23	37.63	27.78
Mexico	Isthmus 33	66.70	88.9	35.31	–45.4	122.15	64.60	47.58	36.81	27.70
Russia	Urals 32	66.30	93.9	34.20	–44.6	119.69	64.74	47.61	35.78	26.16
Total World		**65.62**	**89.8**	**34.57**	**–45.9**	**121.36**	**64.67**	**46.65**	**35.29**	**25.72**
U.S.		63.70	100.6	31.76	–45.9	117.82	63.84	44.79	34.83	25.50

[1]Represents estimated contract prices based on government stated prices, netback deals, and spot market quotations. [2]Average prices (f.o.b.) weighted by estimated import volume.
Source: US Energy Information Administration

FIRST QUARTER WORLDWIDE OIL PRODUCTION

Country	First quarter 2009	First quarter 2008	Change	Change, %
	1,000 b/d			
OPEC	**28,257**	**32,304**	**–4,047**	**–12.5**
Non-OPEC	**41,980**	**41,093**	**887**	**2.2**
Argentina	622	620	2	0.3
Brazil	1,913	1,766	147	8.3
Canada	2,638	2,584	54	2.1
Colombia	636	560	76	13.6
Mexico	2,667	2,911	–244	–8.4
United States	5,292	5,115	177	3.5
Other	2,947	3,209	–262	–8.2
Western Hemisphere	16,715	16,765	–50	–0.3
Norway	2,231	2,205	26	1.2
United Kingdom	1,451	1,484	–33	–2.2
Other	531	581	–50	–8.6
Western Europe	4,213	4,270	–57	–1.3
FSU	12,360	12,291	69	0.6
Other	163	174	–11	–6.3
Eastern Europe & FSU	12,523	12,465	58	0.5
Egypt	657	650	7	1.1
Gabon	240	227	13	5.7
Other	8,071	8,783	–712	–8.1
Africa	8,728	9,433	–705	–7.5
Oman	723	727	–4	–0.6
Syria	383	390	–7	–1.8
Other	19,744	21,929	–2,185	–10.0
Middle East	20,850	23,046	–2,196	–9.5
Australia	479	409	70	17.1
China	3,611	3,771	–160	–4.2
India	646	678	–32	–4.7
Indonesia	860	859	1	0.1
Malaysia	740	773	–33	–4.3
Other	872	928	–56	–6.0
Asia-Pacific	7,208	7,418	–210	–2.8
Total world	**70,237**	**73,397**	**–3,160**	**–4.3**

Source: Oil & Gas Journal

In second quarter 2009, the front-month futures price of oil on the NYMEX averaged $59.79/bbl, down from $123.80/bbl a year earlier.

Meanwhile, the front-month natural gas futures contract during the second quarter was down 67% from a year earlier, averaging $3.81/MMbtu.

The rig count in Canada for June fell to 125 units from 266 a year earlier, according to Baker Hughes Inc. And the U.S. rig count in June averaged about 900, down from 1,902 a year earlier.

G8 Summit: Energy and environment

At a July 2009 meeting in L'Aquila, Italy, G8 leaders promised to speed deployment of renewable energy technologies and to provide more support for carbon capture and storage technology.

The group reaffirmed its commitment to reducing carbon emissions worldwide by 2050, and repeated its pledge to limiting global warming to 2 degrees Celsius above pre-industrial levels.

Key to achieving its ambitious targets will be the creation of attractive carbon markets and a regulatory framework that encourages investment in renewable energy.

U.S., Canadian firms: Terrible first-half 2009

Average commodity prices and rig counts in second quarter 2009 plummeted compared with second quarter 2008, said *OGJ* in a September analysis, crushing the financial results of producers, refiners, and service companies compared to a year earlier.

A sample of U.S. operators recorded a combined 74% decline in earnings for the second quarter of 2009, and for the first half of this year, the group combined for a loss.

Surveyed producers and pipeline operators based in Canada recorded a combined 71% decline in second-quarter 2009 earnings. For the first half, net income sank 54% from a year earlier, as most companies in the group reported poorer results.

Collectively, the sample of service and supply companies posted a 53% decline in net income in second quarter 2009. Revenues for the 21 firms dropped by 21%. Combined first-half earnings were down 40% from the group's first-half 2008 net income.

Global View: 2010 and Beyond

Oil supply, demand and prices

EIA: Short-term Energy Outlook

In mid-January 2010, EIA released its view of expected development in U.S. and global energy markets in 2010 and 2011. It included the following highlights.

- Global oil demand declined in 2009 for the second consecutive year, the first time since 1983 that this had occurred.
- EIA assumes U.S. real GDP growth of 2.0% in 2010 and 2.7% in 2011, and world oil-consumption-weighted real GDP growth of 2.5% and 3.7% in 2010 and 2011, respectively.
- The world oil market gradually tightens in 2010 and 2011, assuming that the global economic recovery continues as projected.
- Non-OECD countries lead 2010 demand recovery, but OECD nations should begin to show significant oil demand growth in 2011 in response to improving economic conditions.

- Projected economic growth in the OECD more than doubles, from 1.2% in 2010 to 2.7% in 2011.
- Expectations of a continued global economic turnaround continue to buttress oil markets. EIA expects that WTI prices will rise to about $80/bbl in 2010 and $84/bbl in 2011.

EIA: International Energy Outlook

In its May 2009 *International Energy Outlook (IEO)*, EIA projected a sharp drop in future world oil output (compared to previous EIA expectations), coupled with greater reliance on unconventional fuels such as oil sands, ultra-deep oil, shale oil, and biofuels.

The 2009 *IEO* projects that global production of conventional oil in 2030 will be 93.1 million b/d. This compares to 107.2 million b/d for 2030 in the 2007 *IEO* report.

Even with wider use of unconventional fuels, the 2009 *IEO* projects a net decline of 11.1 million b/d in the global supply of liquid fuels compared to the *IEO's* 2007 projections.

In 2009, unconventional fuels—including Canadian oil sands, Venezuelan extra-heavy oil, deep-offshore oil, Arctic oil, shale oil, liquids from coal, and biofuels—cumulatively constituted about 4% of the global liquid fuel supply.

The 2009 *IEO* expected this figure to reach nearly 13% by 2030, with production that year of 13.4 million b/d.

On the demand side, the *IEO* highlighted sharply growing demand for energy in Asia and suggested that China is moving closer to overtaking the United States as the leading global energy consumer.

IEA: World Energy Outlook

In its November 2009 *World Energy Outlook (WEO)*, IEA projected in a reference scenario that global energy demand will climb 40% between 2007 and 2030.

It also says that energy resources are adequate to meet the projected demand increase through 2030 and well beyond.

The reference scenario assumes no change in government policies through the period. Fossil fuels remain the dominant sources of energy worldwide. Although oil remains the single-largest fuel in the energy mix in 2030, its share falls to 30% from 34% in 2009.

Southeast Asia's energy demand expands by 76% between 2007 and 2030 in the reference scenario.

IEA also lays out what it calls a "450 Scenario" (see box), which includes aggressive actions it says are needed to limit the buildup of greenhouse gases in the atmosphere and to keep any global temperature rise to around 2 degrees Celsius above pre-industrial levels.

IEA said that investment in upstream oil and gas had already been cut by more than $90 billion as of November 2009, compared with 2008. While demand for oil has dropped sharply, the agency's reference scenario shows oil demand starting to recover in 2010, reaching 88 million b/d in 2015, and then 105 million b/d in 2030.

IEA: Climate Change Scenario

For the first time, IEA included a climate-change policy scenario as part of its annual forecast.

The agency's *2009 World Energy Outlook*, released in November 2009, noted that 60% of global carbon dioxide emissions come from energy, and that action to curb such emissions will be critical.

The "450 Scenario" lays out actions to limit atmospheric buildup of carbon dioxide to 450 ppm.

More details are given in the Climate Change section of the Key Statistics chapter of this book.

IEA: View to 2014

In its medium-term *Oil Market Report* of late December 2009, IEA raised its projection for average global oil consumption during 2009–2014 by 1.9 million b/d compared to its forecast of June 2009.

The projected increase reflects expected stronger economic growth in 2009–2010 than was assumed in June, plus the effects of economic and fiscal stimulus programs, especially among OECD members.

In IEA's high-growth scenario (global GDP increasing at an average rate of 3.2% per year), worldwide oil demand rises to 90.9 million b/d in 2014 from 86.2 million b/d in 2008.

In a lower-growth case (GDP rising 2.2% per year), global oil demand reaches 87 million b/d in 2014.

IEA also raised its projection of refining capacity growth by 1.1 million b/d from the June report, to 8.7 million b/d for 2008-2014. Dominating the growth are China, other Asian nations, and the Middle East.

IEA: Peak oil view

The Economist magazine reported in December 2009 that the chief economist of IEA believes that if no big new discoveries are made, "the output of conventional oil will peak in 2020 if oil demand grows on a business-as-usual basis."

Despite repeated downward revisions in recent years in its forecasts of global oil supply in 2030, IEA had not previously committed itself to a firm prediction for when oil supplies might stop growing.

In its *2009 World Energy Outlook*, released in November 2009, IEA said only that conventional oil is "projected to reach a plateau sometime before" 2030.

ExxonMobil: View to 2030

In December 2009, ExxonMobil published the latest edition of its *Outlook for Energy: A View to 2030.*

A major finding was the expectation that global energy demand will be about 35% higher in 2030 than it was in 2005, requiring trillions of dollars of investment and a commitment to innovation and technology.

The forecast also suggested that efficiency gains between 2005 and 2030 would curb global energy demand growth by about 65%.

It also projects a natural gas supply expansion, particularly in the United States, where production from unconventional reservoirs is rising rapidly.

The report said that oil would remain the largest global energy source in 2030, but that natural gas would move into second place, with gas demand 55% higher in 2030 than in 2005.

The outlook also includes an assessment of how potential carbon emission policies would affect future energy demand and the fuel mix.

The firm said that meeting the energy demands of almost 8 billion people while protecting the environment were "interlocking challenges" that will require accelerated gains in energy efficiency, expanded availability of reliable and affordable supplies, and development and deployment of technology to mitigate the growth of energy-use emissions.

IHS CERA: Oil demand to 2014

IHS Cambridge Energy Research Associates said in September 2009 that it expects world oil demand to grow in 2010 for the first time since 2007, and to reach pre-recession levels by 2012. The analysis was part of the firm's quarterly *World Oil Watch* report.

IHS CERA expects oil demand growth to rise by 900,000 b/d in 2010 and resume its 2007 high of 86.5 million b/d by 2012.

Global oil demand dropped by 2.8 million b/d to reach 83.8 million b/d in 2009. The last time that the world experienced such a severe drop was in the early 1980s, and it took nine years for demand to return to the 1979 high.

Key factors behind the outlook are accelerating oil demand growth from emerging markets, and fewer options for substituting fuels on a global scale.

IHS CERA expects oil demand to increase to 89.1 million b/d in 2014, and anticipates that 83% of the oil demand growth will come from countries that are not members of the OECD.

Near-stagnation of oil demand growth in the industrial countries of the OECD highlights several structural changes, said the firm:

- Decreasing oil intensity associated with economic growth
- Higher fuel-use efficiency
- Displacement of conventional oil with renewable energy sources
- Slower growth in transportation fuel use.

IHS CERA: Oil production to 2030

Global oil productive capacity will grow through 2030 with no evidence of a peak of supply.

That's the view of a report by IHS CERA, *The Future of Global Oil Supply: Understanding the Building Blocks.*

The November 2009 report, which analyzes more than 10,000 projects around the world, predicts that that capacity will grow to as much as 115 million b/d from the current 92 million b/d—a 25% increase.

Post-2030, supply could struggle to meet demand, but this would take the form of a decades-long "undulating plateau" rather than a sharp fall, the report says.

The firm noted that despite recessionary pressures, the first three quarters of 2009 produced discoveries with collective reserves of more than 8 billion new barrels, not including the revisions and extensions important to reserves growth.

Gazprom: Oil price expectations

In June 2009, Gazprom officials said that oil prices likely would hit $85/bbl by the end of 2009, and warned that if capital investment is not restored, prices could rise to $150/bbl within two or three years.

The firm also forecast that investment in exploration and production would decline by more than 20% in 2009, which could lead to a lower production capacity and lower oil supplies within three to five years.

Gazprom officials said "the oil price greatly depends not on fundamental factors, not on the real oil demand and supply, but on the activities of speculating investors."

Natural gas supply, demand and prices

IEA: Gas demand to 2030

In the reference case in its November 2009 *World Energy Outlook (WEO)*, IEA projects that worldwide natural gas demand will climb by 41% between 2007 and 2030.

IEA expects that the boom in North American unconventional gas production together with the recession's impact on demand will prolong the glut of gas supply for the next few years.

The 2009 *WEO* shows that the annual underuse of inter-regional pipeline and LNG capacity could rise from around 60 bcm in 2007 to 200 bcm by 2015. This glut, says IEA, could pressure suppliers to Europe and Asia-Pacific to modify pricing under long-term contracts, to delink gas prices from oil prices, to sell more spot gas, and to cut prices to stimulate demand.

IGU: Gas outlook to 2030

An October 2009 study by the International Gas Union concluded that under a "continuation of current policies" scenario, global gas demand could grow to about 153 tcf (4.3 trillion cu m) per year by 2030, from about 106 tcf/year (3 trillion cu m/year) currently.

The study also concluded that a "green-policy" scenario—global political agreement to put a high cost on carbon dioxide emissions and to encourage renewable energy—would "only be economically successful" in combination with increased use of natural gas. Such a scenario could boost gas use to about 28% of the global fuel mix by 2030 compared to 21% in 2009.

INGAA Foundation: Infrastructure needs

Projected growth in North American natural gas supplies and markets will require billions of dollars of additional investments in pipelines, storage, and other midstream infrastructure through 2030.

So said a report released in October 2009 by the INGAA Foundation Inc., research division of the Interstate Natural Gas Association of America.

The study projected that investments of $6–10 billion/year will be needed over the next 20 years under various market scenarios.

A major focus is the need to move increased gas production from unconventional shale basins and tight sands to the existing pipeline network.

The study projected that the United States and Canada would need 29,000–62,000 mi (46,671–99,780 km) of gas pipelines and 370–600 bcf of additional storage to meet anticipated market requirements.

Expected power-generation and industrial demand growth—as well as the potential to connect massive Arctic gas resources and LNG imports to the grid—also will be key drivers.

Foundation officials also conclude that there is more than 100 years of technically recoverable gas in the United States and Canada.

OPEC Actions in 2009

February

Their economies battered by falling oil prices, OPEC members said they will delay 35 of 150 planned oil drilling projects—representing about 5 million b/d of oil capacity—by at least four years.

The OPEC secretary general said, "This year, our income will be cut by 50%."

The oil minister of the UAE said that at $40/bbl, the price of crude is half that required to attract enough investment in new supply.

March

Meeting in Vienna, OPEC members agree to seek full compliance with the output-quota cuts of 4.2 million b/d announced over several months in late 2008.

The cartel a few days earlier again reduced its forecast for 2009 world crude demand to 84.6 million b/d (a drop of 1.01 million b/d compared to 2008).

OPEC also said it expected demand for its crude to average 29.1 million b/d in 2009.

April

At the Tenth International Oil Summit in Paris, OPEC said it would try to keep prices at the then-current low level of \$40–\$50/bbl throughout 2009 to help sustain global economic recovery.

May

At the G8 Energy Summit, OPEC member Saudi Arabia warned that oil prices could spike to beyond the near-\$150 level reached in mid-2008 within three years. Saudi officials joined others to call for more investment to boost production over the long term, despite the then-current credit crisis.

In late May, just before a brief OPEC meeting, the Saudi oil ministry said that the world economy had strengthened enough to handle oil prices of \$75–\$80/bb. On May 27, crude prices hit their highest level in six months (about \$63/bbl) following those comments.

At the meeting itself, OPEC voted to maintain its 24.85-million-bbl production ceiling for its 11 members other than Iraq.

July

OPEC officials said that global demand for OPEC crude likely would take five years to recover to pre-financial-crisis levels, forcing a sharp reduction in its investment in new production capacity.

The cartel said that average consumption of its oil peaked at 31 million b/d in 2008.

It also said it would invest \$110–\$120 billion in new capacity to 2013, compared to \$165 billion previously planned.

September

Meeting in Vienna, OPEC ministers said they saw no need for the cartel to reduce its production further because oil prices (about \$70/bbl at the time) were at the low end of the group's target price range.

The group was cautiously optimistic about signs of global economic recovery, but said that it would take action if oil prices fell.

November

In a monthly report, OPEC said it expected that worldwide oil demand would drop to 84.31 million b/d in 2009 (down 1.63%) compared to 2008, but that global economic recovery would push that demand back up to 85.07 million b/d in 2009 (an increase of 0.9%).

The group also forecasts that the world economy will grow 2.9% in 2010 after a contraction of 1.1% in 2009. Most of the growth is expected in emerging Asian economies, including China and India.

December

At a December meeting in Angola, OPEC members chose not to discuss the potential impact on production quotas due to renewed oil production from member Iraq.

Development assistance from IOCs, agreed to earlier in the month, could boost Iraq's production from the current 2.5 million b/d to as much as 12 million b/d by 2010.

It is not clear whether other OPEC members will cut their production quotas (and revenue) to accommodate the rise in Iraqi production.

Iraq's production quota was suspended in August 1990 when Saddam Hussein invaded Kuwait.

The group also said it would not alter production levels at least until March 2010, signaling their aim to keep oil prices in the \$70–\$80/bbl range into the new year.

OPEC OIL EXPORT REVENUES: 2009

In January 2010, EIA reported that in 2009, OPEC earned \$573 billion in net oil export revenues, a 41% decrease from 2008. Saudi Arabia earned the largest share—\$154 billion—representing 27% of total OPEC revenues.

Copenhagen Conference

More than 100 world leaders attended the much-anticipated COP15—the 15th Conference of the Parties to the UN Framework Convention on Climate Change—in December 2009 in Copenhagen.

The goal was to devise a legally binding successor to the first phase of the 1997 Kyoto Treaty (which expires in 2012), regarding reduction of carbon dioxide and other greenhouse-gas emissions that contribute to global warming and attendant climate change.

In brief, a hoped-for legally binding treaty was not agreed to. Attendees instead approved the nonbinding Copenhagen Accord, which will lead to continuing talks in 2010.

The accord does include the commitment of substantial funds by wealthy nations to help developing countries cope with the impacts of climate change.

More details are presented in the Climate Change section of the Key Statistics chapter of this book.

Auto markets: China, India

In a December 2009 analysis, the Financial Times reported that momentum in the automobile industry is shifting to Asia.

China car sales nearly doubled during November 2009 compared to a year earlier, while in India, sales rose 61% as the nation's economy rebounded.

China became the world's largest and fastest growing car market, with a total of 1.04 million passenger cars sold in November, said the China Association of Automobile Manufacturers.

In the first 11 months of 2009, some 9.23 million passenger cars were sold in China, up nearly 50% from a year earlier and well ahead of the 6.76 million sold in all of 2008.

In India, between April and November 2009, carmakers sold 1.22 million passenger vehicles, 20% more than a year earlier.

WTI benchmark challenged

In October 2009, Saudi Arabia announced that it would drop the West Texas Intermediate yardstick as the reference for its U.S. oil sales, starting in January 2010.

IEA said in November that the role of WTI as the world's top oil benchmark "looks increasingly precarious." It pointed to "inherent logistical flaws" caused by the contract delivery point at the landlocked pipeline hub of Cushing, OK, that "frequently led to a disconnect from international markets."

Earlier in 2009, a surge in oil inventories in Cushing depressed the value of the WTI against other global benchmarks, throwing the global oil market into disarray.

Saudi Arabia said it will switch to the new Argus Sour Crude Index (ASCI), which was launched in May. It tracks spot sales of Mars, Poseidon, and Southern Green Canyon crudes produced in the U.S. Gulf of Mexico.

North America

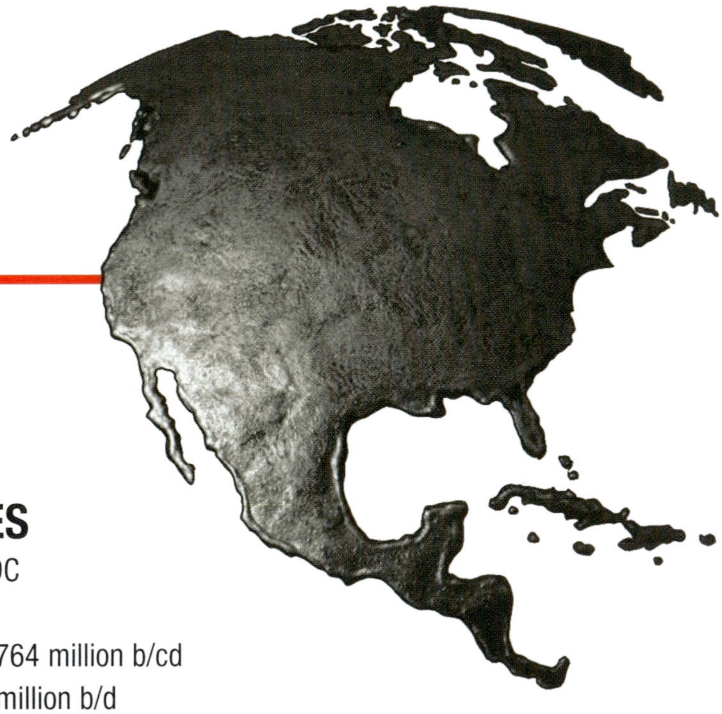

UNITED STATES

Capital: Washington, DC
Monetary unit: Dollar
Refining capacity: 17.764 million b/cd
Oil production: 5.337 million b/d
Oil reserves: 19.121 billion bbl
Gas reserves: 244.656 tcf

Overview

The economy: A snapshot

The Obama Administration took charge in Washington in January 2009, and the nation began to take halting steps toward recovery from the recession that began in the United States in December 2007.

Though the recession statistically ended in June, concerns about unemployment and the impact of increasing government spending dampened economic progress.

- For January, the Bureau of Labor Statistics reported job losses of 750,000. For the first half of 2009, losses topped 600,000/month. By fourth quarter, the average had dropped to 69,000/month.
- The unemployment rate—7.7% in January and 9.5% in June—hit 10.2% in October (the first double-digit rate since 1983). At yearend, the rate was 10.0%. At that level, nearly 16 million people were out of work.
- On March 2, the Dow Jones Industrials Average fell below 7,000 points for the first time in 11 years, pummeled by weakness in banking stocks.
- In August, RealtyTrac estimated that some 13 million primary homes had "negative equity" status (market value lower than the mortgage balance).

In mid-2009, *Oil & Gas Journal* forecast that U.S. GDP would decline 2% for the year, following a drop of 5.5% in the first quarter of 2010.

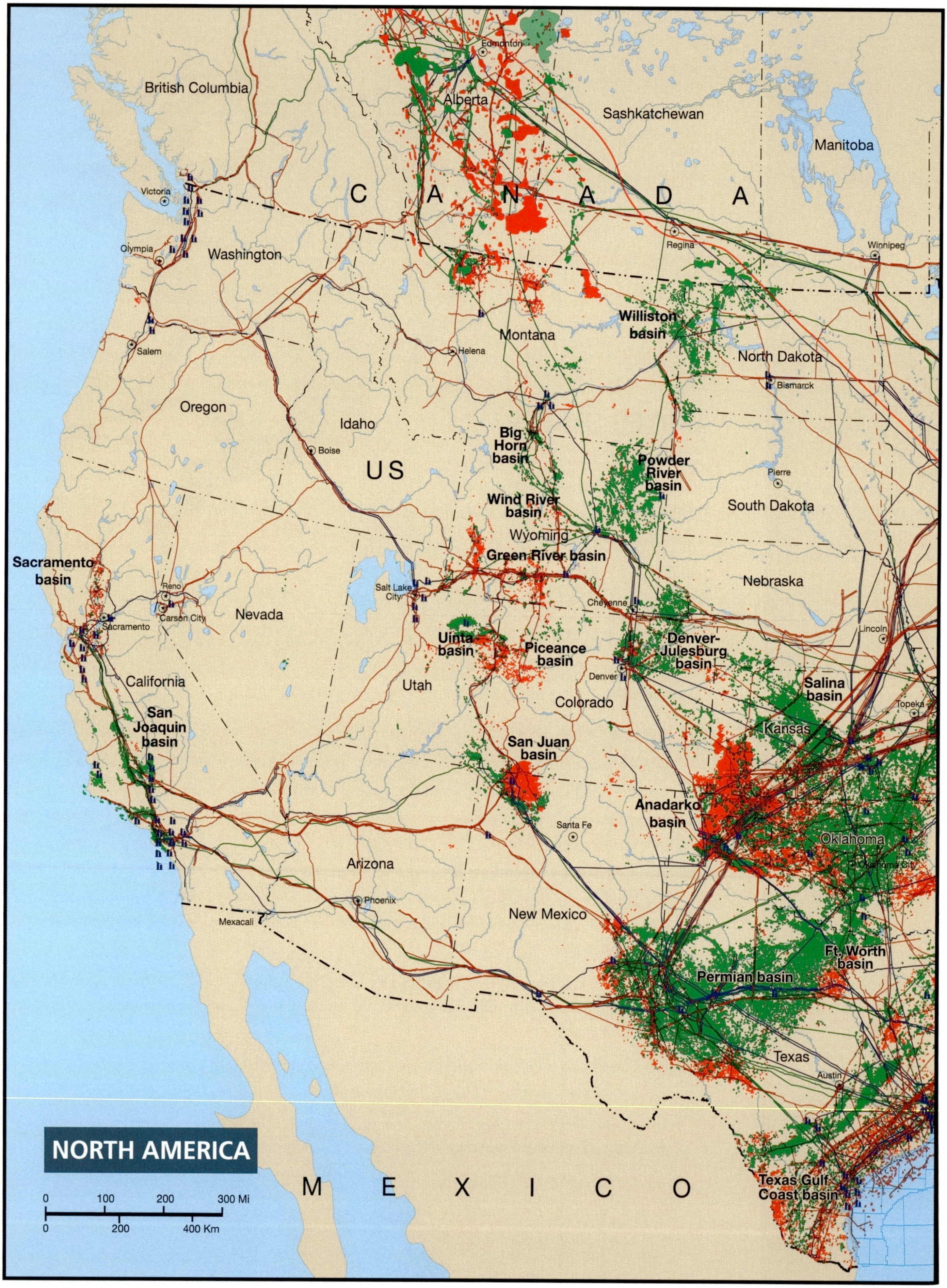
British Columbia
Alberta
Edmonton
Sashkatchewan
Manitoba
C A N A D A
Victoria
Olympia
Washington
Regina
Winnipeg
Williston basin
Montana
Salem
Helena
North Dakota
Bismarck
Oregon
Idaho
Boise
US
Big Horn basin
Powder River basin
Pierre
South Dakota
Wind River basin
Wyoming
Green River basin
Sacramento basin
Reno
Carson City
Nevada
Salt Lake City
Cheyenne
Nebraska
Sacramento
Uinta basin
Piceance basin
Denver-Julesburg basin
Lincoln
Denver
California
Utah
Colorado
Salina basin
Topeka
San Joaquin basin
Kansas
San Juan basin
Anadarko basin
Santa Fe
Oklahoma
Oklahoma City
Arizona
Phoenix
New Mexico
Mexacali
Ft. Worth basin
Permian basin
Texas
Austin
NORTH AMERICA
0 100 200 300 Mi
0 200 400 Km
M E X I C O
Texas Gulf Coast basin

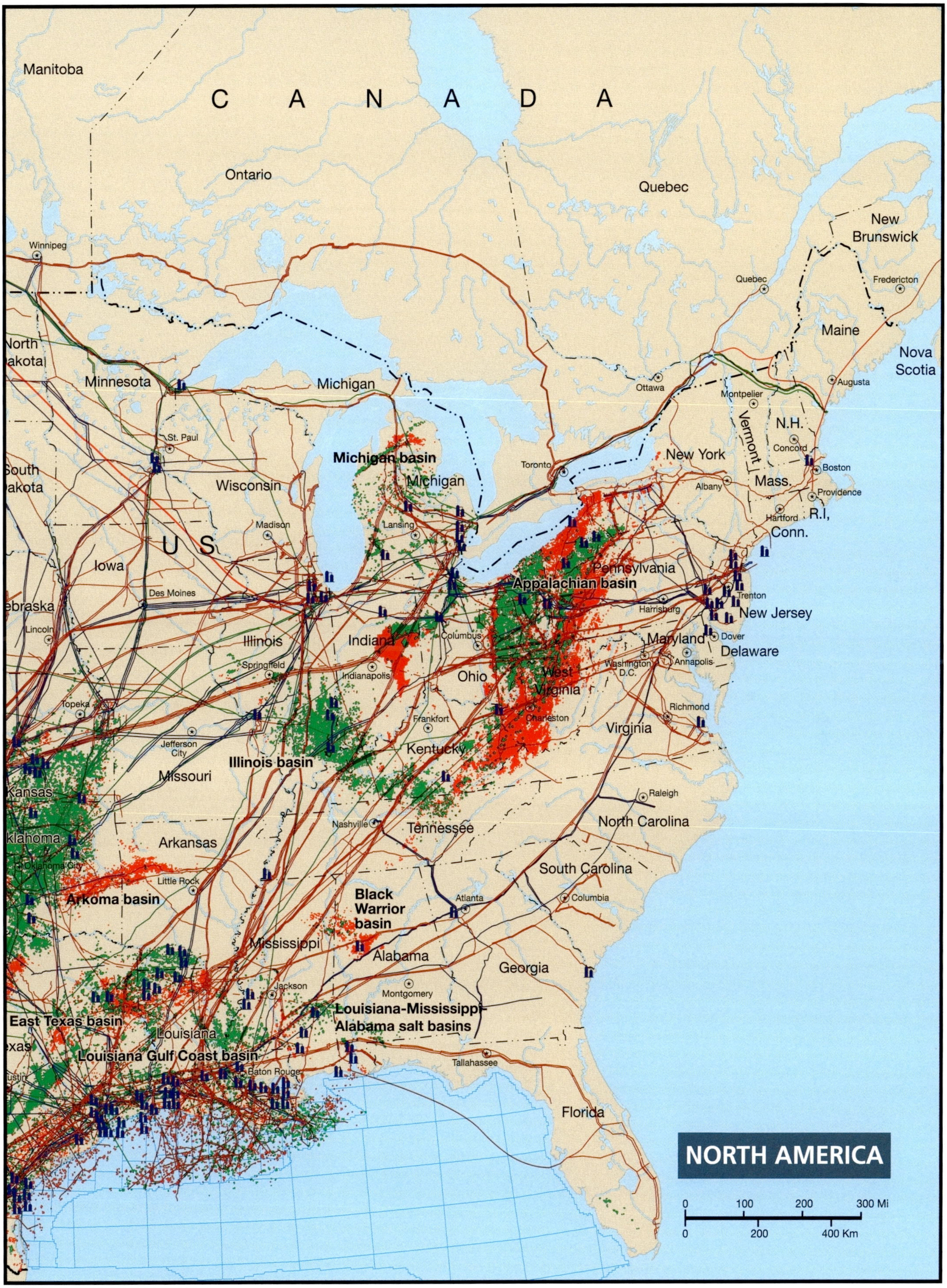
NORTH AMERICA
CANADA
US
Manitoba
Ontario
Quebec
New Brunswick
Nova Scotia
Winnipeg
Quebec
Fredericton
Maine
Augusta
Ottawa
Montpelier
Vermont
N.H.
Concord
Boston
Mass.
Providence
R.I.
Hartford
Conn.
New York
Albany
Toronto
Minnesota
St. Paul
Michigan
Michigan basin
Lansing
Wisconsin
Madison
Iowa
Des Moines
Lincoln
Illinois
Springfield
Indiana
Indianapolis
Columbus
Ohio
Appalachian basin
Pennsylvania
Harrisburg
Trenton
New Jersey
Dover
Delaware
Maryland
Annapolis
Washington D.C.
West Virginia
Charleston
Virginia
Richmond
Frankfort
Kentucky
Topeka
Jefferson City
Missouri
Illinois basin
Kansas
Nashville
Tennessee
Raleigh
North Carolina
Arkansas
Little Rock
Oklahoma City
Arkoma basin
South Carolina
Columbia
Atlanta
Black Warrior basin
Mississippi
Alabama
Georgia
Jackson
Montgomery
Louisiana-Mississippi-Alabama salt basins
East Texas basin
Louisiana
Louisiana Gulf Coast basin
Baton Rouge
Tallahassee
Florida
0
100
200
300 Mi
0
200
400 Km

OGJ FORECAST OF US OIL SUPPLY AND DEMAND

	First half 2009 Volume, 1,000 b/d	First half 2009 % change from 2008	Year 2009 Volume, 1,000 b/d	Year 2009 % change from 2008
DEMAND				
Motor gasoline	8,950	–1.0	9,010	0.2
Dist. 1-4	7,400	–1.0	7,443	0.2
Dist. 5	1,550	–1.0	1,567	0.2
Jet fuel	1,400	–11.8	1,427	–7.3
Dist. 1-4	977	–11.8	992	–7.3
Dist. 5	423	–11.8	435	–7.3
Distillate	3,710	–8.9	3,760	–4.7
Dist. 1-4	3,237	–8.9	3,260	–4.7
Dist. 5	473	–8.9	500	–4.6
Residual	575	–11.0	496	–20.3
Dist. 1-4	433	–11.0	372	–20.3
Dist. 5	142	–11.0	124	–20.3
LPG and ethane	1,900	–7.8	1,847	–5.0
Dist. 1-4	1,855	–7.8	1,796	–5.0
Dist. 5	45	–7.8	51	–5.0
Other products	2,145	–14.2	2,192	–10.8
Dist. 1-4	1,910	–14.2	1,948	–10.8
Dist. 5	235	–14.2	244	–10.8
Total domestic demand	**18,680**	**–6.1**	**18,732**	**–3.9**
Dist. 1-4	15,812	–6.2	15,810	–3.9
Dist. 5	2,868	–5.9	2,922	–3.9
Exports	**1,845**	**1.1**	**1,800**	**–0.1**
Dist. 1-4	1,554	1.1	1,499	–0.1
Dist. 5	291	0.7	301	–0.1
Total demand	**20,525**	**–5.5**	**20,532**	**–3.6**
Dist. 1-4	17,366	–5.6	17,309	–3.6
Dist. 5	3,159	–5.3	3,223	–3.5
SUPPLY				
Domestic production				
Crude & condensate	5,282	3.3	5,275	6.6
Dist. 1-4	3,886	3.3	3,850	6.6
Dist. 5	1,396	3.3	1,425	6.6
NGL	1,770	–4.5	1,770	–0.8
Dist. 1-4	1,709	–4.5	1,705	–0.8
Dist. 5	61	–4.5	65	–0.8
Total domestic production	7,052	1.2	7,045	4.6
Dist. 1-4	5,595	0.8	5,555	4.2
Dist. 5	1,457	2.9	1,490	6.3
Imports				
Crude oil	9,360	–4.0	9,170	–5.8
Dist. 1-4	8,214	–4.0	8,034	–5.8
Dist. 5	1,146	–4.0	1,136	–5.9
Products & unfinished oils	2,900	–11.5	2,975	–5.6
Dist. 1-4	2,736	–11.5	2,805	–5.6
Dist. 5	164	–11.5	170	–5.6
Total imports	**12,260**	**–5.9**	**12,145**	**–5.8**
Dist. 1-4	10,950	–6.0	10,839	–5.7
Dist. 5	1,310	–4.9	1,306	–5.8
Processing gain, loss, etc.	1,345	–28.6	1,383	–26.4
Dist. 1-4	1,103	–28.6	1,134	–26.4
Dist. 5	242	–28.6	249	–26.4
Total new supply	**20,657**	**–5.6**	**20,573**	**–4.3**
Dist. 1-4	17,648	–5.8	17,528	–4.3
Dist. 5	3,009	–3.9	3,045	–4.2
Stock change	**132**	—	**41**	—
Dist. 1-4	282	—	219	—
Dist. 5	–150	—	–178	—
Crude runs to stills	**14,350**	**–3.6**	**14,400**	**–1.7**
Total input to stills	**14,690**	**–3.4**	**14,770**	**–1.7**
Total refining capacity	**17,672**	**0.4**	**17,675**	**0.4**
Refining utilization, %	**83.1**	**–3.8**	**83.6**	**–2.1**
Total industry stocks*	**1,110**	**13.3**	**1,050**	**1.4**
Refined products	760	11.1	720	1.6
Crude oil	350	18.2	330	1.2
SPR crude oil stocks	724	2.5	725	3.3
IMPORT DEPENDENCY				
Total imports % domestic demand	65.6	—	64.8	—
Net imports % domestic demand.	55.8	—	55.2	—

*Million bbl at end of period.

Energy picture: OGJ mid-2009 view

In mid-2009, *OGJ* evaluated the outlook for U.S. energy demand for the year and concluded that it would contract 3% from 2008, to 96.29 quads. Efficiency of energy use was expected to improve to 8,430 Btu/dollar, from 8,520 Btu/dollar in 2008.

Oil and gas were expected to account for 61.45% of the U.S. energy market, just as in 2008.

U.S. oil demand was expected to drop nearly 4% (to 35.837 quads), with gas demand falling by 2% (to 23.333 quads).

U.S. use of gas, coal and nuclear energy was expected to drop, but use of hydroelectric power generation and all other renewable energy sources expected to grow.

U.S. production of crude and condensate was expected to increase, with imports expected to decline almost 6% from 2008.

Plentiful gas supplies were expected in 2009. Inventories were substantial at mid-year due to a surge in production in 2008, and industrial demand was down.

Total U.S. gas production for 2009 was expected to drop about 1% (to 21.228 tcf) as a result of much lower prices than in 2008.

OGJ forecast average U.S. production of crude and condensate to average 5.275 million b/d, up from 4.949 million b/d in 2008. Crude imports were expected to average 9.17 million b/d, down 5.8% from 2008.

U.S. CRUDE, PRODUCTS, AND NATURAL GAS PRICES

Year	Average wellhead crude price, $/bbl	Refiner's acquisition cost of crude, $/bbl	Retail motor gasoline, all types, ¢/gal	Residential heating oil, ¢/gal	Average wellhead natural gas price, $/Mcf
1976	8.19	10.89	59.5	40.6	0.58
1977	8.57	11.96	63.1	46.0	0.79
1978	9.00	12.46	65.2	49.0	0.91
1979	12.64	17.72	88.2	70.4	1.18
1980	21.59	28.07	122.1	97.4	1.59
1981	31.77	35.24	135.3	119.4	1.98
1982	28.52	31.87	128.1	116.0	2.46
1983	26.19	28.99	122.5	107.8	2.59
1984	25.88	28.63	119.8	109.1	2.66
1985	24.09	26.75	119.6	105.3	2.51
1986	12.51	14.55	93.1	83.6	1.94
1987	15.40	17.90	95.7	80.3	1.67
1988	12.58	14.67	96.3	81.3	1.69
1989	15.86	17.97	106.0	90.0	1.69
1990	20.03	22.22	121.7	106.3	1.71
1991	16.54	19.06	119.6	101.9	1.64
1992	15.99	18.43	119.0	93.4	1.74
1993	14.25	16.41	117.3	91.1	2.04
1994	13.19	15.59	117.4	88.4	1.85
1995	14.62	17.23	120.5	86.7	1.55
1996	18.46	20.71	128.8	98.9	2.17
1997	17.23	19.04	129.1	98.4	2.32
1998	10.87	12.52	111.5	85.2	1.96
1999	15.56	17.51	122.1	87.6	2.19
2000	26.72	28.26	156.3	131.0	3.68
2001	21.84	22.95	153.1	125.0	4.00
2002	22.51	24.10	144.1	112.9	2.95
2003	27.56	28.53	163.8	135.5	4.88
2004	36.77	36.98	192.3	154.8	5.46
2005	50.28	50.24	233.8	205.2	7.33
2006	59.69	60.24	263.5	236.5	6.39
2007	66.52	67.94	284.9	259.2	6.37
2008	94.04	94.73	331.7	322.0	8.07
2009*	62.00	64.00	230.0	230.0	4.20

*OGJ estimate.

Source: U.S. Energy Information Administration, for 1976–2008 data

U.S. ENERGY CONSUMPTION AND EFFICIENCY

	GDP (billion 2000 dollars)	Energy consumption (trillion btu)	Energy consumption per GDP, 2000 dollar (Mbtu)	Oil energy consumption (trillion btu)	Oil energy consumption per GDP, 2000 dollar (Mbtu)	Natural gas energy consumption (trillion btu)	Total gas gas energy consumption per GDP, 2000 dollar (Mbtu)	Total oil and natural gas energy consumption (trillion btu)	Oil and gas energy consumption per GDP, 2000 dollar (Mbtu)	Oil and natural gas energy % of total energy
1973	4,341.5	75,708	17.4	34,840	8.0	22,512	5.2	57,352	13.2	75.8
1974	4,319.6	73,991	17.1	33,455	7.7	21,732	5.0	55,187	12.8	74.6
1975	4,311.2	71,999	16.7	32,731	7.6	19,948	4.6	52,679	12.2	73.2
1976	4,540.9	76,012	16.7	35,175	7.7	20,345	4.5	55,520	12.2	73.0
1977	4,750.5	78,000	16.4	37,122	7.8	19,931	4.2	57,053	12.0	73.1
1978	5,015.0	79,986	15.9	37,965	7.6	20,000	4.0	57,965	11.6	72.5
1979	5,173.4	80,903	15.6	37,123	7.2	20,666	4.0	57,789	11.2	71.4
1980	5,161.7	78,122	15.1	34,202	6.6	20,235	3.9	54,437	10.5	69.7
1981	5,291.7	76,335	14.4	31,931	6.0	19,928	3.8	51,859	9.8	67.9
1982	5,189.3	73,234	14.1	30,231	5.8	18,505	3.6	48,736	9.4	66.5
1983	5,423.8	73,066	13.5	30,054	5.5	17,357	3.2	47,411	8.7	64.9
1984	5,813.6	76,693	13.2	31,051	5.3	18,507	3.2	49,558	8.5	64.6
1985	6,053.7	76,493	12.6	30,922	5.1	17,703	2.9	48,625	8.0	63.6
1986	6,263.6	76,722	12.2	32,196	5.1	16,708	2.7	48,904	7.8	63.7
1987	6,475.1	79,156	12.2	32,865	5.1	17,744	2.7	50,609	7.8	63.9
1988	6,742.7	82,774	12.3	34,222	5.1	18,552	2.8	52,774	7.8	63.8
1989	6,981.4	84,886	12.2	34,211	4.9	19,712	2.8	53,923	7.7	63.5
1990	7,112.5	84,654	11.9	33,553	4.7	19,603	2.8	53,156	7.5	62.8
1991	7,100.5	84,522	11.9	32,845	4.6	20,149	2.8	52,994	7.5	62.7
1992	7,336.6	85,866	11.7	33,527	4.6	20,835	2.8	54,362	7.4	63.3
1993	7,532.7	87,579	11.6	33,841	4.5	21,351	2.8	55,192	7.3	63.0
1994	7,835.5	89,248	11.4	34,670	4.4	21,842	2.8	56,512	7.2	63.3
1995	8,031.7	91,174	11.4	34,437	4.3	22,671	2.8	57,108	7.1	62.6
1996	8,328.9	94,176	11.3	35,673	4.3	23,085	2.8	58,758	7.1	62.4
1997	8,703.5	94,766	10.9	36,160	4.2	23,223	2.7	59,383	6.8	62.7
1998	9,066.9	95,183	10.5	36,817	4.1	22,830	2.5	59,647	6.6	62.7
1999	9,470.3	96,817	10.2	37,838	4.0	22,909	2.4	60,747	6.4	62.7
2000	9,817.0	98,975	10.1	38,264	3.9	23,824	2.4	62,088	6.3	62.7
2001	9,890.7	96,326	9.7	38,186	3.9	22,773	2.3	60,959	6.2	63.3
2002	10,048.8	97,858	9.7	38,227	3.8	23,558	2.3	61,785	6.1	63.1
2003	10,301.0	98,209	9.5	38,809	3.8	22,897	2.2	61,706	6.0	62.8
2004	10,675.8	100,351	9.4	40,294	3.8	22,931	2.1	63,225	5.9	63.0
2005	10,989.5	100,485	9.1	40,393	3.7	22,583	2.1	62,976	5.7	62.7
2006	11,294.8	99,875	8.8	39,958	3.5	22,224	2.0	62,182	5.5	62.3
2007	11,523.9	101,554	8.8	39,773	3.5	23,628	2.1	63,401	5.5	62.4
2008	11,652.0	99,275	8.5	37,137	3.2	23,809	2.0	60,946	5.2	61.4
*2009	11,419.0	96,260	8.4	35,837	3.1	23,333	2.0	59,170	5.2	61.5

*Estimated.

Source: U.S. Energy Information Administration

North America

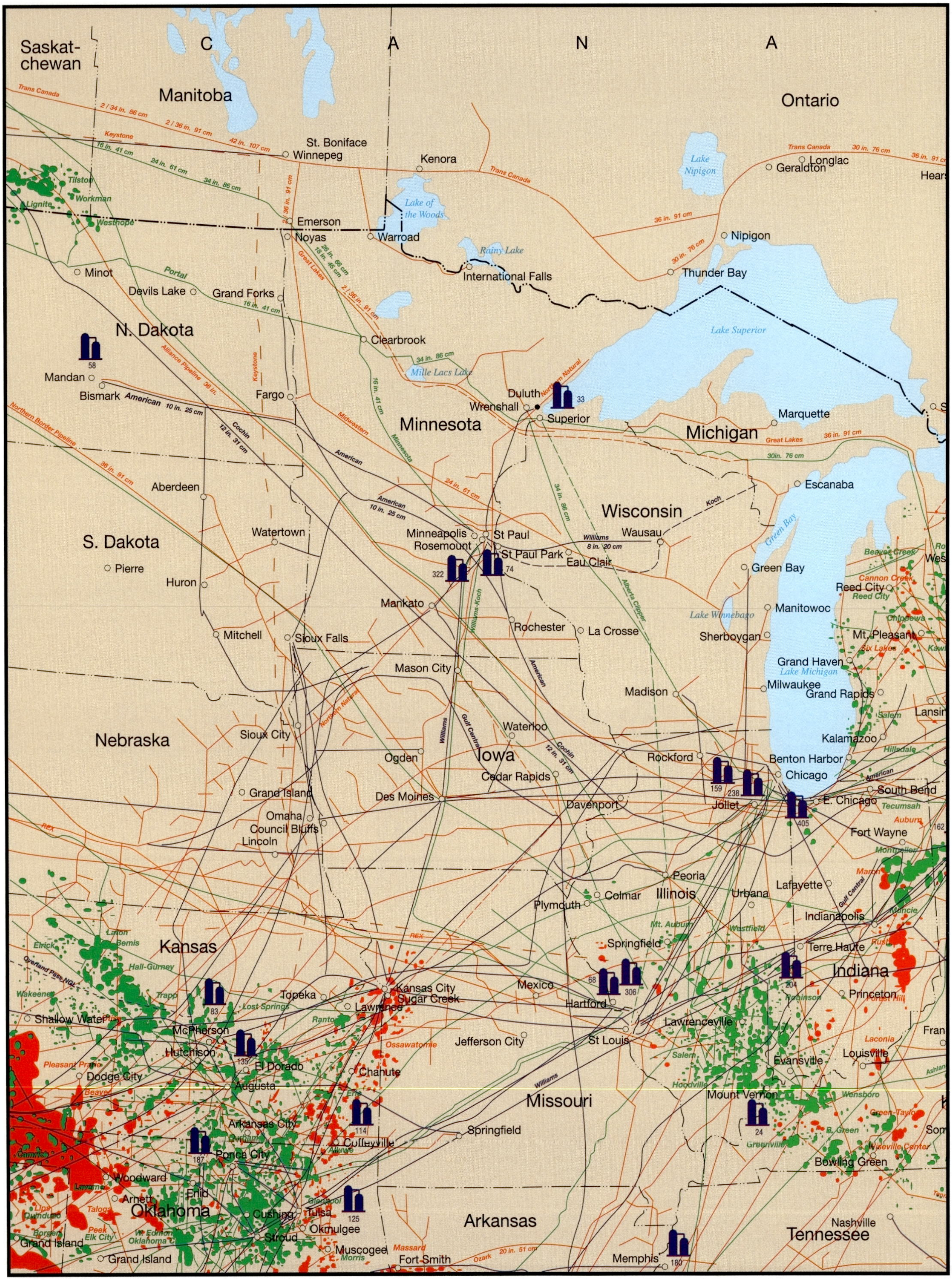

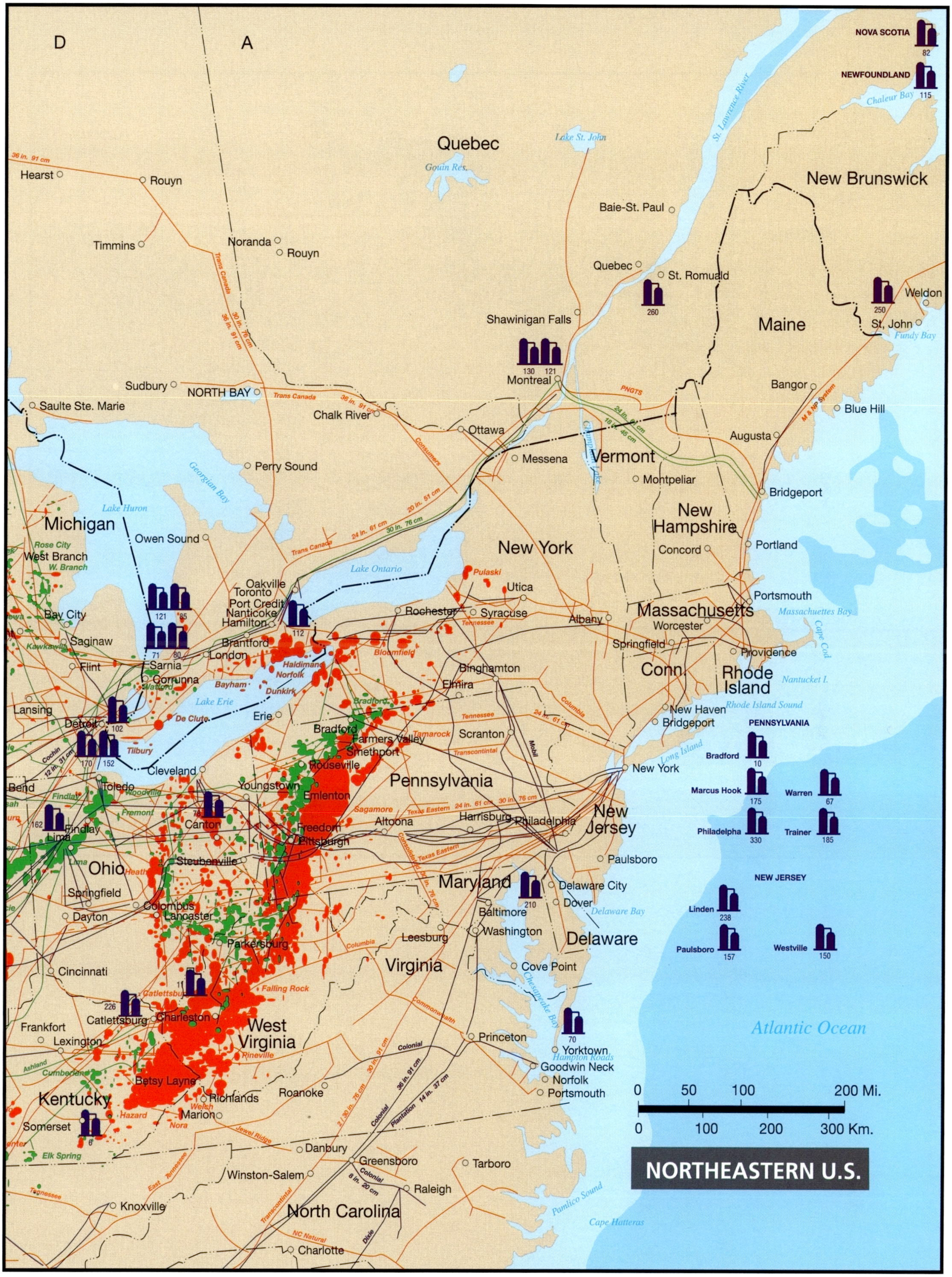

NORTHEASTERN U.S.
Quebec
New Brunswick
Maine
Vermont
New Hampshire
Massachusetts
Rhode Island
Conn.
New York
Pennsylvania
New Jersey
Maryland
Delaware
Virginia
West Virginia
North Carolina
Kentucky
Ohio
Michigan
Atlantic Ocean
NOVA SCOTIA
82
NEWFOUNDLAND
115
PENNSYLVANIA
Bradford
10
Marcus Hook
175
Warren
67
Philadelphia
330
Trainer
185
NEW JERSEY
Linden
238
Paulsboro
157
Westville
150
0
50
100
200 Mi.
0
100
200
300 Km.

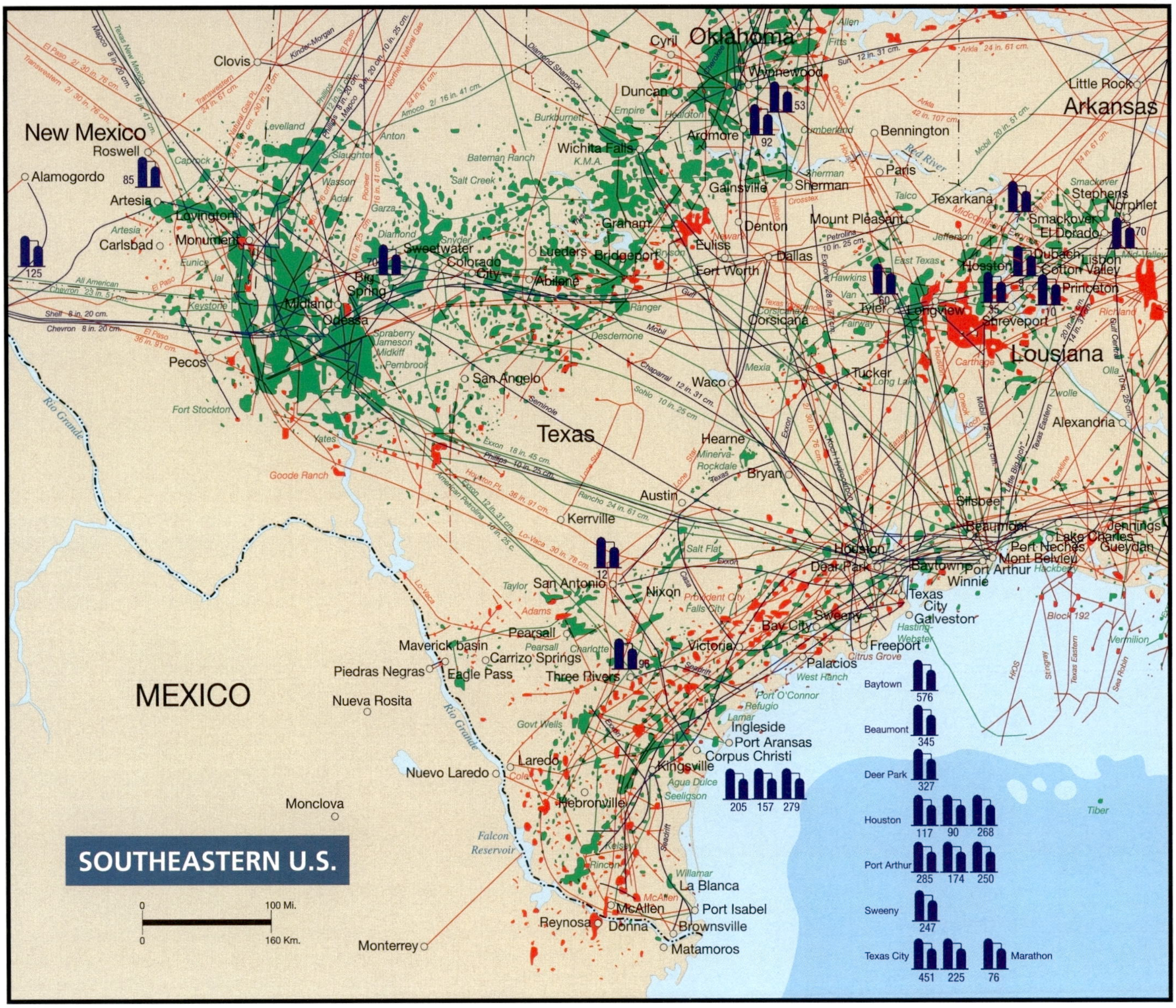

U.S. ENERGY DEMAND

	2008	*2009	Change, %	% share of total energy 2008	*2009
	Trillion btu				
Oil	37,137	35,837	–3.9	37.4	37.2
Gas	23,809	23,333	–2.0	24.0	24.2
Coal	22,421	21,070	–6.0	22.6	21.9
Nuclear	8,455	8,450	–0.1	8.5	8.8
Hydro, other	7,453	7,600	2.0	7.5	7.9
Total	**99,275**	**96,290**	**–3.0**	**100.0**	**100.0**

*OGJ estimate.

Source: 2008 U.S. Energy Information Administration

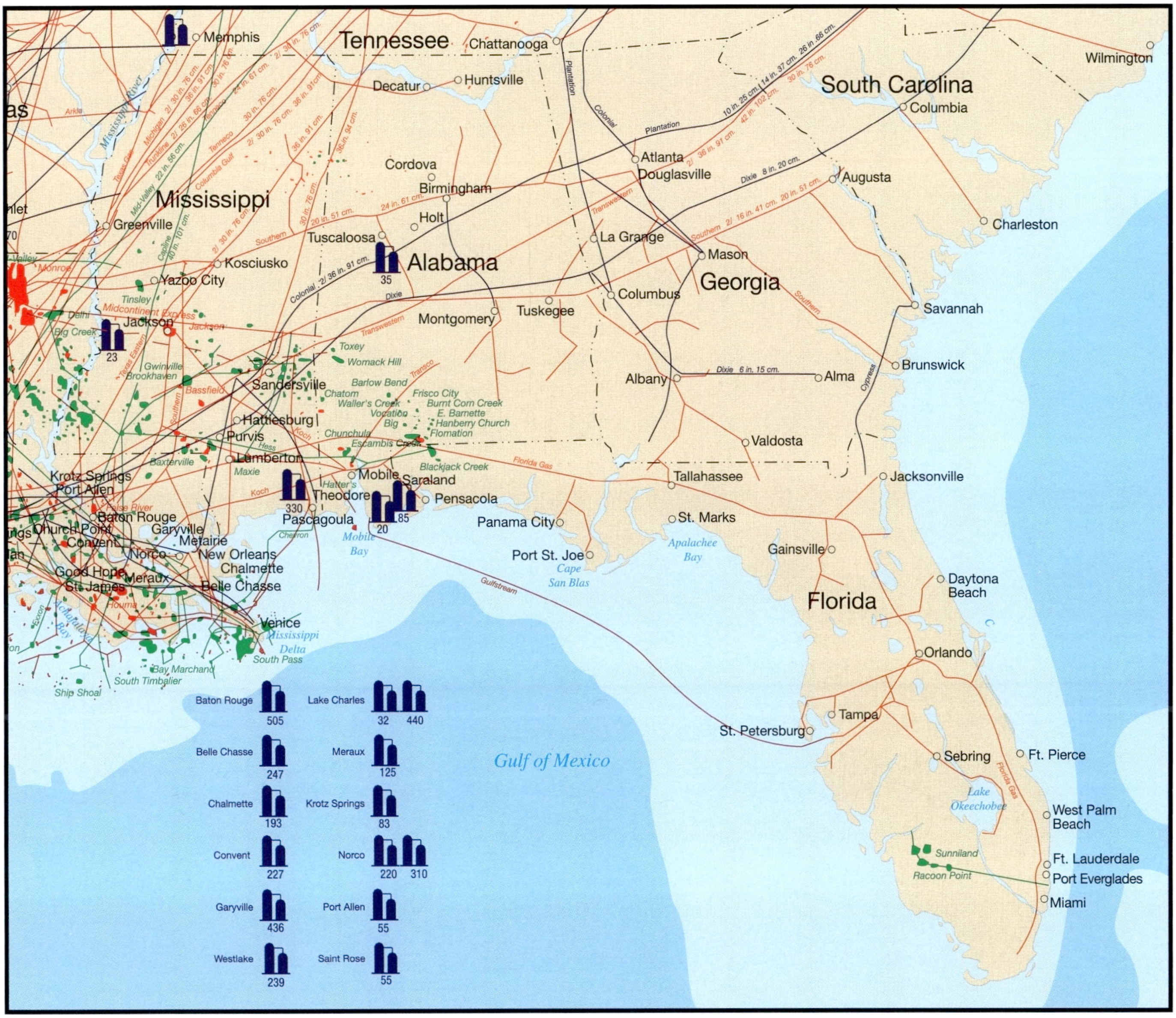

U.S. NATURAL GAS SUPPLY AND DEMAND

	2006	2007 bcf	2008	Change, % 08/07	2009 bcf	Change, % 09/08
Production						
Texas	5,548	6,092	6,969	14.4	6,900	-1.0
Louisiana	1,361	1,364	1,350	-1.0	1,335	-1.1
Federal Gulf of Mexico	2,902	2,799	2,345	-16.2	2,550	8.7
Other states	9,599	9,764	10,778	10.4	10,443	-3.1
Total production	**19,410**	**20,019**	**21,442**	**7.1**	**21,228**	**-1.0**
Imports						
Canada	3,590	3,783	3,586	-5.2	3,227	-10.0
Mexico	13	54	43	-20.4	10	-76.7
LNG	584	771	352	-54.3	528	50.0
Total imports	**4,187**	**4,608**	**3,981**	**-13.6**	**3,765**	**-5.4**
Supplemental gas	66	63	55	-12.7	60	9.1
Losses, etc.*	(818)	(1,014)	(1,298)	28.0	(1,078)	-16.9
Total new supply	**22,845**	**23,676**	**24,180**	**2.1**	**23,975**	**-0.8**
Supply from storage	(436)	193	32	—	(260)	—
Total supply	**22,409**	**23,869**	**24,212**	**1.4**	**23,715**	**-2.1**
Exports	724	822	1,006	22.4	965	-4.1
Total consumption	**21,685**	**23,047**	**23,206**	**0.7**	**22,750**	**-2.0**

*Extraction losses and unaccounted for gas.

Source: 2006, 2007, 2008 EIA; 2009 OGJ.

U.S. REFINERY UTILIZATION

	Crude runs (1,000 b/d)	Input to distillation units (1,000 b/d)	Operable capacity	Utilization rate, %
1987	12,851	12,999	15,643	83.1
1988	13,246	13,447	15,927	84.4
1989	13,401	13,551	15,701	86.3
1990	13,409	13,610	15,623	87.1
1991	13,301	13,508	15,707	86.0
1992	13,411	13,600	15,460	87.9
1993	13,613	13,851	15,143	91.5
1994	13,866	14,032	15,150	92.6
1995	13,973	14,119	15,346	92.0
1996	14,195	14,337	15,239	94.1
1997	14,662	14,831	15,594	95.2
1998	14,889	15,108	15,802	95.6
1999	14,804	15,078	16,282	92.6
2000	15,067	15,296	16,525	92.6
2001	15,086	15,351	16,512	92.6
2002	14,947	15,180	16,700	90.7
2003	15,304	15,503	16,747	92.6
2004	15,517	15,781	16,982	93.0
2005	15,220	15,479	17,128	90.4
2006	15,240	15,598	17,400	89.7
2007	15,143	15,443	17,447	88.5
2008	14,648	15,027	17,607	85.3
2009*	14,400	14,770	17,675	83.6
1990–2009 change				
Volume	1,239	1,160	2,052	–
Percent	7.4	8.5	13.1	–
2008–09 change				
Volume	(248)	(257)	68	–
Percent	-1.7	-1.7	0.4	–

Source: U.S. Energy Information Administration

U.S. OIL IMPORTS

Sources of crude imports[1]

Country	Share of total 2009, %	First quarter average 2009, 1,000 b/d	Change, 09/08, %	Annual average 2008, 1,000 b/d	Change 08/07, %	Annual average 2007, 1,000 b/d
Angola	6.4	612	21.4	504	1.2	498
Ecuador	2.5	241	12.6	214	8.1	198
Nigeria	6.4	607	–34.2	922	–14.9	1,084
Saudi Arabia	11.9	1,128	–25.0	1,503	3.9	1,447
Venezuela	10.8	1,029	–1.0	1,039	–9.5	1,148
Other OPEC	11.9	1,131	–8.3	1,233	21.7	1,013
Total OPEC	**49.9**	**4,748**	**–12.3**	**5,415**	**0.5**	**5,388**
Canada	20.0	1,901	–2.8	1,956	3.6	1,888
China	0.1	5	–54.5	11	83.3	6
Colombia	2.5	235	32.0	178	29.9	137
Gabon	1.1	106	82.8	58	–7.9	63
Indonesia[2]	0.2	21	31.3	16	6.7	15
Mexico	12.7	1,203	1.3	1,187	–15.8	1,409
Norway	0.6	61	103.3	30	–46.4	56
Trinidad and Tobago	0.5	51	121.7	23	–52.1	48
UK	0.6	56	–28.2	78	–22.8	101
Other non-OPEC	11.8	1,122	35.0	831	–9.7	920
Total imports	**100.0**	**9,509**	**–2.8**	**9,783**	**–2.5**	**10,031**

Sources of refined product imports

Country	Share of total 2009, %	First quarter average 2009, 1,000 b/d	Change, 09/08, %	Annual average 2008, 1,000 b/d	Change 08/07, %	Annual average 2007, 1,000 b/d
Algeria	9.0	281	19.1	236	4.0	227
Saudi Arabia	0.7	22	–15.4	26	–31.6	38
Venezuela	5.5	172	15.4	149	–29.7	212
Other OPEC	2.2	70	–45.7	129	12.2	115
Total OPEC	**17.5**	**545**	**0.9**	**540**	**–8.8**	**592**
Canada	19.2	598	11.6	536	–5.3	566
Colombia	1.0	30	30.4	23	27.8	18
Italy	0.6	19	–64.2	53	–3.6	55
Mexico	4.1	127	10.4	115	–7.3	124
Netherlands Antilles	0.4	14	40.0	10	11.1	9
Virgin Islands	10.3	321	0.3	320	–7.5	346
Other non-OPEC	46.9	1,463	–4.7	1,535	–11.1	1,727
Total imports	**100.0**	**3,117**	**–0.5**	**3,132**	**–8.9**	**3,437**

[1]Includes imports for the Strategic Petroleum Reserve. [2]Dropped out of OPEC at end of 2008.

Source: U.S. Energy Information Administration

Obama's energy programs

In January 2009, *OGJ* outlined the major aims of the energy and environmental program proposed by the new federal administration. Readers are directed to the 2009 edition of the *International Petroleum Encyclopedia* for details.

Oil and gas prices

Information about oil and gas price movements during 2009 is given in the chapter titled, World Energy Market: The Big Picture, which precedes this one.

> **Oil imports from Saudi Arabia**
>
> U.S. crude oil imports in March 2009 from Saudi Arabia went below 1 million b/d for the first time since 1990, and were at the lowest level since November 1988, according to Petromatrix.

EIA outlook: Natural gas

In its May 2009 *Short-Term Energy Outlook (STEO)*, EIA said that:

- Domestic gas demand likely will decline 1.9% in 2009 and 0.4% in 2010 as higher prices spur some power plants to switch back to coal.
- Working gas in storage in U.S. reservoirs on May 1 was nearly 2 tcf, more than 360 bcf ahead of the average for the last five years and nearly 500 bcf more than the same week in 2008.
- Henry Hub spot prices were likely to average $4.06/Mcf in 2009, then rise to $5.21/Mcf in 2010.

In its October 2009 *STEO*, EIA said it expected:

- U.S. natural gas inventories to reach a record peak of 3.85 tcf when storage injections end on October 31
- Total U.S. marketed gas production to increase by 1.5% in 2009 and decrease by 3.8% in 2010.
- U.S. LNG imports to increase to about 471 bcf in 2009 and 660 bcf in 2010, compared to 352 bcf in 2008.

Natural gas: WoodMac view

In May 2009, Wood Mackenzie said that weakness in global natural gas markets will delay a recovery in U.S. gas prices and rig counts. It expected the global economic recession to last several more quarters.

US TOTAL NATURAL GAS CONSUMPTION

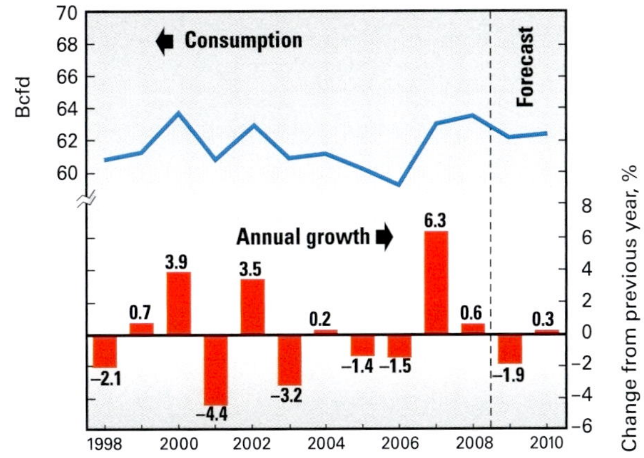

Source: EIA Short-Term Energy Outlook, May 2009

WoodMac said that the U.S. gas market faces three key challenges in the short-to-midterm:

- A combination of low demand and plentiful coal capacity in the electric power market.
- Regarding LNG: global liquefaction capacity and oversupply in Europe and Asia
- Time lags between drilling decisions and supply response.

ANGA study: Gas industry impact

A study released in September 2009 concluded that the U.S. natural gas industry supports nearly 3 million American jobs (622,000 directly, 2.8 million indirectly) and contributed $385 billion to the domestic economy in 2008.

The study was commissioned by America's Natural Gas Alliance and conducted by IHS Global Insight.

API study: Oil, gas industry impact

PricewaterhouseCoopers released in September 2009 a study done for API on the impact of the U.S. oil and gas industry on the American economy.

It concluded that each job in the oil and gas industry supported more than three jobs elsewhere in the U.S. economy in 2007 (the most recent year for which data are available).

The oil and natural gas industry itself employed 9.2 million people full-time and part-time in 2007.

The industry's total value-added contribution to the national economy topped $1 trillion, said the study.

RIK program to end

In September 2009, the U.S. Dept. of the Interior announced plans to eliminate the royalty-in-kind (RIK) program of the U.S. Minerals Management Service.

Officials said the aim it to move to a more transparent and accountable royalty collection operation.

QAEP tax credits

Federal officials unveiled plans in August 2009 to award about $2.3 billion in tax credits to manufacturers of advanced renewable energy equipment. Examples include wind and solar energy systems, battery technology, and fuel cells.

The Qualifying Advanced Energy Project (QAEP) credits are part of the American Recovery and Reinvestment Act (stimulus package).

The first group of credits is to be announced in January 2010.

Cash for clunkers

From mid-2009 through October, the federal government conducted a $3-billion "cash for clunkers" program to spur purchase of more than 700,000 newer, more-fuel-efficient U.S.-made cars.

Most participants traded in old sport-utility vehicles and pick-up trucks for small cars.

Supporters said that buyers would enjoy annual savings in gasoline costs of $700 to $1,000.

But critics said the overall cost to the government of each new car sold was excessive, that buyers just accelerated purchases that they would have made later, and that serviceable older cars were being destroyed unnecessarily.

New FTC rule

A new rule from the U.S. Federal Trade Commission took effect November 4, 2009, prohibiting oil marketplace manipulation.

It prohibits fraud or deceit in wholesale petroleum markets and omissions of material information that are likely to distort petroleum markets, the commission said.

Violators face civil penalties of up to $1 million/day per violation.

Senate, House climate bills

In June 2009, the U.S. House of Representatives passed its version of an energy bill (HR 2454), including cap-and-trade provisions for reduction of carbon emissions.

On September 30, a similar bill, S 1733, was introduced in the U.S. Senate.

More details are given in the Environmental Issues section, later in this chapter.

Energy knowledge: Public survey

More people don't know enough about energy issues to make informed decisions and are still advocating "wishful" solutions.

That was the finding of an early-2009 national survey by Public Agenda, a nonpartisan research organization. Some key findings:

- 51% could not correctly identify a renewable energy source
- 39% could not name a fossil fuel
- 65% overestimated U.S. dependence on Middle Eastern oil

Full survey results are available at: http://www.publicagenda.org/reports/energy.

ACM: New crude benchmark

Platts launched a new oil benchmark in March 2009 to challenge the WTI oil standard in the U.S. market.

WTI, the world's most important oil benchmark, had come under criticism as its price diverged from other global and U.S. references. The IEA said that WTI was sending "misleading" signals.

The new benchmark—the Americas Crude Marker—will reflect the most competitively priced crude oil among four U.S. Gulf Coast sour crudes.

(Editor's note: See the Saudi Arabia section of the Middle East chapter of this book for information about the ASCI index, which also will compete with the WTI.)

Policy

New fuel-economy rules

In May 2009, the Obama Administration set out tougher fuel-economy rules for car manufacturers.

The new corporate average fuel economy (CAFE) standards apply to model years 2012–2016 for all passenger vehicles sold in the United States, including cars, light trucks and SUVs.

By 2016, the CAFE standard becomes 39 mpg for cars and 30 mpg for light trucks and SUVs. This is a 40% improvement over current standards.

The new rules also include for the first time a nationwide standard for emissions of greenhouse gases from vehicles.

Oil into SPR

In mid-January 2009, the U.S. DOE took advantage of low oil prices to buy nearly 10.7 million bbl of oil for the Strategic Petroleum Reserve. It was the first direct purchase of crude for the SPR since 1994.

Additional purchases from May 2009 through January 2010 will add nearly 6.16 million bbl.

EPA: GHG health-threat finding

In December 2009, the U.S. Environmental Protection Agency (EPA) issued a finding based on its study showing that greenhouse gases threaten public health.

The "endangerment" finding is needed before the EPA can regulate carbon dioxide and five other greenhouse gases released from automobiles, power plants, and factories under the federal Clean Air Act.

The U.S. Supreme Court ruled in 2007 that GHGs fit within the CAA definition of pollutants.

In September, EPA proposed regulations that would subject refineries and other large industrial operations to GHG regulation under the CAA while exempting smaller businesses and farms.

CFTC: Regulate emissions markets

In November 2009, the Commodity Futures Trading Commission said that comprehensive regulation of financial derivatives will also need to be a critical component of a well-functioning domestic emissions trading market.

The commission said that regulation was necessary to ensure transparency and compliance, protect the price discovery function, protect against fraud and abuse, and address financial risk, just as in other markets.

New oil, gas taxes

The final federal budget for fiscal 2010, released in May 2009, contained $31.5 billion of new oil and gas taxes over nine years. Collections would begin in fiscal 2011.

Through 2019, total revenues from the taxes are projected to reach $13.3 billion by ending certain tax deductions for oil and gas companies, $8.3 billion from repealing the percentage depletion allowance, and $5.3 billion from placing an excise tax on new Gulf of Mexico production.

Another $3.3 billion would come from repealing expensing of intangible drilling costs, and $1.2 billion from increasing a specific amortization schedule for independent producers.

The budget also reinstates superfund taxes, projected to raise $6.7 billion from 2011–2019.

NGV tax incentives

Legislation to greatly expand the use of natural gas as an alternative to conventional transportation fuel was introduced by the U.S. House in April (HR 1835) and by the Senate in July (S 1408).

The bills would extend for 18 years three critical tax incentives that focus on natural gas as a transportation fuel, the purchase of natural gas vehicles (NGVs), and the installation of commercial and residential natural gas refueling pumps.

New public lands bill

In March 2009, Pres. Barack Obama signed into law a far-reaching public lands bill (HR 146).

The measure won bipartisan support because it combined 160 smaller bills involving historic sites with provisions that add 2 million acres (3,125 sq mi) to the federal wilderness system.

The bill also expands other designations that effectively remove more federal land not just from oil and gas leasing but also from potential use as alternative energy generation sites.

Exploration and Development

California: Major oil find

In July 2009, Occidental Petroleum announced discovery of a major field in Kern County in the San Joaquin basin. The discovery could be California's largest in more than 35 years, Oxy said.

Some analysts estimated it could hold as much as 2 billion bbl, but Oxy said only that the apparent gas-condensate field has an estimated 150–250 million boe recoverable.

Bakken activity

Several companies increased ownership in 2009 in North Dakota Bakken well and production interests under an existing JV with Slawson Exploration Co. Much of the activity centers on the Williston basin Bakken/Three Forks play.

The companies include: Northern Oil & Gas Inc., Lario Oil & Gas Co., and GeoResources Inc.

Slawson expected to drill as many as 45 gross Bakken wells on newly acquired acreage through 2010.

Postulated Lodgepole Expanse

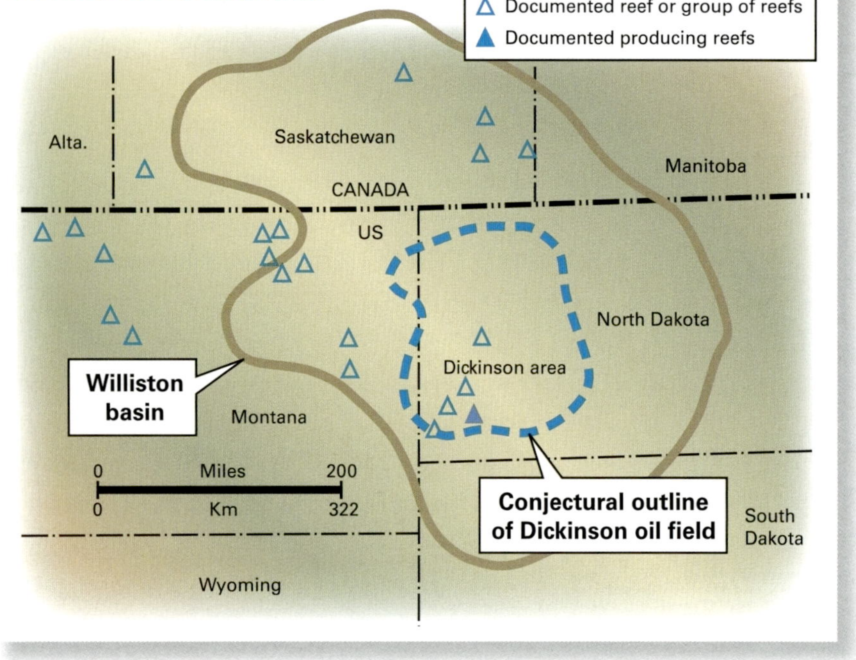

The proposed Tranquillon Ridge project of Plains Exploration & Production Co. could have provided $100 million initially (and up to $2.3 billion of royalty payments) over its 13-year life.

The company planned to directionally drill into state waters from an existing platform in federal waters.

Monterey shale

The emerging Miocene Monterey shale play in California drew increased attention in late 2009.

Venoco Inc. said it plans a big push in 2010, and estimates that the formation may hold the largest shale oil resource in the United States, with 300 billion bbl in place.

At least 2,000 wells have been drilled and logged through Monterey in exploring and producing from deeper formations.

Williston supergiant?

A Calgary-based geosciences consultant estimated in late 2009 that the cluster of Mississippian Lodgepole reef oil fields around Dickinson, ND, could indicate the presence of a supergiant Williston basin field that extends far beyond the Dickinson area.

Exploitation likely will require use of new technologies.

Estimated ultimate recoveries of the Dickinson area Lodgepole formation wells average 1.4 million bbl/well considering only producing wells.

In September 2009, a group led by Armstrong Operating Inc. gauged the first Mississippian Lodgepole oil discovery in the Dickinson area since the late 1990s.

California: Offshore plan

California's state assembly in July 2009 deleted from a budget package a bill that could have authorized the first new oil and gas activity off the Santa Barbara coast in 40 years.

EnCana in Texas, Louisiana

In May 2009, EnCana Corp. described its plans for development of unconventional gas and oil.

The firm has nine gas and four enhanced oil resource plays in the established Jurassic Deep Bossier sand and emerging Haynesville plays in East Texas and North Louisiana.

EnCana is also well established in other North American gas plays, including the Barnett, Eagle Ford, Pearsall, and Niobrara

EnCana Acreage in Gulf Coast Jurassic Trend

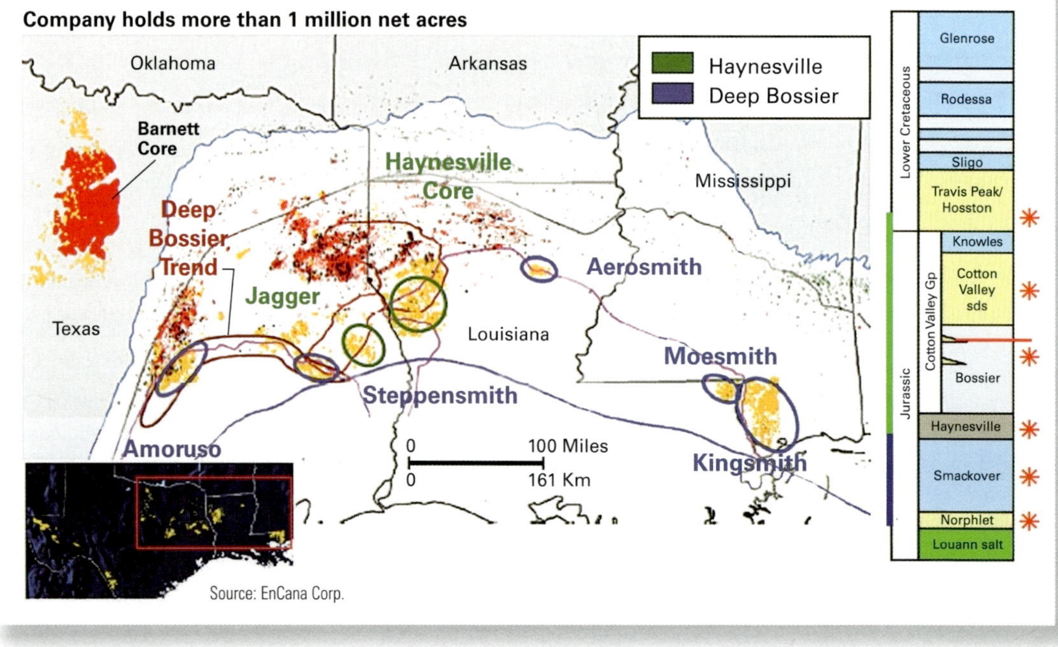

Source: EnCana Corp.

shales in the United States, and the Muskwa and Montney tight sand in British Columbia.

Eagle Ford shale JV

Activity further intensified in 2009 in the development of gas with condensate in the Cretaceous Eagle Ford shale. Companies involved included Petrohawk Energy Corp., Swift Energy Co., St. Mary Land & Exploration Co., and El Paso Corp.

Petrohawk describes the South Texas play "one of the highest-quality shale reservoirs discovered in the United States."

Offshore Florida

In early 2009, Florida state legislators unveiled a proposal that would lift the state's current ban on oil and gas exploration and production in the state waters off Florida's coast.

Such activity could generate at least $1.6 billion/year in state revenues and create more than 19,000 jobs, according to the Associated Industries of Florida (AIF).

Ohio drilling

Since 2004, about 23% of all wells drilled in Ohio have been in areas defined as urban, and in 2008, 318 of the 1,428 drilling permits issued were in such areas.

The rise in the proportion of drilling in urban areas has contributed to a turnaround in an otherwise downward trend in production in the state.

About 8% more permits were issued in 2008 than in 2007. Most (72%) targeted the Clinton sandstone, followed by the Devonian shale, formations below the Cambro-Ordovician Knox unconformity (11%), and the Marcellus shale formation.

For only the third time since 1991, more than 1,000 Ohio wells were drilled in 2008.

Reserves

PGC: U.S. gas resources

Reevaluation of shale plays boosted the Potential Gas Committee's mid-2009 assessment of the U.S. natural gas resource to its highest level ever.

Eagle Ford Shale Drilling Results

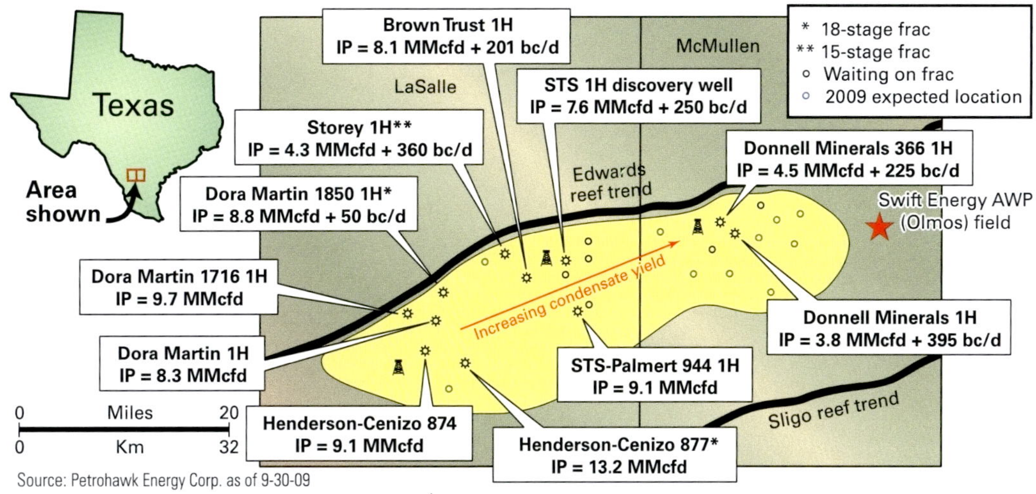

Source: Petrohawk Energy Corp. as of 9-30-09

Ohio Permits and Drilling

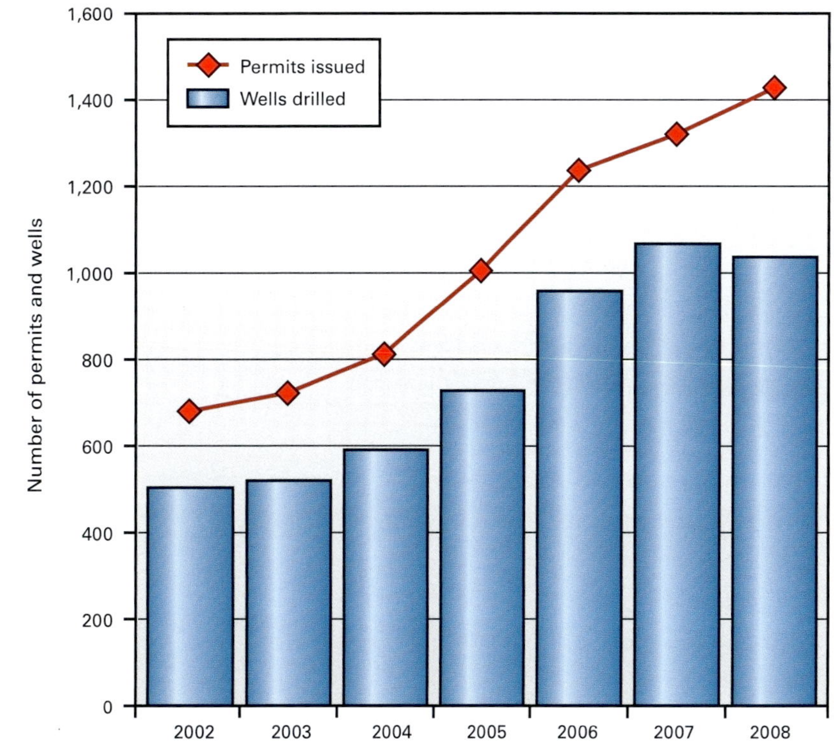

The committee raised its biennial assessment to 1,836 tcf at yearend 2008, from 1,321 tcf at yearend 2006. For 2008, the assessment ascribes 1,673 tcf of gas to traditional reservoirs and 163 tcf to coalbed methane reservoirs.

PGC data, combined with DOE information, projects total U.S. future supply of 2,074 tcf, which is 542 tcf above the 2006 total.

Shale gas accounts for 616 tcf of the new estimate of technically recoverable gas.

The study reevaluated shale plays in the Appalachian, Arkoma, Forth Worth, and Uinta basins as well as several Gulf Coast basins.

USGS: Piceance basin oil shale

The USGS said in early 2009 that western Colorado's Piceance basin has an estimated 1.525 trillion bbl of in-place oil shale. The agency noted that no economic extraction method currently is available in the United States, so it is not known how much of the assessed resource is recoverable.

Oil shale contains kerogen, an organic precursor to oil that must be heated for oil production.

The latest assessment is 50% larger than the 1989 assessment because USGS assessed additional geographic areas and subsurface zones.

DOI: OCS oil, gas assessment

According to an early-2009 report by the Dept. of the Interior, the U.S. OCS holds an estimated 66–115 billion bbl of oil and 362–565 tcf of natural gas. DOI noted that the figures contain significant data gaps.

Estimates of technically recoverable reserves were based on a regional assessment of the entire OCS that was completed by the MMS in 2006.

These estimates include assessments of new areas included in DOI's 2010–15 draft proposed 5-year OCS leasing program—areas in the Atlantic, eastern Gulf of Mexico, Pacific, and Alaska.

EIA: 2007 U.S. oil reserves

In January 2009, the U.S. EIA reported that proved U.S. oil reserves rose by 345 million bbl, or 2%, during 2007 to 21.32 billion bbl, up from 20.97 billion bbl at the beginning of 2007.

The increase was a contrast to the rapid decrease in domestic crude reserves that began in 1970 but which have moderated in the past decade.

During 2007, EIA said, U.S. producers added nearly 2.04 billion bbl of new reserves while producing more than 1.69 billion bbl, resulting in the 345-million bbl net addition.

About 40% of the reserve additions were discoveries, with the rest coming from more intensive development of known producing fields.

Alaska, Texas, and North Dakota accounted for a combined 605 million bbl (the majority) of the year's new reserves.

Elm Coulee: Bakken shale oil

Elm Coulee field, Montana's primary Bakken shale oil producing field, is a giant that is expected to recover more than 200 million bbl.

That was the assessment in third quarter 2009 of the Colorado School of Mines, which noted that the field has more than 600 wells that produce primarily from the middle Bakken.

The Bakken formation depth is 8,500–10,500 ft (2,591–3,200 m), and oil in place is estimated at 5 million bbl/sq mile.

Lease Sales

Lease Sale 208

DOI's MMS reported that Lease Sale 208 in March 2009 saw apparent high bids totaling just over $703 million for 348 tracts in the central Gulf of Mexico.

MMS received 476 bids totaling nearly $994 million from 70 companies.

The tracts offered covered nearly 1.9 million acres (2,969 sq mi) off Louisiana, Mississippi, and Alabama.

Onshore leasing: 2010

In December 2009, the DOI announced that 38 onshore federal oil and gas lease sales have been scheduled for 2010.

DOI noted that significant onshore and offshore acreage has been leased already that has not yet been developed. Offshore, of 7,735 active leases, only 5,211 were producing. Onshore, of 55,385 active leases, about 26,000 were not producing.

One scheduled lease sale (August 2010) would be the first in nearly two years within the National Petroleum Reserve-Alaska (NPR-A).

2009–2010 OCS revenue-sharing

In April 2009, DOI announced that six U.S. coastal states will share nearly $500 million from offshore oil and gas revenues in fiscal 2009 and 2010 to help restore and protect coastal wetlands, wildlife habitat, and marine areas.

Results of Central Gulf of Mexico Lease Sale 208

Top 10 companies based on sum of high bids submitted

Company	Total high bids	Sum of high bids, $
1. Shell Gulf of Mexico Inc.	39	153,631,475
2. BP Exploration & Production Inc.	27	77,466,559
3. Marathon Oil Co.	16	62,441,178
4. Noble Energy Inc.	24	55,442,147
5. BHP Billiton Petroleum (Deepwater) Inc.	28	50,393,232
6. Statoil Gulf of Mexico LLC	9	49,276,000
7. Repsol E&P USA Inc.	20	48,545,024
8. ExxonMobil Corp.	15	24,484,400
9. Chevron USA Inc.	21	22,400,166
10. Ecopetrol America Inc.	26	20,558,066

Companies submitting Top 10 single highest bids

Company	Block	Water depth, m	High bid amount, $
1. Shell Gulf of Mexico Inc.*	Mississippi Canyon 721	800–<1,600	65,611,235
2. Marathon Oil Co.* Repsol E&P USA Inc.	Walker Ridge 578	>2,000	46,471,680
3. Noble Energy Inc.*	Green Canyon 774	800–<1,600	31,786,250
4. BP Exploration & Production Inc.*	Mississippi Canyon 418	800–<1,600	29,194,317
5. Shell Gulf of Mexico Inc.*	Mississippi Canyon 720	800–<1,600	23,511,235
6. Eni Petroleum US LLC* Cobalt International Energy LP Samson Offshore Co.	Walker Ridge 133	1,600–2,000	22,000,222
7. Marathon Oil Co.* Repsol E&P USA Inc.	Walker Ridge 579	>2,000	19,375,173
8. Shell Gulf of Mexico Inc.*	Green Canyon 114	800–<1,600	17,011,235
9. Marathon Oil Co.*	Walker Ridge 134	1,600–2,000	16,226,128
10. Statoil Gulf of Mexico LLC*	Keathley Canyon 698	>2,000	14,289,000

*Denotes the submitter.

Source: U.S. Minerals Management Service

Alabama, Alaska, California, Louisiana, Mississippi, Texas, and their coastal local governments will receive money authorized under the 2005 Energy Policy Act (EPACT).

Funds are allocated based on qualified OCS revenues generated off each state's coast.

Portions of the money will go directly to 67 coastal political subdivisions, DOI said.

Future OCS revenue-sharing

In July 2009, legislation was introduced that would guarantee coastal states a 37.5% share of federal revenue from new oil and gas production off their shores.

That share of rents, royalties, and bonuses would match the share that Alabama, Louisiana, Mississippi, and Texas receive under the 2006 Gulf of Mexico Energy Security Act.

It also would permit leasing of the Destin Dome area near the Florida Panhandle.

Utah leases

In February 2009, DOI canceled 77 federal oil and gas leases in Utah that were had been sold in December 2008. The agency concluded that they were too close to Arches National Park, Dinosaur National Monument and Nine-Mile Canyon.

Wyoming BLM sale

DOI ordered the BLM to remove eight tracts from a February 2009 oil and gas lease sale based on concerns from the governor of Wyoming about the proximity of three parcels to Shoshone National Forest and of five to the Jack Morrow Hills Coordinated Activity Plan boundary.

The lease sale of 137 remaining parcels proceeded as planned.

2010–2015 OCS plan

The MMS released in mid-January 2009 its Draft Proposed Program for 2010–2015, regarding leasing on the U.S. Outer Continental Shelf (OCS). That program was developed by the Bush Administration.

In February 2009, the DOI said that it would:

- extend the public comment period on the plan by 180 days
- assemble a detailed report on conventional and renewable offshore energy resources
- hold four regional conferences to review these findings
- expedite renewable energy rulemaking for the OCS.

EOR using CO_2

U.S. overview

In a late-2009 assessment, *OGJ* pointed to the likely wider U.S. use of carbon dioxide injection for enhanced oil recovery (EOR) as new CO_2 supplies become available.

CO_2 injection in the United States has accounted for recovering about 1.5 billion bbl of oil, and U.S. CO_2 sales for EOR reached an estimated 3 bcfd in 2008.

A 2008 report estimated that another 87.1 billion bbl could be technically recovered using CO_2 injection.

CO_2 source fields provide about 83% of the gas injected in North American EOR projects. Other sources are natural gas plants (10.6%), fertilizer plants (1.4%), coal gas synfuels (4.9%), and ethylene production (0.1%).

US CO_2-EOR SALES

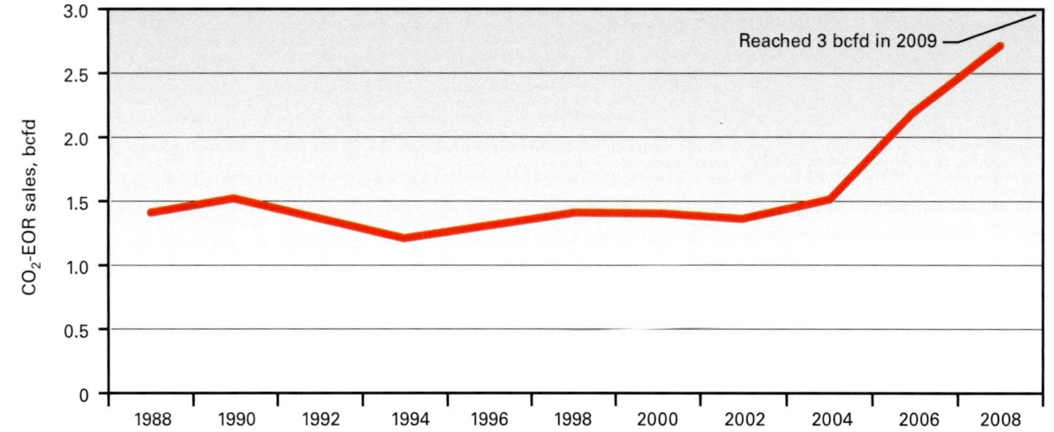

Source: Doll, T., Evans, T., and Melzer, L.S. "North American CO_2 Status," Third Annual CO_2 Conference, Enhanced Oil Recovery Institute, Casper, WY, June 23–24, 2009.

CO_2 SOURCES, PIPELINES

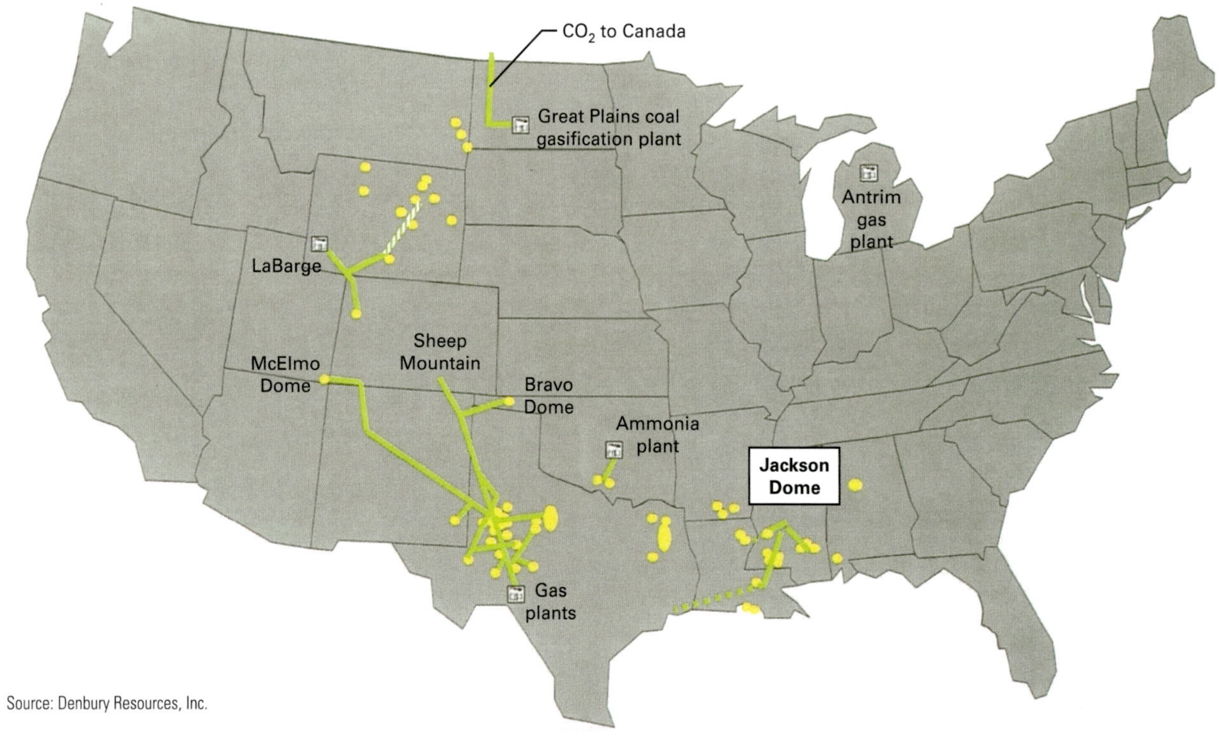

Source: Denbury Resources, Inc.

U.S. CO_2 TECHNICALLY RECOVERABLE RESERVES

	OOIP	Conventionally recoverable	Recoverable with CO_2-EOR
		Billion bbl	
1. Alaska	67.3	22.3	12.4
2. California	83.3	26.0	6.3
3. Gulf Coast (Ala., Fla., Miss. Ala.)	44.4	16.9	7.0
4. Midcontinent (Okla., Ark., Kan., Neb.)	89.6	89.6	10.7
5. Illinois, Michigan	17.8	6.3	1.2
6. Permian (West Tex., NM)	95.4	33.7	17.8
7. Rockies (Colo., Utah, Wyo.)	33.6	11.0	4.2
8. Texas (east-central)	109.0	35.4	17.6
9. Williston (Mont., ND, SD)	13.2	3.8	2.5
10. Louisiana offshore	28.1	12.4	5.8
11. Appalachia (W.Va., Ohio, Ky., Pa.)	14.0	3.9	1.6
Total	**595.7**	**195.7**	**87.1**

Source: Storing CO_2 with Enhanced Oil Recovery, DOE/NETL-402/1312/01-070-08, Advanced Resources International Inc., Feb. 7, 2008.

DENBURY GULF COAST CO_2 PROJECTS

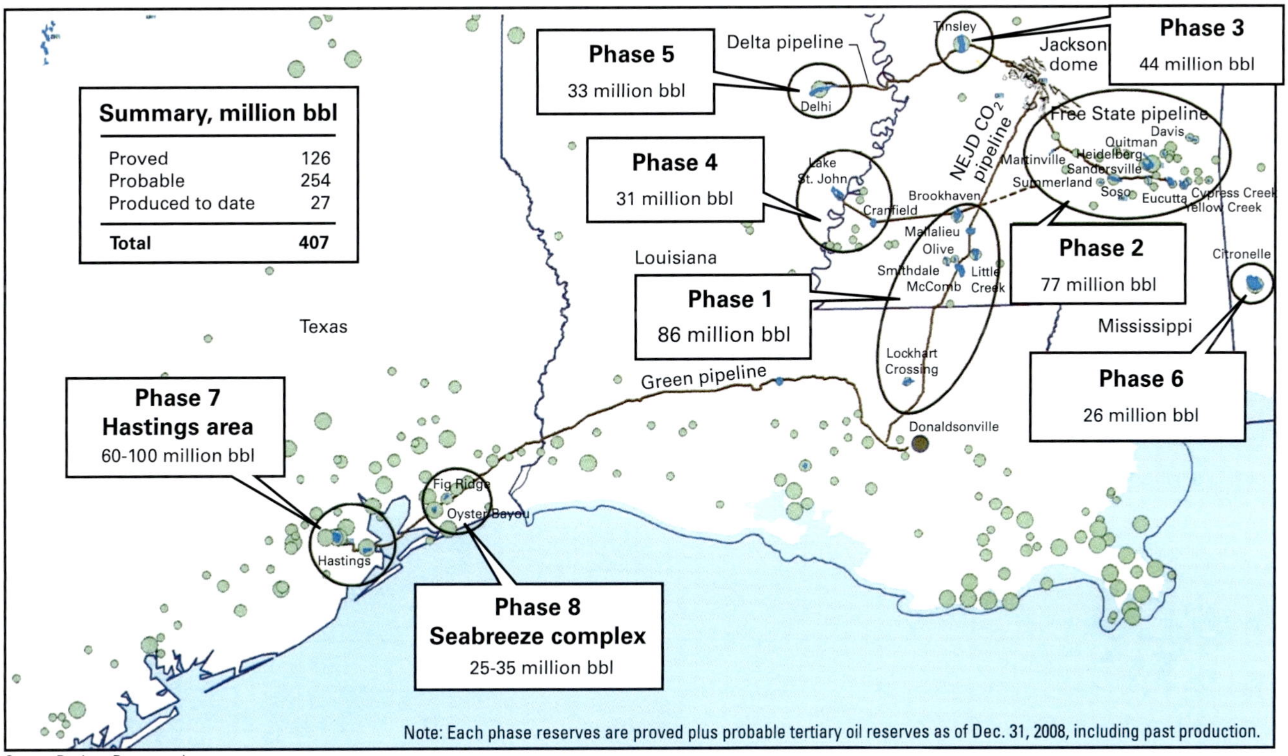

Source: Denbury Resources Inc.

Gulf Coast

Denbury Resources Inc. continued its phased expansion of CO_2 projects. During third quarter 2009, the company reported total production of 24,347 bo/d from all of its CO_2 projects.

From its Gulf Coast CO_2 EOR operations, Denbury expects production to increase to 55,000–65,000 bo/d by 2016.

Also in 2009, the company completed the Delta pipeline and commenced CO_2 injection at Delhi field. It expects first tertiary oil production from Delhi in mid-2010.

Denbury also plans to drill several delineation wells in 2010 to prove probable CO_2 reserves at its Jackson Dome holding in Mississippi. It estimates Jackson Dome reserves at 5.6 tcf proved, 3 tcf probable, and 2 tcf possible.

Rockies

Since 1986, most CO_2 for EOR in the northern Rockies has come from the Wyoming Shute Creek gas processing plant of ExxonMobil Corp. In 2010, a $72-million expansion is to be completed at the plant.

DENBURY QUARTERLY TERTIARY OIL PRODUCTION

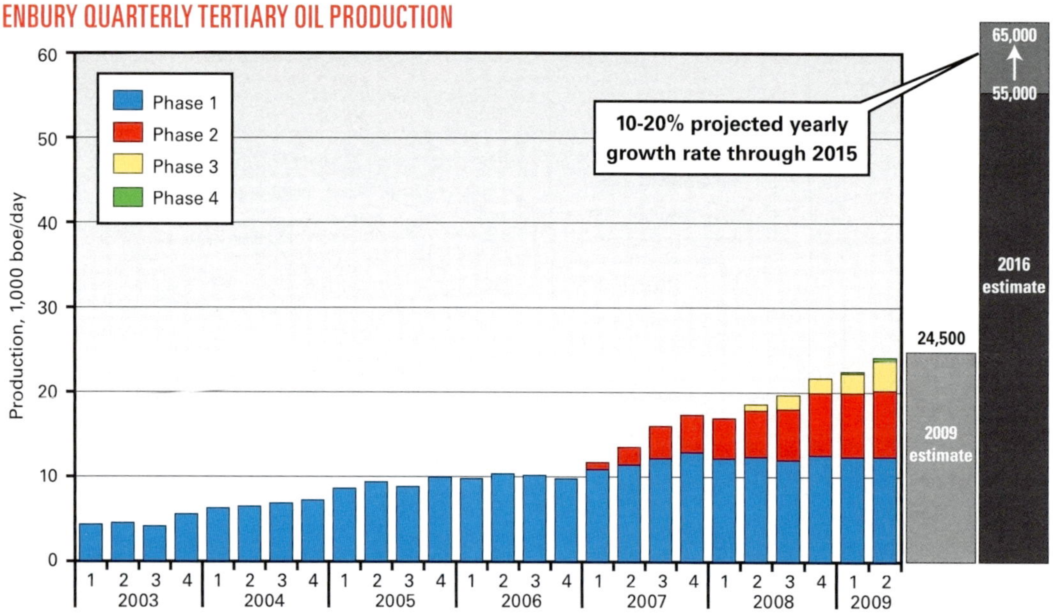

Source: Denbury Resources Inc.

POTENTIAL JACKSON DOME CO_2 SUPPLY

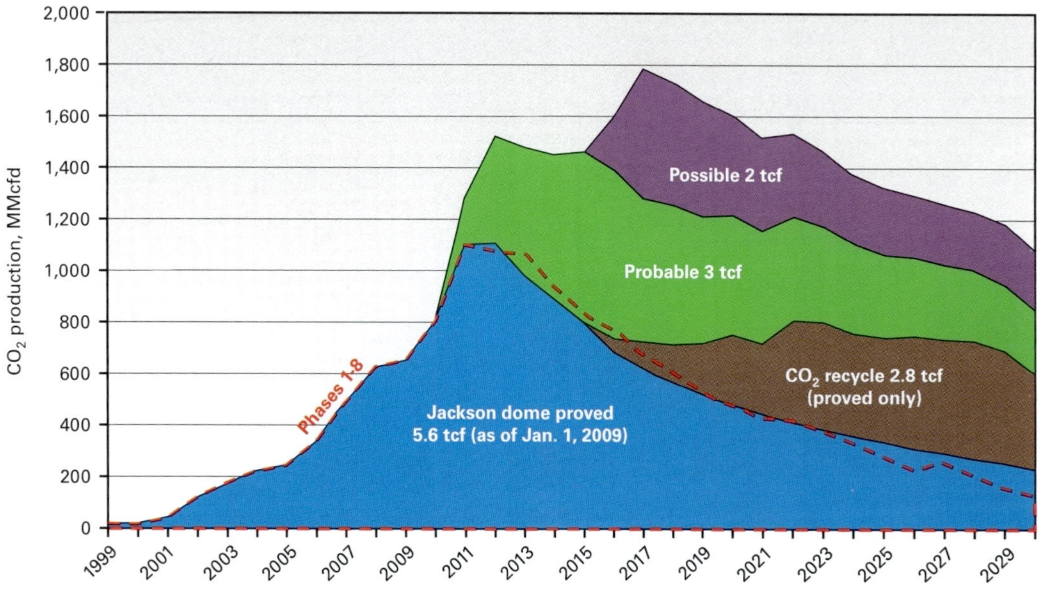

Source: Denbury Resources Inc.

In mid-2009, Encore Acquisition Co. said it plans to buy 50 MMcfd of CO_2 for its Bell Creek EOR project in Montana from the ConocoPhillips-operated Lost Cabin gas plant in Fremont County, WY.

The company would reactivate 275 wells at Bell Creek and drill as many as 75 more.

{Editor's note: In November 2009, Denbury signed a merger agreement with Encore.}

Encore estimates potential recoverable oil from its CO_2-EOR projects total 30 million bbl from Bell Creek, 37 million bbl from Elk basin, 61 million bbl from South Pine, and 136 million bbl from other fields in the Cedar Creek anticline.

Two potential new sources for CO_2 in Wyoming are from a proposed underground coal-gasification project in the Powder River basin and a coal gasification and liquefaction project near Medicine Bow.

And at La Barge, WY, ExxonMobil plans to start up a $100-million pilot project in 2010 to demonstrate its controlled freeze zone single-step process for CO_2 separation.

CO_2-EOR: Permian basin

Wells in the Permian Basin produce about 180,000 bo/d from the use of CO_2 for EOR.

ENCORE CO_2 PROJECTS

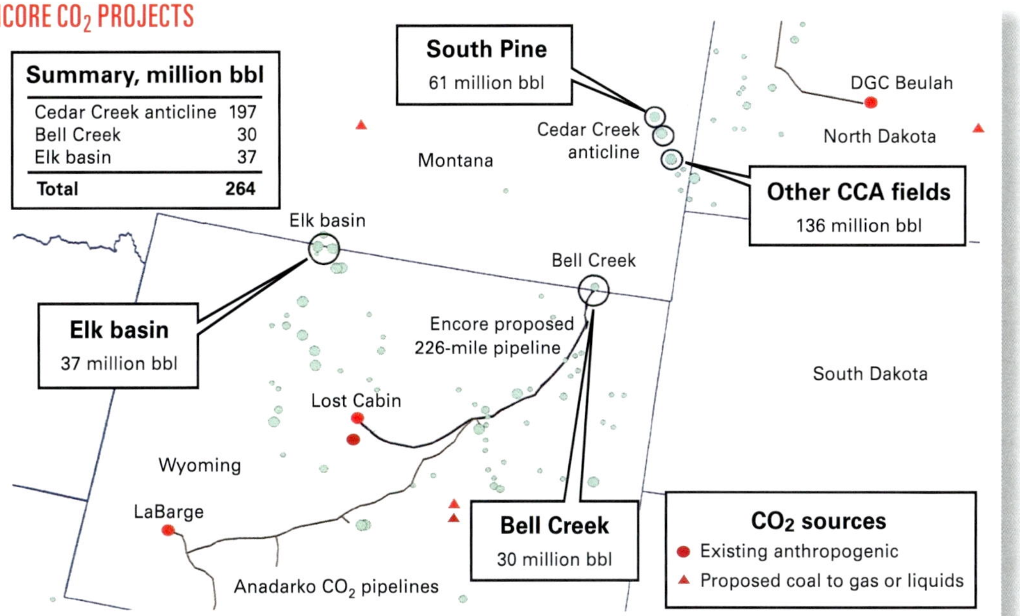

Note: Probable and possible reserves estimates as of Dec. 31, 2008.

Source: Denbury Resources Inc.

PERMIAN BASIN CO_2-EOR

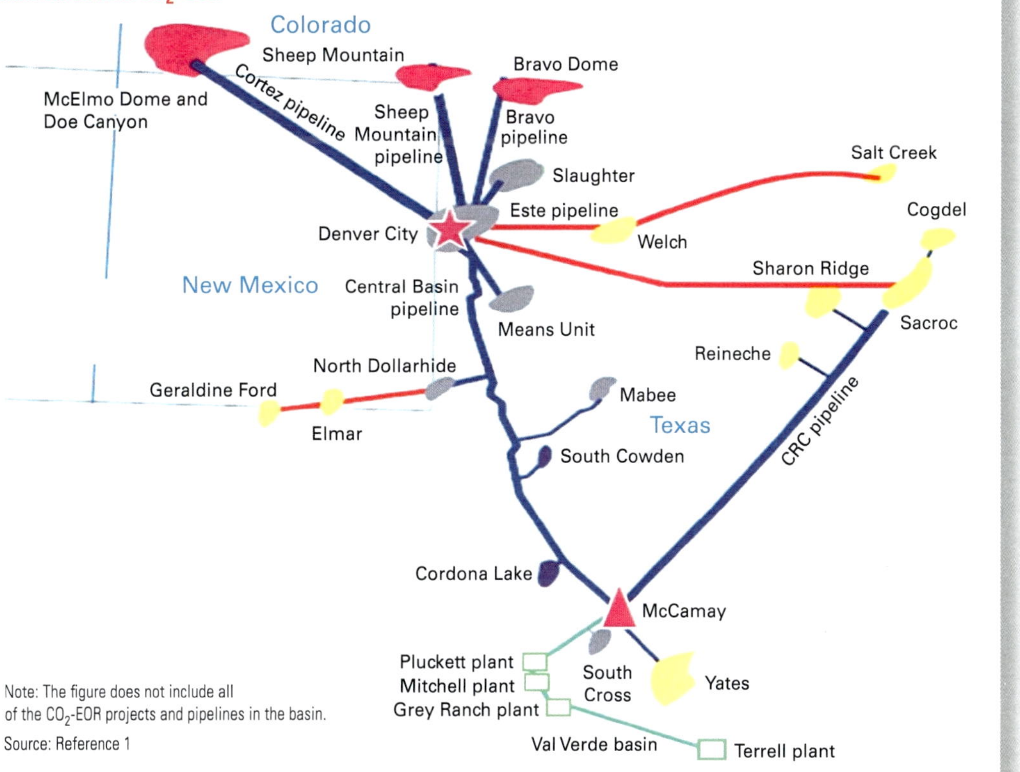

Note: The figure does not include all of the CO_2-EOR projects and pipelines in the basin.

Source: Reference 1

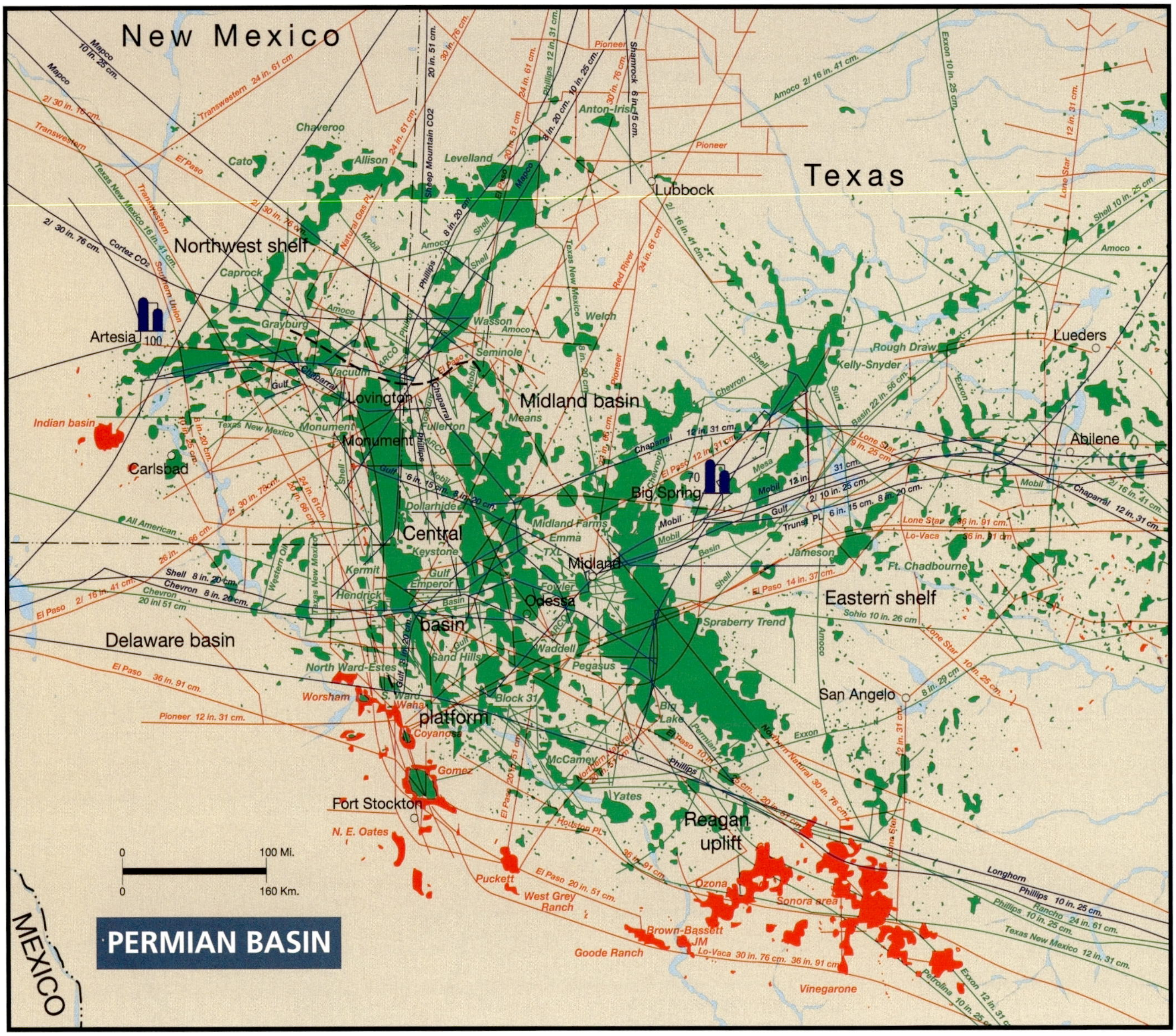

Kentucky

Projects to sequester carbon dioxide and test its use in enhanced oil recovery were under way in early 2009 in western Kentucky.

As much as 8,000 tons of CO_2 was to be injected for six months into the Mississippian Jackson sandstone at 1,870 ft (570 m) in a well in Sugar Creek field in Hopkins County.

Three producing wells were being monitored for changes in production as the result of the CO_2 injection. The geological surveys of Kentucky and Illinois are cooperating in the project.

Also, the Kentucky Geological Survey spud a well in April in Hancock County; CO_2 was to be injected into the well to test the capacity of various formations to serve as permanent storage for the substance.

Drilling and Production

API: New hydraulic frac guidelines

The American Petroleum Institute published a new guidance document in late 2009 outlining industry best-practices for properly drilling and cementing wells that are being hydraulically fractured.

The aim is to ensure that shallow groundwater aquifers and the environment are protected.

Oil-gas price disconnect

In a late 2009 analysis, *OGJ* noted a disconnect between the prices of crude oil and natural gas.

The shrinking share of gas wells and the consequent rise in the share of oil wells in total drilling was a reflection of oil's detachment from gas on a Btu-value basis.

As the price of oil rebounded after its precipitous drop in second-half 2008, it left natural gas prices far behind.

By September 2009, the price of oil had soared to more than $80/bbl on the NYMEX, but gas futures had not moved much higher than $5/MMbtu since the start of 2009.

On a Btu-equivalent basis, oil was trading at three times the price of gas—a very unusual situation.

Oil rig counts and crude prices*

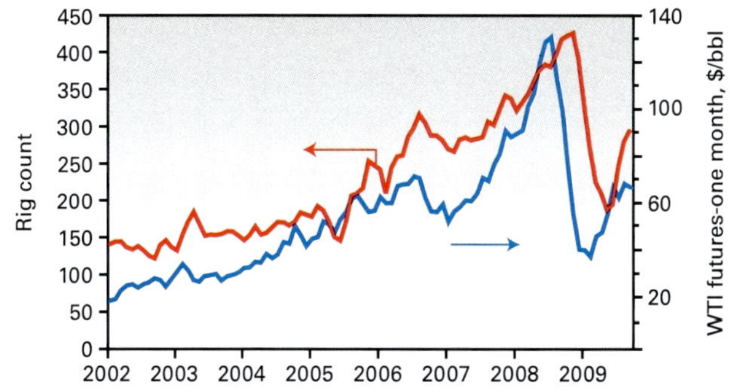

*Data through September 2009. Source: Baker Hughes, Wall Street Journal

Weekly rig counts*

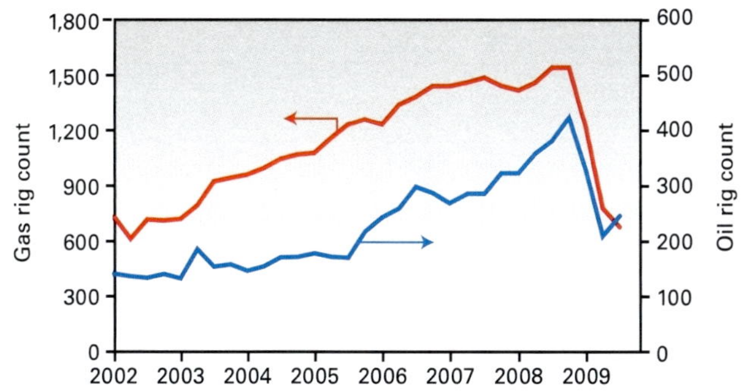

*Data through September 2009. Source: Baker Hughes

Drilling picture: Mid-2009

In a mid-2009 review, *OGJ* concluded that drilling in the United States and Canada could recover modestly by year-end if oil and gas prices stabilized or improved.

How U.S., Canada drilling compare

Average number of active rigs per week,year to date

	Jan. 1–June 27, 2008	Jan. 1–June 26, 2009
Alabama	4.1	3.2
Alaska	7.4	8.5
Arkansas	44.0	47.0
California	38.1	23.5
Land	36.9	22.5
Offshore	1.2	1.0
Colorado	114.0	59.5
Florida	0.1	0.5
Kansas	10.9	17.8
Kentucky	10.0	9.4
Louisiana	149.5	147.0
N. land	52.2	78.2
S. inland waters	20.0	6.4
S. land	25.5	16.5
Offshore	51.9	45.9
Michigan	0.8	0.0
Mississippi	11.3	11.1
Montana	11.3	2.2
Nebraska	0.1	0.2
Nevada	3.2	4.5
New Mexico	73.6	41.8
New York	6.3	2.0
North Dakota	58.7	49.2
Ohio	11.8	7.9
Oklahoma	203.1	109.1
Pennsylvania	20.1	28.4
South Dakota	1.7	0.2
Tennessee	4.8	2.4
Texas	886.8	469.2
Offshore	8.8	3.9
Inland waters	2.5	0.3
Dist. 1	23.9	10.5
Dist. 2	34.3	19.6
Dist. 3	62.2	36.5
Dist. 4	91.6	44.3
Dist. 5	182.8	106.4
Dist. 6	121.1	78.1
Dist. 7B	32.0	13.4
Dist. 7C	60.2	24.3
Dist. 8	128.3	54.4
Dist. 8A	22.5	15.3
Dist. 9	40.5	25.1
Dist. 10	76.2	37.0
Utah	40.3	20.2
Virginia	4.2	3.8
West Virginia	26.4	23.5
Wyoming	71.0	43.5
U.S. total	**1,817.0**	**1,139.0**
U.S. land	**1,732.0**	**1,079.0**
U.S. offshore	**62.4**	**53.4**
Canada total	**341.2**	**210.7**

Source: Baker Hughes Inc.

OIL & GAS JOURNAL WELL FORECAST FOR 2009

State	First half 2009 Total wells	First half 2009 Exploratory wells	First half 2009 Field wells	Full year 2009 Total wells	Full year 2009 Exploratory wells	Full year 2009 Field wells	Full year 2009 Total footage (1,000)
Alabama	175	9	166	360	17	343	1,632
Alaska	60	8	52	125	13	112	848
Arkansas	415	71	344	880	147	733	6,186
California land	895	31	864	1,915	61	1,854	4,927
California offshore	3	–	3	7	–	7	44
Colorado	765	200	565	1,555	400	1,155	10,215
Florida	1	–	1	1	–	1	12
Illinois	130	39	91	295	84	211	726
Indiana	63	19	44	133	40	93	212
Kansas	1,829	132	1,697	3,914	274	3,640	13,656
Kentucky	414	14	400	882	26	856	2,487
Louisiana	1,169	116	1,053	2,534	281	2,253	24,475
North	835	88	747	1,885	228	1,657	17,679
South	122	6	116	235	11	224	2,367
Offshore	212	22	190	414	42	372	4,429
Michigan	177	28	149	365	55	310	626
Mississippi	105	10	95	200	18	182	1,866
Montana	75	10	65	145	18	127	980
Nebraska	17	4	13	36	8	28	184
Nevada	1	–	1	2	1	1	14
New Mexico - East	255	16	239	565	34	531	4,710
New Mexico - West	205	3	202	448	6	442	2,808
New York	20	1	19	43	1	42	171
North Dakota	244	50	194	520	98	422	6,544
Ohio	375	33	342	720	61	659	3,150
Oklahoma	1,072	47	1,025	2,214	93	2,121	17,048
Pennsylvania	2,700	273	2,427	5,724	595	5,129	20,360
South Dakota	4	–	4	8	1	7	32
Tennessee	46	13	33	98	28	70	223
Texas	4,045	357	3,688	8,208	708	7,500	73,062
Dist. 1	131	11	120	255	21	234	1,700
Dist. 2	230	34	196	470	67	403	4,560
Dist. 3	277	34	243	590	70	520	5,296
Dist. 4	340	31	309	625	55	570	6,506
Dist. 5	491	17	474	966	31	935	11,367
Dist. 6	560	67	493	1,135	136	999	12,813
Dist. 7-B	255	11	244	525	22	503	3,418
Dist. 7-C	320	16	304	650	30	620	5,029
Dist. 8	432	25	407	890	52	838	6,268
Dist. 8-A	312	30	282	655	63	592	3,689
Dist. 9	378	12	366	797	27	770	5,181
Dist. 10	297	63	234	610	124	486	6,805
Offshore	22	6	16	40	10	30	430
Utah	277	52	225	585	109	476	4,631
Virginia	240	25	215	495	53	442	1,342
Washington	1	1	–	1	1	–	15
West Virginia	865	104	761	1,790	220	1,570	8,302
Wyoming	955	39	916	2,020	83	1,937	10,779
U.S. total	**7,598**	**1,705**	**15,893**	**36,788**	**3,534**	**33,254**	**222,267**
Western Canada	**4,305**	**903**	**3,402**	**9,854**	**2,037**	**7,817**	**43,759**
Alberta	3,114	604	2,510	6,515	1,244	5,271	26,874
Saskatchewan	650	122	528	2,150	404	1,746	9,333
Brit. Columbia	469	168	301	1,014	367	647	6,862
Manitoba	72	9	63	175	22	153	690
NWT + Yukon Terr.	4	3	1	9	–	9	57
Eastern offshore	3	1	2	7	2	5	47
Eastern land	42	5	37	95	10	85	192

US RIG COUNT TRENDS: 1999-2008*

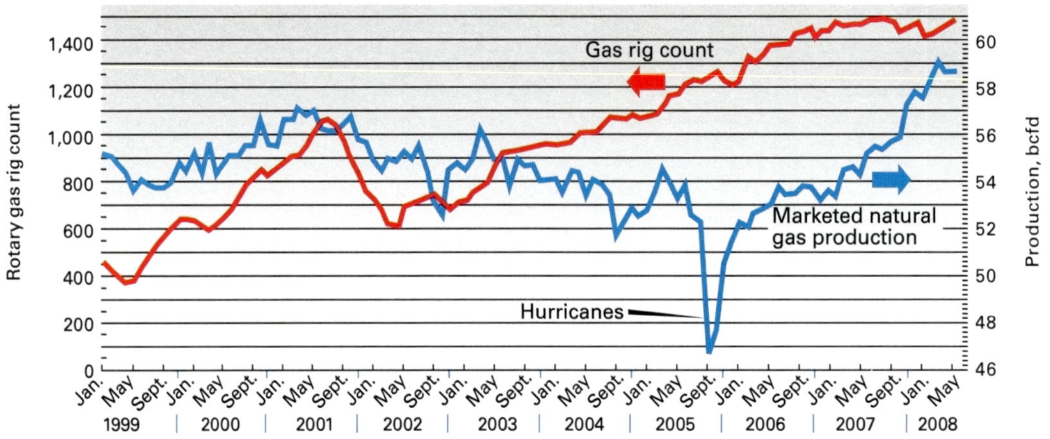

*Average monthly rig count. Sources: Baker Hughes Rig Count, EIA

Highlights of *OGJ's* analysis:

- Operators will drill 36,788 wells in the United States in 2009, down from an estimated 52,097 in 2008.
- All operators will drill 3,534 exploratory wells of all types, down from an estimated 5,474 in 2008.
- The Baker Hughes count of active U.S. rotary rigs will average 1,235 rigs/week in 2009, down from 1,867 in 2008 and 1,768 in 2007.
- Operators will drill 9,854 wells in western Canada, down from an estimated 16,290 in 2008.

A LOOK AT 30 YEARS OF US WELL COMPLETIONS

Year	Total wells[1]	Total footage	Total exploratory wells
[2]2009	43,384	268,721,000	4,291
[2]2008	52,097	325,566,000	5,474
[2]2007	47,057	279,001,000	3,833
2006	49,375	289,959,000	3,696
2005	44,679	254,844,000	3,727
2004	39,051	213,908,000	3,192
2003	30,487	158,221,000	2,593
2002	27,794	145,055,000	2,271
2001	36,061	184,462,000	3,181
2000	31,261	149,848,000	2,517
1999	22,107	109,854,000	2,141
1998	25,822	143,625,000	2,723
1997	30,208	165,480,000	3,353
1996	25,724	138,588,000	3,364
1995	23,061	124,426,000	3,406
1994	23,324	130,654,000	3,788
1993	26,032	138,509,000	3,604
1992	23,921	123,456,000	3,494
1991	28,417	141,391,000	4,399
1990	30,615	149,518,000	5,074
1989	28,363	134,901,000	5,251
1988	32,238	155,164,000	6,350
1987	36,253	163,848,000	6,903
1986	39,015	177,641,000	7,156
1985	70,806	316,778,000	12,208
1984	84,983	368,796,000	15,138
1983	75,738	316,617,464	13,845
1982	83,889	375,382,919	15,882
1981	89,234	406,520,453	17,430
1980	69,486	311,444,837	12,870

[1]Well counts in most recent years subject to reporting lag. [2]Estimated.
Source: 1975–2006 American Petroleum Institute.

New Colorado oil, gas rules

In 1999, Colorado issued 1,000 drilling permits. In 2008, that figure was more than 8,000.

Officials in Colorado said in April 2009 that new oil and gas regulations would allow the industry to continue to grow in a sustainable way, compatible with the state's economy.

Producers remained concerned that the rules will create delays and expenses.

Topics covered by some of the significant new regulations include:

- Maintenance of an inventory by wellsite of chemicals and fuel used in well operations
- Development of comprehensive drilling plans
- Special protection for public water systems
- Assessment of the impact on groundwater of coalbed methane operation
- Control of odors released in drilling operations
- Assessment of effects on wildlife.

Oklahoma: Oil opportunity

The Oklahoma Geological Survey said in early 2009 that the state's thousands of oil fields will ultimately yield only 19% of the more than 84 billion bbl of original oil in place at present decline rates.

It also estimated that every 1% of the remaining oil in place represents 680 million bbl of incremental recovery.

The agency said that 50% recovery is possible in the best blanket sandstone reservoirs, but that recovery was less than 30% in about half the reservoirs studied.

International Activities

U.S. gains from Canadian oil sands

Canadian oil sands development could lead to more than 342,000 new U.S. jobs, said a study by the Canadian Energy Research Institute for the American Petroleum Institute, titled *Canada's Oil Sands and Economic Impact on the USA*.

Demand for U.S. goods and services used for Canadian oil sands production could add $34 billion to the U.S. GDP in 2015 and $42 billion in 2025.

Total-Amso deal

In early 2009, France's Total SA was poised to acquire a 50% interest in American Shale Oil LLC (Amso) in a deal that would enable it to help commercialize Colorado oil shale.

Amso is a unit of IDT, a multinational holding company.

Total and Amso are to pursue joint development of shale oil on Amso's federal research lease acreage in the western part of the state.

Total-Chesapeake deal

Total SA said in early 2009 that it will pay up to $2.25 billion for a 25% interest in the assets of Chesapeake Energy Corp. in the gas-bearing Barnett shale.

BP and Statoil signed similar deals with Chesapeake since mid-2008.

MMS collaborations

MMS signed an MOU in May 2009 with the Norwegian Petroleum Directorate.

The parties aim to share information on resource management and exploration and production technologies and to conduct cooperative research.

In August, MMS also signed an MOU with India's Ministry of Petroleum and Natural Gas Directorate General of Hydrocarbons.

The agreement will aid the exchange of information on developing offshore oil, gas, and mineral resources. Topics of interest include leasing, resource evaluation, and methane hydrate research.

Foreign firms: Interest in U.S. gas

The *Financial Times* reported in August on the growing number of foreign energy companies moving to invest in independent U.S. gas exploration and development companies.

BP and BG Group (UK), Statoil (Norway), and Eni (Italy) all bought into the U.S. gas industry in the preceding 12 months.

Gas Shales: General

New gas shale plays

Advanced Resources International Inc. (ARI) pointed to several new gas shale plays emerging in North America and noted the environmental and land-use issues that will affect their commercial development.

US UNCONVENTIONAL GAS PRODUCTION

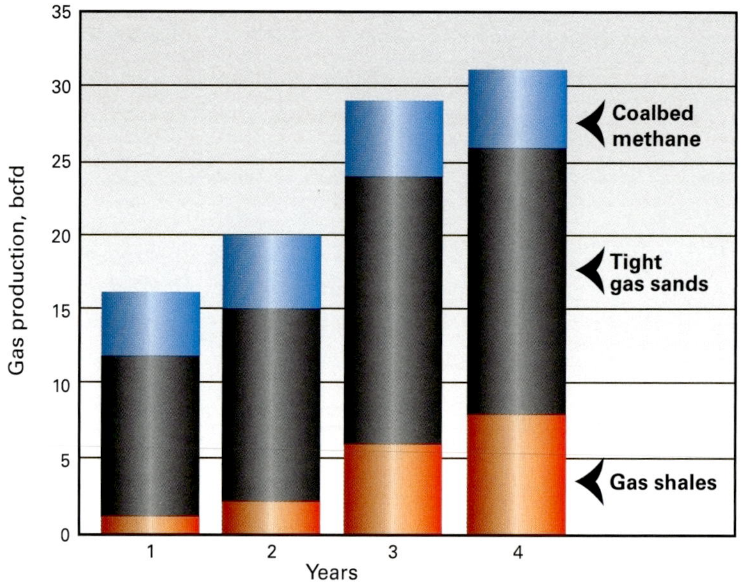

Source: Advanced Resources International Inc.

US SHALE GAS REGIONS

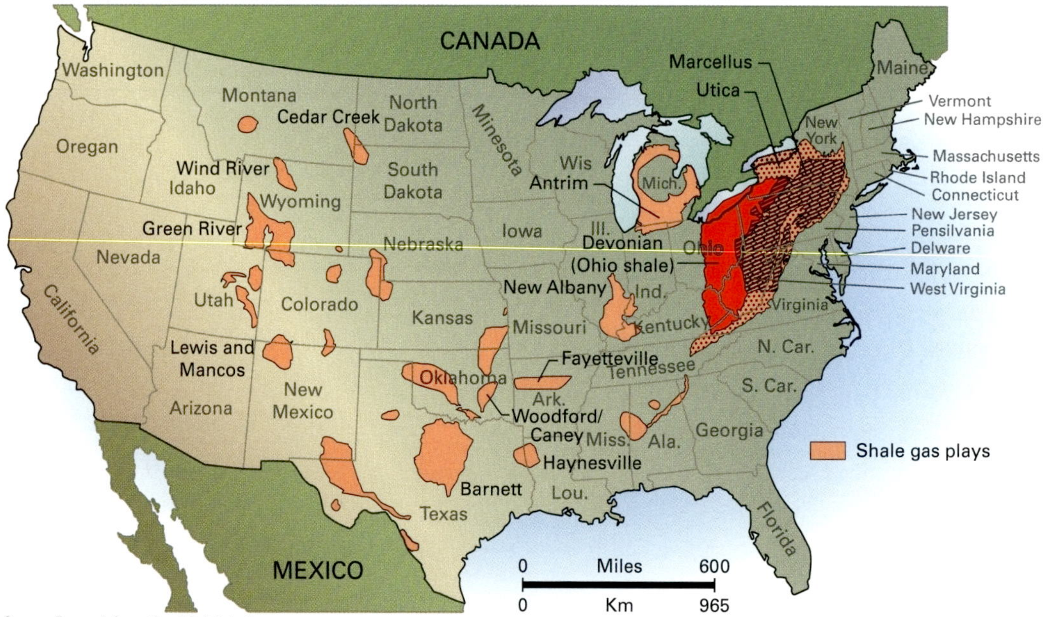

Source: Energy Information Administration

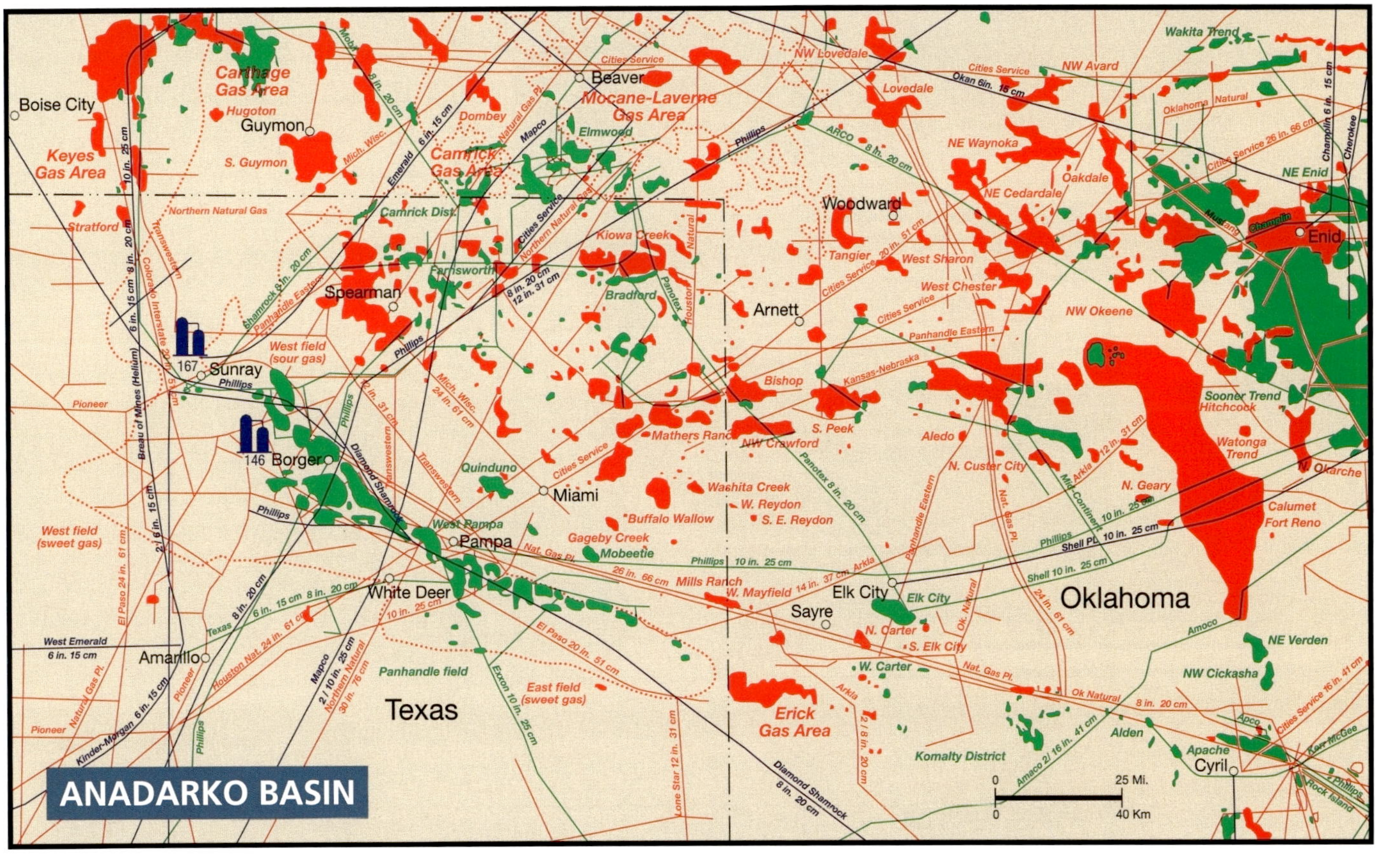

MID-2009 GAS SHALE PRODUCTION

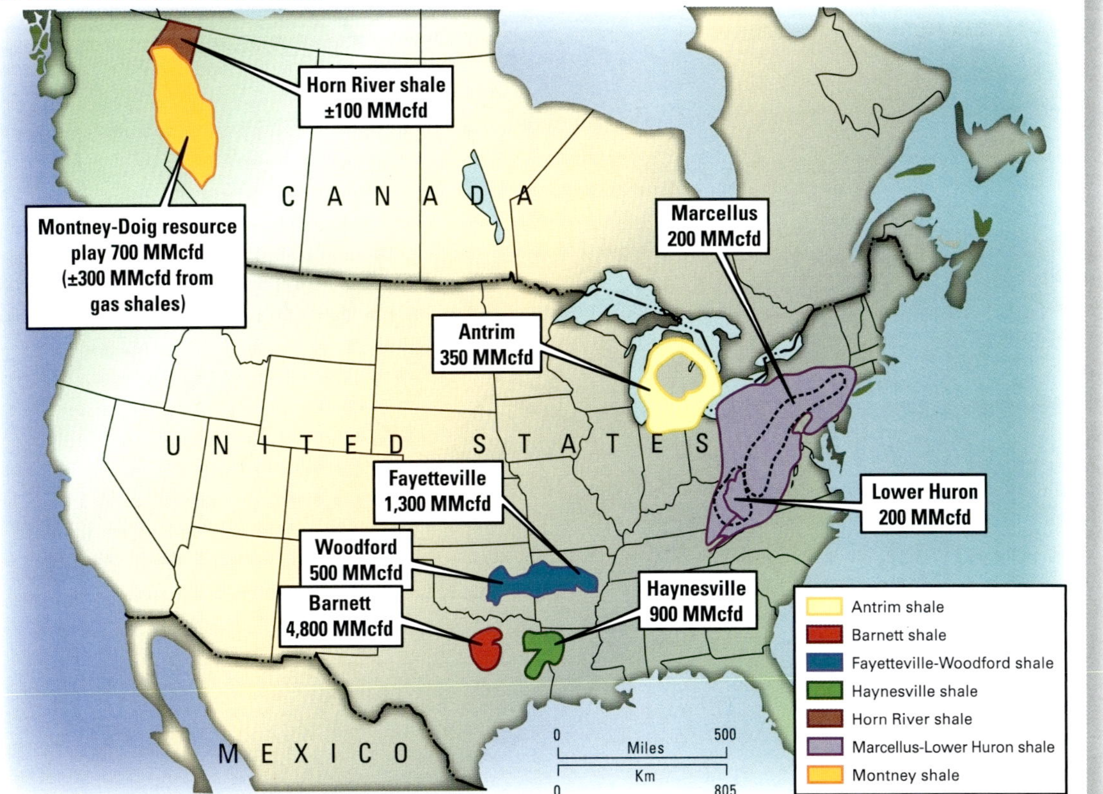

Source: Advanced Resources International Inc.

Eastern US gas shale plays

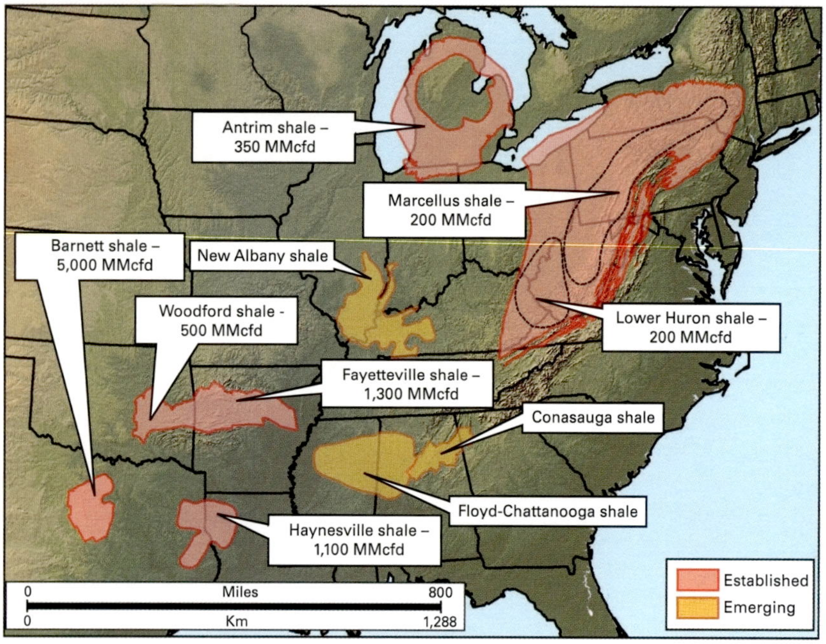

Led by the Barnett (5 bcfed), Fayetteville (1.3 bcfed), and fast-rising Haynesville (1.1 bcfed), shale reservoirs contributed more than 8 bcfd to North American gas supplies in 2009. That figure was just 2 bcfd in 2004.

ARI called two plays potentially noteworthy: the Eagle Ford shale in South Texas, and the Woodford-Anadarko (also called the Cana) shale in western Oklahoma.

Political pressure in recent years has raised calls for more government oversight of hydraulic fracturing in particular and gas shale development in general, due to concerns about environmental impacts.

Eagle Ford gas shale play

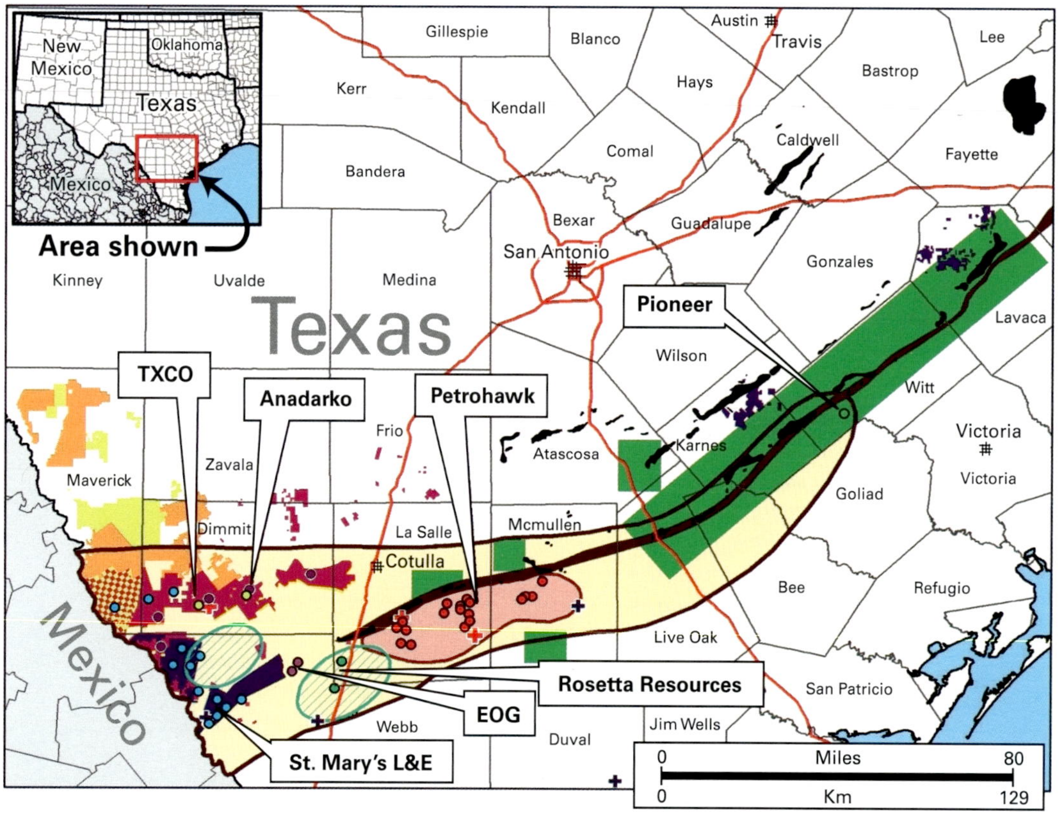

Emerging U.S. Gas Shale Plays

Region	Shale play	Basin	Geologic age	Development stage
Midcontinent	Eagle Ford	Maverick	Cretaceous	Evaluation
Midcontinent	Woodford	Anadarko	Mississippian-Silurian	Evaluation
Midcontinent	Bend	Palo Duro	Pennsylvanian	Exploration
Southwest	Barnett	Permian	Mississippian	Exploration
East	Lower Huron	Appalachian	Devonian	Development
East	Utica	Appalachian	Ordovician	Exploration
Gulf Coast	Chattanooga, Floyd	Appalachian	Mississippian	Exploration
Gulf Coast	Conasauga	Appalachian Thrust Belt	Cambrian	Exploration
Gulf Coast	Tuscaloosa Marine	Gulf Coast	Cretaceous	Exploration
Rockies	Mancos, Manning Canyon	Uinta	Cretaceous	Exploration
Rockies	Pierre, Mancos, Niobrara	Raton	Cretaceous	Exploration
Rockies	Gothic, Hovenweep	Paradox	Pennsylvanian	Evaluation
Rockies	Baxter	Vermillion	Cretaceous	Evaluation
California	Monterey, McClure	San Joaquin	Miocene	Exploration

Deep Shale Fracturing Mixture

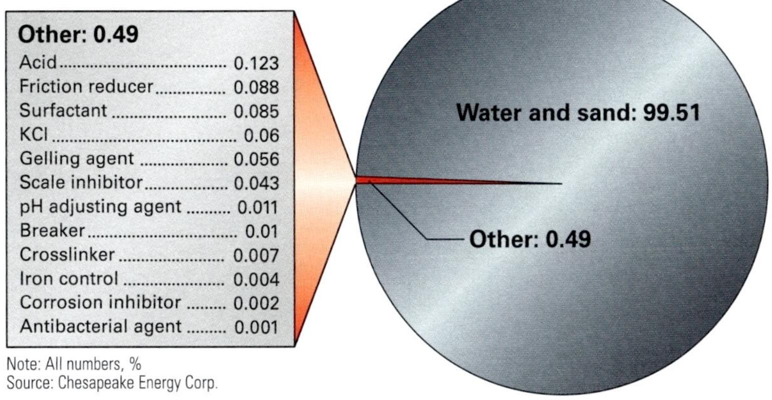

Note: All numbers, %
Source: Chesapeake Energy Corp.

Chesapeake report

In fourth quarter 2009, Chesapeake Energy Corp. reported record gross operated production from its four main U.S. shale gas plays.

The company reported exceeding 1 bcfd in from the Barnett shale, reaching 500 MMcfd from the Haynesville shale, topping 400 MMcfd from the Fayetteville shale, and attaining 100 MMcfd from the Marcellus shale.

In early 2009, Chesapeake said the Haynesville has the potential to become the largest producing field in the United States.

Gas Shales: Marcellus

Newfield-Hess

Newfield Exploration Co. and Hess Corp. announced plans to start drilling in 2010 on a JV exploration project to exploit gas in Devonian Marcellus shale in northeastern Pennsylvania, in Susquehanna and Wayne counties.

Ultra Petroleum

Ultra Petroleum Corp. started production from the Pennsylvania Marcellus shale in mid-2009 at a combined 13 MMcfd from two wells in Tioga County.

It planned to drill more than 32 horizontal Marcellus wells by the end of 2009, and said it has the acreage and resources to drill more than 100 horizontal Marcellus wells in 2010.

ECA

In May 2009, Energy Corp. of America began a 3-year program to develop gas production from Devonian Marcellus shale in Greene County in southwestern Pennsylvania.

It holds 10,000 acres it holds in the county and was already producing 8 MMcfd of gas from wells tapping the Marcellus at a depth of about 8,000 ft (2,438 m).

ECA has drilled 155 Marcellus vertical and horizontal Marcellus wells, and initial potentials of the completed wells have ranged from 500 Mcfd to 3 MMcfd.

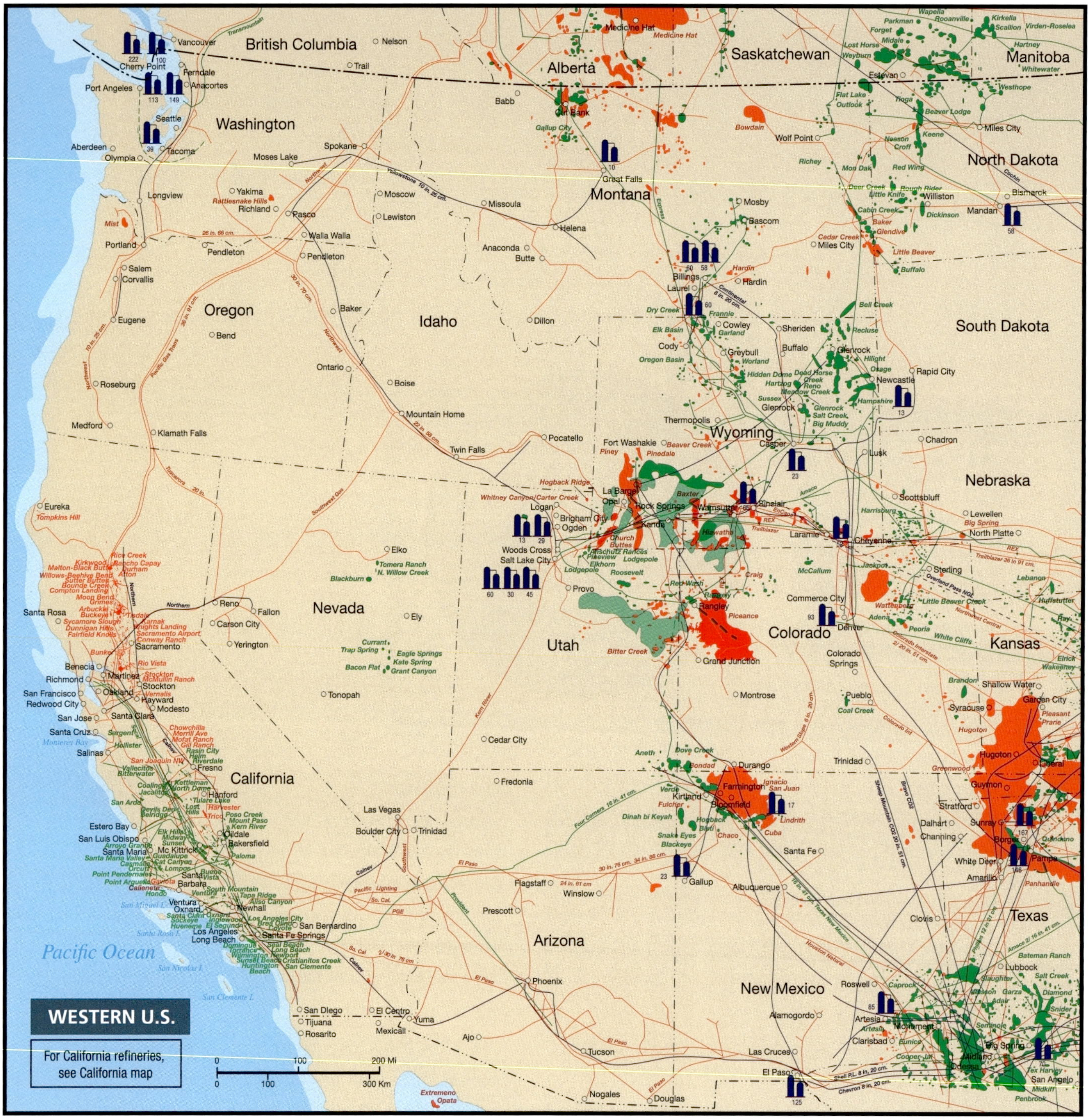

Cabot Oil & Gas

Cabot Oil & Gas Corp. was producing 39 MMcfd of Marcellus shale gas from seven horizontal and 20 vertical wells in northeastern Pennsylvania in mid-2009.

Cumulative recovery is 5.8 bcf since the company's first Marcellus well came on line in 2008, in Susquehanna County.

The most recent horizontal completion, Teel-8H, flowed at a 30-day average of 9.8 MMcfd.

The company planned to drill 18 more horizontal wells by yearend.

Cabot also drilled the vertical Teel-6 well in 2009 with initial flow of 4.2 MMcfd. The $1.4-million well was the first at which the company stimulated the entire column.

Gas Shales: Eagle Ford

Eagle Ford well test

St. Mary Land & Exploration Co. completed its first operated horizontal Cretaceous Eagle Ford shale well in South Texas, where it is running two rigs.

The Briscoe G 1-H in Webb County averaged 4.4 MMcfd of gas and 192 b/d of condensate in seven days (facility constrained). The well took 10 frac stages in a 3,200-ft (975-m) lateral at 7,500 ft (2,286 m) TVD.

Gas Shales: Haynesville/Deep Bossier

BG-Exco: Haynesville

In June 2009, the UK's BG Group made its first move into the U.S. shale gas arena, investing about $1.3 billion in a joint venture with Texas-based Exco.

BG took a 50% stake in 120,000 acres covering gas fields in north Louisiana and east Texas. The majority of the acreage is in the Haynesville shale gas formation.

EnCana: Haynesville

EnCana Corp. announced a $580-million program to drill 50 Haynesville shale gas wells in East Texas and North Louisiana in 2009.

The company said that the 2009 program will enable it to improve its understanding of the play, further evaluate its lands, and retain prospective acreage.

EnCana has committed to supply 150 MMcfd on the proposed Gulf South pipeline expansion and 500 MMcfd on the proposed ETP Tiger pipeline.

Petrohawk: Haynesville

In early 2009, Petrohawk Energy Corp., said it was producing a gross 160 MMcfd of gas equivalent (MMcfed) from 16 operated Jurassic Haynesville shale wells in North Louisiana.

In 2009, the company expected to complete 75–80 gross wells.

LaCrosse Pipeline

Enbridge Inc. was gauging shipper interest in 2009 in its proposed LaCrosse Pipeline, to carry Haynesville shale gas from Carthage, TX, to the Sonat Pipeline southeastern Louisiana.

The 300-mi (483-km) line would interconnect with at least five interstate pipeline along its route.

Enbridge expects to complete LaCrosse by early 2012.

EnCana: Deep Bossier

EnCana reported a 50% increase in production in the Deep Bossier play in East Texas, where it averaged 409 MMcfd of production in first quarter 2009 compared with an average of 334 MMcfd for all of 2008.

Gas Processing

Mont Belvieu

Enterprise Products Partners LP was planning in 2009 to build a 75,000-b/d NGL fractionator at its Mont Belvieu, TX, complex east of Houston, boosting capacity to about 300,000 b/d.

The new unit would accommodate growing NGL volumes from production in the Rockies, the Barnett shale, and the Eagle Ford shale. Completion is expected in 2011.

Willow Creek

Williams Cos. Inc. said in late 2009 that its 450-MMcfd Willow Creek natural gas processing plant in western Colorado had achieved full processing operations, after August start-up.

Located about 25 mi (40 km) northwest of Williams' facilities in Garfield County, the plant was recovering about 20,000 b/d of NGLs.

Meeker II

Enterprise Products Partners LP started operations in early 2009 at its Meeker II natural gas processing plant in Colorado's Piceance basin.

The Meeker II expansion doubles the gas processing capacity at the complex to 1.5 bcfd, with the capability to extract as much as 70,000 b/d of natural gas liquids (NGLs).

Ten of the largest producers in the Piceance basin use the Meeker complex.

In initial operation, inlet gas volume was 750 MMcfd, with 38,000 b/d of NGL extracted.

By yearend, gas volumes were projected to reach 1.1 bcfd, with NGL production of 60,000 b/d.

Additions to U.S. Gas Processing Capacity and NGL Production Capability*

Region	Year	Processing capacity, MMcfd	NGL production (Capability, 1,000 b/d)	Ethane extraction (Capability, 1,000 b/d)
Rockies	2007	1,100	52	26
Texas inland	2007	825	51	20
Midcontinent	2007	—	—	—
Gulf Coast	2007	—	—	—
Total		**1,925**	**103**	**46**
Rockies	2008	1,520	71	35
Texas inland	2008	510	34	13
Midcontinent	2008	360	28	11
Gulf Coast	2008	295	17	7
Total		**2,685**	**106**	**67**
Rockies	2009	700	33	16
Texas inland	2009	655	48	19
Midcontinent	2009	380	31	16
Gulf Coast	2009	100	7	3
Total		**1,385**	**58**	**54**
Rockies	2010	350	30	15
Texas inland	2010	—	—	—
Midcontinent	2010	—	—	—
Gulf Coast	2010	—	—	—
Total		**350**	**30**	**15**
Rockies total, 2007-10		3,670	185	93
Texas inland total, 2007-10		1,990	132	53
Midcontinent total, 2007-10		740	59	27
Gulf Coast total, 2007-10		395	24	10
Grand total		**6,795**	**401**	**182**

*Assumes 100% utilization rates.

LNG

Floating Florida terminal

The Maritime Administration of the U.S. Dept. of Transportation in November 2009 approved plans to build a Gulf of Mexico LNG terminal offshore western Florida.

Port Dolphin Energy LLC has applied for a license to build the deepwater LNG port.

Golden Pass

Construction of ExxonMobil's Golden Pass terminal, in Texas on Sabine Pass, began in 2006.

However, the facility was damaged by Hurricane Ike in September 2008, and the company said that it would likely open in mid-2010, not in 2009 as planned.

Gulf LNG

At Pascagoula, MS, El Paso Corp. continued to build the Gulf LNG terminal in 2009.

It will have an initial sendout capacity of 1.3 bcfd from two storage tanks with combined capacity of 6.6 bcf. The company is targeting start-up for 2011.

Cameron

In June 2009, Sempra Energy's Cameron LNG terminal near Lake Charles, LA, accepted its commissioning cargo. Commercial operations were to start in third quarter.

Cameron lies 18 mi (29 km) north of the Gulf of Mexico in Hackberry, LA. It is the seventh land-based terminal for the United States.

Cameron will be able to send out up to 1.5 bcfd, and has room for expansion.

OceanWay

Woodside Petroleum Ltd. suspended in January 2009 its proposed OceanWay LNG import terminal 28 mi (45 km) off the California coast near Los Angeles.

The firm said its decision reflected rising U.S. domestic gas production and lower gas prices.

The project would have brought Australian LNG to the terminal, regasified it, and sent it ashore.

Broadwater

In April 2009, the U.S. Dept. of Commerce denied an appeal by sponsors of the planned Broadwater offshore LNG terminal in a dispute with New York State.

The decision stated Broadwater had failed to meet its burden of proof that the project "furthers the national interest" to the extent that it "outweighs the…adverse coastal effects."

Broadwater was to be the world's first permanently moored floating LNG terminal, a 1,200-ft (368-m) floating storage and regasification unit (FSRU). It was to be sited near the middle of Long Island Sound.

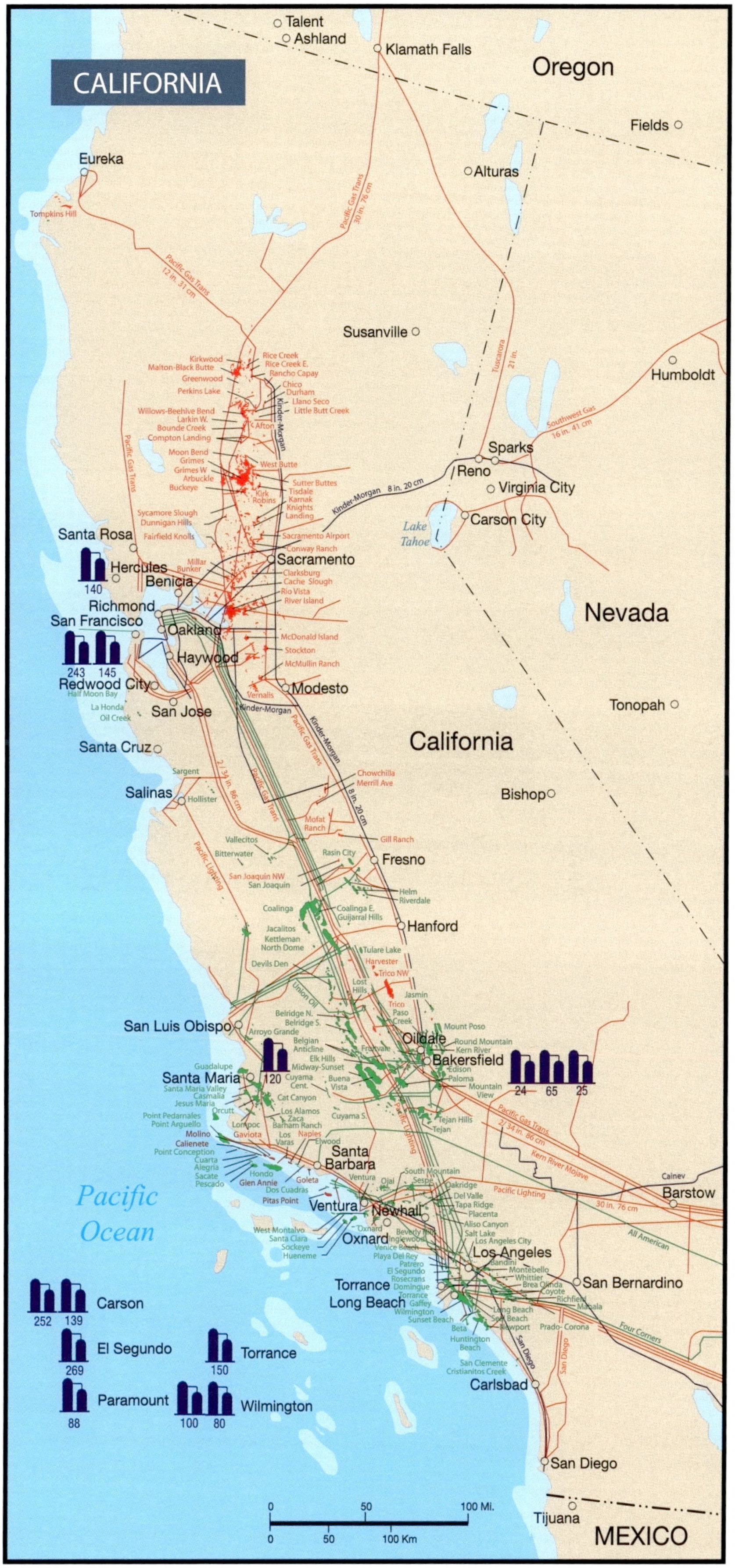

A 22-mi (35-km) pipeline was to carry vaporized LNG ashore for distribution to Long Island, New York City, and Connecticut.

Cove Point

Dominion subsidiary Dominion Cove Point LNG in early 2009 started up its 1.8-MMcfd expansion of the terminal on the Maryland shore of Chesapeake Bay.

The project expands terminal capacity by 80% and LNG storage capacity by nearly 100% to 14.6 bcf (equivalent).

New facilities include two 5.65-MMcf (160,000-cu m) LNG storage tanks, and increased vaporization and associated station facilities adding 800 MMcfd.

A subsidiary of Norway's Statoil has contracted for all the new capacity.

US NATURAL GAS IMPORTS: 1970-2007

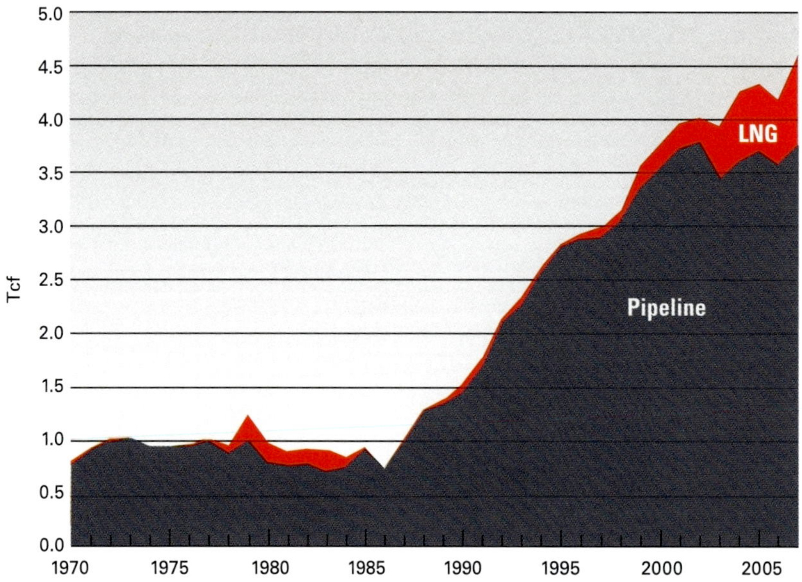

Source: US EIA, Annual Energy Review 2007

US LNG IMPORTS SINCE 1997

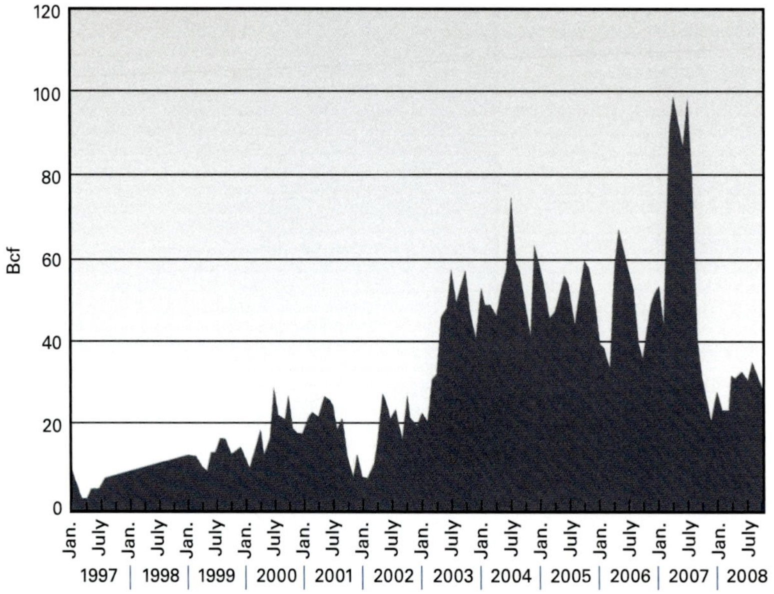

Source: US EIA; http://tonto.eia.doe.gov

US LNG IMPORTS: 3 YEARS

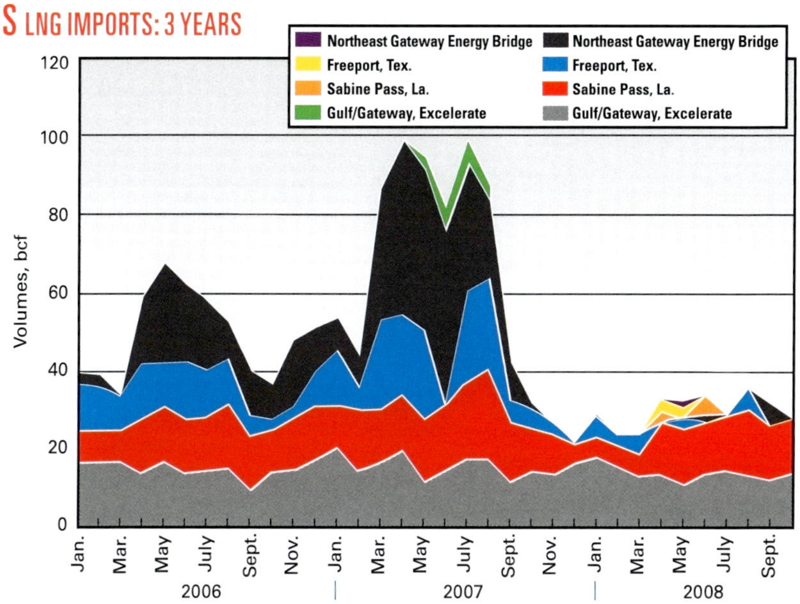

Source: Derived from DOE Office of Fossil Energy data; updated through Dec. 5, 2008

Sparrows Point

In January 2009, the U.S. FERC approved, with conditions, an LNG import terminal and connecting interstate pipeline proposed by AES Sparrows Point LNG LLC and Mid-Atlantic Express LLC.

If built, the terminal will bring 1.5 bcfd to the Northeast. The site is an industrial port southeast of Baltimore, MD.

FERC imposed more than 150 conditions meant to minimize environmental impacts.

Mid-Atlantic Express was approved to build and run a pipeline from the terminal to Eagle, PA.

WoodMac: More LNG to N. America

Imports of LNG into North America are set to increase to 4.2 bcfd in 2014 from 1.7 bcfd in 2009, according to an early-2009 report by Wood Mackenzie.

The report says North American LNG imports will rise despite increased domestic production of shale gas and the current recession that has reduced demand growth.

A significant proportion of the 82 mtpy of new liquefaction capacity coming on stream is already dedicated to Asian markets, said the study. That means no big need for more uncontracted volume in the Pacific Basin, especially as the global economic downturn depresses gas demand.

As a result, a surplus of LNG is expected to flow to the more liquid Atlantic Basin markets, which have available regas capacity.

Pipelines/Transportation

Rockies Express: Final leg

In November 2009, the final 195 mi (314 km) of the Rockies Express-East pipeline was completed, from the Lebanon (OH) Hub in Warren County to Clarington in Monroe County, OH.

Kinder Morgan Energy Partners LP said that the completion made the entire 1,679-mi (2,702-km) Rockies Express Pipeline operational, from northwestern Colorado to eastern Ohio.

Long-term firm commitments have been secured for virtually all of the pipeline's 1.8-bcfd capacity.

Kinder Morgan owns 50% of the line, which it constructed and operates.

ConocoPhillips and a unit of Sempra Energy each own a stake in the line.

ROCKIES EXPRESS PIPELINE

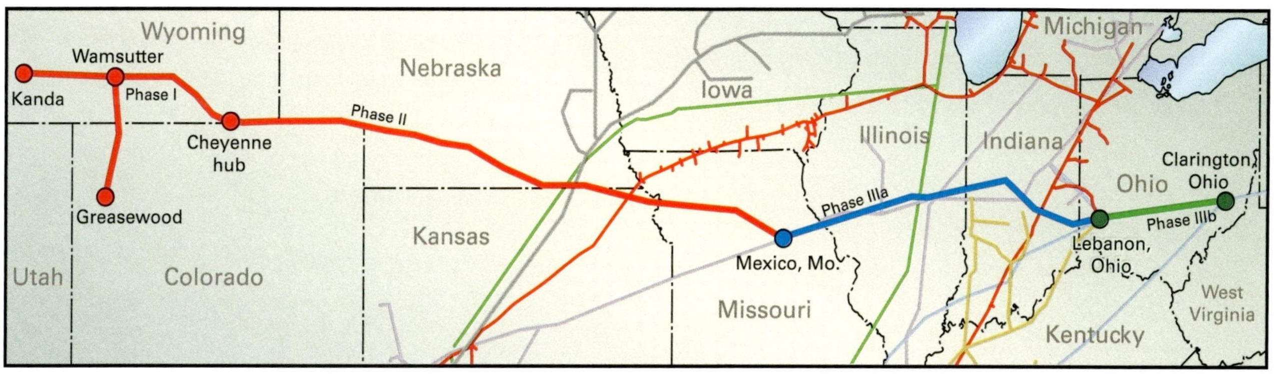

Midcontinent Express

Natural gas service on the roughly 500-mi (805-km) Midcontinent Express Pipeline began in August 2009, between Delhi, LA, and Transcontinental Pipe Line's Station 85 in Butler, AL.

Interim service from Bennington, OK, to Delhi had started in April.

The completed line connects production from the Barnett shale, Bossier sands, and other plays in the region to eastern U.S. markets.

Alberta Clipper

The U.S. Department of State issued a permit in August 2009 for Enbridge Energy LP to build the 1,000-mi (1,609-km) cross-border Alberta Clipper pipeline, from Hardisty, near Alberta's oil sands deposits, to a refinery in Superior, WI.

Initial capacity of the line would be 450,000 b/d.

The department said that added crude oil pipeline capacity between Canada and the United States will advance a number of U.S. strategic interests.

In September, four environmental organizations sued several federal officials, pointing to the negative environmental impact of oil sands production and processing.

They said that the deal contradicts pledges by Pres. Barack Obama to cut global warming pollution, reduce U.S. addiction to imported oil, and invest in a clean energy future.

Keystone

TransCanada Corp. reached agreement in June 2009 with ConocoPhillips to buy its remaining interest in the Keystone pipeline system.

The transaction was expected to close in the third quarter, at which time TransCanada will be sole owner of the project.

The 2,148-mi (3,456-km) line—with the capacity to deliver 435,000 b/d from Hardisty, AB, to Wood River and Patoka, IL—is expected to start up in early 2010.

TransCanada plans to extend the line to Cushing, OK, starting fourth-quarter 2010, raising capacity to 590,000 b/d.

Keystone XL

TransCanada continued work in 2009 on the Keystone Gulf Coast Expansion Project (Keystone XL).

The company anticipates the start of construction in 2010, and intends to start the line in 2012.

Planning began in 2008 on the line, which would provide 500,000 b/d additional capacity from western Canada to existing terminals at Port Arthur, TX, on the U.S. Gulf Coast by 2012.

Keystone XL has secured firm shipping contracts for 380,000 b/d for an average of 17 years.

Texas Independence Pipeline

Energy Transfer Partners LP (ETP) completed construction in 2009 of the 160-mi (258-km) Texas Independence Pipeline (TIP), which increases its gas takeaway capacity in Texas by 1.1 bcfd.

The TIP system will transport gas from Waha, the Bossier sands, and Barnett shale in east and north-central Texas to southeast Texas. It runs from Maypearl to near Henderson, TX.

Rulison expansion

Energy Transfer Partners LP also completed the Rulison gas pipeline expansion project in Colorado in 2009, adding more than 70 MMcfd of capacity.

The expansion also increases the capacity of ETP's South Parachute-Rifle pipeline system and create an outlet for producers to access the Meeker processing plant at the White River hub.

FGT expansion

In September 2009, FERC staff issued a final environmental impact statement for Florida Gas Transmission Co.'s Phase VIII expansion project.

The proposed $2.46-billion project in Alabama and Florida would add 820 MMcfd of capacity to FGT's system. It includes laying 483 mi (777 km) of pipeline, adding 198,000 hp of compression to eight existing stations, and building a new 15,600-hp compression station.

The proposed expansion would start service by mid-2011.

White Cliffs

Mid-2009 saw the startup of the 525-mi (845-km) White Cliffs crude pipeline from Platteville, CO, to storage facilities at Cushing, OK.

White Cliff's current capacity is about 30,000 b/d, expandable to more than 50,000 b/d.

Pacific Connector

FERC issued a final environmental impact statement in May 2009 for the 1.49-bcfd sendout pipeline proposed by Pacific Connector Gas Pipeline LP.

It would extend from the Jordan Cove LNG terminal in Oregon about 234 mi (377 km) southeast to a terminus near Malin.

From there, the proposed line would interconnect with the existing pipeline systems of Gas Transmission Northwest Corp., Tuscarora Gas Transmission Co., and Pacific Gas & Electric Co.

Bakken crude by rail

EOG Resources Inc. said in early 2009 that it would use railcar unit trains to ship crude produced from the Bakken formation in North Dakota to the Cushing, OK, terminal, starting in early 2010.

EOG expects long-term oil netbacks by rail will be better than those receiving by shipping its oil by pipeline through the hub at Clearbrook, MN.

Bakken oil production was under way in Parshall field in North Dakota at mid-year.

Tiger Pipeline

Energy Transfer Partners LP announced plans in early 2009 to construct the $1-billion Tiger Natural Gas Pipeline, working with Chesapeake Energy Corp. as a major shipper.

It will run 178 mi (286 km) from ETP's existing pipeline system near Carthage, TX, through the Haynesville Shale, to a receiving point in Delhi, LA.

Capacity would be 1.25 bcfd initially and could be raised to 2 bcfd. The line is expected to be in service by mid-2011.

Refining and Products

Impact of carbon legislation

A report by Wood Mackenzie in late 2009 said that carbon legislation being considered by Congress was more burdensome than its European counterpart and could threaten the sustainability of the U.S. refining industry.

The study said that the legislation could cost U.S. refiners $100 billion/year within three years.

Delaware City refinery

Valero Energy Corp. decided in late 2009 that it will permanently close its 210,000-b/d Delaware City, DE, refinery, citing financial losses caused by "very poor economic conditions, significant capital spending requirements, and high operating costs."

Earlier in the year, Valero had tried to improve reliability and financial performance and sought a buyer, without success.

New Mexico units

Western Refining Inc. said in late 2009 that it was consolidating operation of its two small refineries in New Mexico to reduce operating costs in a period of shrunken refining margins.

To be consolidated at Gallup are the 23,000 b/d refinery there and the 17,000 b/d facility at Bloomfield, which will be idled.

Total crude throughput will not decline from the plants' combined levels, about 25,560 b/d.

Tulsa refineries

Holly Corp. was moving in October 2009 to buy the 75,000-b/d Sinclair Oil Corp. refinery in Tulsa and integrate the facility with its 85,000-b/d refinery (bought in June from Sunoco) two miles away.

Holly bought the other Tulsa refinery in June from Sunoco Inc.

The integrated complex will have a crude capacity of about 125,000 b/d.

Ethanol and gasoline

The EPA said in early December 2009 that it would postpone, to mid-2010, a final determination on whether to raise the allowable ethanol content in gasoline to 15% from 10%. A biofuels industry group had asked for the increase.

Two congressmen asked the Government Accountability Office (GAO) to explore the impact of such an increase. They expressed concern that ethanol levels above 10% might void vehicle warranties or damage engines.

Ethanol mandates

Federal law requires that use of biofuels grow from 9 billion gal in 2008 to 36 billion gal in 2022. A federal tax credit was put in place to help the domestic ethanol industry.

But in an October 2009 report, the GAO said that the tax credit (45¢/gal) probably won't stimulate production beyond the renewable fuels standard (RFS) levels specified for 2009 unless oil prices climb significantly.

The GAO also noted that the production capacity of the U.S. corn-based ethanol industry is already near the effective RFS limit of 15 billion gal/year.

To meet the RFS, said GAO, domestic biofuels production must increase significantly, with use of other feedstocks likely to be needed.

The U.S. Dept. of Agriculture and DOE are developing advanced biofuels from switchgrass, corn stover, and other cellulosic bio-fuels. A separate tax credit is available for producing advanced cellulosic biofuels.

US OIL PRODUCTION, FOREIGN IMPORTS

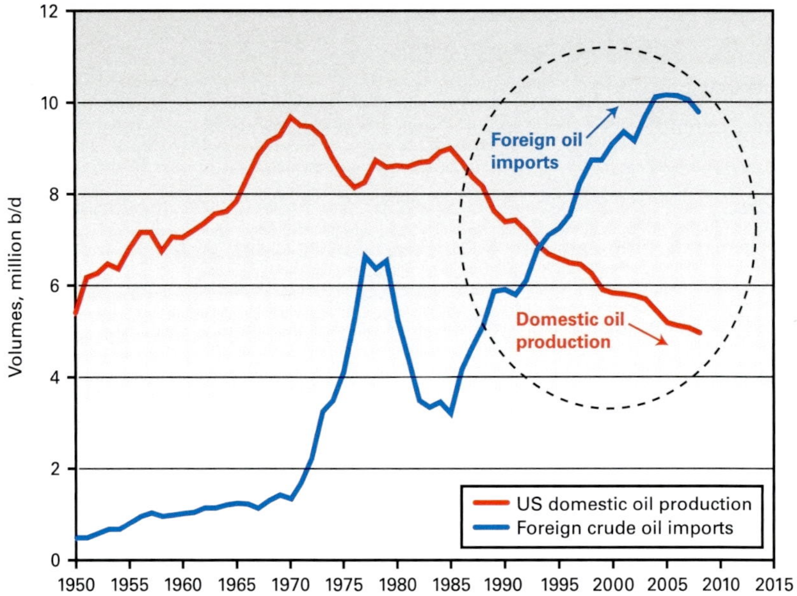

Source: Heritage Foundation; http://www.heritage.org/Research/EnergyandEnvironment/EneryCharts.cfm

US CRUDE IMPORTS: 2007*

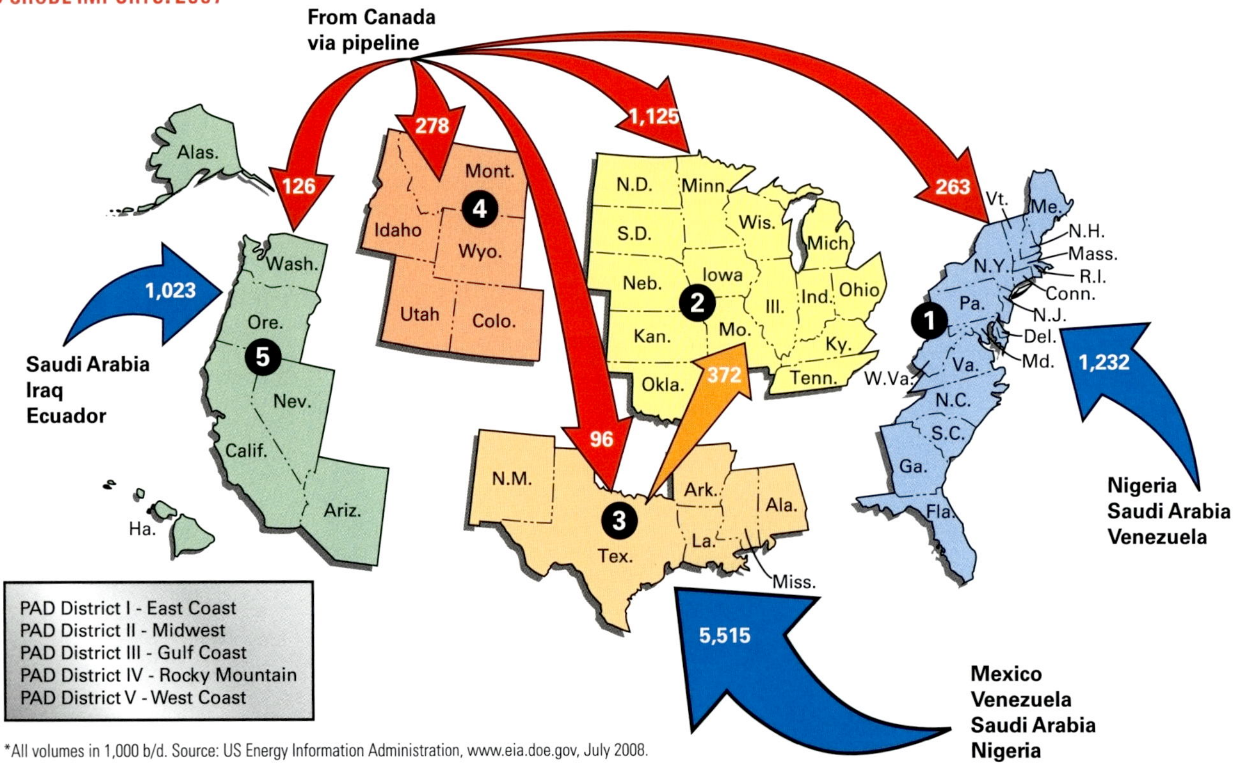

*All volumes in 1,000 b/d. Source: US Energy Information Administration, www.eia.doe.gov, July 2008.

Refining heavy Canadian crudes

A study by Exponent Inc. issued in mid-2009 looked at the impact of changes in the composition of crude oil imported from Canada on the configuration and profitability of U.S. refineries.

A key finding: The past 15 years have seen a movement toward crudes that are heavier and more sour.

Also, refineries adapting to process these crudes have increased requirements for hydrogen, steam, and power, creating opportunities for gasification and cogeneration projects.

US REFINERY CRUDE FEEDSTOCK QUALITY

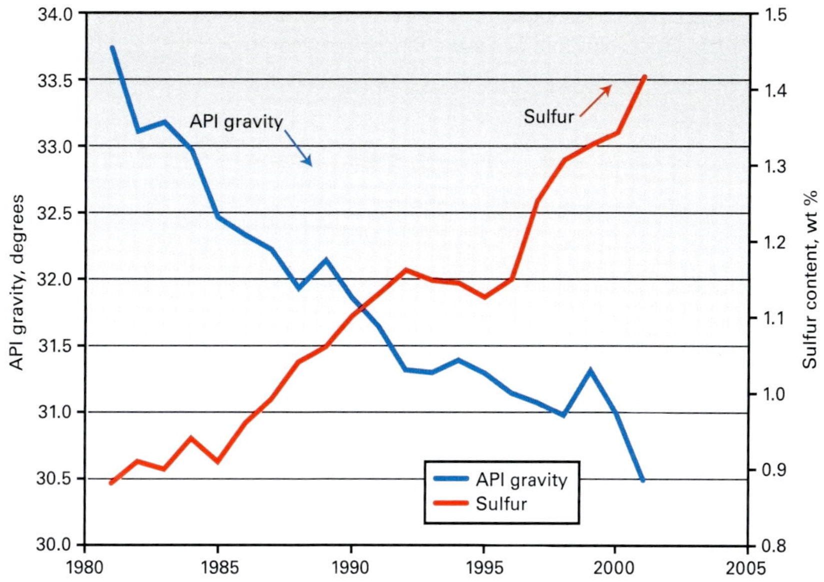

Source: Shore, J., "Refining Challenges: Changing Crude Oil Quality & Product Specifications," Energy Information Administration, World Fuels Conference, Washington, Sept. 25-28, 2002

REFINERY UPGRADES TO PROCESS DILBIT*

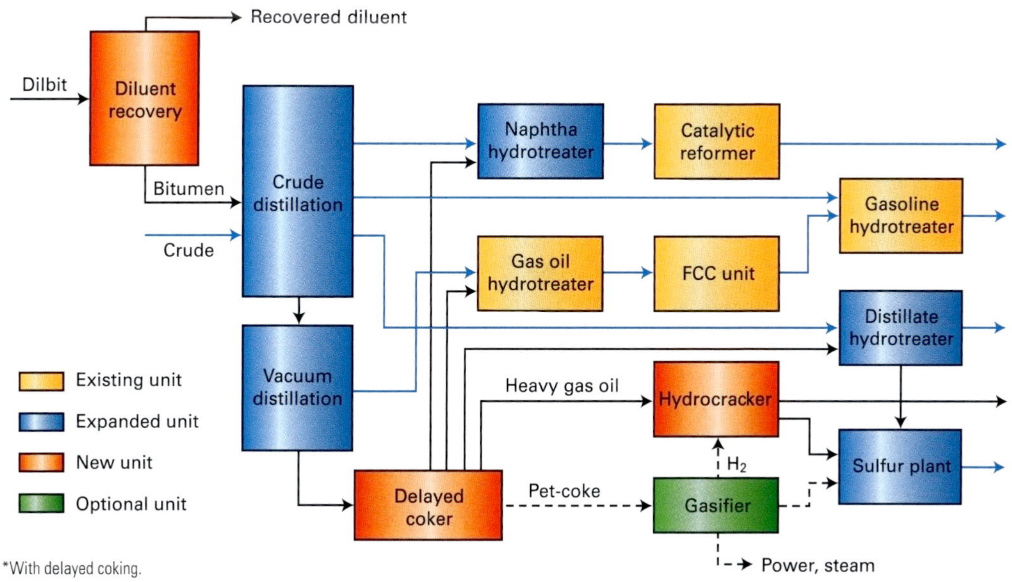

*With delayed coking.

Transporting heavy Canadian crudes

Several projects had been proposed or were under way to expand U.S. distribution of heavy Canadian crude, with an estimated 1 million b/d of pipeline capacity anticipated by 2011. Key projects include:

- Construction of TransCanada's 435,000-b/d Keystone line from Hardisty to Wood River and Patoka by 2010, with a subsequent extension to the U.S. Gulf Coast by 2013.
- Major expansion of Enbridge's main trunkline, the Alberta Clipper project, from Edmonton to Superior, WI, as well as an extension south from Superior to Flanagan, IL (called the Southern Access Expansion and Extension Project).
- Proposed extensions of the Enbridge system to the U.S. East Coast and Gulf Coast.
- Possible interim reversal by Enbridge of its pipeline sections between Portland, ME, and Sarnia, ON, to enable shipment of heavy crude oil down the U.S. Atlantic Coast to the Gulf Coast.
- Intermittent discussion of developing a massive offshore crude oil receiving terminal called the Texas Offshore Port System (TOPS) to meet increasing demand for U.S. crude imports.
- Plans by Sempra and Valero to develop a 500,000-b/d marine terminal near Port Arthur, TX, for crude oil, LPG, and refined products.

Whiting expansion

In October 2009, BP received a setback to its plans for a $3.8-billion expansion of its refinery at Whiting, IN, near Chicago.

CANADIAN, US CRUDE OIL PIPELINES

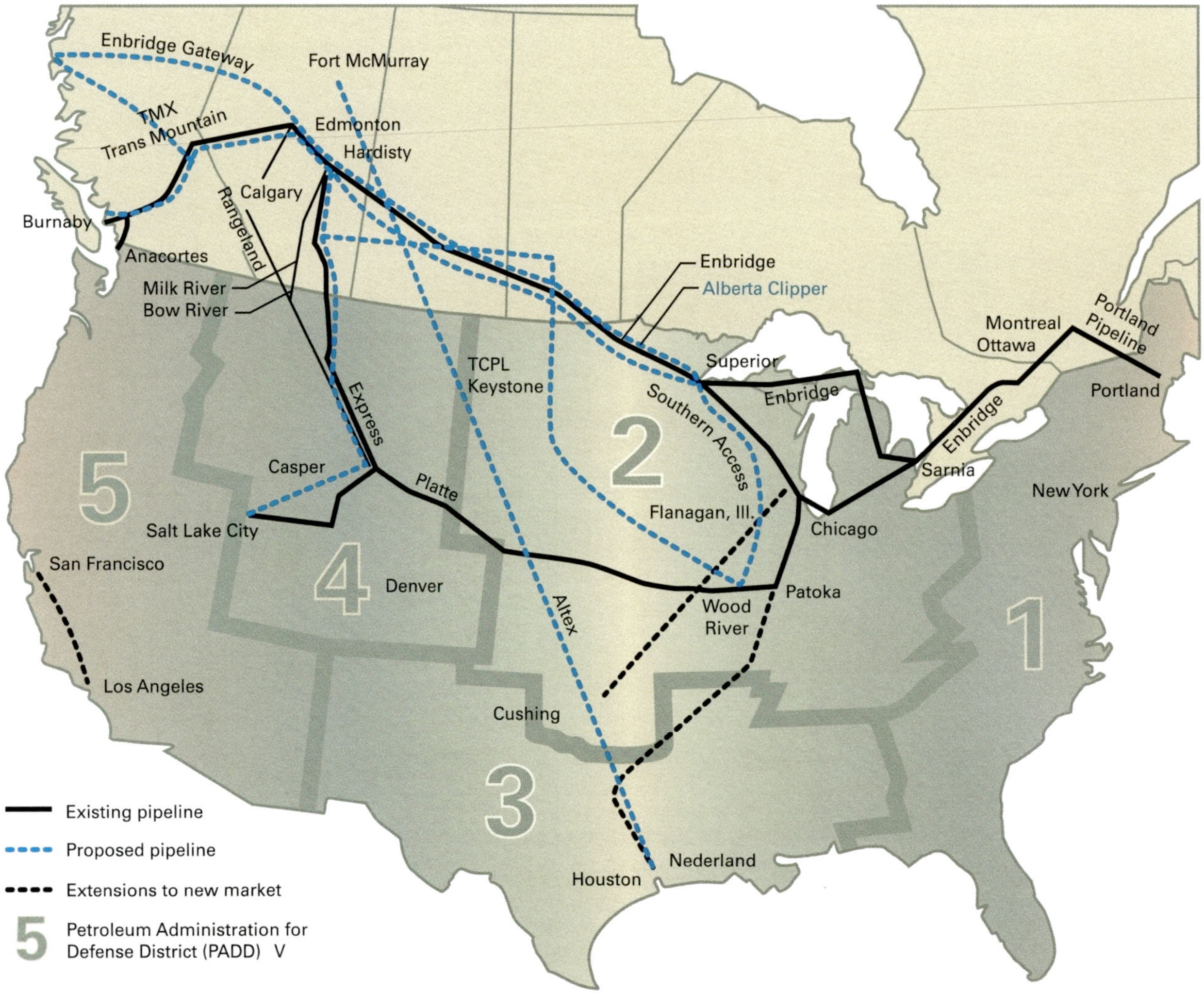

Source: "Alberta's Energy Reserves 2008 and Supply/Demand Outlook 2009-2018," Fig. 2.16, pp. 2-26, Energy Resources Conservation Board, June 2009

The U.S. EPA issued an objection to the operating permit for the refinery that required Indiana environmental officials to rewrite the permit.

Citizens and environmental groups had objected to the permit in 2008 on the grounds that it did not accurately account for the large increases in air pollution that would be caused by the expansion.

The refinery can process 410,000 b/d of crude oil.

BP began a major expansion of the facility in 2008 in order to process Canadian tar sands crude oil. The expansion would make the plant the largest U.S. refiner of tar sands oil in the United States.

Port Arthur delay

Motiva Enterprises LLC decided in early 2009 to delay until first quarter 2012 the completion target for a project that will make its 285,000-b/d refinery at Port Arthur, TX, the largest in the U.S.

Motiva, a joint venture of Shell Oil Co. and Saudi Refining Inc., originally planned to complete the $7-billion expansion in late 2010.

Among other changes, the project will add a single-train crude distillation unit with a capacity of 325,000 b/d and a 150-MW electric power station that will produce 2 million lb/hr of steam.

The expanded Port Arthur facility will be the country's largest refinery producer of sulfur and largest producer of petroleum coke.

Deloitte: Trouble for refiners

In a mid-2009 analysis, Deloitte LLP said that U.S. refining margins are likely to shrink in coming years because stricter fuel economy standards and rising mandates for blending ethanol and other biofuels will reduce gasoline demand.

The study, *A Tsunami of Change Bearing Down on the Refining Industry*, suggests that as much as 1.5 million b/d of U.S. refining capacity could be shut down, though no timeframe was given.

The EIA forecasts a utilization rate for U.S. refining of 78.5% in 2010 due to reduced demand—the lowest level since 1985. EIA expects gasoline demand will decline by 13% by 2030.

Investment needs

Robust worldwide demand growth for middle distillates in 2008 resulted in a disproportionate price increase compared to gasoline. The then-current global refining configuration was misaligned with this demand increase, resulting in a need for significant investment.

These were the conclusions of a December 2008 report (released in early 2009) from Energy Policy Research Foundation Inc., *The U.S. Refining Industry: Background and Perspective for Policy Makers.*

The report also studied historical trends in refinery size, location, complexity, and upgrading.

US REFINER MARGINS

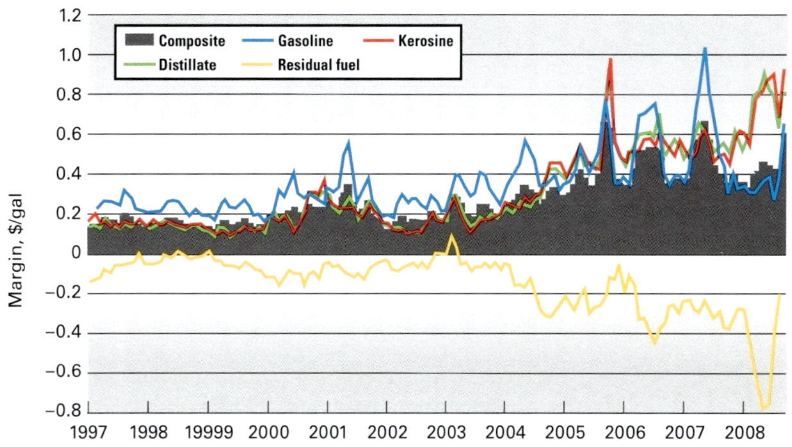

Source: EIA data, EPRINC calculations

US DISTILLATES

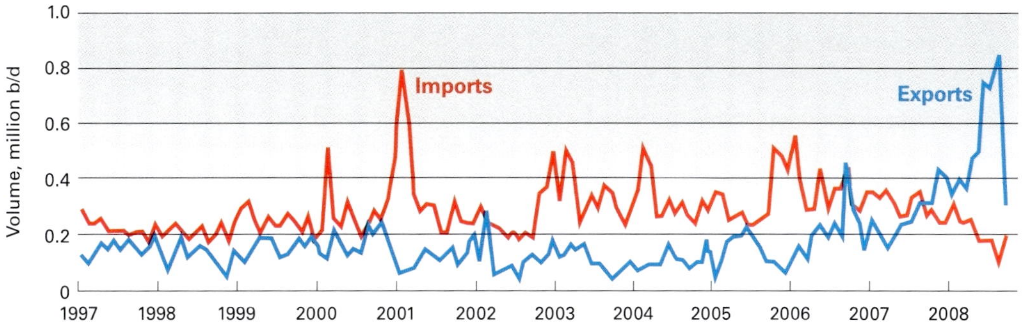

Source: EIA data

US REFINERY UTILIZATION

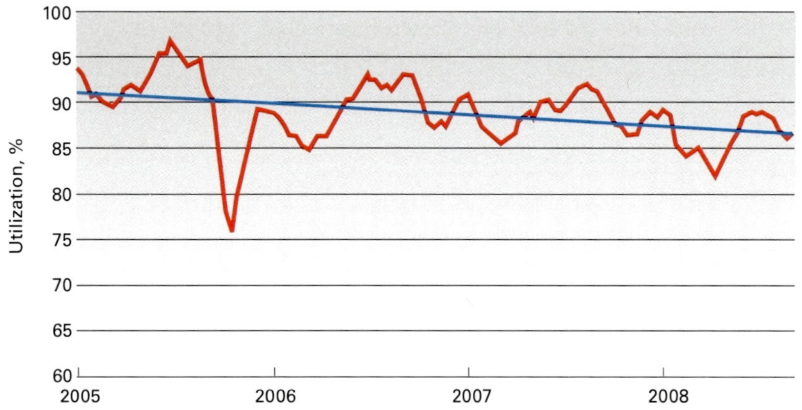

Source: EIA data

SHIFT IN U.S. ETHYLENE CAPACITY

Basic types of ethylene plants	Effective capacity* 2007 Billion lb/year	2007 %	2000 Billion lb/year	2000 %
Purity ethane crackers	4.5	7.2	6.6	22.0
E/P crackers	17.4	28.0	23.2	38.6
Flexifeed crackers	31.4	50.5	27.1	45.1
Heavy feed crackers	8.9	14.4	3.2	5.3
Total effective capacity	**62.2**	**100.0**	**60.1**	**100.0**

*Capacity excludes plants that are mothballed.

Sources: Hodson Report and En*Vantage

Petrochemicals

Ethane capacity requirements

EnVantage Inc. issued a study in early 2009 of the U.S. ethane and ethylene markets.

The firm concluded that the ethylene industry appeared to be healthy and that ethylene producers could handle additional ethane volumes coming onto the market.

However, trends from 1990–2008 revealed signs that it would be difficult for ethylene producers to crack much more ethane in coming years.

For example, by the end of 2007, about 20% of U.S. ethylene plant capacity was at least 35 years old, and this capacity represented 33% of total ethane consumption.

Looking to 2012, the firm sees a base case in which ethylene production averages less than 55 billion lb/year, and a best-case scenario in which production approaches 60 billion lb/year.

The firm suggests that the U.S. gas processing industry should develop a substantial overhang in ethane extraction capability.

US ETHYLENE CAPACITY, PRODUCTION: 1990-2008

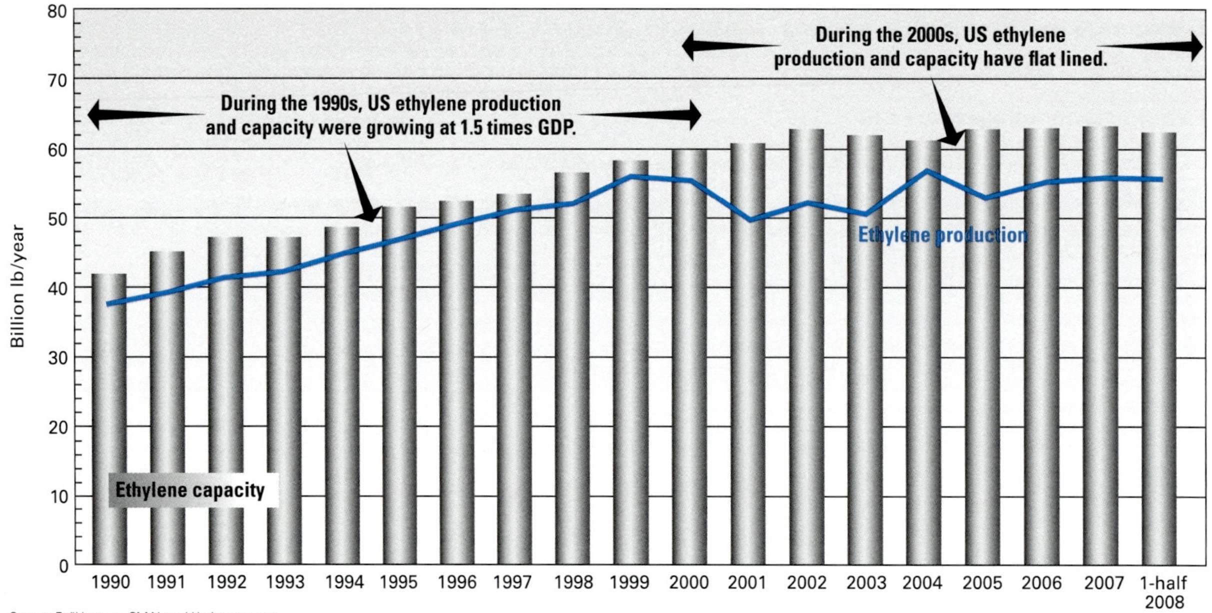

Source: En*Vantage, CMAI, and Hodson reports

US ETHYLENE FEEDSTOCK CONSUMPTION

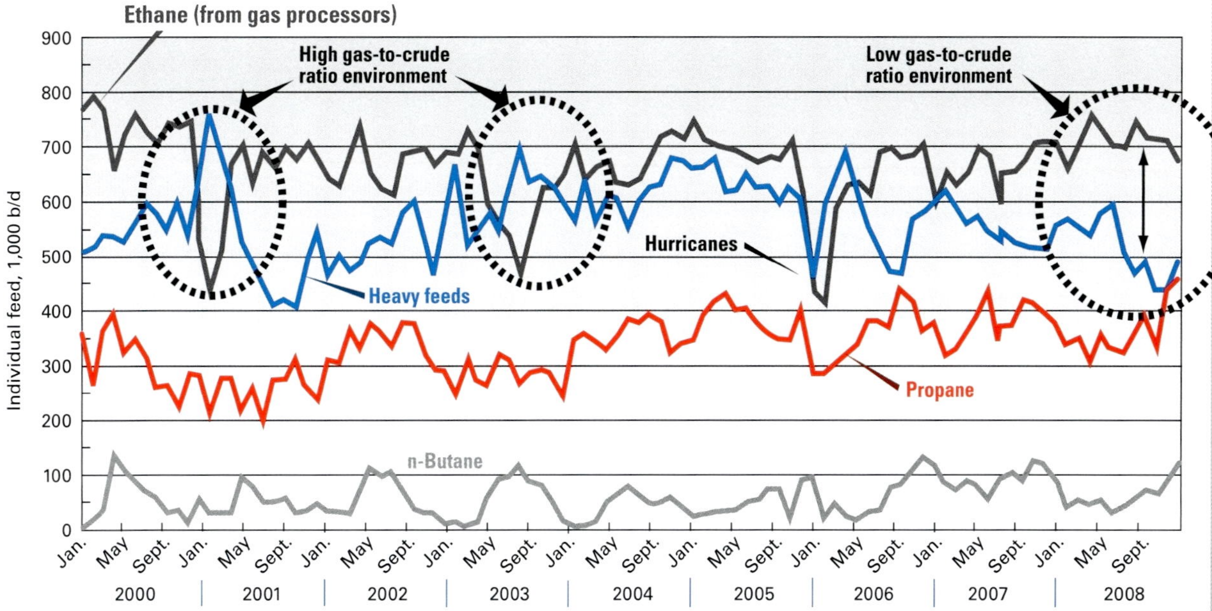

Source: Hodson reports and En*Vantage

US ETHYLENE CAPACITY, PRODUCTION: 2000–12

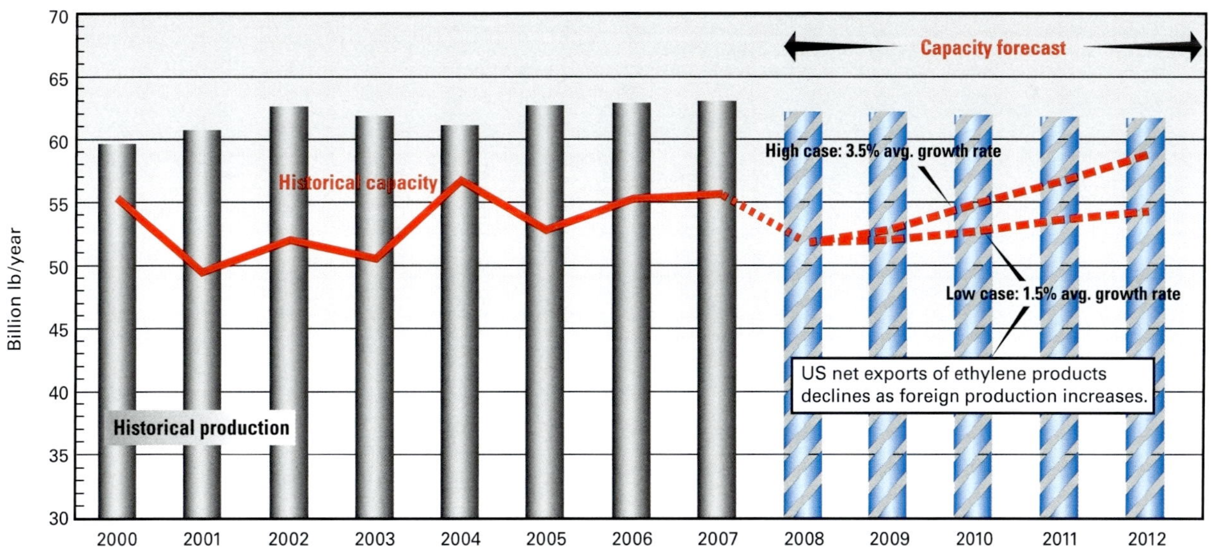

Source: Historical data: CMAI, Hodson reports; forecast: En*Vantage

ETHANE EXTRACTION CAPABILITY, DEMAND

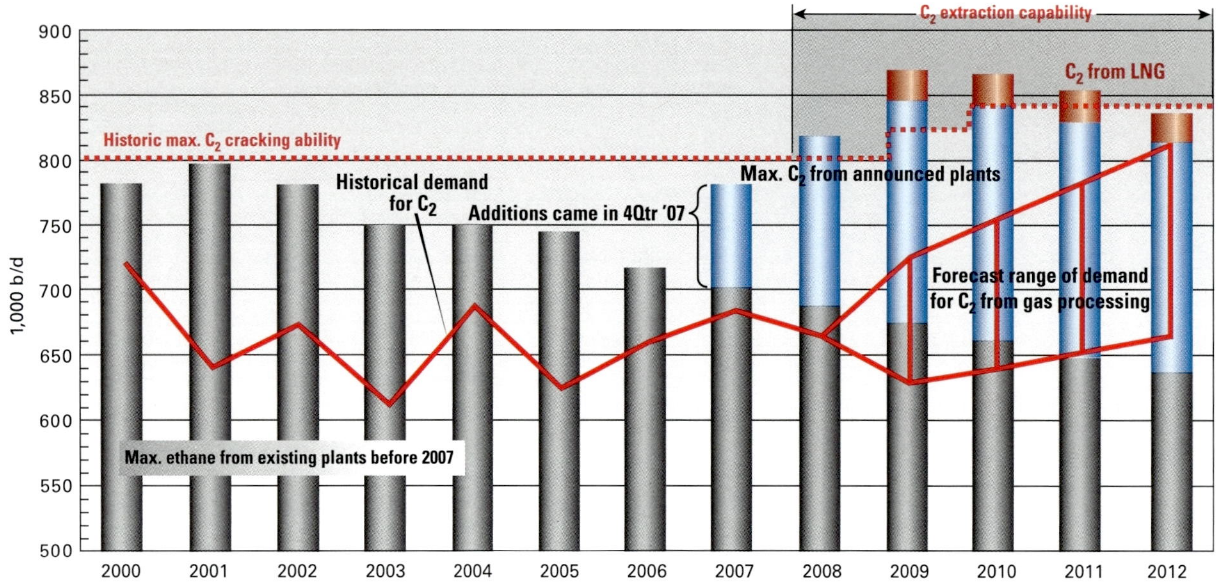

Source: Historical data: CMAI, Hodson reports; forecast: En*Vantage

US ETHANE EXTRACTION CAPABILITY: FORECAST

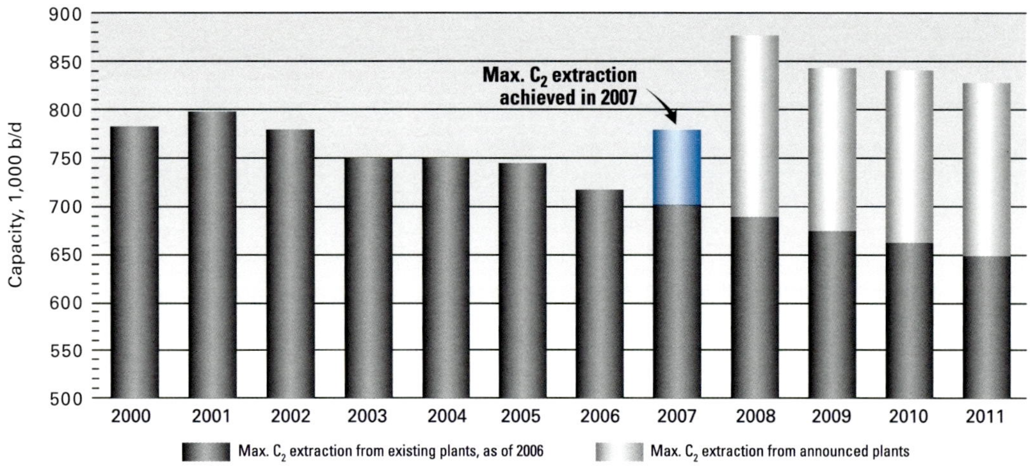

2008 olefins market

Petral Worldwide in early 2009 examined trends in 2007 and 2008 for the U.S. olefins market.

They noted sharp drops in both demand and production in second half 2008, due mainly to a decline in global economic activity, but also to the impact of two hurricanes in the Gulf of Mexico.

Petral noted that olefins producers saw the most dramatic decline ever in feedstock costs as crude oil prices fell from an all-time high of $147/bbl in July 2008 to a low of $31/bbl in late December.

The firm concluded that, in view of OPEC's poor historic record of compliance with its production quota agreements, WTI prices could range from $25/bbl to as much as $50–60/bbl during first half 2009.

US ETHYLENE PLANT FEED SLATE

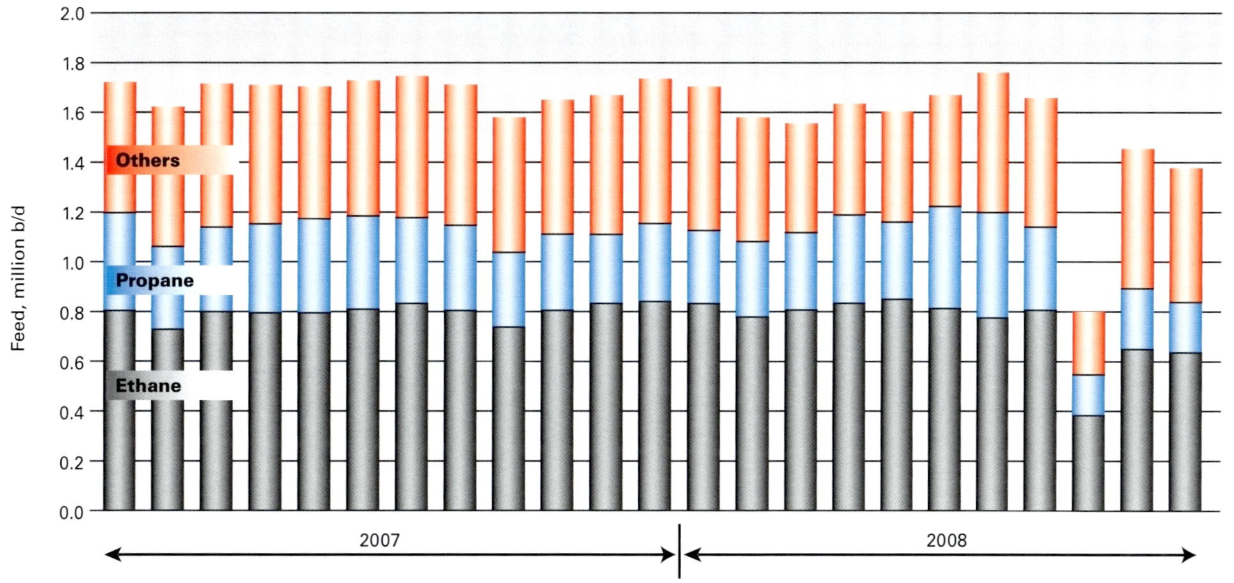

Source: Petral Monthly Olefin Plant Feedslate Survey

US ETHYLENE PRODUCTION

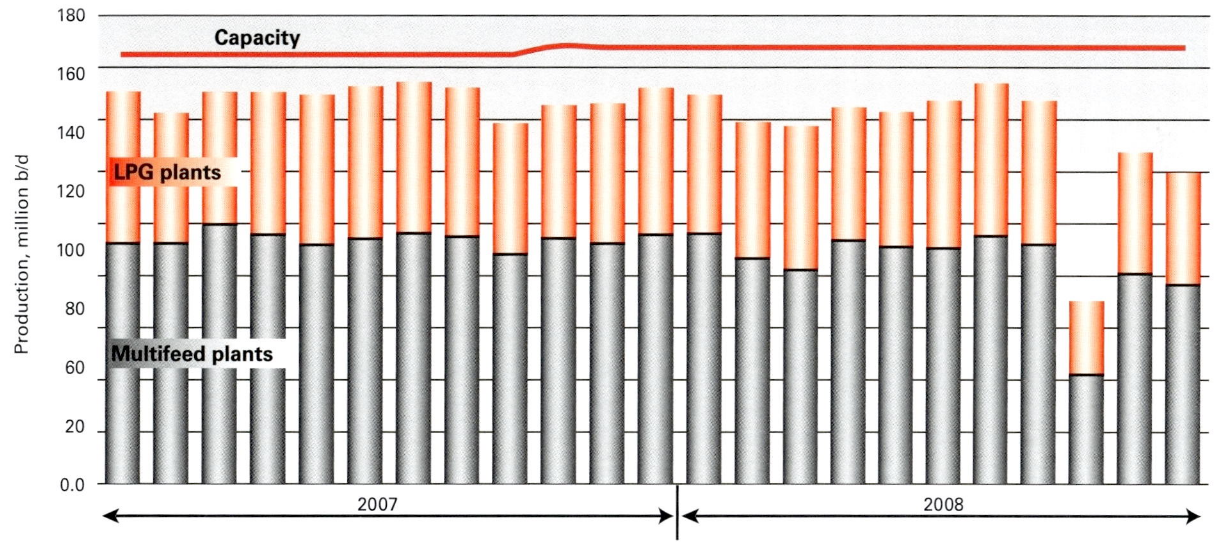

Source: Petral Monthly Olefin Plant Feedslate Survey

Alternative Energy

2008: Boost in renewables use

Americans used more solar, nuclear, biomass, and wind energy in 2008 than in 2007, according to a study in mid-2009 by the Lawrence Livermore National Laboratory (LLNL).

Using data from DOE's Energy Information Administration, LLNL found that:

- Coal and petroleum use dropped
- Natural gas use slightly increased
- Use of geothermal energy remained the same.

Total estimated U.S. energy use in 2008 was 99.2 quads, down from 101.5 quads in 2007.

Of the 99.2 quads used, only 42.15 ended up as useful energy services. The rest was "rejected," as waste heat from power plants, for example.

Wind power outlook

In May 2009, Industrial Info Resources said that 389 wind-farms were in operation in the United States, representing 27,144 MW of generating capacity.

It also said that an additional 46 such facilities were under construction, and that other projects representing 94,000 MW of capacity were being planned and scheduled to begin construction over the next decade.

Notable wind-power projects under development in 2009 included:

- Shepherds Flat Wind Farm (845 MW) near Arlington, OR
- Cape Wind Project (454 MW) in Nantucket Sound, offshore MA
- Pine Tree Wind Project (120 MW) about 100 mi (160 km) southeast of Bakersfield, CA
- Prairie Winds ND 1 (115 MW); some 15 mi (24 km) south of Minot, ND.

Advanced biorefinery support

In early December 2009, the U.S. Dept. of Agriculture and the DOE selected 19 integrated biorefinery projects to receive up to $564 million from the stimulus package to speed construction and operation of pilot, demonstration, and commercial-scale facilities.

Projects in 15 states will evaluate technologies to produce advanced biofuels, biopower, and bioproducts using biomass feedstocks.

Algae-based biofuel

ExxonMobil and Synthetic Genomics set up a $600-million partnership in 2009 to research making biofuels from algae. Head of Synthetic Genomics is Craig Venter, the pioneer of human genome research.

Exxon said it plans to spend $300 million on its internal costs for the venture and up to $300 million more with Synthetic Genomics.

Researchers will aim to produce biofuels resembling gasoline and diesel, or a form of crude oil that could be processed in Exxon's refineries.

Algae can produce almost 2,000 gal of fuel per acre of land used, compared with just 250 gal/acre for corn-based ethanol.

Verenium, BP JV

Verenium Corp. and BP PLC formed a JV (Vercipia Biofuels) in early 2009, to build one of the nation's first commercial-scale, cellulosic-based ethanol plants. It is to be sited in Highlands County, FL.

The project aims to break ground in 2010 and start production in 2012. Construction cost is estimated at $300 million for the plant, which is to produce 36 million gal/year of ethanol.

Environmental Issues

CO_2 EMISSIONS: ACTUAL AND PROJECTED (MPTY)

Nation	1990	2006	2030*	Cuts by 2020 proposed/offered
United States	4,989	5,907	6,414	From 2005 levels: • 20% (S 1733: pending in Senate, Dec. 2009) • 17% (HR 2454: passed by House, June 2009) • 14% (Obama campaign promise)
OECD Europe	4,149	4,429	4,519	From 1990 levels: • 30% if other countries agree • 20% otherwise
Japan	1,054	1,247	1,157	From 1990 levels: • 25%

*2030 levels assume no new emissions reductions made (EIA projections).
Source: U.S. Energy Information Administration

CCS: U.S. plans

DOE has said the United States could have 10 or 12 commercial demonstration projects for carbon capture and storage (CCS) at coal power plants operational by 2016, and be ready for wider commercial deployment by 2019.

US CANDIDATE RESERVOIRS FOR CO_2 SEQUESTRATION

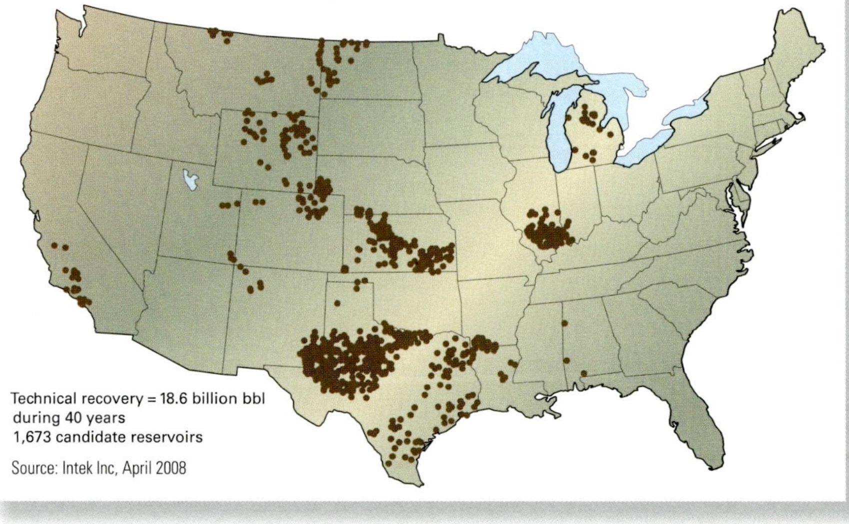

DOE said in 2009 that the government was planning to invest $4 billion in CCS technology development, with another $7 billion coming from the private sector.

As part of the effort, DOE expects to revive the FutureGen project, which envisions development of a an advanced coal-fired power plant that incorporates CCS technology.

CCS: AEP project

American Electric Power and a unit of Alstom SA began sequestering carbon dioxide emissions in a CCS pilot project at AEP's 1,300-MW Mountaineer Power Station in New Haven, WV.

Using Alstom's Chilled Ammonia Process, up to 100,000 metric tpy of CO_2 will be captured and then stored in deep saline aquifers at the Mountaineer site.

After the validation test at Mountaineer, AEP will install a commercial-scale application of Alstom's system on one of its coal-fired units at a location yet to be determined.

CCS: W. Texas, Alabama projects

DOE announced in late 2009 that it will provide $350 million to support the Texas Clean Energy Project. Funds come from the third round of DOE's Clean Coal Initiative.

The 8-year TCEP would capture 90% (2.7 mtpy) of the carbon dioxide created at a proposed 400-MW power plant near Midland-Odessa, TX, and then pipe it to the Permian basin for use in EOR.

A second project will receive $295 million of federal funding over 11 years to retrofit CO_2 capture equipment on an existing Alabama Power Co. plant north of Mobile. The CO_2 would be sequestered in deep saline formations, although some might be used for EOR.

CCS R&D Prize

In November 2009, two U.S. senators proposed a government prize to encourage scientific research and development of carbon capture and storage technology.

The proposal is described in S 2774, the Carbon Dioxide Capture Technology Act.

Climate-change legislation

Prospects for passage of climate-change legislation in 2010 remained uncertain as 2009 ended.

A variety of proposed legislation and other actions were under consideration during 2009.

In February, Congress approved billions of dollars in spending for clean-energy technology as part of the economic recovery bill.

On June 26, HR 2454 (the Waxman-Markey bill) was passed by the House of Representatives.

On September 30, S 1733 (the Kerry-Boxer bill) was introduced in the Senate.

A number of other related actions took place in December 2009.

The U.S. EPA issued a finding that greenhouse gas (GHG) emissions endanger public health and should be regulated under the Clean Air Act.

Sen. Murkowski said that she planned to file a disapproval resolution to keep EPA from trying to regulate GHG emissions under the Clean Air Act.

Senators Kerry, Lieberman, and Graham released a framework for compromise legislation that includes provisions dealing with GHG emissions reduction goals, assistance to consumers and businesses in the transition to a lower-carbon economy, incentives for nuclear power and clean coal, and creation of nonfossil-energy jobs.

Senators Cantwell and Collins introduced the CLEAR Act (Carbon Limits and Energy for America's Renewal). Among other provisions it would set up a climate auction to sell "carbon shares" (with specific uses identified for revenues generated); and establish GHG reduction goals of 20% by 2020 (compared to 17% in the Kerry-Lieberman-Graham framework, HR 2454 and S 1733) and 83% by 2050.

Carbon trading business

The carbon trading market was worth about $126 billion in 2008, about double the amount in 2007, according to the World Bank.

California carbon fee

In September 2009, the California Air Resources Board (CARB) imposed the nation's first statewide carbon fee on some 350 businesses.

Its purpose is to pay for implementing a program required under a 2006 law to bring GHG emissions within the state back to estimated 1990 levels by 2020.

The targeted businesses produce 85% of the state's total GHG emissions, said CARB.

The plan anticipates collections from businesses including refineries, gasoline and diesel fuel importers, large natural gas distributors and consumers, and plants that burn coal and petroleum coke.

They would pay an estimated $63.1 million during fiscal years 2010 and 2011.

Produced-water issues

During 2009, regulatory and public concerns became more visible in the Rocky Mountain West with regard to environmental impacts of oil and gas production in the region.

One major topic was emission of volatile organic compounds (VOCs) and hazardous air pollutants (HAPs) from produced-water surface evaporation ponds.

Western states—including Wyoming, New Mexico, Colorado, Utah, and Montana—generated 2.6 billion bbl of produced water in 2002.

Western Climate Initiative

In 2008, seven Western-state governors and four Canadian province premiers outlined what they called the Western Climate Initiative, aimed at cutting regional greenhouse gas emissions.

WCI's plan is based on an economic model that precludes installation of new electric power capacity in the region except for wind and solar power generation.

In 2009, a report commissioned by the Western Business Roundtable said WCI would prolong the economic recession and weaken already overburdened power grids.

State air quality standards

In January 2009, the federal government granted California and other states the right to raise air quality standards above the national level. The states had asked for authority to adopt more-stringent anti-pollution rules previously but were rebuffed by the Bush Administration.

ROCKY MOUNTAIN REGION OIL, GAS BASINS

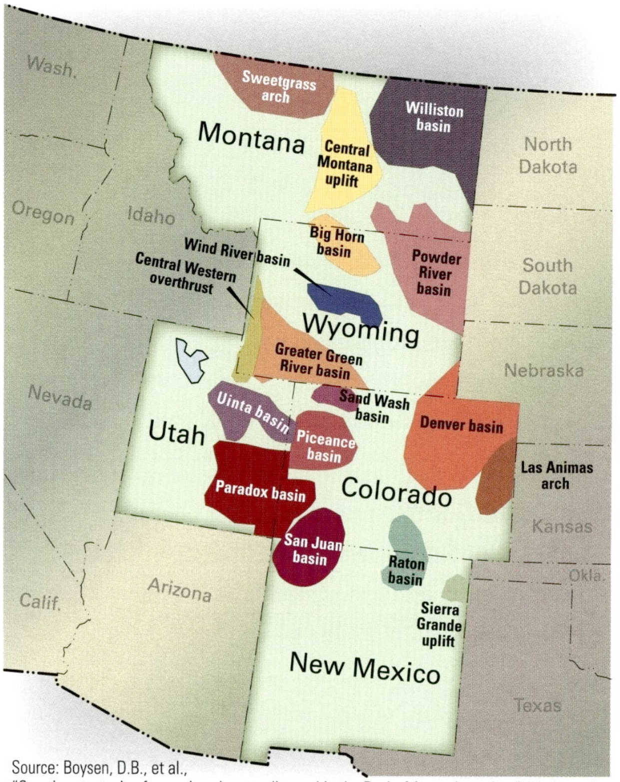

Source: Boysen, D.B., et al., "Creative strategies for produced water disposal in the Rocky Mountain region," 9th Annual International Petroleum Environmental Conference, Albuquerque, NM, Oct. 22–25, 2002.

Company News

ExxonMobil spending

In March 2009, ExxonMobil Corp. said that it would continue investing at "record levels," despite the economic downturn and plunging oil and gas prices.

Exxon was planning to invest $29 billion in 2009, and a total of $25–$30 billion/year through 2014.

Company leaders said they view the current economic situation and severe drop in commodity prices as short-term challenges and that the company has a 30-year investment horizon.

ExxonMobil-XTO deal

In December 2009, ExxonMobil Corp. agreed to buy XTO Energy Inc. in an all-stock deal valued at $41 billion.

The action was seen as part of a strengthened focus by ExxonMobil on unconventional resources.

XTO holds large interests in major U.S. shale, tight gas, and coalbed methane plays as well as the Bakken oil shale. Its resource base is estimated at 45 tcf of gas equivalent.

ExxonMobil also has interests in unconventional resources in Canada, Germany, Poland, Hungary, and Argentina.

Chevron spending

Chevron Corp. announced in December 2009 a $21.6 billion capital and exploratory spending program for 2010, down 5% from projected 2009 spending.

Officials said that 80% of the 2010 program is for upstream exploration and production projects worldwide, with 16% earmarked for downstream businesses.

Chevron-Ecuador suit

In September 2009, Chevron filed a claim before the Permanent Court of Arbitration in The Hague, against the government of Ecuador, citing violations of the country's obligations under the U.S.-Ecuador bilateral investment treaty, investment agreements and international law.

The claim stems from what Chevron calls Ecuador's "exploitation" of a lawsuit filed against Chevron, in which plaintiffs are seeking $27 billion in damages.

The suit accuses Texaco (which Chevron bought in 2001) of leaving behind an environmental disaster when it left Ecuador in 1992.

Chevron said that Texaco spent $40 million on cleanup and that the Ecuador government released it from further responsibility as part of a settlement agreement. Attorneys representing affected Amazonian communities charge that the agreement did not release Chevron from third-party claims.

Hess spending

Hess Corp. said in late 2009 that it has budgeted $3.9 billion in capital and exploratory spending for 2010, a level $700 million higher than it budgeted for 2009.

In 2010, it plans to spend $2.4 billion for production, $600 million for development, and $850 million for exploration.

Devon divestment

Devon Energy Corp. announced plans in November 2009 to divest its Gulf of Mexico and international assets so it can focus on high-growth U.S. and Canadian onshore assets.

It expected to start the divestiture in first quarter 2010 and complete the process by yearend 2010.

The company has assets in Azerbaijan, Brazil, China, and Russia.

Devon's Gulf and international properties comprise 7% of its proved reserves.

BP: Texas City aftermath

In October 2009, the Occupational Safety and Health Administration within the U.S. Dept. of Labor announced a proposed fine of $87.4 million against BP Products North America Inc. The fine is the largest in its history, OSHA said.

OSHA said that BP failed to correct safety problems at its Texas City, TX, refinery following a March 2005 explosion. That explosion killed 15 and injured 170 others at the 446,500-b/cd plant, near Houston.

BP said it would continue to work with OSHA to resolve safety issues.

Denbury-Encore merger

In November 2009, Denbury Resources Inc. acquired Encore Acquisition Co. for $4.5 billion. The combined entity will go forward as Denbury Resources Inc.

Both companies have strategies geared to EOR using CO_2 injection and are also pursuing supplies of CO_2 from industrial sources that need to cut emissions of greenhouse gas through capture and sequestration in the subsurface.

Enterprise-TEPPCO merger

In late 2009, Enterprise Products Partners LP and TEPPCO Partners LP announced completion of the merger of the two partnerships.

With a total value of approximately $30 billion, 48,000 miles of pipelines, and market capitalization of $18 billion, Enterprise becomes the nation's largest publicly traded partnership.

ConocoPhillips asset sale

ConocoPhillips announced in October 2009 that it plans to cut 2010 capital spending by $1.5 billion and to sell $10 billion worth of assets by yearend 2011.

The firm said the asset sales may include oil and natural gas properties, as well as refineries. It plans to use the proceeds to cut debt, aiming to return to a target debt-to-capital ratio of 20–25%.

In early 2009, Conoco announced 1,300 job cuts and said that it would scale back capital expenditures to $12.5 billion in 2009, from $15.3 billion in 2008.

OGJ150 report: 2008 performance

In September 2009, *OGJ* published its review of the performance of the top 150 U.S. oil and gas producers. Shown here are charts summarizing some key business measures.

The complete report is accessible through the *OGJ* Web site at www.ogj.com.

Top 20 in Total Revenue

Rank	Company	Total revenue, $1,000
1	ExxonMobil Corp.	477,359,000
2	Chevron Corp.	273,005,000
3	ConocoPhillips	246,182,000
4	Marathon Oil Corp.	78,569,000
5	Hess Corp.	41,094,000
6	Murphy Oil Corp.	27,513,000
7	Occidental Petroleum Corp.	24,480,000
8	Anadarko Petroleum Corp.	15,723,000
9	Devon Energy Corp.	15,211,000
10	Apache Corp.	12,389,750
11	Chesapeake Energy Corp.	11,629,000
12	XTO Energy Inc.	7,695,000
13	EOG Resources Inc.	7,127,143
14	El Paso Corp.	5,363,000
15	Dominion Energy Inc.	4,312,000
16	Noble Energy Inc.	3,901,000
17	Questar Corp.	3,491,800
18	Williams Cos. Inc.	3,121,000
19	Plains Exploration & Production Co.	2,403,471
20	Pioneer Natural Resources Co.	2,338,287
	Total	**1,262,907,451**

Baker Hughes–BJ Services deal

In third quarter 2009, Baker Hughes said it would buy BJ Services for $5.5 billion in cash and stock.

After the deal, pressure pumping will generate more than 20% of Baker Hughes' revenue, compared to 1% before the merger.

Baker Hughes once owned BJ Services but spun it off in the early 1990s.

M&A report

An early-2009 report by IHS Herold and Standard Chartered Bank said that state-controlled companies continued to buy oil and gas assets in 2008 and early 2009, even as merger and acquisition activity slumped.

Acquisitions by NOCs and sovereign wealth funds reached a record high in 2008, accounting for 15% of the total value of all deals made.

Top 20 in Spending and U.S. Net Wells Drilled

Rank	Company	Capital, exploratory spending, $1,000
1	ExxonMobil Corp.	20,769,000
2	Chevron Corp.	19,666,000
3	ConocoPhillips	19,099,000
4	Devon Energy Corp.	9,375,000
5	Chesapeake Energy Corp.	9,177,000
6	Marathon Oil Corp.	7,146,000
7	Apache Corp.	5,972,846
8	Anadarko Petroleum Corp.	4,801,000
9	EOG Resources Inc.	4,718,860
10	Occidental Petroleum Corp.	4,664,000
11	Hess Corp.	4,438,000
12	XTO Energy Inc.	3,661,000
13	Petrohawk Energy Corp.	3,121,736
14	El Paso Corp.	2,757,000
15	Questar Corp.	2,485,700
16	Forest Oil Corp.	2,338,488
17	Murphy Oil Corp.	2,185,960
18	Newfield Exploration Co.	2,067,000
19	Noble Energy Inc.	1,971,000
20	Southwestern Energy Co.	1,755,888
	Total	**132,170,478**

Rank	Company	U.S. net wells drilled
1	Chesapeake Energy Corp.	1,733.0
2	Anadarko Petroleum Corp.	1,590.8
3	EOG Resources Inc.	1,545.0
4	XTO Energy Inc.	1,247.0
5	Devon Energy Corp.	1,068.9
6	Williams Cos. Inc.	1,050.0
7	Noble Energy Inc.	929.7
8	Chevron Corp.	860.0
9	ConocoPhillips	849.0
10	Occidental Petroleum Corp.	772.4
11	Pioneer Natural Resources Co.	566.0
12	CNX Gas Corp.	552.0
13	Equitable Supply	533.2
14	Range Resources Corp.	490.2
15	El Paso Corp.	450.0
16	ExxonMobil Corp.	442.0
17	Forest Oil Corp.	437.0
18	Quicksilver Resources Inc.	419.7
19	Atlas America Inc.	419.0
20	Exco Resources Inc.	399.5
	Total	**16,354.4**

Top 20 in Liquids Production

Rank	Company	U.S. liquids production, million bbl
1	ConocoPhillips	157.0
2	Chevron Corp.	154.0
3	ExxonMobil Corp.	104.0
4	Occidental Petroleum Corp.	96.0
5	Anadarko Petroleum Corp.	54.0
6	Devon Energy Corp.	41.0
7	Apache Corp.	35.1
8	XTO Energy Inc.	26.2
9	Marathon Oil Corp.	23.0
10	EOG Resources Inc.	20.0
11	Noble Energy Inc.	18.0
12	Pioneer Natural Resources Co.	15.1
13	Kinder Morgan CO_2 Co. LP	15.0
14	Hess Corp.	15.0
15	Whiting Petroleum Corp.	12.4
16	Denbury Resources Inc.	11.5
17	Chesapeake Energy Corp.	11.2
18	Encore Acquisition Co.	10.1
19	Continental Resources Inc.	9.1
20	Cimarex Energy Co.	8.4
	Total	**836.0**

Rank	Company	Worldwide liquids production, million bbl
1	ExxonMobil Corp.	724.0
2	Chevron Corp.	604.0
3	ConocoPhillips	494.0
4	Occidental Petroleum Corp.	171.0
5	Hess Corp.	97.0
6	Apache Corp.	96.9
7	Devon Energy Corp.	81.0
8	Anadarko Petroleum Corp.	80.0
9	Marathon Oil Corp.	77.0
10	Murphy Oil Corp.	38.7
11	Noble Energy Inc.	32.0
12	XTO Energy Inc.	26.2
13	EOG Resources Inc.	22.5
14	Pioneer Natural Resources Co.	18.2
15	Kinder Morgan CO_2 Co. LP.	15.0
16	Whiting Petroleum Corp.	12.4
17	Denbury Resources Inc.	11.5
18	Chesapeake Energy Corp.	11.2
19	Newfield Exploration Co.	11.2
20	Encore Acquisition Co.	10.1
	Total	**2,633.9**

Top 20 in Gas Production

Rank	Company	U.S. gas production, bcf	Rank	Company	Worldwide gas production, bcf
1	ConocoPhillips	896.0	1	ExxonMobil Corp.	2,468.0
2	Chesapeake Energy Corp.	775.4	2	ConocoPhillips	2,095.0
3	Anadarko Petroleum Corp.	750.0	3	Chevron Corp.	1,876.0
4	Devon Energy Corp.	726.0	4	Devon Energy Corp.	940.0
5	XTO Energy Inc.	697.4	5	Chesapeake Energy Corp.	775.4
6	ExxonMobil Corp.	555.0	6	Anadarko Petroleum Corp.	750.0
7	Chevron Corp.	549.0	7	XTO Energy Inc.	697.4
8	EOG Resources Inc.	436.0	8	EOG Resources Inc.	603.5
9	Williams Cos. Inc.	406.0	9	Apache Corp.	592.0
10	Apache Corp.	248.8	10	Williams Cos. Inc.	406.0
11	El Paso Corp.	230.0	11	Marathon Oil Corp.	359.0
12	Occidental Petroleum Corp.	215.0	12	Occidental Petroleum Corp.	307.0
13	Southwestern Energy Co.	192.3	13	Noble Energy Inc.	281.0
14	Newfield Exploration Co.	167.9	14	Hess Corp.	272.0
15	Marathon Oil Corp.	164.0	15	El Paso Corp.	233.0
16	Pioneer Natural Resources Co.	154.3	16	Southwestern Energy Co.	192.3
17	Questar Corp.	151.9	17	Newfield Exploration Co.	167.9
18	Noble Energy Inc.	145.0	18	Pioneer Natural Resources Co.	158.9
19	Ultra Petroleum	138.6	19	Questar Corp.	151.9
20	Exco Resources Inc.	131.2	20	Forest Oil Corp.	141.4
	Total	**7,729.7**		**Total**	**13,467.8**

The survey also showed that "unconventional" sources of oil and gas have become an increasingly important focus of M&A activity. The biggest 2008 deal was the $5.9 billion spent by ConocoPhillips to buy into a coalbed methane project with Origin Energy of Australia.

GULF OF MEXICO

Overview

Hurricane season quiet

In 2009, the Pacific El Nino phenomenon resulted in an especially calm Atlantic hurricane season. Only nine named storms formed in the quietest season since 1997.

The only early landfalls were Bill, which skirted Newfoundland, and Claudette, which formed off Tampa Bay and came ashore in the Florida Panhandle as a tropical storm.

Ida, a rare November system, was briefly a Category 2 hurricane, but then weakened to a tropical storm. It moved across the Gulf Coast from Mobile, AL, through the Panhandle and northern part of Florida before drenching the southeastern coastal United States.

MMS: 2009–18 offshore forecast

In May 2009, the U.S. Minerals Management Service released its 2009–18 forecast for Gulf of Mexico oil and gas production.

MMS expects oil production (including condensate) to peak in 2013 at an average 1.879 million b/d, compared with hurricane-affected production of 1.142 million b/d in 2008.

After the peak, MMS sees the average dropping to 1.735 million b/d in 2018.

For natural gas, MMS sees production averaging 7.03 bcfd in 2009, compared with 6.43 bcfd in 2008. Production then drops to 6.22 bcfd in 2012 before starting an increasing trend; production climbs to 8.27 bcfd in 2018.

MMS sees continued decrease in shallow-water production of both oil and gas, to 82,000 b/d and 900 MMcfd in 2018, respectively.

Deepwater: Key to future gulf production

MMS forecast in May 2009 that Gulf of Mexico oil production will increase substantially in the near term, possibly reaching a record-high 1.8 million bbl/d of oil by 2013.

Deepwater activity is expected to grow. MMS noted that in 2008, operators announced that 15 deepwater discoveries and seven new projects began production in 1,000 ft (305 m) of water or deeper.

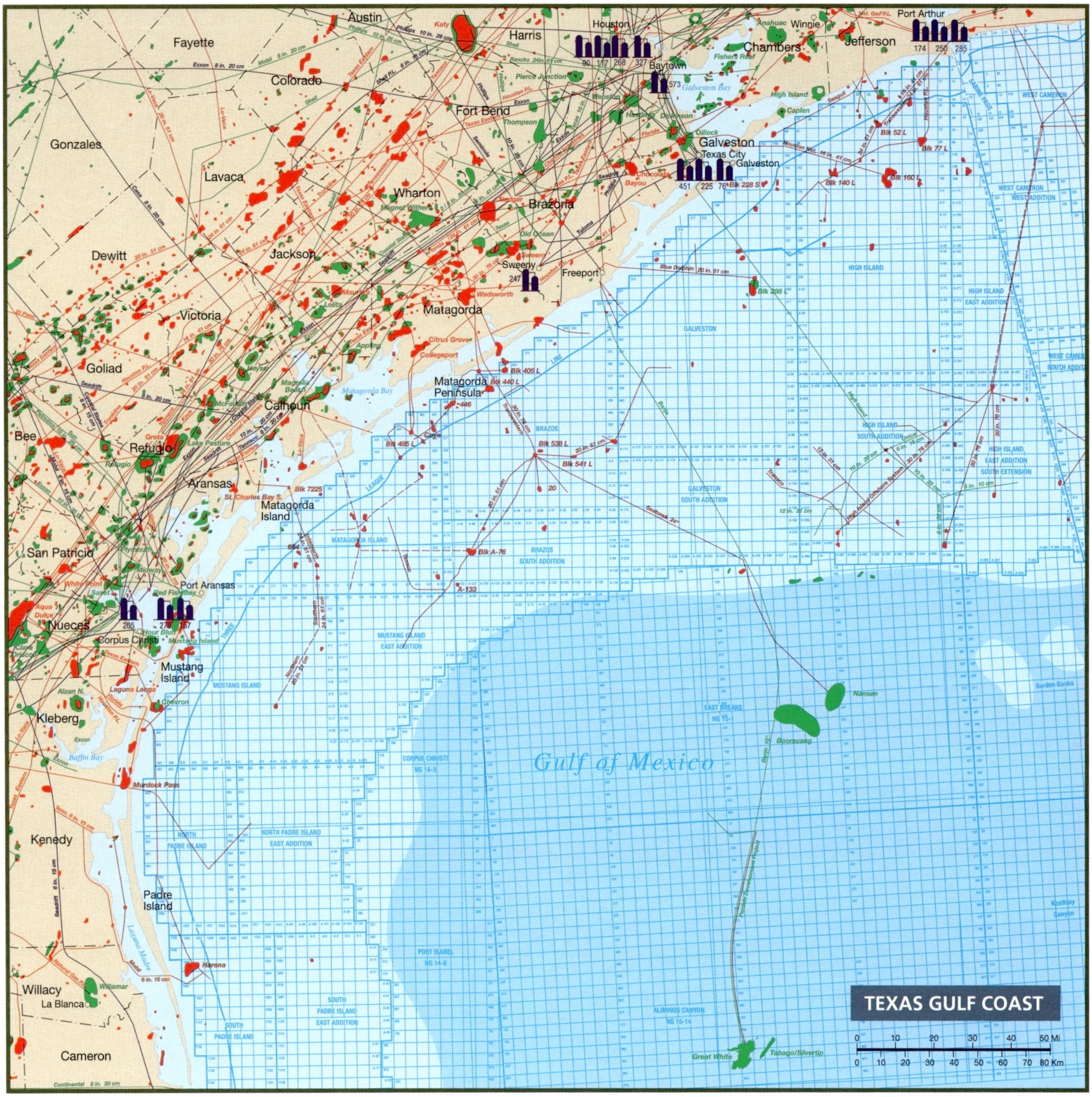

Also in 2008, MMS said that 57% of all GoM leases were in water depths greater than 1,000 ft (305 m) with 141 producing deepwater projects, up from 130 at the end of 2007.

Also, 73% of the tracts receiving bids in the three lease sales held in 2008 were in deepwater.

Exploration and Development

Tiber discovery

BP PLC reported in September 2009 a giant gulf Lower Tertiary deepwater discovery, found with one of the world's deepest wells.

The Tiber well—about 250 mi (402 km) southeast of Houston—went to TD 35,055 ft (10,685 m) on Keathley Canyon Block 102 [almost 6 mi (10 km) below the seabed] in 4,132 ft (1,259 m) of water.

Oil was found in multiple Lower Tertiary (Paleogene) reservoirs.

Unofficial estimates put recoverable oil at 500 million to 1 billion bbl.

BP expects that Tiber will rival its 2006 Kaskida discovery, 45 mi (72 km) to the southeast on Keathley Canyon Block 292. Kaskida is under appraisal through 2010.

BP was working on nine gulf projects: Atlantis Phase 2, Tubular Bells, Kodiak, Freedom, Kaskida, Isabela, Santa Cruz, Mad Dog tiebacks, and Great White.

Other interests in Tiber are Brazil's Petrobras and ConocoPhillips.

Guará: 2 billion bbl?

Also in September, Petrobras reported that the giant Guará field, in Brazil's Santos Basin, contained the equivalent of 1.1–2 billion bbl of recoverable oil and gas—about twice the volumes announced days earlier by BP for its Tiber field, also in the gulf.

BG Group and Repsol are partners with Petrobras in developing Guará, which could begin producing as soon as 2012.

Petrobras points out that the Tupi and Iara fields, part of a group that includes Guará, together hold an estimated 8–12 billion bbl of recoverable oil.

Shenzi field

Oil and gas production began in early 2009 from the Shenzi development in Green Canyon Block 653, said partners Repsol, BHP Billiton, and Hess Oil Corp.

Two new finds were made in the third quarter:

- A new oil pay [555 ft (169 m)] in the existing G104 well, which was drilled to 25,605 ft (7,804 m).
- About 100 ft (30 m) of net oil pay in Shenzi-9 appraisal well, opening additional potential to the Shenzi North Flank.

Shenzi crude export line

Enterprise Products Partners announced completion of the crude pipeline serving Shenzi field in March 2009. Operator of the field is BHP Billiton.

The 83-mi (134-km) line can transport up to 230,000 b/d of from Green Canyon Block 653 to the Ship Shoal 332B junction platform.

Shenzi is 120 mi (193 km) off the coast of Louisiana, in about 4,300 ft (1,311 m) of water.

South Marsh Island Block 230

McMoRan Exploration Co. was setting casing in September 2009 at a proposed 28,000-ft ultradeep shelf prospect.

The company reentered a wellbore on South Marsh Island Block 230 in 20 ft (6 m) of water south of Vermilion Bay. The Davy Jones prospect involves a large ultradeep structure on four blocks.

McMoRan will deepen the well to test Eocene Wilcox, Paleocene, and possibly Cretaceous Tuscaloosa formations.

Mississippi Canyon: Santa Cruz

Noble Energy hit pay in early 2009 with a well on the Santa Cruz prospect in Mississippi Canyon Blocks 519/563.

The well, located in 6,515 ft (1,986 m) of water, was drilled to TD of 18,900 ft (5,761 m).

It found more than 140 ft (43 m) of net gas condensate pay and more than 110 ft (34 m) of net oil pay in multiple reservoirs.

First production is expected in 2011.

Mississippi Canyon: Block 984

Anadarko Petroleum reported in July 2009 that it hit pay with the Vito exploration well on Mississippi Canyon Block 984.

The well encountered more than 250 net ft (76 m) of oil pay in subsalt Miocene sands.

The well was drilled to TD 32,000 ft (9,754 m) in 4,038 ft (1,231 m) of water.

Other interests are Shell Offshore and Statoil USA E&P.

Green Canyon: Mad Dog South

BP PLC said in July 2009 that it drilled a successful appraisal well in a previously untested southern part of Mad Dog field on Green Canyon Block 826.

The 826-5 well, about 100 mi (161 km) south of Grand Isle, LA, was drilled in 5,100 ft (1,554 m) of water.

It found 280 net ft (85 m) of hydrocarbons in the objective Miocene hydrocarbon-bearing sands and also found an oil column of more than 2,200 ft.

Mad Dog field started production in 2005.

Other interests are BHP Billiton and Chevron Corp.

Green Canyon: Samurai

Anadarko Petroleum made a Miocene discovery in mid-2009 at the Samurai prospect in Green Canyon block 432. It was the firm's third deepwater discovery in the gulf in 2009.

The discovery well, drilled to about 31,700 ft (9,662 m), encountered more than 120 ft (37 m) of net oil pay in several high-quality sands.

Other interests are Murphy Oil and Samson Offshore.

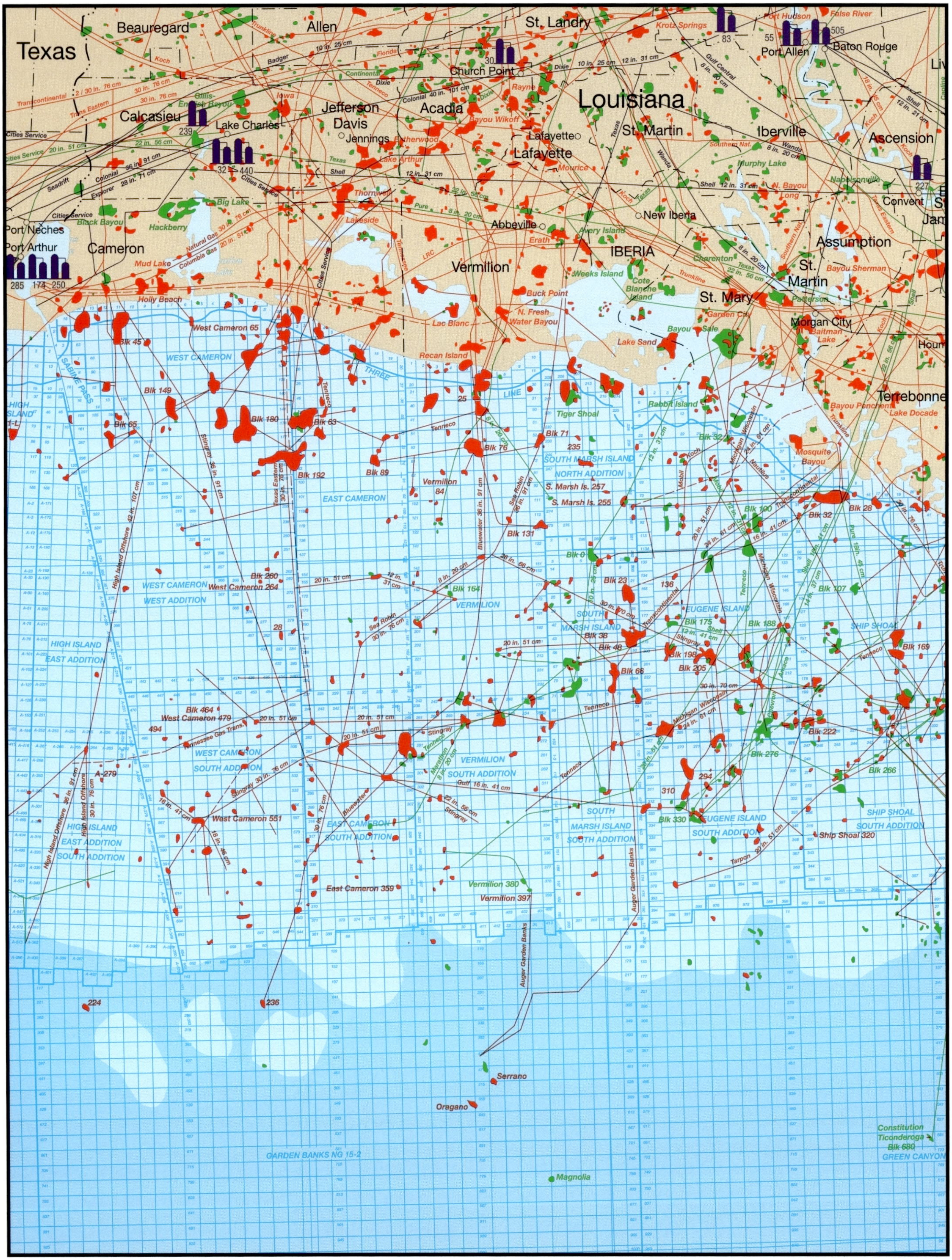

Texas
Beauregard
Allen
St. Landry
Louisiana
Calcasieu
Lake Charles
Jefferson Davis
Jennings
Acadia
Church Point
Lafayette
St. Martin
Iberville
Ascension
Port Allen
Baton Rouge
Port Hudson
False River
Krotz Springs
Convent
Assumption
Port Neches
Port Arthur
Cameron
Hackberry
Big Lake
Black Bayou
Mud Lake
Holly Beach
Abbeville
New Iberia
IBERIA
Vermilion
Weeks Island
Cote Blanche Island
Avery Island
St. Mary
St. Martin
Morgan City
Garden City
Patterson
Lake Sand
Bayou Sale
Buck Point
N. Fresh Water Bayou
Lac Blanc
Recan Island
Terrebonne
Lake Docade
Bayou Penchant
Mosquito Bayou
Tiger Shoal
Rabbit Island
West Cameron 65
WEST CAMERON
SABINE PASS
HIGH ISLAND
THREE
LINE
Blk 45
Blk 149
Blk 65
Blk 180
Blk 63
Blk 192
Blk 89
Blk 76
Blk 71
SOUTH MARSH ISLAND NORTH ADDITION
S. Marsh Is. 257
S. Marsh Is. 255
Vermilion 84
EAST CAMERON
Blk 131
Blk 32
Blk 28
Blk 100
Blk 260
West Cameron 264
WEST CAMERON WEST ADDITION
Blk 164
VERMILION
Blk 23
SOUTH MARSH ISLAND
Blk 38
Blk 48
Blk 66
EUGENE ISLAND
Blk 175
Blk 188
Blk 196
Blk 205
Blk 107
SHIP SHOAL
Blk 169
HIGH ISLAND EAST ADDITION
Blk 464
West Cameron 479
494
A-279
WEST CAMERON SOUTH ADDITION
West Cameron 551
HIGH ISLAND EAST ADDITION SOUTH ADDITION
EAST CAMERON SOUTH ADDITION
East Cameron 359
VERMILION SOUTH ADDITION
Vermilion 380
Vermilion 397
SOUTH MARSH ISLAND SOUTH ADDITION
Blk 330
EUGENE ISLAND SOUTH ADDITION
Blk 222
Blk 276
Blk 266
SHIP SHOAL SOUTH ADDITION
Ship Shoal 320
310
294
224
236
Serrano
Oragano
GARDEN BANKS NG 15-2
Magnolia
Constitution
Ticonderoga
Blk 680
GREEN CANYON
Auger Garden Banks

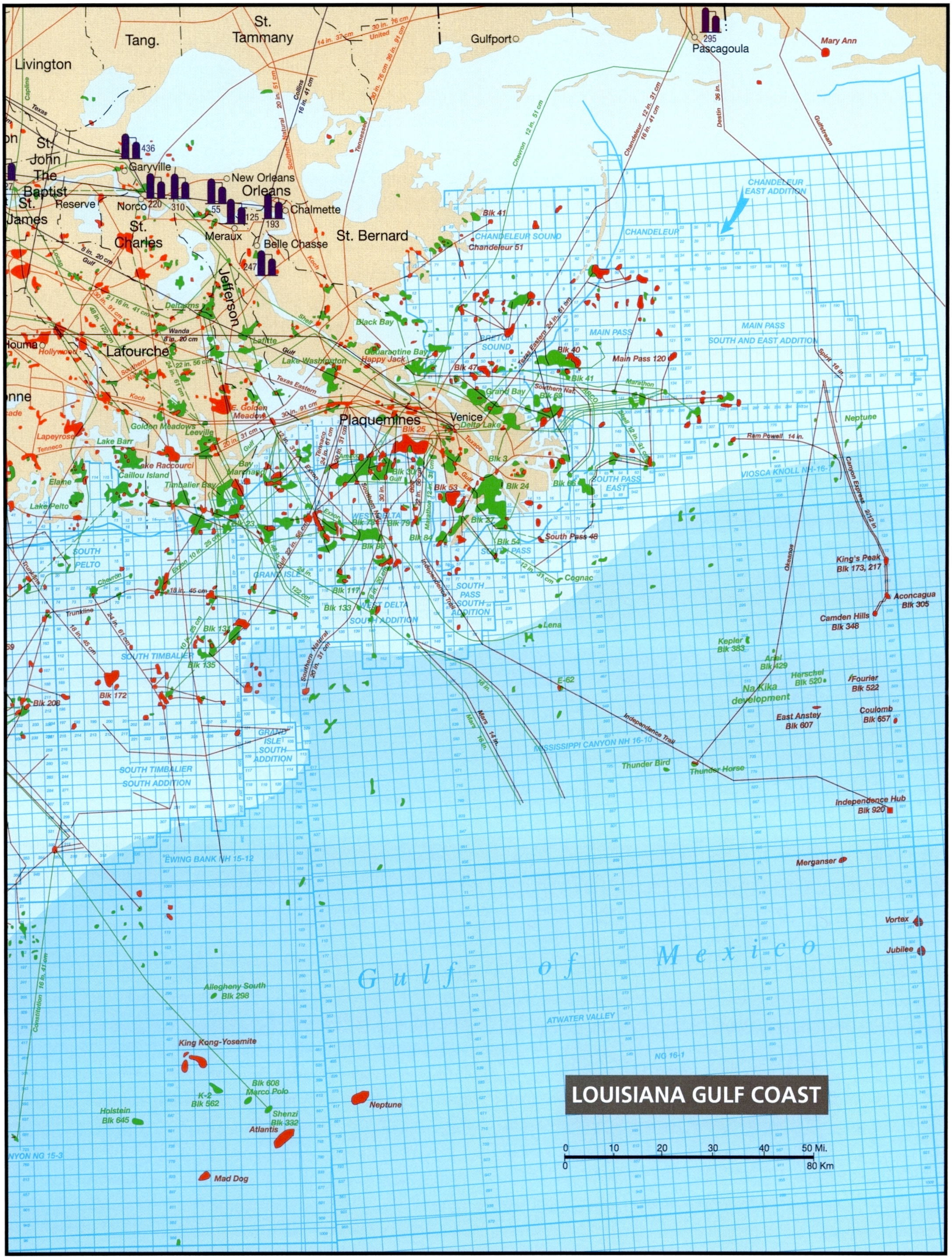
Livington
Tang.
St. Tammany
Gulfport
Pascagoula
Mary Ann
St. John The Baptist
St. James
Garyville
Reserve
Norco
St. Charles
New Orleans
Orleans
Chalmette
Meraux
Belle Chasse
St. Bernard
Jefferson
Houma
Lafourche
Plaquemines
Venice
CHANDELEUR SOUND
CHANDELEUR
CHANDELEUR EAST ADDITION
BRETON SOUND
MAIN PASS
MAIN PASS SOUTH AND EAST ADDITION
SOUTH PASS EAST
VIOSCA KNOLL NH 16-7
SOUTH PELTO
GRAND ISLE
WEST DELTA
SOUTH PASS
SOUTH PASS SOUTH ADDITION
WEST DELTA SOUTH ADDITION
SOUTH TIMBALIER
GRAND ISLE SOUTH ADDITION
SOUTH TIMBALIER SOUTH ADDITION
EWING BANK NH 15-12
MISSISSIPPI CANYON NH 16-10
ATWATER VALLEY
NG 16-1
Gulf of Mexico
Neptune
King's Peak Blk 173, 217
Aconcagua Blk 305
Camden Hills Blk 348
Kepler Blk 383
Ariel Blk 429
Herschel Blk 520
Fourier Blk 522
Na Kika development
East Anstey Blk 607
Coulomb Blk 657
Thunder Bird
Thunder Horse
Independence Hub Blk 920
Merganser
Vortex
Jubilee
Allegheny South Blk 298
King Kong-Yosemite
K-2 Blk 562
Blk 608 Marco Polo
Shenzi Blk 332
Atlantis
Holstein Blk 645
Neptune
Mad Dog
LOUISIANA GULF COAST
0 10 20 30 40 50 Mi.
0 80 Km

Pyrenees, Winter gas finds

Newfield Exploration Co. reported two natural gas discoveries in the deepwater gulf in June 2009, both in the Garden Banks area.

Pyrenees, drilled on Block 293 in 2,100 ft (640 m) of water, encountered 125 ft (38 m) of net pay in three intervals. Newfield described the find as a gas-condensate discovery and said it could be the company's largest find.

Other interests in Pyrenees are Stone Energy Corp., Ridgewood Energy Corp., Arena Exploration, and Deep Gulf Energy.

Winter, drilled on Garden Banks Block 605, in 3,400 ft (1,036 m) of water, found 44 ft (13 m) of net pay in two sands. Other interests include Apache Corp., Deep Gulf Energy, and Royal Offshore.

Caesar Tonga development

The board of Statoil approved development of three discoveries in the gulf in first quarter 2009. Anadarko is the operator for all three: Tonga, Tonga West, and Caesar.

They have been combined into one project, to be a subsea tieback to the *Constitution* spar.

Four initial wells are expected to yield combined average 40,000 b/d, starting in 2011.

Other interests are Royal Dutch Shell and Chevron Corp.

Burgos project

Mexico's Pemex reported discovery of "significant" amounts of natural gas and condensate in February 2009 with its Tsimin-1 wildcat on Mision Block in Burgos field.

The Tsimin-1 well had initial production of 4,400 boe/d, Pemex said.

Pemex also announced first gas flow from the Cali-1 well in its Burgos project, with production starting at 9.1 MMcfd.

The company said development of Burgos will provide an additional 90–110 MMcfd of gas.

Early-2009 discoveries

Operators in the gulf reported a string of discoveries in late January and early February 2009.

Keathley Canyon: Buckskin

A group led by operator Repsol YPF SA found more than 300 ft (91 m) of net pay in Lower Tertiary at the Buckskin No. 1 well, 190 mi (306 km) southeast of Houston in Keathley Canyon Block 872.

The well went to TD 29,404 ft (8,962 m) in 6,920 ft (2,109 m) of water.

Walker Ridge: Shenandoah

Anadarko Petroleum Corp. and partners marked a Lower Tertiary oil discovery with the Shenandoah well in Walker Ridge Block 52.

The well, which encountered net oil pay approaching 300 ft (91 m) in the Wilcox formation, was drilled in 5,750 ft (1,753 m) of water to a TD of 30,000 ft (9,144 m).

Green Canyon: Heidelberg

Another group led by Anadarko gauged a discovery on the Heidelberg prospect in the Middle Miocene.

The well, in 5,000 ft (1,524) m) of water in Green Canyon Block 859, went to TD 28,500 ft (8,687 m) and cut more than 200 ft (61 m) of net oil pay in several Miocene sands.

Garden Banks: Bushwood

Mariner Energy operates Bushwood and Smoothie discoveries.

Bushwood-1, formerly Geauxpher-3, drilled to TD of nearly 25,300 ft (7,712 m) in 2,700 ft (823 m) of water in Garden Banks Block 463. It logged more than 260 ft (79 m) true vertical thickness (TVT) of net gas pay in multiple sands.

Smoothie-2, a deep shelf prospect in 60 ft (18 m) of water in South Timbalier Block 49, went to TD of more than 20,100 ft (6,127 m) and logged more than 200 ft (61 m) TVT of net gas pay in multiple zones.

Block 643: Friesian

Plains Exploration & Production Co. confirmed the Friesian deepwater discovery and made plans to deepen the Friesian-2 well on Block 643.

Deepening the well by about 3,000 ft (914 m)—to 32,500 ft (9,906 m)—the firm aims to test sands that flowed at high rates at the giant Tahiti structure, two blocks west.

Initial production is planned in 2012.

Lease Sales

Lease Sale 207: Western Gulf

In January 2009, MMS accepted high bids valued at $483,959,404 and awarded 313 leases under Western Gulf of Mexico Oil and Gas Lease Sale 207. The original lease sale was August 2008.

Fifty-three companies submitted 423 bids on 319 tracts. The total for high bids submitted on all tracts was $487,297,676.

The highest bid accepted on a tract was $61,110,000 made by Statoil Gulf of Mexico for Alaminos Canyon block 380, in more than 6,562 ft (2,000 m) of water.

The top three companies with the highest number of accepted high bids were:

- Exxon Mobil: 128 bids ($125,264,000)
- Hess Oil Corp.: 22 bids ($14,162,407)
- Chevron: 20 bids ($127,281,322)

Lease Sale 208: Central Gulf

In August 2009, MMS accepted high bids valued at $700 million and awarded 328 leases under the Central Gulf of Mexico Oil and Gas Lease Sale 208. That sale was held March 18.

Seventy companies submitted 476 bids on 348 tracts in the sale.

The total for high bids submitted on all tracts was $703,048,523.

The highest bid accepted on a tract was $65,611,235 submitted by Shell Gulf of Mexico Inc. for Mississippi Canyon Block 721. This tract lies in 2,625–5,250 ft (800–1,600 m) of water and received two bids.

The top five companies with the highest number of accepted high bids were:

- Shell Gulf of Mexico Inc.: 39 for $154 million
- BHP Billiton Petroleum (Deepwater) Inc.: 28 for $50 million
- BP Exploration & Production Inc.: 25 for $77 million
- Ecopetrol America Inc.: 22 for $19 million
- Noble Energy Inc.: 22 for $54 million.

Lease Sale 210: Western Gulf

In August 2009, apparent high bids totaled more than $115 million for 162 tracts in the Western gulf during Lease Sale 210.

MMS received 189 bids totaling just over $145 million from 27 companies. (See two tables on this page.)

The ultradeep water was a big draw. Of the tracts receiving bids, 94 blocks were in 2,625–5,250 ft (800–1,600 m) of water, and 26 were in water deeper than 5,250 ft (1,600 m).

Top 10 companies by sum of apparent high bids

Rank, company	Total high bids	Sum of high bids, $
1. BP Exploration & Production Inc.	37	50,634,191
2. ConocoPhillips	22	15,172,500
3. Petrobras America Inc.	4	10,000,000
4. ChevronUSA Inc.	26	9,075,343
5. ExxonMobil Corp.	17	8,587,021
6. Focus Exploration Inc.	15	4,732,858
7. Anadarko E&P Co.	5	4,367,544
8. Castex Offshore Inc.	3	1,912,000
9. Contango Operators Inc.	3	1,651,000
10. SWEPI LP	3	1,499,426

Source: U.S. Minerals Management Service

The 3,435 blocks offered covered more than 18 million acres in the western gulf's OCS planning area off Texas.

BP Exploration & Production Inc. made 37 bids totaling $50 million, and Chevron USA Inc. placed 26, totaling $9 million. ConocoPhillips Co. placed 22 total high bids totaling $15 million.

BP E&P placed the highest single bid for a block—$28.1 million for Keathley Canyon Block 96, which lies in more than 9,482 ft (3,000 m) of water.

Lease Sale 213: March 2010

The MMS said in November 2009 that Oil and Gas Lease Sale 213 for the Central Gulf of Mexico Planning Area was planned for March 17, 2010.

The proposed sale encompasses some 6,800 unleased blocks covering more than 35.9 million acres offshore Louisiana, Mississippi, and Alabama.

Top 10 tracts by high bid

Rank, company	Block	Water depth, m	High bid, $
1. BP Exploration & Production Inc.	Keathley Canyon 96	800-<2,600	28,133,843
2. Petrobras America Inc.	Keathley Canyon 223	800-<1,600	8,122,000
3. BP Exploration & Production Inc.	Keathley Canyon 180	800-<1,600	5,133,843
4. ConocoPhillips	Alaminos Canyon 476	800-<1,600	3,060,000
5. BP Exploration & Production Inc.	East Breaks 875	800-<1,600	2,933,843
6. ConocoPhillips	East Breaks 406	400-<800	2,550,000
7. Anadarko E&P Co. LP	East Breaks 729	800-<1,600	2,249,351
8. BP Exploration & Production Inc.	Garden Banks 841	800-<1,600	1,933,843
9. Chevron USA Inc.	Alaminos Canyon 774	>2,000	1,462,076
10. Anadarko E&P Co. LP Nexen Petroleum Offshore USA Inc.	Garden Banks 448	400-<800	1,415,200

Source: U.S. Minerals Management Service

MMS said that the sale could result in production of up to 1.3 bbl of oil and 5.4 tcf of natural gas.

The acreage is 3–230 mi (5–370 km) offshore, in water depths ranging from 10 ft (3 m) to more than 11,200 ft (3,400 m). It includes area 181 South in the southeastern part of the Central Planning Area, which contains about 4.2 million acres.

In a policy shift to spur timely development, MMS set shorter initial lease terms for the sale:

- 5 years (not 8) for tracts in water 1,312–2,625 ft (400–800 m) deep
- 7 years (not 10) for blocks in 2,625–5,250 ft (800–1,600 m) of water

Payments to Gulf States

The Dept. of the Interior in March 2009 paid more than $25 million to Gulf of Mexico coastal states and communities as their share of bonuses and rents on energy leases in federal waters off their coasts.

Amounts were: $6.18 million to the state of Alabama; $6.35 million to Louisiana; $5.51 million to Mississippi; and $2.16 million to Texas.

Forty-two counties, parishes and other political subdivisions in those states will receive a total of $5.05 million.

These were the first payments under the 2006 Gulf of Mexico Energy Security Act (GOMESA).

It provides that these states and counties receive 37.5% of the oil and gas qualified leasing revenues from certain OCS areas.

The law directs that the funds be used for specific purposes, including for coastal protection and for carrying out a federally-approved marine, coastal, or comprehensive conservation management plan.

Drilling and Production

Analysis: Marginal GoM fields

Researchers at Louisiana State University analyzed in late 2009 a study by the MMS of marginal oil and gas production in the gulf.

That study estimated marginal production and gross revenue streams associated with the inventory of producing structures in place in early 2007.

Near the end of a structure's life, revenues approach operating cost and the asset becomes marginally profitable. Enhanced-recovery methods can be applied, but doing so is challenging offshore.

Nonetheless, marginal fields are expected to contribute a greater percentage of world supply in the future.

Cumulative hydrocarbon production from the January 2007 inventory of producing assets is estimated to be 1.056 billion bbl of oil and 13.3 tcf of gas.

Deepwater structures do not contribute materially to the structure count in the charts here.

GULF OF MEXICO CUMULATIVE STRUCTURES

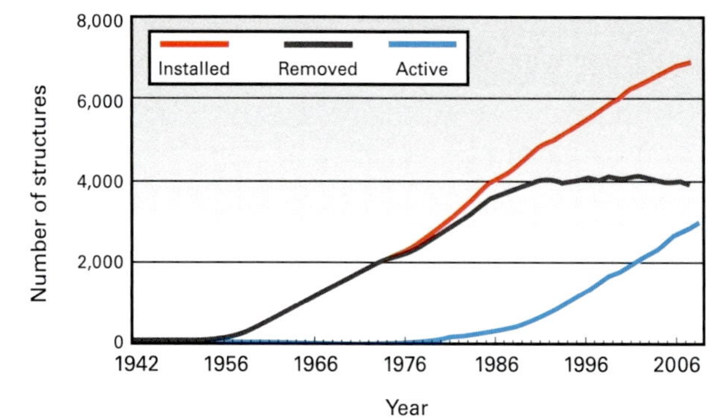

ACTIVE GULF OF MEXICO STRUCTURES

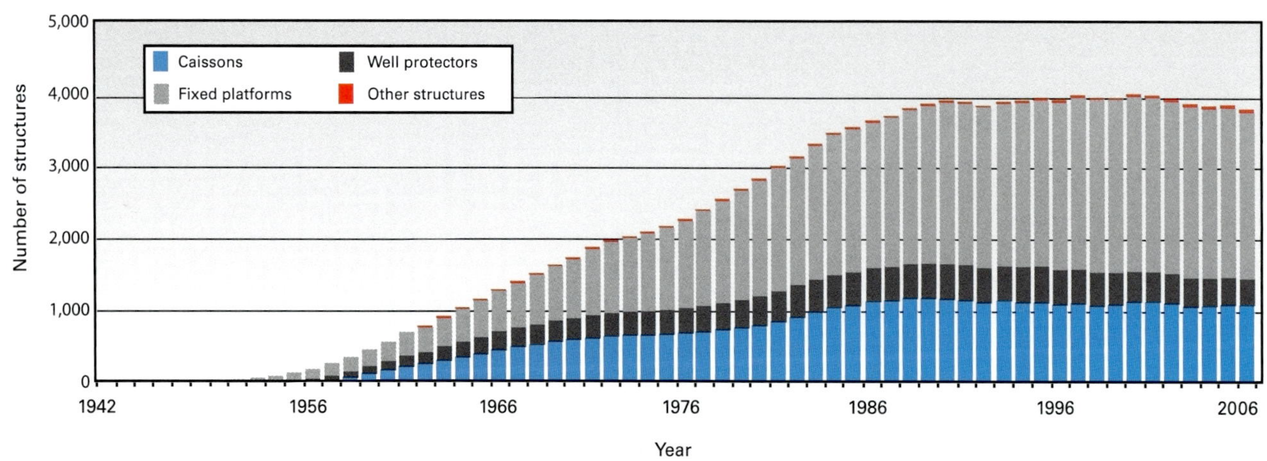

Gulf of Mexico Producing Structures

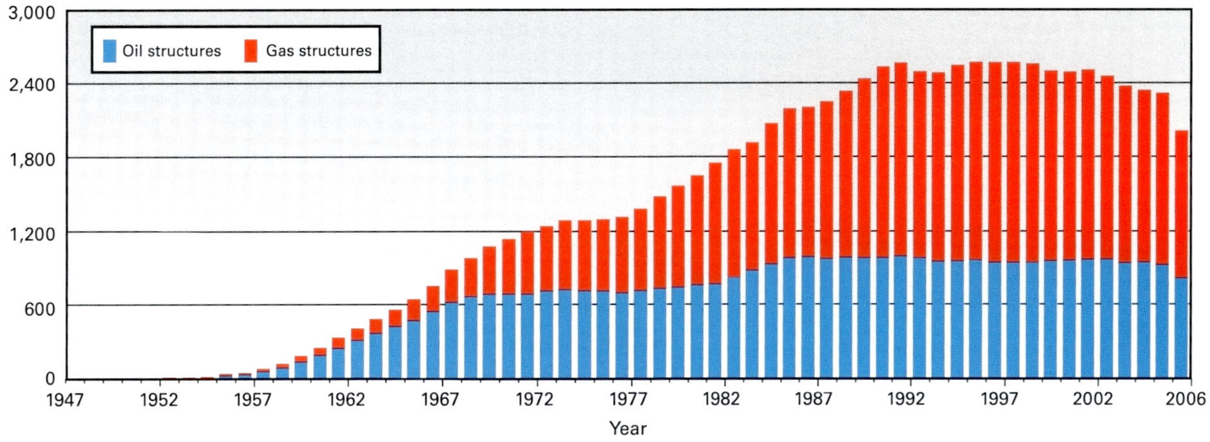

Oil Production from Gulf Structures

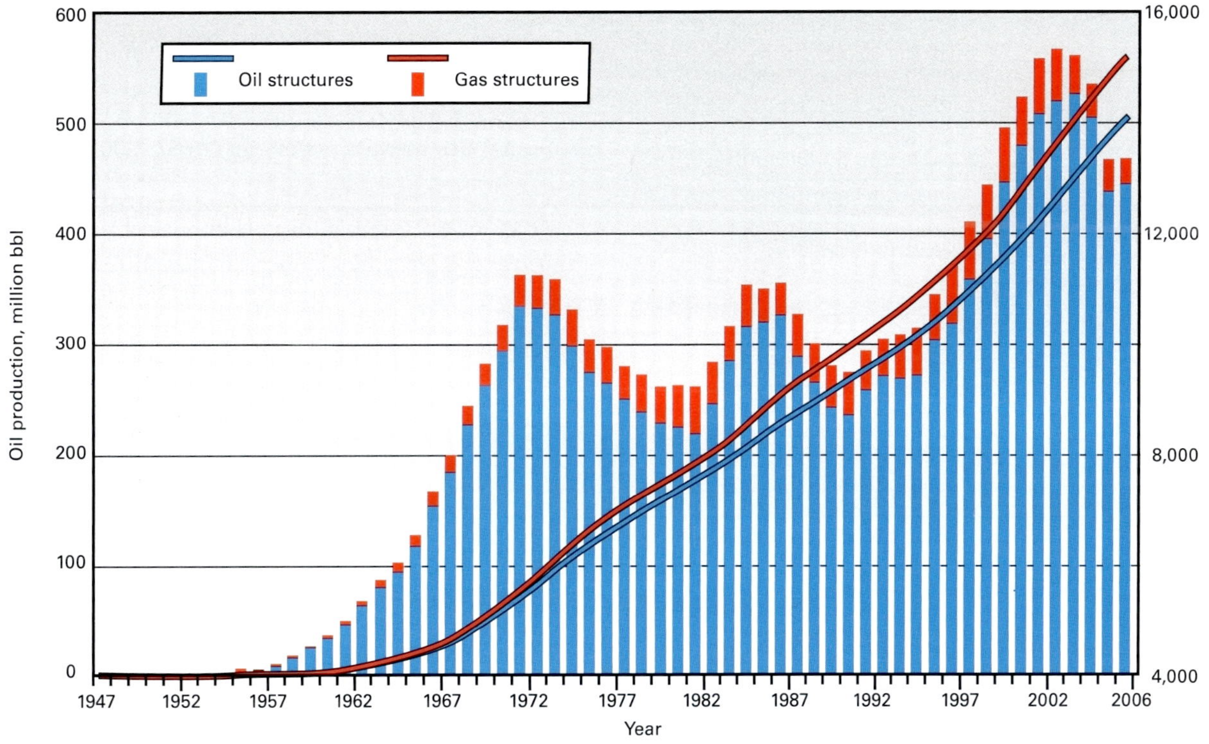

First Gulf of Mexico FPSO

In November 2009, Keppel Shipyard said it was on track to deliver the first FPSO for use in the U.S. Gulf of Mexico.

The *BW Pioneer* will be moored in water 8,530 ft (2,600 m) deep, the deepest for an FPSO, the company said.

Petrobras America has leased the vessel for the Cascade and Chinook fields. Production is scheduled for 1Q 2010.

The FPSO has a storage capacity of about 600,000 bbl of oil, a process capacity of 80,000 b/d of oil, and gas export facilities of 16 MMcf/d.

Gas production from Gulf structures

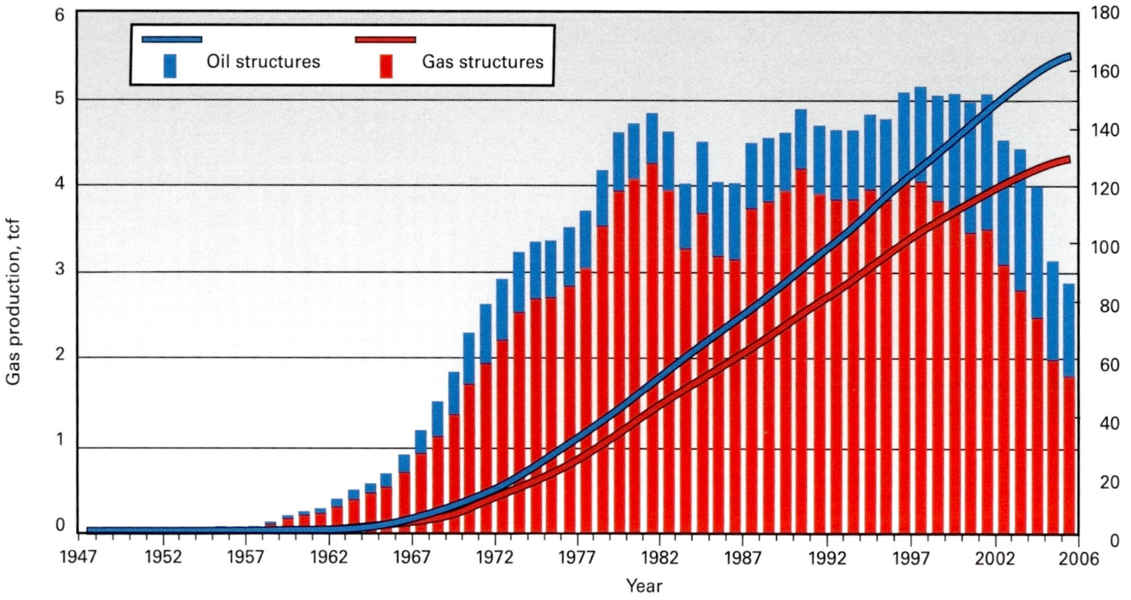

Gulf oil structures by daily oil production

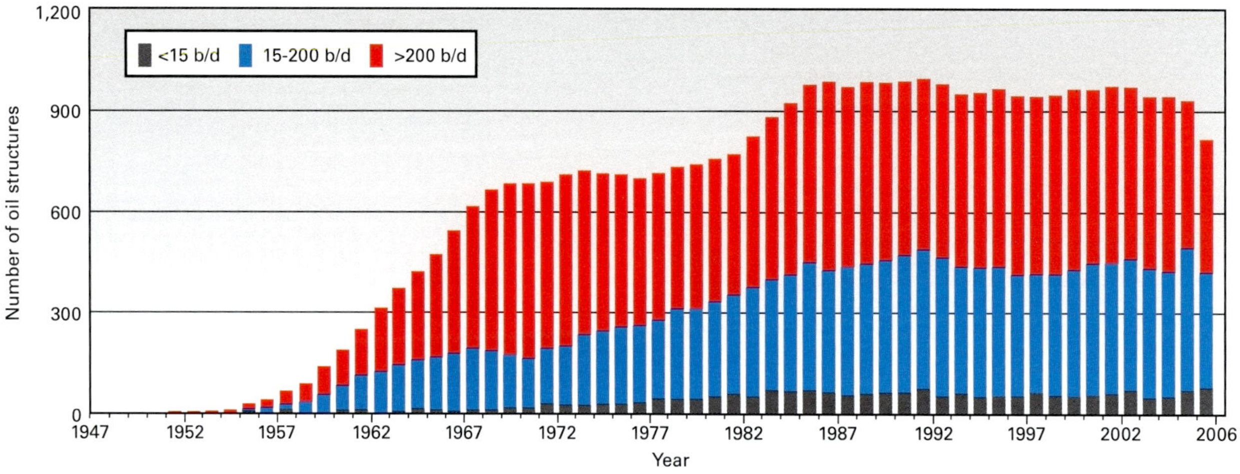

Gulf gas structures by daily gas production

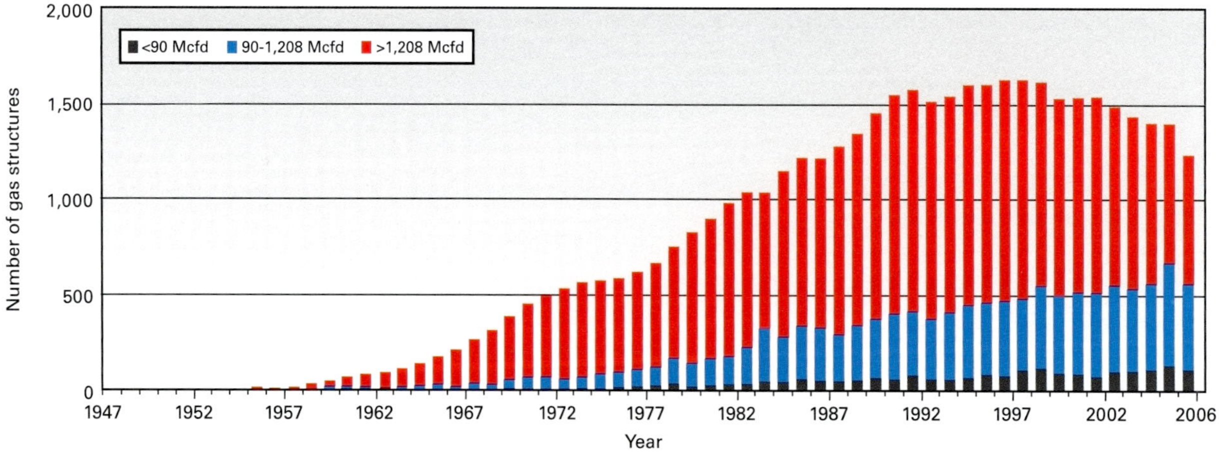

Thunder Hawk

Production started in July 2009 from the Murphy Oil-operated Thunder Hawk field.

The field production unit can process 45,000 b/d of oil and 70 MMcf/d of natural gas.

Thunder Hawk, discovered in 2004 and sanctioned in 2006, is in Mississippi Canyon Block 734.

Geauxpher

Mariner Energy reported in May 2009 the start of production from the Geauxpher development.

Sited in 2,700 ft (823 m) of water, Geauxpher includes two wells with gross production of approximately 115 MMcf/d of natural gas equivalent.

The wells are connected to a platform on Garden Banks Block 72.

Dorado, King South

BP PLC announced in May 2009 that production started from Dorado and King South fields.

Both are subsea tiebacks to BP's Marlin tension-leg platform on Viosca Knoll Block 915.

Dorado is 2 mi (3 km) from the TLP; King South is 18 mi (29 km) from the TLP.

Gulf contribution to U.S. oil

In early 2009, the U.S. EIA forecast that domestic U.S. oil production would increase by 320,000 b/d in 2009, to an average of 5.25 million b/d—the first production increase since 1991.

And the agency expected that two-thirds of the increase would come from three gulf platforms: Atlantis, Thunder Horse, and Tahiti.

The three will account for two-thirds of the total increase, said EIA.

By early 2010, the three are expected to reach their combined designed production capacity of 575,000 b/d of oil.

Production from Atlantis field (Green Canyon Block 699) began in October 2007 and from Thunder Horse (Mississippi Canyon Block 778) in mid-2008.

First production from Tahiti (Green Canyon Block 640) was expected by third quarter 2009.

Thunder Horse: Production up

BP America reported that new wells at the Thunder Horse semisubmersible platform boosted production to more than 300,000 boe/d (from seven wells) in March 2009.

Two more wells were to be added by yearend 2009.

Thunder Horse, the gulf's largest deepwater producing asset, is 150 mi (241 km) southeast of New Orleans, in 6,050 ft (1,844 m) of water on Mississippi Canyon Blocks 778 and 822.

The platform is designed to process 250,000 b/d of oil and 200 MMcfd of gas.

The field was discovered in 1999 in a reservoir almost 16,000 ft (4,877 m) below the seabed, with pressures of 13,000–18,000 psi and temperatures of 88–135°C.

Tahiti: Production starts

Chevron Corp. reported that production began in early May 2009 from Tahiti—deepest producing field in the Gulf of Mexico.

The firm expected daily production to ramp up to about 125,000 bo/d and 70 MMcfd of natural gas by yearend.

Discovered in 2002, Tahiti field is one of the largest in the gulf and contains 400–500 million boe of recoverable resources, according to Chevron. The first phase of the project cost $2.7 billion.

Tahiti lies on Green Canyon Blocks 596, 597, 640, and 641 in 4,100 ft (1,250 m) of water about 190 mi (306 km) south of New Orleans.

The deepest producing well has a depth of more than 26,700 ft, a record for the gulf, said Chevron.

Perdido: Topsides installed

Royal Dutch Shell reported in March 2009 that it completed installation of the 9,500-ton drilling and production platform atop a 555-ft (169-m) cylindrical spar at its Perdido development, in 8,000 ft (2,438 m) of water some 200 mi (322 km) from Houston.

It is the deepest such facility in the world, company officials claimed.

First oil production was expected by third quarter.

The platform can produce 130,000 boe/d from a subsea well drilled to 9,356 ft (2,852 m) below the water's surface. Another well to 9,627 ft (2,934 m) is planned.

Shell operates Perdido; partners are BP PLC and Chevron Corp.

Perdido-to-shore infrastructure

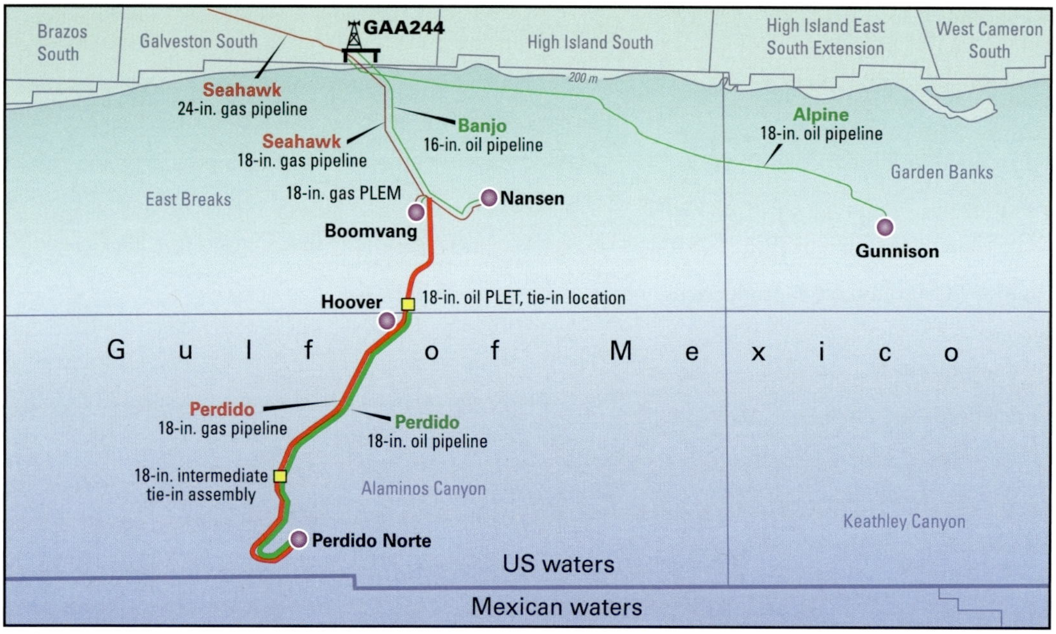

Longhorn gas field onstream

Production began in late 2009 from the Longhorn gas field in the gulf, according to operator Eni. Initial gas production was 200 MMcf/d from four subsea wells connected to the Corral platform.

The platform is in Mississippi Canyon Block 365 about 20 mi (32 km) northwest of the field in 620 ft (189 m) of water.

Daniel Boone onstream

Gas and oil began flowing in October 2009 from the Daniel Boone discovery well in Green Canyon Block 646, said W&T Offshore.

Path of Hurricane Gustav

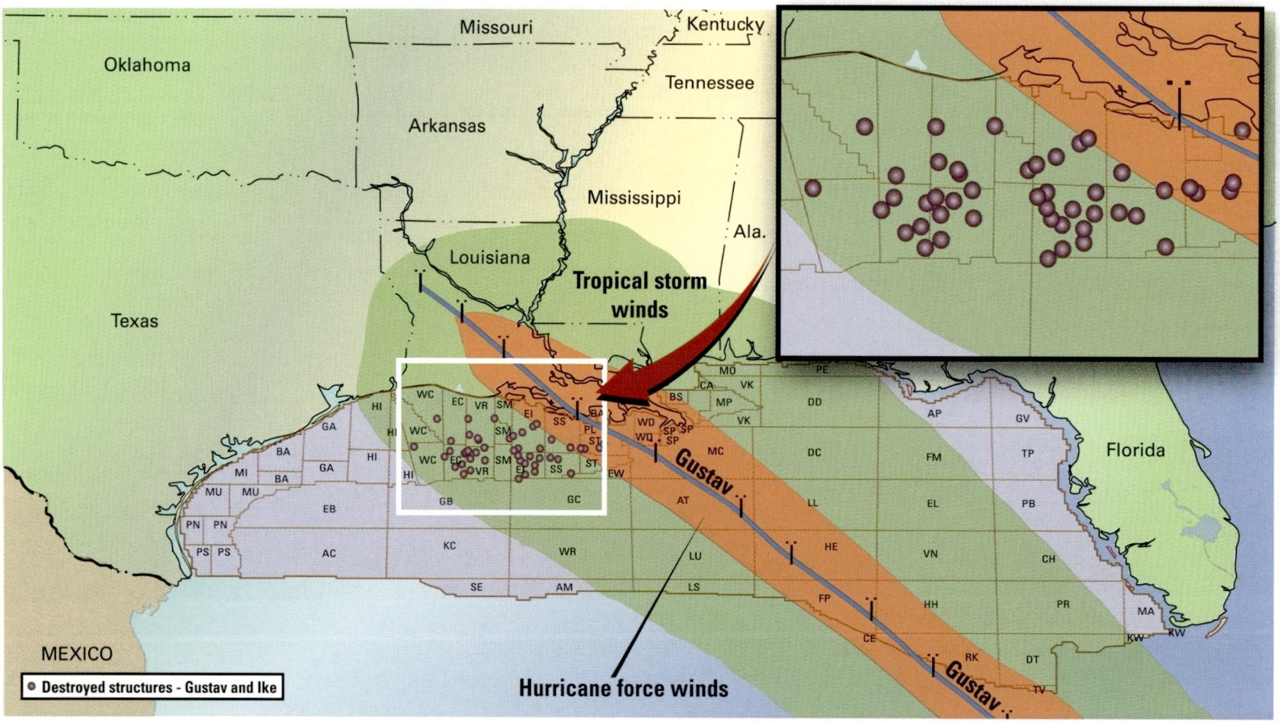

Gross daily production was about 6,000 b/d of oil and 5.7 MMcf/d of natural gas, or 6,950 boe/d.

Viosca Knoll Block 1003

Newfield Exploration Co. was planning to start production in November 2009 from the Fastball discovery in Viosca Knoll Block 1003.

Shenzi milestone

In September 2009, Shenzi field in the deepwater gulf achieved sustained rates of 120,000 b/d of oil from seven wells, said BHP Billiton Petroleum.

Two additional wells are shut in and available for production as needed.

The Shenzi facility is 120 mi (195 km) off the Louisiana coast in approximately 4,300 ft (1,300 m) of water on Green Canyon Block 653.

Recovery from Gustav, Ike seen

Researchers at Louisiana State University in 2009 estimated that the gas structures damaged during Hurricanes Gustav and Ike (August–September 2008) present better economics and redevelopment potential than oil structures.

They also predicted that companies will redevelop 196 million boe, or nearly 95%, of the remaining reserves.

An official government assessment listed 60 destroyed platforms and 31 extensively damaged structures. Remaining reserves from these platforms are estimated at about 234 million boe.

Path of Hurricane Ike

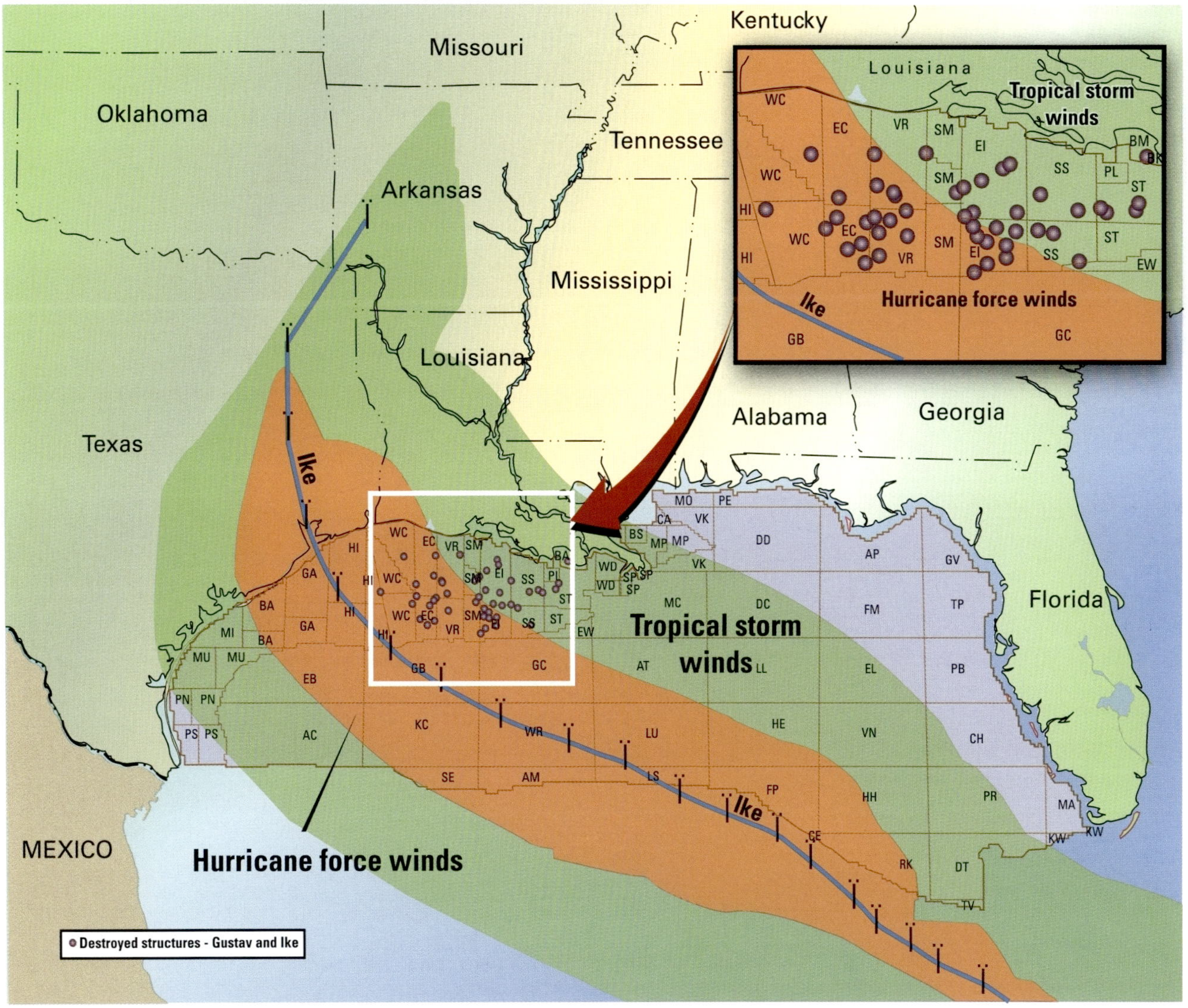

Destroyed structure historic production

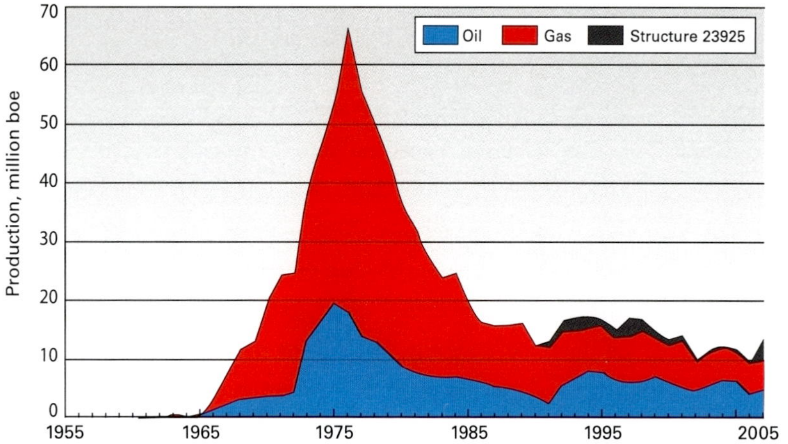

Hurricanes Gustav, Ike destroyed structures

Production type	Caisson	Well protector	Fixed platform	Total
Oil	1	2	14	17
Gas	1	2	31	34
Total	**2**	**4**	**45**	**51**

Source: Minerals Management Service

Structures destroyed

Complex/ structure	Operator	Area	Block	Structure type	Prod. type	Depth, ft	First year on prod.	Last year on prod.	Total number of drilled wells	Total no. of producing wells in 2007	Cumulative prod. — Oil, 1,000 bbl	Cumulative prod. — Gas, bcf	Cumulative prod. — Million boe
305	Energy Partners Ltd.	VR	320	Fixed	Gas	207	1999	2007	4	1	8.5	6.3	1.1
626	Energy XXI GOM LLC	WC	248	Caisson	Gas	80	2000	2002	1	0	0.2	0.9	0.1
806	ERT GOM Inc.	EC	364	Fixed	Gas	373	2000	2007	6	3	170.5	10.2	1.9
927	Merit Energy Co.	EI	288	Fixed	Gas	202	2001	2007	1	1	24.5	5.2	0.9
1055	W&T Offshore Inc.	EI	397	Fixed	Gas	472	2002	2007	19	13	4,269.60	32.5	9.7
1286	Sterling Energy Inc.	EI	268	WP	Gas	190	2003	2007	1	1	3.5	1.6	0.3
1340	Bois D'Arc Offshore Ltd.	VR	122	Fixed	Gas	76	2004	2007	3	3	307.3	11.4	2.2
20068	Mariner Energy Inc.	EI	292	Fixed	Gas	210	1970	2007	17	5	209.6	448	74.9
20135	Conn Energy Inc.	EC	229	Fixed	Gas	115	1972	2005	12	0	0.1	9.3	1.5
20524	Samson LLC	VR	217	Fixed	Gas	121	1967	2005	14	0	572.5	36.9	6.7
20556	Stone Energy Corp.	VR	267	Fixed	Gas	169	1973	2005	12	0	1,641.90	113.8	20.6
20732	Stone Energy Corp.	VR	131	Fixed	Gas	57	1967	2005	2	0	769.7	73.5	13
20739	Apache Corp.	EI	296	Fixed	Gas	214	1973	2007	26	2	7,721.20	276.2	53.8
20940	Chevron USA Inc.	EC	272	Fixed	Gas	182	1973	2005	12	0	4,630.50	126	25.6
20947	McMoran Oil & Gas LLC	SM	49	Fixed	Gas	98	1966	2007	12	7	4,058.80	362.8	64.5
21008	Shell Offshore Inc.	EI	331	Fixed	Gas	246	1972	2003	30	0	26,016.10	170.3	54.4
21009	McMoran Oil & Gas LLC	EI	179	Fixed	Gas	96	1969	2007	7	1	1,112.40	124.5	21.9
21215	Mariner Energy Inc.	EI	266	Fixed	Gas	167	1970	2005	14	0	162.4	226.8	38
21463	Maritech Resources Inc.	SM	48	Fixed	Gas	107	1966	2005	7	0	1,046.70	148.1	25.7
21471	Beryl Oil and Gas LP	ST	196	Fixed	Gas	105	1971	2007	10	2	5,991.80	111.9	24.6
21579	Devon LP	EI	330	Fixed	Gas	244	1973	2007	39	9	38,320.80	384.4	102.4
21764	Stone Energy Corp.	EC	281	Fixed	Gas	175	1976	2007	13	2	1.3	169.1	28.2
21768	Ncx Company LLC	EI	349	Fixed	Gas	320	1975	2002	44	0	8,356.30	266.3	52.7
22012	Stone Energy Corp.	VR	329	Fixed	Gas	220	1979	2005	10	0	1	86.8	14.5
22248	Merit Energy Co.	VR	386	Fixed	Gas	324	1981	2001	14	0	3,191.90	53.6	12.1
22343	Mariner Energy Inc.	EI	267	Fixed	Gas	176	1981	2007	11	2	51.5	62.8	10.5
22444	Beryl Oil and Gas LP	ST	195	Fixed	Gas	100	1981	2007	11	1	178.2	105.7	17.8
23098	SPN Resources LLC	EC	330	Fixed	Gas	255	1985	2007	8	2	1,210.10	21.8	4.8
23308	W&T Offshore Inc.	EI	371	Fixed	Gas	415	1989	2007	24	4	3,229.00	79.8	16.5
23925	McMoran Oil & Gas LLC	EW	947	Fixed	Gas	477	1990	2007	30	11	5,594.20	109.5	23.8
26044	Apache Corp.	SS	291	WP	Gas	235	1994	2007	2	1	98	23.8	4.1
29049	Apex Oil & Gas Inc.	WC	473	Fixed	Gas	130	1996	2006	1	0	6.9	6.5	1.1
80003	Walter Oil & Gas Corp.	EI	390	Fixed	Gas	350	1997	2007	3	1	26.1	39.3	6.6
90004	St Mary Land Exp. Co.	VR	281	Fixed	Gas	176	1997	2007	9	2	820.2	18	3.8
1038	Energy Resource Technology GOM Inc.	EI	302	Fixed	Oil	224	2002	2007	5	4	2,735.70	9.5	4.3
20330	Apache Corp.	EI	175	Fixed	Oil	85	1966	1984	12	0	9,378.70	11.3	11.3
20530	Century Exploration Inc.	SS	154	Fixed	Oil	62	1955	2000	20	0	11,233.60	12.3	13.3
20565	Energy XXI GOM LLC	ST	21	Fixed	Oil	41	1962	1978	2	0	373	0.6	0.5
20670/1	Chevron USA Inc.	EC	272	Fixed	Oil	188	1973	2007	15	3	16,544.90	17.7	19.5
20670/3	Chevron USA Inc.	EC	272	Fixed	Oil	188	1973	2007	16	5	10,879.80	28.3	15.6
20737	SPN Resources LLC	SS	253	Fixed	Oil	187	1967	1979	8	0	986	4.4	1.7
20810	Devon Energy Production Co. LP	EI	119	Fixed	Oil	37	1962	1990	1	0	1,712.90	1.4	1.9
20832	Devon LP	EI	125	Fixed	Oil	40	1972	2004	2	0	799.7	1.3	1
21031	Devon LP	EI	330	Fixed	Oil	254	1973	2007	39	19	72,469.60	118.6	92.2
21583	Chevron USA Inc.	EI	339	Fixed	Oil	260	1973	2007	44	16	32,889.40	69.2	44.4
21781	Unocal Pipeline Co.	SS	208	Fixed	Oil	97	1964	2007	10	2	11,459.10	53.8	20.4
21795	Chevron USA Inc.	EI	339	Fixed	Oil	268	1975	2007	27	6	6,084.10	12.3	8.1
23009	Chevron USA Inc.	EC	272	Fixed	Oil	185	1985	2007	10	2	1,793.50	8.1	3.1
23370	Apache Corp.	VR	284	Fixed	Oil	186	1991	2007	8	5	10,320.60	12.6	12.4
23974	Nexen USA Inc.	EI	258	WP	Oil	155	1992	2002	2	0	2,355.10	6.1	3.4
27008	Maritech Resources Inc.	EC	328	Fixed	Oil	243	1995	2007	28	14	3,896.20	9.8	5.5
Total									**678**	**150**	**315,715.20**	**4,101**	**998.9**

Note: WP = well protector.

Active structure estimated remaining reserves, revenues

Complex/structure	Remaining reserves, 1,000 boe, $60-140/bbl	Discounted revenue at 10%, $1,000 — $60/bbl	$80/bbl	$100/bbl	$120/bbl	$140/bbl
305	0-0	0	0	0	0	0
806	63-81	2,093	2,790	4,329	5,195	6,060
927	0-0	0	0	0	0	0
1055	6,910-6,927	187,857	250,594	313,242	375,891	438,539
1286	0-0	0	0	0	0	0
1340	697-708	21,021	28,271	35,339	42,407	49,475
20068	3,763-3,924	70,759	94,561	118,352	142,062	165,809
20739	243-386	6,904	10,601	13,960	17,327	20,668
20947	91-135	2,659	4,124	5,405	6,722	8,058
21009	0-23	0	0	517	1,088	1,684
21471	39-46	1,627	2,169	2,711	3,724	4,345
21579	1,050-1,139	29,759	40,254	50,745	61,067	71,401
21764	745-844	15,427	20,912	26,313	31,717	37,114
22343	170-270	4,458	6,845	9,014	11,189	13,347
22444	15-43	508	1,132	1,813	2,520	2,941
23098	1,347-1,463	31,957	42,998	53,944	64,892	75,777
23308	18,797-18,884	391,749	522,346	652,939	783,533	914,124
23925	146,956-147,042	2,553,676	3,404,902	4,256,129	5,107,355	5,958,581
26044	0-54	0	1,090	2,234	2,680	3,907
80003	1,946-2,107	38,361	51,530	64,677	77,684	90,755
90004	64-79	2,716	4,040	5,049	6,442	7,515
1038	881-881	36,287	48,383	60,478	72,574	84,670
20670/1	825-896	25,203	33,910	42,543	51,178	59,763
20670/3	1,201-1,293	31,211	41,875	52,477	63,080	73,678
21031	16,094-16,152	427,729	570,317	712,902	855,488	998,073
21583	11,936-11,995	315,107	420,158	525,206	630,254	735,301
21781	1,135-1,362	17,481	23,356	29,208	35,056	40,901
21795	6,043-6,100	143,825	191,786	239,755	287,717	335,675
23009	0-0	0	0	0	0	0
23370	317-375	11,785	16,199	20,643	24,944	29,398
27008	11,723-11,776	301,059	401,427	501,791	602,155	702,517
Subtotal-A	185-366	4,966	9,067	13,061	16,389	20,195
Subtotal-B	31,536-32,901	677,032	907,539	1,138,408	1,368,988	1,598,993
Subtotal-C	201,330-201,716	3,989,219	5,319,962	6,650,248	7,980,563	9,310,890
Total	**233,051-234,983**	**4,671,216**	**6,236,569**	**7,801,717**	**9,365,940**	**10,930,077**

Note: Remaining reserves correspond to oil prices that range from $60/bbl to $140/bbl. Revenue computed based on the assumption that first revenue occurred 1 year after initial investment. Structures with no remaining reserves are uneconomic at the time of the assessment. Subtotal A computed from all structures with prehurricane production less than 100 boe/d and designated as unlikely to be redeveloped. Redevelopment rates from this class are estimated at 10-20%. Subtotal B computed from all structures with prehurricane production greater than 100 boed and less than 1,000 boe/d designated as a possible redevelopment opportunity. Only a percentage of these reserves are likely to be recaptured in the future, estimated at 40-60%. Subtotal C computed from structures with prehurricane production greater than 1,000 boe/d and are considered highly likely to be redeveloped. Most of these reserves are likely to be recaptured, estimated at 90% or greater.

Redevelopment opportunity matrix

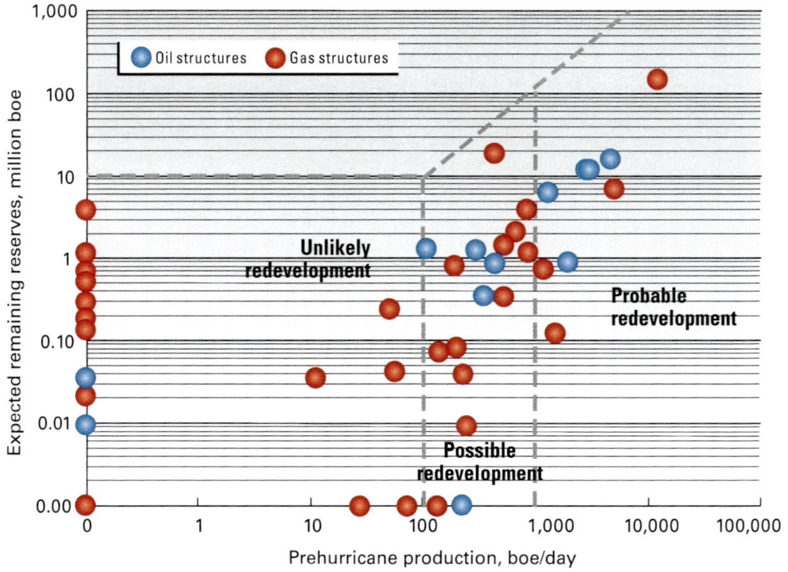

Redevelopment opportunity matrix includes (1) unlikely, (2) possible, and (3) probable redevelopment. The estimated probability of occurrence in each quadrant is $0.10 \le p(1) \le 0.20$, $0.40 \le p(2) \le 0.60$, $0.85 \le p(3) \le 0.95$.

Deepwater action: 2008

EIA reported in early 2009 that the total number of wells spudded in the gulf during 2008 was down about 17% compared to 2007, likely due to disruptions from Hurricanes Gustav and Ike.

However, deepwater discoveries were about on par year-over-year, with operators reporting nine successful wells in 2008:

- Gladden (Newfield Exploration): In water 3,116 ft (950 m) deep, Mississippi Canyon Block 800
- Auduin West (Newfield Exploration): 2,696 ft (822 m), Mississippi Canyon Block 754
- Dalmation (Murphy Oil): 5,900 ft (1,798 m), DeSoto Canyon Block 48
- Freedom/Gunflint (BP PLC): 6,100 ft (1,860 m), Mississippi Canyon Block 948

- Geauxpher (Mariner Energy): 2,700 ft (823 m), Garden Banks Block 462
- GC 448 (LLOG Exploration): Mississippi Canyon Block 72
- MC 72 (LLOG Exploration): Green Canyon Block 448
- Kodiak (BP PLC): 5,000 ft (1,500 m), Mississippi Canyon Block 771.
- Tortuga (Noble Energy): 6,500 ft (1,981 m), Mississippi Canyon Block 561.

History of U.S. Deepwater Discoveries

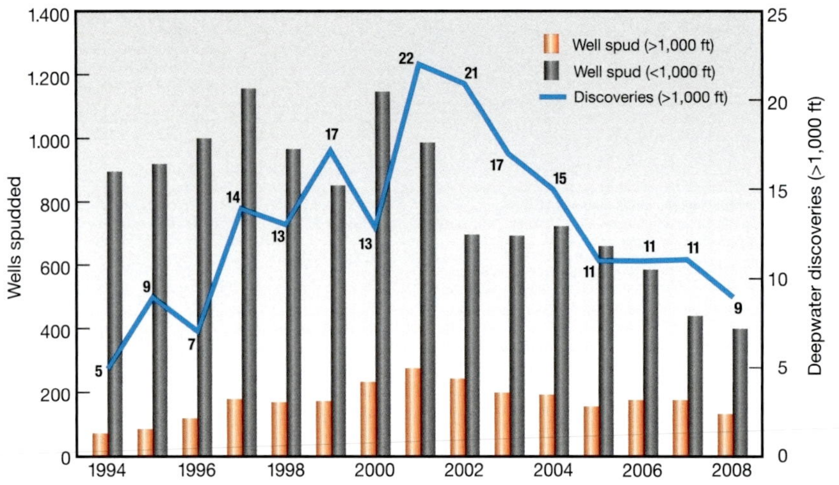

Source: MMS and *Offshore Magazine*

Gulf of Mexico wells spudded

Year	Well type	1–500 ft	501–1,000 ft	1,001–3,000 ft	≥ 3,001 ft	Total
2007	Development	252	2	18	44	316
2007	Exploration	183	3	48	66	300
2007	Relief	2				2
2008	Development	247	1	12	17	277
2008	Exploration	147	4	21	82	254

Note: Data current as of Dec. 1, 2008. Source: MMS.

Gulf of Mexico leases undrilled

Expiration year	1–500 ft	501–1,000 ft	1,001–3,000 ft	≥ 3,001 ft	Total
2008		1	20		21
2009	370	21	86		477
2010	270	31	116		417
2011	278	60	167		505
2012	168	66	276		510
2013	398	78	257		733
2014	45	95	230		370
2015	4	88	214	22	328
2016	58	134	275		467
2017		39	225	15	279
2018	2	194	734	16	946

Note: Data current as of Dec. 1, 2008. Source: MMS.

Pipelines/Transportation

Oil port JV ended

In April 2009, a partnership to build a large offshore crude oil receiving terminal in the gulf dissolved.

Enterprise Products Partners LP and an affiliate of TEPPCO Partners LP jointly withdrew from the Texas Offshore Port System partnership in a disagreement with the third partner, an affiliate of Oiltanking Holding America Inc.

Oil line for Big Foot

Enbridge Inc. signed a letter of intent in late 2009 with Chevron Corp., Statoil and Marubeni Oil & Gas to construct and operate a 40-mi (64-km) oil pipeline from the proposed Big Foot ultradeepwater development in the gulf.

The Big Foot Oil Pipeline, running about 170 mi (274 km) south of the Louisiana coast, will reach depths of up to 5,900 ft (1,798 m). It will transport up to 100,000 b/d to a subsea connection on existing deepwater pipeline infrastructure.

Environmental Issues

Carbon storage evaluation

In October 2009, the University of Texas at Austin announced that it will use $6 million in federal and state grants to identify possible carbon sequestration sites on state-owned property under the Gulf of Mexico.

The U.S. DOE issued a $4.8-million grant and the Texas General Land Office a $1.2-million grant.

ION Geophysical donated UT researchers access to extensive regional seismic data for areas of interest.

Formosa Plastics and subsidiary Neumin Production Co. provided a 3D seismic survey valued at $3.3 million.

Research partners with UT include Sandia Technologies LLC, Los Alamos National Laboratory, and the Environmental Defense Fund.

ALASKA

Exploration and Development

Beaufort Sea exploration plan: Shell

The U.S. MMS approved in October the exploration plan of Shell Offshore Inc. for two Beaufort Sea leases, but insisted on stringent environmental requirements.

Environmental groups protested the action.

Shell obtained the two leases during U.S. OCS Lease Sales 195 and 202, in 2005 and 2007, respectively.

The company was planning initial drilling activity on the leases from July to October.

Shell's plan is limited to the far western area of Camden Bay.

MMS approval followed a sequence of court reviews during 2009 (see box).

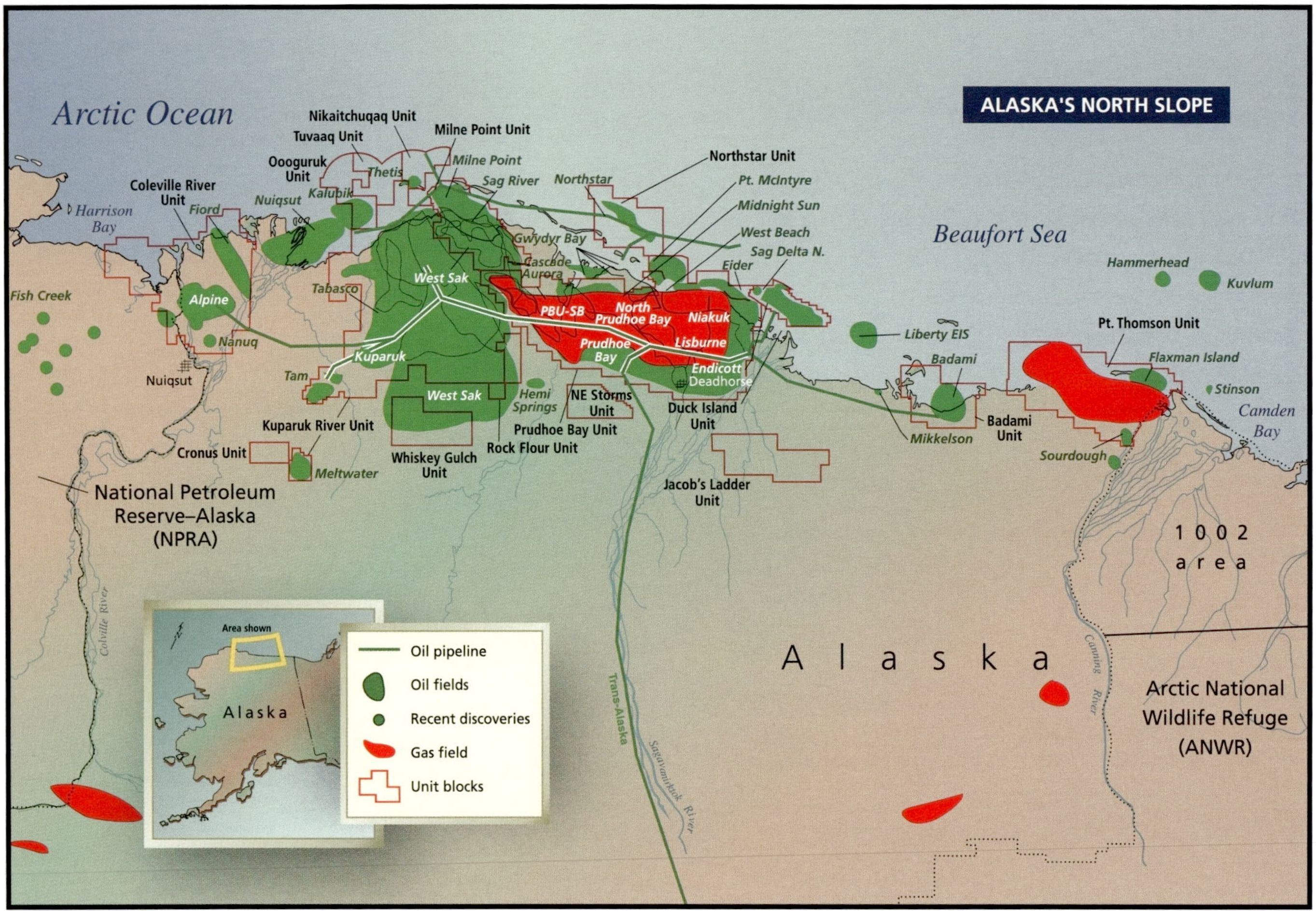

Beaufort Sea leases upheld

A federal appeals court in August 2009 upheld a lower court's decision regarding the actions of MMS in conducting OCS Lease Sale 202 in April 2007.

The appeals court agreed with the finding that the MMS did not need to prepare a supplemental environmental impact statement before conducting that sale, for blocks in the Beaufort Sea off Alaska.

The 9th Circuit Court of Appeals unanimously said that MMS had satisfied its duties under the National Environmental Protection Act.

The North Slope Borough and the Alaska Eskimo Whaling Commission had said in their legal challenge that MMS did not adequately consider seismic and other impacts on wildlife.

Point Thomson

ExxonMobil Production Co. drilled its first well in May and its second well in September 2009 at Point Thomson gas-condensate field on the Alaska North Slope (ANS), 60 mi (97 km) east of Prudhoe Bay.

The first two wells are to be completed to their final depths by yearend 2010.

Point Thomson owners committed $120 million to drilling and development in 2008, and investments of $250 million are expected in 2009.

Other major owners participating in the work include BP Exploration (Alaska) Inc. and ConocoPhillips Alaska Inc.

NPR-A finds

ConocoPhillips and Anadarko Petroleum Corp. reported the discovery and test production in early 2009 from two wells in the National Petroleum Reserve in Alaska (NPR-A).

Pioneer 1, tested in March, and Rendezvous 2, tested in winter 2008, both lie in the Greater Mooses Tooth Unit, about 20 mi (32 km) southwest of the Colville River Unit development on the ANS.

Test production rates ranged from 500–1,300 b/d of oil and 1.5 MMcfd of gas for each well.

Chukchi Sea drilling plans

In December 2009, MMS approved, with conditions, the exploration plan of Shell Gulf of Mexico to drill three exploratory, information-gathering wells in the Chukchi Sea.

The closest proposed drill site is more than 60 mi (97 km) offshore Alaska.

The approval requires close monitoring to ensure that operations are conducted in a safe and environmentally responsible manner, said the Dept. of the Interior.

Drilling and Production

Prudhoe Bay field: Snapshot

- Discovered in 1968 on the ANS
- Producing 400,000 b/d in mid-2009.
- Ultimate recovery (estimated): 14 billion bbl
- Reservoir size: 234 sq mi (606 sq km)
- Anchors 25 other ANS producing fields
- 1,200 active wells, drilled radially from 42 pads
- Six gathering stations send oil to the Trans-Alaska Pipeline
- Associated gas is processed to remove NGLs, then used mostly for reinjection
- Associated gas production: 8.8 bcfd winter, 5 bcfd summer
- Decline rate 2008–2009: 1–2% per year

Prudhoe Bay field: Gas future

Observers suggest that the length of Prudhoe's future depends on continuing to produce oil while beginning to sell natural gas. The ANS has two chief gas reservoirs:

- Prudhoe: estimated 24 tcf of recoverable gas
- Point Thomson [60 mi (97 km) east of Prudhoe]: more than 8 tcf of gas and 200 million bbl of condensate estimated recoverable.

Several options for building a 4- to 5-bcfd pipeline are being evaluated to move gas from the ANS to markets in Canada, the United States and elsewhere.

Also, a site has been identified at Prudhoe Bay on which could be built a 2- to 4-bcfd gas treatment plant.

It is not likely that gas shipments from the ANS would begin before 2018.

Oooguruk project

At the Oooguruk project on the ANS, operations were under way in mid-2009 to drill four horizontal laterals in the Nuiqsut formation, said Pioneer Natural Resources.

Two will be fracture-stimulated producing wells and two will be water injection wells.

An initial water injection well tested at 2,500 b/d of oil and was put on production at a stabilized rate of approximately 1,000 b/d of oil.

Net oil production from Oooguruk was forecast to average 5,000 b/d of oil in 2009 and increase to 10,000–14,000 b/d in 2011.

ANWR drilling?

U.S. Sen. Lisa Murkowski in February 2009 planned to propose use of directional drilling on state land and waters for initial development of oil and gas resources in the Arctic National Wildlife Refuge (ANWR).

That approach, she said, would mean there could be no occupancy, pipelines, construction, or facilities that would impact the refuge or its wildlife.

She estimated that current directional drilling technology would permit only about 10% of ANWR's estimated oil and 80% of its gas to be produced, but that future technology could increase production.

Environmental groups criticized the proposal.

Governor Steps Down

With little advance notice, Alaska Gov. Sarah Palin quit her office at the end of July 2009, turning over the reins of government to Lt. Gov. Sean Parnell.

One factor she cited was media and political attacks on her, and she suggested that she was leaving to allow the state to move forward under Parnell without distraction.

One of her initiatives was a push for construction of a major natural gas pipeline. (See Pipelines section below.) Mr. Parnell said that the pipeline project would be one of his highest priorities.

Ms. Palin was chosen by U.S. Sen. John McCain (AZ) to be his running mate in the 2008 presidential election.

Alaska Natural Gas Pipeline

Background

When oil was discovered at Prudhoe Bay in 1968, natural gas was also discovered. Proved reserves are estimated at 35 tcf.

The trans-Alaska oil pipeline now operates at less than half its capacity as ANS production winds down, and producing gas to replace oil revenue has been under study for several years.

Presented below is a summary of major developments in 2008 and 2009 related to efforts to build a pipeline approximately 1,750 mi (2,817 km) long, south from the ANS.

The line would carry gas to existing pipeline networks in northwestern Alberta, Canada, allowing delivery to a variety of Canadian and U.S. markets.

ANS-to-Alberta pipeline proposals

There were initially three proposals to build a major Alaska natural gas pipeline.

However, Alaska Northwest Natural Gas Transportation Co. told FERC in December 2008 that sponsors concluded that the project was no longer viable. The company has dissolved.

> **Oil spill**
>
> In late November 2009, a ruptured 18-in. pipeline created a major crude oil spill on the ANS.
>
> Initial estimates said that about 46,000 gal of crude and produced water had been released.
>
> Massive ice plugs had formed in the line several weeks earlier, causing BP Exploration (Alaska) Inc. to stop operating it. One plug was more than 1,500 ft (457 m) long.
>
> State officials said that the large breach at the bottom of the pipe indicated that the ice plugs increased pressure to a level that split the line.
>
> The leak was stopped and cleanup began quickly, according to state and federal officials.
>
> The flow line had carried a mix of oil, water and natural gas from wells to the Lisburne Production Center where the components were separated.

Moving forward in 2009 were the other two pipeline proposals by:

- TransCanada [through TransCanada Alaska Co. LLC (TC Alaska)] and ExxonMobil Corp.
- Alaska Gas Pipeline LLC (also called Denali), owned by BP PLC and ConocoPhillips.

TC Alaska, licensee under the Alaska Gasline Inducement Act (AGIA) program, plans a 48-in., 5-bcfd line running about 1,500 mi (2,414 km) from ANS to the Alberta Hub.

Denali plans a 48-in., 4-bcfd line about 2,000 mi (3,208 km) long, from ANS to the Alberta Hub.

Open season (bidding) is expected in mid-2010 to gauge shippers' interest in using the proposed pipelines. Depending on bidding results and on expectations for future U.S. gas supply and prices, the two projects may have to merge to be economically viable.

Construction is expected to take 8–10 years.

FERC: August 2009 report

In its August semiannual report to Congress, FERC noted the following activity since February concerning plans for a natural gas pipeline from the ANS to Alberta:

Denali activity

- The pre-filing process continued.
- Argonne National Laboratory was chosen in May as third-party contractor for the preparation of a required environmental impact statement.

TC Alaska activities

- TC Alaska got FERC approval in May to begin the pre-filing process.
- In June, TransCanada Corp., TC Alaska's parent, announced that it had agreed with ExxonMobil Corp. affiliates to work together on an Alaska gas line, and that the two firms would share some expenses associated with advancing the project.

ANS-to-Valdez pipeline/LNG proposal

TransCanada said in early 2009 that it also would gauge the interest of potential shippers in mid-2010 in another Alaskan gas pipeline project under study.

That line would run approximately 800 mi (1,287 km) from Prudhoe Bay to the port of Valdez, on Alaska's southeast coast.

There the gas would be converted to LNG and shipped to the Lower-48, South America, or Asia.

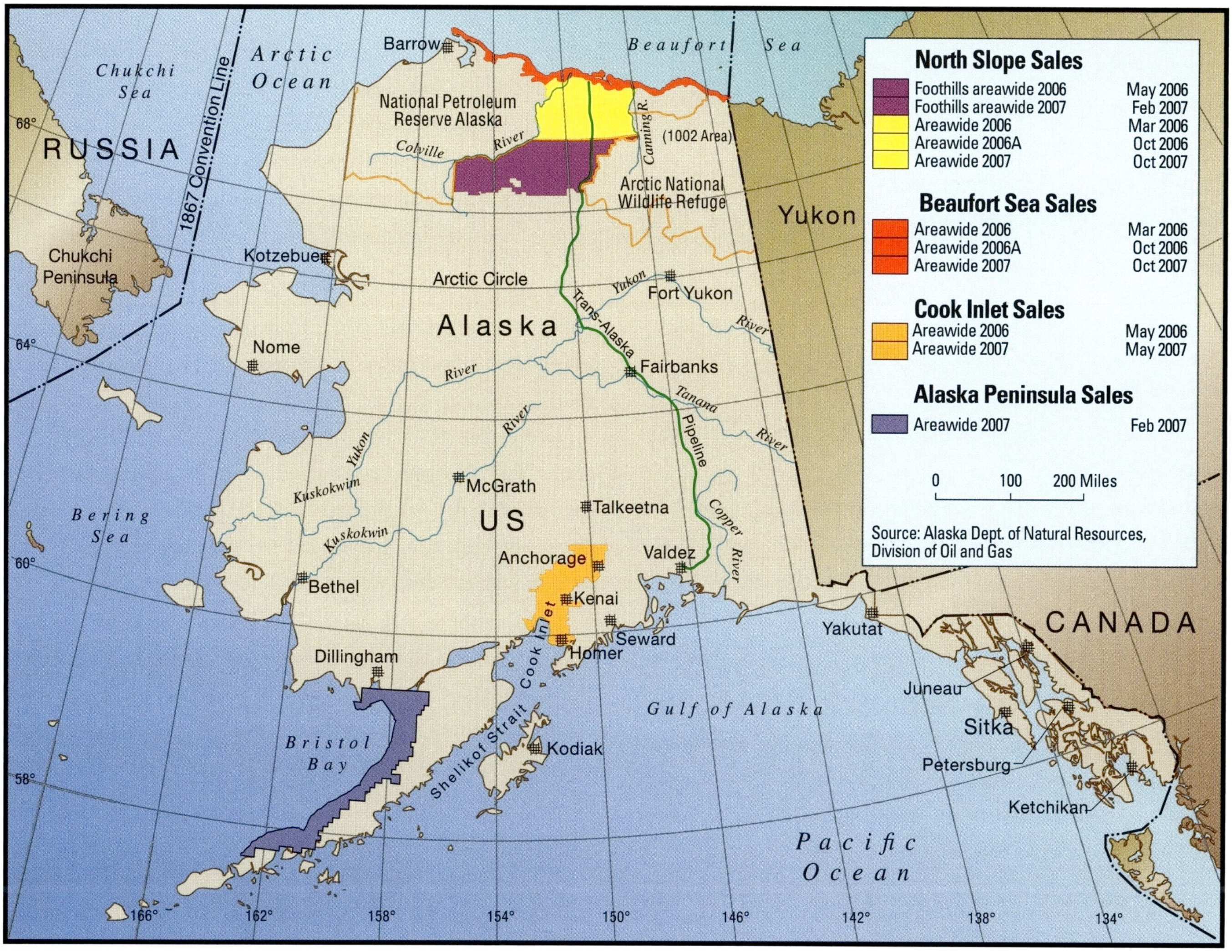

Intrastate Gas Pipelines

Background

During 2009, Alaska's Natural Gas Development Authority continued to develop plans for intrastate gas pipelines that could serve the state's widely scattered communities.

One plan would build a 460-mi (738-km) system from Beluga (west of Anchorage, in southern Alaska) to Fairbanks. It would be used first to transport gas north from Cook Inlet. Later, it would connect to either the TC Alaska or Denali system to carry gas south from the ANS.

Four options for moving gas south from ANS to Cook Inlet were under study, including a pipeline along the Parks Highway or along the Richardson and Glenn highways.

The cost of one such option—an 800-mi (1,287-km), 24-in. line carrying 500 MMcfd—was estimated at $4 billion.

Denali gas plant contract

Officials with the Denali Alaskan gas pipeline project (Alaska Gas Pipeline LLC) let a contract in early 2009 for the design of a major gas treatment plant.

The plant would remove impurities from ANS gas prior to shipment through Denali's proposed 4-bcfd pipeline from northern Alaska to Alberta.

Denali calls it the world's largest gas treatment plant, with process modules of up to 9,000 tons.

The proposed 2,000-mi (1,243-km), 48-in. Denali pipeline would cost an estimated $30 billion and would not start operation before 2018.

CANADA

Capital: Ottawa
Monetary unit: Canadian Dollar
Refining capacity: 2,029,450 b/cd
Oil production: 2.570 million b/d
Oil reserves: 178.092 billion bbl
Gas reserves: 57.906 tcf

Overview

Canada is one of the world's largest producers and exporters of energy.

According to the Energy Information Administration (EIA) of the U.S. Department of Energy, almost all of Canada's energy exports go to the United States, making it the largest foreign source of U.S. energy imports.

Oil

According to *Oil & Gas Journal (OGJ)*, Canada had 178.1 billion bbl of proven oil reserves as of January 2009, second only to Saudi Arabia. More than 95% of these reserves are oil sands deposits in Alberta, which are much more difficult to extract and process than conventional crude oil.

Canada is a net exporter of oil, with 2007 net exports of 1.0 million b/d.

Canada's oil production has steadily increased as new oil sands and offshore projects have come on-stream to replace aging fields in the western provinces.

Canadian oil production comes mainly from three sources:

- the Western Canada Sedimentary Basin (WCSB)
- the oil sands deposits of northern Alberta
- offshore fields in the Atlantic Ocean.

Alberta produces most of Canada's oil because it holds the majority of oil sands deposits and the bulk of the WCSB. According to Statistics Canada, Alberta accounted for 68% of Canada's national oil production in 2007.

Phased development of the Alberta oil sands continues, but volatile oil prices and the global financial crisis in late 2008 and into 2009 hampered progress on some projects.

In February 2009, the Canadian Energy Research Institute (CERI) lowered its production outlook. It expects oil sands to produce 1.9–2.9 million b/d in 2015, and 3.7–5.4 million b/d in 2030.

Three oil projects off the Atlantic coast are all in the Jeanne d'Arc basin: Hibernia, Terra Nova, and White Rose. Investment in the basin has increased in recent years, and both White Rose and Hibernia have announced plans to develop satellite fields.

Other offshore activity was unfolding in 2009 in the Mizzen field in the Flemish Pass basin and in the Hebron-Nevis field.

Natural Gas

Canada had 57.9 tcf of proven natural gas reserves in January 2009, according to *OGJ.* The country produced 6.5 tcf of natural gas in 2006 and is the second largest producer of natural gas in the Western Hemisphere, after the United States.

Natural gas production is concentrated in the WCSB, particularly in Alberta. Even though there have been some new conventional natural gas finds in the WCSB, many analysts predict that conventional natural gas production in the WCSB has reached its zenith.

Future natural gas production could center on coalbed methane (CBM) deposits in the WCSB, Arctic frontier natural gas deposits, the Deep Basin area, and offshore fields.

Offshore Nova Scotia is the center of Canadian gas production on the Atlantic coast. In the Sable Offshore Energy Project (SOEP), the Alma and South Venture fields are the latest brought on stream.

Offshore Newfoundland and Labrador, the Hibernia and White Rose fields contain a combined 4 tcf in recoverable natural gas reserves. There is no current natural gas production at either site, but both ExxonMobil (Hibernia) and Husky Energy (White Rose) have expressed interest in starting gas production.

Basins offshore British Colombia contain an estimated 43.4 tcf of total natural gas reserves, but a federal moratorium on drilling prevents production activity.

The Mackenzie Delta in the NWT holds an estimated 5–6 tcf of recoverable natural gas reserves. Moving gas from the region to southern markets hinges on completion of the Mackenzie Gas Pipeline. There are three large, proven gas fields in the Mackenzie Delta: Taglu, Parsons Lake, and Niglintgak.

Although CBM production is still in its infancy in Canada, it could eventually offset some of the decline in conventional gas production. The Alberta Geological Survey has estimated as much as 500 tcf of CBM gas in place in Alberta alone.

Shale gas also is still in the early stages of development. According to industry sources, the Montney shale formation in British Columbia could contain 50 tcf.

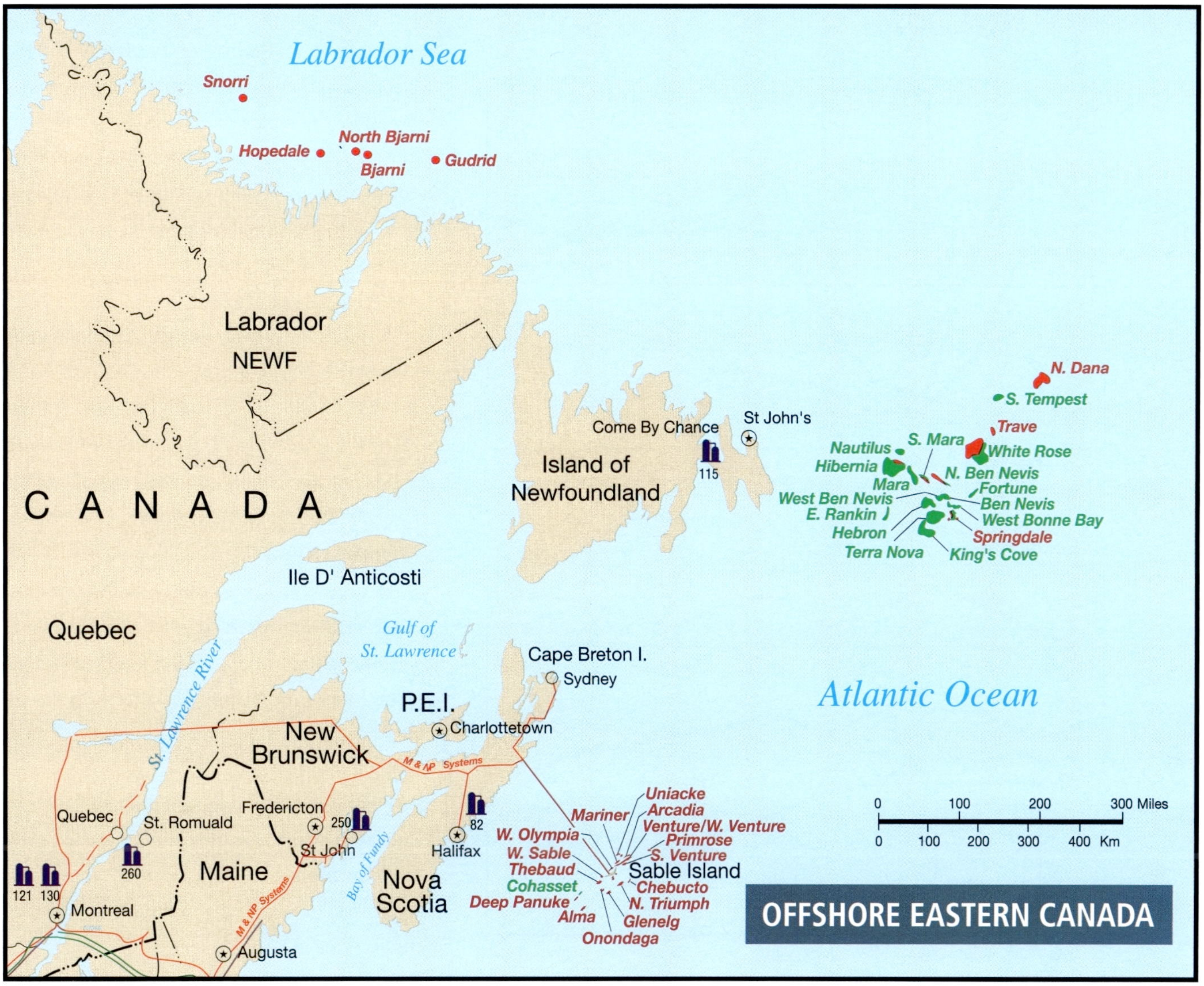

Dawson Creek bombings

In mid-2009, The Royal Canadian Mounted Police were investigating six bombings that occurred between October 2008 and July 2009 at the facilities of EnCana Corp.

All were in the Tomslake area near Dawson Creek, BC.

The company offered a reward of $1 million for help in solving the attacks.

A sociologist at the University of Alberta speculated that the bomber may be reflecting the resentment of local residents unhappy with the disruption caused by the energy boom in the area.

Gas Shales

Overview: OGJ, CAPP

Oil & Gas Journal presented a late-2009 view of the emerging role of natural gas from shale formations as an important potential source of supply growth in Canada.

Two shale gas plays—in the Horn River basin straddling the boundary between northern British Columbia and Alberta, and the Triassic Montney shale to the south—draw greatest current drilling attention. But potential exists elsewhere as well.

Among Canadian plays, only the Montney has reached the commercial stage. (Several U.S. plays are at that stage: the Marcellus, Barnett, Haynesville, Fayetteville, and Woodford shales.)

Shales in the Quebec lowlands, Nova Scotia, and New Brunswick were at the early evaluation-drilling stage in 2009.

Canadian drilling, gas production

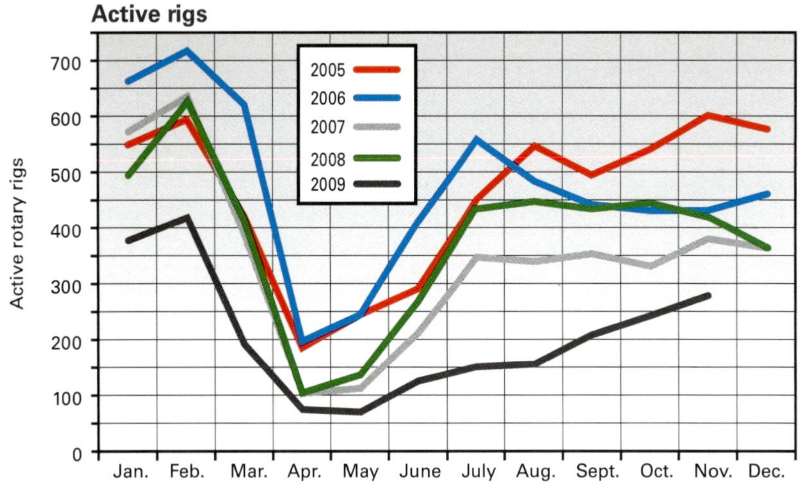

Source: Baker Hughes

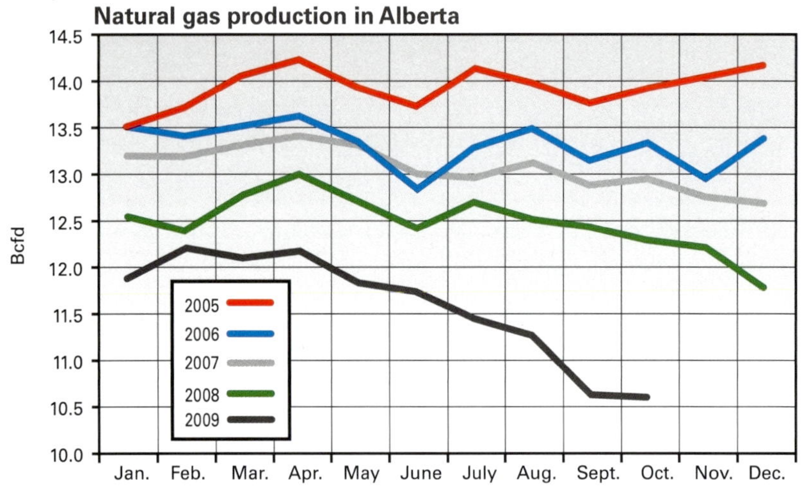

Source: Canadian Association of Petroleum Producers from FirstEnergy Capital data

CAPP's outlook for Canadian gas production

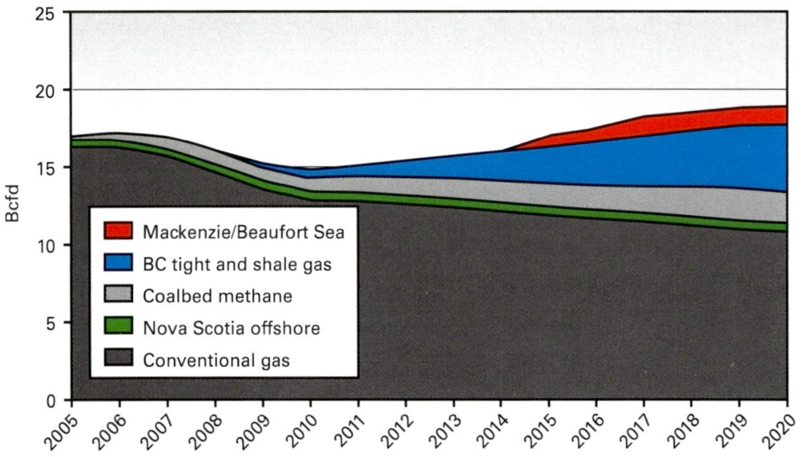

Source: Canadian Association of Petroleum Producers

In a November 2009 analysis, the Canadian Association of Petroleum Producers (CAPP) documented declines during the past several years in Canadian drilling and gas production in Alberta. CAPP sees gas from tight formations and shales in British Columbia as the most important element of an expected recovery and eventual expansion of gas production through 2020.

A June 2009 study by Ziff Energy Group concluded that gas prices {about \$3/Mcf (Can.)} were not high enough to cover the costs of replacing produced reserves in the conventional reservoirs of western Canada. Ziff estimated full-cycle costs for gas production at \$8/Mcf.

Overview: NEB

In a comprehensive November 2009 review, the NEB estimated that total volumes of recoverable gas are 1–10 bcf from each horizontal well drilled into Canadian gas shales. It said amounts are likely to grow as technology improves.

In British Columbia, expansion of gas treatment facilities and pipeline capacity likely will be needed to keep up with anticipated increases in shale gas production. Planned projects include:

- Expansion of a gas processing plant at Fort Nelson (southern Horn River basin) by Spectra Energy.
- Addition of capacity at Cabin Lake by EnCana Corp.
- Extension by Spectra of its raw gas-gathering capacity via a new pipeline from the Fort St. John area to the Taylor area in British Columbia.
- Extension by Nova Gas Transmission Ltd. of its Alberta System into the shale gas area of British Columbia. The proposed Groundbirch Mainline Project would have capacity of 1 bcfd.

Other promising shale gas areas are the Utica play (just north of New England), and the Horton Bluff play in New Brunswick and Nova Scotia.

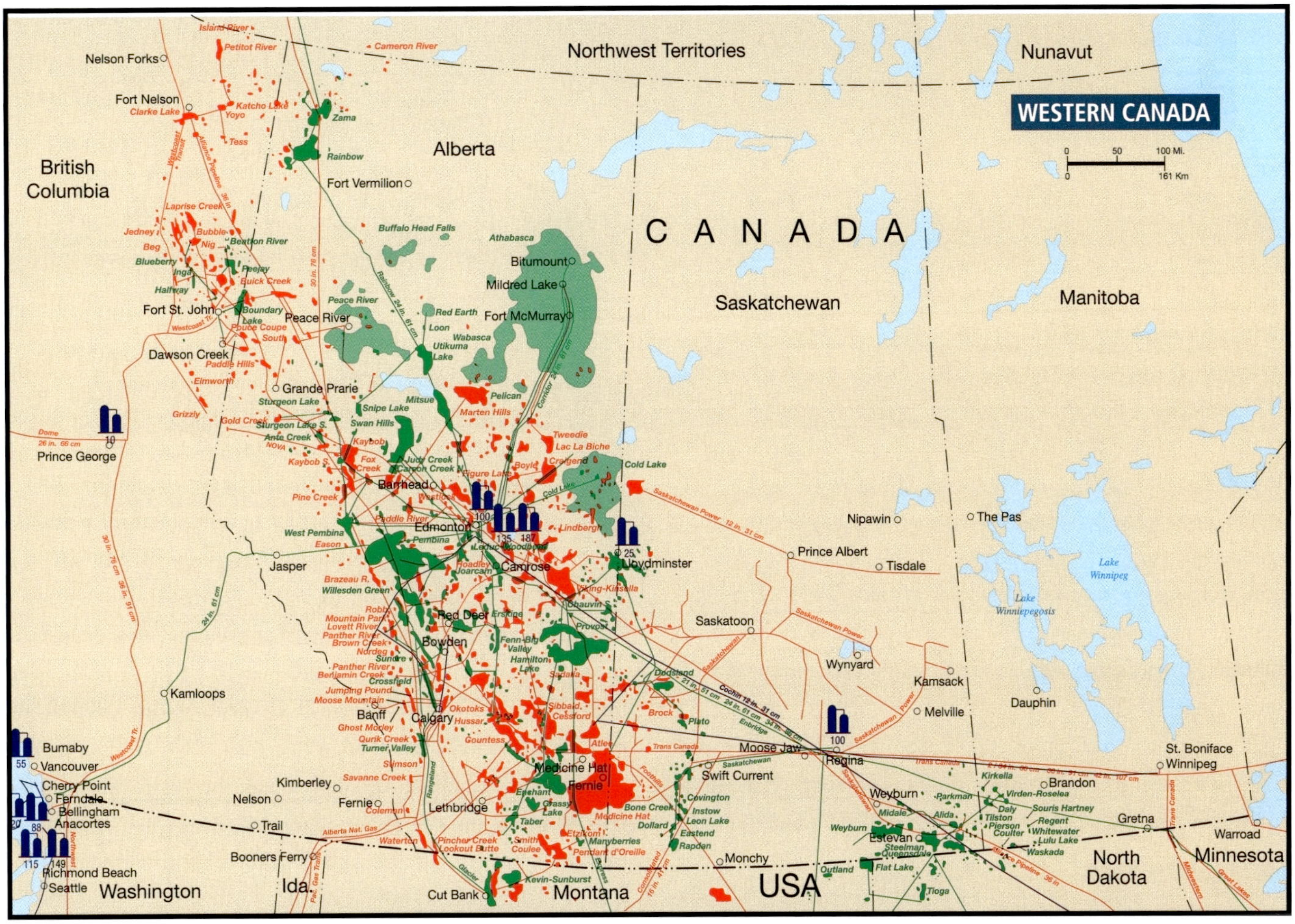

A LOOK AT CANADA'S GAS SHALE PROSPECTS

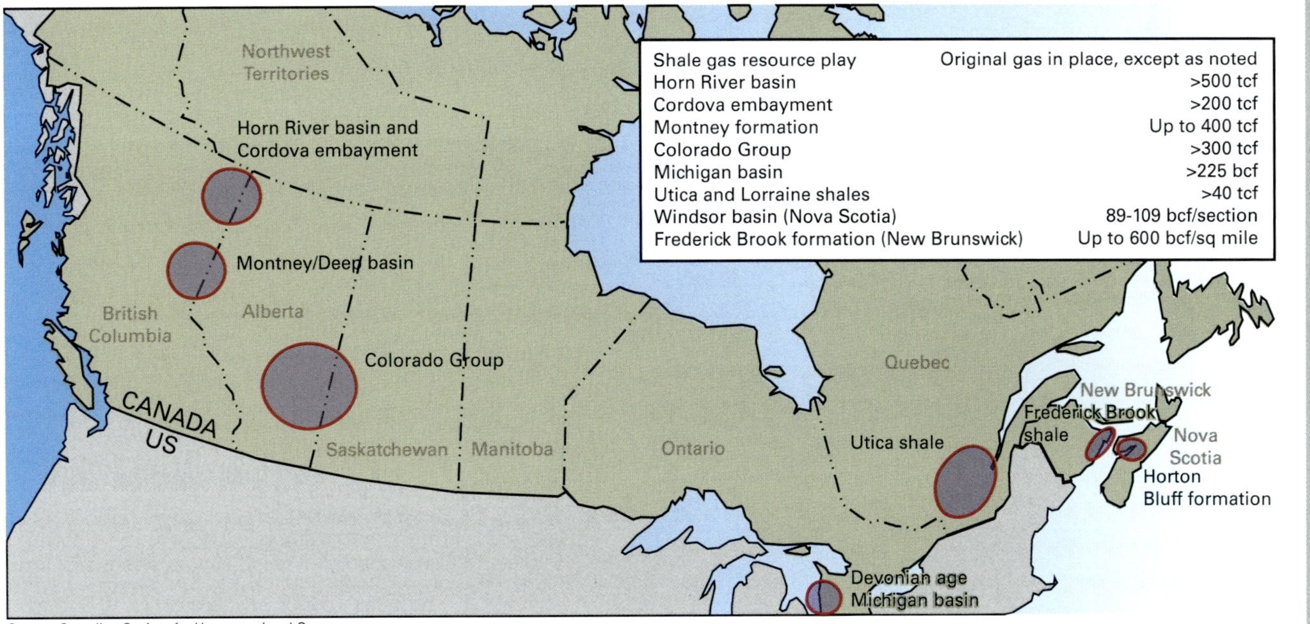

Shale gas resource play	Original gas in place, except as noted
Horn River basin	>500 tcf
Cordova embayment	>200 tcf
Montney formation	Up to 400 tcf
Colorado Group	>300 tcf
Michigan basin	>225 bcf
Utica and Lorraine shales	>40 tcf
Windsor basin (Nova Scotia)	89-109 bcf/section
Frederick Brook formation (New Brunswick)	Up to 600 bcf/sq mile

Source: Canadian Society for Unconventional Gas

A National Energy Board Review of Canadian Gas Shales

	Horn River	Montney	Colorado	Utica	Horton Bluff
Depth (m)	2,500–3,000	1,700–4,000	300	500–3,300	1,120–2,000+
Thickness (m)	150	Up to 300	17–350	90–300	150+
Gas-filled porosity (%)	3.2–6.2	1.0–6.0	Less than 10	2.2–3.7	2
Total organic carbon (%)	0.5–6.0	1–7	0.5–12	2.2–3.7	10
Maturity (R_o)	2.2–2.8	0.8–2.5	Biogenic	1.1–4	1.53–2.03
Silica (%)	44–65	20–60	Sand and silt	5–25	38
Calcite or dolomite (%)	0–14	Up to 20%	—	30–70	Significant–
Clay (%)	20–40	Less than 30	High	8–40	42
Free gas (%)	66	64–80	—	50–65	—
Adsorbed gas (%)	34	20–36	—	35–50	—
CO_2 (%)	12	1	—	Less than 1	5
Gas in place/section* (bcf)	60–318+	8–160	22–62	25–210	72.4–600+
Play area gas in place (tcf)	144–600+	80–700	>100	>120	>130
Horizontal well cost, including frac (million Can. $)	7–10	5–8	0.35 (vertical only)	5–9	Unknown

*In western Canada, a section is 1 sq mile.

Source: National Energy Board compilation from various sources

EIA view

The U.S. EIA sees Canadian gas shale formations as an increasingly important part of U.S. natural gas imports in 2009.

In a special report published in January 2009, EIA cited renewed optimism about unconventional gas recovery in Canada following the successes seen in the Texas Barnett shale and other formations in the Lower 48 states.

EIA estimated 2007 production in the Upper Montney region of British Columbia at 80 MMcfd of gas and expected this to rise significantly in 2009.

BC: Horn River basin

Nexen Inc. said in September 2009 that is boosted shale gas production to 15–20 MMcfd from five wells in the Dilly Creek area of the Horn River basin.

Nexen said its 88,000 acres in that area could support 500–700 wells and are estimated to contain 3–6 tcf of contingent recoverable resource.

Most production comes from three recently completed wells. The company was developing a winter drilling program.

Exploration and Development: East and Maritimes

NB: Green Road well

Corridor Resources Inc. pronounced its Green Road well as a new field discovery in late 2009 after it tested gas from two shaly intervals.

The Green Road G-41 well, 4 km (2.5 mi) north of Elgin, NB, stabilized at 430 Mscfd of gas at 147 psi final flowing wellhead pressure from a black shale at 2,000–2,050 m (6,562–6,726 ft).

Flow from a second zone at 1,850–1,900 m (6,070–6,234 ft) was 3 MMscfd at a final flowing wellhead pressure of 699 psi.

Offshore NL: White Rose area

Husky Energy reported discovery of additional oil resources in the White Rose area in late 2009, some 350 km (217 mi) southeast of St. John's, NL.

Drilling to explore the potential of a second, deeper horizon encountered 55 m (180 ft) of net high quality, oil-bearing reservoir in the Hibernia formation, estimated to hold 30–85 million bbl of light crude oil.

In the 1980s, the E-09 well found the main White Rose pool, also tapping the Hibernia formation. The E-09 discovery contains an estimated 100 to 250 million bbl of light crude.

Offshore NL: Hibernia

In August 2009, federal and provincial governments approved the development-plan application of Hibernia Management and Development (HMDC) for the AA block. The block is estimated to contain 48 million bbl of oil.

Local news sources reported that the company plans to expand the Hibernia project to develop reserves southeast of the Hibernia platform, offshore Newfoundland and Laborador.

The development aims to offset decline from the Hibernia field and extend the facility's life from 2027 to 2036.

Western NL: Parson's Pond area

A group led by provincial energy corporation Nalcor Energy (Newfoundland) was selecting drillsites for as many as three wells in western Newfoundland's Parsons Pond area in late 2009.

The program is designed to test the deeper and larger prospects in the Cambrian Ordovician carbonate platform in the Parson's Pond area.

An independent assessment said that the Nalcor properties could contain an aggregate of 284 million bbl of oil in place.

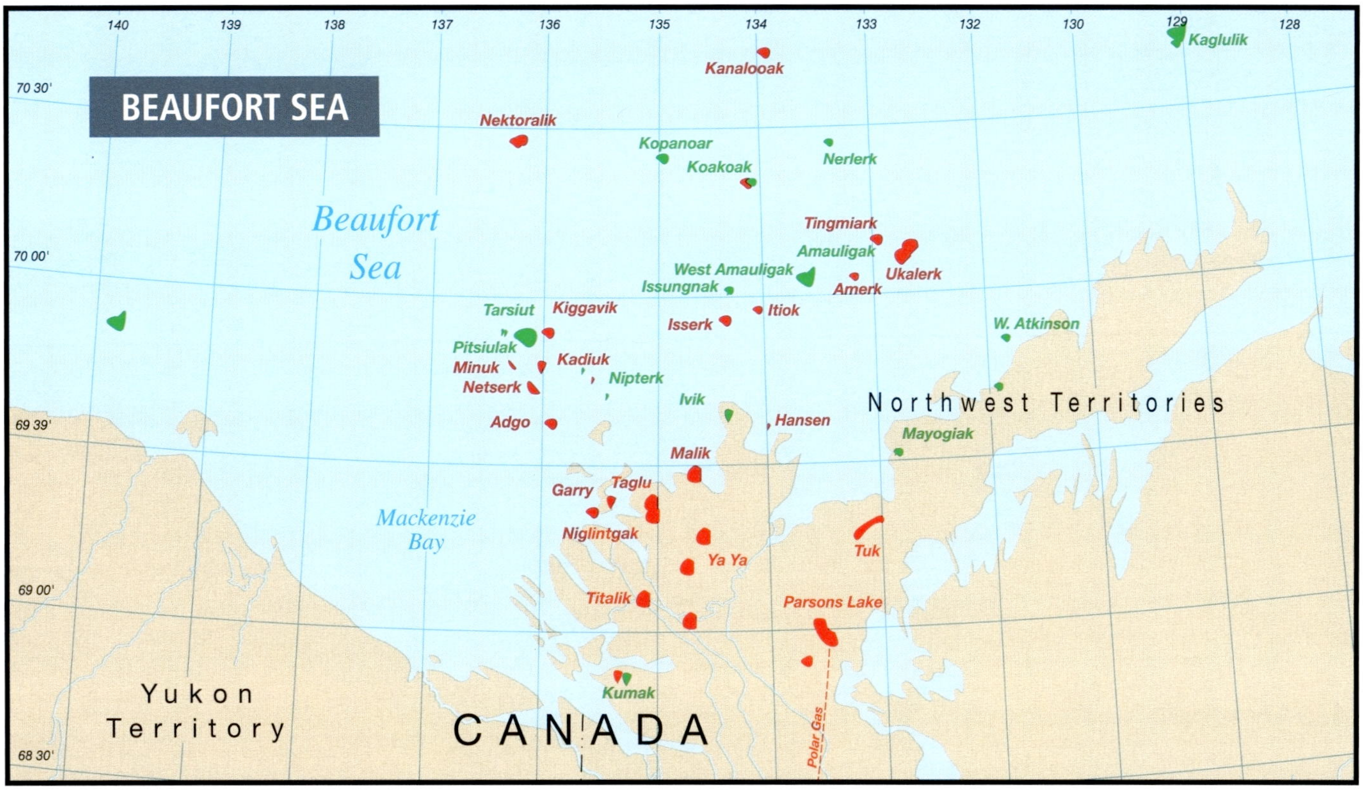

Western NL: Bay St. George basin

Vulcan Minerals Inc., was preparing in mid-2009 to drill in the Bay St. George basin, onshore western Newfoundland, in a JV with Investcan Energy Corp.

Robinson's-1 well is to reach 3,600 m (11,812 ft) on Permit 03-106 about 40 km (25 mi) southwest of Stephenville.

Southern NB: Sussex, Elgin subbasins

In mid-2009, Corridor Resources Inc. reported an updated reserves estimate for the overpressured Carboniferous Frederick Brook shale in the Sussex and Elgin subbasins.

Consulting engineers developed a best estimate of 67.3 tcf of gas in place.

The F-58 well in McCully field had been flowing at a steady 150 Mcfd at 550 psi wellhead pressure for 14 months and had recovered 69 MMcf of gas following frac treatment.

The McCully P-76 well flowed gas at 150 Mcfd for 76 hr at 32 psi final wellhead pressure after fracing.

Depth to the top of the shale is 2,000–3,000 m (6,562–9,843 ft) in much of the area.

Offshore NL Hebron project

Operator ExxonMobil Canada Properties estimated in mid-2009 that the Hebron project on the Grand Banks should start producing by the end of 2017, building to 150,000 b/d by 2020.

Hebron Offshore Project Area

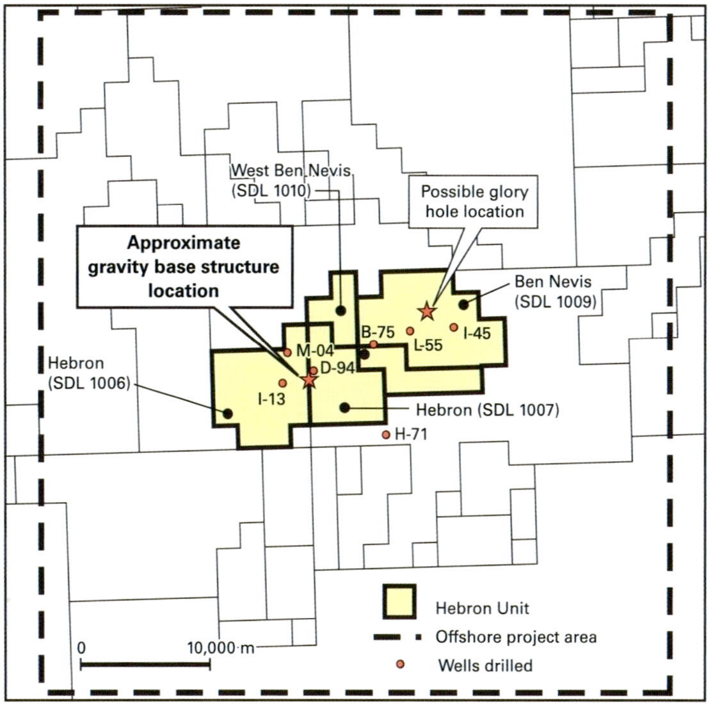

Schematic cross-section of Hebron project area

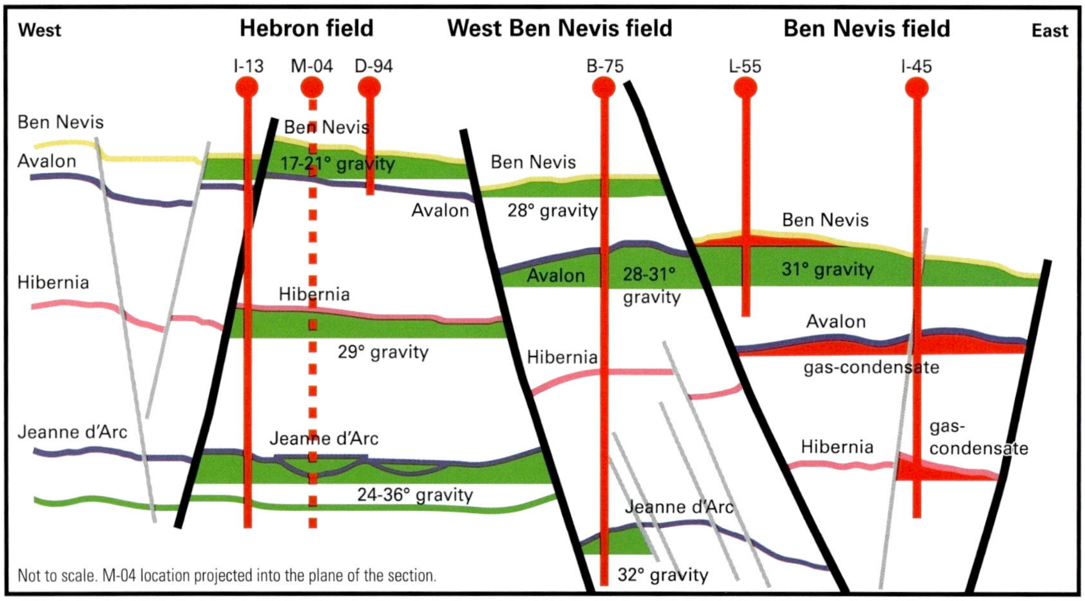

Over 30 years, the project is estimated to recover 566 million bbl of oil.

The project will develop Hebron, Ben Nevis, and West Ben Nevis fields in the Hebron Unit in 88–102 m (289–335 ft) of water in the Jeanne d'Arc basin.

The unit is 340 km (211 mi) east-southeast of St. John's, NL.

Initial development will involve Hebron field only.

Other interests in the Hebron unit are Chevron Canada Ltd., Petro-Canada, and Statoil Canada Ltd.

The government's 4.9% stake is held by Nalcor Energy-Oil & Gas Inc.

Eastern PEI

Ardent Resources Canada Ltd., was planning in mid-2009 to drill two exploration wells on the eastern part of Prince Edward Island, under an agreement with PetroWorth Resources Inc.

By doing so, Ardent Canada will earn a 50% interest in six permits that cover 1,797 sq km (444,000 acres).

The first well was to be drilled by yearend, and the second by mid-2010.

Onshore NS: Windsor Block

The Nova Scotia government finalized in early 2009 a 10-year oil and gas production lease covering the Windsor and Horton groups on the Windsor Block, onshore north of Halifax.

Having already conducted seismic, drilling, completions, and other work, Triangle Petroleum Corp. agreed to drill seven wells by mid-2014.

The company mainly seeks gas in Mississippian–Devonian Horton Group shales.

Offshore NL: Flemish Pass basin

Statoil Canada Ltd. reported a likely deepwater discovery in early 2009 in the Flemish Pass basin.

The discovery came at Mizzen O-16 in 1,100 m (3,609 ft) of water on EL 1049, some 500 km (311 mi) east-northeast of St. John's and 180 km (112 mi) east of White Rose oil and gas field.

Husky Energy Inc. also has an interest in the license.

NB: Stoney Creek field

In January 2009, Contact Exploration Inc., placed the N-78-2328 well at Stoney Creek field onstream at 30 b/d of oil.

The well was drilled to TD 1,115 m (3,658 ft) through the Hiram Brook, Fredericks Brook, and Dawson Settlement formations.

Quebec: St. Lawrence Lowlands

In mid-2009, consulting engineers estimated a prospective resource of 4.28 tcf of natural gas in place in the Ordovician Utica shale in the deep fairway of Quebec's St. Lawrence Lowlands.

The estimate of 150 bcf/sq mi is 66% higher than earlier industry figures, said Questerre Energy Corp. The range of the resource is 2.2–8 tcf.

The evaluation focused on acreage between the Yamaska growth fault and Logan's Line.

It did not include the shallower Ordovician Lorraine interval but did include results from the recently completed Talisman Energy Inc.-operated Edouard-1 exploratory well.

Earlier in the year, Gastem, and Canbriam Energy Inc., took a farmout on Mundiregina permits in the St. Lawrence Lowlands.

The permits are in the Ordovician Utica shale fairway southwest of Gastem's Yamaska permit.

Gastem and Cambrian are committed to shoot seismic and drill six wells by the end of 2010.

Exploration and Development: West/Central

Saskatchewan: East-central

Hunt Oil Co. and Nordic Oil & Gas Ltd. signed a strategic development agreement in mid-2009 regarding gas exploration in a non-producing area of east-central Saskatchewan.

Nordic said its holdings on the northeast flank of the Western Canada Sedimentary basin have gas and oil potential in at least 10 formations, from the Second White Specs to the Winnipegosis sand.

One area of interest is Preeceville, 209 km (130 mi) northeast of Regina.

BC: Horn River basin

ExxonMobil Corp. and its Canadian affiliate reported in mid-2009 that their first four wells in the Horn River basin of northeast British Columbia tested at 16–18 MMcfd.

In addition, TAQA North, a subsidiary of Abu Dhabi National Energy Co. PJSC, acquired 32,000 acres 48 km (30 mi) north of Fort Nelson earlier in the year.

TAQA is among Canada's top 12 oil and gas producers. It also has operations in Montana, North Dakota, and Wyoming.

AB, BC: Progress Energy plans

Progress Energy Resources Corp. was planning to drill 30–35 wells in the Alberta Deep Basin in 2009, run four to six rigs in the Foothills of Northeast British Columbia, and drill four vertical and two horizontal Devonian Montney shale wells in British Columbia in the first half.

Exploration and Development: North

NWT: Mackenzie Delta

MGM Energy Corp. was planning to drill the Ellice A-25 well after reporting better than expected test results from the Ellice J-27 discovery in the western Mackenzie Delta.

The wells are 129 km (80 mi) northwest of Inuvik, NWT.

Ellice A-25 will test a prospect on the Ellice Island anticline.

Ellice J-27's second test, in Taglu at 1,369–73 m (4,491–4,505 ft) MD, flowed at a maximum restricted 22.4 MMcfd. The zone had 9 m (30 ft) of net pay.

Arctic ice patterns

The *Financial Times* put the spotlight in mid-2009 on the early breakup of sea ice in the Canadian Arctic. Analysts pointed out several events:

- September 2007 was the first time in living memory that the entire Northwest Passage was open water from east to west.
- In 2008, despite slightly more ice, a record six private yachts moved through the passage.
- A scientist with the Institute of Ocean Sciences near Victoria, BC, said the extent of northern sea ice in 2009 was about 5% below the long-term average of 12.2 million sq km (4.7 million sq mi).
- A July 2009 forecast by the Canadian Ice Service showed that in the western Arctic, ice break-up was 1–3 weeks early in many coastal areas.

Arctic Interests

In mid-2009, Prime Minister Stephen Harper visited the Canadian Arctic for five days, as part of an effort to protect the country's claims of sovereignty in the region.

Russia and Denmark also are becoming more assertive of their territorial claims to Arctic areas.

Global interest in the Arctic has intensified as warming temperatures raise the prospects of a new northern shipping route and potential hydrocarbon exploration.

Some scientists predict the Arctic could be ice-free in the summer of 2037.

Arctic oil, gas resources

The USGS, working with scientists from Canada, Denmark, Greenland, Norway and Russia, evaluated Arctic gas and oil supplies.

Their June 2009 report showed with 95% probability that there were 44 billion bbl of undiscovered oil and 770,000 bcf of undiscovered gas in the Arctic.

Much of the oil was located in Alaska, with a large portion of the gas in the Russian Arctic.

The study looked only at resources of at least 50 million bbl of oil or 300 bcf of gas.

Lease Sales

Nova Scotia

The Canada-Nova Scotia Offshore Petroleum Board issued Call for Bids NS09-1 in December 2009 for two shallow water parcels in the Sable subbasin. Bids must be received by June 24, 2010.

Newfoundland and Labrador

The Canada-Newfoundland and Labrador Offshore Petroleum Board announced in May 2009 that there would be three separate Calls for Bids for the year.

- NL09-01 (Jeanne d'Arc basin): one parcel.
- NL09-02 (Laurentian Subbasin): two parcels.
- NL09-03 (Western Newfoundland, Labrador offshore region): one parcel.

Sealed bids were due by November 19 for all three calls.

In September, C-NLOPB announced Call for Nominations NL09-01, concerning offshore lands to be considered for a possible call for bids in 2010.

Sealed nominations were to be submitted by November 2, 2009.

Drilling and Production: Big Picture

PSAC view: 2009 and 2010

The Petroleum Services Association of Canada said in late 2009 that it expects the oil and gas industry to drill 8,000 wells in Canada during 2010.

This number was about the same as its forecast for 2009.

PSAC's 2010 breakdown:

- 5,095 wells in Alberta (a 5% decrease from 2009)
- 630 in British Columbia (a 7% increase)
- 1,935 wells in Saskatchewan (a 10% increase)
- 300 wells in Manitoba (a 22% increase).

PSAC sees adequate oil prices in 2010 to sustain conventional drilling activity in areas such as Saskatchewan and northeast Alberta, but continued low gas prices that will reduce by 30% conventional shallow-gas drilling in southeast Alberta.

PSAC based its 2010 forecast on an average $5 (Can.)/Mcf gas price and a $72/bbl West Texas Intermediate oil price.

OGJ view: 2009

In a mid-2009 review, *Oil & Gas Journal* expected that drilling in the United States and Canada could recover modestly in the second half of 2009 if oil and gas prices stabilized or improved.

As oil prices tumbled from highs near $150/bbl in mid-2008, the rig count followed, dropping to 2009 lows. Rig activity seemed to begin to stabilize and pick up again only in June 2009.

U.S. and Canadian rig counts averaged about 37% lower in the first half of 2009 compared to the first half of 2008.

Elsewhere in the world, comparing the same period shows declines of 10% in Africa, 7% in Europe, 5% in the Middle East, 4% in Asia-Pacific, and 2% in the Western Hemisphere outside the United States and Canada. The international rig totals do not include figures for Iran, Iraq, Russia, or China onshore.

Highlights of *OGJ's* midyear drilling forecast for 2009 are given in the United States secion of this chapter.

OGJ expects that operators will drill 9,854 wells in western Canada in 2009, down from an estimated 16,290 in 2008.

Alberta incentives

In early 2009, Alberta offered oil and gas producers royalty incentives in a move designed to revive drilling.

The government reduced royalty for conventional wells spudded between April 1 and March 31, 2010, and for wells coming onto production in that period.

The drilling incentive is a crown royalty credit of $200(Can.)/m [$61(Can.)/ft)] of hole drilled.

Benefits would be greatest for the smallest producers, determined by a sliding scale based on production.

The incentives don't apply to oil sands drilling or production.

Drilling and Production: Oil Sands

Resource estimate: 2008

In a June 2009 report, Alberta's Energy Resources Conservation Board (ERCB) estimated the province's oil sands resource at yearend 2008 as:

- Initial in place: 1,731 billion bbl
- Initial established: 177 billion bbl
- Cumulative production: 6.4 billion bbl
- Remaining established: 170 billion bbl
- 2008 production: 0.477 billion bbl (1.3 million b/d)
- Ultimate potential: 315 billion bbl.

Of the 170 billion bbl remaining established reserves, ERCB considers 80% recoverable with in situ methods and the remaining 34 billion bbl recoverable with surface mining methods.

In 2008, average production from the three main oil sands areas was:

- Athabasca: 721,500 b/d mined; 232,100 b/d in situ.
- Cold Lake: 310,700 b/d in situ.
- Peace River: 42,000 b/d in situ.

About 9,700 wells were producing in the oil sands during 2008. The average well produced 62 bo/d.

New Tailings Rules

In February 2009, the Alberta ERCB announced new rules for regulating tailings reclamation from oil sands operations.

Oil sands mining companies submitted tailings management plans to the ERCB in September in accordance with Directive 074. However, only two of nine management plans submitted would meet the new regulatory requirements starting in 2011.

That was the finding of Pembina Institute & Water Matters Society of Alberta, released in late 2009.

Meeting the rules were plans for the Millennium and North Steepbank mine, and the Fort Hills Energy Ltd. mine—both operated by Suncor Energy Inc.

The institute said liquid tailings lakes in Alberta in 2009 covered about 130 sq km (50 sq mi).

Production forecasts

Several organizations lowered their expectations in mid-2009 for increased Alberta bitumen production, but most still expected a sizable rise during the next decade.

Alberta's ERCB said it expects production to reach 2.7 million b/d in 2018. This rate is a decrease from its 2008 forecast that saw oil sands production increasing to 3.23 million b/d in 2017.

For its current projections, ERCB lowered its crude price assumptions, expecting West Texas Intermediate to average $55/bbl in 2009 and increase to $120/bbl in 2018. The lower oil price scenario ties into the slow down in world economies that has lowered future oil demand expectations.

CAPP estimated that oil sands production under a growth scenario will reach 2.6 million b/d in 2018 and 3.3 million b/d in 2025. Without growth (only projects currently in operations and under construction producing), CAPP expects production to remain relatively constant at 1,957 million b/d in 2018 and 1,987 million b/d in 2005.

CERI estimated in late 2009 that in its "realistic" scenario, Alberta oil sands production would grow to 1.7 million b/d by 2015 and to 4.5 million b/d by 2030.

Oil Sands Development Status

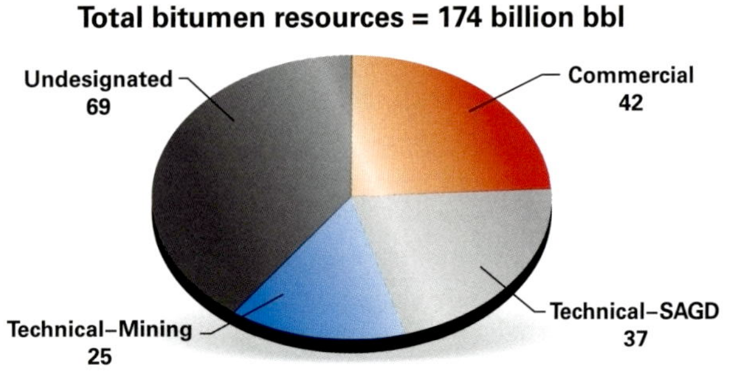

Note: Technical reserves have not received approval or are unlikely to be developed in the near-term. Undesignated reserves have yet to be assigned to a project.
Source: Wood Mackenzie

Oil Sands Regions

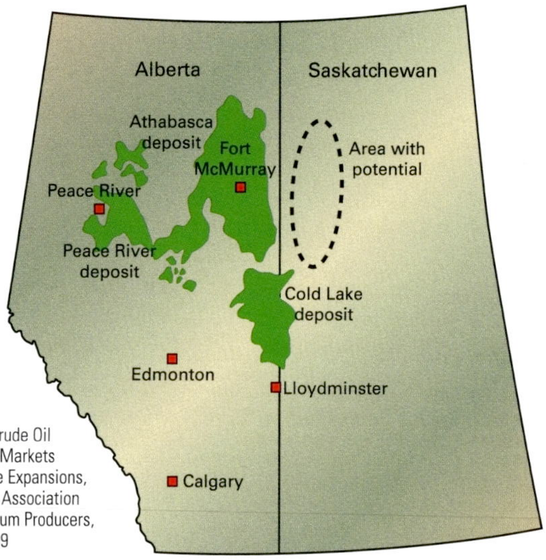

Source: Crude Oil Forecast, Markets & Pipeline Expansions, Canadian Association of Petroleum Producers, June, 2009

Announced Delays in Oil Sands Projects*

Project	Bitumen production rate, 1,000 b/d	Originally stated in-service date	Report
BA Upgrader Phase 1-3	70 x 3	2009, 2010, 2013	Construction has been halted
CNRL Horizon			Cutting spending
Imperial Oil Kearl Phase 1	100	2010	Investment decision in 2009
Korea National Oil Co. Phase 1	10	2010	Delayed for a year
Laricina Energy Ltd.:			
Germain	1.8	2Q 2009	Delayed 12–18 months; cap changed to 600 b/d Delayed 6–9 months; cap
Saleski	1.8	4Q 2009	changed to 1,200 b/d
Opti Long Lake Phase 2	72	2013	Sanctioning subject to various factors
Patch International – Ells River	10	2011	Project on hold
Petro-Can Fort Hills Mine:			
Phase 1	190	4Q 2011	Design for go-ahead will be deferred until some time in 2009
Phase 2	130	2015	
Petro-Can Fort Hills Upgrader 1	165	2012	Decision on go-ahead deferred indefinitely
Petro-Can Fort Hills Upgrader 2/3	175	2014	
Shell Jackpine 1B	100	2012	Suspending expansions for now
Shell Carmon Creek (in-situ) Phase 1-2	50 x 2	2008; 2017	Withdrawn application due to potential changes in design; more info to be provided in early 2009
StatoilHydro Upgrader	250	2014	Withdrawing application; project canceled
Suncor Voyager Upgrader	300 (orig. 350)	2012	Completion delayed by 1 year
Suncor Firebag			
Phase 5	68	1Q 2012	Schedule flexible to market conditions
Phase 6	68	3Q 2012	Schedule flexible to market conditions
Syncrude Stage 4 Expansion	140	2015	Reevaluating plans
Total Joslyn Mine	100	2013	May 2008 onstream date moved to 2014

*As of the end of December 2008. Since compilation of this list, Suncor has suspended its Voyager upgrader and Firebags Phase 3 projects.
Source: Canadian Association of Petroleum Producers

In 2008, it had forecast reference-case production of 3.4 million b/d in 2015 and 5.0 million b/d by 2030.

Wood Mackenzie noted that in 2009, high costs for steel and labor—coupled with the global economic recession and low commodities prices—forced many oil sands firms to suspend or reduce activities.

The firm concluded that cost deflation alone will not be sufficient to restart that activity.

It expected companies to stay in a low-cost, safe mode until there are clear signs of an improved energy market or economic environment.

U.S. ECONOMIC GAINS SEEN

Canadian oil sands development could lead to more than 300,000 new U.S. jobs.

That was one finding from a 2009 CERI study commissioned by the American Petroleum Institute.

"Canada's Oil Sands and Economic Impact on the USA" said more Canadian oil sands production could stimulate economic activity in both countries.

As oil sands production and investment in Canada rises, demand for U.S. goods and services would add an estimated $34 billion to U.S. GDP in 2015 and $42.2 billion in 2025, the study said.

Between 2011 and 2015, some 342,000 positions could be added.

CERI assumed that oil sands output rises from about 1.4 million b/d to 4 million b/d in 2025.

OIL SANDS PRODUCTION FORECAST

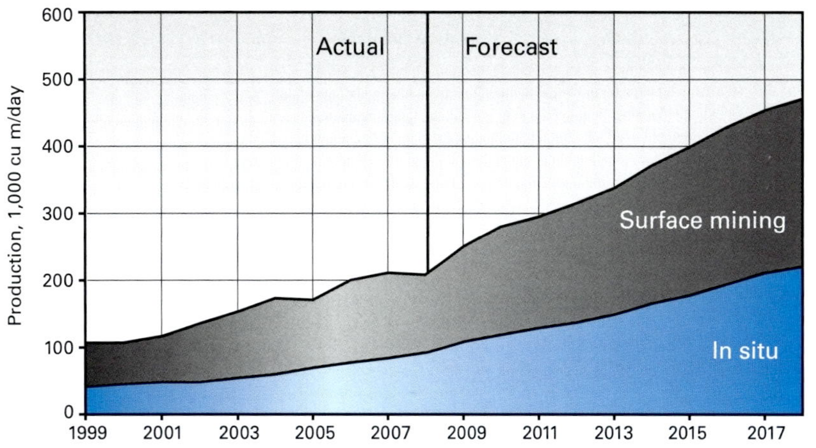

Source: Alberta's Energy Reserves 2008 and Supply/Demand Outlook 2009–2018, ST98–2009, Energy Resources Conservation Board of Alberta, June 2009.

CRUDE BITUMEN, SCO DISPOSITION FORECAST

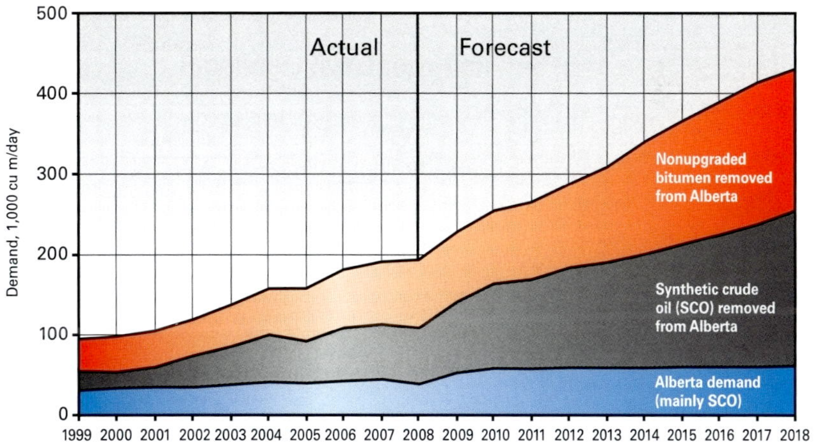

Source: Alberta's Energy Reserves 2008 and Supply/Demand Outlook 2009–2018, ST98–2009, Energy Resources Conservation Board of Alberta, June 2009.

IN SITU PRODUCTION

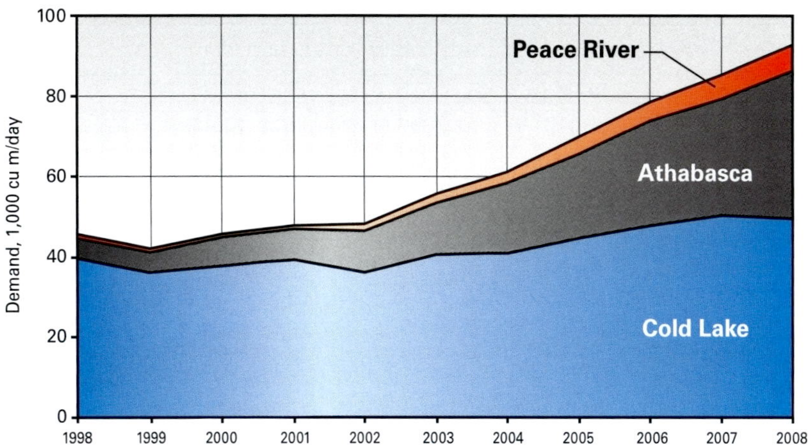

Source: Alberta's Energy Reserves 2008 and Supply/Demand Outlook 2009–2018, ST98–2009, Energy Resources Conservation Board of Alberta, June 2009.

Minable areas under active development

	Area, 1,000 acres	Initial established in place	Initial established reserves	Cumulative production	Remaining established reserves
		million bbl			
Albian Sands	34	4,227	2,636	308	2,334
Fort Hills	47	4,397	2,290	—	2,290
Horizon	70	5,246	3,378	—	3,378
Kearl	49	8,329	5,485	—	5,485
Jackpine	20	2,271	1,396	—	1,396
Suncor	47	6,228	4,322	1,573	2,749
Syncrude	109	13,028	8,215	2,334	5,882
Total	**375**	**43,725**	**27,722**	**4,215**	**23,514**

Source: Alberta's Energy Reserves 2008 and Supply/Demand Outlook 2009–2018, ST98–2009, Energy Resources Conservation Board of Alberta, June 2009.

Mining projects considered in forecast

	Start-up	Capacity, 1,000 b/d	Status
Suncor			
North Steepbank extension	2010		Under construction
Voyageur South Phase 1	TBD*	120	Application
Syncrude			
Stage 3 debottleneck	TBD	47	Announced
Stage 4 expansion	TBD	140	Announced
Alberta Oil Sands Project (Shell)			
Muskeg River expansion and debottlenecking	TBD	115	Approved
Jackpine Phase 1A	2010	100	Under construction
Jackpine Phase 1B	TBD	100	Approved
Jackpine Phase 2	TBD	100	Application
Pierre River Phase 1	TBD	100	Application
Pierre River Phase 2	TBD	100	Application
CNRL			
Horizon Phase 2/3	TBD	135	Approved
Petro-Canada/UTS/Teck Cominco			
Fort Hills Phase 1	TBD	165	Approved
Fort Hills debottleneck	TBD	25	Approved
Kearl Phase 1	TBD	100	Approved
Kearl Phase 2	TBD	100	Approved
Kearl Phase 3	TBD	100	Approved
Total E&P Canada			
Joslyn Phase 1 (North)	TBD	50	Application
Joslyn Phase 2 (North)	TBD	50	Application

*TBD—To be determined. Source: ERCB

Oil sands capital investment

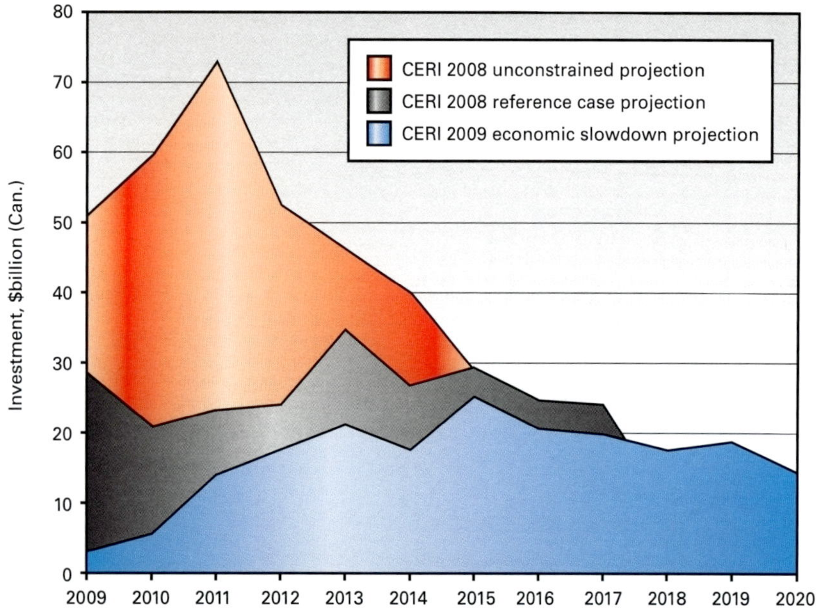

Source: McColl, D., The Eye of the Beholder: Oil Sands Calamity or Golden Opportunity? Cannadian Energy Research Institute, Oil Sands briefing, ISBN: 1-896091-85-7, February 2009.

Firebag project

Suncor Energy Inc. said in November 2009 that it will resume investment in the half-completed third stage of its Firebag in situ development in Alberta. Work had been suspended early in 2009.

The third and fourth stages are to have production capacities of 68,000 b/d each from steam-assisted gravity drainage. Production from the third stage is to start by mid-2011 and from the fourth stage in late 2012.

Total production from the first two stages averaged 54.3 million b/d in July-September.

A total of six stages are planned.

PetroChina: Oil sands JV

Athabasca Oil Sands Corp. (AOSC) entered a series of JV agreements with PetroChina International Investment Co. Ltd. in mid-2009.

PIIC agreed to pay $1.9 billion (Can.) to acquire 60% working interest in AOSC's MacKay River and Dover oil sands projects in northeastern Alberta.

As JV partners, AOSC and PIIC plan to use in situ methods to develop the oil sands projects.

AOSC said it intended to file a regulatory application for the first 35,000 b/d phase of the MacKay River commercial project at yearend.

GHG emissions

Total emissions of greenhouse gases associated with Canadian oil sands exceed those related to average crude oil processed in the United States by 5–15%, said IHS Cambridge Energy Research Associates.

Their 8-month study released in early 2009 evaluated production and processing of raw materials as well as consumption of the oil products.

IHS CERA said 37 stakeholder organizations contributed to the study, including Canadian and U.S. government agencies, oil companies, and environmental and community groups.

It pointed out that the difference in total carbon emissions from oil sands to that of other crude oil sources occurs mainly in the extraction and processing phases—also called "well-to-retail pump" or "well-to-pump."

The study projected that production from Canadian oil sands will reach 2.3–6.3 million b/d by 2035.

In a different 2009 study released in late 2009, CERI estimated that compliance with greenhouse-gas emissions rules could cost the oil sands sector $130 billion over the next 35 years.

LIFE-CYCLE GREENHOUSE GAS EMISSIONS

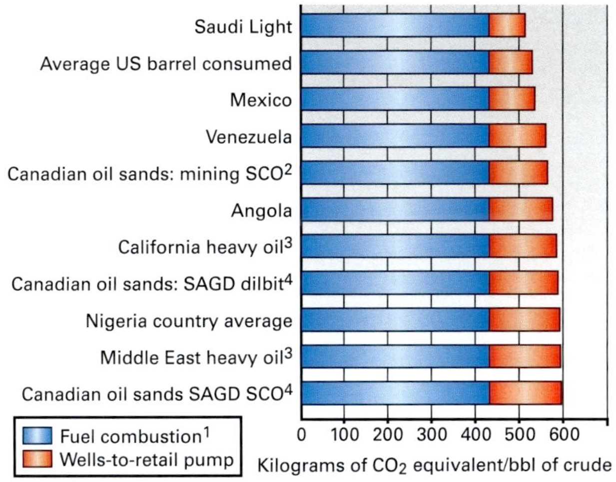

[1]Based on a per-barrel of crude basis assuming an average carbon content. [2]SCO = synthetic crude oil. [3]Assumes steam-assisted gravity drainage (SAGD) is used for production. [4]Assumes steam:oil ratio of 3:1. Dilbit is bitumen diluted with light hydrocarbon liquids.

Source: Cambridge Energy Research Associates based on data from various sources

Kearl go-ahead

ExxonMobil subsidiary Imperial Oil decided in May 2009 to proceed with the $8-billion first phase of its Kearl oil sands project in northeast Alberta.

The decision was in contrast to cutbacks by other oil sands operators concerned about the plunge in crude oil prices that started in mid-2009.

In the first phase, Kearl will extract oil sands through surface mining.

It is due to start producing 110,000 b/d in 2012, slightly more than one-third of total projected output.

Jackpine, Muskeg River challenged

In April 2009, two Canadian environmental groups said Royal Dutch Shell failed to honor its sworn affidavit to cut greenhouse gas emissions at its Jackpine and Muskeg River oil sands mines.

Pembina and EcoJustice estimated Shell's promise was the equivalent of pledging to cut GHG emissions by 900,000 tonnes.

The groups said Canadian regulators would not have approved the Jackpine mine and Muskeg River mine expansion had it not made that promise.

As of early 2009, Shell had begun work at the sites but was not yet producing bitumen.

Shell said it had taken early voluntary action, making its ventures the least greenhouse-gas intensive of all mineable oil sands projects.

Long Lake upgrader

In early 2009, the oil sands upgrader at the Long Lake oil sands facility of Nexen Inc. started production. Full design production of 60,000 b/d of sweet crude is expected by mid-2010.

Total oil sands bid

In early 2009, Total launched a public offer worth $617 million (Can.) [$502 million (U.S.)] to acquire UTS Energy, a Canadian company with oil sands assets.

UTS Energy's main asset is a 20% stake in the Fort Hills project, operated by Petro-Canada.

Total has two other projects in the Athabasca oil sands: Northern Lights and Joslyn.

Drilling and Production: Other than Oil Sands

Saskatchewan: Conventional oil

In 2008, Saskatchewan reported 4,045 oil and gas wells drilled, compared to 3,451 in 2007.

Latest estimates put the province's remaining recoverable reserves of conventional crude at 1.18 billion bbl (611 million bbl heavy oil, the rest medium and light crude).

These figures do not include oil sands or oil shale, classified as unconventional.

EOR production levels, using steam and carbon dioxide, yielded 50,788 b/d in 2008.

Canada's largest CO_2 EOR initiative is the Weyburn project operated by EnCana Oil & Gas Partnership. In 2005, Apache Canada began CO_2 injection in Midale field.

Saskatchewan is Canada's second-largest oil producer and third-largest natural gas producer.

Saskatchewan: Bakken play

One promising play is that portion of the oil-bearing Bakken formation that lies in Saskatchewan.

Production from that area reached 57,000 b/d in December 2008, up from 950 b/d in October 2004.

In mid-2009, the provincial energy ministry attributed rising production to advances in horizontal drilling and hydraulic fracturing. The province has not yet established a Bakken reserve estimate.

BC: Horn River gas plant

EnCana Corp. was making plans in 2009 to build the large Cabin Lake gas processing plant in the Horn River basin to aid in shale development.

The first phase of the plant is slated to begin operation in third quarter 2011. Initial capacity is 400 MMcfd, to be expanded in stages as needed.

LNG

Canaport

In December 2009, an advanced Q-Flex vessel delivered 4.6 bcf of LNG to the Canaport LNG facility in St. John, NB. It was the tenth ship received at Canaport since its commissioning in June.

The terminal has received LNG from Trinidad and Tobago, Egypt and Qatar.

Canaport LNG has a send-out capacity of 1 bcfd, serving markets in Canada and the U.S. northeast through the Brunswick pipeline to Maritimes Northeast Pipeline.

Irving Oil Ltd. and Repsol-YPF SA are partners in the terminal.

Kitimat-Kogas agreement

Kitimat LNG Inc. signed an MOU to supply Korea Gas Corp. (Kogas) with as much as 40% of Kitimat LNG's production, as well as an option to acquire an equity stake in the export terminal in Kitimat, BC.

Kitimat LNG said Kogas plans to purchase 2 mtpy of LNG from the proposed terminal over 20 years at a total purchase value of more than $20 billion.

Kitimat LNG also said it was in discussion with other potential terminal users and investors for terminal capacity, offtake, and equity in the 5-mtpy project.

KSL pipeline project

The government of British Columbia announced plans in early 2009 to facilitate investment by a number of First Nations in the Kitimat to Summit Lake Pipeline Looping (KSL) project.

The $1.2-billion (Can.) KSL system would extend 467 km (290 mi) from Summit Lake to the Kitimat LNG export project. Design capacity is 1 bcfd.

Pipelines

Oil sands pipeline capacity

The global economic downturn, coupled with estimates of excess crude pipeline capacity coming out of western Canada by 2010, slowed development of oil sands export pipeline projects in 2009, said CAPP.

The group also noted that three lines already had been approved and were under construction: Keystone (TransCanada); Alberta Clipper (Enbridge); and Keystone Extension, or XL (TransCanada).

In parallel work, TransCanada had Purvin & Gertz Inc. (PGI) assess the potential supply-demand balance between Western Canadian crude oil and U.S. Gulf Coast refineries.

PGI forecast supply growth between 2007 and 2017 of 1.55–2.42 million b/d.

PGI also noted the following:

- The U.S. gulf refining market was receiving only 65,000 b/d of Canadian crude via the Pegasus pipeline but had coking capacity sufficient to run large amounts of heavy crude.
- U.S. gulf capacity to use heavy crude had been expanding at the same time supplies of heavy crude have been falling
- The U.S. gulf market could absorb a minimum of 500,000 b/d of Canadian crude
- Without Keystone XL, heavy crude oil pipeline capacity out of Western Canada would reach capacity by 2015.

Alberta Clipper status

The U.S. Dept. of State in August 2009 issued a permit authorizing construction of the cross-border Alberta Clipper crude oil pipeline, finding that it would increase diversity of supply, shorten oil delivery time, and increase supply from a major non-OPEC producer.

Enbridge Energy LP said that it expects the 1,609-km (1,000-mi) line to enter service in mid-2010. It is to run from Hardisty, AB to the U.S. border near Neche, ND, then to a refinery near Superior, WI.

Oil sands crude export pipelines

Destination	Project	Route	Capacity, b/d	OD, in.	Completion
East	TransCanada Keystone	Hardisty, Alta., to Wood River-Patoka, Ill.	435,000	—	December 2009
	Keystone Extension	Wood River-Patoka to Cushing, Okla.	590,000 (total, including above)	—	Fourth-quarter 2010
	TransCanada DilBit pipeline	Fort McMurray, Alta., to Hardisty	400,000	—	2012-14
	Enbridge Alberta Clipper	Hardisty to Superior, Wis.	450,000 (expandable to 800,000)	36	July 2010
	Enbridge Southern Access Extension	Flanagan, Ill., to Patoka	400,000	36	2012
	Bow River	Hardisty to Montana	30,000	—	First-quarter 2010
	Enbridge Trailbreaker	Sarnia, Ont., to Montreal	215,000	—	—
	Sunoco	Buffalo, NY, to Philadelphia	400,000	24	—
Gulf Coast	BP-Enbridge Gulf Access	Flanagan to Port Arthur-Nederland, Tex.	400,000	—	2012
	Sunoco	Cushing to Nederland	300,000	—	—
	ExxonMobil-Enbridge	Patoka to Beaumont, Tex.	445,000 (expandable to more than 550,000)	30	—
	TransCanada Keystone XL	Hardisty to Port Arthur-Houston	700,000 (expandable to 1.5 million for full Keystone system)	36	2012
	Altex Energy	Alberta to Port Arthur-Beaumont	425,000 (expandable to 1 million)	36	—
West	Kinder Morgan TMX2	Edmonton to Kamloops, BC	80,000	—	2012
	TMX3	Kamloops to Anacortes, Wash.	320,000	—	2013
	K-M Northern Leg Expansion	Rearguard, BC, to Kitimat, BC	400,000	—	2014
	Enbridge Northern Gateway	Edmonton to Kitimat	500,000	36	—
	TransCanada AB-California	Alberta to California	400,000	—	2016

Source: Canadian Association of Petroleum Producers

Hardisty terminal flows, storage

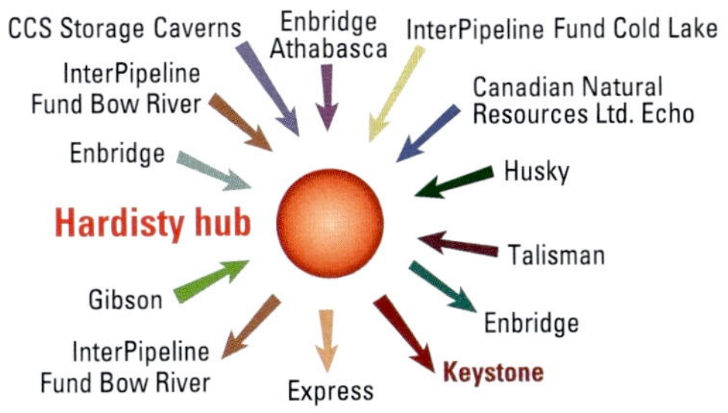

Total storage capacity: 18 million bbl

Major storage operators, capacities; million bbl	
Gibson:	2.8
Husky:	2.3
EMI:	7.5
EPI:	2
CCS:	3
Express:	0.6

Source: TransCanada Keystone Pipeline GP Ltd.

Keystone pipeline system

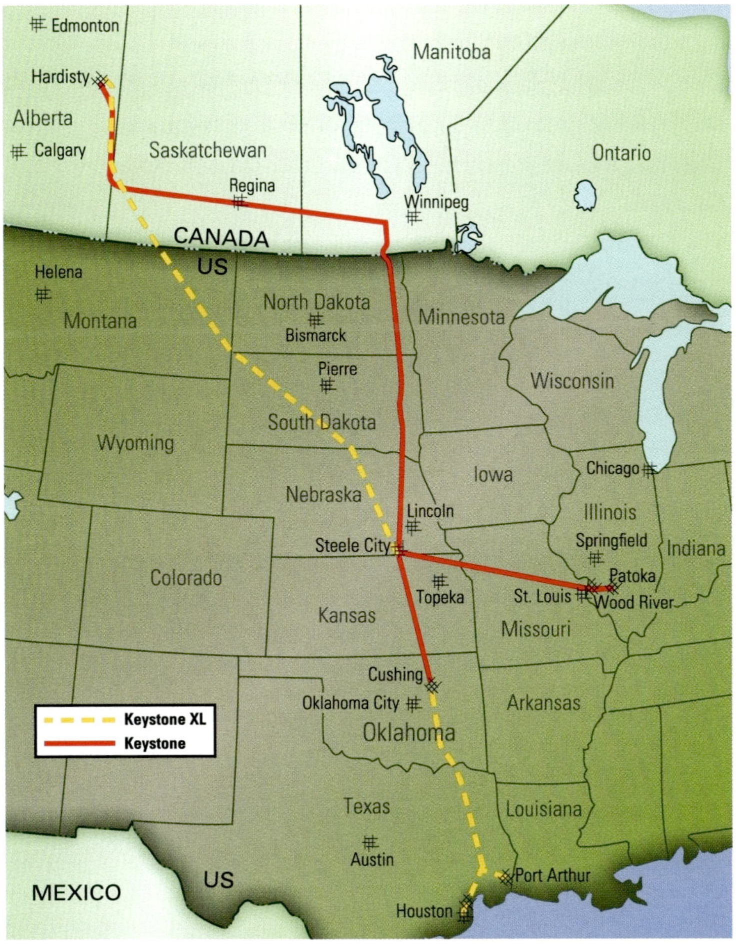

Initial capacity will be 450,000 b/d, but ultimate capacity could be as high as 800,000 b/d.

In September, several groups sued the U.S. secretary of state and other federal officials. They alleged violations of the U.S. National Environmental Policy Act due to inadequate analysis of indirect and cumulative environmental impacts of the pipeline.

Keystone status

The Keystone pipeline project of TransCanada Corp. will span 3,456 km (2,148 mi), transporting crude oil to the U.S. Midwest. In addition to 2,219 km (1,379 mi) of newbuild U.S. line, Keystone includes additions to existing Canadian pipelines and mainline flow reversals.

It was expected to start operation in late 2009 or early 2010, with the capacity to deliver 435,000 b/d from Hardisty to Wood River and Patoka, IL.

TransCanada plans to extend the line to Cushing, OK, starting in fourth quarter 2010, expanding to 590,000 b/d.

In mid-2009, TransCanada reached agreement with ConocoPhillips to buy its remaining interest in the Keystone system.

Keystone XL status

TransCanada continued preparatory and administrative work on the Keystone Gulf Coast Expansion Project (Keystone XL). It will provide 500,000 b/d additional capacity from western Canada to the U.S. Gulf Coast by 2012.

In 2009, Keystone XL secured firm contracts for 380,000 b/d for an average of 17 years from shippers.

Keystone XL consists of 3,187 km (1,980 mi) of pipe, starting in Hardisty and extending to a delivery point near existing terminals in Port Arthur, TX.

TransCanada anticipates beginning construction in 2010, pending regulatory approvals, and intends to start the line in 2012.

Northern Gateway status

Enbridge renewed plans to build the 525,000-b/d Northern Gateway Pipeline from near Edmonton to a tanker terminal near Kitimat, BC, for shipment to China, other parts of Asia, and California.

The 1,170-km (727-mi) project was put on hold in December 2006 to focus on completing the Enbridge Southern Access-Southern Lights project.

Enbridge expects to build Northern Gateway in 2012–2014, pending regulatory approval of filings made in 2009.

The company would also operate the Kitimat terminal.

Enbridge Northern Gateway Pipeline

Source: Enbridge Northern Gateway Pipelines

Alaska natural gas pipeline

In June 2009, TransCanada Corp. and ExxonMobil Corp. agreed to work together on TransCanada's Alaska Pipeline Project natural gas pipeline.

The 2,736-km (1,700-mi) line with 4.5 bcfd capacity would carry gas from Alaska's North Slope to existing pipeline infrastructure in Alberta.

Details are given in the Alaska section of this chapter.

Groundbirch gas line

In mid-2009, Nova Gas Transmission Ltd. (a unit of TransCanada Corp.) said that the Triassic Montney tight sand formation will provide about 90% of the gas to be carried by the proposed 1.66-bcfd Groundbirch pipeline.

Expected in-service date for the 77-km (48-mi) line is November 2010.

The line is to run from Groundbirch, BC, to Gordondale, AB, passing north of the Dawson Creek area.

Nova projects flow at 255 MMcfd in 2010–11 rising to more than 1.4 bcfd by 2021.

Refining and Products

NB: St. John refinery actions

Irving Oil Corp. and BP PLC suspended plans in mid-2009 to build the 300,000-b/d Eider Rock refinery at St. John, NB. The companies said a feasibility study determined that the project "was not viable at a time of global economic recession and dampening forecasts for petroleum product demand in North America."

However, Irving Oil said in September that it would spend $220 million to upgrade its existing 250,000-b/d refinery at St. John.

Environmental Issues

Quest CCS project

In late 2009, the Alberta and federal governments promised $865 million for the Quest carbon capture and storage (CCS) project near Edmonton.

Officials acknowledged that it will be at least 2015 before any carbon dioxide is piped underground.

The governments have teamed with Shell Canada, Chevron Canada and Marathon Oil Sands on the project. It is expected to eventually collect up to 1.2 mtpy of carbon dioxide from the Scotford oil sands upgrader, pipe it to a wellhead, and inject it more than 2 km (1.2 mi) underground.

The CO_2 would be sealed in rock formations by impermeable rock or clay caps.

Total project cost is expected to be $1.35 billion, with $485 million coming from the companies.

It would be the first initiative to draw on Alberta's $2-billion CCS fund and Ottawa's $1-billion pool for such projects.

Project Pioneer CCS initiative

The government of Alberta signed a letter of intent in late 2009 with TransAlta Corp. and its partners Capital Power and Alstom for Project Pioneer at the Keephills 3 plant west of Edmonton.

The project is expected to sequester 1 mtpy of CO_2 starting in 2015.

The initiative will retrofit a coal-fired electricity plant using $431 million from the Alberta CCS fund.

The federal government is also contributing $343 million for the project through the Clean Energy Fund and the federal econENERGY Technology Initiative.

Company News

EnCana split

In November 2009, EnCana Corp. split into two energy companies: Cenovus Energy Inc., an integrated oil company; and EnCana Corp., a pure play natural gas company.

Cenovus Energy focuses on oil projects and established natural gas and crude oil production in Alberta and Saskatchewan. The company also owns refineries in Illinois and Texas.

EnCana Corp. has a portfolio of shale and other unconventional natural gas developments in basins from northeast British Columbia to east Texas and Louisiana.

The company had planned the move in 2008, but the action was halted due to uncertainty in the global debt and equity markets.

KNOC to buy Harvest Energy Trust

In December 2009, Korea National Oil Corp. was poised to buy Harvest Energy Trust for $1.8 billion (Can.) plus the assumption of $2.3 billion (Can.) in debt.

At mid-month, Harvest unit-holders approved the KNOC takeover bid; the deal was expected to close by yearend.

Harvest reported production of 52,745 boe/d as of August 31, 2009, and reserves of 154.3 million boe as of December 31, 2008.

A Harvest subsidiary owns the 115,000 b/cd Come by Chance refinery in Newfoundland.

By acquiring Harvest assets, KNOC will have more than 380 million bbl in reserves and up to 126,000 b/d in production capacity in Saskatchewan and Alberta.

Suncor, Petro-Canada merger

In August 2009, Suncor Energy Inc. and Petro-Canada merged in a $19.3-billion (Can.) transaction. The resulting company, called Suncor Energy Inc., has approximately 7.5 billion boe of proved and probable reserves with existing production of 680,000 boe/d.

With the merger, Suncor became the largest holder of oil sands properties for both mined and in-situ resource recovery. It also holds assets in every major oil development project on Canada's East Coast.

International holdings involve oil and natural gas production from the North Sea, North Africa, and Latin America.

Ontario Nuclear Reactor Plan

In June 2009, the province of Ontario suspended a $26-billion (Can.) ($22.4-billion U.S.) plan to build what would have been North America's first nuclear reactors in 30 years.

Officials pointed to cost overruns and uncertainty about over the future of the government-owned nuclear agency—Atomic Energy of Canada Ltd.—that likely would build them.

Latin America

ARGENTINA

Capital: Buenos Aires
Monetary unit: Peso
Refining capacity: 627,075 b/cd
Oil production: 610,000 b/d
Oil reserves: 2.520 billion bbl
Gas reserves: 14.070 tcf

Exploration and Development

Incentive program: Unconventional gas

Argentinean officials approved in September 2009 the first contract under its "Gas Plus" incentive program to encourage development of unconventional natural gas resources.

Apache Corp. will drill as many as 48 wells in two Neuquen basin fields by 2013 and supply 50 MMcfd of gas at a price of $5/MMBtu. The contracts take effect in January 2011.

Apache earlier reported that of the 83 wells it drilled in 2008 in Argentina (in the Neuquen, Austral, and Cuyo basins), 72 were productive.

Falklands: Consortium focus

A consortium led by Madrid-based Repsol YPF SA plans oil and gas exploration in international waters off the Falkland Islands beginning in early 2010.

No details were given, but the area of interest appears to be the Colorado Marina basin.

Argentina and Britain both claim large areas of the seabed around the islands. Reserves offshore have been estimated at 18 billion bbl.

Britain lodged a claim for 1.2 million sq km (7.5 million sq mi) in May 2009. Argentina filed a claim for the same area in April with the UN Commission on the Limits of the Continental Shelf.

South Falklands basin

In early 2009, Borders & Southern Petroleum PLC identified three key play fairways and two promising prospects in the offshore South Falklands basin.

The company holds six frontier licenses, 150 km (93 mi) southeast of the islands, in 200–3,000 m (656–9,842 ft) of water.

ARUBA
CURACAO
GRENADA
BARBADOS
Bridgetown
TRINIDAD and TOBAGO
Port of Spain
COSTA RICA
Pto. Limon
Colón
Panama
PANAMA
Santa Marta
Barranquilla
Cartagena
Maracaibo
Tibu
Caracas
Lechoso
Jose
Puerto la Cruz
Punzan
San Roque
Grico
Penal-Barrackpore-Wilson
Teak
Balata-Bouvallius
Southeast Galeota
Radix
E. Queens Beach
Quarry-Coora-Quinam
Hato
Sinco
Palmita
Silvestre
S. Silvestre
Ciudad Bolivar
Barrancabermeja
S. Cristobol
Georgetown
Tambaredjo/Jossikreek
VENEZUELA
GUYANA
Cantimplora
Volcanera
Cupiagua
Cusiana
Bogota
Apiay Complex
Chichimene
Castilla
Rubiales
Tulua
Cali
Dina
Tello
Andalucia
San Francisco
SURINAME
Karanambo
Roraima
Tumaco
Esmeraldas
COLOMBIA
Quito
ECUADOR
Para
Guayaquil
Amistad
Tumbes
Zorritos
Piedra Redonda
Talara
Iquitos
Concordia
Manaus
Amazonas
Jurua
Southwest
Rio Urucu
Coari
Leste do Urucu
Nova Olinda
Paripi
Bayovar
Maquia
B R A Z I L
Pucallpa
Aguaytia
Aguas Calientes
Porto Velho
Acre
PERU
Camisea
Fields
Rondônia
Lima
La Pampilla
Conchan
Pisco
Paracas
Mato Grosso
BOLIVIA
Trinidad
Lago de Titicaca
Pirin
La Paz
Surubi
Pto. Villarroel
Cochabamba
Sicasica
GCC
Santa Cruz
Oruro
Colpa
Tita
Rio Grande
Sucre
Corumba
Arica
Tatarenda
Camiri
Boyuibe
Monteagudo
Vuelta Grande
Villemontes
Caigua
Buena Vista
Sanandita
Tigre
Madre Jones
PARAGUAY
Tocopilla
Toro
Campo Duran
Bermejo
Ramos
Lomitas Tranquitas
Villa Elisa
Mejillones
CHILE
Rio Pescado
Caimancito
Jujuy
Salta
ARGENTINA
Asuncion
Antofagasta
0 100 200 300 400 Miles
0 200 400 600 Km
SOUTH AMERICA

SOUTH AMERICA
0 100 200 300 400 Miles
0 200 400 600 Km
Atlantic Ocean
Georgetown
Tambaredjo/ Jossikreek
Paramaribo
Cayenne
GUYANA
SURINAME
FRENCH GUYANA
Belem
Amazon River
Sao Luis
Atum
Curima
Espada
Pescada
1-CES-8
Xareu
Fortaleza
Canto do
Fazenda Belem
Ubarana
Agulha
Amaro
Serraria Lorena
Guamare
Faz. Pocinhos
Estreito
Manaus
Nova Olinda
Rio Tapajos
Rio Teles Pires
Rio Araguaia
Rio Tocantins
BRAZIL
Recife
Maceio
Aracaju
Salvador
Catu
269
Rio Sao Francisco
Cuiaba
Brasilia
Senador Canedo
Ethanol
Sao Mateus
Atlantic Ocean
Corumba
Campo Grande
Aquidauana
PARAGUAY
Rio Grande
Parana
145
Betim
Belo Horizonte
Vitoria
Jubarte
RJS-150
Frade
Roncador
Campos
Duque de Caxias
232
7
6
55°
45°
35°
10°
5°
0°
15°
20°

SOUTH AMERICA
BRAZIL
PARAGUAY
URUGUAY
ARGENTINA
CHILE
Atlantic Ocean
Falkland Islands
(Islas Malvinas)
Oruro
Arica
Sucre
Santa Cruz
Tita
Rio Grande
Camiri
Tatarenda
Corumba
Monteagudo
Vuelta Grande
Boyuibe
Buena Vista
Caigua
Sanandita
Tigre
Madre Jones
Toro
Campo Duran
Bermejo
Ramos
Rio Pescado
Lomitas Tranquitas
Caimancito
Jujuy
Salta
Metan
Tocopilla
Mejillones
Antofagasta
Campo Grande
Aquidauana
Villa Elisa
Asuncion
Formosa
Resistencia
Corrientes
Paso de los Libres
Uruguaiana
Sao Gabriel
Apucarana
Irati
Perimbo
Tremembe
Betim
Belo Horizonte
Campos
Duque de Caxias
Rio de Janeiro
Maua
Sao Paulo
Cubatao
Santos
Sao Sebastiao
Merluza
Tupi
Guara
Curitiba
Sao Francisco do Sul
Florianopolis
Laguna
Canoas
Porto Alegre
Tramandai
Rio Grande do Sul
Ethanol
Cordoba
San Lorenzo
Mendoza
Lumunta
Tupungato
Cacheuta
Concon
Valparaiso
Santiago
Refugio
Colorados
Barrencas
El Sosneado
San Luis
Campana
Buenos Aires
La Plata
Montevideo
Dock Sud
Las Flores
Cnl. Pringles
Bahia Blanca
Punta Alta
Puerto Rosales
Concepcion
Colonia Catriel
El Medanito
Guanaco
Plaza Huincul
Challaco
Cerro Bandera
Barda Gonzales
San Antonio
Golfo San Matias
Rivadavia
Anticlinal Grande
Cerro Dragon
Canadon Grande
Canadon Seco
Comodoro Rivadavia
Cerro Tortuga
Comodoro
Caleta Olivia
Pico Truncado
Santa Cruz
Rio Gallegos
Punta Arenas
Tierra Del Fuego
Rio Grande
Ushuaia
3 20 145 356 14 181 232 47 163 242 193 7 32 181 17 98 38 87 110 37 4 8 189 106 113 25 29 16
0 100 200 300 400 Miles
0 200 400 600 Km

South Falkland Basin Licenses

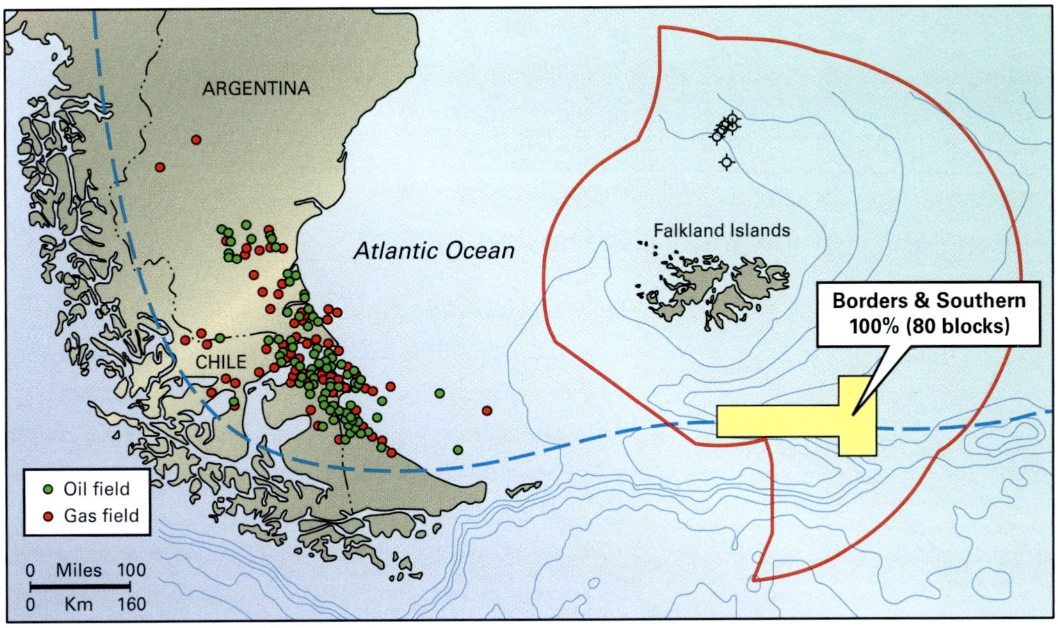

Source: Borders & Southern Petroleum PLC

The firm said the Darwin and Stebbing prospects have the best chance of success and will be the preferred targets for initial tests.

Cuyana basin

In early 2009, Otto Energy Ltd. finalized a farmout from Oromin Explorations Ltd. on the Santa Rosa Block in Argentina's Cuyana basin.

The block lies east of the basin's oil and gas producing area. Oromin was planning to start drilling by mid-year.

Pipelines/Transportation

Strait of Magellan

Plans were under way in mid-2009 to build a gas pipeline through the Strait of Magellan.

The 38-km (24-mi) undersea line is to join Cape Espiritu Santo and Cape Virgenes in waters outside the eastern mouth of the Strait, linking gas fields in the south with the nation's main gas network.

Completion was expected by yearend 2009.

Refining and Products

La Plata refinery

Repsol YPF SA was making plans in early 2009 to upgrade the delayed coking unit at the 189,000-b/d La Plata refinery in Buenos Aires Province.

The improved coker was expected to be ready for operation completed by yearend 2009, for converting heavy residue feed into high-value transport fuel.

Alternative Energy

Shown are selected examples of significant initiatives on wider use of renewable energy resources.

First solar power plant

Power company Energia Provincial Sociedad del Estado (EPSE), San Juan Province, opened a bidding process in mid-2009 for construction of the first solar power plant in Argentina.

To be built in the Ullum Department in San Juan Province, it will have a capacity of 1.2 MW.

About 9,000 solar panels will be installed, aiming for initial operation by third quarter 2010.

World's largest windfarm

Madrid's Grupo Guascor said that it plans to start building a $2.4-billion wind power project in Patagonia, in Santa Cruz Province in mid-2010.

The facility will produce 600 to 900 MW, using 700 turbines—about double the size of the largest windfarm operational in Europe. Target date for start-up is yearend 2013.

BAHAMAS

Capital: Nassau
Monetary unit: Bahamian Dollar
Refining capacity: 0
Oil production: 0
Oil reserves: 0
Gas reserves: 0

Offshore exploration interest

In mid-2009, BPC Ltd. and Norway's Statoil ASA formed a JV to explore for oil and gas off the Commonwealth of the Bahamas, but were awaiting approval of license applications.

The government could approve the license applications by yearend, BPC said.

The territory of interest is between four other Bahamas blocks wholly held by BPC southwest of Andros Island, and six blocks in the Florida Straits off northwestern Cuba, operated by Repsol YPF SA.

BPC also has identified 22 exploration leads on its fully owned Bain, Cooper, Donaldson, and Eneas licenses, 225–425 km (140–264 mi) southeast of Miami, and its Miami license, 85–150 km (53–93 mi) east of Miami.

Only five wells have been drilled in the Bahamas since 1947, the last one 50 km (31 mi) off Cuba in 1986.

Cay Sal area

In mid-2009, Statoil reached agreement with BPC Ltd. to become the operator of the Zapata, Islamorada, and Falcones licenses in the Cay Sal area of the southwestern Bahamas.

Approval and award of the licenses by the Government of the Commonwealth of The Bahamas was pending.

The license has an initial three-year term. Seismic and satellite data indicate the presence of several large structures in the license areas, said Statoil.

Bahamas Land, Offshore Exploration Wells

Well	Year of discovery	Operator	Total depth, m	Age at total depth
Andros Island-1	1947	Superior	4,446	Early Cretaceous
Cay Sal-1	1959	Bahamas California	5,763	Jurassic or Early Cretaceous
Long Island-1	1970	Bahamas Gulf	5,351	Jurassic or Early Cretaceous
Great Isaac-1	1971	Bahamas California	5,440	Jurassic
Doubloon Saxon-1	1986	Tenneco	6,626	Early Cretaceous

BPC PROSPECTS, LEADS OFF BAHAMAS

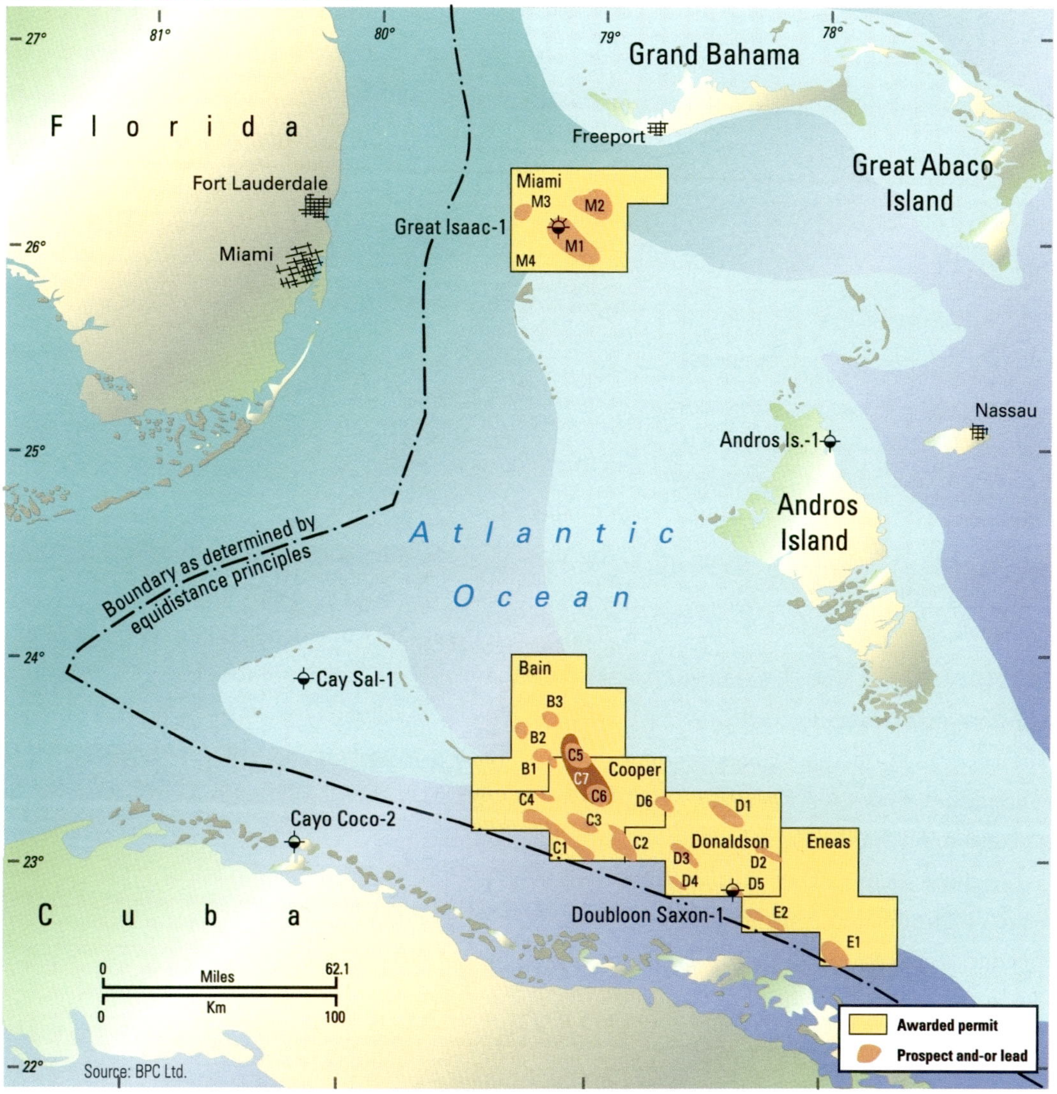

BELIZE

Capital: Belmopan
Monetary unit: Belizean dollar
Refining capacity: 0
Oil production: 0
Oil reserves: 6.700 million bbl
Gas reserves: 0

Belize has the lowest population density in Central America. But as a popular tourist destination, the country consumed 7,400 b/d of oil in 2007—more than twice the level used in 1998, according to the U.S. EIA.

As for production, Belize Natural Energy Ltd. struck oil in the country in 2006, and as of February 2009, production from BNE's seven wells averaged 4,300 b/d—about 3,100 b/d short of domestic needs.

In 2006, under the PetroCaribe program, Venezuela began to supply Belize with 4,000 b/d of products.

In 2007, Belize sought more help from its neighbor, including construction of a small refinery, without success.

The country then went ahead to establish a 2,500-b/d refinery (now operated by Blue Sky Belize Ltd.) and also signed a production-sharing agreement with Taiwan's state-owned CPC in January 2009.

Development plans were being considered in mid-2009.

BOLIVIA

Capital: La Paz
Monetary unit: Boliviano
Refining capacity: 41,200 b/cd
Oil production: 40,000 b/d
Oil reserves: 465.000 million bbl
Gas reserves: 26.500 tcf

Bolivia has the second-largest natural gas reserves in South America (after Venezuela) but remains the continent's poorest country.

Pres. Evo Morales nationalized the country's oil and gas sector in May 2006, and state-owned Yacimientos Petroliferos Fiscales Bolivianos (YPFB) now runs all oil and gas pipelines and domestic sales and exports.

The nationalization action ordered YPFB to hold a majority stake in several foreign-owned refineries.

In early 2009, the *Financial Times* observed that the country was struggling to secure long-term investments for its hydrocarbons sector. The nationalization raised questions about its reliability as a supplier and uncertainty about export demand.

The government said it expected to boost oil and gas investment to $530 million in 2009. Investment peaked at $581 million in 1999, but dropped to $149 million in 2007.

The dozen foreign companies that operate in Bolivia were awaiting details of a hydrocarbons law required by the new constitution.

YPFB CORRUPTION SCANDAL

In February 2009, a scandal emerged regarding corruption at state-run YPFB.

The head of Catler Uniservice was murdered—the firm that won an $86-million contract from YPFP to build an NGL separation plant in Rio Grande.

During the resultant investigation, Pres. Morales and other government officials claimed that the U.S. Central Intelligence Agency was involved, but no evidence was provided.

The case also implicated former YPFB president, Santos Ramirez.

In February 2009, Pres. Evo Morales signed an agreement with Russia's Gazprom to develop Bolivia's gas reserves until 2030. The agreement includes a joint project with France's Total.

Itau field

In mid-2009, Total declared the giant Itau gas-condensate field commercial, ten years after its discovery in Block XX Tarija West. Production is to start in mid-2010 at an initial 50 MMcfd.

Gas will be processed at Petrobras Bolivia's plant in adjacent San Alberto field.

The Total Group also was planning to spud an appraisal well by yearend 2009 at Incahuasi gas-condensate field, 129 km (80 mi) north of Itau.

GCC pipeline

In May 2009, the third stage of the 130-MMcfd Gasoducto Carrasco-Cochabamba (GCC) pipeline began.

The $170-million line will extend 250 km (155 mi) from the gas-producing region of Carrasco to the city of Cochabamba.

GCC is intended to help resolve gas supply restrictions for the industry in Cochabamba, La Paz, Oruro, and Potosi due to the insufficient capacity of the Gasoducto al Altiplano (GAA) line.

BRAZIL

Capital: Brasilia
Monetary unit: Real
Refining capacity: 1.908 million b/cd
Oil production: 1.950 million b/d
Oil reserves: 12.802 billion bbl
Gas reserves: 12.862 tcf

Overview

According to the U.S. Energy Information Administration (EIA), Brazil is the third-largest energy consumer in the Western Hemisphere, behind the United States and Canada.

Offshore exploration and production contributes 81% of Brazil's oil and gas production.

Oil

State-owned Petroleo Brasileiro SA (Petrobras) controls more than 95% of crude oil production in Brazil. Rio de Janeiro state contains about 80% of Brazil's total production.

About 90% of Brazil's proved oil reserves are offshore, with 64% of the concession areas in deep and ultradeep waters, according to Petrobras. These reserves consist of mostly heavy grades.

Petrobras has aggressively expanded production in recent years and is expected to continue to do so.

Fields expected to start and/or ramp up in Brazil in coming years include Parque Da Conchas (BC-10), Cachalote, Frade, Marlim Leste Jubati, Marlim Leste P-53, Marlim Sul P-51, and Pinauna.

Pres. Luiz Inacio Lula da Silva has encouraged Petrobras to invest in overseas oil and gas projects

Natural gas

The Campos and Santos basins off the southeast coast hold the majority of Brazil's gas reserves. Natural gas production has grown slowly in recent years, however, mainly due to a lack of domestic transportation capacity and partly due to low domestic prices.

Brazil's largest natural gas production is offshore in the Campos Basin in Rio de Janeiro state. Most onshore production occurs in Amazonas and Bahia states.

Petrobras has predicted that gas use will reach 134 MM cu m/d (4.73 bcfd) in 2012, almost triple the level of 2006.

Gas is used mostly as a substitute for fuel oil in industrial and power-generating applications. The importation of natural gas has helped fuel rapid growth in domestic consumption.

Alternative fuels

Brazil is the world's second-largest ethanol producer after the United States, and is the world's leading exporter of ethanol made from sugar cane.

EIA has estimated that Brazil's ethanol production could reach 530,000 b/d in 2009.

Most of the nation's cars run on pure ethanol or an ethanol-gasoline mixture.

Biofuel production is anticipated to grow by around 60,000 b/d in 2010.

Join OPEC?

In April 2009, Luiz Inácio Lula da Silva seemed to reverse earlier decisions, saying Brazil was again considering an invitation to join OPEC.

In March, an official with the Mines and Energy Ministry said the country was officially invited to join OPEC, but that it should enter the organization only when it effectively becomes an oil exporter.

Brazil, which consumes all of its own production and imports additional oil, could not implement OPEC output cuts.

In September 2008, Brazil declined an invitation to join OPEC, citing plans to export refined oil, not crude from its deep water reserves.

Brazil's New Licensing System

Cost recovery mechanisms	• Contractors will recoup costs of exploration, development, operation, and abandonment through "cost oil," limited per period by a ceiling established in their contract. • Costs above ceiling can be carried and recovered in succeeding years.
Production shares before and after cost recovery	• Petrobras will be the operator of all production-sharing contracts, with a minimum participation of 30% of the costs incurred and profits obtained. • "Cost oil" will be shared by contractors, and "profit oil" will be shared among government and contractors according to the auction result.
Royalties	• Royalties are still under consideration by the Brazilian Congress. • The model is expected to follow that of most other fiscal systems, with royalties taken as a percentage of gross revenues (calculated from the percentage of production).
Taxes	Taxes applied will depend on existing legislation.
Uplifts	A portion of capital costs can be recovered through cost recovery, determined on a case-by-case basis by regulation or by contract.
Adjustments to price changes	• Adjustments on "profit oil" shares between the government and contractors can be previewed in contracts to protect against future price fluctuations. • Adjustments will be approved by the national energy council.
Terms of contract auctions	• The winner of auctions for exploration and production-sharing contracts will be the company or group of companies whose proposal offers the greatest percentage of "profit oil" to the Brazilian government. • Upon assigning a contract, the contractor must pay a bonus.
Existing concession contracts	• Concession contracts already established in the presalt region will be fully honored by the Brazilian government. • The new sharing model will be effective only for those areas which have not yet been auctioned, representing about 72% of the whole presalt area, or 107,000 sq km.

Source: Brazil's Energy Research Corp. (EPE)

Presalt Resource Development

New development rules

In August 2009, Pres. Lula da Silva proposed new legislation to control development of Brazil's potentially huge oil reserves in the subsea presalt layer, about 270 km (168 mi) offshore.

The plan also aims to establish a strong petrochemical industry to refine the oil for export as value-added products.

Ratification of the new development model by Brazil's Congress was expected by yearend. It includes:

- a new production-sharing system (replacing a concession approach) for contracts
- a new public company for presalt contract agreement and administration
- a new social fund for investment in education and mitigating poverty in Brazil.

The move to a production-sharing system for the award of new contracts reflects efforts to change in Brazil's standing from an oil importer to a self-sufficient global energy producer.

Under the proposed system, Petrobras will be operator of all presalt E&P contracts; interested parties can seek contracts through a partnership agreement.

The production-sharing system will apply to new contracts signed for fields in 72% of the presalt area, while previously awarded contracts (which involve 28% of the presalt region) will remain unchanged.

A new oil company, Petro-Sal, totally government-owned, will oversee production-sharing agreements.

Reduced resource estimate

In September 2009, Credit Suisse reduced its estimate for presalt oil reserves in the Santos basin, to 28.2 billion bbl from 50 billion bbl.

The previous estimate was made when potential reserves of 8 billion and 4 billion bbl were announced for the Tupi and Iara prospects, respectively.

Concession vs. PSA

In a concession, oil companies are given any oil they produce in return for bearing exploratory and operational risks and for paying royalties and other fees to the government.

In a production-sharing agreement (PSA), the oil remains government property and oil companies are given a share of it as payment for their services.

The revision was called necessary because the region is becoming better defined as more wells are drilled. In July, two wells, on Blocks BM-S-22 and BM-S-52, turned up dry.

Credit Suisse analysts called the Guara prospect, on BM-S-9 block, "the most promising area in the block."

New research center

The Federal University at Rio de Janeiro said in September 2009 that it plans to open a $10-million facility for research into development of presalt hydrocarbons offshore.

Petrobras and Schlumberger are partnering with the university to build the facility on Fundao Island.

Officials plan initial research on improved characterization of deep reservoirs using such technologies as electromagnetics, nuclear magnetic resonance, and seismic evaluation.

Exploration history

Petrobras discovered the Tupi field in 2007 in the Santos basin in a presalt zone. It is the biggest oil discovery since the Kashagan field off Kazakhstan, and the largest in the Americas since 1976.

Oil in the presalt zones appears to be lighter and sweeter than most of Brazil's existing production.

Additional presalt discoveries include Carioca, Iara, and Guara. Preliminary industry estimates indicate recoverable presalt oil and gas reserves could approach 56 billion boe.

In June 2009, the 4-BRSA-711-RJS (4-RJS-647) well confirmed reserves estimates of 5–8 billion boe of oil and natural gas in the Tupi area.

Petrobras expects to invest $28.9 billion in the presalt play by 2013. EIA estimates that large-scale development of presalt reserves likely will not occur until well after 2010.

Development challenges

Although Brazil's presalt play is seen as an exciting new world-class exploration province, Wood Mackenzie noted in early 2009 that a number of uncertainties about development remain:

- Water depths are at or beyond today's production technologies.
- Waxy or heavy crudes will require special equipment.
- High CO_2 content at Tupi and other fields will require corrosion-resistant equipment.
- Carbonate reservoirs are heterogeneous, and quality is difficult to predict.
- Monetizing gas will require pipelines, LNG, or CNG infrastructure.

Bidding in 2010

The Ministry of Mines and Energy said in May 2009 that Brazil will invite IOCs to bid for concessions in the country's large presalt oil fields in 2010.

U.S. Ex-Im Bank financing

In August 2009, acting through the Export-Import Bank, the United States agreed to raise support by $8 billion to $10 billion for development of presalt oil reserves.

The bank had approved in May a $2-billion preliminary commitment to encourage purchase of U.S. goods and services by Petrobras, which plans to invest $174 billion in development through 2013.

Petrobras also negotiated a 10-year, $10-billion bilateral credit line with the China Development Bank.

Santos basin: Abare Oeste

Petrobras discovered oil and gas in September 2009 on its Abare Oeste prospect in the Santos basin.

Abare West is the fourth well drilled in the BM-S-9 offshore block, which also includes the promising Carioca, Guara and Iguacu discoveries.

Traces of oil and natural gas were discovered in 2,163 m (7,096 ft) of water and at a total depth of 5,150 m (16,896 ft).

Petrobras is lead operator of Block BM-S-9; other interests are BG Group PLC and Repsol YPF.

BG Group announced a separate find at Abare Oeste shortly after the Petrobras announcement.

Abaré Oeste well

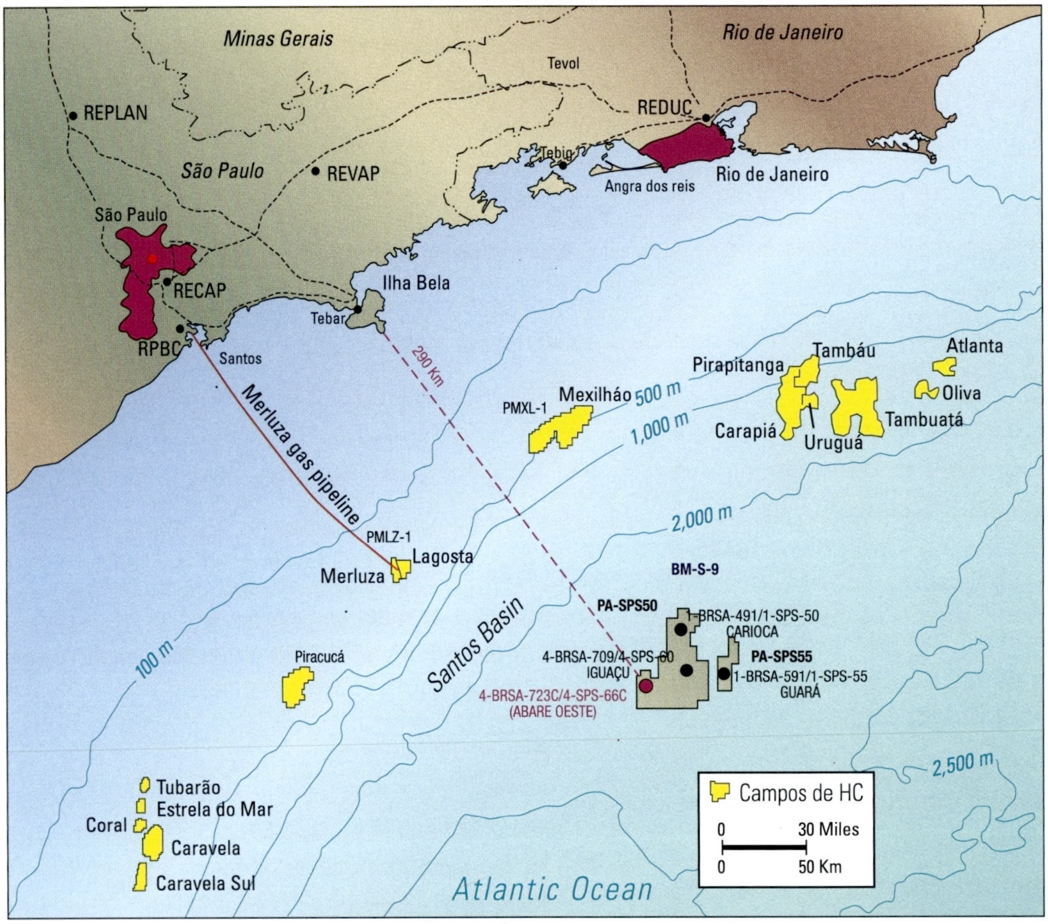

Presalt: A description

Brazil's presalt region is approximately 274 km (170 mi) off the Atlantic coast. Stretching from Espírito Santo to Santa Catarina State, it is some 805 km (500 mi) long and 200 km (125 mi) wide.

"Presalt" oil deposits lie under a thick layer of salt below the seabed.

Water depth typically exceeds 2,000 m (6,560 ft), overlying a further 3,000 m (9,840 ft) of sand and rock, and a layer of salt that also can be more than 2,000 m (6,560 ft) thick.

As of late 2009, Petrobras estimated 5–8 billion bbl of producible oil in Tupi field; 3–4 billion bbl in Iara; and 1.1–2 billion bbl in Guara.

Santos basin: Tupi EWT

In mid-2009, Petrobras started production of oil—in an extended well test (EWT)—from the presalt layer of the Santos basin, on Block BM-S-11.

The action, in the Tupi field, aimed to gather technical data on such issues as fluid movement during production; underwater offloading; and optimum geometry for final wells.

The Tupi pilot project was expected to start up after the conclusion of the EWT, with a capacity to produce and process 100,000 bbl of oil.

Santos basin: Guara

Under limited flow tests, the Guara presalt discovery well 1-SPS-55 (1-BRSA-594) produced at about 7,000 bo/d, said Petrobras in third quarter 2009. Initial production could approach 50,000 b/d.

The well, drilled in mid-2008 on Block BM-S-9, is about 55 km (34 mi) southeast of the Tupi 1-RJS-628A (1-BRSA-369A) discovery well. Water depth is 2,141 m (7,024 ft).

Others with interests in the block are BG Group and Repsol YPF SA.

Santos basin: Azulao well

ExxonMobil Corp. reported in early 2009 discovering traces of oil in the Azulao-1 well, on offshore Block BM-S-22 in the presalt layer of the Santos basin, near the Tupi and Carioca areas.

Others with interests in the well are Hess Corp. and Petrobras.

Other Exploration and Development

Campos basin: Wahoo No. 2

In late 2009, Anadarko's Wahoo No. 2 appraisal/exploration well (also called Wahoo North) in the Campos basin encountered more than 90 ft (295 m) of high-quality net oil pay.

Anadarko said the pay was in the same presalt interval as the original Wahoo discovery.

The company plans to drill up to two exploration/appraisal wells on block BM-C-30 in 2010.

Campos basin: Block BM-C-43

OGX Oil & Gas Co. said in late 2009 that its 1-OGX -1-RJS well in offshore Block BM-C-43 encountered an oil column of more than 200 m (656 ft) and net pay of 57 m (187 ft).

The firm estimates recoverable oil at 500 million to 1.5 billion bbl.

The well, in the southern part of the Campos basin, was drilled to 2,347 m (7,700 ft) in 140 m (459 ft) of water.

Santos basin: Piracuca field

Petrobras and partner Repsol YPF SA filed a declaration of commercial viability in April 2009 for a light oil and gas discovery made on Block BM-S-7, Piracuca field, in the Santos basin.

The field lies in 200 m (656 ft) of water off Sao Paulo state, 200 km (124 mi) southeast of Santos.

The partners estimated recoverable volumes at 88.5 million cu m or about 550 million boe.

Santos basin: Iguaçu well

Well 4-BRSA-709 (4-SPS-60), known as Iguaçu, found another accumulation of light oil in the BM-S-9 concession in the Santos basin, said Petrobras in April 2009.

The BM-S-9 concession comprises two evaluation areas:

- Carioca: 1-BRSA-491-SPS (1-SPS-50)
- Guará: 1-BRSA-594-SPS (1-SPS-55)

Iguaçu is in the Carioca area, some 340 km (211 mi) off the coast of São Paulo State in 2,140 m (7,021 ft) of water.

Potiguar basin

In late 2008 and early 2009, several companies reported traces of hydrocarbons at several test wells in the Potiguar basin:

- Petrobras: well 3-BRSA-695RN on Block POT-T-520; and well 3-BRSA-700DRN on Block BT-POT-8, off Rio Grande do Norte State.
- Galp Energia and Petrobras: onshore well 1-GALP-26RN on Potiguar's POT-T-354 Block
- Galp Energia and Petrobras: an onshore well on Sergipe-Alagoas' SEAL-T-412 Block.
- Petrobras and Starfish Oil & Gas SA: at Well 1-STAR-BRN on Block POT-T-794 onshore.

Drilling and Production

Petrobras: Five-year plan

Petrobras released its 2010–13 strategic plan in January 2009.

It includes development of presalt discoveries as well as continued development of discovered heavy oil and gas accumulations.

The company projects its oil production will grow to 2.68 million b/d in 2013, from 2.05 million b/d in 2009. Combined gas and oil production increases to 3.32 million boe/d in 2013 from 2.51 million boe/d in 2009.

PETROBRAS'S 2010-13 PLAN

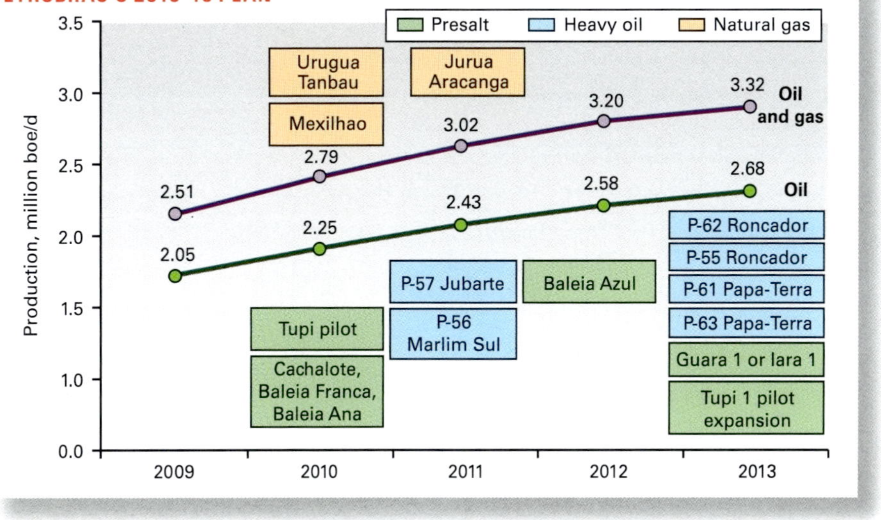

Petrobras slated five large gas production projects to start in 2009.

- Camarupim: 6 MM cu m/d
- Canapu: 2 MM cu m/d
- Lagosta: 1.5 MM cu m/d
- Manati expansion: 2 MM cu m/d
- Urucu system: bringing about 6 MM cu m/d of gas from the middle of Amazonas to Manaus.

In 2009, Petrobras was aiming to add seven rigs to the two drilling in the presalt in 2008. It plans to add six more in 2010, and another in 2011, for 16 drilling rigs for the deepwater presalt.

During 2009–13, Petrobras expects to start production from ten large oil projects and three large gas projects, including the Tupi pilot (late 2010), Cachalote, Baleia Franca, and Baleia Ana.

The company estimates that production from the presalt will reach 219,000 b/d of oil and 7 MM cu m/d of gas in 2013, rising to 582,000 b/d of oil in 2015, and 1.8 million b/d in 2020.

Campos basin: BC-10 field

In mid-2009, JV partners reported the start of production from the FPSO Espirito Santo in the BC-10 field development in the Campos basin.

JV members are Royal Dutch Shell, ONGC Campos, and Petrobras.

The FPSO, moored in 1,789 m (5,869 ft) of water, is designed to process 100,000 b/d of oil and treat up to 45 MMcf/d of gas for injection or export.

Campos basin: Parque das Conchas

Production began in July 2009 from the multi-field Parque das Conchas project in the Campos basin, according to Royal Dutch Shell.

In the two-phase project, initial production was drawing from the Abalone, Ostra, and Argonauta B-West fields. The second phase will focus on the Argonauta O-North field.

Campos basin: Frade field

The Brazilian subsidiary of Chevron Corp. began crude oil production in June 2009 from Frade field, the company's first operated deepwater development in Brazil.

The project is expected to achieve peak production of 90,000 b/d of crude oil and NGLs in 2011.

The field is in 3,700 ft (1,128 m) of water, some 230 mi (370 km) northeast of Rio de Janeiro. It contains an estimated 200–300 MMbbl of recoverable oil.

Besides Chevron, interests include Petrobras and Frade Japão Petróleo Ltda., a Japanese joint-vehicle company of INPEX, Sojitz, and JOGMEC.

Espírito Santo: Cangoá, Camarupim fields

In June 2009, Petrobras announced initial gas production from the Cangoá and Camarupim gas fields, off the northern coast of Espírito Santo.

The projects are part of the Gas Production Anticipation Plan (Plangás), launched in 2006 to leverage the domestic production of natural gas.

Well 7-CMR-1H-ESS at Camarupim field is anchored in 790 m (2,592 ft) of water.

Cangoá field was producing through well 7-CAN-1D-ESS, 42 km (26 mi) from the mouth of the Doce River, in 60 m (197 ft) of water.

The field is capable of producing about 400,000 cu m/d of natural gas.

$830-million drillship

Petrobras and Mitsubishi Corp. agreed in February 2009 to form a 50-50 JV to construct and operate a drillship with the capacity to drill in water 3,050 m (10,000 ft) deep.

The ship, to be delivered in mid-2010, will cost $830 million.

The JV will charter the vessel to a firm that will provide drilling services for the exploration and development of Petrobras's deepwater oil and gas fields in Brazil and overseas.

Carioca find

Repsol YPF SA reconfirmed in January 2009 that it will spend €500 million in 2008–12 to develop the Carioca presalt discovery.

Interests in the Santos basin BM-S-9 Block that contains the Carioca find are Repsol YPF, Petrobras, and BG Group.

The ANP estimated in 2008 that Carioca could contain 33 billion boe, possibly making it one of the biggest oil finds ever.

Daily production record

Petrobras reported a new daily oil-production record in March 2009, due mostly to the start up of P-51 and Cidade de Niteroi in the Marlim Sul field, and P-53 in Marlim Leste, all in the Campos basin.

The 2,012,654 boe figure beat its previous record, from December 2007.

Officials expected yearend 2009 production of 2.8 MMb/d, climbing to 3.7 MMb/d by 2013, and perhaps 5.8 MMb/d by 2020.

Campos basin: Jabuti well

In February 2009, the Jabuti well (7-MLL-54HP) began producing, about 120 km (75 mi) off the coast of Rio de Janeiro in the Campos basin.

Petrobras reported initial flow of 35,000 b/d from the well, in water depth of 1,413 m (4,636 ft).

Jabuti represents the first commercial light oil accumulation ever discovered in deepwater carbonatic reservoirs, proving that these formations can still accumulate and retain oil at such depths.

International Activities

Russian oil, gas expertise

On a July 2009 visit to Moscow, Brazilian and Russian officials expressed interest in sharing information about each country's technology for developing oil and gas fields.

Deepwater oil production technology was of particular interest.

Oil to China for loans

Under a February 2009 agreement, Petrobras pledged to supply China with between 100,000 and 160,000 bbl/day of oil in return for Chinese loans of up to $10 billion.

The funds would help Petrobras and its partners develop presalt fields.

Petrobras said it signed a contract with a subsidiary of Sinopec and an MOU with PetroChina.

In May, Petrobras also finalized a deal with the China Development Bank for a 10-year, $10 billion bilateral credit line.

LNG

Summary of terminals

Aiming to reduce its dependence on gas from Bolivia, Brazil has constructed two LNG regasification terminals, and in 2009, planning was under way for several more.

The Petrobras Bahia de Guanabara LNG terminal became operational in August 2009. It can regasify up to 14 million cu m/d, sending the gas primarily to power stations in southeastern Brazil.

The Petrobras Pecem LNG terminal was completed in 2008 in the state of Ceara. It can regasify 7 million cu m/d. The gas fuels three other power plants (two in Ceara State, one in Rio Grande do Norte State).

Petrobras was conducting environmental studies in September regarding another LNG terminal. Possible locations are São Francisco do Sul, Aracruz, and Ipojuca. Completion in 2013 was estimated.

Five private Brazilian firms agreed in October 2009 to consider construction of a 6-million cu m/d regasification terminal by 2013 in Barcarena, Pará. Initial LNG likely would come from Trinidad and Tobago.

Pipelines

Urucu-Coari-Manaus line

In November 2009, Petrobras began commercial operation of the Urucu-Coari-Manaus natural gas pipeline.

The 661-km (411-mi) line is part of Brazil's Growth Acceleration Program to bring gas from the Amazonian Solimoes basin to market. Branch lines run to Coari, Codajas, Anori, Anama, Caapiranga, Manacapuru, and Iranduba.

Plans call for boosting the initial capacity of 4.1 million cu m/day to 5.5 million by September 2010.

Refining/Petrochemicals

Abreu e Lima accord

Petrobras and Venezuela's PDVSA completed negotiations in November 2009 for joint construction and operation of the Abreu e Lima refinery, in Brazil's Pernambuco state.

The refinery will be able to process 230,000 b/d of heavy oil, supplied equally by the two firms. The refinery's main product will be low-sulfur diesel. No start-up date was specified.

The refinery was conceived to process crude from a PDVSA-Petrobras JV in Venezuela's Orinoco belt, but agreement on terms has taken years.

In October, Petrobras said it had resolved all outstanding issues, but that PDVSA would have to pay Petrobras at least $400 million when it signs the final agreement.

In September, Petrobras estimated the project cost at $12 billion, nearly three times the original figure.

Refining first presalt crude

In October 2009, Petrobras began refining 264,000 bbl of oil from Tupi field—its first load from the presalt region—at the Henrique Lage refinery in Sao Jose dos Campos, Sao Paulo State.

Petrobras said Tupi's oil is paraphinic, with low sulfur content.

The Tupi reservoir is more than 3,000 m (9,842 ft) under the seabed, beneath 2,000 m (6,562 ft) of salt, in water 2,140 m (7,021 ft) deep, some 300 km (186 mi) offshore São Paulo State.

Comperj project

Petrobras said in January 2009 that its $8.5-billion planned investment in the Rio de Janeiro petrochemical complex known as Comperj would be maintained in its five-year strategic plan.

Comperj is part of Brazil's Program for Accelerated Growth (PAC), an economic expansion program outlined by Pres. Lula da Silva.

Construction on the 150,000-b/d refinery began in March 2008; it is due to come online in 2012.

Alternative Energy

Shown are selected examples of significant initiatives on wider use of renewable energy resources.

Renewables role

In May 2009, the government announced plans to increase emphasis on renewable resources even as it continues oil and gas development.

At an earlier international conference, North and South American countries discussed a proposal to generate 50% of their energy from renewable sources by 2050.

Resources expected to get more attention include biodiesel and ethanol made from sugarcane.

Brazil's national energy policy aims to produce 23.3 billion l./year of ethanol and to export 5 billion l. For biodiesel, the goal is to reach production of 3.3 billion l./year by 2010.

Ethanol, electricity from sugarcane

Late 2009 saw the creation of LDC-SEV Bioenergia, in Sao Paulo, the second-largest sugarcane processor in the country.

Two companies combined assets to create the new firm: Louis Dreyfus Commodities Brasil SA, and Santelisa Vale Bioenergia SA.

LDC-SEV Bioenergia will operate 13 plants that process 40 mtpy of cane to produce sugar, ethanol and electric power.

First wind-only energy auction

A unit of the Ministry of Energy and Mines was poised at yearend 2009 to hold its first exclusive auction for wind-power reserve energy.

More than 400 participants—chiefly from the states of Rio Grande do Norte, Ceará, and Rio Grande do Sul states—submitted applications.

Officials said they expected to award rights to develop 1 to 2 GW of capacity.

Company News

Petrobras: Investment plans

Petrobras said in February 2009 that it planned to invest $174.4 billion through 2013 for exploration and production, despite the current turmoil in financial and commodity markets. More than 500 projects are planned.

A major focus will be development of its huge Tupi discovery in the Santos basin, estimated to hold reserves of 5–8 billion boe. Early production is expected to start at yearend 2010.

Production from two other promising presalt resources—Iara and Guara—is anticipated in 2012 and 2013, respectively.

Petrobras expects that 90% of its spending will be in Brazil, and 10% abroad.

Its domestic oil production is expected to jump to 2.68 million b/d in 2013, with worldwide production of 3.6 million boe/d.

Of the total earmarked for E&P, some $29 billion will be allocated to Santos and Campos basin presalt projects alone, including funds for the Parque das Baleias, off Espirito Santo.

The company also said it expects to finish expanding its Cenpes research center in Rio de Janeiro in early 2010.

Petrobras: Funding

In July 2009, Petrobras announced a $13.3-billion funding agreement with the Brazilian Development Bank (BNDES) to support its four-year business plan.

The company also had sought funding from other governments, including China and the United States.

One major project is the development of the Tupi field in the Santos basin.

Petrobras wants to boost its oil and gas production from the current 2.2 million b/d to 3.3m b/d by 2013, and to 5.7 million b/d by 2020.

Petrobras: Report card

In a review of company performance in mid-2009, *Oil & Gas Journal* said that Petrobras "appears to be doing everything right," including:

- Going partially public
- Getting loans from China
- Developing expertise and a strong reputation for deepwater exploration and development
- Applying that deepwater expertise in the Gulf of Mexico and foreign waters

- Being a profitable business partner for U.S. service companies and IOCs
- Developing oil supplies for the domestic market rather than for export.

At the end of 2008, Petrobras's domestic proved reserves of oil, condensate, and natural gas totaled 14.093 billion boe, up 1.2% from 2007.

Proved reserves of oil and condensate equaled 11.969 billion bbl. Proved gas reserves were 337.6 bcm.

These figures did not include presalt discoveries in the Santos basin still being evaluated.

Petrobras: Investigation

In July 2009, a parliamentary committee began an investigation of alleged fraud, corruption, over-billing and tax avoidance by Petrobras.

Charges also were leveled against the ANP—the government regulator for petroleum and gas.

Petrobras denied the allegations and supporters of Pres. Lula da Silva dismissed the inquiry as politically motivated.

OGX: Plans

OGX Petroleo e Gas Participacoes SA—Brazil's largest independent in terms of offshore exploration acreage—said in February 2009 that it was planning to drill or participate in 51 wells from 2009–2012, in Brazilian land and offshore basins.

The company holds 1.7 million acres of offshore rights on 22 blocks in four basins.

Largest Windfarm

The 38-MW Monte Redondo Wind Park—the largest windfarm in Chile—began operation in October 2009.

Located in Ovalle (Comquimbo region), the $100-million facility is run by a subsidiary of GDF Suez.

Nineteen 2-MW turbines generate electricity at the site.

A power distribution firm in Santiago has committed to buy 100 GW/year for 14 years, starting in January 2010.

Officials said the project will cut CO_2 emissions by 54,000 tpy.

CHILE

Capital: Santiago
Monetary unit: Peso
Refining capacity: 226,800 b/cd
Oil production: 2,500 b/d
Oil reserves: 150.000 million bbl
Gas reserves: 3.460 tcf

Overview

Chile has one of South America's most robust and open economies. But in 2008, its real GDP grew by only by an estimated 3.2%, down 30% from 2006 and 2007, according to the *World Factbook* of the U.S. Central Intelligence Agency.

The country has limited domestic energy resources and must import the bulk of the energy it uses. Argentina historically has been a key source of energy imports in recent years.

Exploration and Development

Fell Block, Monte Aymond field

GeoPark Holdings Ltd. was evaluating a gas-condensate development well in early 2009 on its Fell Block.

It also said it expected Monte Aymond-33 well (Monte Aymond field) to come on line by mid-2009. That well, TD 2,391 m (7,844 ft) tested at 3.12 MMcfd of gas, 22.6 b/d of condensate, and 33.9 b/d of water.

The company was completing its $57-million 2008 capital program on that block. It said that more than 90% of its output in 2008 (77% of it gas) came from the 440,000-acre Fell Block.

Magallanes basin

Apache Corp. said in October 2009 that it planned to drill three wells on its onshore concessions in the Magallanes basin by yearend.

The plan includes first spudding the Esmeralda A-1X well on the Rusfin concession, which surrounds the depleted Lago Mercedes oil field, and then the Carmen A-1X and B-1X on the Lenga concession to the north.

Separately, GeoPark said it also planned to drill on the Tranquilo and Otway Blocks in the basin.

LNG

GNL Quintero facility

BNamericas reported that Chile's first LNG regasification terminal, operated by the GNL Quintero consortium, was up and running in October 2009, sending out 4.5 MMcmd of natural gas.

Interest-holders are BG Group, Endesa, Enap and local gas distributor Metrogas.

The plant—the first land-based LNG regasification facility built in South America—has a sendout capacity of 10 MMcmd, which could be expanded to 20 MMcmd.

First LNG was delivered in June from Trinidad and Tobago, but shipments from Equatorial Guinea and Egypt also were planned for 2009.

GNL Quintero has a long-term LNG contract for the equivalent of 6.5 MMcmd and can dock the largest LNG tankers now operating.

GNL Mejillones facility

A JV of Suez Energy International (a unit of GDF Suez SA), and Chilean copper company Codelco, was developing the GNL Mejillones LNG regasification facility—the second such plant in Chile.

Initial construction was to be complete by yearend 2009, with delivery of a first LNG cargo from Trinidad and Tobago planned for January 2010.

The plant initially will use a floating storage unit, but plans call its eventual replacement with a land-based storage and regasification facility.

COLOMBIA

Capital: Bogota
Monetary unit: Peso
Refining capacity: 285,850 b/cd
Oil production: 660,000 b/d
Oil reserves: 1.355 billion bbl
Gas reserves: 3.955 tcf

Overview

Colombia is an important petroleum producer, although political unrest and stagnant reserves have led to decreased exports in recent years. According to the U.S. Energy Information Administration (EIA), the country aims to boost hydrocarbon exploration to preserve its status as a net oil exporter in the longer term.

Longstanding civil conflict has taken a toll on the country's energy sector. The frequency of attacks against energy infrastructure has declined in recent years, but pipelines and power lines still suffer damage from insurgent groups.

Oil

Colombia's oil production has declined steadily since 1999, when it peaked at 830,000 b/d, due to a lack of sizable new reserve discoveries. Average production was 578,000 b/d in 2008.

Much of the country's crude is lighter and sweeter than that of other major Latin American producers.

All oil production is overseen by state-owned Ecopetrol in contracts of association with foreign companies. Colombia's oil licensing agency, Agencia Nacional de Hidrocarburos (ANH) has said Ecopetrol aims to raise output to 700,000 b/d by 2015 and to 1 million b/d by 2020.

To attract oil companies, Ecopetrol has cut its mandatory stake to 50–55% in any exploration contract with a private-sector company—down from the 70% required in previous decades.

Natural gas

Colombia's gas reserves are spread across 18 basins, seven of which have active production. Most reserves are in the Llanos basin, although the Guajira basin accounts for most of current production.

The government's Natural Gas Mass Consumption Plan aims to increase domestic natural gas use and establish Colombia as the "gas hub" for the Andean region.

Exploration and Development

Rubiales oil field extension

Pacific Rubiales Energy Corp. said in mid-2009 that it appears to have extended Rubiales heavy oil field in the Llanos basin into the Quifa Block to the southwest.

The firm was planning to drill four more wells surrounding the field in 2009.

Niscota Block

A subsidiary of Ecopetrol logged an indicated gas-condensate discovery in mid-2009 on the Niscota Block (Llanos basin), 300 km (186 mi) northeast of Bogota.

Talisman Energy Inc., whose Colombian subsidiary also has an interest in the block, said the Huron-1 exploration well was drilled to 18,275 ft.

One reservoir tested at 3,400 b/d of condensate. Gas volume was not disclosed.

Caguan basin

Emerald Energy PLC and Canacol Energy Ltd. were delineating an apparent large heavy-oil discovery in the Caguan basin, 322 km (200 mi) south of Bogota in early 2009.

Five wells had been drilled as of April, with two others in the works for the southern part of Capella, an accumulation on the Ombu E&P contract area.

Emerald Energy estimated gross proved and probable reserves at 14.8 million bbl, with estimated initial oil in place of 245 million to 1.1 billion bbl.

The vertical Capella-1well went to TD 1,159 m (3,802 ft) in mid-2008, tapping two sandstones.

Orinoco trend in Llanos basin

ExxonMobil Corp. said that it planned to explore an extension of Venezuela's Orinoco heavy oil belt in far eastern Colombia in 2009.

The company was awarded a technical evaluation agreement in 2008 on Block CPE-3 covering 25,900 sq km (10,000 sq mi) in an unexplored area of the eastern Llanos basin, on trend with the Orinoco belt.

The region is 643 km (400 mi) east-northeast of Bogota and about the same distance south-southwest of Caracas.

Corcel Block

Petrominerales Ltd. reported in March 2009 that production exceeded 13,000 b/d from two wells on the Corcel Block, east of Villavicencio in the Llanos basin.

The firm was awarded blocks 25 and 31 north of Corcel in late 2008.

It planned to spud an exploration well on the Guatiquia Block, just southwest of Corcel, by mid-year.

Lease Sales

Pacarana Block

In July 2009, ANH awarded the Pacarana Technical Evaluation Area in the Caguan-Putumayo basin to Canacol Energy Ltd.

The Pacarana Block covers 4,694 sq km (1,812 sq mi) south of the Ombu E&P contract area, where Emerald Energy PLC and Canacol are appraising the 2008 Capella heavy oil discovery.

Capella could hold as much as 100 million bbl of recoverable oil.

Drilling and Production

Costayaco field

Gran Tierra Energy Inc. said in mid-2009 it was on track to a 19,000 b/d production plateau at Costayaco field by yearend.

Tests at the Costayaco-8 development well yielded 2,640 b/d from the Caballos formation and 2,211 b/d from the Villeta Upper T sandstone.

The company also was building the location for Costayaco-9, just southwest of Costayaco-8, and finaliz

Pipelines/Transportation

Ocensa oil line

Colombia's Ecopetrol entered into an agreement in early 2009 with Enbridge Inc. to buy the latter's interest in the Oleoducto Central SA oil pipeline.

The 845-km (525-mi) Ocensa system can transport as much as 650,000 b/d of crude. It extends from Cuisiana and Cupiagua fields to the port of Covinas.

The deal boosts Ecopetrol's stake in Ocensa to 60%.

CUBA

Capital: Havana
Monetary unit: Peso
Refining capacity: 301,400 b/cd
Oil production: 50,000 b/d
Oil reserves: 124.000 million bbl
Gas reserves: 2.500 tcf

Overview

Cuba uses more than 150,000 b/d of oil. As a member of Petrocaribe—a group of 18 nations that buy oil from Venezuela under preferential terms—it receives about 95,000 b/d from that country. That is more than all other members combined.

By some estimates, Cuba has allocated $1.7 billion to its energy industry since 2004, working with a variety of international partners.

Existing oil production comes from onshore and near-shore fields. The oil is heavy and high in sulfur.

Canadian firms Sherritt International Corp. and Pebercan Inc. together have been producing about 60% of all Cuban output, in collaboration with Cupet (Unión Cubapetróleo) the national oil company.

Repsol-YPF SA, ONGC Videsh, Petrobras, Statoil AS, and Petrovietnam all were making plans in 2008 to explore for oil.

Cuban officials continued to seek additional foreign investment, aiming to raise oil and gas production to 100,000 boe/d by 2009.

On 10 of Cuba's 59 Gulf of Mexico oil and gas exploration blocks, more than 25 oil fields have been discovered and prepared for drilling. The largest, Varadero, has reserves estimated at 3 billion bbl.

There is also an additional projected reserve of 2.8 billion bbl in the southeast section of the Gulf of Mexico, just inside Cuba's Exclusive Economic Zone.

Cuba's oil fields; CUPET-PDVSA plans

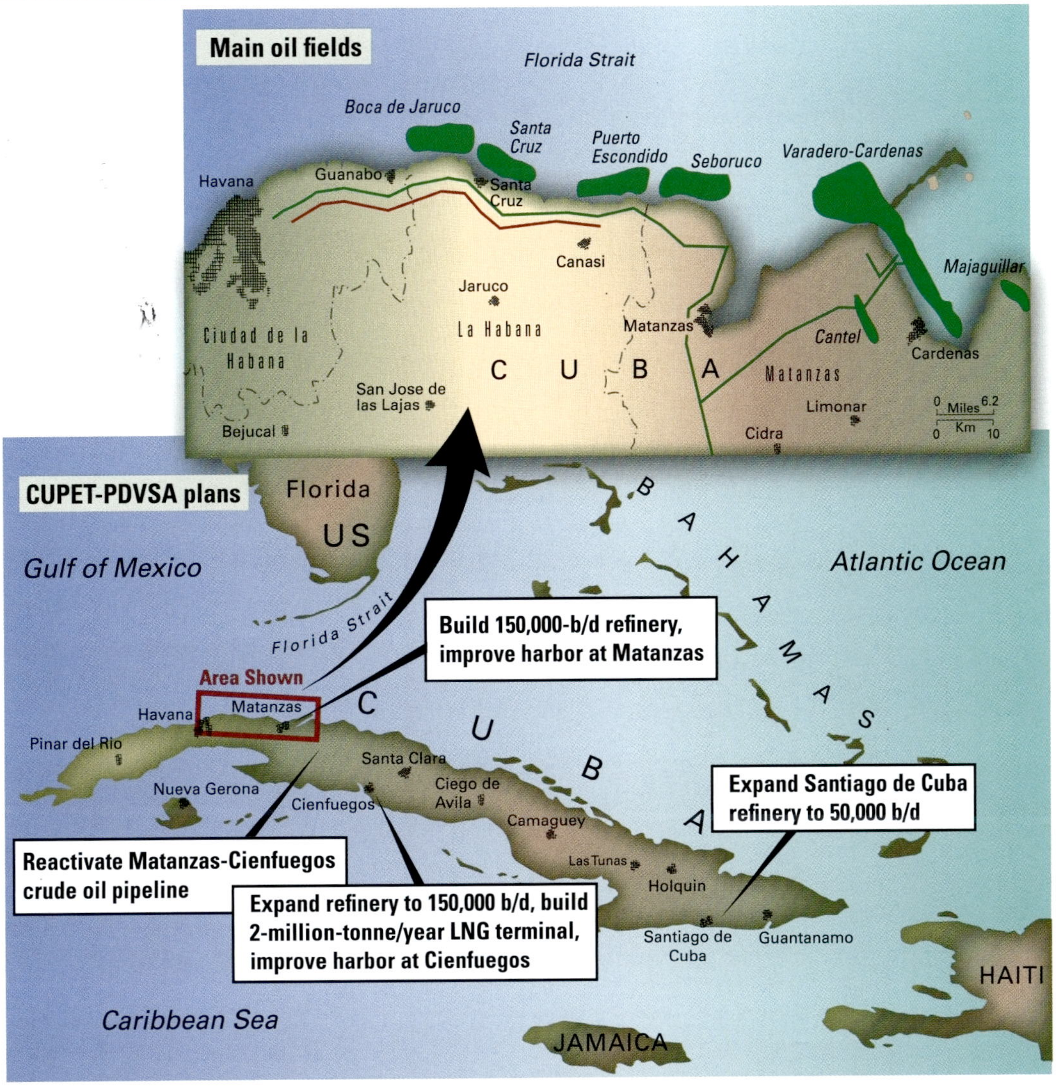

International Activities

U.S.-Cuba offshore cooperation?

One part of an early-2009 proposal within a Brookings Institution project supported offering licenses to U.S. companies to develop offshore oil and gas (as well as renewable resources) with Cuba.

The proposal went to U.S. Pres. Barack Obama, prior to his decision in April to ease travel, financial and communications restrictions on Cuba.

More than a dozen former ambassadors, foreign policy specialists and academics prepared the proposal as part of a larger project called "U.S. Policy Toward a Cuba in Transition."

ECUADOR

Capital: Quito
Monetary unit: U.S. dollar
Refining capacity: 176,000 b/cd
Oil production: 470,000 b/d
Oil reserves: 6.500 billion bbl
Gas reserves: 282.000 bcf

Overview

Ecuador has the third-largest oil reserves in South America. The country's oil production was estimated at 500,000 b/d in 2007, making Ecuador the fifth-largest producer in South America.

New Energy Minister

Pres. Rafael Correa, vowing to crack down on IOCs operating in Ecuador, named Germanico Pinto to be oil and mining minister in June 2009.

Germanico is expected to take a much firmer approach toward IOCs than his predecessor, Derlis Palacios.

The state-owned petroleum firm is Petroleos de Ecuador (Petroecuador), and its production arm is Petroproduccion.

About a dozen foreign oil companies had contracts and assets in Ecuador in 2009, including Brazil's Petrobras, Repsol-YPF SA (Madrid), EnCana Corp. (Calgary), Agip SPA (Milan, Italy), Perenco SA (Paris), and Occidental Petroleum Corp. (Los Angeles).

The largest refinery in Ecuador is the 110,000-b/d Esmeraldas facility on the Pacific coast.

Exploration and Development

Pungarayacu field

Ivanhoe Energy Inc. expected to start drilling by yearend 2009 at giant Pungarayacu heavy oil field on Block 20. The field covers 648 sq km (250 sq mi). The wells will go only to about 450 m (1,500 ft).

Based on data from 27 wells drilled in the 1980s, original oil in place has been estimated at 4.3 to 12.1 billion bbl (most likely 6.4 billion bbl). No recovery factor was estimated.

Crude oil is commonly held to be of 10–11° gravity, but Ivanhoe Energy said it might be lighter.

Four blocks targeted

Petroamazonas—a subsidiary of Petroecuador that operates oil assets previously owned by Occidental Petroleum Corp.—expected to develop oil fields on four blocks in 2009.

The unit aims to drill 14 wells in Panacocha field, with initial production set at 5,000 b/d starting in second quarter 2010, eventually rising to 25,000 b/d.

Thirteen wells would be drilled in Panayacu-Quinde field, with output potential of 10,000 b/d.

Four wells will be drilled in Paka Sur field, with initial output of 3,000 b/d possibly online by yearend.

Pending financing, 14 wells also may be drilled on Block 31, with production starting in early 2011.

Drilling and Production

Sacha field

Rio Napo, a 70:30 JV of Petroecuador and Venezuela's PDVSA, began operating Sacha oil field in Ecuador's Amazon region in late 2009.

Output of 70,000 b/d of oil is expected by 2014–2015, about 40% higher than the current 49,780 b/d. Sacha field holds 491 MMbbl of oil reserves.

The Rio Napo JV is one of a series of accords reached between Ecuador and Venezuela since Pres. Rafael Correa took power.

International Activities

China, Petroecuador agreement

In mid-2009, Petroecuador agreed to sell 2.88 MMbbl of oil per month to PetroChina International Co. Ltd. over two years, for a $1-billion advance payment.

The total monthly volume would comprise 2.16 MMbbl of Oriente crude and 720,000 bbl of Napo crude, with price of each shipment determined by market conditions.

Petroecuador said it would invest the prepayment in the oil industry and in public works.

In addition to the $1-billion advance, Quito also was negotiating with Beijing on a separate $1-billion credit with a four-year maturity.

Deals with Angola, China

In April, Petroecuador signed an agreement with Angola's Sonangol concerning cooperation on several blocks in Ecuador, including Blocks 28 and 29 in the Amazon region and offshore Block 42.

Ecuador also signed an E&P contract in October 2009 for its Oglan field with a joint company of Petroecuador and China's Sinopec.

Sinopec joined a handful of other Chinese state oil companies that have partnered with Petroecuador, including CNPC, Andes Petroleum, Petroriental, and CPEB Changqing Petroleum.

LNG

New plant: El Oro State

In April 2009, Petrocomercial signed a $49-million contract for construction of an LNG plant in El Oro State, designed to process 200 metric tons/day, equivalent to 10 MM cu m/d.

Natural gas for the plant will come from Guayaquil Gulf and will in turn supply the states of Pichincha, Guayas and El Oro.

The plant is expected to come online in fourth quarter 2009.

Gas Processing

LPG JV

Petrocomercial awarded a major contract in October 2009 for an LPG transport and distribution project. Other venture partners include several Quito-based firms, plus Ismocol (Colombia) and Thailand's Shi Asia.

Construction plans call for a pump station in Monteverde, a storage and distribution terminal in El Chorrillo, and a 123-km (76-mi) pipeline between the two locations.

Refining and Products

Refinery at El Aromo

Construction of a $4-billion oil refinery, as part of a major petrochemical and refining complex, was under development in late 2009 by Petroecuador and Venezuela's PDVSA.

A new entity, Refineria del Pacifico—owned 51% by Petroecuador and 49% by PDVSA—was created to manage the project.

Located in the coastal town of El Aromo, the plant is described as the largest oil refinery on South America's Pacific Coast.

It will process 300,000 b/d of Ecuadoran crude and export derivatives, starting in late 2013.

Unlike the country's three existing refineries—Esmeraldas, La Libertad, and Shushufindi—the new facility will be able to process heavy crudes.

Company News

Chevron: $27-billion lawsuit

Chevron Corp. ramped up its 15-year battle over allegations of pollution made by environmentalists who have filed what has been described as the biggest environmental lawsuit in history.

Filed in Ecuador, the legal action seeks $27 billion in potential damages.

In September 2009, the company filed a claim against the government before the Permanent Court of Arbitration in The Hague, citing violations of terms of the U.S.-Ecuador bilateral investment treaty, investment agreements and international law.

Chevron also demanded the removal of the judge in the case, claiming political interference and corruption in the proceedings.

Chevron bought Texaco in 2001, and the plaintiffs charge Chevron is therefore responsible for Texaco's alleged failure to follow appropriate environmental standards in handling of certain materials.

They say that Texaco's actions contaminated waterways and caused high levels of associated illness.

Chevron responds that it met its obligation when it cleaned up many well sites as part of a $40-million remediation agreement with Ecuador in 1995.

The company contends that PetroEcuador—Texaco's consortium partner, which still operates in parts of the country—is responsible for the remaining sites.

Tax fight

Perenco said in July 2009 that it would temporarily suspend its operations in Ecuador because of a long-running battle with the government over taxes, according to Reuters news agency.

Ecuador raised the windfall oil tax to 99% from 50% in 2007 as a way to pressure foreign oil companies to rework their production agreements, but it later reduced the tax to 70%.

Perenco sued Ecuador in 2008, disputing the legality of the tax, saying the levy violates its contract. In reaction, Ecuador announced in 2009 its intention to withdraw from the World Bank's International Center for Settlement of Investment Disputes (ICSID).

By July, Ecuador had seized control of oil fields operated by Perenco. It says the company owes more than $350 million in windfall taxes. It also was trying to sell 2 MMbbl of oil confiscated from the firm.

The ICSID accepted a request from Perenco in May to block any forceful collections until the center rules on the legality of the tax.

Repsol YPF SA also has asked ICSID to resolve the tax issue, and Occidental Petroleum Corp. has applied to the center for $1 billion in compensation for oil fields seized in 2006.

GUATEMALA

Capital: Guatemala City
Monetary unit: Quetzal
Refining capacity: 0
Oil production: 14,000 b/d
Oil reserves: 83.070 million bbl
Gas reserves: 0

Guatemala and Belize are the only oil-producing countries in Central America.

Most of Guatemala's oil is produced from fields in its northern jungles, adjacent to the Mexican border.

Perenco produces the majority of the oil in the country. Domestic refining capacity is meager and almost all of their production is exported to the United States. The country must import petroleum products.

Peten basin

A subsidiary of Quetzal Energy Ltd. was redeveloping the A7-2005 license in mid-2009, aiming to drill three development wells in the Atzam field and six in the Tortugas field in the Peten (Chapayal) basin.

PetroLatina has spud the Atzam-3 well in Guatemala's Atzam oil and gas field in License A7-2005 in the southern Peten (Chapayal) basin.

Projected depth for the Atzam-3 well was 1,280 m (4,200 ft), being drilled just east of the productive Atzam-2 well. The reworked Atzam-2 produced 102,168 bbl of oil from March 2008 until May 2009. Previous operators produced 90,000 bbl before the well watered out.

JAMAICA

Capital: Kingston
Monetary unit: Dollar
Refining capacity: 36,000 b/cd
Oil production: 0
Oil reserves: 0
Gas reserves: 0

Jamaica consumed about 27.8 million bbl of petroleum and petroleum products in 2007, according to the Petroleum Corporation of Jamaica (PCJ), the state energy company. The country depends on imported oil to meet over 90% of its energy needs.

Jamaica is among several Caribbean countries that signed the Petrocaribe Initiative in 2005, which allows it to buy crude oil and refined products from Venezuela under favorable financing terms.

The similar San Jose Accord enables Jamaica to receive oil and refined products from Mexico.

Offshore survey

CGGVeritas completed an extensive multi-client, long-offset 2D survey offshore Jamaica in mid-2009.

It was done with PCJ to complement the second official offshore Jamaica licensing round, expected to be announced in early 2010.

The long-offset survey aims to delineate the tectonic setting and prospectivity of the North and South Nicaragua Rise area offshore, as well as its relation to the more well-known Walton basin (Pedro Bank area).

CGGVeritas said that initial data review indicated large basins and thick sections of Cretaceous and possibly Jurassic strata on the South (Lower) Nicaragua rise.

MEXICO

Capital: Mexico City
Monetary unit: Peso
Refining capacity: 1,540 million b/cd
Oil production: 2.605 million b/d
Oil reserves: 10.404 billion bbl
Gas reserves: 12.702 tcf

Overview

Oil

The oil sector generates approximately 40% of Mexico's government revenues.

In March 2009, the U.S. EIA forecast that Mexico will produce 2.9 million b/d of oil for the year and 2.7 million b/d in 2010.

Mexico's energy secretary in September estimated the 2010 figure at 2.5 million b/d.

Cantarell

Cantarell field in the Gulf of Campeche, off the southeast coast, is one of the largest fields in the world. It produced 2.12 million b/d of crude at its peak in 2004.

But state-owned Petroleos Mexicanos (Pemex) has warned that Cantarell is in long-term decline and that production is expected to drop by about 14% per year.

Output of 550,000 b/d was anticipated for December 2009 (21% of national oil production), falling to an estimated 255,000 b/d in 2017 (8.4% of national production).

KMZ

Another major oil production center in the Gulf of Campeche is Ku-Maloob-Zaap (KMZ), adjacent to Cantarell. The largest source of new production growth, it overtook Cantarell as Mexico's biggest producer, with record output of 814,000 b/d in April 2009.

The KMZ complex produced 550,700 b/d of crude in 2007 and 702,000 b/d in 2008. Pemex expects KMZ production to peak at 820,000 b/d in 2009, but fall back to 810,000 b/d in 2010.

MEXICO–WEST
US
Texas
MEXICO
Nogales
Nogales
Cananea
Magdalena
Sonora
Hermosillo
Guaymas
Navojoa
El Paso
Ciudad Juarez
Rio Grande
Celulosa de Chuhuahua
Chihuahua
Chihuahua
Delicias
Cuidad Camango
Jimenez
Coahuila
Ulua
Buena Suerte
Baja California
Gulf of California
Topolobanpo
Sinaloa
Culiacan
Durango
Durango
Gomez Palacio
La Paz
Zacatecas
Zacatecas
San Luis Potosi
Mazatlan
Chicura
Canuch
Jayabo
Nayarit
San Blas
Tepic
Aguas.
Aguascalientes
Lagos
Guanajuato
Guadalajara
Salamanca
Lago de Chapala
245
Puerto Vallarta
Jalisco
Morelia
Colima
Manzanillo
Colima
Texoman
Michoacan
Lázaro Cárdenas
Pacific Ocean
30°
25°
20°
15°
115°
110°
115°
100 0 100 200 Miles
100 0 100 200 300 Km

MEXICO-EAST
CANTARELL COMPLEX
Tunich
Tabay
Zazil-Ha
Maloob
Zaap
Bacob
Kutz
Lum
Ku
Bacab
Cantarell
Tartunich
Ixtoc
Luch
Batab
Nohoch
Och
Abkatun
Alux
Uech
Pich
Chac
Caan
Kay
Kax
Takin
Chem
Gulf of Mexico
Bay of Campeche
Gulf of Tehuantepec
Texas
Coahuila
US
MEXICO
Nuevo Leon
Tamaulipas
Zacatecas
San Luis Potosi
Aguas
Guanajuato
Queretaro
Hidalgo
Tlax.
Veracruz
Michoacan
D.F.
Puebla
Morelos
Guerrero
Oaxaca
Tabasco
Chiapas
Campeche
Yucatan
Quintana Roo
BELIZE
GUATEMALA
HONDURAS
EL SALVADOR
Corpus Christi
Nuevo Laredo
Laredo
Nuevo/Laredo
Lampazos
Ulua
Buena Suerte
Monclova
La Presa
Arcos
Sta. Rosalia
Lojitas
Canor
Culebra
Carretas
McAllen
Reynosa
Mision-Trevino
Brasil
Fortuna
Brownsville
Matamoros
Cadereyta
275
Primavera
Cabeza
S. Bernardo
Benavides
Ramirez
18 de Marzo
Chapul
Huizache
Pobladores
Monterrey
San Fernando
Lerma
Ciudad Victoria
190
Barcodon
Tamaulipas-Constituciones
Ciudad Madero
Ebano Panuco Fields
Tampico
Arenque
Cabo Nuevo
Arrecife Medio
Isla de Lobos
Tuxpan
Golden Lane Fields
Tiburon
Esturon
Marsopa
Bagre
Escuala
Salamanca
Queretaro
315
Tula
Pachuca
Poza Rica
Pnenche
Morelia
Mexico City
Toluca
Tlaxcala
Jalapa
Veracruz
Remudauera
Cuernavaca
Puebla
Miralejos
Capite
M. Piancho
Mecayucan
Angastura
Mirador
Coapo
Novillero
Veinte
Mata Verde
Cacuite
Macuile
Azcapotzelco
Cardenas
Minatitlan
185
Sinán
Citam
Yum
Bolontikú
Kix
May
Dos Bocas
Ciudad del Carmen
Fontera
Ciudad Pemex
Villahermosa
Chilpancingo
Oaxaca
Acapulco
Tuxtla
Tehuantepec
Juchitan
Salina Cruz
330
Merida
Campeche
Belmopan
Xan
Caribe
W. Chinaja
1 San Diego
Rubelsanto
Yalpemech
Guatemala
Sonsanate
22
San Salvador
25°
20°
15°
100°
95°
90°
100 0 100 200 Miles
100 0 100 200 300 Km

CENTRAL AMERICA
Gulf of Mexico
Straits of Florida
Great Bahama Bank
Santa Cruz
Puerto Escondido
Yumuri
Seboruco
Varadero-Cardenas
Boca de Jaruco
Guanabo
Tarara
La Luz
Havana
122
Matanzas
Matanzas
Bacuranao
Motembo
2
76
Cabaiguan
Cienfuegos
CUBA
Cristales
Jatibonico
102
Santiago
30°
Yucatan
Cozumel
MEXICO
Quintana Roo
JAMAICA
36
Kingston
Chetumal
Caribbean Sea
BELIZE
Belmopan
Gulf of Honduras
15°
Puerto Cortes
HONDURAS
Tegucigalpa
22
San Salvador
La Union
NICARAGUA
20
Managua
Bluefields
Lake Nicaragua
10°
COSTA RICA
24
Limon
San Jose
Punta Cahuita
Panama City
PANAMA
Golfito
David
La Palma
100 0 100 200 Miles
100 0 100 200 300 KM
85°
80°
COLOMBIA

Chicontepec

The Chicontepec basin, with 29 fields northeast of Mexico City, is potentially a large contributor to future production. Pemex says it contains possible reserves of 17.7 billion boe.

Pemex had planned to invest $11 billion 2009–2013 at Chicontepec to boost production to 700,000 b/d by 2017. But in late 2009, the government cut its 2010 budget by 60% from the 2009 level, citing poor results and execution problems.

Through early November 2009, Chicontepec oil production averaged just 29,000 b/d, against a yearly goal of 58,000 b/d. Expectations are for 43,000 b/d in 2010.

Pemex has predicted that the declines in Cantarell production can be offset by new wells coming online in deepwater Gulf of Mexico and in the Chicontepec basin.

But many experts say both areas would require at least a decade of exploration and development before significant production begins.

Southern Mexico

Important onshore production centers in southern Mexico include Bellota-Jujo and Samaria-Luna.

Natural gas

According to Pemex, the southern region contains the largest share of proven gas reserves. However, the northern region will likely be the center of future reserves growth because it contains almost ten times as much probable and possible natural gas reserves as the south.

Mexico is a net importer of natural gas. In 2007, Mexico produced 1.98 tcf while consuming 2.4 tcf, with imports coming in both by pipeline from the United States and as LNG.

In 2007, gas consumption was more than 90% greater than it was a decade ago. Gas is now more widely used for power generation, to free up oil for export and other uses.

Pipeline network

While Mexico's oil and upstream gas activities are restricted to Pemex, natural gas transmission, distribution and storage has been open to the private sector since 1995.

The extensive Pemex oil pipeline network connects major production centers with domestic refineries and export terminals. More than 450 pipelines span 4,665 km (2,900 mi), chiefly in the south.

Pemex also operates more than 9,170 km (5,700 mi) of natural gas pipelines in Mexico. The company has 11 natural gas processing centers that produce NGLs, condensates, and LPG.

Pemex also operates most of the country's natural gas distribution network, which includes 12 active connections with the United States.

A subsidiary of TransCanada Pipelines Ltd. (Calgary) operates the 130-km (80-mi) Tamazunchale Pipeline, which connects the Pemex gas processing facility in Naranjos to a gas-fired power plant near Tamazunchale.

Pemex snapshot

In an August 2009 review, *Oil & Gas Journal* described Pemex as the sole producer of crude, natural gas, and petroleum products in Mexico and the country's most important company.

It is the biggest enterprise in Mexico, the most significant source of income for the government, and the third-largest exporter of crude to the United States, behind Canada and Saudi Arabia.

Drilling activity in 2009 was twice the 2007 level, but most observers say that Pemex needs better technology to explore ultradeep waters of the Gulf of Mexico and significant funding to support exploration and development.

The 2009 goal for total oil production was 2.65 million b/d, but output reached only 2.56 million b/d in July. Of that July production:

- 58.1% was heavy crude, 31.5% was light crude, and 10.4% was super-light crude
- 76.8% was from marine regions, 19.4% from the southern region, and 3.8% from the north.

At the start of 2009, Pemex claimed proved reserves of 14.3 billion boe of oil and gas, including 10.2 billion boe developed. It also had 14.5 billion boe of probable reserves and 14.7 billion boe possible.

Proved natural gas reserves were 17.6 tcf, including 11.5 tcf of associated gas. Proved, probable, and possible gas reserves totaled 60.4 tcf.

2008 oil production

Pemex said that its total oil production in 2008 declined by 9.2% to just under 2.8 million b/d. This was the lowest since 1995, when it reached 2.62 million b/d, according to the energy ministry.

Pemex said that the decrease was due largely to continuing depletion of Cantarell field and to Hurricane Ike, which disrupted production in November in the Gulf of Mexico.

Pemex said that depletion of Cantarell was partly offset by increased output at Ku-Maloob-Zaap.

KMZ field reached a record production level of 802,000 b/d on December 29.

Hedging oil prices

According to the *Financial Times*, Mexico took out a $1-billion insurance policy in early December 2009 against oil prices falling in 2010.

Mexican officials were reported to have hedged all net oil exports for 2010 by buying protection against oil prices falling below $57 a barrel.

Earlier in 2009, another hedging move at $70/bbl yielded netted the country more than $5 billion as oil prices dropped between January and June.

Competitiveness review

In July 2009, the Mexico Competitiveness Institute (IMCO) called for major policy action to confront Mexico's decline in the world economy.

IMCO studied the relative competitiveness of 48 countries and reported that the country had fallen two places (to 32nd) in the three years ended in 2007.

Mexico was the only country that scored zero on IMCO's index regarding energy policy liberalization.

Regarding oil policy, IMCO recommended the following:

- Open the upstream sector to competition from private investment
- Liberalize current restrictions on payments to Pemex contractors.

Observers noted that the IMCO data were from 2007, well before oil reforms of 2008 and their implementation in 2009.

Those reforms allow Pemex to create incentive-based service contracts with private companies and give the company more autonomy.

Ex-Im Bank loans

In early 2009, the U.S. Export-Import Bank approved $900 million worth of 10-year direct loans to Pemex to support purchase of U.S. goods and services for use in hydrocarbon project development.

- $600 million is directed to projects at 18 onshore and offshore oil and gas exploration sites in the Bay of Campeche.
- $300 million is specifically for purchase of materials for work at the Cantarell field.

The bank said the financing will help ensure that U.S. products would be competitive.

Engineering services, oil field equipment, offshore platforms, drilling and upgrade services, and upgrade/rehabilitation services are expected to come from suppliers in several different states.

Since 1998, Pemex has been the bank's largest borrower, getting $8.3 billion of financing to buy U.S. goods and services for oil and gas exploration, development, and processing projects.

How Mexico Compares Regarding Government Effectiveness[1]

	Finland[2]	Japan[2]	Oil producers: Brazil	China	Norway	US	UK	Avg. oil	Mexico	Var. oil, %
Regulation quality	92.4	74.8	43.7	37.8	85.9	86.0	97.8	70.2	56.0	-20
Effectiveness	90.5	70.9	24.9	33.7	96.1	80.1	84.8	63.9	32.8	-49
Promotion of competition	100.0	50.0	50.0	25.0	75.0	100.0	100.0	70.0	25.0	-64
Public policy re competitiveness	100.0	47.3	39.5	75.8	64.0	84.7	66.6	66.1	47.3	-28
Liberalization of energy sector	100.0	100.0	100.0	66.7	100.0	100.0	100.0	93.3	0.0	-100
Gov't. independence of interest groups	100.0	33.3	66.7	33.3	100.0	66.7	100.0	73.3	33.3	-55
e-Government	69.6	73.5	36.6	24.6	95.7	90.7	76.6	64.8	40.5	-38
Control of corruption	100.0	61.7	22.0	10.6	96.2	68.3	80.7	55.6	18.9	-66
Accountability	97.6	80.5	64.6	0.0	98.7	85.3	94.3	68.6	51.4	-25

[1]Shows index values of selected countries. Mexico's score is 20% or more lower than selected oil competitors. [2]Finland and Japan shown for reference. Finland scored 100 in 23 of 75 indexes. Data: IMCO 2009.
Sources: Regulation—World Bank Governance Indicators; Effectiveness—World Bank Governance Indicators; Competitiveness—IMCO; Energy liberalization—IMCO; Interest groups—Economist Intelligence Unit (EIU); e-Government—UN E-Government Survey; Corruption—World Bank Governance Indicators; Accountability—World Bank Governance Indicators.
Mexico Energy Intelligence, July 24, 2009. Data: IMCO 2009 (imco.org.mx) Table: MEI

Where Mexico Stands Relative to Three Countries with Open Energy Markets*

	World Economic Forum rankings of 134 countries: Mexico	Brazil	Norway	US	Comparative advantage (No. of country rankings): Br>Mex	Nor>Mex	US>Mex
Institutions							
Transparency of gov't. policy-making	94	101	8	28	-7	86	66
Strength of auditing standards	71	60	11	20	11	60	51
Efficacy of corporate boards	82	42	9	12	40	73	70
Percentile score of boards	39	69	93	91			
Infrastructure							
Quality of electric supply	87	58	10	16	29	77	71
Technological readiness							
Availability of latest technologies	92	58	6	5	34	86	87
Firm-level technology absorption	92	42	9	3	50	83	89
Innovation							
Capacity for innovation	67	27	13	6	40	54	61
Quality of scientific research institutions	79	43	22	1	36	57	78
Company spending on R&D	71	31	19	3	40	52	68
University-industry research collaboration	84	50	17	1	34	67	83
Gov't. procurement of high tech products	104	84	16	4	20	88	100
Availability of scientists and engineers	105	57	18	6	48	87	99
Utility patents	56	58	19	2	-2	37	54
Average of innovation	81	50	18	3	31	63	78
Percentile ranking	40	63	87	98			

*Shows only indirect measures of efficiency of energy policy and markets. Mexico's scores are much lower than favorite policy models on key measures.
Sources: Data—World Economic Forum 2009; Table—Mexico Energy Intelligence; Date—July 26, 2009

Exploration and Development

Deepwater plans

In May 2009, Pemex said that it was accelerating its deepwater exploration and production strategy to help reverse Mexico's declining production.

Pemex said it shot extensive 3D and 2D seismic and drilled eight wells in its gulf deepwater basins from 2000–2008, discovering two nonassociated gas fields and two extra-heavy oil fields.

Two deepwater rigs were working offshore Mexico in early 2009. Exploratory drilling is to expand northward from Cantarell. Five rigs are to be working by 2012.

The company said it has identified five deepwater geological trends over a 500-sq km (193-sq mi) area, in water at least 500 m (1,640 ft) deep.

Offsetting the Cantarell downturn

Mexico's daily *El Universal* reported in late 2009 that Pemex and the Secretaria de Energia (Sener) were preparing risk contracts to offer to IOCs and domestic oil companies in order to accelerate the search for oil and gas.

The action is driven mostly by concern about the decline of production at Cantarell.

The awarding of risk contracts and the purchase of materials will be aimed at developing key suppliers for priority areas, to cut costs and delays in constructing rigs, drilling, and laying pipelines. Contract awards are expected in early 2010.

GOM: Leek-1 well

Pemex reported in October 2009 that it found oil in the deep waters of the Gulf of Mexico at the Leek-1 well. It was the second deepwater discovery by Pemex for the year, following the earlier Holok well.

No precise location was given, but Pemex said the water depth was 700 m (2,297 ft).

Drilling and Production

Chicontepec budget

In November 2009, Mexican officials reportedly slashed funding for the Pemex Chicontepec oil development in 2010 by more than 60% from 2009 levels.

The technically challenging initiative is also called the Tertiary Gulf Oil Project.

Mexico's daily *El Universal* said "poor results, technical limitations, and execution problems" at Chicontepec prompted the decision.

Pemex said that Chicontepec will produce less oil in 2009 than expected and will likely produce only 43,000 b/d in 2010. The 2009 target was 58,000 b/d, but production averaged only 29,000 b/d from January 1 to November 6.

Cantarell: EOR

In October 2009, Pemex was studying enhanced oil recovery (EOR) methods to extract more oil from the tight reservoir matrix of Cantarell field. Injection of steam or carbon dioxide was being evaluated.

The company will aim for 60% oil recovery from the field.

To date, Pemex has produced about 12.2 billion bbl from Cantarell and has achieved a 41% recovery for the Akal portion of the field.

MEXICO CHICONTEPEC FIELD DIVISIONS AND WELLS*

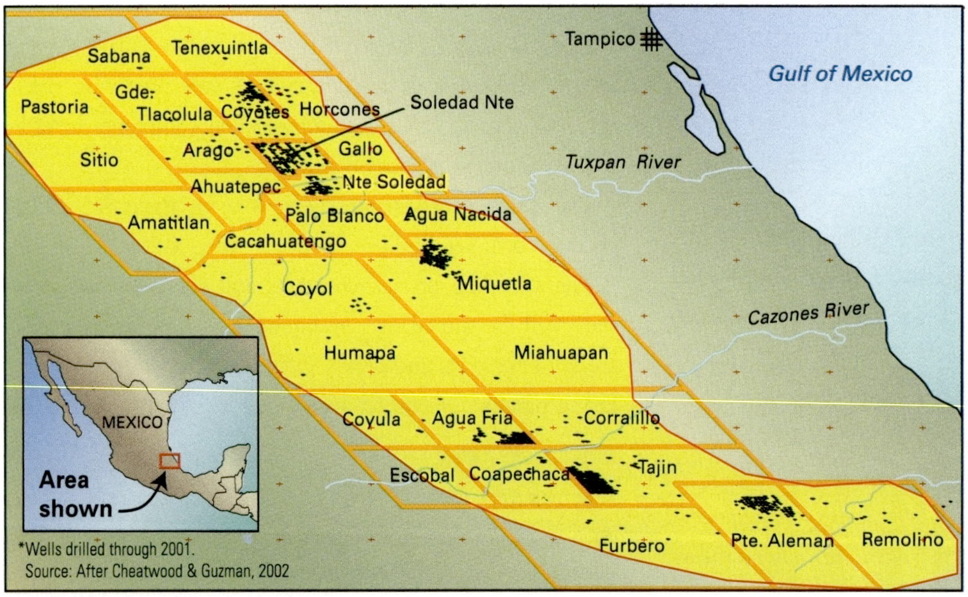

*Wells drilled through 2001.
Source: After Cheatwood & Guzman, 2002

Kambesah field

Pemex said it plans to start oil and gas production in 2010 from small Kambesah field in Campeche Sound, according to press reports in mid-2009.

Reserves weren't specified, but output is expected to peak at 13,700 b/d of oil and 9.3 MMcfd of gas.

Kambesah, near Ixtoc field, was discovered in late 2008.

LNG

Manzanillo terminal

The LNG terminal being built at the Pacific coast port of Manzanillo is on track for startup in 2011.

Two other LNG regasification plants are already operating in Mexico—at Altamira on the Gulf of Mexico, and Ensenada on the Pacific Coast.

National plans call for raising LNG imports to 1.99 bcfd by 2017, from January 2009 levels of 503 MMcfd.

Pipelines/Transportation

Central Mexico gas line

In October 2009, Pemex was seeking bids for a $600-million natural gas pipeline project in central Mexico.

The plan calls for a 230-km (143-mi) pipeline from Tamazunchale to San Luis de la Paz, and a 56-km (35-mi) line from San Luis de la Paz to San Jose Iturbide.

The aim of the project is to increase transmission capacity by 400 MMcf/d.

Guadalajara Pipeline

TransCanada Pipelines Ltd. was chosen in May 2009 to build, own, and operate the 310-km (193-mi), $320-million Guadalajara Pipeline.

It will extend from an LNG regasification plant being built near Manzanillo on the Pacific Coast to Guadalajara and carry up to 500 MMcfd of natural gas.

The targeted in-service date is 2011.

State-owned utility Comision Federal de Electricidad (CFE) has a 25-year contract for all gas shipped through the line to fuel power plants.

LNG from Peru will supply most of the gas for the pipeline.

Refining and Products

Tula and Salamanca facilities

In late 2009, Pemex chose a site at Tula, Hidalgo State, to construct a $9-billion refinery. It will be the first in 30 years to be built in Mexico. An added $859 million will be spent on associated pipelines.

The company also will invest $3 billion to modernize another facility, at Salamanca, Guanajuato State.

The 300,000-b/d Tula facility is expected to come on stream in 2015.

The expanded Guanajuato facility is to ready by yearend 2014.

Petrochemicals

Cangrejera complex

In May 2009, Pemex was reviewing bids from parties interested in participating in a project to expand the capacity of the Cangrejera petrochemical complex.

Investment of $504 million over three years is expected, to set up three new plants and to double paraxylene production capacity to 488,000 tpy.

Wind power

A law passed in 2008 in Mexico encourages wider use of renewable energy resources.

Based on that law, the Energy Ministry has called for increasing alternative energy use to meet 26% of Mexico's energy needs by 2012. Currently, renewable resources provide about 23% (19% from hydropower).

Wind power is of particular interest. The goal is to generate 2,500 MW by 2012 (compared to about 300 MW in 2009).

One major project is the 250-MW La Ventosa windfarm, in the southern province of Oaxaca.

The windfarm will be the largest in Latin America.

Another 500-MW facility (with 333 turbines) was under development in mid-2009 in the La Rumorosa zone of Baja California, near the U.S. border, by Union Fenosa.

Construction is slated to begin in mid-2010, with startup in 2012.

Hurricane Ida

In early November 2009, Hurricane Ida ravaged parts of Central America.

Floods and landslides killed at least 120 in El Salvador and thousands were left homeless in Nicaragua.

The storm, with winds of 160 km (100 mi) per hour, then swept across Mexico's Caribbean coast and into the southern Gulf of Mexico.

NICARAGUA

Capital: Managua
Monetary unit: Córdoba
Refining capacity: 20,000 b/cd
Oil production: 0
Oil reserves: 0
Gas reserves: 0

Sandino basin

Norwood Resources Ltd. was planning to further develop its San Bartolo-1 discovery well in the Sandino basin in mid-2009.

Six targets for coring and testing were identified, at interval depths ranging from 1,905–1,914 m (6,250–80 ft) to 2,313–2,338 m (7,590–7,670 ft).

The six reservoirs contain nontraditional volcaniclastic, feldspathic sandstones with 10–35% hydrophilic clays. Permeabilities appear to range from less than 10 md to 20–30 md.

Russian accord

The Nicaraguan government, following the lead of Venezuela, signed an agreement in mid-2009 allowing exploration within its borders by a Russian consortium.

The concessions include the Caribbean and Pacific, both offshore and on land.

Carrying out the exploration will be the Russian National Oil Consortium, comprising Rosneft, Gazprom, Lukoil, TNK-BP, and Surgutneftegaz.

Bolivian hydrocarbons adjacent to NW Paraguay blocks

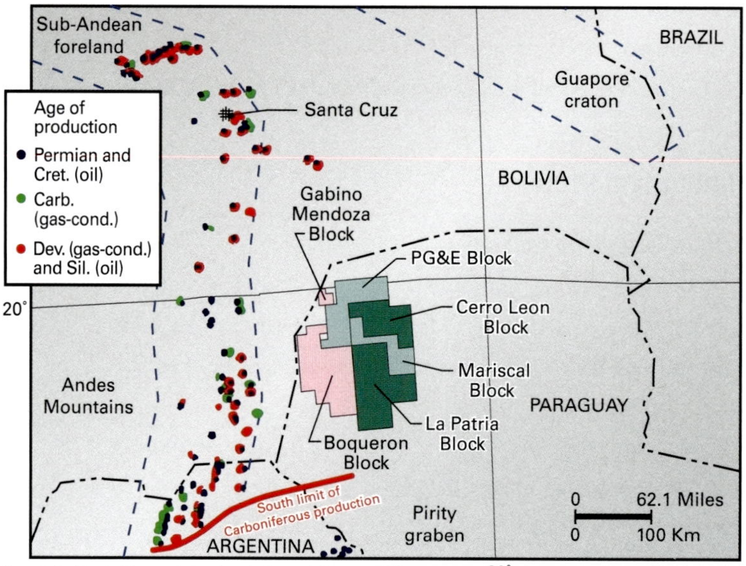

Source: Updated with permission from LeRoy and Carlson, 2002, used by permission of Paraguay Vice-Ministry of Mines & Energy. Modified from Lindquist, 1998.

PARAGUAY

Capital: Asunción
Monetary unit: Guarani
Refining capacity: 7,500 b/d
Oil production: 30 b/d
Oil reserves: 0
Gas reserves: 0

Landlocked Paraguay has a market economy marked by a large informal sector. A large percentage of the population, especially in rural areas, is engaged in agricultural activity.

Accurate economic measures are difficult to obtain.

According to the *World Factbook* (produced by the U.S. Central Intelligence Agency), per-capita real income has stagnated at 1980 levels.

The economy rebounded between 2003 and 2008, however, as growing world demand for commodities combined with high prices to support Paraguay's commodity-based export expansion.

The estimated real growth rate in GDP was 4.3% in 2006, 6.8% in 2007, and 5.8% in 2008.

Despite having some of the most liberal petroleum legislation in the world, Paraguay's limited prospects and severe lack of infrastructure in the vast Chaco plain have discouraged drillers.

Paraguay's potential

In early 2009, Quincy Oil & Gas Corp. analyzed the hydrocarbon industry in Paraguay. Factors creating an optimistic outlook include higher regional gas prices, plus increasing demand, particularly in Argentina, Brazil, and Chile.

Paraguay consumes 30,000 boe/d as refined products, including gasoil, gasoline, lubricants, and aviation fuel, all imported.

CDS Oil & Gas Group PLC was planning a drilling program in 2009, and any commercial output could be processed at the government-owned 7,500-b/d Petropar refinery at Villa Elisa, near Asunción.

If exploratory drilling by CDS or others uncovers significant commercial volumes, export pipelines would need to be built.

Paraguay near-term work programs

Company	Properties/projects	Comments
Quincy Oil & Gas Corp.	2 prospecting blocks covering an aggregate of 1,107,000 ha. Independent prospective resource estimate of 112 bcf of conventional gas and 1.2 tcf unconventional gas.	Aeromag, geochem, and stratigraphic drilling planned for 2009.
CDS Oil & Gas Group PLC	2 concession laws and 1 prospecting permit covering an aggregate of 1,547,990 ha. Published prospective recoverable resource of 6.4 million bbl and 142 bcf.	Announced plans to drill in 2009.
Aurora Petroleos SA and Boreal Petroleos SA, both co-owned by Artemis Petroleum PLC and Pantera Petroleum Inc	Concession laws on 5 separate blocks totaling 1.6 million ha. Published prospective resource of 129 million bbl	2009 program not confirmed.
Crescent Global Oil SA and Pirity Hydrocarburos SRL owned by Crescent Global Oil LLC and Petro Victory LLC	2 exploration and production permits and 1 prospecting permit for an aggregate of 2.33 million ha.	Plan an aggregate of 2,800 km of 2D and 3D seismic in 2009 and drilling in 2010.
Amerisur Resources PLC	2 exploration and production permits covering 1.6 million ha. Estimated unrisked prospective resource of 708 million boe on the San Pedro Block and 248 million boe on the Curupayty Block.	Planning geochemical survey and seeking farmout participants.

Oil and Gas Exploration in Paraguay

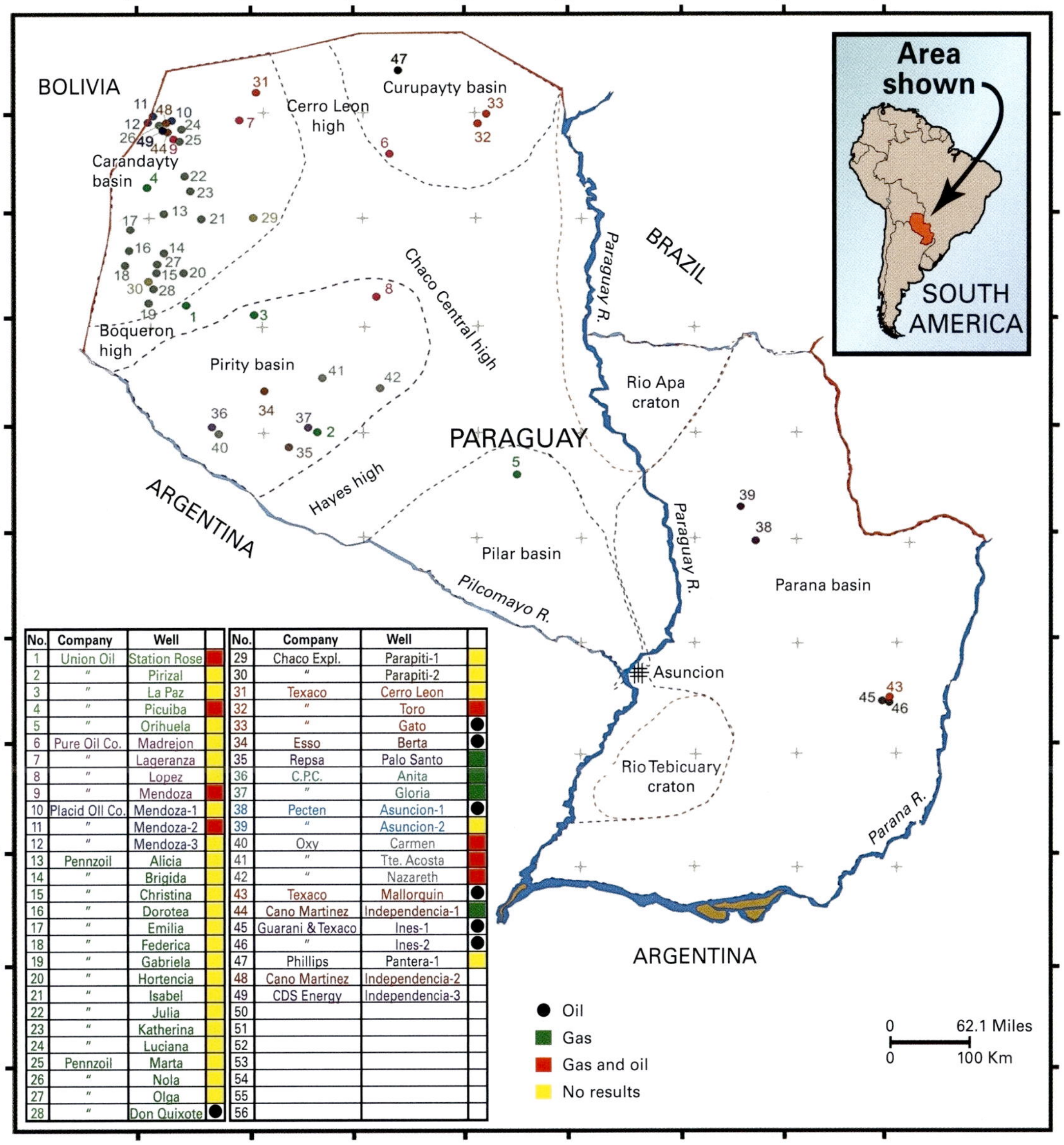

No.	Company	Well		No.	Company	Well	
1	Union Oil	Station Rose	Gas and oil	29	Chaco Expl.	Parapiti-1	No results
2	"	Pirizal	No results	30	"	Parapiti-2	No results
3	"	La Paz	No results	31	Texaco	Cerro Leon	No results
4	"	Picuiba	Gas and oil	32	"	Toro	Gas and oil
5	"	Orihuela	No results	33	"	Gato	Oil
6	Pure Oil Co.	Madrejon	No results	34	Esso	Berta	Oil
7	"	Lageranza	No results	35	Repsa	Palo Santo	Gas
8	"	Lopez	No results	36	C.P.C.	Anita	Gas
9	"	Mendoza	Gas and oil	37	"	Gloria	Gas
10	Placid Oll Co.	Mendoza-1	No results	38	Pecten	Asuncion-1	Oil
11	"	Mendoza-2	Gas and oil	39	"	Asuncion-2	No results
12	"	Mendoza-3	No results	40	Oxy	Carmen	Gas and oil
13	Pennzoil	Alicia	No results	41	"	Tte. Acosta	Gas and oil
14	"	Brigida	No results	42	"	Nazareth	Gas and oil
15	"	Christina	No results	43	Texaco	Mallorquin	Oil
16	"	Dorotea	No results	44	Cano Martinez	Independencia-1	Gas
17	"	Emilia	No results	45	Guarani & Texaco	Ines-1	Oil
18	"	Federica	No results	46	"	Ines-2	Oil
19	"	Gabriela	No results	47	Phillips	Pantera-1	No results
20	"	Hortencia	No results	48	Cano Martinez	Independencia-2	
21	"	Isabel	No results	49	CDS Energy	Independencia-3	
22	"	Julia	No results	50			
23	"	Katherina	No results	51			
24	"	Luciana	No results	52			
25	Pennzoil	Marta	No results	53			
26	"	Nola	No results	54			
27	"	Olga	No results	55			
28	"	Don Quixote	Oil	56			

The Quincy analysis noted that hydrocarbons have been produced for more than 75 years in Bolivia, just to the west, and for more than 25 years in Argentina, a few miles south of Paraguay.

The first exploration well drilled for oil in Paraguay was in 1947, but only 48 wells were drilled by several international oil companies and two private Paraguayan companies between that year and 2005.

In 2002, Earth View Associates estimated the area-wide recovery potential for the Paraguayan Chaco basin, east of the current production in Bolivia, at more than 4 billion boe.

PERU

Capital: Lima

Monetary unit: New sol

Refining capacity: 192,950 b/cd

Oil production: 106,000 b/d

Oil reserves: 447.382 million bbl

Gas reserves: 11.800 tcf

Overview

Oil is the dominant fuel source in Peru, with hydroelectricity second, according to the U.S. EIA. However, production since

2004 from the Camisea project has led to increased use of natural gas.

An LNG liquefaction plant being built at Pampas Melchorita will allow Peru to become an important regional exporter of natural gas.

Oil

Crude oil production in Peru has been in decline for the last decade. The country produced 113,000 b/d of total oil liquids in 2007, of which 68% was crude, said EIA.

In 2007, Peru consumed about 170,000 b/d of oil, importing about 57,000 b/d mostly from Ecuador and other South American countries.

Argentina's Pluspetrol controls over half of Peru's crude oil production. Other major producers include Occidental Petroleum, Brazil's Petrobras, and Petro-Tech Peruana.

Peru mostly privatized the former state-owned oil company, Petroperu, in 1993. The company still controls the country's only crude oil pipeline and most refineries.

Petroperu said that it wants to become an oil producer again, aiming to produce enough oil to meet at least half of domestic fuel demand by 2015.

In 2008, Petroperu reached exploration agreements with Discover Petroleum (Norway), Reliance Industries Ltd. (India), CNPC (China) and Pluspetrol.

Production of crude oil (mostly a heavy, sour type) is concentrated in the north. The largest oil blocks are Block 1-AB along the border with Ecuador, Block 8 in the northeastern Amazon, Block X in the northwest, and Block Z-2B off the northwest coast.

The first two account for almost two-thirds of total production.

Natural gas

Peru had proven natural gas reserves of 11.9 tcf in 2008, the fifth-largest amount in South America. Officials want to increase gas use and reduce reliance on oil and hydroelectricity.

Continued ramp-up of the Camisea project will likely boost natural gas production in coming years.

The government was planning to invest $35 million in 2008 to expand and upgrade the country's gas distribution networks.

Conflict over Development

In a re-emergence of conflict that first arose in August 2008, AIDESEP, an umbrella-group of jungle-Indian peoples, led protests in mid-2009 against government policies that grant development rights in Indian communal lands

The national government argues that investment in development that could turn the country into an oil exporter, benefiting all Peruvians.

For weeks during the summer, tens of thousands of Amazonian Indians blocked roads and ports and seized hydroelectric plants and oil and gas pumping stations. Their aim was to overturn decrees supporting oil exploration, commercial farming and logging in parts of the jungle.

Peru has 70 million hectares (173 million acres) of rainforest, about 70% of which has now been granted or offered as concessions for oil and gas exploration. Several companies, some foreign, operate many of the concessions.

In May, the government signed 13 oil and gas concessions with foreign companies, including Reliance Industries (India). At least a dozen more concessions were to be auctioned by November.

One area in dispute is Lot 67, in Loreto in the northeastern jungle, where Perenco plans to invest up to $2 billion by 2013 to extract heavy oil. The area includes part of a proposed Indian reserve.

Construction continued in 2009 on construction of the Peru LNG liquefaction plant—at the coastal town of Pampa Melchorita. Completion is slated for 2010, with export of up to 4.4 mtpy.

Exploration and Development

Block Z34

In late 2009, Gold Oil PLC was seeking partners to help develop Block Z34.

The company believes the offshore block has significant oil potential because it is west of four oil fields in northern Peru that have already produced a total of 1.6 billion bbl.

Of six drilling leads identified, some are in water 61–91m (200–300 ft) deep, but the most promising are in much deeper water.

The company believes it has confirmed the extension of the neighboring Z2b field. That field, producing about 14,000 b/d, is operated by Petro-Tech Peruana, a JV between Colombia's Ecopetrol and the Korean National Oil Company.

Gold Oil estimates it will need at least $60 million to develop Z34 and drill three exploration wells.

Lease Sales

New contracts

Petroperu signed 13 new contracts with IOCs in May 2009 for oil and gas exploration, valued at $650 million. The action brought to 92 the total number of oil contracts in effect.

Those submitting winning bids included Faulkner-Suits Exploration Inc.; Cepsa; Golden Oil Corp.; Petrolifera Petroleum Ltd.; KEI Pty.; Olympic Peru Inc.; Emerald Energy PLC; Talisman Energy Inc. and Colombian Ecopetrol SA; and Pan Andean Resources PLC.

A consortium of Pluspetrol, Petroperu, Reliance Exploration & Production DMCC (India), and Chinese-owned Sapet Development Peru Inc. also won a contract.

The companies will drill 35 wells during the exploration phase. If commercial volumes of oil and gas are discovered, the average royalty rate paid to the government will be more than 30%.

November auction plan

Reuters news agency reported in August 2009 that Petroperu was planning an auction in October or November of 17 blocks for oil and gas exploration.

The auction would be Peru's first since 2008, when several top government officials were fired or forced to resign after being accused of steering concessions to favored bidders.

Drilling and Production

BPZ Energy projects

In early 2009, BPZ Energy Inc. was testing the latest of six wells drilled in the 80-million-bbl Corvina field off the northwest coast. It controls 2.4 million net acres in northwest Peru, in blocks Z-1, XIX, XXII, and XXIII.

About 50 drilling prospects have been mapped in previously explored but undeveloped fields.

BPZ's Z-1 BLOCK, OFF PERU

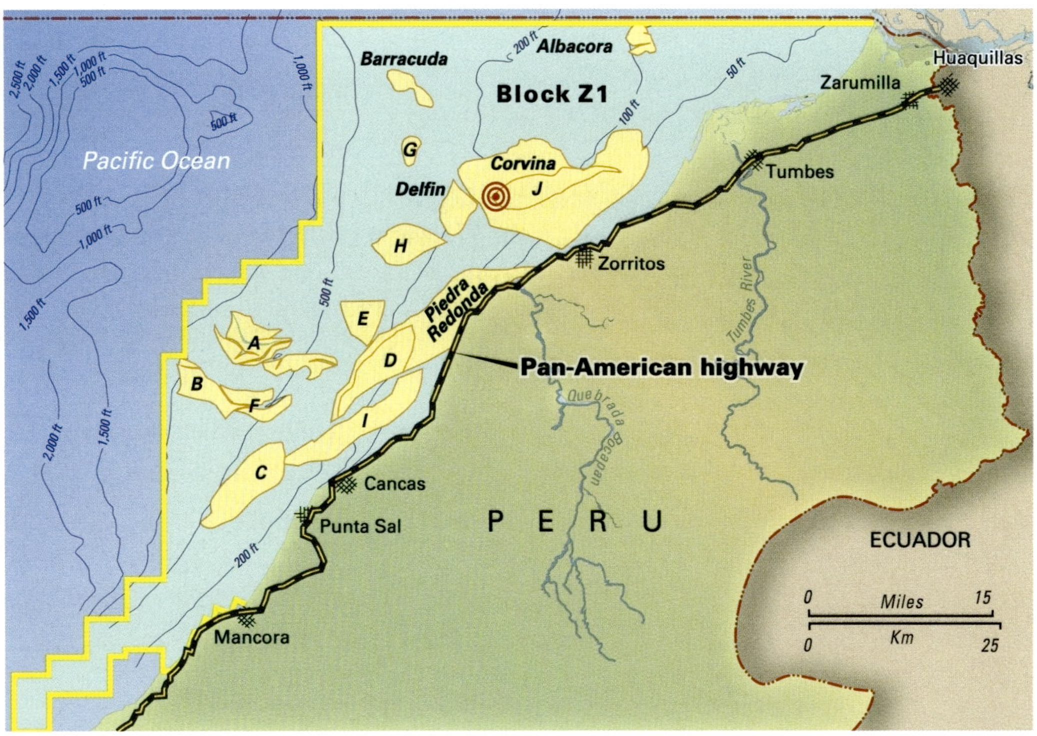

The firm also announced in January a new contract with the Peruvian government to deliver about 17 MMbbl of crude from Corvina field to the state-owned Talara refinery over seven years.

BPZ also was moving forward with its plan to build an integrated Corvina gas-to-power project, requiring 40 MMcfd of natural gas. Delivery of gas turbines for the power plant was expected in fourth quarter 2009, with start-up scheduled for mid-2010.

Other BPZ gas development projects in the works in Peru include pipelines to carry gas from Corvina field to two towns in Ecuador: Arenillas (75 MMcfd) and Guayaquil (150 MMcfd).

International Activities

Algeria interest

Algeria's Sonatrach said in early 2009 that it will join Petroperu in hydrocarbon exploration and production activities.

Peru's Energy and Mines Ministry had earlier asked for more assistance from Libya, particularly in the transport of offshore hydrocarbons and in building and operating petrochemical and NGL plants.

Sonatrach already is a shareholder in both Transportadora de Gas del Peru (TGP)—which moves gas and liquids from Camisea gas fields to Peru's Pacific coast—and in Consorcio Camisea, which operates Camisea fields.

LNG

Pampa Melchorita plant

Completion of construction of the Peru LNG liquefaction plant—at the coastal town of Pampa Melchorita, 169 km (105 mi) south of Lima—is slated for 2010.

The $3.8-billion facility is not only the largest industrial project ever to be undertaken in Peru; is also South America's first baseload LNG export facility.

The single-train facility will have a capacity of 4.4 million mtpy.

The project will include marine export facilities and a 408-km (254-mi) pipeline to transport feed gas from Ayacucho to Pampa Melchorita.

CB&I has been performing engineering, procurement, fabrication, and construction since early 2007.

Pipelines

Camisea lines

Two major upgrades were under way in early 2009 on the pipeline system that carries gas from the Camisea fields to the Pacific coast:

- An increase in the capacity at a compressor station in the Ayacucho region
- Construction of a loop along the coast to increase capacity to 450 MMcfd from 314 MMcfd.

Both projects were to be completed in second-half 2009.

Even so, to meet rapidly growing gas demand, officials said discussions also had begun for a new pipeline that could increase capacity to 1 billion cfd.

TRINIDAD and TOBAGO

Capital: Port of Spain
Monetary unit: Dollar
Refining capacity: 168,000 b/cd
Oil production: 107,000 b/d
Oil reserves: 728.300 million bbl
Gas reserves: 15.400 tcf

Exploration and Development

New fiscal regime

Trinidad and Tobago announced in September 2009 that, before the next offering of exploration blocks in first-quarter 2010, it will put in place a new fiscal regime.

The goals are to make the country more competitive and to encourage exploration for gas, particularly smaller fields in shallow water.

Officials expect that the government will offer six blocks for bid, including NCMA 2, NCM 4, and NCM 4B.

Galeota block

State-owned Petrotrin LLC signed an exploration and production contract in April 2009 with Bayfield Energy (Galeota) Ltd. for exploration of Galeota block and revitalization of Trintes oil field.

The work includes seismic and exploration wells as well as production enhancement facilities.

Discovered in the 1960s, Trintes oil field has so far produced about 22 million bbl of oil.

Seismic surveys suggested that a second field similar to Trintes might be found in the southwest quadrant of the block.

The Galeota exploration program acquire 120 sq km (46 sq mi) of 3D seismic and drilling of at least seven exploration wells to depths ranging from 1,981–3,658 m (6,500–12,000 ft).

Galeota block lies 11 km (6 nautical mi) east of Galeota Point in southeastern Trinidad.

Reserves

Updated gas reserves

A 2008 study described by Ryder Scott Company Petroleum Consultants in third quarter 2009 indicated that Trinidad and Tobago has proved gas reserves of 15.4 tcf and another 30 tcf in unrisked exploration potential.

The 2008 study also showed 8.5 tcf of probable reserves and 6.3 tcf of possible reserves.

Analysts said that 1.5 tcf of proved gas had been used up and the challenge to the country was to replace it.

Government officials confirmed that Trinidad and Tobago produces about 4.1 bcfd of gas and that there is a critical need for the nation to encourage exploration as wells as appraisal and development drilling.

They also said changes are planned to the taxation regime for exploration, particularly in marginal fields and in the deep water.

Trinidad and Tobago is the largest exporter of LNG to the U.S.

Drilling and Production

Savonette field

BP Trinidad & Tobago (BPTT) started its first natural gas production in October 2009 from Savonette field. It is in 88 m (290 ft) of water, about 80 km (50 mi) offshore to the southeast, in the prolific Columbus basin.

Gas from Savonette will supply Atlantic LNG's liquefaction plant for export as LNG to both international and domestic markets.

Production from the field is expected to average 600 MMscfd of gas, plus associated condensate, from four wells. The Savonette platform is designed to handle as much as 1 bcfd of gas.

Pipelines/Transportation

Gas pipeline extension to Barbados

Construction was under way in mid-2009 on a natural gas line from Trinidad to sister island Tobago. Startup is slated for first quarter 2010.

At the same time, planners were considering extending the line to Barbados.

Earlier in the year, Barbados passed hydrocarbon legislation, hoping to discover oil and gas of its own.

Barbados

In January 2009, Barbados awarded rights to two blocks—Carlisle Bay and Bimshire—to BHP Billiton. The Barbados energy ministry said it still planned to open as many as 26 blocks for exploration.

Barbados produces just 1,100 b/d of its own oil from onshore wells. Barbados National Oil Co. Ltd. sends the crude to Trinidad and Tobago for refining, then brings it back to Barbados.

Production by BNOC meets just 15% of Barbados' fuel needs; the other 85% is imported.

Like Trinidad and Tobago, Barbados has chosen not to join the PetroCaribe plan, under which Venezuela's PDVSA offers low-priced crude under long-term payment plans.

Company News

Niko Resources: Block 2AB

In mid-2009, Niko Resources Ltd. obtained a 26% interest from Centrica PLC and became operator of offshore Block 2AB, its first block in the country.

The block covers 1,605 sq km (620 sq mi) near giant Angostura oil and gas-condensate field off northeastern Trinidad.

Other interests in Block 2AB are Voyager Energy and state-owned Petrotrin.

Centrica: Block 5(c)

In mid-2009, Centrica PLC increased its position in LNG development in Trinidad by acquiring a 45% interest in gas development Block 5(c), off the southeast coast.

The company agreed to pay $142.5 million cash to Canadian Superior Energy Inc. and will use the estimated 650 bcf of recoverable gas to supply customers in the UK, North America, or other Atlantic Basin markets starting in 2014.

The block is close to existing gas pipeline infrastructure and LNG export facilities.

Centrica will join operator BG Group, which has a 30% stake.

Centrica also has an existing interest in Block 2AB.

URUGUAY'S 2009 BLOCK OFFERING

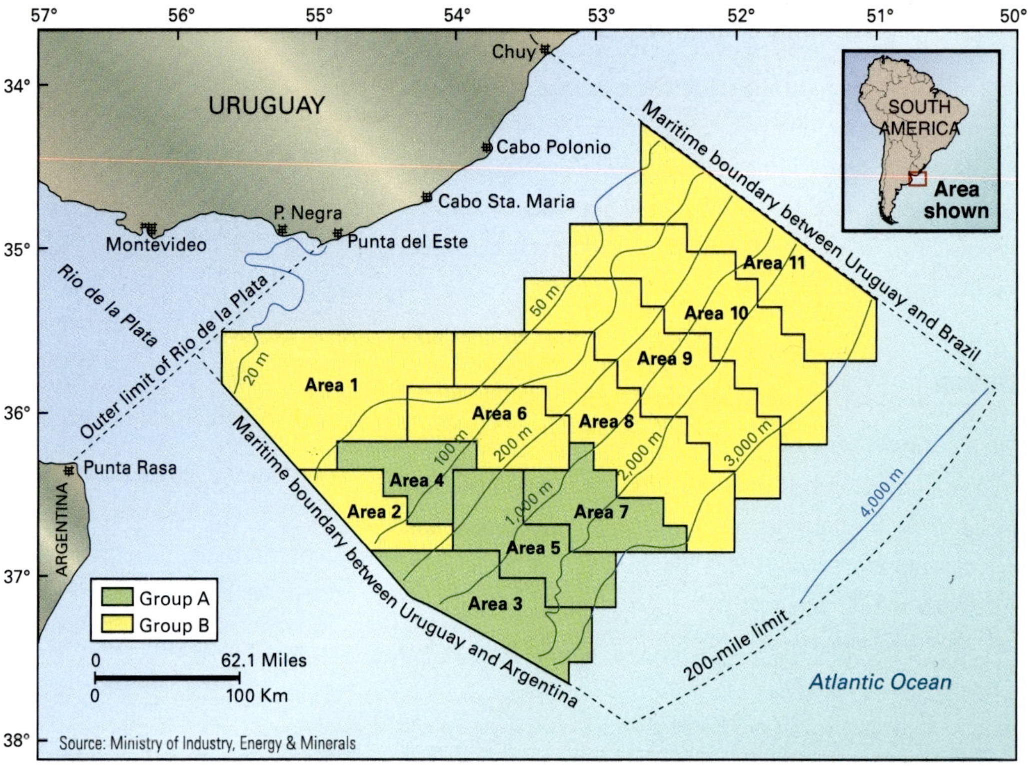

URUGUAY'S BASINS

URUGUAY

Capital: Montevideo
Monetary unit: lUruguayan peso
Refining capacity: 50,000 b/cd
Oil production: 950 b/d
Oil reserves: 0
Gas reserves: 0

Lease sales

In July 2009, national oil company ANCAP offered 11 offshore blocks in the Punta del Este, Pelotas, and Oriental del Plata basins. It was Uruguay's first licensing round.

The blocks lie in 1,000–3,000 m of water.

Interested companies included BHP Billiton Ltd., Galp Energia, SGPS SA, PDVSA, Petrobras, Pluspetrol SA, and YPF SA.

Punta del Este basin

In October, ANCAP accepted the bids it received for two blocks in the Punta del Este basin.

ANCAP selected Argentina's YPF SA to operate Block 3 and Brazil's Petrobras to operate Block 4. YPF, Petrobras and Portugal's GALP Eneregia have interests in each block.

Block 3 covers 5,500 sq km (2,124 sq mi) in 100–2,000 m (328–6,562 ft) of water, 93–186 km (150–300 mi) off Montevideo.

Block 4 covers 3,000 sq km (1,158 sq mi) in 50–200 m (164–656 ft) of water, 201–241 km (125–150 mi) off Montevideo. It is north of Area 3.

VENEZUELA

Capital: Caracas
Monetary unit: Bolivar
Refining capacity: 1.282 million b/cd
Oil production: 2.170 million b/d
Oil reserves: 99.377 billion bbl
Gas reserves: 175.970 tcf

Overview

Venezuela is the fourth-largest producer among the Organization of Petroleum Exporting Countries (OPEC). A report by Business Monitor International in 2008 concluded that Venezuela will provide about 28% of Latin America's oil supply by 2011.

Oil production has fallen 25% since Pres. Hugo Chávez took office in 1999, largely as a result of his dismissal of more than 20,000 experienced engineers and executives with the state oil company, Petroleos de Venezuela SA (PDVSA). They had joined a 2002–03 strike urging that Pres. Chávez be removed from office.

The government's unpredictability and inflexibility are seen by BMIas a serious deterrent to investment by foreign companies in Venezuela.

BMI's *Venezuela Oil & Gas Report* projected that export capability would be about 2.6 million b/d by 2011, with gas production of 48 bcm that year. Gas export potential exists beyond 2011.

The report pointed out that recent changes in licensing, production-sharing, and taxation have reduced foreign involvement.

At the same time, BMI said, the country has considerable untapped hydrocarbons potential, with a high reserves-to-production ratio and significant output growth potential.

Assessments by reputable sources indicate likely substantial reserves in the Orinoco extra-heavy crude belt. In the Orinoco Magna Reserve project, the government expects to certify more than 250 billion bbl of crude oil reserves. If correct, that would give Venezuela the largest hydrocarbons reserves in the world.

More nationalization

The Venezuelan government continued actions to nationalize or otherwise control energy-sector assets in the country during the first half of 2009.

- In February, Pres. Chávez won a referendum that allows him to run for office indefinitely, prompting action to consolidate his power over opponents and the energy sector.
- In March, government troops took over key oil ports in the country.
- On May 7, Congress issued a law that reserved to the state goods and services related to "primary hydrocarbon activities."
- Pursuant to the new law, in May and early June, the government seized the facilities of 76 private oilfield service companies in the country. Seized property included oil rigs, terminals, boats, and other installations, including two major gas operations.
- In May, the national assembly was considering a law to put all petrochemical activity under state control.
- Also in May, the government seized operational assets owned by transnational Williams Companies Inc., with a combined capacity of 2 bcf/d for natural gas injection and compression.

At the end of 2008, PDVSA owed contractors an estimated $14 billion, according to the National Assembly, and some observers saw the seizures as a way to eliminate those debts.

PDVSA: Sliding?

PDVSA is the largest employer in Venezuela. It accounts for one-third of the country's GDP, half of the government's revenue, and 80% of Venezuela's exports earnings.

But PDVSA's revenue fell 52% in the first half of 2009 compared with the same period in 2008 due to a sharp drop in world oil prices, according to company data.

In the third quarter, the oil sector's portion of Venezuela's gross domestic product fell 9.5% compared with a year earlier, its third straight year-over-year decline.

And in mid-2009, troubles were arising on several other fronts as well.

Dissident union leaders said that many of the 100,000 workers employed by PDVSA and some contractors were unhappy about overdue pay; delay in renegotiating a contract with oil workers; and government efforts to consolidate all unions into a pro-government, socialist organization.

Industry analysts pointed to several other disturbing actions and trends:

- In May 2009, national troops seized the assets of many foreign and domestic oil service companies (see above).
- Company revenues that exceeded $120 billion in 2008 were expected to drop to about $50 billion in 2009.
- It is estimated that PDVSA will need to spend $3 billion/year just to maintain present production from existing fields, many of which have decline rates of 25% per year or more.
- The government pulls billions from the company for nonoil investments.
- PDVSA claims Venezuela's production capacity tops 3.4 million b/d, but OPEC reported that oil production was only 2.24 million b/d in May 2009.
- Oil exports to the United States have been falling since peaking at 1.77 million b/d in 1997.

The late-2009 recovery in oil prices to around $70/bbl provided some respite, but sustaining social spending in a volatile global energy market environment may require more government borrowing.

Stake in Deltana Platform

PDVSA said in November 2009 that it had acquired from ConocoPhillips a stake in the Deltana Platform—a joint natural gas project with Chevron Corp.

The platform is in the Atlantic Ocean, between Trinidad and Tobago and the mouth of the Orinoco River.

Gas output is expected to reach 750 MMcfd, flowing through a 300-km (186-mi) pipeline to the Gran Mariscal de Ayacucho Industrial Complex, being built in Sucre State.

Bond issues, buybacks

Declining oil revenues in 2009 hurt PDVSA's ability to pay contractors and suppliers. By the middle of the year, PDVSA owed them a reported $8.145 billion.

From 2007 to late 2009, PDVSA issued about $12 billion worth of bonds to raise money, according to a Reuters analysis. This included about $3 billion in July 2009 and about $3.2 billion in October.

In the last half of 2009, company sources said, PDVSA bought back between $300 million and $600 million of its dollar-denominated bonds to take advantage of their fall in market price.

The company also was in loan talks with several foreign private financial institutions in December.

Exploration and Development

Perla field

Italy's Eni SPA said in October 2009 that it made a "world class" natural gas discovery with an exploration well drilled in Perla field on the Cardon IV block in the Gulf of Venezuela.

Eni estimated that the field has a reserve potential higher than the 6 tcf previously estimated.

The Perla 1X well was drilled 50 km (31 mi) offshore, finding a 240-m (787-ft) hydrocarbon column. The field is 209 km (130 mi) northeast of Maracaibo City.

Production tests high-quality gas with a capacity of 600,000 cu m/day and 500 b/d of condensate.

Repsol YPF and PDVSA also have interests in the project.

Orinoco: Carobobo Blocks

Venezuela says the Orinoco Belt region holds 272 billion bbl of recoverable reserves, including proven reserves of 80 billion bbl.

And within the belt, the government has delineated seven Carabobo Blocks, which could collectively produce more than 800,000 b/d.

In August 2009, Venezuela said it planned to complete the bidding round for the blocks by yearend.

Some 20 companies—including BP, Chevron, CNPC, Ecopetrol, Eni, Galp Energia, Petronas, and Statoil, and several Japanese and Indian organizations—showed interest in the tender.

In June, Venezuela and Russia signed an energy cooperation deal to develop two oil fields in the Orinoco region. One aspect would aim to develop the Carabobo 1 North and Carabobo 1 Center fields.

The Carabobo project aims to turn the tar-like Orinoco crude into oil for exports and produce around 200,000 b/d. Initial investment of $10–20 billion per area is likely.

Orinoco: Block Junín 6

In September 2009, PDVSA signed an agreement with a consortium of Russian companies (integrated by Rosneft, Lukoil, Gazprom, TNK-BP, and Surgutneftegaz) to develop Block Junín 6 in the Orinoco Belt.

The area has production potential of more than 400,000 b/d of extra-heavy oil.

Work is expected to start in 2012.

Orinoco: China deal

In September 2009, Venezuela signed a $16-billion investment deal with China spanning three years, to raise oil output by several hundred thousand barrels per day in the Orinoco Belt.

Few details of the deal were released.

REPSOL-ENI'S PERLA GAS DISCOVERY OFF VENEZUELA

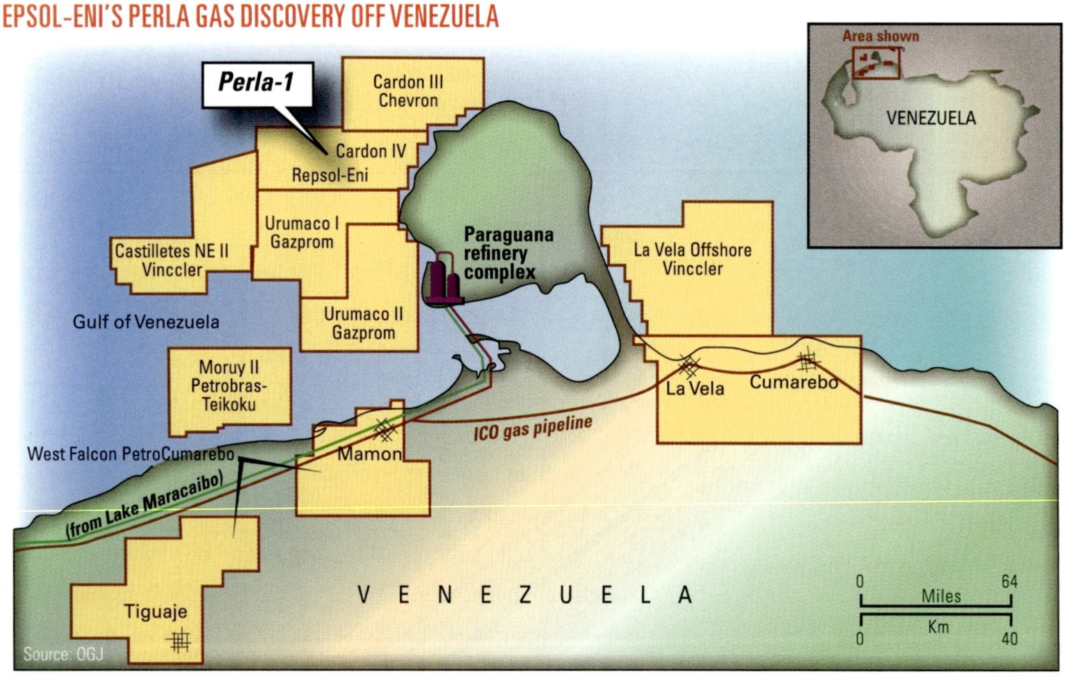

Orinoco Oil Belt location

The Orinoco Oil Belt occupies the southern strip of the eastern Orinoco River Basin in Venezuela. It is located south of the Guárico, Anzoátegui, Monagas and Delta Amacuro States, and it follows the line of the Orinoco River. The belt is approximately 600 km (370 mi) from east to west, and 70 km (43 mi) from north to south, encompassing an area of about 55,314 sq km (21,357 sq mi).

Drilling and Production

Sacha field

Rio Napo, a 70:30 joint venture of state oil company Petroecuador and PDVSA, began operating the Sacha oil field in Ecuador's Amazon region in late 2009.

Output of 70,000 b/d of oil is targeted for 2014–2015.

Further details are given in the Ecuador section of this chapter.

LNG

Cigma facility

In November 2009, work was proceeding on construction of three LNG trains at the Gran Mariscal de Ayacucho Industrial Complex (Cigma) in the eastern state of Sucre.

PDVSA said it had signed JV accords with several IOCs and foreign companies (notably from Japan and Russia) since 2008 for the three trains, each with a design capacity of 4.7 mtpy.

Plans called for commissioning all three by 2014.

Some concerns were expressed about the pace of construction of pipelines to carry gas to Cigma, and about the ability of PDVSA to produce enough natural gas to supply the LNG plants.

Venezuela discussed in 2008 an initiative with Russia's Gazprom to liquefy for export some of the gas produced on Venezuela's coastal shelf.

International Activities

Japanese investment

Venezuela and Japan agreed in April 2009 to cooperate on 12 hydrocarbon projects.

Japan is expected to invest $33.5 billion in Venezuela's hydrocarbon sector through 2014, including projects for petrochemical production, oil refining, LNG, and oil exploration.

Some $8 billion is targeted to develop oil blocks in the Orinoco belt, including Junin 11 Block.

That block has estimated reserves of 6 billion bbl.

An additional $6 billion will focus on oil exploration in the coastal region.

The two nations also created a $4-billion investment fund, as part of a larger package of investment in oil, petrochemicals, and LNG production.

Other initiatives address offshore gas-field development in the Proyecto Mariscal Sucre region. Terms call for potential transport of gas to Japan after 2013.

Oil to China

Venezuelan and Chinese leaders agreed in April 2009 to speed up the start date for a long-planned increase in Venezuelan oil exports to China, to 2010 rather than 2013.

Pres. Chavez said the goal of shipping 1 million b/d to China is reachable because Venezuela plans to divert much of the current 1.5 million b/d it sends to the United States.

Reports vary on just how much oil Venezuela actually ships to China. In May 2008, Venezuela said oil shipments were 500,000 b/d. The Chinese reported only 80,000 b/d of crude and 300,000 b/d of products.

Also discussed was a jointly owned refinery in China's Guangdong Province to process 20 mtpy of Venezuelan heavy crude.

Relations with Iran

Venezuela and Iran developed closer commercial and political ties in 2009.

- The two agreed to invest $760 million in each other's energy sectors.
- Iran signed a deal to import gasoline from Venezuela in the event that sanctions are imposed by Western nations unhappy with Iran's nuclear program development.

- Iran will acquire a 10% stake in developing Dokobuki oil field in Venezuela
- Iran is partnering with Venezuela to develop Block 7 of the Ayacucho oil field.
- PDVSA will invest 10% in the Phase 12 development of the South Pars gas field in Iran.
- Venezuela agreed to invest in the exploration of 12 blocks in Iran.

Venezuela and Iran also established a 50:50 JV to explore for oil and distribute petroleum and oil products in Central Asia, Central and South America, and Africa (but not in Iran or Venezuela).

Panama to join PetroCaribe

Venezuela decided in early 2009 to allow Panama to join the PetroCaribe fuel assistance program.

Launched by Pres. Hugo Chavez in 2005, PetroCaribe now includes 18 countries in and around the Caribbean Sea.

Member countries pay 60% of the cost of Venezuelan oil at the time of purchase and can defer the remaining 40% as financing for development projects, repayable over 25 years at a 1% interest rate.

Refining and Products

Abreu e Lima accord

Petrobras and Venezuela's PDVSA completed negotiations in November 2009 for joint construction and operation of the Abreu e Lima refinery, in Brazil's Pernambuco state.

The refinery will be able to process 230,000 b/d of heavy oil, supplied equally by the two firms.

Details are given in the Brazil section of this chapter.

Stake in Dominican refiner

In late 2009, after months of delay, PDVSA completed its purchase of a 49% stake in the state refiner of the Dominican Republic (DR), Refineria Dominicana de Petroleo SA (Refidomsa). The DR firm owns a 34,000-b/d facility in Haina.

The deal would allow the DR to become an oil distribution center for the Caribbean and possibly Central America.

Terms call for the DR to buy 30,000 b/d of oil from Venezuela, in addition to the 50,000 b/d it already receives under the PetroCaribe accord.

Refinery in Syria

Venezuela and Iran signed an MOU in September 2009 to build a 140,000-b/d oil refinery at Furqlus, near Homs, in the western region of Syria.

Syria and Malaysia's Al Bukhari Group (Kuala Lumpur) also will participate.

Details are given in the Syria section of the Middle East chapter of this book.

Other action in Latin America

Several other nations in Latin America saw activity in the oil and gas sector in 2009.

Costa Rica

Capital: San Jose

Refinery joint venture

CNPC (Beijing) and Refinadora Costarricense de Petroleo S.A. (Recope) (San Jose) agreed to set up a joint-venture refinery.

Exploration policy

In May 2009, Costa Rican Pres. Oscar Arias reversed earlier government statements, declaring his opposition to any oil and gas exploration in his country.

However, the country was still in negotiation with Venezuela for admission to the PetroCaribe program under which Caracas provides oil at discounted prices.

Refineries

Also in May 2009, Costa Rica affirmed the validity of a 25-year agreement made with Chinese officials to help expand Costa Rica's Moin refinery to 60,000 b/d by 2013 from the current 25,000 b/d.

Additionally, China's CNPC and Costa Rica's Recope were evaluating plans to set up a $6-billion Costa Rican refinery with the capacity to process 200,000 b/d.

Panama

Capital: Panama City

Pipeline start-up

Tesoro Corp. announced in September 2009 the shipment of the first oil through the reversed 130-km (81-mi) Petroterminal de Panama Trans-Panamanian Pipeline, from the Atlantic side of the isthmus to the Pacific.

The reversal will enable Tesoro to source a broader range of crude oils to supply its Pacific Rim refining system.

Very large crude carriers of 2-million bbl capacity will be able to transport West African and other crudes to the port of Chiriqui Grande, Bocas del Toro, on the Caribbean for the journey across the isthmus of Panama.

Crude will be piped to the port of Charco Azul on the Pacific Coast where it will be received by tankers and shipped to U.S. West Coast refineries.

Refinery at Puerto Armuelles

In early 2009, Panama was completing studies for construction of a 350,000-b/d refinery at Puerto Armuelles, about 550 km (342 mi) west of Panama City.

In 2007, the Panamanian government agreed to cooperate with Qatar Petroleum and Occidental Petroleum Corp. on the project, then valued at $7 billion, and the two companies commissioned Foster Wheeler Ltd. to conduct the studies.

Since then, Tesoro Corp. entered the country with a throughput agreement that will allow the company to transport crude oil through a pipeline owned by Petroterminal de Panama (PTP).

Suriname

Capital: Paramaribo

Refinery expansion

Staatsolie Maatschappij Suriname NV let a project management contract in late 2009 to Aker Solutions for its refinery expansion project in Suriname.

The project would double the refinery's processing capacity to 15,000 b/d, producing diesel, gasoline, fuel oil, bitumen, and sulfuric acid. The expansion is to be completed in 2013.

Europe

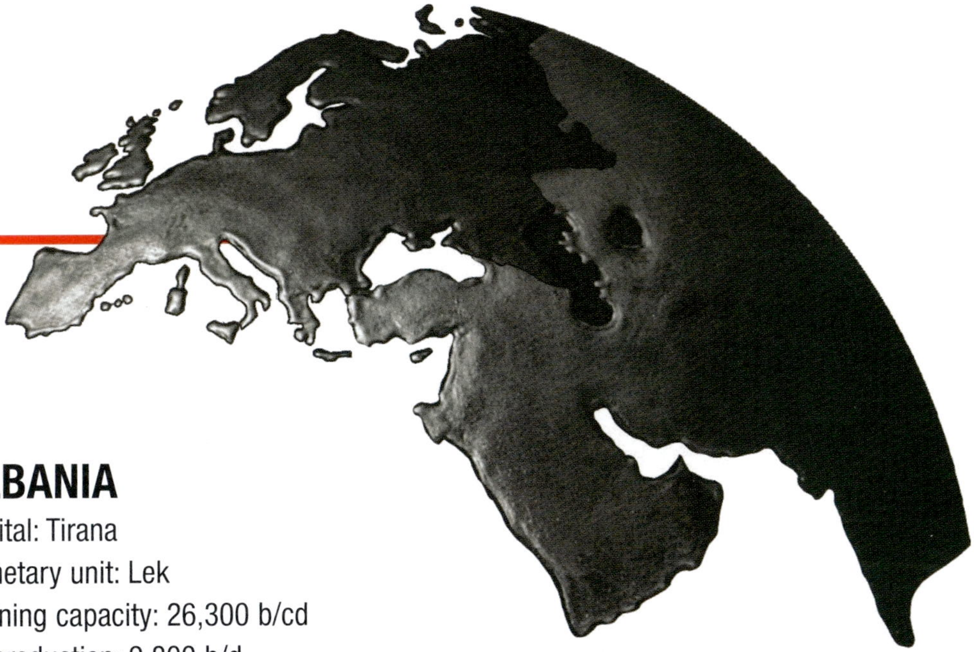

ALBANIA

Capital: Tirana
Monetary unit: Lek
Refining capacity: 26,300 b/cd
Oil production: 9,800 b/d
Oil reserves: 199.140 million bbl
Gas reserves: 30.000 bcf

The oil and gas sector in Albania is administered by the Ministry of Industry and Energy under which the National Petroleum Agency is responsible for oil and gas licensing activities. The Ministry of Economy commercially supervises them.

The Albanian Petroleum Corp. (APC) is the parent to Albpetrol, responsible for exploration and production. Armo, responsible for refineries at Ballsh and Fier and Servcom, supports Albpetrol in development and drilling.

Limited exploratory drilling is under way in the country although results generally have been poor.

Trans-Adriatic Gas Pipeline

Albania said in early 2009 that it will work with Croatia to lobby for the major Trans-Adriatic Pipeline (TAP) project.

Energy trading company Elektrizitats-Gesellschaft Laufenburg (EGL) and Statoil agreed in early 2008 to a JV to build and operate the Trans-Adriatic Pipeline (TAP).

A section of the line would run through Albania, transporting Caspian and Middle Eastern natural gas to the heart of Europe.

Lezhe energy park

Planning moved ahead in 2009 for a $1.3-billion energy park in Lezhe, near the Adriatic coast.

It is to include a 140-MW power generation plant fueled by biomass or biodiesel, as well as two 234-MW wind farms in the same region.

Electricity from the park is expected to be sold to Italy and other international markets, in addition to addressing local power needs. Cables would carry power under the Adriatic Sea between Albania and Apulia, in southeastern Italy.

Albania makes wide use of hydroelectricity for power, but still imports more than 40–50% of the power it needs.

EUROPE
0 50 100 150 Miles
0 100 200 Kilometers
North Sea
Irish Sea
Celtic Sea
Bristol Channel
Lyme Bay
Bay of Biscay
Gulf of Cadiz
Mediterranean Sea
IRELAND
N. IRELAND
ENGLAND
WALES
NETHERLANDS
BELGIUM
LUX.
GERMANY
DENMARK
FRANCE
SWITZERLAND
SPAIN
PORTUGAL
CORSICA
SARDINIA
MINORCA
MAJORCA
IBIZA
MOROCCO
ALGERIA
TUNISIA
Corrib
Morecambe
Formby
Waterford
Kinsale Head
Weymouth
Londonderry
Belfast
Sligo
Ballina
Galway
Dublin
Ballough
Foynes
Limerick
Arklow
Tralee
Wexford
Cork
Waterford
Whitegate
Bantry
Moffat
Carlisle
Newcastle
Darlington
Teesside
Middlesbrough
Barrow
Heysham
Lancaster
Bradford
Eastham
Manchester
Liverpool
Sheffield
Killingholme
Grimsby
Lincoln
Theddlethorpe
Bacton
Ellesmere Port
Aberystwyth
Birmingham
Milford Haven
Pembroke
Neath
Gloucester
Cambridge
Ipswich
Coryton
Oxford
London
Bath
Exeter
Plymouth
Southampton
Fawley
Portsmouth
Dunkerque
Antwerp
Brussels
Amsterdam
Rotterdam
Groningen
Emden
Leer
Bremen
Wilhelmshaven
Cuxhaven
Heide
Kiel
Brunsbuttel
Schwerin
Harburg
Hamburg
Lingen
Loningen
Verden
Salzbergen
Beckedorf
Meerdf.
Buckeburg
Herringhausen
Salzgitter
Gelsenkirchen
Duisburg
Hamm
Dortmund
Essen
Dusseldorf
Cologne
Rheinland
Eisenach
Jena
Erfurt
Suhl
Bonn
Verviers
Frederica
Kaergard
Kalundborg
Skaelskor
Notre Dame de Gravenchon
Port Jerome
Cherbourg
Le Havre
Gonfreville L'Orcher
Gargenville
Petit Couronne
Plestin Les Greves
Brest
St. Malo
Quimper
Lorient
Vannes
Laval
Donges
Nantes
Angers
Cholet
Les Sables
Poiters
Chauvigny
La Rochelle
Niort
Royan
Le Verdon
Pauillac
Ambes
Bordeaux
Limoges
Chazelles
Bergerac
Tulle
Roussiness
Orleans
Blois
Briare
Vierzon
Paris
Maubeuge
Charlesville
Grandpuits
Luxembourg
Metz
Frankfurt-am-Main
Karlsruhe
Vohburg
Ingolstadt
Munich
Bourbonne
Epinal
Wallbach
Dijon
Besancon
Dole
Zurich
Zug
Hochst
Chur
Bern
Geneva
Annemasse
Martigny
Aosta
Lugano
Mandello
Macon
Bourg
Les Ancizes
Feyzin
Lyon
Vienne
Issoire
St. Etienne
Chamberry
Grenoble
Le Puy
Valence
Brescia
Padua
Milan
Biella
Chivasso
Cremona
Rovigo
Turin
Asti
Busalla
Parma
Aquata
Genoa
Savona
Pistola
Leghorn
Villefranche
Toulouse
Castres
Carcassonne
Pamiers
Lourdes
Narbonne
Fos-sur-Mer
Berre l'Etang
Marseille
Toulon
L'Isle
Manosque
Nice
Monaco
Bastia
Muro
Aleria
Zonza
Perpignan
La Junquera
Gerona
Palamos
Mataro
Barcelona
Tarragona
Vert-La-Gravelle
Coulommes-Vaucourtois
Villeperdue
Chaunoy
Marolles
Valence en Brie
Villemar
Aufferville
Donnemarie
Flacy
Cisy-les-Nobles
Joigny
Chateaurenard
Velye-Germinon
Bussey-Lettree
Dommartin-Lettre
Coudemagnes
Grandville
Forcelles
St. Martin-de-Bossenay
Trois Fontaines
Revigny
Gallician
Lavergne
Cazaux
Lugos
Mothes
Parentis
Ychoux
N. Mimizan
Lucats
Cabeil
Lacq sup
Gaviota
Bayonne
Pont d'As
Lacq
St. Marcet
Proupiary
Rousse
Meillon
St. Faust
El Serrablo
Ayolvengo
Otomin
Ferrol
La Coruna
Aviles
Santander
Vizcaya
Bilbao
Logrono
Burgos
Vitoria
Huesca
Barbastro
Monzon
Balaguer
Tarrega
Zargoza
Fayon
Zamora
Vallodolid
Porto
Madrid
Chipirón NE
Dorado
Angula
Salmonete
Montanaza
Tarraco
Casablanca
Castellon
Amposta
Vinaroz
Castellon de la Plana
Vallencia
Cabo Ruivo
Lisbon
Setubal
Sines
Puertollano
Almendralejo
Cordoba
Huelva
Sevilla
Rota
Cadiz
Gulf of Cadiz
Granada
Malaga
Almeria
Cartegena
Murcia
Algeciras
Porto Torres
Maddalena
Alghero
Olbia
Nuoro
Arbatax
Iglesias
Porto Scuso
Cagliari
Algiers
Bejaia
Skikda
Tunis
Oran
Arzew

Nuclear plant plans

In April 2009, Albania and Croatia said they plan to sign an agreement to jointly develop a nuclear power plant.

The $5.3-billion facility would generate 1,500 MW. A design team had been formed, but the construction schedule was not specified.

Croatia's state-owned power company is expected to construct the plant, with a major share of the power output going to Croatia.

The plant is to be built near Skadar, Albania, close to the border with Montenegro and about 30 km (19 mi) from that country's coast.

Some in Montenegro are concerned that that region is seismically active and that waste heat from the plant could disrupt animal and plant life in and near Skadarsko Lake.

Bosnia-Herzegovina and Montenegro might join the project because power from the plant would pass through those nations to reach Croatia.

AUSTRIA

Capital: Vienna
Monetary unit: Euro
Refining capacity: 208,600 b/cd
Oil production: 18,400 b/d
Oil reserves: 50.000 million bbl
Gas reserves: 570.000 bcf

Austria-Slovakia pipeline

OMV AG announced plans in mid-2009 to open the long-delayed pipeline between Vienna and Bratislava, capital of Slovakia, by 2010.

The pipeline, which would give OMV direct access for the first time to Russian oil, will connect OMV's Schwechat refinery with the Druzhba pipeline, which originates from southeastern Russia.

OMV now relies on costly shipments through the port of Trieste, Italy.

Plans for the pipeline were first discussed in 2003, but the parent firm (Yukos Oil Co.) of one of the project's partners filed for bankruptcy in 2004.

OMV also has expressed interest in developing crude oil assets in Russia as well as Kazakhstan.

Hydropower projects

Tiroler Wasserkraft AG (TIWAG) said in July 2009 that it plans to build a new pumped-storage hydropower plant in the Alps to extend the capacity of its existing facilities.

The Kuehtai II project would include a new reservoir and two turbines to generate a total of 180 MW.

TIWAG owns and operates 45 hydropower plants in eastern Austria.

Construction is slated to start in early 2011, with completion set for September 2015.

Also, Verbund-Austrian Hydro Power AG announced investment plans of $730 million to extend, modernize or construct elements of two of its hydropower plants in Austria: Reisseck II and Limberg II.

> Austria gets more than two-thirds of its electricity from hydropower stations located in the Alps or the country's rivers.
>
> In mid-2009, the Association of Austrian Electricity Companies and the Austrian Energy Ministry announced investments through 2010 that total $3.2 billion for the construction of new power plants and transmission grids, focusing on renewable resources.

European gas balance

Demand for natural gas is falling in Europe—because of high prices in 2008 and a deteriorating economic outlook—but the region also faces a potential of oversupply of gas, said Wood Mackenzie.

The firm pointed to large number of LNG projects expected to come on stream soon. By 2011, as much as 110 bcm (3.88 tcf) of LNG likely will enter the market.

The firm also said it expects that national oil companies who reduced their dependence on international oil companies as energy prices climbed in 2008 may need new help from the IOCs if global economic weakness continues.

Russia-Ukraine gas dispute

At the start of 2009, Ukraine and Russia were still far apart on a gas supply deal for the year ahead. The uncertainty had clear implications for European gas supply security.

Gazprom was seeking to get Ukraine to pay higher gas prices, while Ukraine was insisting on higher gas transit fees while threatening to end a long-term contract (signed in 2006) governing the movement of gas across its territory.

At the end of 2008, Gazprom was asking $250/M cu m of gas and Ukraine was offering about $200. But in early 2009, Gazprom was insisting on $450, compared to the average $179.50 price in 2008.

Russian officials also offered to boost gas transit payments to Ukraine, from $1.70/M cu m to $3.40, if Ukraine agreed to pay a "market price" for supplies.

The *Financial Times* said the ongoing dispute also was rooted in:

- Gazprom's resolve to end decades of energy subsidies for its former Soviet neighbors.
- Ukraine's acceptance of gas prices rising to European levels, but over the three-year adjustment period Moscow agreed to in October 2008.
- Russia's continued resentment at being dependent on Ukraine for access to pipelines built and controlled by Russia before the end of Soviet rule.
- The role of RosUkrEnergo in Ukrainian gas trade—an obscure Swiss-based company owned jointly by Gazprom and Ukrainian business people.

On January 1, 2009, gas flow destined for Ukraine was cut off while transit deliveries to the EU were sharply reduced. Negotiations and lawsuits ensued, and on January 19, the heads of Gazprom and Naftohaz signed a 10-year agreement on gas supplies to Ukraine.

Gas flow was fully restored on January 21.

At yearend 2009, the threat of another disruption surfaced regarding oil flow to Slovakia, Hungary, and the Czech Republic, but payment issues were resolved.

BELGIUM

Capital: Brussels
Monetary unit: Euro
Refining capacity: 797,629 b/cd
Oil production: 0
Oil reserves: 0
Gas reserves: 0

Vitol-Petroplus deal

In late 2009, a division of Vitol Group agreed to acquire Petroplus Refining Antwerp and Petroplus Refining Antwerp Bitumen from Petroplus Holdings AG for $25 million.

Closing, subject to regulatory approvals, was expected by yearend.

The Belgium refining assets include a bitumen processing plant with 875,000 tpy capacity.

Olefins-to-polypropylene

In February 2009, Total SA reported progress on a Belgian research project aimed at diversifying petrochemicals feedstock away from naphtha and gas.

Norsk Hydro and UOP have developed a methanol-to-olefins (MTO) process that converts methanol to ethylene and propylene. Working with UOP, Total has developed a technology that takes the heavier olefins from the MTO unit and converts them into propylene.

The pilot plant at Feluy to assess this olefin cracking process is Total's second-largest current research project after its Lacq CO_2 capture-and-storage pilot plant.

Thornton Bank wind farm

In mid-2009, a unit of Electricité de France commissioned Belgium's largest wind farm and the first such facility offshore. At the Thornton Bank facility, 28 km (17 mi) offshore near Zeebrugge, a first 30-MW section became operational in June.

The $1-billion project will consist of sixty 5-MW turbines built by REpower Systems AG. The first phase of work installed six 5-MW turbines in 2008.

In four more phases, total capacity will be brought to 300 MW by 2013.

Belgian utilities are required to generate or source a percentage of electricity from renewable sources.

BULGARIA

Capital: Sofia
Monetary unit: New lev
Refining capacity: 115,240 b/cd
Oil production: 1,000 b/d
Oil reserves: 15.000 million bbl
Gas reserves: 200.000 bcf

Drilling and Production

Kavarna, Kaliakra gas fields

In late 2009, the Bulgarian government issued a Certificate of Commerciality for the Kavarna gas field in the Black Sea, according to Melrose Resources PLC.

The company has applied for a similar certificate for the Kaliakra field, also in the Black Sea.

The Kavarna field, holding estimated reserves of 680 million cu m (24 bcf), is expected to be onstream by July 2010.

First gas from the Kaliakra field, which contains 1.64 bcm (57 bcf) of reserves, is expected by October 2010.

Pipelines/Transportation

Nabucco, South Stream deliberations

In late 2009, Bulgarian officials said that they would give priority to the Nabucco gas pipeline project (backed by the EU) over the rival South Stream gas pipeline project backed by Russia.

Belene Nuclear Plant

The Bulgarian government said in August 2009 that it would cancel construction of the new nuclear plant at Belene due to lack of funding.

Officials said they cannot afford the loans needed to take on the government's 51% stake in the 2,000-MW plant in northwest Bulgaria.

The plant was important in Bulgaria's plans to secure its future power generation, remain a leading energy exporter in the Balkans, and meet its EU targets for reducing carbon dioxide emissions.

Bulgaria, which currently gets virtually all of its gas from Russia, wants to diversify its energy sources. A dispute over gas delivery in January 2009 between Russia and Ukraine severely affected Bulgarian industry and households.

The Nabucco pipeline would bring gas from the Caspian Sea to Western Europe, across Azerbaijan, Georgia, Turkey and Bulgaria. In July 2009, the EU and Turkey signed an agreement to build the $11-billion pipeline.

The South Stream project would carry Russian gas beneath the Black Sea, then through Bulgaria, and either across Greece to Italy or northwest, across the Balkans toward Slovenia, Hungary and beyond.

The construction of both Nabucco and South Stream is still in the planning stage.

Bulgaria said it would still partner with Russia on other projects. One would carry Russian gas between the Bulgarian Black Sea port of Burgas to a terminal in Alexandroupolis in northeastern Greece.

Help from Greece

In January 2009, Greece pumped natural gas to Bulgaria during a European energy crisis caused by the halt of natural gas supplies from Russia.

Details are given in the Greece section of this chapter.

Spur to tap ITGI gas pipeline

Greek, Bulgarian and Italian companies agreed in July 2009 to build a 160-km (99-mi) pipeline that would help to end Bulgaria's reliance on Russian natural gas. A completion date of 2012 was set, but observers believe 2015 may be more likely.

The $167-million Greece-Bulgaria Interconnector (GBI) would provide Bulgaria access to the Italy-Turkey-Greece Interconnector (ITGI) pipeline carrying gas from Azerbaijan to Turkey, Greece and Italy.

The Turkey-to-Greece section of the ITGI was inaugurated in 2007, and completion of the Greece-to-Italy segment is planned for 2012.

The ITGI presents a challenge to Nabucco, the proposed pipeline from Turkey to Austria, which will be competing for the same Azeri gas.

The GBI would connect to the ITGI in Greece and would have a capacity of 3–5 bcm/year (106–177 bcf/year).

CROATIA

Capital: Zagreb
Monetary unit: Kuna
Refining capacity: 250,317 b/cd
Oil production: 15,700 b/d
Oil reserves: 73.350 million bbl
Gas reserves: 1.080 tcf

Adria LNG Project

In late 2009, German utility company RWE AG withdrew from the Adria LNG consortium. That project aims to build an LNG regasification terminal at the island of Krk in Croatia.

Saying it chose to focus on expanding its presence in the LNG business of northern Europe, RWE offered its shares to the other Adria consortium members.

Those others are E.ON AG, OMV AG, Total SA, and natural gas company Geoplin d.o.o. Ljubljana.

Adria LNG said that RWE's withdrawal would not affect pursuit of the project. The final investment decision is expected in 2010. Project costs are estimated at $1.2 billion, with construction scheduled to start in 2010 and first operation in 2014.

Capacity initially will be 10 bcm/year, to be ramped up to 15 bcm/year in a second stage.

Hungary-Croatia gas pipeline link

In early 2009, the European Commission proposed that a planned pipeline linking Hungary's and Croatia's gas networks should receive additional subsidies, according to Hungarian officials.

Details are given in the Hungary section of this chapter.

New gas sources

In early 2009, government officials of Croatia and Bosnia were discussing possible new sources of natural gas, in a bid to counterbalance overdependence on Russian supplies.

The talks were said to have focused on possible importation of gas from some Arab countries.

Nuclear plant

In April 2009, officials from Albania and Croatia said that they plan to sign an agreement for the joint development of a nuclear power plant near Skadar, Albania.

Details are given in the Albania section of this chapter.

CZECH REPUBLIC

Capital: Prague
Monetary unit: Koruna
Refining capacity: 183,000 b/cd
Oil production: 4,500 b/d
Oil reserves: 15.000 million bbl
Gas reserves: 140.000 bcf

CEZ: Shift to gas-fired power

Czech energy group CEZ AS, responsible for 75% of the Czech Republic's power output, underscored in early 2009 its plans to move toward wider use of gas-fired power generation.

The group announced that it will build an 800-MW plant in Melnik, about 24 km (15 mi) north of Prague.

The plant will replace three CEZ-owned coal-fired units 2015, helping the group meets its emission targets. CEZ has relied on coal for 50% of its power generation and on nuclear energy for about half of the rest.

The Melnik plant will be the group's gas-fired power plant in the Czech Republic.

Oil supply from Russia

In late December, Russia warned that it was prepared to cut off oil supplies to Europe through the Druzhba pipeline over a dispute with Ukraine over transit fees.

The Czech Republic as well as Slovakia and Hungary—the three countries that would be most affected by such action—each said that they had enough oil to meet at least three months of domestic demand.

Russia's threats to European energy supplies have become an annual affair, with halts to oil shipments in 2007 and 2008 having little lasting impact on the crude futures market.

DENMARK

Capital: Copenhagen
Monetary unit: Krone
Refining capacity: 174,400 b/cd
Oil production: 265,000 b/d
Oil reserves: 1.060 billion bbl
Gas reserves: 2.165 tcf

Overview

New energy policy

In 2009, Denmark was in the second year of implementing its 2008–2011 energy policy.

The policy includes several initiatives to reduce the country's dependence on fossil fuels and enhance its ability to meet its obligations to the integrated climate and energy proposal presented by the European Commission in January 2008.

The EC set for Denmark the target of generating 30% of its energy from renewable sources by 2020.

Denmark's plans include development of two offshore wind farms by 2012 and offering more generous subsidies for development of energy from wind, biogas and biomass.

UN climate conference

The largest UN climate change conference in history was held December 7–18, 2009, in Copenhagen, with delegates from 192 nations in attendance.

It was the 15th session of the Conference of the Parties (COP) to the United Nations Framework Convention on Climate Change (UNFCCC).

Details are in the Climate Change Issues section in the Key Statistics chapter of this book.

Exploration and Development

Gita: Major discovery

In April 2009, Maersk Olie og Gas A/S announced that its recent Gita gas discovery in the Danish North Sea could hold resources of 250–600 MMboe, based on a new estimate from partner Noreco.

The well encountered gas in Mid-Jurassic sandstones in a hydrocarbon-prone area of the Central Graben in the Danish sector. The vertical well was drilled to TD 5,162 m (16,936 ft) below sea level.

The Gita prospect straddles the licenses 9/95 Maja and 9/06 Gita.

Slick mapping

In the second half of 2009, Infoterra conducted an oil-slick mapping and interpretation project over Baffin bay for Nunaoil, the National Oil Company of Greenland.

The firm acquired about 480 radar satellite images over the 240,000-sq-km (92,665-sq-mi) area.

> **GREENLAND ACTION**
>
> Greenland, an autonomous country within the Kingdom of Denmark, is viewed as one of the most promising new frontiers in the hunt for oil.
>
> Cairn Energy reported in 2009 that it had raised £196 million ($310 million) to fund exploration efforts in Greenland through a deal with Petronas of Malaysia. The two firms will evaluate the six exploration blocks off the coast of Greenland where Cairn is the operator.
>
> Cairn's Greenland licenses cover an area equivalent to 250 North Sea blocks, on which only six wells have been drilled to date. The company expects to start a drilling program in 2011.

Alternative Energy

Wind energy outlook

Capitalizing on abundant wind energy is a key element of Denmark's energy planning.

Role of wind power

In 2007, just under 20% of Denmark's electricity came from wind power. The country's Energi21 national energy plan aims to boost that figure to 40% by 2030.

In 2009, more than 5,000 wind turbines were running, with total generating capacity of 3,200 MW.

As of May 2009, eight offshore wind farms (with a combined capacity of 424 MW) were operating, and three more being built or in an advanced planning stage.

DONG Energy, operator of both offshore Horns Rev facilities (see below), said it plans to inaugurate five more wind farms by mid-2010 and to triple its energy production from renewable sources by 2020.

Two Danish wind-energy industry groups devised an ambitious plan in 2009 that they say could produce as much as 50% of the country's electricity from wind systems by 2020.

They envision 3,800 MW from onshore and 2,550 MW from offshore turbines.

Offshore wind systems

The three largest offshore wind farms in Denmark in September 2009 were the 209-MW Horns Rev II and 160-MW Horns Rev I projects on the west coast of Jutland, and the 158-MW Nysted unit near Lolland.

Horns Rev II generated first power in September. Full operation is expected in early 2010.

Two more offshore facilities were in the active planning stage:

- The 200-MW Roedsand II system being built by E.ON Sverige AB, for start-up in 2011
- A 400-MW facility off Anholt Island, for which construction bids were invited in mid-2009.

Second-generation bioethanol plant

In late 2009, a second-generation bioethanol plants opened in Kalundborg. With the capacity to produce up to 5.4 million l./year (1.5 MMgal/year) of cellulosic ethanol, it is the largest such plant in Europe.

The plant was built by a subsidiary of DONG Energy and uses about 30,000 tpy of straw as feedstock.

The first year's output has already been bought by Statoil, which will mix the ethanol into the Danish fuel supply in 2010.

Environmental Issues

Copenhagen climate plan

Copenhagen aims to reduce its CO_2 emissions to zero and become the first carbon-neutral capital city by 2025, according to a new climate plan introduced in early 2009.

An interim goal is to cut CO_2 emissions by 20% by 2015.

Initiatives to reach these targets include:

- development of a new, larger-capacity geothermal power station
- use of hydrogen-powered and electric cars
- energy-efficient refurbishing of educational institutions
- increased use of wind turbines
- use of biomass in power stations.

NORTHERN EUROPE
NORWAY
DENMARK
SCOTLAND
IRELAND
N. IRELAND
WALES
ENGLAND
NETHERLANDS
BELGIUM
LUX.
GERMANY
SWITZERLAND
FAROE I.
SHETLAND I.
ORKNEY I.
Norwegian Sea
North Sea
Irish Sea
Celtic Sea
Donegal Bay
Galway Bay
Firth of Forth
Firth of Clyde
Moray Firth
Troll
Midgard
Åsgard
Tyrihans
Draugen
Ormen Lange
Njord
Agat
Åsgard Transport
Heimdal
Boa
Gekko
Kneler
Corrib
Morecambe
Formby
Galley Head
Seven Heads
Helvick
Ardmore
Kinsale Head
Waterford
Weymouth
Mongstad
Kollsnes
Sotra
Bergen
Oslo
Kopervik
Karsto
Stavanger
Sola
Slagen
Vall oy Tonsberg
Arendal
Mandal
Goteborg
Skagen
Tjeldbergodden
Trondheim
Nyhamna
Arhus
Fredericia
Kaergard
Kalundborg
Esbjerg
Skaelskor
Heide
Kiel
Cuxhaven
Brunsbuttel
Schwerin
Hamburg
Harburg
Wilhelmshaven
Emden
Groningen
Salzbergen
Meerdf
Roxforrde
Amsterdam
Rotterdam
Gelsenkirchen
Dortmund
Duisburg
Essen
Dusseldorf
Cologne
Rheinland
Bonn
Eisenach
Suhl
Frankfurt-am-Main
Antwerp
Brugge
Brussels
Liege
Verviers
Lille
Dunkerque
Calais
Maubeuge
Grandpuits
Luxembourg
Metz
Karlsruhe
Vohburg
Ingolstadt
Baden-Wurttemberg
Munich
Reichstett
Vendenheim
Zurich
Sullom Voe
Lerwick
Kirkwall
Flotta
Wick
Stornoway
Nigg Bay
Inverness
Mallaig
St. Fergus
Cruden Bay
Aberdeen
Dundee
Finnart
Ardyne Point
Grangemouth
Edinburgh
Glasgow
Ardrossan
Moffat
Londonderry
Belfast
Sligo
Ballina
Galway
Balloughо
Dublin
Arklow
Foynes
Limerick
Tralee
Wexford
Waterford
Cork
Whitegate
Bantry
Carlisle
Newcastle
Port Clarence
Darlington
Teesside
Middlesbrough
Scarborough
Barrow
Heysham
Lancaster
Bradford
Leeds
Hull
Eastham
Manchester
Liverpool
Grimsby
Ellesmere Port
Lincoln
Theddlethorpe
Nottingham
Bacton
Aberystwyth
Birmingham
Leicester
Coventry
Northampton
Cambridge
Milford Haven
Pembroke
Llandarcy
Gloucester
Neath
Oxford
Ipswich
London
Bath
Reading
Fawley
Southampton
Portsmouth
Bournemouth
Exeter
Plymouth
Notre Dame de Gravenchon
Cherbourg
Le Havre
Port Jerome
Gonfreville L'Orcher
Rouen
Gargenville
Petit Couronne
Reims
Paris
Plestin Les Greves
Brest
St. Malo
Quimper
Rennes
Laval
Lorient
Vannes
Donges
Nantes
Orleans
Blois
Vert-La-Gravelle
Couldmmes-Vaucourtois
Velye-Germinon
Bussey-Lettree
Chaunoy
Villeperdue
Soudron
Dommartin-Lettre
Courdemagnes
Marolles
Valence en Brie
Donnemarie
Grandville
Trois Fontaines
Forcelles
Villemar
Aufferville
Flacy
Gisy-les-Nobles
Joigny
St. Martin-de-Bossenay
Bourbonne
Chateaurenard
2 Lines
8 lines
0 50 100 150 Miles
0 100 200 Kilometers
15°
10°
5°
0°
60°
55°
50°
203
116
70
106
83
268
78
107
10
89
244
191
81
373
404
305
357
115
21
137
227
327
101
310
85
262
110
206
12
100
221
221
272
27
71
104
210
163
330
141
223
338
154
229

FINLAND

Capital: Helsinki
Monetary unit: Euro
Refining capacity: 258,500 b/cd
Oil production: 0
Oil reserves: 0
Gas reserves: 0

Nord Stream Pipeline permits

In November 2009, Finland and Sweden granted permits that will allow the Nord Stream Pipeline to be routed through the Exclusive Economic Zones (EEZ) of the two countries.

The three other needed EEZ permits were granted by Denmark in October, and by Germany and Russia in December.

The Baltic Sea pipeline will carry natural gas 1,223 km (760 mi) from Russia to the European Union.

Construction of the two parallel lines that comprise Nord Stream is planned to start in 2010, with the first pipeline operational in 2011 and the second in 2012.

Gazprom holds 51% of the joint venture. German energy companies E.ON Ruhrgas AG and Wintershall AG each hold 20%, and Dutch company Nederlandse Gasunie NV holds the remaining 9%.

Second renewable-diesel plant

Neste Oil Corp. started up its second plant to produce renewable diesel in July 2009 at the company's main refinery at Porvoo. The new plant can produce 170,000 tpy of a fuel called NExBTL—chemically similar to fossil diesel—using bio-based inputs.

Neste's first 170,000-tpy NExBTL plant was commissioned in Porvoo in 2007.

It is also building plants in Singapore and Rotterdam, each capable of producing 800,000 tpy, that are to be completed in 2010 and 2011, respectively.

Wind power plans

As of May 2009, Finland had 119 wind turbines in operation, generating 144 MW of power. The country aims to generate 2,000 MW of wind energy by 2020.

However, reports in mid-2009 from the Finnish Wind Power Association indicate that some 1,700 wind farm projects—combined capacity nearly 7,000 MW—are already in the process of seeking permits. They include both onshore and offshore systems.

It is estimated that if all the proposed and planned projects are realized, wind power could surpass hydroelectric power, which accounts for almost 19% of the total electricity generated in Finland.

CCS: Meri-Pori power plant

An ambitious carbon capture and storage (CCS) project that began in early 2008 in Finland was expanded in 2009 to integrate transportation and carbon-dioxide-use options.

The two Finnish companies that launched the project—Fortum Oyj and Teollisuuden Voima Oyj—were joined in early 2009 by Copenhagen's A.P. Møller-Mærsk A/S.

The heart of the project is the extraction of an estimated 1.2 mtpy of carbon dioxide from flue gases at the 565-MW coal-fueled Meri-Pori power station.

The CO_2 will be shipped in liquid form to the North Sea coast of Denmark in special tankers from Maersk. There, it will be injected into depleted oil and gas fields to enhance recovery.

Funding is expected from the CCS demonstration program of the EU, with details to be finalized in 2012. The start of operations in 2015 is anticipated.

FRANCE

Capital: Paris
Monetary unit: Euro
Refining capacity: 1.984 million b/cd
Oil production: 18,300 b/d
Oil reserves: 101.150 million bbl
Gas reserves: 250.000 bcf

Exploration and Development

Shale gas interest

Several companies were seeking permits in mid-2009 for shale gas prospects in southeast France in Languedoc Roussillon, the Cevennes mountains region, and the Savoie area near the Swiss border.

They included Total SA, Mouvoil SA, Bridge Oil Ltd., and Diamoco Energy.

Permits of interest are the Ales, Bassin d'Ales, Plaine d'Ales, Montelimar, Cevennes, and Navacelles.

French authorities planned to assess the competing proposals for overlapping shale acreage onshore north of the Gulf of Lion and make an award in October 2009.

European Shale-Gas Database

The GFZ German Research Centre for Geosciences designed and launched in mid-2009 what it describes as the "first and biggest and most comprehensive" study on shale gas in Europe.

It couples interdisciplinary reservoir and regional-scale research with compilation of an international data base on black shale in 20 European nations. Participants in the database effort include Geoforschungszentrum GFZ, IFP, and TNO.

The project started in May 2009 and will initially run to mid-2012, with research focusing on gas-in-place and producibility.

Study of the processes contributing to shale gas formation will include evaluation of Alum shale (Sweden), Posidonia shale (Germany), Lower Carboniferous (Germany and The Netherlands), and Barnett shale and Marcellus shale (both in the United States).

Lease Sales

Mozambique Channel

In early 2009, the French government awarded two 5-year exploration permits to mid-sized oil companies to explore offshore Juan de Nova.

The tiny island, a French possession, lies between Madagascar and Mozambique.

Details are given in the Madagascar section of the Africa chapter of this book.

International Activities

Accords with Kazakhstan

In late 2009, Kazakhstan and France signed a number of agreements involving oil and gas.

Under one agreement, French companies will participate in siting a major export pipeline between the offshore Kashagan oil field and Baku. The oil then will flow from Baku to Europe.

Details are given in the Kazakhstan section of the Former Soviet Union chapter of this book.

Total-Novatek deal

In mid-2009, Russia's Prime Minister Vladimir Putin approved an exploration JV deal between Total and Novatek, Russia's largest independent gas producer.

The JV will develop the Termokarstovoye gas field in the Yamal-Nenets region of the Siberian Arctic.

Total will take a 49% stake in the Novatek subsidiary that holds a license at the field.

The unexplored 47.3-bcm (1.67-tcf) field has been called a "field of national importance" to Russia, in which foreign investors require special government permission to operate.

Total already holds a 25% stake in the initial development phase of Shtokman, the gas exploration project of Gazprom. Since 2006, Total has been one of the few foreign majors to flourish in Russia.

Total-Santos deal

In mid-2009, GDF Suez bought 60% of the interests of Santos Ltd. in the Petrel, Tern, and Frigate gas fields in Australia's Bonaparte Gulf.

The deal is the first entry of GDF Suez into Australian petroleum exploration and production.

The offshore fields lie 250–300 km (155–186 mi) west of Darwin.

The JV (Bonaparte LNG) will develop the fields using a floating LNG facility capable of producing 2 mtpy of LNG, and will also market the gas.

GDF Suez said all LNG production will be shipped to the Asia-Pacific region.

Contingent reserves in the fields total 220 million boe, of which 42.48 bcm (1.5 tcf) is estimated to be dry gas.

Total-Chesapeake deal

Total SA moved in early 2009 to purchase an interest in the U.S. shale-gas assets of Chesapeake Energy Corp.

Details are given in the United States section of the North America chapter of this book.

Stakes in Kazakh oil field

In late 2009, Total and GDF Suez entered a partnership with Kazakhstan's state-owned KazMunaigas (KMG) to develop a large offshore oil and gas field. Total will acquire a 17% stake in the Khvalynskoye gas and condensates deposit from KMG and GDF Suez will take an 8% interest.

The two French firms expect to invest about $1billion in the field, where production of up to 9 bcm/year (317.8 bcf/year) of gas will begin in 2016.

Khvalynskoye straddles the offshore border between Kazakhstan and Russia. Total said that all gas production from Khvalynskoye would be supplied to Russia.

Toreador subsidiaries

Toreador Resources Corp. entered into two definitive agreements in late 2009 to sell separately its Turkish subsidiary, Toreador Turkey Ltd., and its Hungarian subsidiary, Toreador Hungary Ltd.

The company's Turkish unit will be purchased by private Norwegian energy company Tiway Oil.

Toreador's Hungarian unit will be acquired by private Austrian company Rohol-Aufsuchungs AG.

Toreador expected both deals to close by mid-October, with total combined cash proceeds of $16.4 million.

LNG

Fos Faster Terminal

In mid-2009, Koninklijke Vopak NV established a JV with Royal Dutch Shell PLC to set up an LNG terminal at Fos-sur-Mer, near Marseille.

The project, known as the Fos Faster LNG Terminal SAS, will be part of the overall capacity expansion plans of the Grand Port Maritime de Marseille. The port complex is the largest in France and one of the largest in Europe.

Terminal capacity will be about 8 bcm/year (282.5 cm/year). A tentative start-up date for the facility is 2015.

GDF Suez: New subsidiaries

In early 2009, GDF Suez began setting up two new wholly owned subsidiaries.

Elengy will manage the firm's LNG tanker operations as well as three methane terminals in France: Fos Tonkin; Montoir-de-Bretagne; and Fos-Cavaou (to be commissioned in early 2010).

GDF Suez plans to raise the unloading capacity at Montoir-de-Bretagne from the current 10 bcm/year (353.1 bcf/year) to 16.5 bcm/year (582.6 bcf/year).

The other subsidiary, Storengy, will oversee operation of the firm's facilities for underground storage of natural gas. GDF Suez operates 12 such sites in France and has a presence in Canada, Germany, and the UK.

Pipelines/Transportation

Galsi extension?

In mid-2009, French officials were talking with Algeria about a possible extension to the planned 840-km (522-mi), 8-bcm/year (282.5 bcf/year) Galsi (Gasdotto Algeria Sardegna Italia) pipeline.

The current planned pipeline route runs under the Mediterranean Sea from Algeria to Sardinia, and then on to Italy. The extension would bring gas to the French island of Corsica, just north of Sardinia.

Pending a final investment decision slated for mid-2010, first delivery of Algerian gas into Italy could start in 2013.

Alternative Energy

Minervino, Grighine wind farms

EDF Energies Nouvelles SA, part of Electricité de France, commissioned a new wind farm in Italy in mid-2009.

The facility at Minervino comprises 16 turbines supplied by REpower Systems A.G.

As of mid-year, a subsidiary of EDF Energies Nouvelles had completed five wind projects in Italy, with combined gross wind energy capacity of 266 MW.

EDF Energies Nouvelles said that Italy's largest planned wind project, the 98.9-MW Monte Grighine wind farm in the western part of Sardinia, was slated for commissioning by yearend 2009.

Small-scale photovoltaic initiative

In mid-2009, the Commission de Regulation de l'Energie announced the start of a bidding process to build a small solar photovoltaic power plant in every region of France by 2011.

The aim is to increase the installed capacity of renewable energy generation capacity by 300 MW.

The 300 MW of new capacity would come from 27 units designed to produce 5–10 MW each.

The tender also includes regions outside the metropolitan area, such as Corsica, Guayane, Guadaloupe, Martinique, Mayot, and Reunion.

Until 2008, about 70 MW of solar photovoltaic power was connected to the grid, chiefly in the Languedoc-Roussillon, Rhone-Alpes, and Pays de la Loire regions.

Between 2006 and 2008, projects for about 1,600 MW were authorized by the ministry.

One firm announced plans in mid-2009 to build a 104-MW solar power plant in the village of Barben.

Nuclear support

France obtains most of its power from 58 nuclear reactors at 19 nuclear power plants.

In late 2009, the French government pledged €1billion ($1.4 billion) to fund next-generation nuclear development. The support is part of a €35 billion ($50 billion) stimulus package designed to rejuvenate the French production market.

The government is also actively seeking foreign energy companies to invest in new French power plants.

The government has majority stakes in leading utility company Electricité de France, which runs most of the nuclear plants, and in Areva SA, which manufactures nuclear reactors.

The government also announced that the country's Atomic Energy Commission will now be known as the Commission of Atomic Energy and Alternative Energy, to reflect future research into renewable energy sources.

Environmental Issues

Carbon permits market: caution

In the run-up to the late-2009 UN climate-change conference in Copenhagen, the French finance ministry expressed concern about unregulated "dark pools" in the carbon trading market and called for creation of a single European regulator to prevent future abuses.

One official said carbon permits fall somewhere between commodities and financial instruments, creating a regulatory challenge. Following establishment in 2005 of the European Union Emission Trading Scheme, tax fraud was uncovered in the UK, Denmark, and France.

One analyst of the evolving carbon market said that the world market for greenhouse gas credits could rise to as much as $3 trillion by 2020 if the United States joins the trading system.

Carbon tax on imports?

In September 2009, French Pres. Nicolas Sarkozy renewed his call for a European carbon tax on imports to the bloc.

He previously drew charges of protectionism after calling for European import tariffs on products from countries that do not abide by international targets on carbon emission reductions.

France will become the largest economy to levy a carbon tax of €14–16/tonne of CO_2, starting in 2010. Finland, Sweden, and Denmark have already introduced carbon taxes.

A rate of €14 ($20) per tonne translates into 3 cents/l. of gasoline.

The government has promised to cut greenhouse gas emissions by 75% from 1990 levels by 2050.

First CCS project

Total SA received administrative authorization in mid-2009 to start injecting CO_2 in France's first carbon capture and storage (CCS) pilot project.

Injection into the nearly depleted Rousse reservoir at Lacq gas field in southwestern France is planned to be conducted through mid-2011.

CO_2 will be captured from exhaust gases at a boiler at Lacq's steam generating plant and then compressed and sent through a pipeline for injection into the Rousse at a depth of 4,500 m (14,764 ft).

The €60-million pilot is being carried out in cooperation with Institut Francais du Petrole and Bureau de Recherche Geologique et Miniere (Mining and Geological Research Office).

E10 gasoline

In March 2009, French officials announced the introduction of "green" E10 unleaded gasoline at all service stations, incorporating 10% ethanol instead of the previous 5%.

The four major oil companies providing gasoline in France (Total, BP, Esso, and Shell) will all provide the new fuel blend, though some delay in nationwide introduction was anticipated.

The oil trade group UFIP estimated that 75% of all service stations would be selling it by yearend.

The E10 can be used only by automobiles sold after 2000.

The government said that introducing E10 would cut carbon dioxide emissions by 1 mtpy by 2010.

GERMANY

Capital: Berlin
Monetary unit: Euro
Refining capacity: 2.411 million b/cd
Oil production: 55,500 b/d
Oil reserves: 276.000 million bbl
Gas reserves: 6.200 tcf

Overview

Gas reserves

Germany has the third-largest natural-gas reserves in the EU, after The Netherlands and UK. Most of its 175.6 bcm (6.2 tcf) of reserves are in the northwestern part of the country between the Elbe and Weser rivers. Substantial deposits have also been found in the North Sea region.

However, environmental restrictions on exploration and production have made Germany a net importer of natural gas. In 2007, imports totaled 88.35 bcm (3.12 tcf).

The country is the world's third largest consumer of natural gas, after the United States and Russia.

Gas Processing

Crystal Gas Plant

Work progressed in late 2009 on the Crystal Gas Plant project in Etzel, Germany.

Electricité de France SA, in coordination with EnBW Energie Baden-Wurttemberg AG, awarded Technip an EPC management contract for the plant.

The project covers gas compression and treatment facilities to store gas in underground salt caverns until needed to meet periods of peak demand in Germany and The Netherlands.

The storage facility is expected to begin operations by 2011.

Pipelines/Transportation

Nord Stream permit

At yearend 2009, Germany issued approvals for its 31-km (19-mi) section of the Nord Stream pipeline that will pass through the country's Exclusive Economic Zone.

The line will run beneath the Baltic Sea to carry gas from Vyborg, Russia, to Greifswald, Germany, where it will enter the European gas network.

Details about approvals from four other nations are given in the Finland section of this chapter.

Largest gas pipeline system

Plans were under way in late 2009 to merge five existing natural gas pipeline networks to create Germany's largest pipeline system.

The combined network will comprise 14,800 km (9,196 mi) of pipelines and more than 400 pipeline grids. The network will potentially carry more than 50% of the high-heating-value natural gas (H-gas) transported in Germany.

The merging systems are those of:

- E.ON Gastransport GmbH
- Bayernets GmbH, the network company of Bayerngas GmbH
- GRTgaz Deutschland GmbH, a subsidiary of GDF Suez SA
- Eni Gas Transport Deutschland SpA
- GVS Netz GmBH, a JV of Energie Baden-Wurttemberg AG and Eni.

The combined operation will be managed by NetConnect Germany GmbH & Co. KG.

Storing Nord Stream gas

In early 2009, Gazprom announced plans to build two underground storage units in northeastern Germany to store natural gas from the proposed Nord Stream pipeline project.

Up to 15 bcm (530 bcf) of gas will be stored in bedrock about 700 m (2,279 ft) underground.

The facility, expected to be the largest gas storage site in Western Europe, is slated to be commissioned by 2012.

Other gas storage facilities have been proposed for Bernburg (510 million cu m; 18.0 bcf) and Gronau-Epe (200 million cu m; 7.06 bcf).

According to CEDIGAZ, Germany has more than 20.17 bcm (712.2 bcf) of natural gas storage capacity, which is the largest in the EU and fourth-largest worldwide.

Refining and Products

Rheinland refinery

Royal Dutch/Shell awarded an EPC management contract to Technip in late 2009 for the first phase of its "Connect" project in Germany.

By connecting two existing refineries in Godorf and Wesseling, Shell will create the largest refinery in Germany: the Rheinland refinery.

In the first phase, to be implemented at Wesseling, workers will modify selected process units as well as build new facilities. The units there also will be used to remove sulfur from gas oil produced in Godorf.

Wilhelmshaven: Delay of upgrade

In late 2009, ConocoPhillips confirmed that it will delay the upgrade of its 260,000-b/d Wilhelmshaven refinery, citing unfavorable market conditions.

The firm had planned to add a coker, hydrocracker, and hydrogen units.

Alternative Energy

Desertec solar-power consortium

Desertec Industrial Initiative (DII) GmbH—a sustainable-energy initiative of a consortium of 12 European companies—proposed plans in 2009 for a massive solar power generation project in the Sahara Desert.

The facility, built on 17,000 sq km (6,564 sq mi) of desert, would supply electricity to Europe starting in 2015 and meet 15% of the continent's electricity requirement by 2050. It would also send power to the Middle East and North Africa.

Project design is to be completed by 2012. Investment requirements are estimated at €400 billion (US$593.6 billion) through 2050.

The first stage of the project will install fields of concentrating solar collectors that generate steam in special-design boilers to drive turbines.

Solar power

In December 2009, Germany was poised to bring online what it called the world's biggest solar power project, located at Lieberose in the eastern state of Brandenburg.

The €160-million facility, whose 700,000 photovoltaic solar modules cover an area about the size of 210 football fields, will produce 53 MW.

Earlier, in August, a €33-million demonstration solar-thermal "power tower" went online near Cologne, capable of generating 1.5 MW.

Officials said that Germany's solar industry employs 280,000 people today and could employ 500,000 by 2019.

The European Photovoltaic Industry Association forecasts that the solar photovoltaic (PV) capacity in Germany could reach 17,000 MW by 2013.

Offshore wind farms

Officials in Hamburg granted approval in 2009 for construction of the $900-million Veja Mate 400-MW offshore wind farm in the German North Sea.

It would be about 90 km (56 mi) northwest of the island of Borkum, near the Germany-Netherlands border. Full operation is scheduled for September 2012.

Construction was already under way on the 400-MW BARD Offshore 1 wind farm in the same general area. It is to be completed by yearend 2010.

A total of 25 offshore wind farm projects in the North Sea and Baltic Sea (combined capacity 7,400 MW) have already been approved by the agency.

6-MW wind turbine

REpower Systems AG announced in early 2009 the launch of what it claims is the world's most powerful wind turbine. The firm said that the 6M Series turbine generates 6.15 MW.

The first test installation of three of the new turbines took place in March at the onshore Westre wind farm in Germany near the German-Danish border.

The 6M has a rotor diameter of 126 m (413 ft) with a hub height of approximately 100 m (328 ft).

Ocean-current technology

Voith AG and Siemens formed a JV in 2009 with RWE Innogy, a subsidiary of RWE AG, to develop, manufacture, and market ocean current technologies.

Over the next few years, the partners said they will invest more than $37 billion in the venture.

Voith subsidiary Voith Hydro developed an ocean current turbine that was to be deployed and tested by yearend 2009 in the first 110-kW prototype offshore South Korea.

RWE Innogy said it plans to build a 10.5-MW ocean current power station off the Welsh coast in 2012.

Nuclear power future

Discussions regarding the future of nuclear power began within the German government following the September 2009 federal election. A key issue was whether to reverse an earlier government decision to shut down Germany's nuclear power reactors by 2022.

Coming to power as part of the re-election of Chancellor Angela Merkel was a pro-nuclear coalition that was seen as favoring reversal. Observers speculated that the new government would try to claw back billions of euros of windfall profits from German utilities in exchange for a reversal decision.

A majority of Germans were said to favor a phase-out of nuclear power.

Environmental Issues

Draft CCS legislation

The German government passed a draft law in early 2009 that will allow major energy producers to forge ahead with developing carbon capture and storage (CCS) systems at coal-burning power plants.

The draft law includes strict safety rules and requires that companies sign insurance agreements or make significant security deposits.

They also will be responsible for operating the CCS facilities for 30 years after decommissioning power plants that use the facility before being able to transfer management to the federal government.

The government plans to review the viability of CCS in 2015.

Verein der Kohlenimporteure e.V., the German Coal Importer Federation, released a report in April suggesting that CO_2 emissions from Germany's coal-fired power stations could be cut by as much as 90 mtpy by 2050 through the use of CCS technology.

Carbon-capture projects

In August 2009, RWE AG and its partners launched Germany's first carbon-capture pilot plant at the coal-fired power plant in Niederaussem, Bergheim.

RWE is working with BASF SE and Linde AG. The pilot plant is expected to have a capture efficiency of 90%.

And in third quarter 2009, a pilot CO_2-capture plant near the E.ON AG coal-fired Staudinger power plant, near Hanau, started operation. The facility is a joint project of E.ON and Siemens AG.

The pilot uses a post-combustion process developed by Siemens that the partners said would remove more than 90% of the CO_2 from the plant's flue gas. The pilot project will run until 2010.

About 50% of Germany's electricity comes from coal-fired plants, and the country is the biggest emitter of CO_2 in Europe.

Legislation to speed up the development of CCS technology was postponed in early 2009 by the German government, but is expected to be introduced in early 2010.

GREECE

Capital: Athens
Monetary unit: Euro
Refining capacity: 423,000 b/cd
Oil production: 1,600 b/d
Oil reserves: 10.000 million bbl
Gas reserves: 35.000 bcf

Epsilon field

Aegean Energy SA and its 95%-owned subsidiary Kavala Oil SA were preparing in third quarter 2009 to drill what would be the deepest and longest well ever in offshore Epsilon field.

The drilling site was in the Gulf of Kavala about 8 km (5 mi) west of Thassos Island.

The well was expected to reach TVD of 5,500 m (18,044 ft) and vertical depth of 2,900 m (9,514 ft).

Aegean Energy said that its PNA-H3 horizontal well, drilled earlier in 2009 on Prinos North field, and the Epsilon well together are expected to increase production to 5,000 b/d from 1,300 b/d.

Aegean is implementing a 5-year, $200-million investment plan—described by the company as the biggest the sector has seen in Greece in more than 25 years. The firm expects to drill at least two more exploration and appraisal wells in first quarter 2010.

LNG expansion

State-owned gas supplier, DEPA, said in mid-2009 that the company was planning to develop LNG terminals as a medium-term strategy to avoid delays in securing long-term supplies of natural gas delivered by pipeline.

DEPA is already expanding the country's only LNG terminal at Revythousa, in the Gulf of Megara, to supply 5 bcm (176.6 bcf/year).

It plans to build a similar-sized facility near the north Aegean port of Kavalla, and a third terminal is under consideration for the port of Astakos in western Greece.

DEPA planned to sign a deal in June with RasGas of Qatar to purchase LNG on the spot market. It already has similar agreements to buy spot cargoes from Gaz de France, British Gas, and Eni.

Gas to Bulgaria

In January 2009, Greece piped natural gas to Bulgaria amid the European energy crisis caused by the halt of natural gas supplies from Russia.

In mid-month, state-run DEPA started sending gas under a deal to provide 2.5 million cu m/d (88.3 MMcf/d) for a week using an existing pipeline link.

Although Bulgaria is heavily reliant on Russian gas, Greece was been less affected by the cuts because it also imports gas from the Caspian Sea via Turkey and has stores of LNG.

South Stream participation

In May 2009, Greece joined Serbia and Bulgaria in signing on to Russia's planned South Stream project to ship up to 60 bcm/year (2.12 tcf/year) of gas by pipeline beneath the Black Sea to a Bulgarian port.

Greece-Bulgaria Interconnector

In mid-2009, Greek, Bulgarian and Italian companies signed an agreement to build a 160-km (99-mi) pipeline that would end Bulgaria's reliance on Russia for its natural gas.

Details are given in the Bulgaria section of this chapter.

HUNGARY

Capital: Budapest
Monetary unit: Forint
Refining capacity: 161,000 b/cd
Oil production: 13,700 b/d
Oil reserves: 26.566 million bbl
Gas reserves: 286.000 bcf

Exploration and Development

Mako Trough

Results were disappointing from late-2009 drilling by ExxonMobil Corp., MOL Rt., and Falcon Oil & Gas Ltd. in Hungary's Mako Trough.

Two fracs in the Foldeak-1 well to test the Szolnok formation yielded excessive levels of carbon dioxide and sulfur, with high water flow rates.

The partners were pondering additional tests at the well to better understand the resource.

Northeastern region

Several operations were under way in August 2009 in the northeastern part of Hungary.

A group led by Ascent Resources PLC spudded the PEN-105 well in the Peneszlek area, projected to 1,500 m (4,921 ft). The well was to test a structure that previously tested gas from Miocene volcaniclastic reservoirs in the PEN-12 well drilled in 1982.

In addition, Hungarian Horizon Energy Ltd. spudded the Gorbehaza-1 well in the Nyirseg North permit.

The well was projected to 1,300 m (4,265 ft) targeting Pannonian clastic and Miocene volcaniclastic reservoirs.

Also, gas production began in August 2009 from the 2008 Hajdunanas discovery in the Pannonian basin, in a project led by Hungarian Horizon Energy Ltd., a subsidiary of Aspect Energy LLC.

Evaluation estimated a preliminary 340 million cu m (12 bcf) of gas in place in Pannonian sands.

The discovery is on the Hernad I and II licenses.

Lovaszi, Ujfalu fields

The Hungarian subsidiary of Ascent Resources PLC and Hungary's MOL RT were planning in mid-2009 to explore 88 sq km (34 sq mi) in southwestern Hungary around Lovaszi and Ujfalu oil and gas fields.

Ascent was to shoot 3D seismic and fund the drilling of two wells.

Ascent will also shoot 3D seismic over 90 sq km (35 sq mi) in the adjoining part of Slovenia that includes the Petisovci redevelopment project. The area is highly prospective for tight gas.

Pipelines/Transportation

South Stream gas pipeline

The Hungarian Development Bank and Russia's Gazprom entered into a JV in early 2009 to build the portion of the South Stream gas pipeline that will run across Hungary.

The pipeline is scheduled to start operation by yearend 2015, transporting gas from Russia and Central Asia to Austria and Italy.

The $12.7-billion, 31-bcm/year (1.09 tcf/year) pipeline will begin in Dzhubga on the Russian Black Sea coast and will run about 900 km (560 mi) underwater to Varna in Bulgaria.

From there, the northern route of the pipeline will take it through Serbia and Hungary to Austria, while a southern leg will be routed through Greece to Italy.

The planned Nabucco pipeline—to deliver gas through a 3,300-km (2,050-mi) pipeline from the shores of the Caspian Sea through Turkey, Bulgaria, Romania, and Hungary to Austria and Germany—is viewed as a competitor to the South Stream pipeline.

Nabucco conference

In late January 2009, high-ranking officials from the countries in the Nabucco gas pipeline consortium—Austria, Bulgaria, Germany, Hungary, Romania, and Turkey—as well as potential supplier nations, transit countries, and other interested parties, met in Hungary to discuss the path forward.

Observers noted some progress but concluded that the conference failed to deliver assurance that the Nabucco line will be built. The project has been under discussion since 2002.

Hungary-Croatia gas link

In early 2009, the European Commission proposed that a planned pipeline linking the gas networks of Hungary and Croatia should get additional subsidies, according to Hungarian officials.

The EC earmarked a €20 million contribution (about 10% of total cost) for the project in an updated proposal.

The updated project list also includes €25 million for a gas interconnector between Hungary and Slovakia, and €30 million for a gas interconnector between Hungary and Romania.

Gas storage facility

In mid-2009, the European Bank for Reconstruction and Development (EBRD) agreed to provide a $277-million loan to Hungary's MOL Magyar Olaj-es Gazipari NyRt for development of a natural gas storage facility.

The project will convert the existing Szoreg-1 reservoir in the Algyo oilfield in the south into an underground gas storage facility in an effort to secure energy supplies.

Design capacity is 1.2 bcm (42.4 bcf) for strategic reserves and 700 million cu m (24.7 bcf) for commercial use. Construction was well under way, and commissioning is slated for April 2010.

Total cost of the project is estimated to be about $848 million.

Company News

Surgutneftegaz-MOL-OMV deal

In March 2009, Russia's Surgutneftegaz oil production company took a 21.2% stake in MOL, from Austria's OMV for €1.4 billion. OMV had been trying for two years to take over MOL.

The transaction raised concern among Hungarian leaders, but observers said that the acquisition was not likely to change the level of EU dependence on Russian oil and gas.

REPUBLIC OF IRELAND

Capital: Dublin
Monetary unit: Euro
Refining capacity: 71,000 b/cd
Oil production: 0
Oil reserves: 0
Gas reserves: 350.000 bcf

Exploration and Development

Offshore west coast

In mid-2009, Serica Energy PLC in partnership with Germany's RWE announced it had struck oil off Ireland's Atlantic west coast. It was described as the first discovery in Irish waters in 28 years.

The discovery was made on the Bandon exploration prospect in PEL 01/06 in the Slyne basin, off County Mayo, 45 km (28 mi) southwest of undeveloped Corrib gas field.

The Corrib field, 85 km (53 mi) off northwest Ireland, holds an estimated with 22.7 bcm (800 bcf) of recoverable gas. Shell and partners are investing €900 million (£776 million) to develop the field, whose reserves might be sufficient to meet up to 60% of Ireland's gas needs.

Spanish Point

Bergen Oilfield Services planned to start work in mid-2009 on a 3D seismic survey in the Porcupine basin off western Ireland. Providence Resources commissioned the survey over the Spanish Point discovery.

Celtic Sea reserves

In mid-2009, Island Oil & Gas upgraded its reserves estimates for the Old Head of Kinsale and Schull gas fields in the Celtic Sea.

Island estimated proven in-place gas at Old Head and Schull at 2.83 bcm (101 bcf), more than double the 1.36 bcm (48 bcf) estimated in 1997.

Natural gas

LNG project

Shannon LNG Ltd. has been working since 2006 to build an LNG receiving and re-gasification terminal on a site between Tarbert and Ballylongford.

The firm said that construction was expected to start in late 2009, despite some speculation from a U.S. source about possible delay. Planners said the facility could be ready to accept its first shipments of gas by 2012.

Corrib gas pipeline

Ireland's planning authority, An Bord Pleanala, told Royal Dutch Shell PLC in November 2009 that it did not accept plans for the proposed Corrib high-pressure gas pipeline.

The authority said that parts of the pipeline route were unacceptable because of its proximity to houses, its impact on local land value, and insufficient plans to upgrade the road network in the area.

Partners have until February 2010 to provide more information on the safety of the pipeline and its effect on the peat bogs along the route.

Local opponents, who have been protesting the gas line for a decade, urged a complete rethinking of the safety and economics of the project.

Alternative Energy

Wave energy

Sweden's Vattenfall AB created a JV (Tonn Energy) with Irish firm Wavebob to develop technology to extract power from wave energy. The multi-million dollar venture also will get funds from Ireland's Investment and Development Agency.

Work is under way at a site off the coast of County Mayo on the west coast.

Ireland has said that it wants to send 500 MW of ocean-derived energy into the national grid by 2020.

Pre-commercial trials of a full-scale energy converter are planned by 2013, to be followed by commercial electricity generation by 2015.

Wind energy

The Irish Wind Energy Association estimated in mid-2009 that the wind-energy sector will support more than 10,760 direct and indirect jobs and that about one-third of the €14.75 billion invested in the sector would be retained in the local Irish economy.

Ireland aims to get 40% of its electrical power from renewable sources by 2020, compared to the 20% target set by the EU for its members.

Ireland had about 1,320 MW of wind-power capacity in place in early 2009.

Renewable-energy jobs

Ireland's Electricity Supply Board unveiled in early 2009 a $1.3-billion/year plan to create 3,700 "green" jobs through renewable power initiatives and to sustain 1,300 existing jobs.

The ESB previously announced a $14-billion fund for investing in a range of renewable energy projects in other European countries.

ITALY

Capital: Rome
Monetary unit: Euro
Refining capacity: 2.337 million b/cd
Oil production: 80,000 b/d
Oil reserves: 423.678 million bbl
Gas reserves: 2.466 tcf

Overview

Italy is Europe's third largest producer of oil and gas. Most new activity in the past decade has focused onshore at fields such as the prolific Val d'Agri in the south.

Offshore, the dominance of Eni SpA has been virtually unchallenged since Royal Dutch Shell (The Hague, TheNetherlands) and BG Group (London) exited their licenses earlier this decade.

Eni's offshore oil production comes mainly from the Vega field off southern Sicily, and Aquila off the coast of Puglia. It also produces gas from numerous fields in the northern Adriatic, although expansion has been constrained for environmental reasons.

In recent years, it has made further small gas discoveries such as Argo offshore Sicily, and Benedetta in the Adriatic.

However, there are numerous other "stranded" fields in Italian waters that the government is eager to monetize.

Exploration and Development

Elsa discovery

Vega Oil and partner Petroceltic were laying plans in late 2009 to drill an appraisal well in 2010 on the Elsa discovery in the central Adriatic Sea.

An independent reserve audit of Elsa, in permit BR268RG, suggested best-estimate contingent recoverable resources of 104 MMbbl.

Sambuca prospect

In mid-2009, AuDAX Resources Ltd. invited bids for drilling in first half 2010 on the Sambuca prospect in the Sicily Channel off Tunisia, in the Pelagian basin.

Sambuca is in the contiguous G.R15.PU exploration permit northwest of Pantelleria Island off Italy and the Kerkouane permit off Tunisia.

It appears to be one of the largest undrilled structures in the Mediterranean Sea with a mean unrisked potential of 270 MMboe recoverable.

EASTERN EUROPE
Copenhagen
Malmo
BORNHOLM
RUGEN
Reinkenhagen
Kolberg
Rostock
GERMANY
Stettin
Schwedt
Berlin
Vestrup
Rudersdorf
Roxforde
Magdeburg
Staakow
Guben
Cottbus
Frankfut-an-der-Oder
Grunberg
Leipzig
Dresden
Jena
Zeitz
Litvinov
Kralupy
Pardubice
Kolin
Prague
Waidhaus
CZECH REP.
Brno
Trebio
Znojmo
Pribor
Zukow
Stawno
Vacenovice
Breclav
Hodonin
Ianzhot
Stefanov
Malacky
Wysok
Linz
Vienna
Bratislava
Munich
Salzburg
AUSTRIA
Graz
Gdansk
Kaliningrad
Krasnoborskoye
Ushakovskoye
Kaunas
Vilnius
Grodno
Plock
Warsaw
Poznan
Lodz
Radom
Pulawy
POLAND
Kielce
Sandomierce
Rozwadow
Trzebina
Grobla
Partinia
Lubaczow
Jaroslaw
Rudki
Jaslo
Czechowice
Gorlice
Smikov
Strazske
SLOVAKIA
Uzhgorod
Miskolo
Salushsk
3 Lines
2 Lines
Baranovich
BELARUS
Davydovskoye
Minsk
Vishanskoye
Mozyr
Pinsk
Kobrin
Brest
Ovruch
Kovel
Lutsk
Rovno
Dubno
Brody
Lvov
Ternopol
Dashava
Kradobnensk-Kalusz-
Dolina
Stryj
Starynj
Bitkov
Kosov
Nadvornaya
Kosmachskoye
Chernortsy
Krasnoilsk
Unecha
Kursk
Kharkov
Lisichansk
Kiev
UKRAINE
Berdichev
Staro-Konstantinov
Vinnitsa
Kremenchug
MOLDOVA
Kishinev
Ugeny
Iasi
Odessa
Kherson
HUNGARY
Budapest
Nedudvar
Hajduszoboszlo
Biharnagybajom
Szandaszolos
Pusztafoldvar
Battonya
Szeged
Algyo
Ulles
Nagylanizsa
Lovaszi
Budafap
Nagylengyel
Legrad
ROMANIA
Sarmasel
Teleac
Zaul-de-Cimpie
Bogota
Tauni
Miercurea
Cucacesti
Saros
Singeorgiu
Noul-Sasesc
Bazna
Zemes
Bacau
Stanesti
Victorovca
Belmaolia
Dranbanesti
Frumusita
Valeni
Izmail
Valenii de Munte
Boldesti
Moreni
Ploiesti
Cimpina
Pitesti
Bucharest
Gura-Ocnitei
Bilteni
Arad
Timisoara
Resita
Ticleni
Turnu-Severin
Leonardi
Budisterni
Giurgiu
Gigen
Frasinu-Mislea
Delta
Midia
Lebada
Navodari
Constanta
Black Sea
Tulcea
Maribor
Petisovci
SLOVENIA
Lendava
Ljubljana
Zagreb
Paluzza
Belluno
Trieste
Koper
Rijeka
Porto Marghera
Gulf of Venice
Pula
Ferrara
Bologna
Ravenna
Ivana
CROATIA
Gospic
Sisak
Lipovljani
Banja Luka
BOSNIA & HERZEGOVINA
Bosanski Brod
Sarajevo
Split
Pecs
Babocsa
Novi
Knezevac
Mokrin
Srpska Crnja
Medja
Boka
Pancevo
Belgrade
Velebit
Kikinda
Nikolinki
Mramorak Selo
SERBIA AND MONTENEGRO
Aleksinac
Nis
Pristina
Dubrovnik
Shkoder
ALBANIA
Tirana
Durres
Diviaka
Seman
Bubullima
Povelca
Frakulla
Vlore
Finiq
Pekisht
Kucova
Patos-Marinza
Cakran
Ballsh
Gorisht
Drashovice
Skopje
MACEDONIA
Thessaloniki
GREECE
Trikala
Larissa
Arta
Delfi
Gulf of Corinth
Elefsina
Athens
Aspropyrgos
Argos
Tripoli
Prinou
S. Kavalla
Kavala
Komotini
Alexandroupolis
Agean Sea
BULGARIA
Sofia
Vraca
Chiren
Dolni Dabnik
Gorni Dabnik
Devetak
Stara Zagora
Plovdiv
Burgas
Otmanli
Varna
Kavarna
Tjulenovo
Staro Orjachovo
Tolloukhin
S.Eregli
Cst.Eregli
Eregli Deep
Aliphi
S. Akcakoca
Ayazli
Cst. Akcakoca
Eregli
Istanbul
Izmit
Bandirma
Bursa
Burhaniye
TURKEY
Izmir
Soke
Rodhos
Khania
Rethimnon
Iraklion
CRETE
Florence
San Marino
Falconara
Ancona
ITALY
Marta
Rome
Bellante
Cellino
Allano
Chieufi
Rospo
Lanciano
San Stefano
Cupello
Vasto
Valle Cupo
S. Salvo
Portocannone
Monte Stillo
Ripi
Petroliard-Farnesina
Terracina
Candela
Gaeta
Naples
Pietralunga
Grottole
Ferrandina
Cerro Falcone
Monte Alpi/Enoc
Tempa Rossa
Nova Siri
Giovinazzo
Bari
Aquila
Taranto
Brindisi
Pomarico
Pisticci
Rotondella
Adriatic Sea
Tyrrhenian Sea
Ionian Sea
Crotone
Luna
San Leonardo
C. Cimiti
Vibo Valentia
Rosarno
Lavinia
Bagnara
Messina
SICILY
Palermo
Milazzo
Gaglioano
Mazara del Vallo
Marinella
Bronte
Cisina
San Nicola
Catania
Ardizzone
Marsala
Porta Enpedocle
Gela
Commarata
Pone
Drillo
Perla
Ragusa
Vega
Augusta
MALTA
Mediterranean Sea
0 50 100 150 Miles
0 50 100 150 200 Kilometers

The main drilling objective is the Miocene Birsa sandstone at 1,440 m (4,724 ft).

West Sicily seismic

In early 2009, Northern Petroleum PLC completed 2D seismic surveys in six West Sicily offshore thrust belt licenses G.R17.NP, G.R18.NP, G.R19.NP, G.R20.NP, G.R21.NP, and G.R22.NP.

The survey vessel acquired 2,463 km (1,530 mi) of data during the first part of the work program which was funded by Shell Italia E&P.

Po basin: AleAnna

AleAnna Resources LLC was awarded the Ponte del Diavolo exploration permit in northern Italy's Po basin in early 2009. The site is just west of its Corte dei Signori permit, where it earlier finished shooting 3D seismic. Drilling at Corte dei Signori was anticipated to start by yearend.

Po basin: Po Valley Energy

Po Valley Energy Ltd. said in July 2009 that log analysis of its Sillaro-2 appraisal well in the Po basin confirmed the main pay zones in the Sillaro-1 discovery.

Test data also highlighted three possible attractive shallower targets.

Po Valley Energy was planning to perforate several production zones to complement existing production capacity in Sillaro-1 and run short flow-tests mid-year.

Cascina San Pietro permit

Po Valley Energy Ltd. reported above-expected test flow rates from its Bezzecca-1 appraisal well, which tapped the deepest Miocene horizon in the Cascina San Pietro permit.

The well, east of Milan, flowed 2.2 MMcfd of gas under restricted test conditions from the target interval at about 1,930 m (6,332 ft).

Continued testing was planned at intervals as shallow as 1,860 m (6,102 ft).

Drilling and Production

$1.5-billion FPSO contract

Eni SpA awarded Saipem SpA a $1.5-billion, 20-year contract in late 2009 for the supply and operation of an FPSO vessel.

The vessel will have a storage capacity of 700,000 b/d and a production capacity of 12,000 b/d of oil. Production activities will start in 4Q 2011.

For the first eight years of the contract, the FPSO will operate in the Aquila field in the Adriatic Sea offshore Italy.

Guendalina field

The Environment Ministry granted approval in May 2009 to develop the offshore Guendalina gas field in the northern Adriatic, about 25 km (16 mi) off Ravenna.

Final award of the production concession was expected in the third quarter, and first production is expected by late 2010.

Eni SpA operates the field; Mediterranean Oil & Gas PLC has a 20% interest.

Reserves are 538 million cu m (19 bcf) proved, and 85 million cu m (3 bcf) probable, with output expected to be 0.566 million cu m/d (20 MMcfd).

Pantelleria Island

Northern Petroleum (UK) Ltd. was pursuing 2D seismic operations in six thrust belt licenses off Pantelleria Island, Sicily, in mid-2009.

The island is 100 km (62 mi) southwest of Sicily and 70 km (43 mi) east of the Tunisian coast.

The project expects to acquire more than 2,500 km (1,553 mi) of data in a work program funded by Shell Italia as part of a farm-in agreement.

The licenses are G.R17.NP, G.R18.NP, G.R29.NP, G.R20.NP, G.R21.NP, and G.R22.NP plus C.R147.NP.

Offshore licenses, Pantelleria Island, Sicily

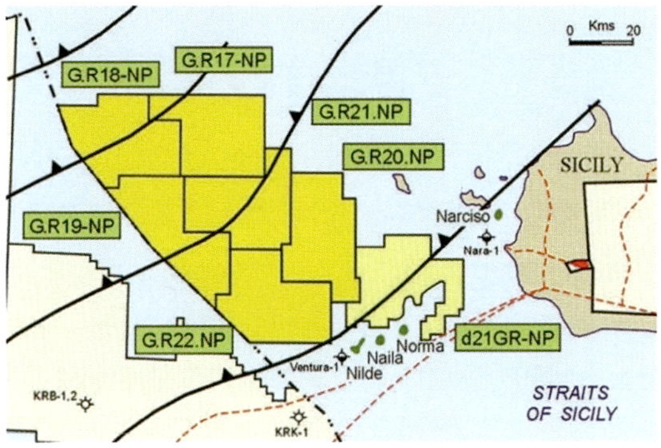

International Activities

Eni-KazMunaiGas deal

In late 2009, Eni SpA and KazMunaiGas signed a preliminary agreement that could result in up to $50 billion in investments in Kazakhstan's oil and gas sectors.

Details are given in the Kazakhstan section of the Former Soviet Union chapter of this book.

Gazprom stake in SeverEnergia

Eni SpA and Enel SpA said in mid-2009 that they will assign to Gazprom a 51% stake in SeverEnergia, a Russian holding company with exploration and production licenses. The price-tag is $1.5 billion.

SeverEnergia was formerly known as EniNeftegaz and was two production companies belonging to Russia's now-defunct Yukos Oil Co.

By 2013, SeverEnergia hopes to produce 150,000 boe/d, focusing on west Siberian fields. SeverEnergia has the Arcticgaz, Urengoil, and Neftegaztechnologia E&P licenses, which hold an estimated 5 billion boe.

The parties also agreed to produce first gas by June 2011 from the Samburskoye field.

LNG

Adriatic terminal

In October 2009, the terminal of Adriatic LNG was inaugurated, offshore Porto Levante. At full capacity, expected by yearend, the $3-billion terminal will be able to deliver 8 bcm/year (282.5 bcf/year) of natural gas, representing about 10% of Italy's current requirements.

It received its first cargo in August. Gas from Qatar will supply the plant for the next 25 years.

The terminal is the world's first offshore gravity-based LNG structure.

Panigaglia terminal

The Panigaglia regasification terminal, Italy's first, is at Fezzano di Portovenere (SP). Owned by Eni SpA, it has a capacity of about 4 bcm/year (141.2 bcf/year). Plans were under way at mid-year to expand capacity to 8 bcm/year (282.5 bcf/year) at a projected cost of €450 million.

MedGas terminal

In early 2009, officials announced that startup of the planned MedGas LNG terminal—to be built at Gioia Tauro, near Calabria—likely would be delayed from 2013 to 2014.

Backers of the project to build the 12-bcm/year (423.7-bcf/year) regasification facility were expecting final approval to proceed by yearend.

Porto Empedocle terminal

Enel SpA was poised in late 2009 to choose the company that will perform EPC work on its Porto Empedocle LNG facility in Sicily.

Trieste terminal

In July 2009, the federal environmental officials formally granted approval to build a planned 8-bcm/year (282.5-bcf/year) LNG facility at the port of Trieste in northern Italy.

The plant is to be built by Germany's E.ON, which was awaiting regional approvals in October. It will be managed by Spain's Gas Natural SDG SA.

Pipelines/Transportation

Spur to tap ITGI gas pipeline

Greek, Bulgarian and Italian companies agreed in July 2009 to build the 160-km (99-mi) Greece-Bulgaria Interconnector (GBI) would provide Bulgaria access to the Italy-Turkey-Greece Interconnector (ITGI) pipeline carrying gas from Azerbaijan to Turkey, Greece and Italy.

Completion of the Greece-to-Italy segment of the ITGI is planned for 2012.

More details are given in the Bulgaria section of this chapter.

Alternative Energy

First offshore wind farm

The Environmental Ministry in September 2009 approved the environmental-impact evaluation needed for construction of the first offshore wind farm in Italian waters.

The 162-MW San Michele wind farm will take a year to build, near the southern city of Termoli.

Hydrogen power plant

In August 2009, Enel SpA began operating a 16-MW hydrogen-fired unit at the Andrea Palladio power plant near Venice.

Enel called the modified GE 10 gas turbine the world's first such unit, achieving "zero emission" electricity production. The hydrogen comes from a nearby petrochemical plant.

Solar power infrastructure

In mid-2009, Italian energy transmission operator Terna SpA announced plans to invest about €320 million ($450 million) by 2010 on solar power projects in the country.

Terna has also proposed to set up a new company to oversee the development of projects with a total power-generating capacity of 70–100 MW.

Italy currently has some 37,000 solar power installations, generating about 450 MW. Total installed photovoltaic capacity is expected to reach 1,300 MW by 2010.

$58 billion for renewables

The Italian Observatory of Renewable Energy in mid-2009 forecast that more than $58 billion will be invested by 2020 in the renewable energy sector of the country.

One driver for such investment is the fact that Italy imports about 85% of its energy.

The Ministry of Economic Development reported that domestic supplies of natural gas have fallen sharply over the past decade and met only 12% of consumption in 2008.

Likewise, crude pumped from domestic oilfields satisfied just 7% of national needs.

Monte Grighine wind farm

EDF Energies Nouvelles, half-owned by France's EDF, decided in mid-2009 to take a 50% stake in Italy's largest planned wind project, the 99-MW Monte Grighine wind farm in the western part of Sardinia.

The share was bought from Denmark's Greentech Energy Systems A/S.

Construction of Monte Grighine began in 2007 but progress has been delayed by a lack of funding.

The deal also gave EDF the option to take a 50% equity stake in every project from Greentech's wind-power project portfolio in Italy and Poland.

Nuclear JV

A joint venture established by Enel SpA and Electricité de France in mid-2009 will assess the feasibility of building at least four nuclear power plants in Italy.

Italy ended domestic nuclear power production in 1987, a year after the accident at the Chernobyl nuclear power plant in Ukraine. But public opposition now appears to be softening.

In July, the senate approved a law that, among other things, authorizes the return of nuclear power and calls for the creation of a nuclear safety agency.

MALTA

Capital: Valletta
Monetary unit: Lira
Refining capacity: 0
Oil production: 0
Oil reserves: 0
Gas reserves: 0

Area 4 development

Leni Gas & Oil PLC proceeded with plans in 2009 to develop the 5,700-sq km (2,200-sq mi) Area 4 offshore Malta, focusing on Blocks 4, 5, 6, and 7.

The company, with JV partner Mediterranean Oil & Gas, had signed a production sharing contract with the government in 2008.

The $2.5-million project aims to gain improved understanding of the area and to increase the chance of successful identification of sites with the highest potential for drilling in 2010–11.

For four prospects and five leads in Area 4, total most likely hydrocarbon potential has been estimated at 5 billion bbl of oil in place; most likely total prospective recoverable oil potential is estimated at 1.47 million bbl.

Malta-Sicily electricity interconnect

In mid-2009, EU ministers approved a proposal to build an electricity connection grid from Malta to Sicily. Funding will come from the €4-billion ($5.64-billion) energy infrastructure development program initiated by the EU as part of an economic recovery package.

Establishment of formal construction contracts is expected by early 2010.

Malta has no domestic energy sources, and its electricity grid is not linked to any other power network in the EU.

THE NETHERLANDS

Capital: Amsterdam
Monetary unit: Euro
Refining capacity: 1.206 million b/cd
Oil production: 25,500 b/d
Oil reserves: 100.000 million bbl
Gas reserves: 50.000 tcf

Exploration and Development

L8-D field

The Dutch subsidiary of Cirrus Energy said in early 2009 that initial results from its L11-13 offshore appraisal well appeared to be encouraging.

The well was drilled to TD of 17,717ft (5,400m) in the L8-D field. The well was designed to test the southern extension of a large, well defined fault block which contains the L8-D field.

In October, the company reported the start of production from its L11b-A06 well on the L8-D field at a constrained gross rate of 32.8 MMcf/d. Production began less than a month after start-up of the M7-A field.

West Netherlands basin

In October 2009, the government awarded to a Dutch subsidiary of Northern Petroleum PLC the Engelen exploration license in the West Netherlands basin.

The 4-year license covers the Kerkwijk South gas prospect, mapped from seismic and on trend with the 1988 Kerkwijk gas discovery.

Engelen is adjacent to the Utrecht license, where two undeveloped discoveries will be evaluated in tandem with Engelen.

Drilling and Production

Groningen field turns 50

June 2009 marked the 50th anniversary of the discovery of the massive Groningen natural gas field in the northern part of The Netherlands.

It is one of the largest gas fields in Europe and—in 1959—was the third-largest in the world based on its known reserves. Gas from Groningen was exported to Western Europe, positioning The Netherlands as an important supplier.

However, the oil crisis in the 1970s, the liberalization of the European gas market, and the rise of other large gas producers in other countries led the Dutch government to devise a 'small fields' policy, which exploited smaller fields in the hopes of slowing Groningen field's depletion rate.

To date, the field's 300 wells have yielded more than 60% of its original producible gas reserves, estimated at 2.7 tcm (95.3 tcf).

The JV of ExxonMobil Corp. and Royal Dutch Shell PLC that manages Groningen said that the field should contribute to Europe's gas supply security for at least another 50 years.

LNG

LionGas terminal

Dutch company 4Gas was poised in late 2009 to complete FEED studies for the LionGas LNG terminal, to be built in the Port of Rotterdam.

The terminal, expected to come onstream in 2012, initially will have a regasification capacity of 9 bcm/year (317.8 bcf/year). A second phase of development will boost that capacity to 12 bcm/year (423.7 bcf/year).

JV for French terminal

In mid-2009, Koninklijke Vopak NV established a JV with Royal Dutch Shell PLC to set up an LNG terminal at Fos-sur-Mer, near Marseille, France.

Details are given in the France section of this chapter.

Pipelines/Transportation

First salt-cavern gas storage

In 2009, work was under way to develop the first-ever salt-cavern natural gas storage facilities in The Netherlands. The project is funded by NV Nederlandse Gasunie and utility firm NV Nuon.

Completion is anticipated in mid-2010.

The storage facility, being developed near the Groningen gas field, will consist of four salt caverns, each with a working-gas volume of 45 MMcm.

Total investment is estimated at $496 million for the project, which is projected to become operational by mid-2010. A JV of Gasunie and Nuon, Zuidwending VOF will, operate and run the facility.

Gas storage facility

In August 2009, Gazprom agreed to make use of a gas storage facility in The Netherlands that is being developed by a consortium led by the Abu Dhabi National Energy Co. (TAQA).

Gazprom's export division is to provide cushion gas to pressurize what will become Europe's largest gas storage facility. In return, Gazprom is to receive working capacity at the facility.

Located in a depleted gas field north of Amsterdam, the facility will provide 4.1 bcm (144.8 bcf) of storage when it begins commercial operation, expected in 2013.

Gazprom said the facility would help it supply customers with gas from the planned Nord Stream pipeline from Russia to Germany.

Refining and Products

Pernis upgrade

Royal Dutch Shell PLC said in mid-2009 that it plans to proceed with development of a $500-million hydro-desulfurization plant at its Pernis refinery in Rotterdam.

The 400,000-b/d refinery is Shell's largest such facility in Europe and one of its largest globally.

The new HDS plant will allow the refinery to process heavier crude oils.

Most of the output from the new plant will be sold in the German market as domestic heating oil.

The company said it is considering investing as much as $1.3 billion in the refinery by 2016.

Shell refineries offered

Indian conglomerate Essar bid in August 2009 for three European refineries offered by Royal Dutch Shell as part of the latter's downstream restructuring plans.

One plant is the Stanlow facility in the UK; the others are in Harburg and Heide, Germany.

Other buyers in other countries also expressed interest in the plants, which have an estimated combined value of about $2.4 billion.

Shell has said it is moving to shed smaller refineries and to invest instead in large, integrated complexes, such as those in Port Arthur, TX, and Rotterdam.

Total-Dow-Lukoil deal

Total SA acquired a 45% share in the 147,000-b/d Vlissingen refinery in The Netherlands from Dow Chemical in mid-2009 and entered an agreement to sell the stake to Lukoil.

Details are given in the Russia section of the Former Soviet Union chapter of this book.

Petrochemicals

RIL bid for LyondellBasell

In late 2009, it was reported that India's Reliance Industries Ltd. offered to buy a controlling stake in LyondellBasell, the bankrupt Dutch-based U.S. petrochemical firm, for about $10 billion, all cash.

The deal would be one of the largest offshore acquisitions by an Indian group.

The action was seen as reflecting the view of a range of Indian conglomerates, in fields from telecom to IT services, who are also seeking to expand globally.

Reliance generates sales of more than $30 billion/year from petrochemicals, oil and gas, and retailing.

In 2008, LyondellBasell reported just over $50 billion in revenues, with employees in 19 countries.

However, in early 2009, it filed for bankruptcy protection for its U.S. operations and one European holding.

Alternative Energy

Tromp Binnen wind farm

The renewable-energy unit of RWE AG received permission in mid-2009 from the Dutch government to build the 300-MW Tromp Binnen wind farm off the northern coast of The Netherlands.

RWE Innogy said the facility, about 75 km (47 mi) off the coast near Callantsoog, will be one of the largest in the North Sea, and the farthest from shore

Largest biodiesel plant

Finland's Neste Oil Corp. began construction in May 2009 of a $940-million renewable diesel plant in Rotterdam. The 240 MM gal/year plant will produce Neste's proprietary NExBTL fuel from palm oil.

The firm says it will be the largest renewable fuels plant in Europe when it starts up in 2011.

Environmental Issues

CCS: Barendrecht

In late 2009, the Dutch government gave Royal Dutch Shell PLC approval to more ahead with its plan to capture CO_2 from a refinery and store it in a depleted gas field under the small town of Barendrecht.

Local residents and officials and environmental groups opposed the plan.

The CO_2 would come from Shell's Pernis oil refinery.

The first stage of the project, spanning three years, will capture 800,000 tons of CO_2, compress it, and pipe it to a site for injection into a depleted gas field almost 2 km (1.2 mi) beneath the town.

The Dutch government said it may be possible to store up to 10 million tons of CO_2 under Barendrecht.

CCS: Limburg, Groningen

Dutch energy provider NV Nuon said in early 2009 that it will invest $53 million on a large-scale CCS test project at its Buggenum power plant in Limburg Province.

Construction of the project was expected to start in 2009. A two-year test is scheduled to start in late 2010 to evaluate a pre-combustion carbon capture process.

It is to be a prelude to a larger-scale CCS plant at the proposed Nuon Magnum power project, near Groningen. That 1,300-MW plant will gasify a combination of coal and biomass.

The company expects that with CO_2 storage, emissions at Nuon Magnum will be cut by about 2 mtpy.

Carbon capture

A subsidiary of Siemens AG joined with TNO (Netherlands Organization for Applied Scientific Research) in a project to improve and commercialize a carbon-capture technology, developed by Siemens, based on the use of amino acid salt.

Aiming to enhance the technology and make it ready for market entry, the team will build a pilot plant by 2014 in the Frankfurt Hoechst Industrial Park to demonstrate the feasibility and effectiveness of the proprietary process developed by Siemens.

Company News

Shell: Refining assets

In August 2009, Royal Dutch Shell PLC announced plans to sell about 8% of its refining assets in 2009–10 because of declining demand and oversupply in the market.

That includes its refining assets at Heide and Harburg in Germany and at Stanlow in the UK, as well as its service stations in Greece and in other countries.

Nigeria settlement

In June 2009, Royal Dutch Shell agreed to pay $15.5 million to the Ogoni people of the Niger Delta.

The action aims to settle a court case brought in the United States by nine plaintiffs, including relatives of a noted environmentalist and writer and people claiming environmental damage due to Shell's actions.

Ken Saro-Wiwa was executed by Nigeria after what was seen by many as a sham trial. He campaigned against Shell's activities in Nigeria, and plaintiffs accused Shell of complicity in his death.

Shell denies any wrongdoing and called its payout a humanitarian gesture. It said that part of the money will support a trust fund for the Ogoni.

NORWAY

Capital: Oslo
Monetary unit: Kroner
Refining capacity: 319,000 b/cd
Oil production: 2.025 million b/d
Oil reserves: 6.680 billion bbl
Gas reserves: 81.680 tcf

Exploration and Development

Ormen Lange delineation

Norske Shell, operator of the Ormen Lange field, completed drilling appraisal well 6305/5-3 S in production license 209 in the North Sea in late 2009 to delineate the northern extent of the field.

Results will be incorporated into a major re-interpretation of the field aimed at guiding future development and long-term production.

Continental shelf limits

The UN Commission on the limits of the Continental Shelf submitted its final recommendation in early 2009 for the limits on the Norwegian continental shelf to the north.

This establishes the extent of the seabed (235,000 sq km; 90,734 sq mi) for which Norway is responsible regarding the management of natural resources.

Data for UN use was collected by the Norwegian Petroleum Directorate.

Some of the area identified is beyond the 200 nautical mi traditionally set as the limit of a continental shelf.

However, most of the area beyond 200 nautical mi is at a sea depth of more than 2,500 m (8,202 ft) and on oceanic crust. The presence of oil or gas reserves in these areas is unlikely, said the NPD.

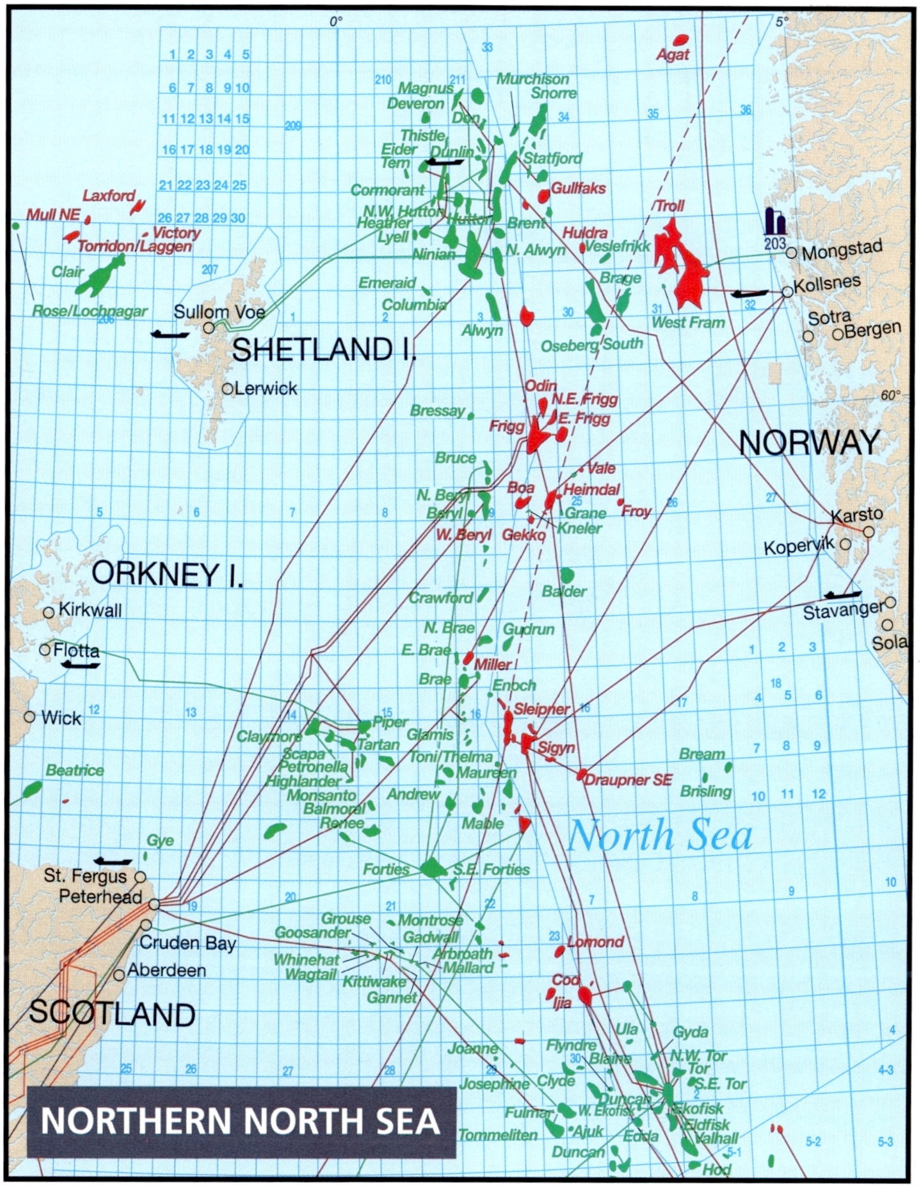

Discoveries noted

Several firms reported discoveries in exploratory drilling offshore Norway in 2009. Drilling location and estimated recoverable volumes are listed:

- Marathon Petroleum Norge: Marihøne prospect in PL340 (20–30 MMboe).
- Statoil: Frigg Delta prospect (18–35 MMboe)
- ExxonMobil: Jetta prospect, south of Jotun field (tbd)
- Norwegian Energy Co. ASA (Noreco): Tasta prospect, PL 412 (34–91 MMboe)
- A/S Norske Shell: Gro prospect, PL 326 (353 bcf to 3.5 tcf of gas)
- Statoil: Titan prospect, Tampen area (5.6–12.5 MMboe)
- Statoil: Gyrid prospect, PL 348, southern Norwegian Sea (20–30 MMbbl of oil)
- Statoil: Fulla prospect, close to Lille Frigg field (63–107 MMboe)
- Statoil: Asterix prospect, west of Sandnessjoen [16 bcm (565 tcf) of gas]
- Statoil: Pan/Pandora (34/8-14 discovery), south of Vislund field (7.5 MM cu m oil equivalent)
- Statoil: Dompap field (25–50 MMbbl)

SOUTHERN NORWEGIAN SEA GAS FIELDS, DISCOVERIES

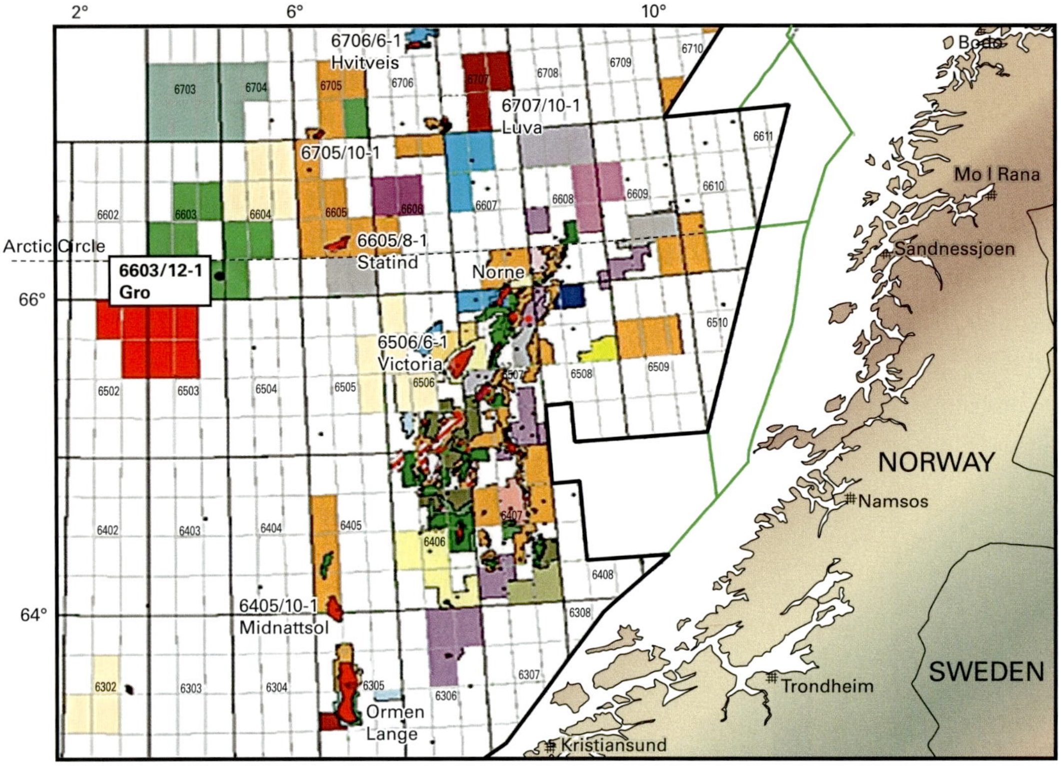

After Norwegian Petroleum Directorate

Jan Mayen

The Norwegian government initiated an opening process in third quarter 2009 for petroleum activities around Jan Mayen in the Arctic Ocean.

The Ministry of Petroleum and Energy (MPE) was preparing a proposed study program, including environmental assessment.

Following studies and consultations, a white paper will be submitted to the parliament, which will make a final decision on opening the area.

The first licensing round was held in early 2009 for the Icelandic shelf, including blocks in the Icelandic portion (Dreki area) of the Jan Mayen Ridge.

Åsgard potential

Statoil reported in mid-2009 that Wells 6506/12-12 S and 6506/12-12 A had confirmed oil, gas, and condensate in the northeastern part of the Smørbukk deposit on the Åsgard field in the Norwegian Sea.

The company estimated proven resources of 15–25 MMboe. Hydrocarbons were found in the Mid/Lower Jurassic.

Goliat field

In mid-2009, parliament gave the go-ahead for Eni Norge AS to develop Goliat field in the Barents Sea with a circular floating production platform designed by Sevan Marine ASA.

With reserves of 180 MMboe, Goliat is the first oil development in its area (Blocks 7122/7,8,9,10 and 7123/7. Production is scheduled to start in late 2013.

Oselvar field

Authorities approved the development plan of DONG Energy for the Oselvar field in the Norwegian North Sea using subsea wells linked to BP's Ula platform.

Partner Noreco estimates reserves of 52 MMboe, with further exploration potential.

Lease Sales

20th licensing round

In April 2009, the Ministry of Petroleum and Energy awarded new production licenses based on results from the 20th licensing round for areas on the Norwegian shelf. That round was launched in mid-2008.

Thirty-four 34 companies received offers to participate in 21 new production licenses.

A total of 79 blocks had been offered: 51 in the Norwegian Sea and 28 in the Barents Sea.

Reserves

Barents Sea shelf

According to an assessment done in 2008 and reported in mid-2009 by the U.S. Geological Survey, more than 76 billion boe may be technically recoverable on the Barents Sea shelf.

The shelf lies entirely north of the Arctic Circle and covers 1.76 million sq km (680,000 sq mi) off northern Norway and Russia. Most lies in water less than 500 m (1,640 ft) deep.

The mean estimate of undiscovered, conventional, technically recoverable resources in four geologic provinces on the shelf included 11 billion bbl of oil, 380 tcf of natural gas, and 2 billion bbl of NGLs.

Most of the undiscovered petroleum appears to be in the East Barents basin province.

Continental shelf

The Norwegian Petroleum Directorate released in early 2009 its overview of the estimated quantities of recoverable petroleum resources on the Norwegian continental shelf. Recoverable resources consist of both the original marketable and the remaining petroleum resources.

The estimates are based on annual reports from operating companies, NPD evaluations for fields and discoveries, and NPD estimates of undiscovered resources.

The overview also shows the geographical distribution of resources for the North Sea, the Norwegian Sea, and the Barents Sea.

Total recoverable petroleum resources on the Norwegian shelf as of yearend 2008 were estimated at 13 bcm of oil equivalents—essentially the same as in NPD's previous account.

The current report shows 25 discoveries made in 2008: 12 in the North Sea, 4 in the Barents Sea, and 9 in the Norwegian Sea. Expected volume of the new discoveries is estimated at 50 MMcm of oil and 72 bcm of gas.

At yearend 2008, 11 fields were being developed: Alve, Gjøa, Morvin, Rev, Skarv, Tyrihans, Vega, Vega Sør, Volund, Yme, and Yttergryta.

Drilling and Production

Start of production

Production began in 2009 from a number of prospects and fields in Norway, including:

- Rev gas and condensate field, Norwegian North Sea: Talisman Energy Norge AS (January)
- Alve gas and condensate field, Norwegian Sea: Statoil (March)
- Tune South satellite gas field, Norwegian North Sea: Statoil (July)
- Tyrihans South and North oil and gas fields, Norwegian Sea: Statoil (July)
- Volund oil field, North Sea: Marathon Petroleum (September)

Statoil: Shutdowns, upgrades

Statoil conducted 21 production shutdowns of facilities on the Norwegian continental shelf during 2009 to improve safety and efficiency. Many installations also were to be upgraded to expand capacity, make technical improvements, and extend economic life.

The turnarounds, scheduled to be done by October, affected the following installations:

Oseberg, Statfjord, Gullfaks, Snorre, Troll, Kvitebjørn, Visund, Sleipner, Grane, Huldra, Veslefrikk, Brage, Heimdal, Volve, Glitne, Snøhvit, Norne, Åsgard, Kristin, Njord, Sture, and Kårstø.

In April, the company also temporarily halted operations on Snøhvit and at Hammerfest LNG on Melkøya for planned maintenance.

International Activities

MMS, NPD agreement

The Norwegian Petroleum Directorate signed an MOU in May 2009 with the U.S. Minerals Management Service.

Objectives of the agreement are to establish and build relationships; promote cooperative resource management activities in exploration and production; promote sharing of technical information, exploration and production strategies, and technical solutions; and conduct cooperative research.

Gazprom, Statoil Arctic cooperation

In mid-2009, Gazprom and Statoil signed an MOU pledging to cooperate through 2012 on exploration, development, and production operations in the Arctic areas of Russia and Norway.

The agreement replaces a similar one made in 2005 among Gazprom, Statoil, and Norsk Hydro before Statoil acquired the latter's assets.

The two companies also will work on technologies to find, produce and transport oil and gas. Russian officials pointed to the Shtokman field as a starting point for development of such technologies.

Environmental Issues

CO_2 storage in North Sea

In May 2009, Norway and the UK announced that they will study possible storage of captured carbon dioxide under the North Sea.

The CO_2 would come from carbon capture systems being evaluated by several European countries. In several of those countries, energy companies cannot build new gas or coal-fired power plants without a plan for carbon capture and storage (CCS).

The Norway-UK study will assess how quickly the North Sea could be needed for CO_2 storage and the related preparations that countries must make.

In April, ScottishPower PLV published a report claiming that all of Europe's CO_2 emissions well into the next century could be stored in saline aquifers beneath North Sea and in depleted oil and gas fields.

Carbon storage: Norwegian Sea

In early 2009, FEED work was about to start on a CO_2/amine project at the Åsgard B offshore production platform in the Norwegian Sea by Statoil.

The project aims to maximize absorption and secure storage of CO_2 at the facility, which is in Haltenbank, south of Heidrun field.

Statoil has been operating a CCS project, at the Sleipner gas field for the past 13 years. The company extracts nearly 2,600 tons of CO_2 and stores it 1,000 m (3,281 ft) below the sea surface.

Mongstad carbon capture plan

In early 2009, Statoil submitted its master plan for carbon capture at its Mongstad facility, north of Bergen, to the Ministry of Petroleum and Energy and the Ministry of the Environment.

When the Mongstad energy project was established, Statoil agreed to prepare such a plan.

The plan confirms that carbon capture is possible and describes two alternatives for doing so.

Iceland: Hydrocarbon potential

An analysis in 2009 by Global Geo Services (GGS) Spectrum said that the key to the hydrocarbon potential of Iceland's northern waters is the Jan Mayen Ridge (JMR).

The slice of continental crust is around 500 km (311 mi) long and runs south from Jan Mayen Island into Icelandic waters in the Dreki area. Water depth is 800–2,000 m (2,625–6,562 ft).

The JMR likely has many geological similarities with east coast Greenland and offshore mid-Norway. There is a proved hydrocarbon system on the Norwegian side.

The main hydrocarbon source rocks are expected to be of Early Jurassic and Upper Jurassic age, with other source potential in the Cretaceous and Tertiary.

ICELAND LEASE BLOCKS: 2009

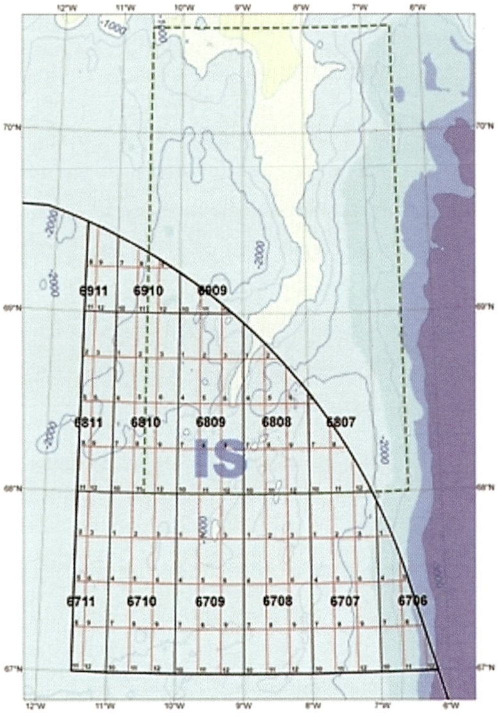

At full capacity, Statoil says, the facility can store three million metric tpy of CO_2, removed from natural gas in the gas fields Sleipner (North Sea), Salah (Algeria), and Snøhvit (Barents Sea).

EXPLORATION AREAS OF INTEREST: ICELAND, NORWAY

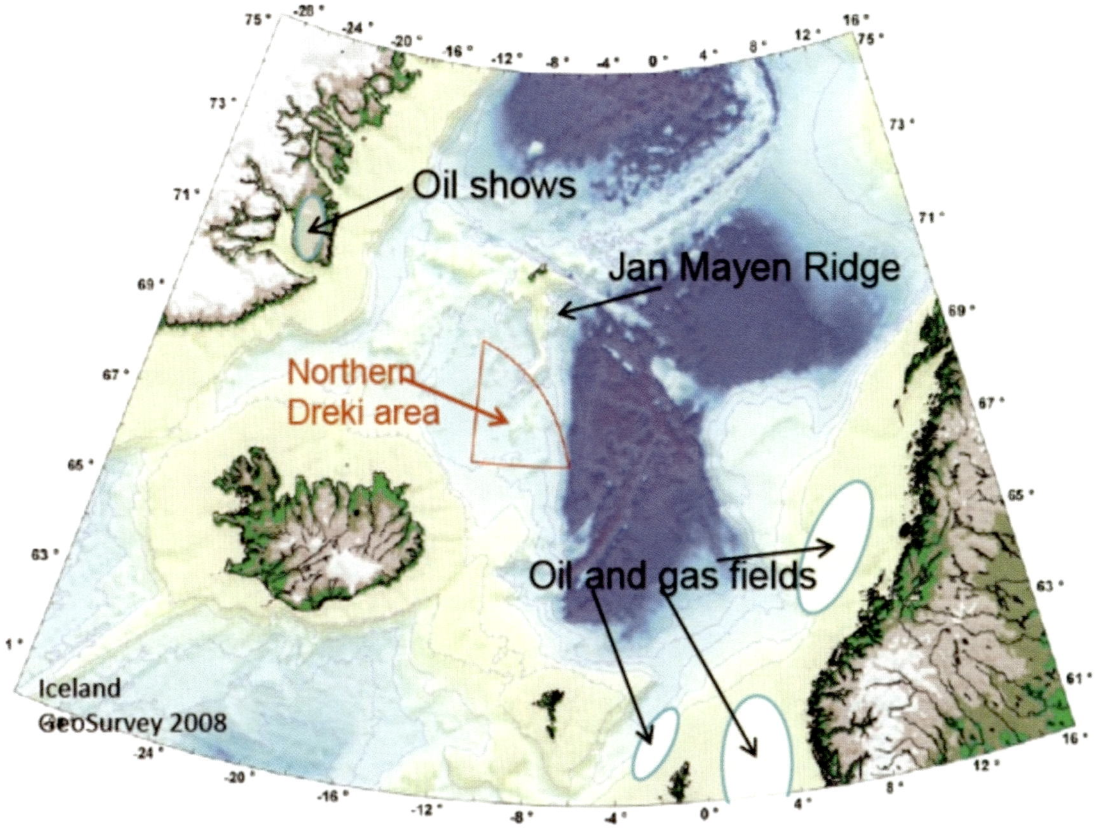

Iceland: First licensing round

In mid-2009, Iceland's National Energy Authority (NEA) said it received bids from two companies for exploration rights to the northern Dreki area. The bids were in response to the country's first-ever licensing round, launched in January.

One bid was from Aker Exploration, which merged in 2009 with Det norske oljeselskap. The other was a combined bid from Sagex Petroleum and Lindir Exploration (later withdrawn by Sagex).

Areas of interest are Blocks IS6708/1, IS6708/2, IS6808/11, and IS6909/11.

Selection of winning bids was scheduled for late 2009.

NEA must consult with Norwegian authorities before it can grant permission to companies to explore the area of shared sovereignty.

Iceland was planning to offer a total of 100 offshore oil and gas licenses in 2009.

Company News

DNO-Aker Exploration merger

Det norske oljeselskap completed its merger in December 2009 with Aker Exploration.

The merged company, which will operate 32 licenses and have a portfolio of 70, will continue to operate under the name Det norske oljeselskap.

Statoil name, visual identity

In late 2009, StatoilHydro changed its name to Statoil ASA and adopted a new visual identity.

The company says its new logo is meant to symbolize a guiding star.

POLAND

Capital: Warsaw
Monetary unit: Zloty
Refining capacity: 492,950 b/cd
Oil production: 12,000 b/d
Oil reserves: 96.375 million bbl
Gas reserves: 5.820 tcf

Shale gas potential

In late 2009, Poland awarded Chevron Polska Exploration & Production a license to explore the region around Zamosc in the southeast.

In the same time frame, the prospect of gas shale development led to deals with other companies, based in Canada, Austria, Italy, the United States, and the UK.

Areas of interest include the Starogard, Slupsk, and Slawno oil and gas concessions, the Baltic basin region near the Port of Gdansk, and the Mazowieckie and Lubelskie Provinces.

In 2008 and 2009, Poland granted 30 concessions for unconventional gas prospecting in the country.

Fences area

In late 2009, Polish Oil & Gas Co. and FX Energy Inc. reported first production from Rotliegend sandstone at 3,000 m (9,842 ft) at the Roszkow well in the Fences area of western Poland.

Initial flow rate was about 15 MMcfd.

Four other wells in the Fences concession were awaiting pipeline connections, of which three were expected to go on production later in 2010.

LNG sale

In mid-2009, Qatar Liquefied Gas Co. III (Qatargas) signed a 20-year agreement to sell 1 mtpy (1.4 bcm/year) of LNG to Poland's PGN under a contract said to be worth about $550 million/year.

The LNG will be received at a $650-million terminal being built at Swinoujscie, scheduled for completion in 2015.

Plock petrochemicals expansion

In early 2009, PKN Orlen announced plans to invest more than $1.3 billion to double the capacity of its 800,000-tpy petrochemical plant in Plock in a five-year project.

The expansion will increase production of both teraphthalic acid and paraxylene by 600,000 tpy.

Nuclear power roadmap

The government unveiled plans in mid-2009 to begin generating nuclear power by 2021.

The initiative aims to help reduce reliance on coal-fired power plants, which currently produce 93% of the nation's electricity. It will also help cut greenhouse gas emissions in line with the EU target.

The plan includes training programs and nuclear energy research to be conducted under the direction of a new National Atomic Energy Agency.

PKN Orlen: Diversification

In late 2009, state-controlled PKN Orlen SA said it plans to diversify beyond its role as a refiner by entering the power generation sector.

The firm was considering building a gas-fired power plant in northern Poland or in the Czech Republic.

PORTUGAL

Capital: Lisbon
Monetary unit: Euro
Refining capacity: 304,172 b/cd
Oil production: 0
Oil reserves: 0
Gas reserves: 0

Sines LNG terminal

REN Atlántico Terminal de GNL awarded an EPC contract in late 2009 for expansion of its LNG facility at Sines. The project will add a storage tank to increase receiving capacity from 6 million metric tpy to 9 million metric tpy.

The $265-millon project is expected to be completed in 2012.

Largest photovoltaic power plant

In December 2009, a Portuguese subsidiary of Spain's Acciona Energy inaugurated operation of the world's largest photovoltaic solar power plant.

The 46-MW, $355-million Central Solar de Amareleja is sited in the Alentejo Region in southern Portugal. Its 262,000 photovoltaic modules generate enough power for 30,000 homes.

ROMANIA

Capital: Bucharest
Monetary unit: New leu
Refining capacity: 516,557 b/cd
Oil production: 90,000 b/d
Oil reserves: 600.000 million bbl
Gas reserves: 2.225 tcf

Histria XVIII Block

In third quarter 2009, Romania's largest oil company, Petrom, started production at its Delta 6 and Lebada Vest wells on the Histria XVIII Block in the Black Sea.

The company expected output from Delta 6 to reach a maximum of 3,400 boe/d in the fourth quarter.

Petrom said that output from the two new wells accounts for 10% of its daily offshore production.

Petrom operates five producing commercial fields: Lebada Est, Lebada Vest, Sinoe, Pescaru, and Delta.

New crude terminal

Rompetrol Group NV commissioned in early 2009 the new Midia Marine Terminal, sited in the Black Sea about 9 km (6 mi) off the city of Constanta.

The $100-million crude oil facility can transfer 24 mtpy of crude oil from tankers.

Its location cuts 33 km (21 mi) off the previous transport distance to the Rompetrol Rafinare refinery.

A floating docking system speeds loading and unloading of carriers up to 160,000 DWT capacity.

Europe's largest onshore wind farm

In third quarter 2009, the GE Energy unit of General Electric Co. signed a contract to provide 101 wind turbines for the second phase of Europe's largest onshore wind power project.

The wind farm is at Cogealac and Fantanele, in the Dobrogea region of Romania.

The Cogealac section will have a capacity of 253 MW and is expected to start up in 2011.

Earlier, in March, GE Energy agreed to provide 139 of its 2.5-MW turbines for the 347-MW Fantanele section, which is to be operational by mid-2010.

The capacity of the completed complex will be 600 MW, making it almost twice the size of any other European onshore wind facility.

The estimated €1 billion for the complex is coming from CEZ Group.

SLOVAKIA

Capital: Bratislava
Monetary unit: Koruna
Refining capacity: 115,000 b/cd
Oil production: 1,000 b/d
Oil reserves: 9.000 million bbl
Gas reserves: 500.000 bcf

Reverse-flow tactic

In early 2009, Slovakia resolved an imminent gas-delivery crisis by reversing the flow of the Bratstvo (Brotherhood) pipeline from west to east, to carry some 15 million cu m (530 MMcf) of gas from the Czech Republic back to Slovakia.

To prevent another gas shortage in winter 2009–2010, Slovakia bolstered its gas stores during 2009 and contracted for additional supplies. The government also acted to further improve 'reverse-flow' operation of the Bratstvo link and to upgrade a pipeline to Austria (to be completed in 2010).

The European Commission made available some €1.4 billion ($2 billion) to develop reverse-flow projects and pipeline interconnections as part of its €5-billion economic recovery plan.

Vienna-Bratislava oil line

Austria's OMV AG announced plans in mid-2009 to open the long-delayed pipeline between Vienna and Bratislava (capital of Slovakia) by 2010.

Details are given in the Austria section of this chapter.

Druzhba pipeline

Druzhba (Friendship) is the world's longest oil pipeline at more than 4,000 km (2,485 mi).

Originating in southeastern Russia, it reaches Belarus, the Czech Republic, Germany, Hungary, Poland, Slovakia, Slovenia, and Ukraine.

It was built to supply Russian oil to most of the former member states of the Soviet Union.

Germany receives about 500,000 b/d of crude from the pipeline; Poland 360,000; Czech Republic 100,000; Slovakia 76,000; Lithuania 300,000; and Belarus 400,000 b/d.

Jaslovske Bohunice nuclear plant

In late 2009, the European Commission granted approval to proceed with plans to build a new nuclear power plant in western Slovakia.

The development JV includes Czech power company CEZ Group and Jadrova a Vyradovacia Spolocnost (JAVYS), a Slovak nuclear-plant operating and decommissioning company.

The plant, to be sited at Jaslovske Bohunice, will generate between 600 and 1,600 MW, based on a feasibility study to be completed in mid-2010.

Construction is to start in 2013, with cost estimated at €4–6 billion ($5.9–8.9 billion). Commissioning would take place in 2020.

SPAIN

Capital: Madrid
Monetary unit: Euro
Refining capacity: 1.272 million b/cd
Oil production: 1,900 b/d
Oil reserves: 150.000 million bbl
Gas reserves: 90.000 bcf

Saggas LNG expansion

Work continued in early 2009 on an expansion project at the Saggas LNG regasification facility and storage terminal in Sagunto, 50 km (31 mi) north of Valencia.

The project, to be completed by mid-year, was to add a fifth vaporizer and a third storage tank.

A second expansion was also being discussed.

New refinery in Badajoz

Steel fabricator Grupo Alfonso Gallardo was planning in mid-2009 to build a $3-billion, 110,000-b/d crude oil refinery in Badajoz, aiming to select a FEED contractor by yearend.

The product slate will focus on diesel refined products.

Construction is expected to begin in 2011, targeting start-up in 2014 to produce a product slate focusing on diesel products.

Role of renewables

By the end of 2009, renewable sources were expected to account for almost 25% of Spain's electricity production, well above the EU average in 2008 of less than 8%.

Spain is Europe's second largest producer of wind and solar energy after Germany.

The high interest in renewables use reflects the fact that Spain imports about 80% of its energy and that it is struggling to meet its international commitments to reduce CO_2 emissions.

Federal incentives paid to spur investment in clean energy totaled €950 million in 2008.

Europe's largest solar site

When completed, the Andasol Solar Power Station in the province of Granada in southern Spain will include three units, each using parabolic-trough CSP technology to produce 50MW of electric power.

In 2009, Andasol 1 went online, Andasol 2 was undergoing testing, and Andasol 3 was under construction. The full facility is scheduled to be operational in 2012.

The three units in the complex have a combined collector area equal to about 210 football fields, making it the largest solar energy site in Europe.

One Spanish group and five German firms are developing the Andasol complex.

Solar-tower power plant

In early 2009, Abengoa Solar began operation of its 20-MW PS20 Solar Tower Power Plant at the Solucar Platform near Seville.

The facility is the world's largest solar tower power plant and the second in commercial use.

A 10-MW PS10 solar power plant, also owned by Abengoa Solar at the same location, became operational in 2007.

Chinese talks with Repsol

In mid-2009, Chinese firms CNPC and CNOOC were reported to be in talks with Repsol YPF about possible asset purchases and joint ventures worth billions of dollars.

One element of discussion was the possible sale of Repsol's 75% stake in Argentina's YPF to CNPC, parent of listed PetroChina.

CNOOC was said to be discussing a proposal under which Chinese group might invest at least $15 billion in a JV involving some of Repsol's main exploration and production assets outside Argentina.

Other reports said that CNOOC also was interested in a minority stake in YPF.

Enel: Endesa stake

In February 2009, major Italian utility Enel SpA announced its acquisition of 25.01% of Spain's Endesa SA from Acciona Energia, a unit of Acciona SA.

With the acquisition, Enel will own 92.06% of Endesa and will become the second-largest power company in Europe, with total installed capacity of 94,300 MW in 22 European countries.

The value of the acquisition was put at €11 billion ($13.9 billion).

Nuclear plant reprieve

In mid-2009, the central government decided to let Spain's oldest commercial nuclear power plant keep operating until 2012, contradicting an earlier pledge to start a phase-out of nuclear power.

The 466-MW Santa María de Garoña plant in northern Spain provides only about 1.4% of the country's electricity, but concerns about job losses and associated economic impact spurred the decision.

SWEDEN

Capital: Stockholm
Monetary unit: Krona
Refining capacity: 437,000 b/cd
Oil production: 0
Oil reserves: 0
Gas reserves: 0

Brunnsviksholme LNG terminal

AGA Gas AB awarded a $32-million contract in early 2009 to NCC Construction to build an LNG terminal and related facilities in Brunnsviksholme, near the eastern port city of Nynashamn.

AGA Gas expects that the 350,000-tpy terminal—the first of its kind in Sweden—to be completed by mid-2011. It is said to represent the single largest investment made by the firm in Sweden.

LNG will come from the Skangass liquefaction plant in Stavanger, Norway, and from Central Europe.

Nord Stream permits

In November 2009, Sweden and Finland granted permits that will allow the Nord Stream Pipeline to be routed through the Exclusive Economic Zones of the two countries.

The line will run beneath the Baltic Sea to carry gas from Vyborg, Russia, to Greifswald, Germany, where it will enter the European gas network.

Further details are given in the Finland section of this chapter.

Skanled on hold

In mid-2009, the partners in the Skanled initiative suspended the pipeline project that was to deliver Norwegian gas to Sweden and Denmark because of the global economic uncertainty and increased commercial risk.

The pipeline was scheduled to start operations in 2012.

Norway's Gassco AS, consortium leader, said future gas demand was not clear, but that the project might be relaunched if commercial conditions improve.

Design capacity of the Skanled line is 24 million cu m/d (847.4 MMcfd). Operation is dependent on receiving terminals in Norway, Sweden, and Denmark.

Brunsmo wind project

GE Energy, part of General Electric Co., was poised to deliver by yearend 2009 a total of 57 wind turbines for the planned Brunsmo wind power project of Arise Windpower AB.

Total generating capacity of the facility, near Karlskrona in southern Sweden, will be 142 MW.

Arise Windpower intends to bring 300 large wind turbines on stream by 2014, producing a total of 2 terawatt-hrs/year of electric power.

Sweden aims to get 50% of the country's total energy from renewable sources by the year 2020.

Largest land-based wind farm

Government officials were considering in mid-2009 a proposal to build the world's largest land-based wind farm by 2020 at Markbygden in northern Sweden, west of Pitea.

Plans call for erection of 1,101 turbines over an area covering 500 sq km (193 sq mi).

Construction cost is estimated at €5 billion ($6.8 billion) for the facility, which would have a total generating capacity of about 4 GW.

The project is led by Markbygden Vind AB, which is owned by the Swedish power group Svevind and German turbine manufacturer Enercon.

Permanent nuclear waste storage

Sweden announced in mid-2009 that the world's first permanent disposal site for spent nuclear fuel will be located in Forsmark, 200 km (124 mi) north of Stockholm.

The Swedish Nuclear Fuel and Waste Management Co. said that initial site work is expected to start in 2013, construction in 2015, and facility operation by 2024.

An application to build the disposal site will be sought in 2010 from Sweden's nuclear authorities.

TURKEY

Capital: Ankara
Monetary unit: Lira
Refining capacity: 714,275 b/cd
Oil production: 45,000 b/d
Oil reserves: 262.200 million bbl
Gas reserves: 215.000 bcf

Exploration and Development

Samsun and 3921 Blocks

In mid-2009, Turkey's government approved a deepwater exploration agreement in the Black Sea between state-owned TPAO and ExxonMobil Exploration and Production Turkey.

The firms will jointly explore prospects in the Samsun Block and in the eastern portion of 3921 Block.

Water depths are up to 1,980 m (6,500 ft).

Western Black Sea

At mid-year 2009, Sherritt International Oil & Gas Ltd. was planning to spud the Durusu-1 exploration well in the Thrace basin, in the western Black Sea off Turkey.

The well, in 80 m (262 ft) of water 20 km (12 mi) off the coast, will evaluate the Tertiary Danisman and Osmancik formations that produce gas onshore to the west.

South Akcakoca: West Ayazli

The partners in the phased South Akcakoca development in the Turkish Black Sea were drilling the nearby West Ayazli prospect in third quarter 2009.

According to Stratic Energy Corp., the well—targeting a Cretaceous interval—was expected to encounter gas in the Kusuri formation. First gas is anticipated in late 2010.

South Akcakoca Phase 1 involved developing the Akkaya, Ayazli, and East Ayazli fields.

Cyprus, Libya agreements

In mid-2009, Libya was planning to sign agreements with Cyprus concerning the supply of oil and natural gas, according to Libyan officials.

In April, Libya's National Oil Corp. said it was considering investing in oil and natural gas exploration off southern Cyprus.

Pipelines/Transportation

Samsun-Ceyhan line

A joint agreement was signed in late 2009 by officials from Italy, Russia, and Turkey for construction of the 550-km (342-mi) Samsun-Ceyhan oil pipeline.

It will run across Turkey, between the port of Samsun on the Black Sea and the port of Ceyhan on the Mediterranean Sea.

The pipeline is to transport Russian and Kazakh crude across Turkey without having to ship it across the congested Bosporus Strait.

The initial capacity of 1 MMb/d later will be expanded to 1.5 MMb/d.

Companies involved in the project will be Eni SpA, Rosneft Oil Co., JSC Transneft, and Çalik Enerji, a unit of Turkish conglomerate Çalik Holding AS.

Nabucco agreement

In July 2009, the prime ministers of Turkey, Bulgaria, Romania, Hungary, and Austria signed an agreement under which the five nations will allow the 3,300-km (2,050-mi), $10-billion Nabucco pipeline to pass through their territories.

An invitation was extended to Iran and Russia to be part of the pipeline project, although Russia has always declined to participate.

The proposed line would transport an expected 31 bcm/year (878 MMcf/year) gas from Central Asia to Austria and Germany via Turkey, Bulgaria, Romania, and Hungary.

A continuation of the Baku-Tbilisi-Erzurum pipeline, Nabucco is expected to be completed in 2014.

The potential suppliers of gas to the pipeline are Azerbaijan, Iran, Iraq, Turkmenistan, and Uzbekistan.

In 2009, Iraq offered to supply enough gas to fill half the capacity of the pipeline by 2015, confident of expected production from its rehabilitated fields.

Refining and Products

Cooperation with Russia

Turkey and Russia were moving forward in late 2009 with plans to build a refinery as part of an earlier agreement to jointly build the Samsun-Ceyhan pipeline.

Under a broad protocol signed in August by the two nations, a working group was to be established to examine the pipeline project, possible construction of a "major refinery," and potential joint sales of petroleum products, including in third countries.

Agreements with Iran

Iran and Turkey agreed in late 2009 to promote several oil and gas projects and initiatives in the region.

The scope of the agreement includes:

- transportation of 35 bcm/year (1.24 tcf/year) of gas from Iran to Europe by way of Turkey
- the development of Iranian gas fields, including the giant South Pars
- augmenting participation of Turkish firms in Iran's petrochemical ventures
- development of a $2-billion crude oil refinery in northern Iran.

Alternative Energy

Renewable-energy JV

Germany's Energie Baden-Wuerttemberg AG entered into an MOU in early 2009 with Turkish conglomerate Borusan Yatirim ve Pazarlama AS to set up wind power facilities and hydropower plants in Turkey.

A JV is to be established to plan the investment of $1.2 billion from 2009–2013, focusing chiefly on renewable energy resources.

The agreement envisions an initial aggregate generating capacity of about 1,000 MW, growing to 2,000 MW in the longer term.

Company News

Heritage-Genel Enerji merger off

In late 2009, Heritage Oil called off its proposed $6-billion merger with Turkey's Genel Enerji, following its agreement to sell its Ugandan oil fields to Eni SpA.

In June, the deal looked solid, as limited oil began to flow from Kurdistan—identified as an important region of focus by Heritage as it discussed the merger.

Heritage had said that the merged company would aim to be a major player among the several small firms drilling in a patchwork of license areas in Kurdistan.

UNITED KINGDOM

Capital: London
Monetary unit: Pound
Refining capacity: 1.866 million b/cd
Oil production: 1.305 million b/d
Oil reserves: 3.084 billion bbl
Gas reserves: 10.312 tcf

Overview

Mid-year tough for offshore industry

Cost-cutting measures, dwindling backlogs of work, and pricing pressures were taking a toll in mid-2009 on the UK offshore oil and gas industry, particularly the supply chain.

That was the conclusion of a report commissioned by industry group Subsea UK and compiled by Douglas Westwood.

The outlook for 2010 was more optimistic with a growth in exploration and production expenditure expected by more than half of companies surveyed. Thirty-two percent indicated that spending would be up by 20% or more.

Analysis by Deloitte LLP showed a 78% drop in exploration drilling in first quarter 2009 compared with a year earlier due to low oil prices, high costs, and financing difficulties.

The firm also said in mid-year that drilling on the UKCS may fall by 50% in 2009 and drop again in 2010 despite new incentives and tax changes introduced by the Chancellor in April.

It projected investment in drilling at new and existing fields at £3.5–4.5 billion in 2009 and £2.5–4 billion in 2010, and estimated that government tax receipts from the North Sea would drop by almost half in 2009.

In a related mid-year report, Parliament's Energy and Climate Change Committee issued a report warning that urgent government action was needed to prevent a decline in North Sea oil and gas revenue that could lead to the loss of 50,000 jobs.

Committee members called for more effective tax breaks, action to ease the credit crunch, and moves to ensure new developers can access existing platforms and pipeline to get oil ashore from fields too small to justify building their own facilities.

A tax change called the Value Allowance was made in early 2009 to help stimulate marginal field development in the offshore sector.

Energy security concerns

In August 2009, a former UK energy minister now advising the government on international energy issues said that an "interventionist" approach by the government will be needed if security of energy supply is to be guaranteed.

His report was published by the Dept. of Energy and Climate Change (DECC).

In October, energy regulator Ofgem said that investment of up to £200 billion would be needed over the next 10–15 years to maintain reliable energy supplies in the UK while cutting carbon emissions.

The agency also said that a late-2009 drop in natural gas prices would not last.

And a related October report by the advisory Committee on Climate Change said that UK greenhouse gas emissions need to fall more sharply than seen in 2003–2007 if deeper long-term cuts are to be achieved. More-aggressive regulation may be needed, it said.

Exploration and Development

Small fields seen as key

Industry association Oil & Gas UK said in late 2009 that to sustain a reasonable level of self-sufficiency in oil and gas, the industry will have to develop a series of smaller fields.

The group said that discoveries on the UK shelf now average around 15 MMboe, and the typical size of new UK field development projects is around 25 MMboe.

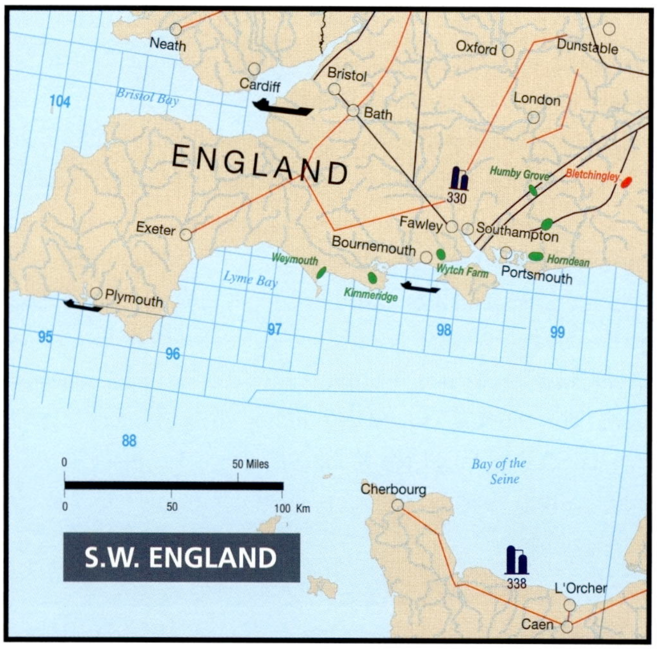

Discoveries noted

Several firms reported discoveries in exploratory drilling offshore UK in 2009.

- OMV AG: Tornado exploration well; oil and gas; west of the Shetlands
- DONG Energy: Glenlivet discovery; gas; west of the Shetlands
- Venture Production: well 49/4c-7z; gas; northern section of Kew field
- Talisman Energy: Godwin structure; oil; North Sea Block 22/17
- Dana Petroleum: SE Rinnes well 210/24a-12; oil; Northern North Sea Block 210/24a
- Nexen Inc.: Hobby exploration well; oil; North Sea Block 20/1N
- GDF Suez and First Oil Expro: Juliet well; gas; southern basin of UKCS.

UKCS challenges

In a mid-2009 report, industry group Oil & Gas UK said that exploration activity in the UK North Sea dropped by more than 50% since mid-2008, with serious implications for energy security in 2010.

The group said that investment in the industry, which was £4.8 billion in 2008, could dip below £3 billion in 2010. It said that about £5 billion/year is required to maintain exploration.

UKCS PROJECTED RESERVES/RESOURCES

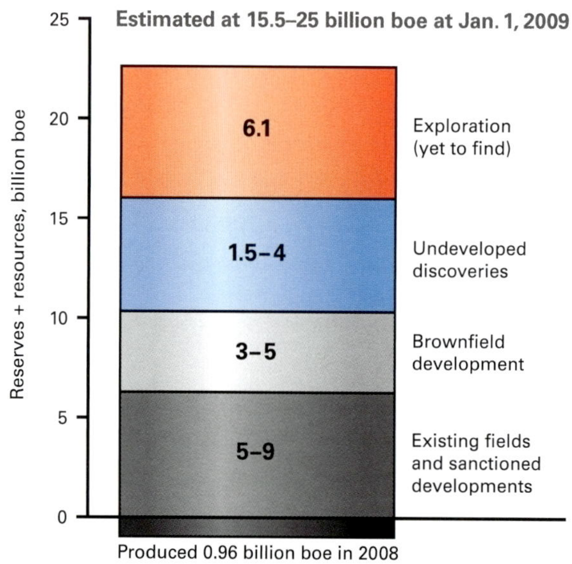

Source: Oil & Gas UK

UKCS FIELD DEVELOPMENTS 2009–10

Field name	Sector	Field type	Operator	Development status	Development type	Production start date	Total recoverable reserves, MMboe
Affleck	Central North Sea	Oil and gas	Maersk	Under development	Subsea	2009	21
Shelley	Central North Sea	Oil	Oilexco	Under development	FPSO	2009	18
Lybster	Central North Sea	Oil	Caithness Pet.	Under development	Ext. rch drlg	2009	3
Rita	Southern Gas basin	Gas	E.ON Ruhrgas	Under development	Subsea	2009	7
Ettrick	Central North Sea	Oil	Nexen	Under development	FPSO	2009	41
Jacky	Central North Sea	Oil	Ithaca	Under development	Subsea	2009	5
Don area	Northern North Sea	Oil and gas		Under development	Subsea	2009	52
Care and Es	Southern Gas basin	Gas		Under development	Subsea	2009	20
Bardolino	Central North Sea	Oil and gas	Shell	Probable development	Subsea	2009	12
Ptarmigan	Central North Sea	Oil	Oilexco	Probable development	Subsea	2009	7
Topaz	Southern Gas basin	Gas	RWE Dea	Probable development	Subsea	2009	4
Athena	Central North Sea	Oil	Ithaca	Probable development	Subsea	2010	29
Babbage	Southern Gas basin	Gas	E.ON Ruhrgas	Probable development	Subsea	2010	32
Ensign	Southern Gas basin	Gas	Venture	Probable development	Fixed platform	2010	14
Rochelle	Central North Sea	Oil	Endeavor	Probable development	Subsea	2010	5
Huntington	Central North Sea	Oil	E.ON Ruhrgas	Probable development	Subsea	2010	114
Fiddich	Central North Sea	Gas-condensate	Talisman	Probable development	Subsea	2010	19
Jacqui	Central North Sea	Oil and gas	ConocoPhillips	Probable development	Subsea	2010	13
Loirston	Northern North Sea	Oil and gas	ExxonMobil	Under development	Ext. rch drlg	2010	3
Cygnus	Southern Gas basin	Gas	GDF Suez	Probable development	Subsea	2010	9

Source: Wood Mackenzie

In the next 2 years, 20 fields are slated to come on stream, and Oil & Gas UK said that another 30 fields are competing to attract investment.

South Falkland basin prospects

Borders & Southern Petroleum PLC identified several play fairways and prospects in early 2009 based on seismic data shot in 2008 on its Falkland Islands acreage.

Details are given in the Argentina section of the Latin America chapter of this book.

Falkland Islands

Madrid's Repsol YPF SA led a consortium in 2009 that plans oil and gas exploration in international waters off the Falkland Islands beginning in early 2010.

Argentina and Britain both claim large areas of the seabed around the islands.

Details are given in the Argentina section of the Latin America chapter of this book.

Falklands claim

Reserves subsea off the Falkland Islands near Argentina are estimated at 18 billion bbl.

In April 2009, Argentine officials presented voluminous documentation to the UN staking the country's claim to 1.7 million sq km (656,000 sq mi) of seabed, including the UK Overseas Territories of the Falklands, South Georgia, and South Sandwich Islands, plus a large portion of Antarctica.

The initiative aims to expand Argentina's maritime territory to 425 mi from shore, based on a claimed extension of the nation's continental shelf.

UK officials were preparing to submit a counter-claim in mid-2009.

Drilling and Production

Prolonging production

A 3-year government study under way in 2009 was examining ways to prolong the production life of the North Sea oil and gas industry.

In a project funded at up to $2.7 million, Senergy Ltd. will support the work of DECC as it seeks to secure maximum recovery from existing oil and gas developments as well as opportunities to access new reserves in technically challenging fields.

UK North Sea Safety

A study released in August 2009 by the Health and Safety Executive (HSE) found substantial improvement in the safety of UK offshore oil and gas industry operations.

For 2008–2009, both the combined fatal and major injury rate and the number of major hydrocarbon releases were at their lowest since the HSE began regulating the industry in 1991.

Decommissioning costs

A late-2009 analysis by Oil & Gas UK concluded that under federal legislation, the vast majority of the 470 installations in use on the UKCS will have to be removed to shore for dismantling and disposal.

In addition to platforms and subsea equipment, this includes some 10,000 km (6,214 mi) of pipelines and about 5,000 wells, and many terminals.

The group projects that the cost to decommission oil and gas installations on the UKCS by 2030 will be £19 billion ($32.2 billion), rising to £23 billion ($38.9 billion) by 2040 for existing facilities.

New facilities could add another £2–3 billion ($3.4–5.1 billion), increasing the total to about £25 billion ($42.4 billion).

Cumulative UKCS decommissioning costs, 2007–2040

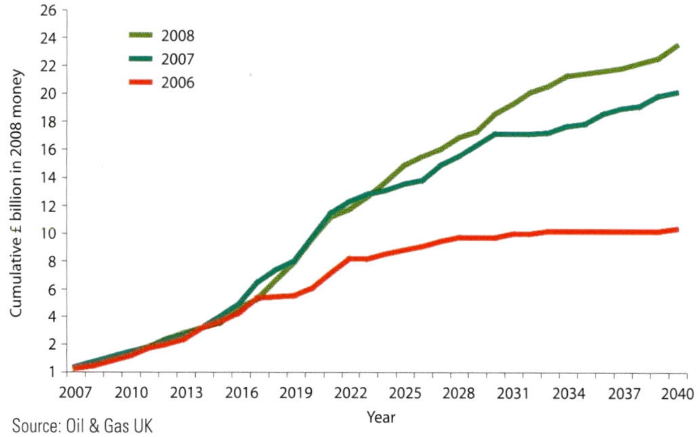

Source: Oil & Gas UK

Current cessation of production (COP) dates for existing UKCS fields

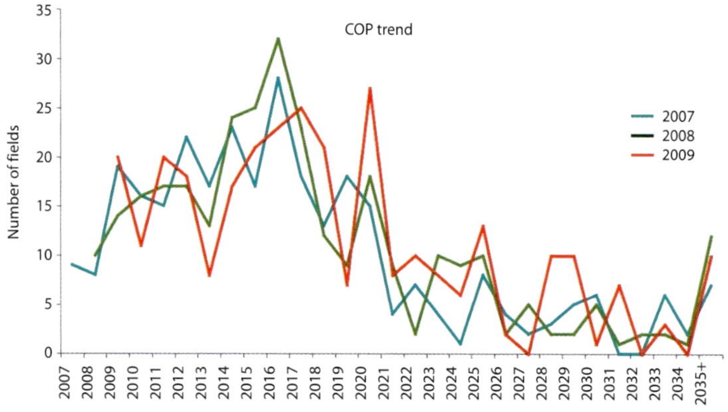

Source: DECC

Babbage platform

E.ON Ruhrgas AG was installing its first operated gas production platform in the UK southern North Sea in late 2009.

It is will process 2 million cu m/d of gas from the Babbage field starting in early 2010.

The field, 80 km (50 mi) off the coast, is expected to produce more than 5 bcm (177 bcf) of gas.

E.ON and partners plan to drill five wells in the field through 2011.

Affleck field

Maersk Oil UK Ltd. brought Affleck oil field on stream in mid-2009.

Initial flow from the field, located in the Central UK North Sea, was reported to be 2,000 b/d.

Shelley field

Premier PLC began producing oil at unspecified rates from the Paleocene Forties reservoir at Shelley field in the UK North Sea in mid-2010.

The firm is using an FPSO from the UK unit of Sevan Marine ASA.

The field is about 40 km (25 mi) south of the Premier-operated Balmoral complex.

Don Southwest field

Petrofac Energy Developments, a division of Petrofac, started oil production from two wells on Don Southwest (Don SW) in the UK northern North Sea in July 2009, adding to initial output from the nearby West Don field that began in April.

Total peak production from both fields is expected to exceed 40,000 b/d.

Forties field

Apache Corp. reported in mid-2009 that the Forties Charlie 6-3 well was on stream at 10,500 b/d of oil—the highest initial rate in the field since 1994.

It is the seventh development well brought into production at Forties in 2009.

Total production from the field was more than 70,000 b/d, the company says.

Jacky field

Oil production began in early 2009 from the Jacky field, said Ithaca Energy Inc.

Independent evaluation assigned gross proven and probable reserves of 5.1 MMbbl to the field.

As part of the same field activity, Ithaca re-started production from the Bravo facility in the Beatrice field. Together, Jacky and Beatrice production are expected to average about 10,000 b/d of oil gross in 2009.

Rita field

In early 2009, the UK unit of E.ON Ruhrgas started natural gas production from Rita field, on Blocks 42/21b and 44/22c in the UK North Sea. No output figure was given.

Seven Seas field development

DECC gave consent to Centrica Resources in early 2009 to develop the Seven Seas gas field in the UK Southern North Sea. First gas is scheduled for winter 2009–2010.

Hydrocarbons will be processed at the West Sole Alpha (WSA) platform and sent to the Dimlington terminal through the existing WSA export line.

Grouse field

In early 2009, Venture Production PLC brought Grouse oil field on stream.

The field, on Block 21/19 in the UK Central North Sea, is expected to produce 10,000 b/d of oil and 3.25 MMscfd of natural gas in 2009.

LNG

South Hook

The first commissioning cargo from Qatargas 2 arrived in March 2009 at the South Hook regasification terminal near Milford Haven, Wales, which is majority-owned by Qatar Petroleum.

South Hook is expected to deliver as much as 2 bcmd (70.6 bcfd) into the UK natural gas grid—meeting about 20% of current UK gas demand. Full operation was to start by yearend 2009.

Qatargas officials call the South Hook LNG terminal the largest LNG receiving terminal in Europe, with a capacity of 15.6 mtpy and using the largest-diameter storage tanks in the world.

Dragon

In September 2009 the 4.4-mtpy Dragon LNG terminal at Milford Haven, South Wales, began commercial operation, following commissioning that had started in July.

BG Group PLC and Malaysia's Petronas have agreements governing capacity rights for a 20-year term, allowing them each 2.2 mtpy of throughput.

Dragon LNG is near the South Hook LNG terminal.

Pipelines/Transportation

Brent system operator

TAQA Bratani Ltd. assumed operatorship in mid-2009 from Shell UK Exploration & Production for the Brent System pipeline in the UK North Sea—the first change in management in almost 30 years.

The pipeline delivers 100,000 b/d of oil from 20 North Sea fields.

TAQA is the UK arm of the Abu Dhabi National Energy Co.

Refining and Products

Teesside refinery

Petroplus Holdings AG decided in mid-2009 to sell the 177,000-b/d Teesside refinery in northeast England. Production ended in May, and Petroplus said that if the plant was not sold by yearend 2009, it would be converted into a storage site.

PetroChina: UK role?

The *Financial Times* reported in June 2009 that PetroChina was contemplating an investment in the Ineos refinery at Grangemouth, Scotland—its first move into the refining sector in Europe.

Ineos, a private chemicals group, was said to be talking with several companies about Grangemouth.

PetroChina has both oil production and refining businesses and has been moving into refining outside China. In May it paid $1billion for a large stake in Singapore Petroleum Co., which has part-ownership in a large refinery in Singapore.

Alternative Energy

UK renewable energy strategy

The UK government released in mid-2009 its new Renewable Energy Strategy white paper, showing how it intends to meet an

EU goal of generating 15% of its energy from renewable sources by 2020.

Wind and tidal/wave power figure prominently in the strategy, although energy from biomass and hydroelectric plants also will play an important role.

Wind energy

Development of wind power projects was a strong element of UK efforts in 2009 to make wider use of renewable resources. Presented here are highlights of some major initiatives.

Crown Estate: Wind-power boost

The Crown Estate, which controls use of the UK coast and seabed out to 23 km (14 mi), announced in mid-2009 that it will allow construction of up to 2 GW of additional offshore wind farm installations.

This will allow deployment of hundreds more wind turbines under Rounds 1 and 2 of the government's wind-power initiative. Round 1 solicitations were made in 2001 and Round 2 in 2003.

The Round 3 solicitation in mid-2008 drew 40 bids; winners are to be announced in January 2010.

Funds for offshore wind farms

The government announced in early 2009 that contracts worth more than £100 billion ($162 billion) would be available to the winners of Round 3 in its offshore wind power development initiative.

Observers said that the decision by the Crown Estate will set off a development program similar to the opening of the North Sea to oil and gas production in the 1970s.

Nine projects will be supported, with a combined capacity of 25 GW. The largest is on Dogger Bank, 100 km (62 mi) off the east coast, with a planned capacity of 10 GW.

Industry body RenewableUK noted that the UK aims to get 30 GW of electricity from wind by 2020.

London Array

Planning was under way in 2009 by the London Array consortium for what it says will be the world's largest offshore wind farm. Consortium members are Dong Energy A/S, Masdar, and E.ON AG.

Located in the Thames River Estuary about 20 km (12 mi) offshore, the facility is to generate 1 GW when completed.

Contracts totaling almost €2 billion ($2.9 billion) were awarded to build the 630-MW Phase I by 2012.

Offshore-wind 'supergrid'

Nine countries agreed in late 2009 to build an offshore 'supergrid' in the North Sea and Irish Sea.

The North Seas Countries' Offshore Grid Initiative would allow countries with significant offshore wind energy facilities to transfer energy easily among the grids of participating countries.

Signing the agreement were the UK, Germany, France, Belgium, The Netherlands, Luxembourg, Denmark, Sweden, and Ireland.

A strategic work plan is to be prepared in early 2010, leading to creation of a formal MOU for signature later that year.

The European Wind Energy Association said that more than 100 GW of offshore wind projects were being planned, which could provide more than 10% of the electricity needed by the EU.

Lincs

Centrica PLC announced in October 2009 that it had final investment approval for the £725-million ($1.2-billion), 270-MW Lincs wind power project, 8 km (5 mi) offshore Skegness.

Construction is to start in 2010, and commercial operation is expected in late 2012.

Rhyl Flats

The 90-MW Rhyl Flats wind farm off the north coast of Wales began generating power in July 2009, according to the unit of RWE Innogy that owns the plant.

A total of 25 turbines will be operational when construction is completed by early 2010.

The company also has approval to build a 750-MW wind farm off the north Welsh coast, and holds a major stake in the planned 500-MW Greater Gabbard wind farm off England's east coast.

Sheringham Shoal

Construction was under way in late 2009 on the 315-MW Sheringham Shoal wind farm, being built offshore the Norfolk coast.

The $1.48-billion plant, scheduled to be operational in late 2011, is a project of Statkraft AS and Statoil. Its 88 turbines will be built in an area spanning 35 sq km (14 sq mi),

Tidal and wave energy

Another segment of the renewables resources sector active in 2009 focused on technology to capture energy from ocean waves and currents. Presented here are highlights of some major initiatives.

Marine Renewables Proving Fund

In September 2009, the government announced its new Marine Renewables Proving Fund, to provide support for developing wave and tidal energy technologies.

It will offer £22 million ($36 million) in grants to accelerate commercial rollout of such systems.

The Carbon Trust, a not-for-profit established to promote the move to a low-carbon economy, estimates that wave energy could become a £2-billion/year market for the UK by 2050.

Severn "tidal fence"

In mid-2009, DECC launched the Severn Embryonic Technology Scheme (SETS) to investigate the possible generation of power from the massive tidal flow that occurs in the River Severn Estuary.

The enormous tidal flow in the estuary would pass through a line of underwater turbines to generate 1,300 MW. Construction cost of one proposed "tidal fence" was estimated to be at least $5 billion.

Other renewable-resource projects

Beyond wind and waves, other renewable energy resources were under evaluation in 2009.

Pumped storage

Scottish and Southern Energy PLC announced plans in mid-2009 for the UK's first large-scale hydroelectric pumped-storage facility in more than 30 years. Each of two units in the Scottish Highlands would have a capacity of 300–600 MW.

Largest ethanol plant

December 2009 saw the startup of Ensus I, Europe's largest fuel ethanol production facility, in Teesside. The $500-million facility will produce 400 million l./year (105 MM U.S. gal/year) of ethanol from wheat feedstocks.

Cellulosic ethanol JV

BP PLC and Verenium Corp. formed a JV in early 2009 to develop one of the first commercial-scale, cellulosic-based ethanol plants in the United States. Details are given in the United States section of the North America chapter of this book.

Nuclear power

In late 2009, the government announced plans to build ten new nuclear power stations that will provide about 25% of the country's power by 2025.

The ten sites deemed "potentially suitable" for the new plants are: Bradwell, Braystones, Hartlepool, Heysham, Hinkley Point, Kirksanton, Oldbury, Sellafield, Sizewell, and Wylfa.

Observers noted, however, that a major overhaul of the current planning system will be needed in order to build the plants in the time frame projected.

Environmental Issues

Clean energy initiatives

In mid-2009, the government was moving ahead with a subsidy program for new offshore wind farms and considering one for pilot "clean coal" power plants that can capture and store their carbon dioxide (CO_2) emissions.

More controversial was possible action to provide financial guarantees for nuclear power projects.

The government has set a target of cutting UK CO_2 emissions by 80% by 2050, an ambitious goal that implies that all electricity production will have to be emissions-free.

New UK energy bill

The government introduced a bill in late 2009 that would provide financial incentives for energy companies to implement CCS technologies. Funded by a tax on consumer electricity bills, it would support up to four, commercial-scale CCS demonstration projects.

Other provisions would give energy regulator Ofgem increased powers, push a low-carbon mandate, and protect vulnerable households from electricity-price increases.

Smart Energy Meters

Under a plan announced in December 2009, UK energy suppliers will be responsible for the installation of 47 million 'smart meters' (gas and electric) in 26 million households by 2020.

The meters will facilitate more efficient flow of electric power.

The plan was developed by the Dept. of Energy and Climate Change (DECC).

In mid-2010, the government will release final details for the CCS incentive mechanism, which will become operational April 2011.

Carbon capture and storage

A number of projects were being carried out or planned in 2009 on technology to capture emissions of CO_2 from fossil-fuel power plants and store the CO_2 to prevent release to the atmosphere.

Ferrybridge

Scottish and Southern Energy PLC said in November 2009 that it plans to build the largest CCS demonstration plant in the UK at the Ferrybridge coal-fired power station in Yorkshire.

Operation of a 5-MW pilot plant could begin in 2011.

Aberthaw

In late 2009, RWE npower, a subsidiary of RWE AG, was poised to seek approval for a CCS project at the company's 1,500-MW coal-fired power plant in Aberthaw, Wales. The first phase would build a 3-MW demonstration plant, starting in 2011.

Teesside

Progressive Energy Ltd. and Centrica PLC were working in 2009 on the initial stages of a project to build the 850-MW Eston Grange facility in Teesside. It would combine an integrated gasification combined-cycle plant with a pre-combustion CCS concept.

CCS required for power plants

The government said in April 2009 that no new coal-fired power plants can be built unless they incorporate working CCS technology.

In addition, the government will demand that a full-scale retrofit to CCS be done within five years of the technology being independently judged as technically and commercially proven.

Call for carbon trading

The government's 2009 Global Carbon Trading Report called for creation of a global carbon-trading network to lower global emissions and reduce the high cost of implementing renewable energy projects.

It also noted that the EU Emissions Trading System (ETS) is already operating, and others are planned in 35 countries in the developed world.

Company News

EDF-Centrica-British Energy

In mid-2009, Electricité de France SA sold a 20% stake in the UK's nuclear operator, British Energy, to Centrica PLC in a deal worth $3.4 billion.

The decision secures for Centrica a chance to participate (with EDF) in what is seen as an upcoming push to build new nuclear plants in the UK.

British Energy, the leading supplier of nuclear energy in the UK, controls eight nuclear reactors with a combined capacity of 8.7 GW.

Heritage-Genel Enerji

In late 2009, independent Heritage Oil called off its proposed $6-billion deal to merge with Turkey's Genel Enerji in the wake of its agreement to sell its Ugandan oil fields to Eni, the Italian energy group.

Details are given in the Turkey section of this chapter.

Dragon-ENOC

Dragon Oil agreed in late 2009 to become a full subsidiary of Emirates National Oil Company (ENOC), Dubai's national oil company, in a $1.14-billion transaction.

ENOC, based in Dubai, said that the takeover of Dragon would be a "major step" to its goal of becoming an integrated oil and gas producer.

Centrica-Venture

In August 2009, Centrica PLC increased its stake to more than 58% in Venture Production, all but securing its £1.3-billion ($2.1-billion) proposed takeover of the North Sea oil and gas producer.

BP: Downplaying renewables

In mid-2009, BP PLC was moving to trim its role in alternative energy development, according to reports in the *Financial Times*, focusing more clearly than ever on oil and gas.

BP invested a peak of $1.4 billion in alternative energy projects in 2008, but that figure was expected to drop to $0.5–1.0 billion in 2009.

BP: Iraq

BP won approval in mid-2009 to return to Iraq to rehabilitate the giant Rumaila oil field.

Details are given in the Iraq section of the Middle East chapter of this book.

BG-Exco

BG Group made its first move into the U.S. shale gas sector in mid-2009, investing about £790 million ($1.3 billion) in a JV with Texas-based Exco.

BG took a 50% stake in 120,000 acres covering gas fields in north Louisiana and east Texas.

Most of the acreage is in the Haynesville shale gas formation, a significant gas resource.

Oilexco

In January 2009, the UK arm of Oilexco filed for administration, marking the first significant failure of a UK oil company following the 2008 plunge in the price of crude and the global economic crisis.

Oilexco was the most active driller of appraisal wells in the North Sea in recent years.

Scotland Activity

Orkney oil storage

Scapa Flow is a natural anchorage sheltered by the Orkney Islands, just off northern Scotland.

Over six days in early 2009, a ship-to-ship transfer of North Sea Forties oil was made there from a VLCC on long-term storage anchor to three receiving vessels.

Earlier, ConocoPhillips loaded three cargoes of crude into a supertanker that it chartered for long-term oil storage in Scapa Flow.

Observers estimated that some 20 MMbbl of North Sea oil was being stored on supertankers in European waters, creating an available source of supply that helped constrain North Sea crude prices.

Whitelee wind farm

In May 2009, a 322-MW, $413-million wind farm was inaugurated in Whitelee, Scotland, 15 km (9 mi) from Glasgow. It was called the largest onshore wind farm in Europe at the time.

Scottish officials approved plans to expand the plant's capacity to 452 MW, and the developer of the facility, ScottishPower Renewables (Glasgow), said it was evaluating eventual expansion to 600 MW.

Wave/tidal turbines

- Scottish and Southern Energy PLC agreed in early 2009 to set up a JV with Aquamarine Power, aiming to develop sites that could produce 1 GW of power from wave or tidal energy by 2020.
- In August, Aquamarine deployed a 315-kW demonstrator of its Oyster® wave energy converter at the European Marine Energy Centre (EMEC) in Orkney. A hydraulic system on the seabed delivers high-pressure water to an onshore turbine that generates electricity.
- In late 2009, Atlantis Resources Corp. was making plans to test what it called the world's largest tidal turbine in 2010. The 1-MW AK-1000 would be sited on the seabed near the EMEC.

CCS pilot: Longannet

In October, Scottish and Southern Energy PLC completed a successful 1-MW CCS pilot project at the 2,304-MW Longannet coal-fired power plant in Fife. A new process was said to have cut by one-third the energy needed for CO_2 separation. A commercial-scale plant at the site is scheduled to start up in 2014.

Subsea carbon storage

The North Sea off Scotland could sequester millions of tons of carbon emissions in saline aquifers and depleted oil and gas fields, said a report published in mid-2009.

Opportunities for Carbon Dioxide Storage Around Scotland was prepared by the Scottish Centre for Carbon Storage, the government, and industry partners.

The CO_2 storage capacity is comparable to that of Norway and greater than that of The Netherlands, Denmark, and Germany combined, the study said.

Middle East

BAHRAIN

Capital: Manama
Monetary unit: Dinar
Refining capacity: 253,650 b/cd
Oil production: 29,500 b/d
Oil reserves: 124.560 million bbl
Gas reserves: 3.250 tcf

Bahrain has one of the most varied economies in the Persian Gulf region and has become a major regional financial center, but its economy still depends heavily on oil.

Oil

Petroleum revenues currently make up about two-thirds of government revenue and export earnings, according to the U.S. Energy Information Administration (EIA).

Bahrain's two major industries are energy-intensive refining and aluminum smelting, so oil prices dramatically affect its economy.

Relatively high oil prices of recent years helped Bahrain's real GDP to grow. In its *World Factbook*, the U.S. Central Intelligence Agency projected a growth rate of 6.3% for 2008.

Bahrain has said it plans to invest $15 billion for oil and gas exploration and aims to increase oil production to 75,000 b/d by 2014 and to 100,000 b/d by 2030.

Natural gas

Bahrain also is planning to increase natural gas output from the present 1.7 bcfd to 2.5 bcfd, chiefly to keep up with growing demand for electric power.

The country's gas reserves are in decline. Talks with Iran about building an export pipeline south from that country stalled in 2008. And Qatar continues to observe a moratorium on new gas developments in its giant North Field, worried about the impact they would have on future output.

In early 2009, Royal Dutch Shell signed an agreement with the government to study gas import options, including construction of an LNG terminal.

Bahrain Petroleum Co. (BAPCO), as part of a $100-million venture, completed in late 2009 drilling three of eight planned gas wells in the Khuff reservoir.

PERSIAN GULF
Caspian Sea
TURKMENISTAN
AFGHANISTAN
I R A N
IRAQ
KUWAIT
SAUDI ARABIA
BAHRAIN
QATAR
UNITED ARAB EMIRATES
OMAN
Persian Gulf
Strait of Hormuz
Gulf of Oman
0 50 100 150 Miles
0 100 200 Km
35°
30°
25°
50°
55°
Astara
Ardebil
Hashtpar
Bandar-e Pahlavi
Rasht
Lulebur
Qazvin
Karaj
Tehran
Shahr-e Rey
220
Saveh
Hamadan
Qum
Arak
170
Isfahan
284
Shahreza
Yazd-e Khvast
Lordegan
Nazarabad
Ramhormuz
Ahwaz
Shushtar
Andimeshk
Jelow Gir
Basra
150
Khorramshahr
Abadan
350
Bandar-e Khomeine
Bandar-e Mah Shahi
Gachsaran
Ganaveh
Kazerun
Shiraz
40
Marv Dasht
Ardakan
Firuzabad
Bushehr
Deyyer
Asaluyeh
Bandar-e Maqam
Bandar-e Lengeh
LAVAN I.
30
Bandar 'Abbas
232
Qotbabad
Baft
Mashiz
Kerman
Bafq
Yazd
Ardakan
Torbat-e Haydari
Mashhad
Sabzevar
Shahrud
Damghan
Shahi
Neka
Behshahr
Gorgan
Shah Pasand
Dasht
Bojnurd
Quchan
Ashkoad
Sarakhs
Khangiran
Dauletabad Donmez
Okrem
Husan Kui
Korpedzhe
S. Bugdaili
Kamyshldzha
N. Okarem
Okarem
Keimir
S. Okarem
Akpatlaukh
Chikishlyar
Veyzenhar/ Malah kuh
Sarkan
Samand
Abu Ghirab
Qaleh Nar
Lab-i-Safid
Lali
Karun
Palangan
Zeloi
Masjid-I-Suleiman
Par-E-Siah
Nur-Buzurgan
Al Hafayan (Hafayat)
Majnoon
Sausangerd
Ahwaz
Naft Safid
Haft Kel
Doudrou
Kuh-I-Rig
Ramin (Mulla Sani)
Mansuri
Marun
Ab Teymour
Shurom
Kuh-E-Bangestan
Parsi
Karanj
Jufeyr
Nahr Umr
Ramshir
Rumaila
Zubair
Tuba
Mutriba
Sabriyah
Bahrah
Raudhatain
Ar Rawdatayn
Hendijan
Rag-E-Safid
Binak
Bahrgansar
Radak-Milaatun
Nargesi
Sarvestan
Nowrooz
Abuzar
Doroud (Darius) Kharg
200
Al Kuwait
Riquah
466
Abdullyah
Mina al Ahmadi
Burgan
Soroush
Minagish
Mina Abdullah
Dorra
Umm Gudair
270
Mina Su'ud
Hout
Wafra
Fawaris
DIVIDED ZONE
Hamur
Lulu/Esfandiar (Iran)
Marjan/Foroozan
BB
Bushgan
Kuh-i-Kaki
Dalan
Aghar
Kuh-i-Mand
Bandubast
Kangan
Nar
Varavi
Hasbab
Khafjil
Safaniya
S. Wafra
Fa
Pars
Salsal
Manifa
Sharar
Kurayn
Karan
Assaluyeh
Lamard
Tabnak
G
Watban
Ras al Ju'aymah
Jana
Fadhi
F
Bakr
Dhib
Jaladi
Fadhili
290
Al Jubail
Berri
Abu-Safah
South Pars
BH
Ras Tanura
Samin
Qatif
550
Ad Dammam
Abqaiq
254
Manama
Awali
Al Rayyan
North
Alpha
W
Resalat
Reshadat
Maydan Mahzan
Bul Hanine
Abu Al Bukhoosh
Salman
Nasr
139
Al Uqayr
Al Mubarraz
Dukhan
Umm Bab
200
Doha
Idd Al Shargi
Umm Said
Belbazem
Mandoos
Zakum
Umm Al Dalkh
Jarnain
Bu Jufair
Falah
Dalama
Hair Dhalma
Ghasha
Bu Tini
Hail
Fateh
Zol
Taweelah
Abu Dhabi
150
Ruwais
350
Bida Al Qimzan
Tarif
Bab
Habshan
Bu Hasa
Zubaya
Arjan
Rumaitha
Sahil
Hawaila
Asab
Shah
Qusahwira
Zayqat
Khurais
Ghawar
Harmaliyah
Haradh
Qirdi
Abu Jifan
Farhah
Mazalij
As Salamiyah
Dilam
Ad Dilam
Sahba
Midrikah
Tinat
Lughfah
Raghib
Jawb
Naim
W. Namak
Sarkhun
Suru
S. Gashu
HD
Gavarzin
Salakh
Hanquan
Saleh
Bukha
W. Bukha
Limah
Baih
Umm al Qaiwain
Sharjah
Sirri A
Sajaa
82
71
Mo'Ayyid
S. Juwaiza
Khubai
Dubai
120
Al Fujairah
Jebel Ali
Margham
TFP
Fujayrah
Mahdah
Suhar
Al' Ayn
Al Khaburah
Muscat
Mina al Fahal
85
Shams
Safah
Lekhweir
Mender
Deleel
Nadir
Al Sahwa
Hafar
Shibka
Nati
M. Hurayma
Habiba
Abeer
Baqlah

Awali fields

Occidental Petroleum Corp., Mubadala Development Co., and the National Oil & Gas Authority of Bahrain (NOGA) formed a joint operating company in late 2009 to revitalize the Awali onshore oil fields.

The joint entity, Tatweer Petroleum-Bahrain Field Development Co., will be operator for the fields.

The Awali fields are reported to have proven resources of 125 million bbl.

In the near term, eight drilling rigs and eight work-over rigs will be employed to boost production.

Bahrain-Saudi oil pipeline

Bahrain and Saudi Arabia were finalizing in late 2009 the details for construction of a new crude oil pipeline between the two nations. The 100-km (62-mi), $350-million line will carry 350,000 b/d and start operation by late 2011.

The new pipeline will replace an existing line with a capacity of 230,000 b/d.

Expansion of Sitra refinery

BAPCO announced in early 2009 that it will invest some $2 billion to augment the capacity of its Sitra refinery, the nation's largest. Built in 1936, it is also one of the oldest in the GCC region.

The expansion would raise capacity from the current 250,000 b/d to 350,000–400,000 b/d.

About one-sixth of the refinery's feed comes from oil fields in Bahrain, the rest from Saudi Arabia.

CO_2 recovery plant

In December 2009, Gulf Petrochemical Industries Co. (GPIC) (Sitra, Bahrain) inaugurated operation of a CO_2 recovery plant at its complex on the island of Sitra. The $55-million plant is said to be the largest in the world.

The CO_2 emitted in flue gases is recovered using the Kansai-Mitsubishi Carbon Dioxide Recovery (KM CDR) Process.

The new recovery plant will reduce greenhouse gas emissions by more than 90%, said GPIC, capturing 450 tpd of CO_2.

Recovered CO_2 will be used as feedstock to produce methanol and ammonia in the complex.

GCC: $18.3 TRILLION IN OIL, GAS RESERVES

An economic report released in late 2009 by the Dubai International Financial Centre estimated that the six nations of the Gulf Cooperation Council—Bahrain, Kuwait, Oman, Qatar, Saudi Arabia, and the United Arab Emirates—have combined oil and gas reserves worth $18.3 trillion.

The estimate was based on conservative prices for oil ($50/bbl) and natural gas ($9/MMBtu).

IRAN

Capital: Tehran
Monetary unit: Rial
Refining capacity: 1.451 million b/cd
Oil production: 3.725 million b/d
Oil reserves: 137.620 billion bbl
Gas reserves: 1,045.670 tcf

Overview

Despite steep production declines and problems with oil field development in recent years, Iran's crude and condensate production capacity is expected to reach 4.7 million b/d by 2012, almost all from South Pars field, according to FACTS Global Energy.

Iran has very large gas reserves, but FGE sees a ceiling of 20–30 mtpy on Iranian gas exports. Domestic demand has been kept high by subsidy pricing, oil field needs for gas injection, large gas-based petrochemical projects, and political opposition to gas exports.

In late 2008, Iran was importing 20–23 MMcmd (700–810 MMcfd) of gas from Turkmenistan, or about 5% of its total consumption.

Iran's only gas exports were to Turkey, in the range of 20 MMcmd (700 MMcfd).

Iran must import about one-third of its domestic liquid fuel. In 2007, the government used rationing to curb demand. In 2009, the oil ministry took initial steps to reduce fuel price subsidies to cut use.

Strait of Hormuz

In December 2009, a report by the U.S. Office of Naval Intelligence noted that the government of Iran had given the Islamic Revolutionary Guard Corps command over naval operations in the Persian Gulf and Strait of Hormuz.

ONI described the move as part of a strategy to block international access to vital sea lanes in the event of a war. Oil movements through the strait account for 40% of all seaborne oil traded in the world, according to the U.S. Energy Information Administration.

Nuclear development

Diplomatic sparring continued in 2009 between Iran and other nations concerned about its nuclear development program.

In mid-October, Iran began negotiations with the United States, France and Russia on a draft proposal from the International Atomic Energy Agency. It called for rapid transfer of two-thirds of Iran's current stock of low enriched uranium out of the country, for conversion into medical isotopes.

However, at the end of the month, talks broke down over the rate of transfer.

Iran did let IAEA inspectors visit its previously secret uranium enrichment plant at Qom. But its unwillingness to reveal more about its nuclear program set the scene for possible tougher sanctions.

Iran has been inviting outside participation in its oil sector, but lack of confidence in the country's political environment and concerns about its economy remain a deterrent.

Development of Iran's energy sector also has been hindered by trade embargoes imposed by the United Nations and the United States, trade sanctions and obsolete technology.

In addition, Iran continues to face strong Western opposition to its nuclear development program.

Fuel subsidies

Iran's parliament approved in October 2009 the introduction of steep cuts in energy and food subsidies.

Such subsidies cost about $100 billion/year, and economists had observed that the Iranian government cannot afford to maintain them at current levels. The subsidies cover oil, gasoline, natural gas, and electricity as well as water, bread, transport and telephone services.

Exploration and Development

South Pars: Phase 11

At mid-year 2009, China was preparing to enter a cooperative program with Iran on the Phase 11 development of the South Pars gas field in the Persian Gulf.

Total SA had an MOU for Phase 11, but NIOC chose to seek new partners due to continuing delays.

NIOC aims to produce 2 bcm/d of gas and 70/000 b/d of condensate from Phase 11.

Binaloud oil field

India's ONGC Videsh announced a discovery in early 2009 in the large Binaloud oil field in the Persian Gulf, according to Iranian Offshore Oil Co (IOOC).

The field, in the Farsi Block, could hold reserves of 1 billion bbl of heavy crude, said IOCC.

The block also holds an estimated 12.8 tcf of recoverable gas.

Darkhowain oil field

Italy's Eni said in early 2009 that it will conduct third-phase development of Darkhowain oil field to raise its output to 160,000 b/d from the current 100,000 b/d.

Drilling and Production

Salman gas field

Gas flow began in October 2009 from the first phase of the Salman field development in the Persian Gulf, according to Petro Iran Co.

The field's KPP gas production platform can produce 300 MMcf/d. Output will gradually increase to 700 MMcf/d, according to Shana, the official news agency of the National Iranian Oil Co.

South Pars output boost

In October 2009, Iran was preparing to raise its natural gas output by 1 bcf/d as a new platform came on stream in the

South Pars field. The new installation is part of the $5-billion South Pars Phases 6, 7, and 8 development project.

Gas from these phases will be injected into the Agajari oil field for pressure support.

$2 billion for domestic oil, gas projects

In third quarter 2009, the Iranian government announced plans to invest about $2 billion on development and expansion of oil and gas projects in the country.

Priority will be given to the NGL 3100 project and the South Gesho project.

South Gesho gas field has reserves of 5 tcf and is estimated to yield 14.2 million cu m/day.

2005–2009 investment

NIOC reported in mid-2009 stated that in the last four years, about $1.7 billion was invested in Iran's oil and gas sector, with about $820 million from domestic sources.

During the period, 17 oil and gas fields and 22 reservoirs were discovered, and seven new contracts for exploration and block development were signed with European and Asian companies.

International Activities

Iran-Oman

Iran and Oman restarted talks in late 2009 about export of gas from Iran.

In 2008, the two countries signed an MOU regarding the Kish gas field in the Persian Gulf, which has in-place gas reserves of 1.36 tcm (48 tcf). Iran would develop the field by 2012, while Oman would provide the investment, estimated at $12 billion.

Phase 1 of the project is expected to produce 2 bcf/d of gas starting in mid-2011. Approximately 65% of the gas would be used by Iran, with the remainder exported to Oman.

Venezuela, Uganda, Syria

In 2009, Iran and Venezuela signed agreements to invest in each other's energy sector and to cooperate on oil and gas development in both nations.

Iran and Venezuela also signed an MOU to build a 140,000-b/c oil refinery at Furqlus in Syria.

And in mid-2009, Uganda said it will build a domestic refinery with Iranian assistance rather than export raw crude oil.

For details, see the Venezuela, Uganda, and Syria sections of this book.

Presidential Election and Aftermath

On June 12, 2009, Mahmoud Ahmadinejad won reelection as president in a contest hotly disputed by his main challenger, Mir-Hossein Moussavi, and others.

Observers noted that the results appeared to be at odds with the huge groundswell of support that formed around Mr. Moussavi in the final week of the campaign.

Although thousands of supporters of Pres. Ahmadinejad publicly celebrated his claimed 62.6% of the vote, many more followers of Mr. Moussavi protested the outcome and took to the streets.

But Ayatollah Ali Khamenei, Iran's supreme leader, declared the contest over.

Protest actions escalated in the weeks after the election in what some observers called the worst internal turmoil in Iran in 30 years. News organizations reported that at least 17 protesters were killed.

Significant protests recurred sporadically during the rest of 2009, and early 2010 brought reports that the government executed several people who had taken part in protests.

Turkey

In late 2009, Iran and Turkey agreed to promote several oil and gas projects in the region. The agreement includes:

- Transportation of gas from Iran to European markets by way of Turkey.
- Development of Iranian gas fields (including South Pars).
- Wider participation by Turkish firms in Iran's future petrochemical ventures.
- Development of a $2-billion crude oil refinery in northern Iran.

The planned 3,300-km (2,050-mi) Nabucco pipeline likely will play a vital role in shipping gas out of Turkey and into Europe, in competition with Russian gas supplies.

LNG

LNG, field-development deal

In December 2009, India's ONGC and a group of partners agreed to invest $10 billion in an LNG project and the development of two large gas fields in Iran.

In return, India is to receive at least 6 mtpy of LNG on a long-term basis.

The $4.32-billion LNG facility at Tombak Port on the southern coast will convert gas procured from two fields—South Pars, and Farzad B in the Farsi Block—into LNG for tanker shipment.

The partnership also will acquire a stake in the Phase 12 (SP-12) development of South Pars gas field.

Full development of SP-12 and Farzad-B is expected to take 7–8 years.

Pipelines/Transportation

IPI gas pipeline

In mid-2009, Iran and Pakistan were moving ahead on a long-awaited gas pipeline between the two countries.

The $7.6-billion project—called both the Iran-Pakistan-India (IPI) Pipeline and the Peace Pipeline—has been delayed by disagreements over transit fees that have put India's participation in doubt.

Pakistan also had been criticized by the United States for economic interaction with Iran.

Officials have said that India would have the option to join the project in the future, and that a link to China might also be considered.

The line would carry about 21.2 million cu m/d (750 MMcfd) gas some 2,100 km (1,309 mi) from Iran's South Pars field.

The pipeline would cover 1,100 km (684 mi) in Iran and 1,000 km (621 mi) in Pakistan, and possibly be extended 675 km (419 mi) further to Rajasthan, India.

As of mid-2009, about 700 km (435 mi) of pipeline from the South Pars gas field had been built, with the remaining 400 km (249 mi) inside Iran under construction.

Completion of the entire line is anticipated by 2012.

Turkmenistan gas to Iran

Turkmenistan was making plans to open a new pipeline in late 2009 to carry natural gas to Iran.

Details are given in the Turkmenistan section of the Former Soviet Union chapter of this book.

Caspian oil export pipelines

In mid-2009, Iran was preparing to issue a tender for construction of a new oil export pipeline by 2012 from the Caspian Sea to the Persian Gulf.

The 970-mi (1,561-km) pipeline would transport 1 million b/d from the Iranian port of Neka on the southern Caspian coast to Jask in the Gulf of Oman.

Information agency Shana said that the tender would be open to international companies.

Pakistan-China oil pipeline

China and Pakistan were reportedly working on a proposal in mid-2009 for a trans-Himalayan pipeline that would carry Middle Eastern crude to western China.

The line would enable China to import oil from the Gulf region without having it travel through the dangerous Strait of Malacca.

Once the oil reached China, it still would have to be shipped thousands of miles further east to major energy markets on the coast.

The pipeline would run from the southern Pakistan port of Gwadar and follow the Karakoram highway.

Gwadar is considered the likely terminus of proposed major pipelines that would carry gas from the South Pars fields in Iran, from Qatar, or from the Daulatabad fields in Turkmenistan, for export.

Refining and Products

Refinery-sector status

In early 2009, FACTS Global Energy reported on Iran's refining sector. Highlights from the report, *Iran's Gasoline Imports and U.S. Politics: An Update on Iran's Refining Buildup*, include the following:

- Iran is embarking on a large refining expansion and upgrading program. By 2012–13, Iran will not need any gasoline imports and may be an exporter after that.
- The government intends to push more gasoline into the non-regulated price market to cut domestic consumption.

- When rationing was imposed in mid-2007, reduced imports saved the country about $2–3 billion/year.
- Gasoline imports for 2008 averaged 112,000 b/d, about 40% lower than 2007.
- It is not clear that Iran can find enough capital and contract labor to build six new refineries that have been announced: Hormuz, , Fars, Shahriar, Caspian, Anahita, and Khuzestan.
- Crude distillation expansions are under way in Lavan, Arak, Tabriz, and Isfahan refineries.
- Distillate treating unit capacity will increase tenfold from 56,000 b/d. This will enable Iran to produce distillate products at Euro Spec 4 and 5 levels by the middle of the next decade.
- The Persian Gulf Star Project is under way. The Bandar Abbas refinery will process 360,000 b/d of South Pars condensate, focusing on gasoline production. Completion by 2013 is expected.

Chinese gasoline

In September 2009, Chinese state oil companies began providing 30,000–40,000 b/d of gasoline to Iran—a level approaching one-third of total imports.

The action runs counter to U.S.-led efforts to restrict fuel supply as a way to dampen Iran's nuclear ambitions.

Petrochemicals

JVs with foreign firms

Iran announced plans to set up five petrochemical projects in joint ventures with foreign companies between April 2009 and March 2010.

State-owned National Petrochemical Co. was expected to provide $2.5 billion of the estimated $8 billion required, with the rest coming from bonds and foreign investments.

Iran's plans call for boosting export of petrochemical products by 30%, to 39 million tons, in 2010.

Iran also said it intends to build 47 petrochemical plants by 2015, adding some 43 mtpy of production capacity. This would boost Iran's production to more than 5% of global output and almost 36% of total output from the Middle East.

IRAQ

Capital: Baghdad
Monetary unit: Dinar
Refining capacity: 637,500 b/cd
Oil production: 2.400 million b/d
Oil reserves: 115.00 billion bbl
Gas reserves: 111.940 tcf

Overview

Rebuilding

Prime Minister Nouri al-Maliki and Iraq's Council of Representatives (national parliament) continued efforts to establish an effective central government, ensure security, rebuild infrastructure, and reestablish social and commercial activities in the country in 2009.

Sectarian violence and attacks against U.S. and coalition forces continued in Iraq in 2009, but at a much reduced level. March marked the sixth anniversary of the U.S.-led invasion to depose then-president Saddam Hussein.

Withdrawal of U.S. combat troops from the country is scheduled to begin in July 2010.

Oil and gas status

Iraq holds the world's third largest proven reserves of crude (some 115 billion bbl) and an estimated 3.17 tcm (112 tcf) of natural gas.

It seeks to boost oil production to 6 million b/d by 2018. However, decades of wars, UN sanctions, violence and sabotage have impaired the country's ability to ramp up production.

Work continued in 2009 on formulation of a federal law regarding the development of the country's oil sources, the sharing of oil revenues, and the appropriate role of foreign companies in oil development.

Auctions in June and December led to deals with several IOCs for developing dozens of oil fields, reopening the country to foreign oil firms for the first time in more than 30 years.

A long-standing dispute continued between the federal government and the Kurdistan Regional Government (KRG) regarding oil contracts. The KRG separately has signed exploration and production contracts with several foreign firms, drawing criticism from Baghdad and stirring controversy in the international industry.

In June, the oil ministry said such agreements were permissible, but that the Iraqi constitution requires the KRG to

Middle East

TURKEY
SYRIA
IRAQ
IRAN
LEBANON
ISRAEL
JORDAN
WEST BANK
GAZA STRIP
KUWAIT
SAUDI ARABIA
Red Sea
Lake Van
Lake Uromiyeh
Buhayrat ath Tharthar
Bahr al-Milh
Aksaray
Mersin
Ceyhan
Dortyol
Batman
Akpinar
Firat
Bolukyala
Cendar
Cukurtas
Cemberlitas
Adiyaman
Kucukpirin
Pyanko
Besikli
Kakus S.
Kahta
Zeynel
Mehmetdere
Kosh
Kayakoy
Hazro S.
Selmo
Yatir
Barbes
Papur
Beycayit
Dodan
Kurkan
Kastel
Celiki
Karaali
Kurtalan
Mollcabir
Magrip
Ambarcik
Garza - Germik
Camurlu
Kozluca W.
Kiztepe
Derik
Zurabeh
Hamzeh
Awbah
Kotbah
Barde
Karatchok
Suedie
Mushora
Ain Zalah
Butmah
Bulgurdag
Shaikh Mansour
Shaikh Sulaiman
Al Hawl
Jeribe
Tishreen
Alaounet
Salhieh
Hasaw
Jibbisah
Ghouna
Gbebeh
Sasan
Alan
Mosul
Demir Dagh
Irbil
Al Qawayr
Al-Quway
Bai Hassan
Kirkuk
Qayyarah
Chamchamai
Khabbaz
Al Hadr (Hatra)
Ash-Sharqat
Jambur
Halabia
Kor Mor
Hamrin
Bayji
Pulkhana
Injana
Kifri
Chia Surkh
Saddam
Tikrit
Gilabat
Khanaqin
Emman Hassan
Naft Khaneh
Naft-e Shah
Sumar
Sar-e Pol-e Zahab
Kermanshah
Samarra
Balad
Hit
Fallujah
Tang-E Bijar
Baghdad
E. Baghdad
Ilam
Mehran
Samand
Veyzenhar/ Malah kuh
Sarkan
Halush
Jelow Gir
As Suwayrah
Ahdab
Karbala
Al Hashimiyah
Al Kut
Al Hayy
Abu Ghurab
Ad Diwaniyah
Kifl
Gharraf
Abu
Sukhayr
An Nasiriyah
Ar Rifai
Buzurgan
Majnoon
Ash Shatrah
As Samawah
Samawah
Suq ash Shuyukh
West Qurnan
Nahr Umr
Zubair
Jalibar
Az Zubair
Rumaila
Safwan
Mutriba
Ar Rawdatayn
Abu Khaimah
Raudhatain
Rafha
Al Qaysumah
Shanpur
Tabriz
Bostanabad
Uromiyeh
Naqadeh
Tibni
Ammala
Sfayeh
Derro
Wadi Vbaid
Attalah N.
Al Mahash
An Nishen
Markada
Habari
Wahab
Al Kharrata E.
Omar N.
El Isba
Sijan
Ash Shola
Omab
Thayyem
Saban
Sajjannah
Tanak
Al Ishara E.
Tayyam E.
Sarhit
El Ward N.
Al Ahmar
Ratka
Mayadin
Akash
Akkas
Al-Haditha
Ash Shaer
Najib
Doubayat
Tadmur
Arak
Al-Hail
Soukhne
Baniyas
Tartus
Homs
Tripoli
Beirut
Sidon
Damascus
Tamar
Haifa
Nazareth
AGP
Meged
Tel Aviv-Yafo
Rihab
Amman
Jerusalem
Kochav
Hebron
Heletz-Brur
Kidod
Zohar
Gurim
Hamza
Ar Rutbah
Risha
Wadis Sirhan
Badanah
Al Jawf
Sakakah
Kahf
Elat
Taba
Al Bad
Midyan
Ash Sharmah
Duba
Al Waih
Marsa Alam
Umm Cajj
Medina
Yanbu al Bahr
Badr Hunayn
Riyadh
Dumah
As Salamiyah
Dilam
Ad Dilam
Raghib
Hilwah
Al Hillah
Al Hawtah
Naim
95
22
133
107
18
130
90
12
25
310
110
7
2
40
100
235
400
120
35°
30°
25°
36°
40°
44°
0
100
200 Miles
0
100
200
300 Km
MIDDLE EAST

coordinate with the federal government on all oil-related matters.

As of yearend 2009, no federal oil law had been passed.

Oil and gas outlook

Business Monitor International Ltd. (BMI) noted that Iraq has ambitious targets for its oil sector.

However, the firm sees major risks ahead, including attacks on oil installations, decrepit infrastructure, Iraq's OPEC production quota, and the degree of success of new energy policy to stimulate investment by IOCs.

BMI estimated that average oil production would grow to 3.15 million b/d in 2012, and then rise steadily to 4 million b/d by the end of 2018.

Natural gas production is expected to climb to 35 bcm (1.23 tcf) by 2018.

Investment needs

The Center for Global Energy Studies estimated in April 2009 that Iraq will need to invest $28–43 billion to raise its oil production capacity to 6 million b/d.

The estimate includes repair of pipelines, storage, and loading facilities to return capacity to 3.5–4 million b/d.

Analysis of fields, reserves

A study by Petrolog & Associates in early 2009 identified 80 significant Iraqi fields that hold 115 billion bbl of proved reserves. Key findings included these:

Location of Oil Structures in Iraq

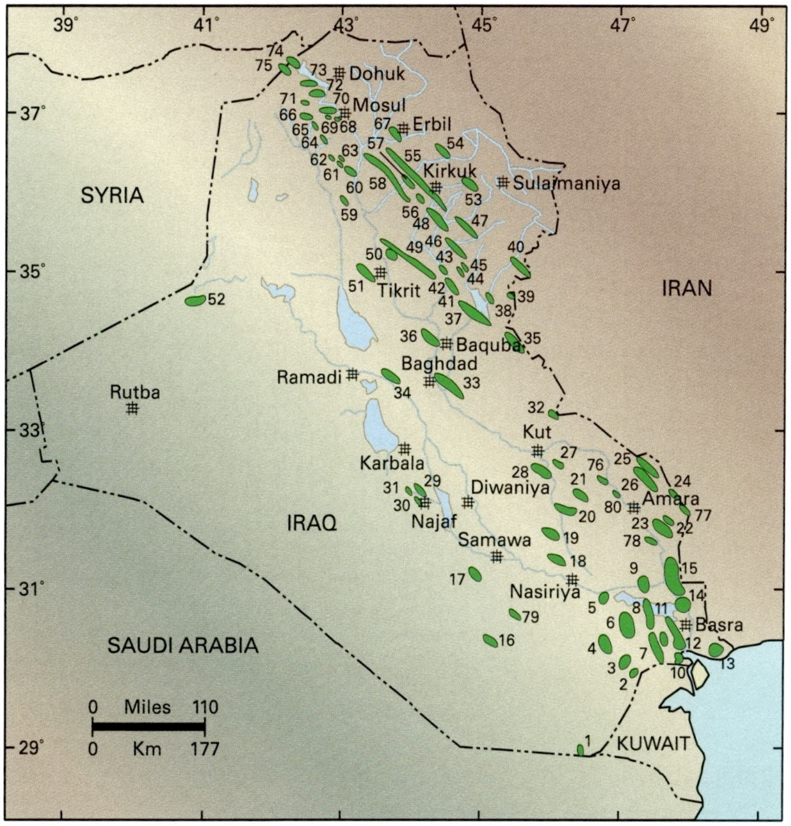

1 Abu Khaima
2 Jerishan
3 Rachi
4 Luhais
5 Subba
6 Ratawi
7 Rumaila S.
8 Rumaila N.
9 W. Qurna
10 Safwan
11 Tuba
12 Zubair
13 Siba
14 Nahr Umr
15 Majnoon
16 Salman
17 Samawa
18 Nasiriya
19 Gharraf
20 Abu Amud (Rafidain)
21 Dujaila
22 Noor
23 Halfaya
24 Jabal Fauqi
25 Abu Ghirab
26 Buzurgan
27 Dhafiriya
28 Ahdab
29 Kifl
30 W. Kifl
31 Marjan
32 Badra
33 E. Baghdad
34 Falluja
35 Naft Khaneh
36 Balad
37 Mansuriya
38 Nau Duman
39 Jaria Pika
40 Chia Surkh
41 Kashm al Ahmar
42 Injana
43 Judaida
44 Gilabat
45 Qamar
46 Pulkhana
47 Kor Mor
48 Jambur
49 Hamrin
50 Saddam
51 W. Tikrit
52 Akkas
53 Chemchemal
54 Taq Taq
55 Kirkuk
56 Khabbaz
57 Bai Hassan
58 Qara Chauq
59 Sadid
60 Qaiyarah
61 Majmah
62 Najman
63 Qasab
64 Adaiyah
65 Ibrahim
66 Sasan
67 Demir Dagh
68 Qalian
69 Atshan
70 Alan
71 Gusair
72 Butmah
73 Ain Zalah
74 Mushorah
75 Sufaya
76 Amara
77 Huwaiza
78 Rifae
79 Diwan
80 Kumait

- Only seven oil fields have been fully developed and another 15 partly developed.
- The seven largest fields—Rumaila South and North, Kirkuk, East Baghdad, Majnoon, West Qurna, and Zubair—contain two-thirds of total reserves.
- Iraq may have the greatest endowed oil reserve base in the world, with total proved reserves of 145.5 billion bbl of oil and potential reserves in excess of 215 billion bbl.
- Evaluation of 157 exploration wells showed that almost every field has multiple reservoirs. The drilling success rate is almost 80% and the discovery rate 70%.
- Iraq thus far has produced only 8% of its oil source base, though it has over 19% of the world reserves and over 31% of the total production of the Middle East major producers.
- Iraq can develop a production plateau of 10 million b/d, but doing so likely would take 15–20 years.

Cumulative Exploration Success Rates

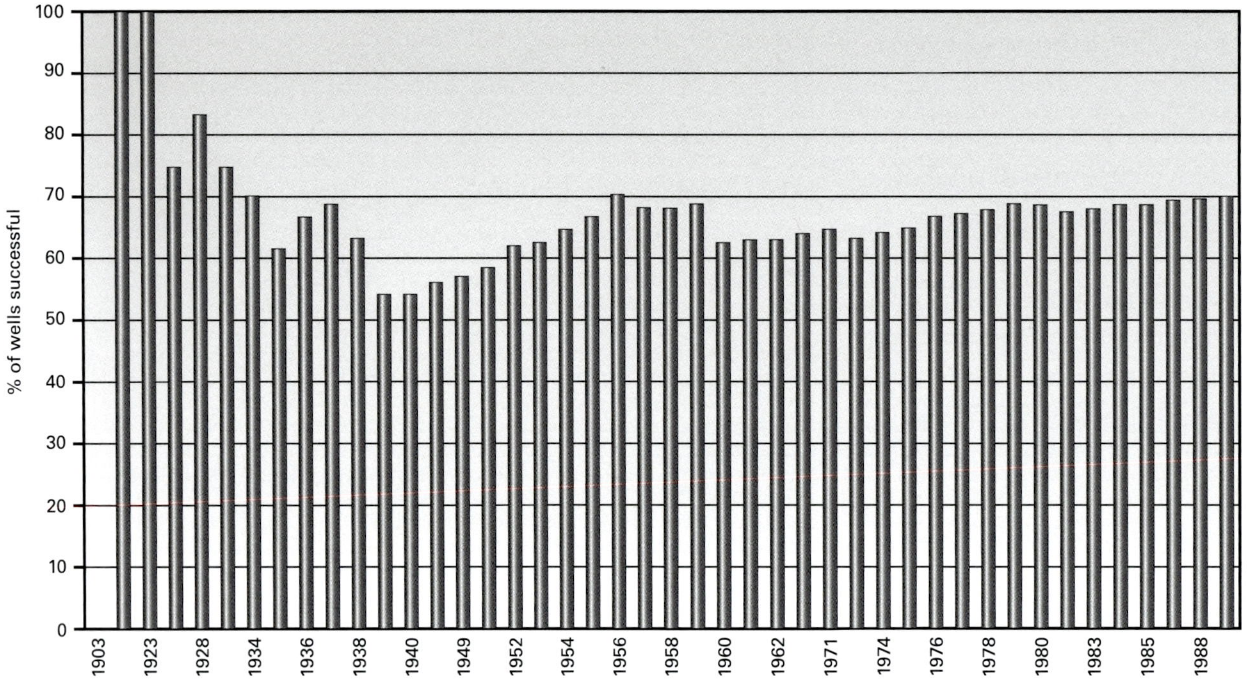

Proved and Potential Oil by Country or Region

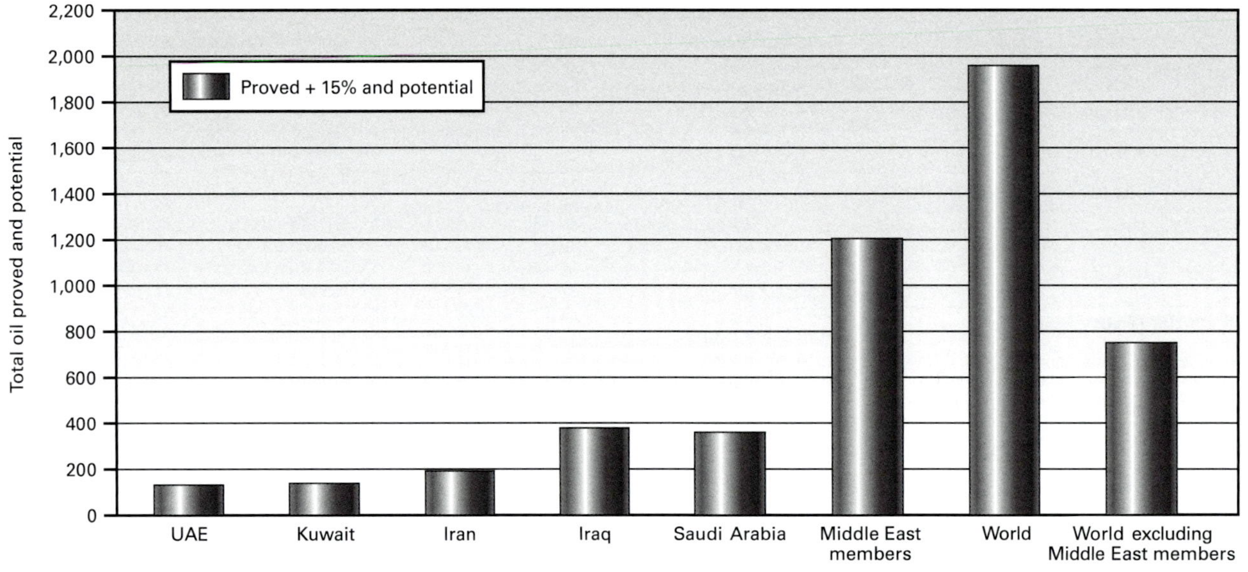

Iraq Oil in a Global Context

Country or region	Production: Billion bbl	Production: % total	Proved + 15% & potential	Oil resource base total, billion bbl
UAE	24.0	15.1	135	159 (55.5 from peak)
Kuwait	35.8	20.3	140	176 (55.2 from peak)
Iran	58.3	23.5	190	248 (65.7 from peak)
Iraq	30.5	8.0	380	410 (174 from peak)
Saudi Arabia	106.4	22.6	365	471 (129 from peak)
Middle East majors	255.0	17.4	1,209	1,464 (477 from peak)
World	1,016.0	34.1	1,964	2,980 (474 from peak)
World excluding Middle East majors	761.0	50.2	754	1,515 (–3.5 from peak)

Source: Petrolog & Associates

Field Development by IOCs

Auction of development rights

Iraq seeks to increase its oil output from the current 2.4 million b/d to at least 6 million b/d by 2018.

The oil ministry said in April 2009 that the oil industry needs $50 billion in investment over the next 5–6 years and much of it is expected to come from IOCs. To that end, officials took important action in 2009.

In June, Iraq allowed the first western oil group back into Iraq in more than 30 years. BP and partner CNPC will repair and develop Rumaila, Iraq's biggest oil field.

In addition, Iraq held two auctions for the rights to develop dozens of oil fields.

In the first round, in June 2009, just one of eight offered blocks found a bidder. The second round, in mid-December, saw 45 companies bid for 45 fields, leading to these outcomes:

- Royal Dutch Shell, in partnership with Petronas, won the right to develop the large Majnoon field, north of Basra near the Iranian border. Current production is 46,000 b/d, but Shell has pledged to increase that to 1.8 million b/d.
- CNPC, China's largest oil and gas producer, with Petronas and France's Total won a contract to develop Halfaya field. Current production is 3,000 b/d.
- Russia's Lukoil and Norway's Statoil won rights to the major West Qurna-2 field.
- Iraq accepted earlier bids from an ENI-led group for the Zubair field, and from ExxonMobil and Shell for West Qurna-1.
- A consortium led by Japan Petroleum Exploration Co. (Japex) won the rights to develop Gharaf oil field.

Further Auction Details

Estimates from 2003 put original oil in place for Majnoon field at 38 billion bbl and original reserves at 12 billion bbl. Corresponding figures for Halfaya are 16 billion bbl and 4.6 billion bbl.

At yearend 2009, Iraq still had no oil law in place to guarantee and protect foreign oil investment, and none was likely to be enacted until after elections set for early 2010.

Also, some observers questioned the ability of the government to build or repair needed infrastructure and to keep it secure.

Others noted that OPEC would eventually have to reduce the production quotas of many of its members to accommodate increased production by Iraq.

Exploration and Development

Block 8

ONGC Videsh Ltd. (OVL) was poised in early 2009 to sign a renegotiated exploration agreement with Iraq. The agreement was originally signed in 2000 during the Saddam Hussein regime.

The deal will allow OVL begin exploring and developing Block No.8, in southern Iraq, bordering Saudi Arabia. The block, is estimated to have reserves of 645 million bbl (54 million bbl recoverable).

The company is expected to invest $186 million in exploration and $1.45 billion for development.

Kurdistan well: Shaikan-1

Gulf Keystone Petroleum International Ltd. (GKP) reported in mid-2009 that its Shaikan-1 well in Iraqi Kurdistan on Shaikan Block found oil.

The well was drilled 85 km (53 mi) northwest of Erbil in an underexplored area near Dihok. It found oil in the Middle Jurassic Sargelu formation at a depth of 1,450–1,510 m (4,757–4,954 ft).

Flow tests indicated 5,000–8,000 b/d. The company said the oil's properties are comparable to oil produced at Tawke field to the northwest of Shaikan Block.

Giant Kurdistan find

In May 2009, Heritage Oil Corp. reported discovery of a giant oil field in Iraqi Kurdistan with 2.3–4.2 billion bbl of oil in place (50–70% recoverable).

The Miran West-1 wildcat showed a restricted flow of 10,000–15,000 b/d from a 500-m (1,640-ft) gross interval in the Shiranish, Kometan, and Qamchuqua formations.

A second interval of 710 m (2,329 ft) produced oil with low sulfur, no water, and a low gas-oil ratio.

Heritage said it could start trucking production from Miran West-1 by the end of 2009.

Khor Mor field and Nabucco

OMV of Austria and Hungary's MOL agreed in May 2009 to invest, with partners, $8 billion to develop a major gas field in Iraqi Kurdistan. The two firms each bought a 10% stake in Pearl Petroleum, which is developing the Khor Mor field.

The field will supply the planned 3,300-km (2,050-mi) Nabucco pipeline, which would bring gas from Central Asia to Europe.

Pearl—jointly owned by Crescent Petroleum and Dana Gas, two UAE-based exploration and production companies—already has spent $605 million developing Khor Mor.

MOL and OMV also each hold a 16.67% stake in Nabucco.

Observers note that Nabucco will face a challenge to find gas suppliers and will also compete with the South Stream pipeline being built by Gazprom and Eni.

Drilling and Production

Tawke and Taq Taq

Modest volumes of Kurdish oil began flowing May 30, 2009, through a pipeline from Tawke and Taq Taq oil fields in Kurdistan to Iraq's main northern export pipeline, which runs to the Turkish port of Ceyhan.

An initial rate of 10,000 b/d was anticipated to rise later to 50,000 bbl/d.

Tawke was described as the first newly developed oilfield to come on stream since 2003 and the first to come on stream anywhere in Iraq for more than 30 years.

Tawke was developed by Norway's DNO International. Taq Taq was a JV, run at the time by Addax Petroleum and Turkey's Genel Enerji.

IOSC JV

In early 2009, Mesopotamia Petroleum Co. Ltd. (MPC) signed an agreement with state-owned Iraqi Drilling Co. to form a new JV aimed at increasing oil and gas production.

The deal was described by MPC as the Iraqi oil ministry's first JV agreement of its type signed with a foreign company since the 2003 fall of Saddam Hussein.

The JV company will drill several wells for the nation's oil companies and international operations.

About 60 wells were planned for the area around Basra in 2009.

Policy change: IOC stakes

In an effort to attract investment, Iraq in early 2009 reversed earlier plans and decided to let IOCs hold as much as 75% stakes in oil drilling projects.

However, IOCs still will not be allowed to book oil reserves—a sign the government intends to keep close control over the nation's oil and gas.

Pipelines/Transportation

Securing pipelines

Most of Iraq's 7,000-km (4,350-mi) domestic oil pipeline network has been badly damaged from attacks by terrorists and criminals, said a senior Iraqi official in early 2009.

He estimated that 80% of Iraq's pipelines linked to refineries and to industrial and power plants have suffered "between 50–100% total destruction."

Starting in 2008, the Oil Police Directorate upgraded its coordination with security personnel of the ministry of the interior and the ministry of defense. A total of 22 battalions are to be in the field by 2012.

Refining and Products

First Kurdistan refinery

In June 2009, the Kurdistan Regional Government of Iraq opened its first oil refinery in Irbil.

Processing crude oil from the Khurmala Dome oilfield, initial production was 20,000 b/d.

The rate was expected to increase to 40,000 b/d by yearend and eventually to 75,000 b/d.

Products will include heavy fuel oil, diesel, kerosene, paraffin, aviation fuel, and gasoline.

KRG officials have plans to build more refineries, increasing capacity to 200,000 b/d by 2011.

Karbala refinery

Iraq's State Company for Oil Projects awarded a $25-million FEED contract in early 2009 to Technip SA for a new refinery to be constructed in Karbala.

Total capacity will be 140,000 b/d. The facility will produce LPG, gasoline, diesel oil, jet fuel, fuel oil and asphalt, mostly for the domestic market.

The project is scheduled for completion by mid-2010.

Iraqi oil, gas infrastructure

Company News

DNO

In October 2009, Norway's DNO got approval to restart operations after the government lifted a suspension related to a controversial sale of DNO shares.

The company said it would not resume export production, focusing instead on sales within Iraq until a payment mechanism was agreed to with the central government in Baghdad.

DNO started exporting more than 40,000 b/d oil a day from Kurdistan in June but had not received payment as of October. The central government had not set up a formula for distributing oil revenues.

Parliamentary elections

Iraq's third election of members of parliament since 2003 had been scheduled for December 2009.

But in mid-May and again in December 2009, Iraqi officials announced that the election would be delayed. At yearend, the specified date was March 7, 2010.

Sinopec-Addax

In mid-2009, Sinopec agreed to a $7.2-billion takeover of Switzerland-based Addax Petroleum, which has interests in Iraqi Kurdistan.

Analysts with Dealogic called the transaction China's largest ever outbound investment in the oil and gas sector.

Addax has a long history in Iraqi Kurdistan, operating the Taq Taq oil field and refinery.

Heritage-Genel

A merger in mid-June 2009 valued at $5.5 billion brought together Heritage Oil and Genel Enerji, the privately held Turkish oil company that is its JV partner developing oil fields in Iraqi Kurdistan.

The USGS estimates that the region could contain 40 billion bbl of oil.

Approval of the merger, creating a new firm called HeritaGE Oil, was expected by fourth quarter 2009.

ISRAEL

Capital: Jerusalem
Monetary unit: Shekel
Refining capacity: 220,000 b/cd
Oil production: 0
Oil reserves: 1.940 million bbl
Gas reserves: 1.075 tcf

Overview

In the Eastern Mediterranean region, Israel has the largest and most developed economy. According to the U.S. CIA World Fact Book, growth of real GDP for 2008 was estimated to be 4.2%.

Oil

Israel produces only a minimal amount of oil. The country imported approximately 319,000 b/d of crude oil in 2007 to meet domestic needs, primarily from Egypt, the North Sea, West Africa, and Mexico.

However, in recent years, Israel also has increased imports from Russia and the Caspian, including Azeri oil transported through the new Baku-Tbilisi-Ceyhan (BTC) pipeline.

More than 20 small independent firms are active in hydrocarbon exploration throughout the country. Israel's Petroleum Commission has estimated that the country could contain 5 billion bbl of oil reserves.

Natural gas

Gas production in 2008 was estimated at 1.2 bcm (42 bcf). However, Israel plans to dramatically increase the share of natural gas in its fuel mix, especially for electricity generation, so demand is expected to reach 7.9 bcm (282 bcf) by 2010.

Supply options being developed include imports from Egypt's Nile Delta and production from domestic offshore reserves.

Israel took part in preliminary discussions in 2008 about construction of the subsea MedStream pipeline that would deliver gas from Turkey. And in 2009, initial planning was under way for an LNG regasification facility in Israel.

Several energy companies have discovered small but commercially viable quantities of natural gas off the coast of Israel and the Gaza Strip. The offshore gas reserves of Israel and the Palestinian Authority are under concession to two main groups:

- Yam Thetis consortium, comprising Noble Energy, Avner Drilling, and Delek Group.
- A partnership of British Gas (BG) with Isramco and others.

Gas Sales Protest

The Egyptian-Israeli consortium Egyptian Mediterranean Gas Co. finalized in August 2009 a revised agreement to supply Egyptian natural gas to Israel's Dorad Energy.

EMG is to sell 12.5–16 bcm (441–565 bcf) of gas to Dorad Energy over 17–22 years at a reported cost of $2.1–3.3 billion.

But critics in Egypt raised two objections:

- that the gas was being sold at prices considerably lower than those on the international market.
- that Egypt should not be selling anything to Israel.

Exploration and Development

Tamar appraisal

In mid-2009, Noble Energy Inc. assessed its Tamar-2 appraisal well offshore Tel Aviv in the Matan license. The well, drilled to TD 5,084 m (16,680 ft), is in 5,530 ft (1,685 m) of water.

Noble said that reservoir thickness and quality were consistent with those encountered by its Tamar-1 discovery well and that pressure data confirmed continuous, high-quality reservoirs.

Based on the assessment, the firm raised its gross mean resource for Tamar by 26% to 6.3 tcf.

Production of first gas is expected by 2012.

Dalit discovery

A group led by Noble Energy Inc. reported in early 2009 that its Dalit deepwater discovery appears capable of producing 200 MMcfd.

The discovery well was drilled to 3,658 m (12,000 ft) in 1,372 m (4,500 ft) of water on the Michal license, 48 km (30 mi) off Haifa.

Noble estimated Dalit's gross mean resource at 14 bcm (500 bcf) of gas from log and test data.

Dalit is the Noble's second subsalt, lower-Miocene discovery in the Levantine basin, the other being the Tamar prospect.

Others with interests in both Dalit and Tamar prospects are Isramco Negev 2, Delek Drilling, Avner Oil Exploration, and Dor Gas Exploration.

TAMAR AND OTHER EAST MEDITERRANEAN HYDROCARBONS

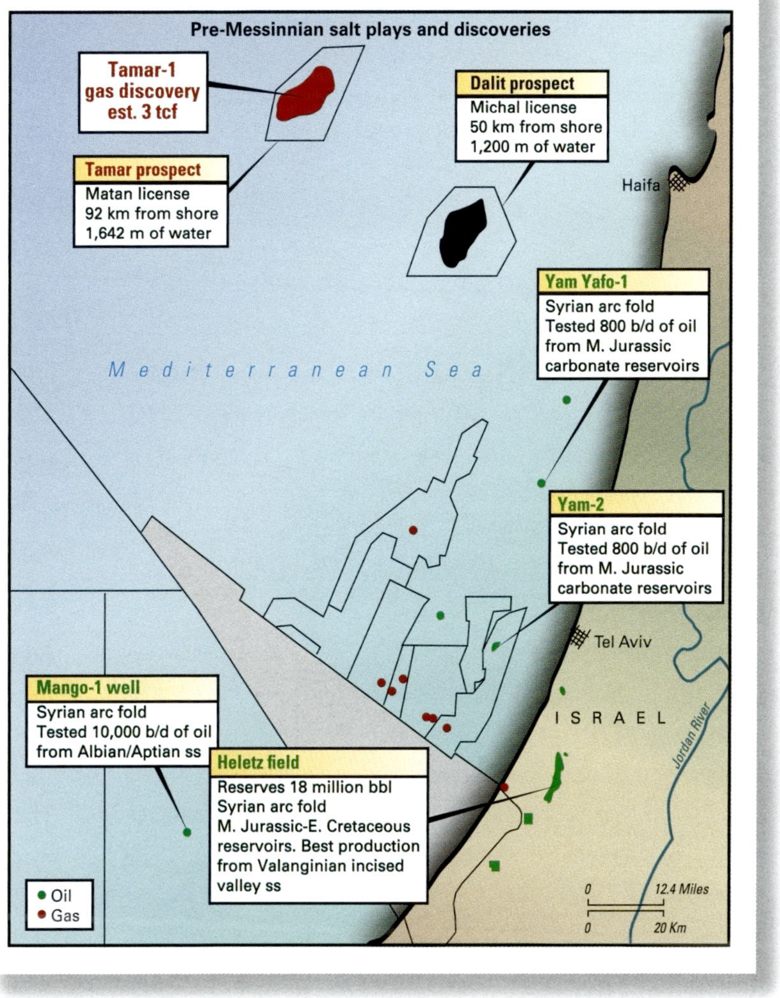

Lease Sales

Gabriella license

In September 2009, Israeli officials awarded Adira Energy Corp. (a subsidiary of AMG Oil Ltd.) the 97,000-acre Gabriella license, 10 km (6 mi) offshore between Netanya and Ashdod.

The license, in 80–120 m (262–394 ft) of water, contains the Yam Yafo-1 exploratory well in an area where two wells tested oil or gas-condensate from Jurassic carbonates in the early 1990s.

Issachar-Zebulun permit

In June 2009, Israel awarded Zion Oil & Gas Inc. an exploration permit adjacent to and east of Zion's Asher-Menashe license.

The permit extends Zion's oil and gas rights from the Mediterranean at Caesarea, across the Carmel Mountains to Megiddo, and to the Jordan River just south of the Sea of Galilee.

Zion named the permit Issachar-Zebulun.

Drilling and Production

Tamar gas sale

Dalia Power Energies, a private Israeli power company, signed a letter of intent in late 2009 to buy natural gas from the Tamar gas field of Noble Energy Inc.

Dalia Power has a license to build a power plant in Israel, with startup planned for 2013.

Sale of gas from Tamar is expected to begin in 2012.

Noble Energy and its partners will sell 5.7 bcm (200 bcf) under a 17-year supply agreement.

Yam Tethys gas field conversion

In mid-2009, a subsidiary of Delek Group Ltd. was reported to be considering conversion of its offshore Yam Tethys gas field, in the vicinity of Ashkelon, into an underground storage facility.

Gas production from the field is expected to be exhausted by 2013.

LNG

Offshore terminal plan

In August 2009, the National Infrastructure and Finance ministries issued a prequalifying tender for an offshore LNG receiving terminal, to be built at a location yet to be determined.

Officials expected bids to be processed in third quarter 2009, and said final tender documents for the $400-million project would be released in March 2010, with the winning bidder chosen by yearend 2010.

The committee set a target date of October 2013 for start-up of the new terminal, which will have a receiving capacity of 4 bcm/year.

The government said buying LNG internationally would provide a backup to domestic supplies, but that early imports would likely be small because of offshore Tamar and Dalit gas discoveries.

Alternative Energy

Solar projects in the Negev

Israeli officials in early 2009 issued tenders for proposed solar power projects in the Negev region.

The 250-MW project in Ashalim would be the country's first national solar power project.

It would comprise two solar thermal plants and a photovoltaic unit.

The tenders pertain to operation and maintenance of the plants as well as construction.

Construction cost was estimated at $600 million.

Israel aims to generate 10–20% of its electricity from solar power by 2020.

JORDAN

Capital: Amman
Monetary unit: Jordanian dinar
Refining capacity: 90,400 b/d
Oil production: 0 b/d
Oil reserves: 1.000 million bbl
Gas reserves: 213.000 bcf

Overview

Jordan faces several challenges in meeting its energy needs.

- It has almost no indigenous energy resources.
- It is dependent on imported energy (97% imported in 2008).
- The cost of energy imports is high (21% of GDP in 2007).
- Primary energy demand is growing sharply.

An integrated law for energy and mines was to be issued by yearend 2009, and a new energy-sector regulatory commission created. The aim is to improve the performance and flexibility of the sector.

The government also is evaluating alternate sources of energy.

In 2008, it initiated a long-term venture with Royal Dutch Shell for the exploration of potential deep shale oil deposits in Jordan. It is reported that the venture will cost approximately $20 billion.

Exploration and Development

Risha concession

In late 2009, BP PLC announced plans to partner with Jordan's National Petroleum Co. (NPC) to exploit the onshore Risha concession in the northeast.

The concession borders Syria, Iraq, and Saudi Arabia. It includes Risha gas field, 274 km (170 mi) northeast of Amman, where the first successful wells were drilled in the late 1980s.

Risha field produces 30 MMcfd of gas from more than 30 wells. Proven gas reserves total 180 bcf.

The gas reservoir appears to be tight sandstones at 26,246–27,886 m (8,000–8,500 ft).

Oil shale evaluation

In mid-2009, Jordan signed a concessionary agreement with Jordan Oil Shale Co. BV, a subsidiary of Royal Dutch Shell PLC, to explore for oil in the country's shale deposits.

The Shell subsidiary will invest up to $540 million in preliminary exploration and assessment.

The project will assess oil shale resources in central, southern and northwestern regions. Jordan estimates that 40 billion tonnes of oil shale exist in 21 locations near the Yarmuk River, Buwayda, Bayt Ras, Ruwayshad, Karak, Madaba, and Ma'an districts.

The government expects production of oil shale within 12–20 years of the conclusion of the agreement.

In April, Jordan approved MOUs between Inter Rao of Russia and Aqaba Petroleum for oil shale exploration and mining in areas near Al Attarat in the south.

Refining and Products

Zarqa refinery expansion

The government announced in November 2009 that it will invest $2.2 billion to expand the capacity of the refinery owned by Jordan Petroleum Refinery Co. Ltd. (JPRC) in Zarqa.

The project will expand capacity from the present 105,000 b/d to 131,000 b/d.

The upgrade will include changes to comply with the latest European emissions standards and enable production of unleaded gasoline and low-sulfur diesel.

Alternative Energy

Wind, solar initiatives

In late 2009, Jordan was pursuing a range of sustainable-energy projects.

- Two tenders were launched for developing wind energy projects for electricity generation at Al-Kamshah and Al-Fujeij, Shoubak.
- Other wind projects are under development at Al-Harir, Ma'an and Wadi Araba.
- Planning was under way for the country's first concentrating solar power (CSP) project, with funding from the European Union expected by yearend 2009.

Jordan is aiming to increase its use of domestic resources from 4% in 2007 to 25% in 2015.

KUWAIT

Capital: Kuwait City
Monetary unit: Dinar
Refining capacity: 936,000 b/cd
Oil production: 2.010 million b/d
Oil reserves: 101.500 billion bbl
Gas reserves: 63.000 tcf

Overview

Oil, gas outlook: BMI

In its October 2009 *Kuwait Oil & Gas Report*, Business Monitor International (BMI) outlined Kuwait's oil and gas sector and its role in the region.

Underlying the analysis is the BMI expectation that Kuwaiti real GDP will fall by 1% in 2009, but then to grow at a 2.1% rate in 2010; 2.5% in 2011; 3.4% in 2012; and 3.7% in 2013.

Oil

BMI forecasts Kuwait's crude production to increase to 2.90 million b/d in 2013, subject to OPEC quotas. Its oil demand is expected to rise from 300,000 b/d in 2008 to 330,000 b/d in 2013.

Kuwait will account for 10.35% of Middle East regional oil supply by 2013, but for just 2.71% of demand.

Natural gas

The nation's gas production should reach 19.3 bcm (681bcf) in 2013, up from 12.8 bcm (452 bcf) in 2008. Consumption is expected to reach 25.3 bcm (893 bcf) in 2013, requiring imports of 6 bcm (212 bcf).

Kuwait contributed 3.29% to 2008 regional gas production; that figure will drop to 3.16% by 2013.

In 2008, Kuwait consumed 3.27% of the region's gas. Its market share is forecast at 4.93% by 2013.

International Activities

Oil and gas JVs

Kuwait's Noor Financial Investment Co. formed two joint venture companies in early 2009 with a Geofizika, a subsidiary of Russia's Gazprom.

One venture will be formed in Russia, the other in Kuwait, but both will focus on oil and gas production, repair and maintenance services.

Middle East Regional Oil, Gas Data

In an October 2009 study, Business Monitor International (BMI) reported that Middle Eastern regional oil production in 2008 averaged 26.29 million b/d and was projected to rise to 28.01 million b/d by 2013.

Regional oil demand was 11.25 million b/d in 2008 and should grow to 12.17 million b/d by 2013.

Exports averaged 15.04 million b/d in 2008 and are forecast to reach 15.84 million b/d by 2013.

BMI says Iraq has the greatest production growth potential, followed by Qatar.

Regarding natural gas, the Middle Eastern region in 2008 consumed 391.5 bcm (13.8 tcf), with demand of 512.8 bcm (18.1 tcf) expected for 2013 (31.0% growth).

Regional production of 389.5 bcm (13.8 tcf) in 2008 should grow by 56.7% to 610.4 bcm (21.6 tcf) in 2013, implying that net exports will rise to 98 bcm (3.5 tcf) in that outyear.

In October 2009, China's Sinopec, Kuwait Petroleum Corporation (KPC), and the government of Guangdong Province signed an MOU for a JV project to build the Sino-Kuwait Guangdong Oil Refinery & Petrochemical Complex.

The start of construction is expected in early 2010 at Donghai Island, Zhanjiang City.

A total investment of about $8.8 billion is anticipated for a facility with a total refining capacity of 300,000 b/d and ethylene production capacity of 1 million mtpy.

The complex will primarily process crude oil imported from Kuwait.

IFC aid

Kuwait Energy Co. reported in fourth quarter 2009 that it had secured $50 million from the International Finance Corp. (part of the World Bank) for exploration and development in Egypt and Yemen.

KEC has six onshore assets in Central Yemen and one off Al Mukalla in the Gulf of Aden.

The funding also will support an environmental and social action plan to begin in September 2009.

LNG

Mina Al-Ahmadi GasPort

The first LNG regasification terminal in the Persian Gulf—Kuwait's offshore Mina Al-Ahmadi GasPort—received its first cargo in September 2009.

The cargo came from the North West Shelf LNG plant of Woodside Petroleum in Australia.

The Mina Al-Ahmadi GasPort, owned by Kuwait National Petroleum Co. (KNPC), is 40 km (25 mi) south of Kuwait City.

An "Energy Bridge" regasification vessel (EBRV) owned by Excelerate Energy will be stationed at Mina Al-Ahmadi to deliver regasified LNG at a baseload rate of up to 500 MMcfd.

Refining and Products

Al Jour refinery

In August 2009, press reports said that KNPC might retender contracts for construction of the 615,000-b/d Al Jour refinery near the border with Saudi Arabia.

The company had cancelled a March tender offer despite initial spending of $1 billion on the project. Parliament objected to soaring cost estimates and to alleged irregularities in the tender process.

The refinery was scheduled to begin operations in 2012, producing low-sulfur fuel oil for power plants.

Petrochemicals

Olefins II complex

The five-year expansion of the Olefins II petrochemicals complex at Shaiba was completed in mid-2009, doubling its capacity. The facility is about 40 km (25 mi) south of Kuwait City.

Dow deal revisited

The government in December 2008 cancelled the $17.4-billion K-Dow petrochemicals JV involving Dow Chemical Co. and state-owned Petrochemicals Industries Co. as the global economy deteriorated.

Dow had aimed to use proceeds from the deal to pursue a $15-billion takeover of specialty chemicals group Rohm and Haas Co.

In early 2009, the Kuwait Investment Authority said that it would consider increasing its support for the takeover beyond the original $1 billion level if the terms of the deal could be changed.

OMAN

Capital: Muscat
Monetary unit: Riyal
Refining capacity: 85,000 b/cd
Oil production: 800,000 b/d
Oil reserves: 5.500 billion bbl
Gas reserves: 30.000 tcf

Saiwan East

A group of seismic specialists announced plans in late 2009 to evaluate three heavy oil zones and a deeper light oil formation, all in late Neoproterozoic and early Paleozoic rocks, in eastern Oman.

Two wells have delineated the accumulations on the Saiwan East structure on the eastern flank of the Ghaba salt basin, 306 km (190 mi) southwest of Muscat.

Block 47

RAK Petroleum PCL was planning in early 2009 to drill the Zad–2 well on Block 47.

The firm acquired subsidiaries of Indago Petroleum Ltd., which hold rights to several blocks in Oman.

The Zad-2 well will test the Cambrian Amin sandstone.

Al Ghubar-Qarn Alam Block

Harvest Natural Resources Inc. signed an exploration and production sharing agreement in 2009 with the government for gas-condensate development from the Al Ghubar-Qarn Alam block.

The newly created block had been part of the Block 6 concession.

Ghubar-Qarn Alam is in the Ghaba salt basin near Barik, Saih Rawl, and Saih Nihayda gas-condensate fields, which support Oman's exports of LNG.

West Bukha field

Oil and gas production began in early 2009 from West Bukha field in Block 8 in the Strait of Hormuz.

Heritage Oil reports initial flow rates of approximately 10,000 b/d of oil and 30 MMcf/d of associated gas from the Mishriff-Maudud and Thamama reservoirs.

Production is from two wells drilled in 90 m (295 ft) of water approximately 25 km (16 mi) from the Musandam Peninsula.

Oman-Iran: Kish field

Oman and Iran restarted talks in late 2009 about export of gas from the Kish field in Iran.

Details are given in the Iran section of this chapter.

Musandam project

Oman Oil Co. opened bidding in third quarter 2009 for an EPC contract to build the Musandam oil processing and pipeline project. The contract has an estimated value of $300–500 million.

The plant will process 45 MMscfd of gas and about 20,000 b/d of oil. It will be linked to the offshore West Bukha field by pipeline.

Al Duqm complex

Oman Oil Co. and Abu Dhabi's International Petroleum Investment Council signed an MOU in fourth quarter 2009 to study the possible development of a refinery and petrochemical complex at Duqm Port, in the Al Wusta region of Oman.

No timetable for the estimated $7-billion project was given.

The complex would be one element in a major plan to develop the Duqm Port region in order to diversify Oman's economy. Development would include a new airport, new hotels, and a free trade zone.

Largest paraxylene plant

Production at a major paraxylene plant was on track to start by yearend 2009.

The facility is at the $1.6-billion aromatics complex in the Sohar Industrial Port region.

The plant, called the largest greenfield paraxylene facility in the world by its operator, has a production capacity of 814,000 tpy.

QATAR

Capital: Doha
Monetary unit: Riyal
Refining capacity: 338,700 b/cd
Oil production: 765,000 b/d
Oil reserves: 25.410 billion bbl
Gas reserves: 899.325 tcf

Drilling and Production

North field study

Qatar officials expect results in 2012 from a study analyzing the outlook for continued production from the major North gas field, offshore in the Persian Gulf.

The single largest nonassociated gas field in the world, it holds an estimated 900 tcf. It will produce 25 bcfd by 2011 once several pending LNG, GTL, and domestic gas projects come on stream.

In 2005, Qatar imposed a moratorium on further developments to ensure the optimal productive life of the reservoir. Several projects were put on hold.

The government said it wants to develop the reserves in such a way that they last 100 years.

Gas Processing

Pearl GTL project

In October 2009, the energy ministry inaugurated the main control room at the large Pearl GTL plant in Ras Laffan and said construction of the full plant was on track for completion in late 2010.

The two-train, $18-billion facility will use the Shell Middle Distillate Synthesis (SMDS) process to convert 1.6 bcfd of gas into 140,000 b/d of GTL products and 120,000 b/d of ethane and NGLs.

Royal Dutch Shell is building the plant in partnership with Qatar Petroleum.

LPG projects: China, Vietnam

Qatar Petroleum announced in late 2009 that by 2015, the company will invest about $9.8 billion in new LPG projects in Vietnam and China to capitalize on growing Asian demand.

- A JV is planned with China's CNOOC and another petrochemicals partner to construct a $5.8-billion, 3.8-mtpy LPG processing facility in Hainan Province. Chinese approval is expected in 2010.
- Advanced discussions were under way with several parties in late 2009 concerning a consortium to build a $4-billion LPG processing facility in Vietnam. Details were not released.

Qatar is seeking to increase LPG output to 12 mtpy by 2010, and has identified Asia as the primary future growth market.

LNG

Qatar owns about one-third of the world's gas reserves, including those in North field, the single largest nonassociated gas field in the world. Qatar's gas production is expected to reach 23 bcfd by 2013.

LNG plays a major role in the nation's economy and energy sector. Qatar leads the world in LNG exports, and LNG production has been rapidly expanding in recent years. Plans call for raising production capacity from about 31 mtpy in 2009 to 77 mtpy by 2012.

A current moratorium on new projects in the North field was expected to remain in place until 2013 or 2014. (See the Drilling and Completion section on this page).

The nation's LNG industry is split into two operating companies—Qatargas and RasGas.

Qatargas is a series of joint ventures between Qatar Petroleum and several major oil companies, including Shell, ConocoPhillips, Total, and ExxonMobil.

RasGas is a series of joint ventures involving ExxonMobil, state-owned Qatar Petroleum, and several (mainly Asian) investors.

LNG to Italy, Poland

RasGas Co. Ltd. was preparing to deliver its first LNG cargo to Italy in mid-2009 from its newly completed 7.8-million tpy Train 6.

Also in mid-2009, Qatar Liquefied Gas Co. III (Qatargas) signed a 20-year agreement to sell 1 mtpy of LNG to Poland. The value of the contract was put at $550 million/year.

Poland will receive LNG at a terminal being built at Swinoujscie, to be completed in 2015.

Qatargas expects its LNG production to rise to 42 mtpy by the end of 2010 from 10 mtpy in 2008.

More LNG to China

Qatargas boosted exports of LNG to China in late 2009, signing supply agreements with CNOOC and PetroChina International Co. Ltd.

Qatargas intends to supply a further 3 mtpy to CNOOC starting in 2013, with possible sale of another 2 mtpy to be discussed in the future. CNOOC got its first LNG cargo from Qatar in October 2009.

The firm also expects to start sending 2 mtpy of LNG to PetroChina in 2012, in addition to the 3 mtpy already supplied by under a previous agreement.

Qatargas-2: Trains 4 and 5

Qatargas-2 partners Qatar Petroleum and ExxonMobil Corp. completed the 7.8-mtpy Train 5 at Ras Laffan in September 2009.

Production from the 7.8-mtpy Train 4 began mid-year after resolution of technical and labor difficulties.

Ras Laffan III: Trains 6 and 7

Ras Laffan LNG Co. Ltd. III started up the 7.8-mtpy Train 6 in mid-2009 at the LNG facility operated by RasGas Co. Ltd. at Ras Laffan Industrial City.

Ras Laffan III is a JV of Qatar Petroleum and ExxonMobil.

Train 7, also with 7.8-mtpy capacity, was scheduled to start up by late 2009.

Both trains will be supplied by gas from North field.

Qatargas-4 project

In late 2009, Royal Dutch Shell announced that commissioning of the $8-billion Qatargas-4 LNG project would likely be delayed until late 2010, with a first LNG shipment expected in early 2011.

The delay was attributed to difficulties meeting work schedules.

Qatargas-4 is a JV of Shell with Qatar Petroleum. The plant will process 1.4 bcfd of natural gas to produce 7.8 mtpy of LNG and 420 MMcfd of natural gas liquids.

Refining and Products

Al Shaheen refinery

In October 2009, Qatar Petroleum was preparing to seek bids to build the Al Shaheen refinery. Contracts are to be awarded in first quarter 2010. The project budget is now targeted at $10 billion.

FEED work was completed in August 2008, but later that year, Qatar Petroleum decided to reconsider the project because of falling demand, lower oil prices, and the reduced price of raw materials.

Observers now estimate startup no earlier than 2014.

Construction of the 250,000-b/d refinery complex at the Mesaieed Industrial City will include pipelines to bring feedstock from the offshore Al Shaheen oil field, 180 km (112 mi) north of Doha.

Current oil production from the field is about 320,000 b/d.

Petrochemicals

Singapore JVs

In late 2009, Qatar Petroleum International signed agreements to acquire two Singapore chemical joint ventures from Royal Dutch Shell PLC.

The transactions were expected to be completed by yearend. Terms were not revealed.

The two JVs will be called Petrochemical Corp. of Singapore Pte. Ltd. (PCS) and The Polyolefin Co. (Singapore) Pte. Ltd. (TPC).

Other shareholders in the two are Japanese consortia led by Sumitomo Chemical Co. Ltd.

PCS owns and operates two naphtha steam crackers for olefins production.

TPC owns and operates units producing low-density polyethylene (LDPE) and polypropylene.

LDPE plant

A new facility to produce 300,000 tpy of LDPE was under development by Qatar Petrochemical Co. (Qapco) in 2009.

The plant, being built as part of Qapco's existing PE production facility in Mesaieed Industrial City, is scheduled for completion in December 2011.

Ras Laffan complex

Qatar was poised in September 2009 to commission the Ras Laffan Olefins complex.

The 1.3-mtpy plant includes a condensate refinery (first in the country) with a capacity of 146,000 b/d.

Ras Laffan Olefins Co. Ltd. is developing the project at the Ras Laffan Industrial City.

Olefins units will convert ethane from gas projects in the North Field to a variety of products for Asian and European markets.

Raw condensate from Al Khaleej Gas, QatarGas, and RasGas will be converted to LPG, kerosene jet fuel, naphtha and gas oil.

SAUDI ARABIA

Capital: Riyadh
Monetary unit: Riyal
Refining capacity: 2.080 million b/cd
Oil production: 7.920 million b/d
Oil reserves: 259.900 billion bbl
Gas reserves: 263.000 tcf

Overview

Oil

Saudi Arabia is noted for its dominant role in oil production. The country was producing about 8.3 MMb/d in 2009, in line with quotas established by OPEC.

However, by yearend 2009, the country was poised to achieve what the *Financial Times* called "a formidable feat of engineering"–an increase in its oil production *capacity* to a record 12.5 MMb/d.

FT also said that the size and complexity of the five-year, $100-billion program undertaken to achieve that milestone "are unmatched in the industry."

The production restraint of Saudi Arabia and several other Arabian Gulf nations kept oil prices in the $70/bbl range in late 2009, close to the target range desired by both OPEC and Saudi Arabia.

Natural gas

Saudi Arabia also is a major natural gas producer, holding the world's fourth largest proven reserves. However, it is struggling to meet gas demand that is growing at 7% per year.

Citing forecasts from Saudi Aramco, the U.S. Energy Information Administration expects Saudi gas demand to reach 14.5 bcfd by 2030, up from about 7.1 bcfd in 2007.

Much of Saudi Arabia's gas is "associated;" i.e., tied to oil production. The petroleum ministry announced a $9-billion program in 2006 to add 50,000 bcf of non-associated reserves in 10 years through new discoveries, according to EIA.

Expanding its search for gas, Saudi Aramco will begin drilling in the Red Sea in 2012 (for the first time) and will continue exploration in the challenging Empty Quarter. No commercial discoveries had been made in the quarter as of late 2009.

The Karan field—the nation's first non-associated offshore field—was discovered in 2006, and Aramco began awarding contracts to develop the field in 2009.

Karan is expected to produce 400 MMscfd in 2011 and 1.8 bcfd when completed.

Other gas 2009 gas and oil discoveries are described in sections below.

Saudi Arabia also said it wants to boost gas processing capacity to 9 bscfd (from 6.2 bscfd) by 2015.

WTI benchmark

Starting with January 2010 sales, Saudi Aramco said, it will switch from WTI to an index of Gulf Coast sour crudes as the benchmark for pricing its oil for sale in the U.S. market.

Aramco instead will use Argus Media Ltd.'s Argus Sour Crude Index (ASCI) in pricing its Extra Light, Arab Light, Arab Medium, and Arab Heavy crudes.

That index is based on daily spot sales of Mars, Poseidon, and Southern Green Canyon (Gulf of Mexico) crudes.

Observers saw the move as providing "insurance" against volatility caused by WTI price dislocations.

Exploration and Development

Arabian Gulf discoveries

Saudi Aramco said in early 2009 that it made discoveries at several wells in the Arabian Gulf.

- At the Arabiyah-1 well over the Khuff B reservoir, open-hole tests at a depth of 3,200 m (10,500 ft) measured gas flow at 41.1 MMcf/d.
- Rabib-1 flowed 40.6 MMcf/d in a cased hole test over the Jauf reservoir at 5,221 m (17,130 ft).
- Hasbah-16 flowed 62.1 MMcf/d from an open hole test over the Khuff B reservoir at 12,700 ft (3,871 m).
- Jurayd-101 flowed oil from three reservoirs that had not previously proven to be oil productive.

Eastern Province finds

The Saudi oil ministry reported in January 2009 that Saudi Aramco had discovered five more oil fields and three gas fields in the Eastern Province.

Four of the oil fields are on land and one is in the Persian Gulf. They include Jaouf-11, Ramthan-9, Nayashin-1, Jareed-101, and Khorsaniya-114.

The three gas field, all offshore, are Arabiya-1, Rabib-1, and Hisbah-16.

Test production results were released for three of the wells.

Drilling and Production

Wafra test

Chevron Corp. started a $340-million pilot test in mid-2009 of steamflooding in the Neutral Zone between Saudi Arabia and Kuwait.

The test site is the Wafra field, an Eocene heavy-oil carbonate reservoir. Most fields in the Neutral Zone, Saudi Arabia and Kuwait are carbonates.

Steamflooding tends to dissolve minerals in carbonate rock, clogging reservoir pores and well passages. But Chevron was confident it could make Wafra the world's first commercial conventional steamflood in a carbonate reservoir.

Manifa field

Saudi Aramco said in late 2009 that it had made significant progress on its shallow-water Manifa heavy-oil development project. The company calls Manifa "the largest single offshore crude oil project in Saudi Aramco's history."

Startup is now slated for 2013, following delays related to falling oil prices in 2008.

The project is expected to produce 900,000 b/d of Arabian Heavy crude, 90 MMscfd of associated gas, and 65,000 b/d of condensate.

Khurais field

In mid-2009, Saudi Arabia announced the start of production at the giant Khurais oil and gas field.

The field, with the capacity to produce 1.2 MMb/d of Arabian Light crude, was described as the final element in Saudi Aramco's plan to raise its production capacity to 12.5 MMb/d by yearend 2009.

The $10-billion project is 145 km (90 mi) east of Riyadh. The field is estimated to hold at least 12 billion bbl of recoverable oil, making it the country's second-largest oil field.

Khurais began production in 1970 and had yielded 111 million bbl by mid-1979. It was later shut in, and the project to restart it began in 2005.

Gas Processing

Wasit project

Preliminary engineering for Saudi Arabia's largest gas plant—Saudi Aramco's $2.3-billion Wasit gas development program—is scheduled to be completed by first quarter 2011.

The project aims to produce and process up to 2.5 bscfd of gas from the Hasbah and Arabiyah offshore non-associated sour-gas fields for use in power generation and industrial plants.

Startup is scheduled for 2014.

Shaybah NGL program

Saudi Aramco said work was slated to begin in October 2009 on its NGL facility to process output from the Shaybah oil field.

The field is in the Empty Quarter near the border with Abu Dhabi. It came online in 1998, with estimated reserves of more than 14 billion bbl of oil and 25 tcf of gas.

A contract for FEED and project management services was awarded to KBR.

Pipelines/Transportation

Bahrain-Saudi Arabia oil line

In late 2009, details were being finalized for construction of a new pipeline to carry 350,000 b/d of crude oil from Saudi Arabia to Bahrain.

Details are given in the Bahrain section of this chapter.

Refining and Products

Jizan refinery

In January 2010, the oil ministry received two bids to build a new oil refinery at Jizan. Expected capacity will be 250,000 to 400,000 b/d.

With an estimated cost of at least $10 billion, Jizan would be the first Saudi oil refinery to be 100% privately owned.

Oil storage in Japan

Saudi Aramco was considering a proposal from Japan in late 2009 for joint use of that country's oil storage facilities on the island of Okinawa.

Spare capacity was triggered by oil refiners either closing down or idling capacity.

The deal also would give Aramco increased access to expanding oil markets in China and Asia.

Petrochemicals

Yanbu, Jubail, Ras Tanura

According to the *Financial Times*, Saudi Aramco sees the export of refined oil products and plastics as more profitable than shipping crude oil, and so aims to build more refineries and petrochemical plants.

A final investment decision is expected in late 2010 on a combined refinery and petrochemical project at Ras Tanura.

Work is to start in 2013 on a 400,000-b/d refinery at Jubail Saudi.

And a $6-billion refinery at Yanbu Industrial City is now targeted for completion in 2014, after delays in 2007 due to low energy prices.

SABIC-Sinopec JV

Construction of the $3-billion Tianjin Petrochemical Integrated Project was expected to be completed in September 2009. The project is a JV of Saudi Basic Industries Corp. (SABIC) and China's Sinopec.

Located in Tianjin, China, the project will produce about 3.2 million metric tpy of petrochemicals.

The project boosts plant output of ethylene from 200,000 metric tpy to 1.2 million metric tpy, and raises crude processing capacity from 5 million metric tpy to 12.5 million metric tpy.

Sipchem-Hanwha JV

In mid-2009, Saudi International Petrochemical Co. (Sipchem) formed a JV with South Korean petrochemical company Hanwha Chemical Corp.

The JV will build a $1.1-billion petrochemical facility at Sipchem's site in Jubail Industrial City.

Production is to start in 2013.

CONCEPT FOR EU-MENA[1] POWER TRANSMISSION GRID[2]

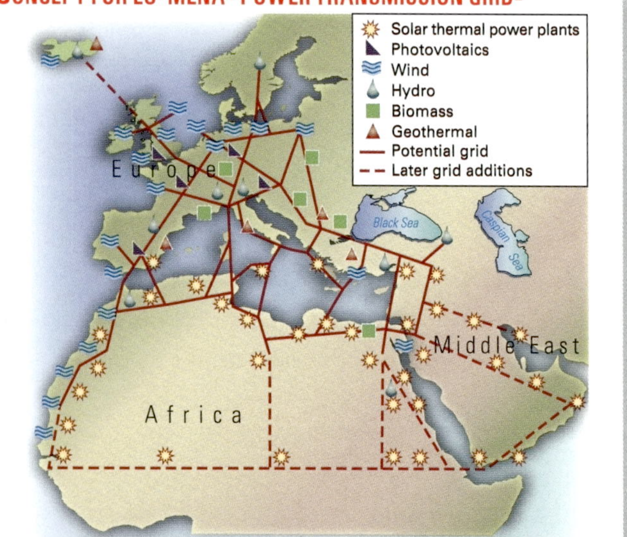

[1]Possible sustainable facilities to supply power to Europe, the Middle East, and North Africa (EU-MENA). [2]High voltage, direct current electric power grid with power transmission losses to Eurpoe less than 15%.
Sources: Trans-Mediterrranean Renewable Energy Cooperation (www.desertec.org), the DESERTEC WhiteBook, various studies

Petro Rabigh expansion

Saudi Aramco signed an agreement in early 2009 with Sumitomo Chemical Co. to develop Phase 2 of their $10.3-billion Petro Rabigh petrochemical JV. Completion date is 2014.

The project focuses on the refinery and chemicals complex at Rabigh, on the Red Sea coast.

A first phase upgraded the 400,000-b/d complex to produce higher-quality petrochemicals. Construction for Phase 2 was planned to start by yearend 2009.

Total investments were estimated at ¥300–500 billion, with funds to come from both firms.

Alternative Energy

EU-MENA "super-grid"

The Middle East-North Africa (MENA) region would be a key element in a transcontinental 'super energy grid' concept under consideration by the EU in 2009.

The ambitious project would capture solar and wind power for export as electricity to Europe.

The Trans-Mediterranean Renewable Energy Cooperation (TREC) initiative is a concept developed by a consortium led by the Club of Rome, the German Aerospace Bureau, and several universities in Europe and the Middle East.

TREC proposes a super-grid that spans the MENA region, with a Saudi Arabian solar plant at the heart of the project.

Some of the energy from a mix of concentrated solar thermal power (CSTP) plants and wind turbines would move through power lines to Europe, and some would run seawater desalination plants in MENA.

The long-term goal is to produce 100 GW of exportable power by 2050.

SYRIA

Capital: Damascus
Monetary unit: Syrian pound
Refining capacity: 239,865 b/cd
Oil production: 365,000 b/d
Oil reserves: 2.500 billion bbl
Gas reserves: 8.500 tcf

Syria's proven oil reserves were about 2.5 billion barrels at the end of 2009.

Since the 1960s, Syria has produced heavy crude oil from its northeastern fields. However, production slumped considerably, from about 600,000 b/d in 1992 to 365,000 b/d at the end of 2009, due to obsolete technology and infrastructure.

Syria's two refineries, at Homs and Baniyas, have a combined capacity of 239,865 b/d.

The country, which has witnessed very little international investment in recent years, is now welcoming international firms to develop its oil and gas sector. (See refinery reports below.)

Palmyride basin

In October 2009, Stratic Energy Corp. spudded the Al Tayr-101 well on the Jebel Nasrani prospect.

That prospect is in Block XVII in the southwestern Palmyride basin.

The well is projected to reach 3,000 m (9,842 ft) to tap Triassic Kurrachine carbonates.

Blocks returned

Groundstar Resources Ltd. agreed in September 2009 to a request from the Ministry of Petroleum & Mineral Resources to return Blocks XIV and XVI, which had been awarded to Groundstar in 2008.

Groundstar analyzed seismic and related well information on and adjacent to the blocks and identified several leads. However, deteriorating credit market made it impossible to secure a partner.

Groundstar said it expects to participate in upcoming bid rounds for exploration and EOR projects.

Abu Khashab: New refinery

China and Syria signed an agreement in late 2009 to build a $1.5-billion, 100,000-b/d refining complex at Abu Khashab, northeast of Deir ez-Zor. It which will process heavy Syrian crude, and is expected to be completed by 2011.

Furqlus: New refinery

Venezuela and Iran signed an MOU in September 2009 to build a 140,000-b/d oil refinery at Furqlus, near Homs, in western Syria.

Syria and Malaysia's Al Bukhari Group also will participate.

Syria will supply 50% of the crude feed; Venezuela and Iran will provide 30% and 20%, respectively.

Venezuela and Iran established a JV company, Venezuela Iran Oil and Gas Company (VENIROGC) in Tehran to oversee investments in the project.

UNITED ARAB EMIRATES

Capital: Abu Dhabi
Monetary unit: Dirham
Refining capacity: 773,250 b/cd
Oil production: 2.271 million b/d
Oil reserves: 97.800 billion bbl
Gas reserves: 214.400 tcf

Overview

Oil

Abu Dhabi has about 95% of the UAE's proven oil reserves and about 8.5% of the world's reserves.

Experts forecast Abu Dhabi will be able to keep pumping crude for another 100 years, and the emirate has been steadily working to expand production.

The current capacity is about 2.9 million b/d and the plan is to reach 3.7 million b/d by 2017.

The emirate produces about 2.2 million b/d, in line with OPEC direction.

Natural gas production

Abu Dhabi is the largest gas producer in the UAE at 4 bscfd, but the country imported about 740 MMscfd in 2009 from the massive North field offshore Qatar, through the Dolphin Pipeline.

According to a late-2009 analysis by FACTS Global Energy, the country likely would buy more if the current Qatari gas moratorium were lifted. (See the Qatar section of this chapter for details.)

Abu Dhabi gas infrastructure

Gas demand also is increasing rapidly for petrochemical and industrial projects.

The emirate also currently re-injects some 1.4 bscfd of natural gas into its oil fields to maintain production levels.

Increased use of alternatives to gas for generating power is under study, including construction of Abu Dhabi's first oil-fired electric power stations.

Renewables, nuclear

Abu Dhabi is also looking beyond hydrocarbons to meet energy needs.

Initiatives include the ambitious Masdar Initiative, a multi-billion-dollar project to develop renewable energy technology, particularly solar power.

However, a major push in the UAE's plan—led and funded by Abu Dhabi—is to develop a nuclear program by 2017 (see box). The UAE's first-ever nuclear plants could be producing 4,000 MW by 2020.

The UAE is also a net importer of gas, bringing in about 2 bscfd from Qatar.

Significant UAE gas reserves are associated gas; their development depends on crude production.

Also, much of the associated gas is re-injected to maintain oil-reservoir pressure or to produce NGLs.

Another complication: Much of UAE non-associated gas is sour, making it expensive and technically challenging to develop.

Abu Dhabi has put in place a number of programs to bring additional gas online. These include:

- new exploration
- construction of a pipeline that will enable up to 700 MMcf to be produced offshore and piped to the mainland
- development of the sour gas in the Shah onshore field (see separate item below).

Natural gas use

Abu Dhabi uses 2.3 bscfd of gas. Almost 60% goes for electric power production, the rest to industry.

Power generation is fueled almost totally by gas, and gas demand for that sector is expected to grow at 8–9% per year from 2008 to 2020.

Post-oil economy

The *Financial Times* observed in September 2009 that several of Abu Dhabi's sovereign investment funds have been leading an initiative to move beyond an oil-based economy.

Those funds include the Abu Dhabi Investment Authority (ADIA); the Advanced Technology Investment Co. (ATIC); Mubadala; and the International Petroleum Investment Co. (IPIC).

The groups are investing in such areas as manufacture of semiconductors and automobiles; renewable energy development; banking and commercial finance; and even private-sector spaceflight.

Drilling and Production

Shah gas development

ADNOC chose ConocoPhillips in 2009 to lead a $10-billion project to develop ultra-sour gas reserves in Abu Dhabi's Shah field. Located 180 km (112 mi) southwest of Abu Dhabi city, it will be the most expensive gas project in the UAE.

The project will included one the world's largest sulfur-removal plants as well as sulfur processing and exporting facility at Ruwais Industrial City.

First gas production is expected in 2015.

Shah's reserves are expected to produce roughly 1 bscfd of raw gas; however, after gas treatment, only 500–600 MMscfd of dry gas likely would be available for the domestic market or re-injection into oil wells, according to FGE.

Gas and products projects

Abu Dhabi has several new projects pending that could produce large amounts of gas, condensate, or other gas liquids from 2010 to 2015.

OGD-III

Abu Dhabi's Onshore Gas Development III (OGD-III) was working to finish building by yearend 2009 a gas processing plant in the Habshan Zone of Bab field.

The plant will process 1.3 bscfd of raw gas and produce 1.2 bscfd of dry gas, 4.3 mtpy of NGLs, and 130,000 b/d of condensates.

AGD-III

The Asab Gas Development II (AGD-II) will install two gas treatment and NGL recovery units with a total capacity of 743 MMscfd. Startup was scheduled for December 2009.

It is designed to recover 2.3 mtpy of NGL and to re-inject residual gas back into the reservoir.

A third NGL extraction train was being built at the Al-Ruwais refinery, west of Abu Dhabi City, to treat NGLs from the OGD-III and AGD-II gas projects.

IGD

Abu Dhabi's $7-billion Integrated Gas Development (IGD) project is designed to produce 2 bscfd of high-pressure gas from the offshore Umm Shaif and Khuff reservoirs.

It is expected that 1 bscfd of gas will be treated release to the domestic market and the other 1 bscfd will be re-injected into Umm Shaif oil field.

The IGD project is expected to be completed in 2014.

Gas Processing

Ruwais, Habshan gas projects

Abu Dhabi Gas Industries Ltd. awarded more than $9 billion in contracts in mid-2009 for work on its integrated gas projects at Ruwais and Habshan.

The projects are scheduled for completion by third quarter 2013.

At Ruwais, a fourth NGL train and storage tanks will be built. The facility currently gets feedstock from NGL plants at Asab, Bu Hasa, Habshan, and Bab, as well as LPG from the Takreer refinery.

The Habshan project will be built at a new location called Habshan 5. Workers will install four gas processing trains (total capacity 2 bcfd) and four new sulfur-recovery units.

Habshan 5 will produce 900 MMscfd of sales gas, 12,000 tpd of NGL, and 5,000 tpd of liquid sulfur.

Pipelines/Transportation

Port of Fujairah

A multi-million-dollar EPC contract was awarded in third quarter 2009 for a major expansion of the Port of Fujairah. Topside facilities will be built for four berths at Oil Terminal 2.

A deepwater facility at the mouth of the Straits of Hormuz, the port is the second-largest bunkering facility in the world, after Rotterdam.

Some 10,800 vessels used the port's offshore anchorage facilities in 2008.

Oil storage in Japan

Nippon Oil Corp. reached agreement with ADNOC in late 2009 for the UAE firm to store oil in an effort to bolster Japan's national reserves.

Details are given in the Japan section of the Asia-Pacific chapter of this book.

Taweelah-Fujairah gas pipeline

In early 2009, a subsidiary of Russia's Gazprom started work on the $418-million Taweelah-Fujairah natural-gas pipeline (TFP) project. This is the first involvement of a Russian company in a large-scale energy project in the UAE.

The 240-km (149-mi) pipeline will transport gas from the existing natural-gas receiving facility in Taweelah on the Persian Gulf coast to a power and desalination plant at Fujairah on the coast of Oman.

TFP will be the longest overland gas pipeline in the UAE.

The pipeline is part of the $7-billion Dolphin Gas project.

Petrochemicals

Borouge 2 and 3: Abu Dhabi

Borouge, a JV of ADNOC and Austria's Borealis AG, is a major manufacturer of polyolefins.

Its current production capacity is 600,000 tpy of polyethylene, but an expansion project called Borouge 2 to increase capacity to 2.1 million tpy is scheduled to begin in mid-2010.

Also, in June 2009, the company awarded a $1.1-billion contract and chose a project manager (Bechtel Group) to build the Borouge 3 facility in Ruwais, Abu Dhabi.

That complex, designed to produce 2.5 mtpy of polyolefins, is expected to be operational in late 2013.

Taweelah complex

Abu Dhabi was moving forward in early 2009 with plans to build what it calls the world's largest fully integrated grassroots petrochemical complex.

The plant will be in the new Mina Khalifa Industrial Zone in the Taweelah area.

Masdar City Complex

The Masdar Carbon-Neutral City Complex is to be an ambitious 6-sq km (2.3-sq mi) residential and commercial township project, 17 km (11 mi) off Abu Dhabi.

When completed in 2018–19, it is expected to house 45,000 people and provide office facilities to more than 1,500 companies. Cost of the project, initiated in 2006, is estimated at $22 billion.

It is to be self-sustaining, using solar and geothermal energy and waste-conversion technologies.

Nuclear power deal

In December 2009, a South Korean consortium won a $20.4-billion contract to develop a civilian nuclear program for the UAE.

The group outbid competing consortia from France and a U.S.-Japanese team.

The contract involves the design, construction and operation of four 1,400-MW units.

Plans call for the first nuclear plant to start producing electricity in 2017, the other three in 2020.

The operator will be the new Abu Dhabi National Chemicals Co., also called Chemaweyaat.

Production from the first $15-million phase of the complex is scheduled for 2014.

The full complex will have a combined production and export capacity of more than 6 mtpy.

Alternative Energy

Masdar: Green energy focus

The Abu Dhabi state-controlled company called Masdar plans to invest about $15 billion on renewable energy projects through 2015.

As of mid-2009, some $3 billion had already been invested, and $10 billion is to be available by 2016.

The initiative will include wind, solar and hydrogen-based power projects in Abu Dhabi.

A signature initiative is construction of a Carbon-Neutral City Complex (see box).

In 2009, Masdar connected a 10-MW, $50-million photovoltaic power plant to the grid, consisting of more than 87,000 panels.

Domestically, Masdar also plans to capture 6.5 mtpy of carbon dioxide at four sites across Abu Dhabi and sell it to ADNOC for use in EOR.

The company (through subsidiaries and with partners) also will continue to invest in renewable energy projects in Europe. Examples include a windfarm in London, a photovoltaics manufacturing plant in Germany, and three solar power plants in Spain.

Environmental Issues

Power plant with CCS

In early 2009, plans for a $2-billion hydrogen power and carbon capture and storage (CCS) power plant in Abu Dhabi were moving ahead. The plant is slated for completion in late 2014.

The project is a JV of Abu Dhabi's Masdar clean energy fund and Hydrogen Energy International LLC (also in Abu Dhabi). HEI is itself a JV between BP PLC and the Rio Tinto Group.

The 420-MW plant is called the world's first large-scale CCS project.

It will extract carbon dioxide and hydrogen from natural gas, before the latter is burned in a gas turbine to produce electricity. The extracted carbon dioxide will be used for EOR.

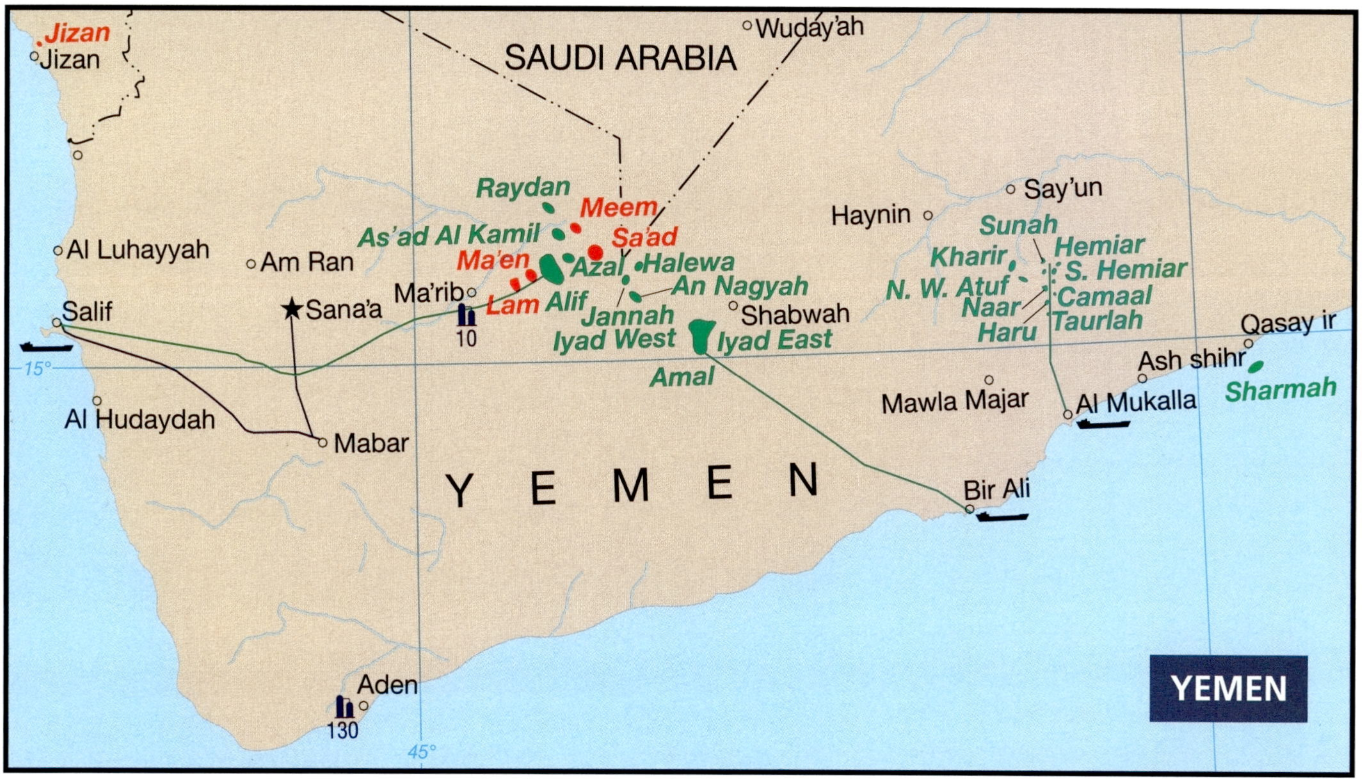

YEMEN

Capital: Sana'a
Monetary unit: Rial
Refining capacity: 140,000 b/cd
Oil production: 270,000 b/d
Oil reserves: 3.000 billion bbl
Gas reserves: 16.900 tcf

Kuwait: IFC aid

Kuwait Energy Co. secured $50 million in 2009 from the International Finance Corp. (part of the World Bank) for exploration and development in Yemen and Egypt and to support related environmental and social management activities. Details are given in the Kuwait section of this chapter.

"Gas City" envisioned

In September 2009, Gas Cities Ltd., based in Sharjah, UAE, reached agreement with the government of Yemen to study construction of a "gas city" in the country.

The aim is to develop an integrated industrial city based on primary gas-processing industries, from which other high-employment industries would benefit.

Yemen's gas city would include fertilizer and methanol factories as well as a power plant. Self-sufficient for all utilities, it would send any excess power into the Yemen national electricity grid.

In the first phase of the agreement, the government and Gas Cities Ltd. will investigate a suitable feedstock profile and find a viable location.

Gas Cities Ltd. is a 50:50 JV between Dana Gas PJSC and Crescent Petroleum, established in 2008.

The venture plans to build at least four gas cities in the Middle East, North Africa and South Asia.

In 2008, work began on the first such complex, at Chemchemal in Iraqi Kurdistan, and in May 2009, the JV announced plans to build a gas city in Egypt.

Yemen LNG: first production

In October 2009, Total SA announced that the 6.7-mtpy Yemen LNG plant produced its first LNG from the first of two planned trains. The second train was under construction.

The $4.5-billion project takes gas from Block 18 in the Marib region of central Yemen. A 320-km (199-mi) pipeline connects the field to the LNG plant, which is at the port of Balhaf on the southern coast.

No customer was specified for the initial production.

Three gas sales agreements were signed in 2005 with Kogas, GDF-Suez, and Total Gas & Power Ltd.

Africa

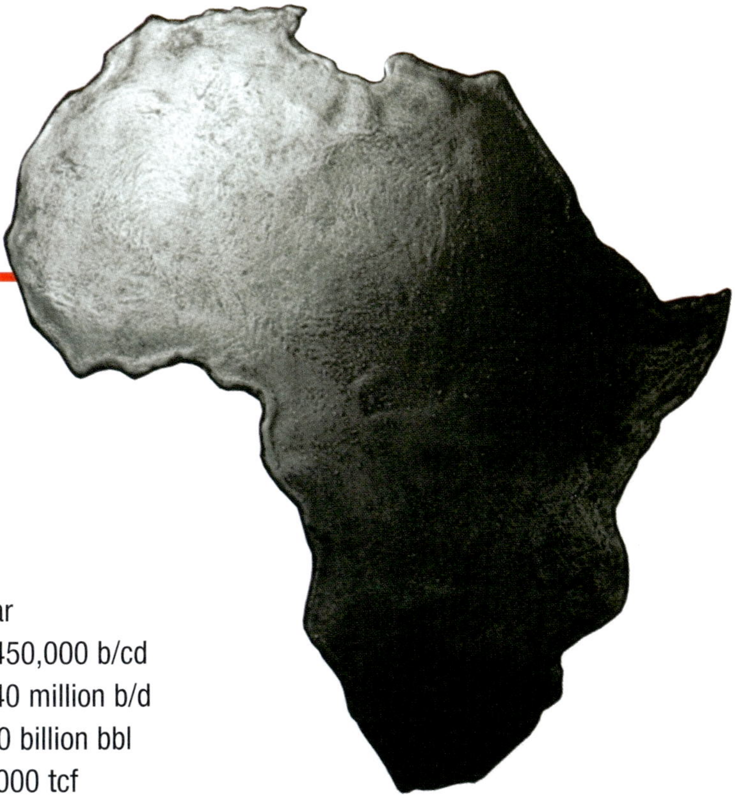

ALGERIA

Capital: Algiers
Monetary unit: Dinar
Refining capacity: 450,000 b/cd
Oil production: 1.240 million b/d
Oil reserves: 12.200 billion bbl
Gas reserves: 159.000 tcf

Exploration and Development

Timimoun basin

In October 2009, Sonatrach, Total SA, and Cepsa reported that the Algerian National Oil and Gas Development Agency had approved their development plan for the Tout gas license in west-central Algeria, in the Timimoun basin near Adrar.

Development was to start by yearend to develop 10 fields and drill 40 production wells.

Gas production is slated to begin in 2013 and to peak at 4.5 bcm/year (159 bcf/year).

Developers also will build a connection to the Sonatrach pipeline that will carry gas from fields in west-central Algeria to Hassi R'Mel.

The 10 fields of interest are Hassi Ilatou, Hassi Ilatou Cambrien, Hassi Ilaout Nord Est, Gour Nefrat Gedinnen, Gour Nefrat Ordovicien, Bou Hadid, Bou Hadid Ouest, Oued Hamou, Oued Zine, and Sbaa.

Illizi basin

Sonatrach reported an oil discovery in mid-2009 in the Illizi basin with its Ain Antar Is-1 well.

The well was drilled on Block 244A in the Tinrhert area.

It was the seventh discovery of the year in Algeria, five of which were made by IOCs.

Gazprom: El Assel license

In early 2009, the Dutch arm of Gazprom secured from Sonatrach the rights to explore the onshore El Assel concession in the Berkine basin in Algeria. Gazprom will have a 49% stake.

Provisional estimates put recoverable oil reserves of the area at about 30 million tons (220 MMbbl)

WEST AFRICA
Atlantic Ocean
Mediterranean Sea
Ionian Sea
Gulf of Guinea
Bight of Biafra
MOROCCO
ALGERIA
TUNISIA
LIBYA
WESTERN SAHARA
MAURITANIA
MALI
NIGER
CHAD
SENEGAL
GAMBIA
GUINEA-BISSAU
GUINEA
SIERRA LEONE
LIBERIA
IVORY COAST
BURKINA FASO
GHANA
TOGO
BENIN
NIGERIA
CAMEROON
CENTRAL AFRICAN REPUBLIC
EQUATORIAL GUINEA
GABON
CONGO (FORMER ZAIRE)
CABINDA
ANGOLA
NAMIBIA
BOTSWANA
REPUBLIC OF SOUTH AFRICA
SICILY
GREECE
CRETE
Athens
Rabat
Mohammedia
Casablanca
Kechoula
Essaouira
Safi
Djebel-Jeer
Sidi Rhalem
Sidi-Kacem
Las Palmas
Sidi Ifni
Tarfaya
Oran
Arzew
Algiers
Skikda
Bejaia
Bizerte
Tunis
Gela
El Golea
Hassi Messaoud
Bel Rhazi
Touat
Ilatou
SBAA
Azzene
Reggane
In-Salah
In Amenas
Wafa
Amasralad
Le Camp
Murzuk
Tripoli
Sider
Tobruk
Farrud
Umm Farud
Amal
Nafoora
Augila
Messla
Sarir
Defa
Chinguetti
Nouakchott
Bamako
Bissau
Conakry
Freetown
Monrovia
Ouagadougou
Niamey
Lake Chad
N'Djamena
Koudalwa
Kano
Kaduna
Abuja
Makurdi
Komé
Bolobo
Miandoum
Doba
Bangui
Porto Novo
Lagos
Lome
Accra
Abidjan
Espoir
Panthere
Bellier
Foxtrot
Jubilee
Saltpond
Yaounde
Kribi
Ceiba
Libreville
Sao Tome
Cape Lopez
Port Gentil
Tschengue
Rembo-Kotto
Asswe
Sette-Cama
Gamba-Ivinga
Mossaka
Brazzaville
Kinshasa
Pointe Noire
N'Kossa
Emeraude
GCO
Essungo
Mavanga
Sulele
Guntala
Maleva North
Luanda
Benfica
Tobias
Lobito
Windhoek
Walvis Bay
Gaborone
Luderitz
Johannesburg
Kudu
Ibhubesi
Cape Town
Cape of Good Hope
Mossel Bay
Sable
F-A
Oribi
3 Lines
2 Lines
0 300 600 Mi.
0 200 400 600 Km.

EAST AFRICA
0 300 600 Miles
0 200 400 600 Km
Mediterranean Sea
Red Sea
Gulf of Aden
Indian Ocean
Mozambique Channel
Persian Gulf
Ionian Sea
Aegean Sea
SICILY
TURKEY
GREECE
CRETE
CYPRUS
SYRIA
LEBANON
ISRAEL
JORDAN
IRAQ
IRAN
PAKISTAN
Kuwait
SAUDI ARABIA
OMAN
YEMEN REPUBLIC
TUNISIA
LIBYA
EGYPT
NIGER
CHAD
SUDAN
ERITREA
DJIBOUTI
ETHIOPIA
SOMALIA
NIGERIA
CAMEROON
CENTRAL AFRICAN REPUBLIC
CONGO (FORMER ZAIRE)
UGANDA
KENYA
GABON
CONGO
RWANDA
BURUNDI
CABINDA
TANZANIA
MALAWI
ANGOLA
ZAMBIA
MOZAMBIQUE
ZIMBABWE
NAMIBIA
BOTSWANA
MADAGASCAR
SWAZILAND
LESOTHO
REPUBLIC OF SOUTH AFRICA
Athens
Nicosia
Beirut
Damascus
Baghdad
Amman
Jerusalem
Manama
Doha
Abu Dhabi
Riyadh
Hilwah
Ghawar
Tripoli
Tunis
Bizerte
El Kala
Annaba
Gela
Hassi Messaoud
Mellitah
Wafa
Sider
Tobruk
Alamein
Abu-Qir
Alexandria
Cairo
Suez
Mango
Ramadan
Ras Gharib
Midyan
Geisum
Asyût
Farrud
Umm Farud
Amal
Nafoora
Augila
Abu Gharadig
Defa
Sarir
Messla
In Amenas
Murzuk
Le Camp
Port Sudan
Jizan
Sana'a
Ma'bar
Aden
Assab
Djibouti
Khartoum
Kassala
Asmera
Kosti
Sheraf
Munga
Adar
Muglad
Heglig
Tabaldi
Unity
Abu Gabra
Melut
Bentiu
Addis Ababa
Calub
Lake Chad
Lake Tana
Kano
N'Djamena
Bolobo
Doba
Miandoum
Komé
Makurdi
Yaounde
Kribi
Bangui
Mogadishu
(Idle)
Lake Albert
Kampala
Lake Victoria
Nairobi
Mombasa
Libreville
Tschengue
Rembo-Kotto
Asswe
Sette-Came
Gamba-Ivinga
Mossaka
Kigali
Bujumbura
Brazzaville
Kinshasa
N'Kossa
Cabinda
Emeraude
GCO
Essungo
Mavanga
Guntala
Maleva North
Sulele
Luanda
Benfica
Tobias
Lake Tanganyika
Dodoma
Zanzibar
Dar es Salaam
Songo-Songo
Kilwa
Lobito
Ndola
Lake Malawi
Lilongwe
Lusaka
Lake Kariba
Harare
Mutare
Beira
Pande
Temane
Antananarivo
Walvis Bay
Windhoek
Thohoyandou
Gaborone
Pretoria
Johannesburg
Maputo
Mbabane
Luderitz
Kudu
Sasolburg
Maseru
Durban
Ibhubesi
Umtata
Cape Town
Mossel Bay
Cape of Good Hope
Sable
Oribi
F-A

El Assel, which includes three blocks covering about 3,000 sq km (1,158 sq mi), is Gazprom's first project in Algeria.

Regane, Ahnet basins

Repsol YPF said in early 2009 that several natural gas discoveries in the Regane and Ahnet basins in the Sahara Desert could prove to be important future resources.

Exploration well KLS-1 tapped the Kahlouche Sud structure at 3,720 m (12,205 ft) in the North Reggane concession. On test, gas flowed at 629,000 cu m/d (22.2 million cfd).

Discovery well OTLH-2 in the M'Sari Akabli concession flowed gas on test at 249,000 cu m/d (8.79 MMcfd) at a depth of 1,280 m (4,200 ft), and 110,000 cu m/d (3.88 MMcfd) at 1,340 m (4,396 ft).

Lease Sales

2009 licensing round

After failing to generate interest in a December 2008 offering, Algeria launched a licensing round in mid-2009 for 25 blocks in basins where previous discoveries were made by Repsol-YPF SA and Statoil.

Officials said it would seek bidders with experience in developing gas from tight sands.

Bids were to be submitted by December 20, 2009, with contracts to be signed in January 2010.

E.ON Ruhrgas AG, BG Group PLC, Eni SPA, and Gazprom filed successful applications for the licensing round.

Drilling and Production

El Merk hub

Two initial contracts were awarded in mid-2009 to build facilities for the El Merk production hub in Block 208 in the Berkine Basin, about 250 km (155 mi) southeast of Hassi Messaoud.

The hub will be operated by a JV between Anadarko Petroleum Corp. and Sonatrach.

The El Merk facility will handle hydrocarbons from the El Merk oil field, serving as a central processing and distribution unit for the region.

It is expected to produce 29,000 b/d of condensate, 98,000 b/d of oil, 600 MMcfd of NGLs, and 31,000 b/d of LPG.

Gas Processing

Gas handling plant: Qartzites de Hamra

Sonatrach awarded a $1.1-billion, 39-month contract in mid-2009 to Canada's SNC-Lavalin Group Inc. to build natural gas handling facilities in Qartzites de Hamra in eastern Algeria.

SNC will build a complex to serve four fields: Rhourde Nouss Central, Rhourde Nouss Southwest, Rhourde Adra, and Rhourde Adra South.

To be completed in 2012, the project will process and treat about 10 bcm/d (353 MMscfd).

Residue gas will be piped to Arzew, where Sonatrach is expanding an LNG production plant.

Gas treatment: MLE field

In March 2009, Sonatrach and First Calgary Petroleums Ltd. awarded a $1.8-billion EPC contract to Saipem SpA to provide gas gathering and processing equipment for the Menzel Ledjmet East (MLE) field in the Berkine basin, 350 km (217 mi) southeast of Hassi Messaoud.

Saipem will develop facilities to treat gas extracted from the MLE field and others in the Central Area Field Complex, as well as gas gathering systems, a gas pipeline to Gassi Touil, and an oil pipeline to Point Kilometrique 0 (PK0).

All facilities are to be operational by 2012.

LNG

Skikda delay

The oil ministry announced in early 2009 that the Skikda LNG plant will not come online until 2013, two years later than its earlier announced completion date. No reasons were given for the delay.

New train at Arzew

In early 2009, Daewoo Engineering and Construction won a $208-million contract from Sonatrach to build a 4.7-mtpy LNG train in Arzew. It will be the third train at the facility.

A consortium led by Saipem SpA is leading the project, which will cost an estimated $4.5 billion.

Daewoo also will build gasoline, propane and butane storage tanks and an export terminal. Construction is scheduled for completion by 2012.

The Rhourde Nouss and Gassi Touil fields will feed the plant.

Pipelines/Transportation

Pipelines to Europe

The majority of Algeria's gas production is exported to Europe and provides about 10% of Europe's requirements. Algeria is second only to Russia as an exporter of gas to Europe.

- The TransMed pipeline, also known as the Enrico Mattei pipeline, runs undersea from Tunisia to Sicily and was upgraded in 2008 to raise its capacity to 30.2 bcm/year (1.07 tcf/year).
- The Maghreb-Europe Gas pipeline, also known as the Pedro Duran Farrell pipeline, runs to Cordoba in Spain and was completed in 1996.
- The Medgaz pipeline, to run undersea from Algeria to Spain, is to deliver 8 bcm/year (282 bcf/year) of gas in early 2010.
- The Galsi pipeline will carry 8 bcm/year (282 bcf/year) from Algeria through Sardinia to Italy and perhaps to Corsica, starting in 2013.

Algeria was planning to increase exports of gas and LPG by 25% to 77.5 bcm (2.74 tcf) in 2009, and to 85 bcm (3.0 tcf) in 2010.

Medgaz

Medgaz announced in March 2009 completion of the offshore section of the pipeline, and reported in November that 97% of the overall project had been completed.

In December, the company began a testing sequence expected to lead to start-up of pipeline operation in first quarter 2010.

With an initial capacity of 8 bcm/year (282 bcf/year), the Medgaz line will carry natural gas 210 km (130 mi) from Beni Saf on the Algerian coast, along the seabed to Spain, to a connection point in Almería.

Galsi

In June 2009, the Ministry of Energy announced that construction of the Galsi (Gasdotto Algeria Sardegna Italia) gas pipeline will start in 2010.

The 1,490-km (926-mi) line will transport 8 bcm/year (282 bcf/year) from Algeria through Sardinia to the island of Corsica through four sections:

- a 640-km (398-mi) onshore section from Hassi R'Mel to El Kala
- a 280-km (174-mi) offshore section from El Kala to Cagliari
- a 300-km (186-mi) onshore section to Olbia in Sardinia
- a 270-km (168-mi) offshore section from Olbia to Tuscany.

The line is a joint development of Sonatrach, Edison SpA, Enel SpA, and Hera SpA.

France has expressed interest in receiving Galsi gas by tapping the segment between El Kala and Tuscany to add a spur to French-controlled Corsica.

Originally scheduled for 2012, the opening of the line has been delayed to 2013. The final investment decision is expected by June 2010.

GK3

Sonatrach awarded EPC contracts in mid-2009 worth a total of $1.4 billion to Saipem SpA and Petrojet to build the 784-km (487-mi) GK3 gas pipeline, connecting the Hassi R'Mel gas field with El Kala on the northeastern coast of Algeria.

Petrojet will develop two sections, from Hassi R'Mel to Chaiba and on to Mechtatine.

Saipem will build the third, from Mechtatine to Tamoulka and then further to Skikda and El Kala.

El Kala will be the departure terminal for the planned Galsi pipeline, which will run from Algeria to Italy by way of Sardinia.

Trans-Sahara Gas Pipeline

The planned $10-billion Trans-Sahara Gas Pipeline aims to transport as much as 30 bcm/year (1.06 tcf/year) of gas from Nigeria to Europe, passing through Algeria and Niger.

Although no date has been set for the start of work on the 4,100-km (2,548-mi) line, its first shipment of gas is scheduled for 2015.

In early 2009, Total SA said that it was ready to participate, and Gazprom also has announced plans to invest in the TGSP through a JV called NiGaz Energy Co. Ltd. with state-owned Nigerian National Petroleum Corp.

In mid-2009, Algeria, Niger, and Nigeria agreed to start the process of constructing the line.

Refining and Products

Refinery overview

As of mid-2009, Algeria had a total refining capacity of 450,000 b/d from four Sonatrach-owned facilities (all operated by Naftec SpA):

- Hassi Messaoud: 30,000 b/d
- Algiers: 60,000 b/d
- Arzew: 60,000 b/d
- Skikda: 300,000 b/c.

A fifth refinery is planned for Tiaret (300,000 b/d), to start up in 2014.

Skikda is to be modernized by 2012 to raise its capacity to 18 mtpy (132 million bbl/year).

Expansion work is slated to be completed in 2010 at both Arzew and Algiers.

Algeria officials have said the country intends to invest about $28 billion from 2009–2013 to overhaul the downstream sector, including refining and petrochemicals.

Sonatrach is expected to contribute about $8 billion toward the effort.

Alternative Energy

Desertec consortium

Desertec Industrial Initiative (DII) GmbH—a sustainable-energy initiative of a consortium of 12 European companies—proposed plans in mid-2009 for a massive solar power generation project in the Sahara Desert.

Details are given in the Germany section of the Europe chapter of this book.

Hybrid gas/solar power plant

Work on Algeria's first commercial hybrid power plant began in fourth quarter 2009 in Hassi R'Mel, 418 km (260 mi) south of Algiers.

The plant will generate 150 MW. Of this total, 125 MW will come from a combined cycle system and 25 MW from a facility using almost 180,000 sq m (1.94 million sq ft) of concentrating solar collectors.

The plant is scheduled to be completed by late 2010.

Algeria also has a goal of supplying Europe with electricity generated by solar systems in the Sahara Desert. By 2012, undersea cables would carry the power from Algeria to Sicily and Spain.

Environmental Issues

In Salah: CO_2 storage

Krechba is one of eight gas fields comprising the In Salah gas project in central Algeria that includes an industrial-scale CO_2 storage project. It has been in operation since 2004.

The facility at In Salah is being used as an experimental and demonstration project to learn about CO_2 geological storage in deep saline formations.

Gas from several fields, which has a CO_2 content of 5–10%, is processed to remove the CO_2 from the production stream to meet export specifications.

KRECHBA, ALGERIA, CO_2 INJECTION, STORAGE

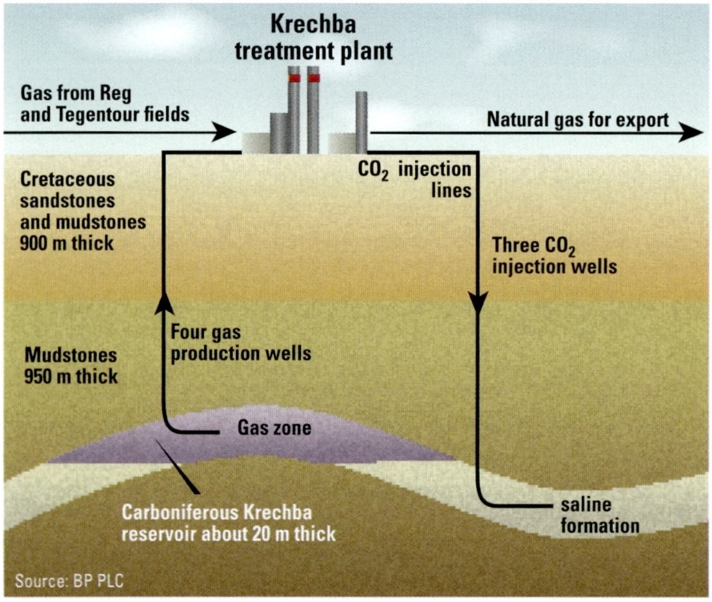

Rather than vent that separated CO_2, BP and its JV partner Sonatrach invested an incremental \$100 million in technology to compress, dehydrate, transport, and inject that CO_2 into a deep saline formation.

That formation is a 20-m (66-ft) thick Carboniferous sandstone, 1,900 m (6,233 ft) below ground.

By the end of 2008, more than 2.5 million tonnes of CO_2 had been stored underground.

A Joint Industry Project was set up to monitor the CO_2 storage over an initial 5-year period.

Project details can be found at http://www.insalahco2.com.

Company News

Sonatrach plans

Despite declines in oil and gas prices in the first half of 2009, Sonatrach said that it would maintain its investment program. That includes expansion in traditional areas such as refining and LNG as well as more speculative ventures in desalination, power generation, and petrochemicals.

It estimated a 2008–2012 investment of \$45 billion, but said that 2009–2013 spending would total \$64 billion, boosted by the 60 discoveries it made in 2008. Its budget includes \$1.5 billion/year for exploration.

IHS Global Insight said Sonatrach faces a risk of delay in its upstream gas exploration and production because of low industry interest in its December 2008 licensing round.

IHS said that the setback could hinder development in the southwest—a region from which gas will be needed to meet future export commitments and domestic needs.

ANGOLA

Capital: Luanda
Monetary unit: Kwanza
Refining capacity: 39,000 b/cd
Oil production: 1.790 million b/d
Oil reserves: 9.500 billion bbl
Gas reserves: 9.600 tcf

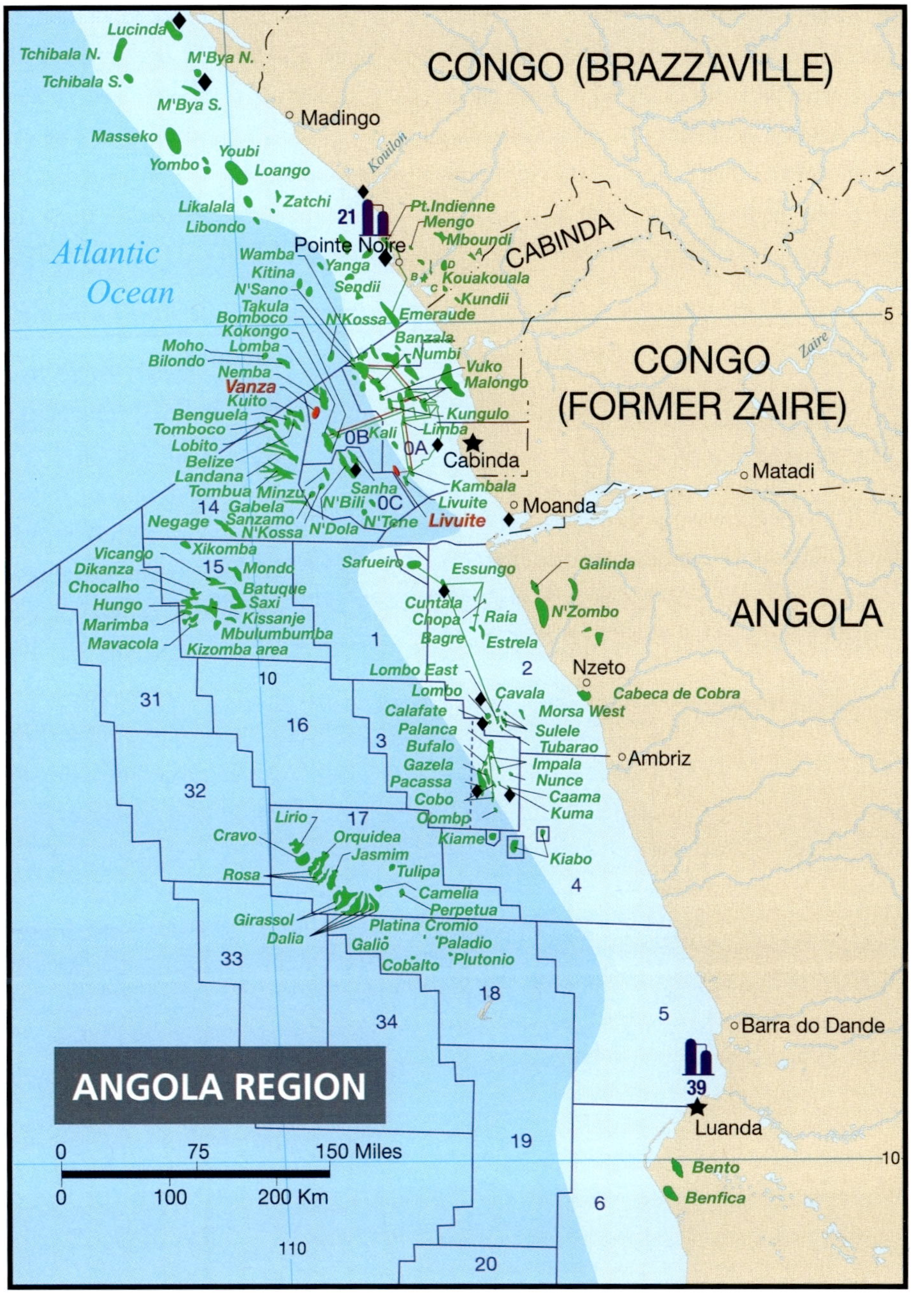

Overview

Angola has overtaken Libya as the region's second most prolific oil producer. (Nigeria is first.)

The country's production is predominantly from deep water overlaying thick salt layers.

In mid-2008, there were 21 deepwater fields on production.

In 2009, Tombua and Landana fields (Block 14) and Mafumeira Norte (Area A, Block 0) began production. The Pazflor field on Block 17 is expected to produce first oil in 2011.

Exploration and Development

Block 32 deal

In mid-2009, a subsidiary of Marathon Oil Corp. signed a $1.3-billion agreement with China's CNOOC and Sinopec regarding development of offshore Block 32.

CNOOC and Sinopec were expected to form a JV to acquire the Marathon unit's undivided 20% participating interest in the production-sharing contract and joint-operating agreement in the block.

Marathon will retain a 10% working interest in the block, which has already yielded 12 discoveries.

The companies expected to close the transaction by yearend.

Discoveries reported

A number of significant discoveries were made offshore Angola in 2009, including these:

- Petrobras: Manganês-01 well; oil; Block 18/06
- Eni Angola: Cabaça Norte-1 well; oil; Block 15/06
- BP PLC: Tebe discovery; oil; ultra deepwater Block 31
- BP PLC: Oberon-1 well; oil; Block 31
- BP PLC: Leda well; oil; Block 31
- CABGOC; exploration well 79-3XST1; oil and gas; Block 0; Greater Vanza Longui Area
- Maersk Oil Angola; Chissonga-1 well; oil; Block 16

Eni-Sonangol partnership

Based on an MOU signed in 2008, Eni SpA made three agreements in early 2009 with state oil company, Sociedade Nacional de Combustiveis de Angola (Sonangol).

Eni will evaluate setting up an onshore electric power plant that will use associated gas recovered in oil production operations.

Sonangol and Eni will study highly prospective Angolan onshore basins and their production potential.

The two companies will bolster educational projects and training of Angolan professionals.

Eni has been in Angola since 1980 and currently operates deepwater Block 15/06.

Cabinda

Cabinda is a tiny enclave of Angola but geographically separate from it.

A 2009 overview by the *Washington Post* noted that the province contributes the majority of the oil revenues that make up about 42% of Angola's GNP and 90% of its budget.

Cabinda generates about 60% of the country's oil, and some of Angola's most productive oil fields lie in Block 14 and Block 0 offshore Cabinda.

On Block 14, Kuito oil field came on line in 1999, and by 2000, was producing 100,000 b/d.

Block 0 has been in operation for more than 20 years, producing more than 2 billion bbl since 1968.

In 1998, the block established a single-day production record of 510,451 bbl.

Chevron Corp. (formerly ChevronTexaco) has been very influential in the development of Cabinda.

Asia seeking African oil?

A study released in mid-2009 by the UK's Royal Institute of International Affairs argued that "fears expressed in western capitals about an Asian takeover in the Nigerian and Angolan oil sector...were highly exaggerated."

The authors noted that except for Japan, Asian oil companies only acquired equity participation in both countries in the last five years.

Drilling and Production

Tombua-Landana

In third quarter 2009, Chevron Corp. subsidiary Cabinda Gulf Oil Co. Ltd. (CABGOC) started oil production from the $3.8-billion Tombua-Landana project on Block 14, about 80 km (50 mi) off Angola.

CABGOC expects that the two fields will reach peak production of 100,000 bo/d in 2011.

Landana and Tombua contain an estimated 350 million bbl of recoverable oil.

The Angola LNG project, under construction at Soyo, will process the Tombua-Landana gas along with associated gas from other fields on Blocks 0, 14, 17, and 18. Completion is expected in 2012.

Mafumeira Norte

CABGOC brought the Mafumeira Norte oil project, 24 km (15 mi) offshore, on stream ahead of schedule in mid-2009. Its 14 wells will deliver a maximum of 30,000 b/d and 30 MMcfd of natural gas in 2011.

The project is the first-phase development of Mafumeira field in Area A of Block 0.

Block 17: Pazflor development

Operator Total SA reported the start of drilling in March 2009 at the giant Pazflor oil field development. It will be the third development center on offshore Block 17, following Girassol and Dalia, both on production.

Production is slated to begin in 2011.

Pazflor lies in 600–1,200 m (1968–3937 ft) of water about 150 km (93 mi) offshore and 40 km (25 mi) northeast of Dalia field.

The project aims to bring four fields into production: Perpetua, Hortensia, Zinia, and Acacia.

LNG

First LNG plant

In early 2009, Angola LNG Ltd. was completing site preparation work at a 5.2-mtpy production facility in Soyo, in Zaire Province. It will be the country's first gas liquefaction facility.

First LNG from the project is expected by early 2012.

Refining and Products

Project Sonaref

Sonangol was proceeding in mid-2009 with its project to build a 200,000 b/d, $8-billion refinery, called Project Sonaref, near the port city of Lobito. KBR Inc. won the contract in 2008 to design the refinery.

Commissioning is scheduled for late 2011.

CAMEROON

Capital: Yaounde
Monetary unit: CFA franc
Refining capacity: 37,000 b/cd
Oil production: 74,000 b/d
Oil reserves: 200.000 million bbl
Gas reserves: 4.770 tcf

Etinde permit

Bowleven PLC announced plans in third quarter 2009 to drill the Isongo F-2 and F-3 appraisal wells on the IF field within the MLHP-7 block of the Etinde permit by second quarter 2010.

The Isongo-F-1r exploration well in early 2009 averaged 3,371 b/d of oil.

Bowleven said the discovery changed the thinking about Cameroon's petroleum systems, which previously ascribed oil potential only to the northern part of the Rio del Rey basin.

The Etinde permit covers three shallow water blocks off northern Cameroon, spanning parts of the Rio del Rey and Douala basins. Both have proven petroleum systems and discovered hydrocarbon resources.

The IF field lies on Block 7 and is estimated to hold 53 million bbl of contingent resources, according to independent analysis.

Bowleven also will reprocess 3D seismic and acquire new 3D seismic over the IE and IF discoveries.

If drilling is successful, the firm estimates that first oil could be produced in 2012, likely using an FPSO vessel.

Logbaba field

Victoria Oil & Gas PLC spudded a well in mid-2009 to develop Logbaba gas-condensate field in the city of Douala.

Logbaba has independently assessed proved and probable reserves of 104 bcf.

The development plan calls for drilling one new well to 3,048 m (10,000 ft); building a 340,000-cu m/d (12-MMcfd) gas processing plant; and laying 15 km (9 mi) of pipeline to industrial users.

Production is to start in 2010.

CHAD

Capital: N'Djamena
Monetary unit: CFA franc
Refining capacity: 0
Oil production: 145,000 b/d
Oil reserves: 1.500 billion bbl
Gas reserves: 0

Timbre field

An ExxonMobil Corp. subsidiary was planning to start production from Timbre field in Chad in 2009.

Waterfloods and production optimization were continuing to maximize recovery, the company said.

Chad's oil production, shipped through the Chad-Cameroon pipeline, averaged 127,000 b/d in 2008, the majority coming from Kome, Miandoum, and Bolobo fields.

Oil pipeline

The China National Petroleum Corp. (CNPC) started work in mid-2009 on a 300-km (186-mi) oil pipeline in Chad.

The line will carry crude from the Koudalwa oilfield in the Chari-Baguirmi region to the Djarmaya refinery, 40 km (25 mi) north of N'Djamena.

The project, scheduled for completion by 2013, is a joint venture between CNPC and state oil company Société des Hydrocarbures du Tchad (SHT).

Construction of the Djarmaya refinery began in late 2008 and is slated for completion in 2011. It was designed to produce 700,000 tpy of gasoline and diesel, 20,000 tpy of kerosene and 60,000 tpy of LPG.

Refinery in Chad

CNPC signed agreements in August 2009 with the government of Chad for construction of a JV refinery in that country.

The refinery, to be built 40 km (25 mi) north of N'Djamena, has a total planned crude processing capacity of 53,000 b/d. Initial operation is slated for 2011, serving regional markets.

Oil will be sourced from Zone H of the Sedigui oil field.

The project's 40-MW power plant also will supply electricity to N'Djamena.

The project funders will be a subsidiary of CNPC and SHT.

CHAD-CAMEROON PIPELINE

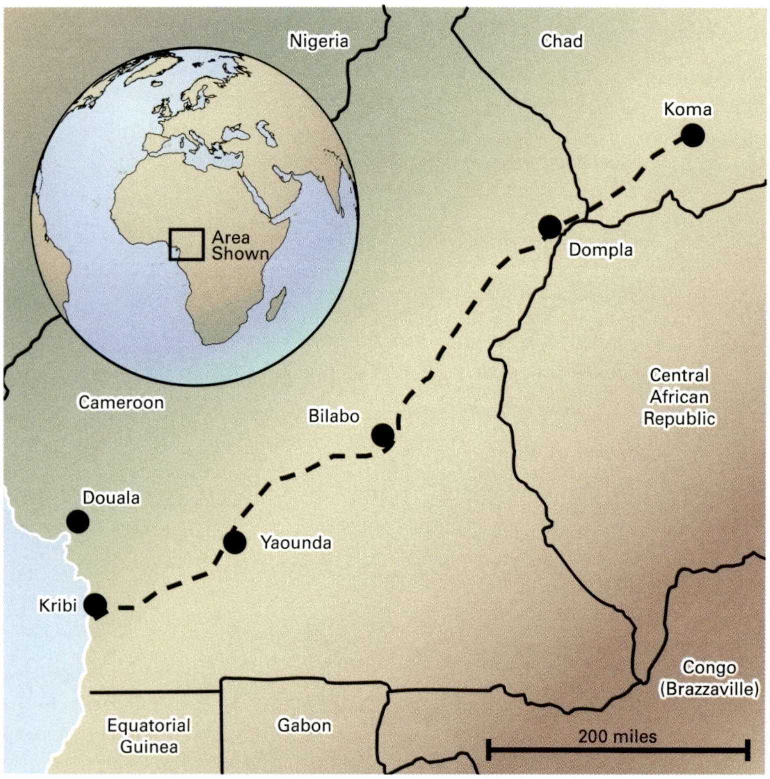

OUTLOOK FOR GROWTH

The International Monetary Fund warned in early 2009 that economic growth in sub-Saharan Africa likely would drop by 50% compared to the average of the past decade, to 3.25% in 2009, due to the impact of the global economic crisis and declining oil prices.

Separately, the World Bank said that developing countries likely would face a financing shortfall of $270–700 billion in 2009 as private-sector creditors avoid emerging markets.

Africa's fortunes, said the IMF, depend heavily on the recovery of global growth through stimulus packages and financial-sector reforms in more-developed economies.

CONGO (BRAZZAVILLE)

Capital: Brazzaville
Monetary unit: CFA franc
Refining capacity: 21,000 b/cd
Oil production: 240,000 b/d
Oil reserves: 1.600 billion bbl
Gas reserves: 3.200 tcf

Mer Profonde Sud Block

Murphy Oil Corp. reported an oil discovery in mid-2009 at the Turquoise Marine-1 prospect on the offshore Mer Profonde Sud (MPS) Block.

The discovery well, which found more than 41 m (136 ft) of net oil pay, lies 27 km (17 mi) from Murphy-operated Azurite field.

The discovery was the second for Murphy on the MPS block.

Moho-Bilondo field

Total E&P Congo gauged a third deepwater oil discovery in the northern part of the offshore Moho-Bilondo license in early 2009. The company had two successful exploration wells there in 2007.

The Moho Nord Marine-4 well, 75 km (47 mi) offshore, flowed 8,100 b/d from the Albian F formation. TD was 4,239 m (13,907 ft).

The company started production in the Phase 1 development of Moho-Bilondo field in the southern part of the license in April 2008. It expects a plateau of 90,000 b/d of oil equivalent from 14 wells in 2010.

Total noted that it has drilled half of all exploration wells in Congo, begun production on 15 fields, and discovered 65% of initial listed reserves in the country.

Azurite field

Murphy Oil Corp. started oil production in mid-2009 from the offshore Azurite field on the Mer Profonde Sud Block through what it says is the industry's first floating drilling, production, storage, and offloading facility (FDPSO).

The FDPSO has storage capacity of 1.3 million bbl of oil and can process 40,000 b/d of oil.

CONGO (FORMER ZAIRE)

Capital: Kinshasa
Monetary unit: Zaire
Refining capacity: 0
Oil production: 25,000 b/d
Oil reserves: 180.000 million bbl
Gas reserves: 35.000 bcf

Nganzi Block

SOCO International PLC completed a 360-km (224-mi) 2D seismic survey in late 2008 over the coastal Nganzi Block. The company said that initial interpretation of results in early 2009 led to identification in early 2009 of several large structures.

More detailed interpretation continued, with expectations of initial drilling in second half 2010.

Block V

In 2008, the government granted to Dominion Petroleum Ltd. a production sharing contract for a new exploration concession on Block V.

In August 2009, the company said that it planned to complete seismic surveys in the block by yearend.

Adjacent to the border with Uganda, the block covers 7,105 sq km (2,743 sq mi). It includes part of Lake Edward and adjoins Block 4B in Uganda, where Dominion has been active.

Both blocks are part of the Albertine Rift system where major oil discoveries were made in 2006–2007.

Relations with Uganda

In early 2009, Congo and Uganda agreed to upgrade diplomatic relations and boost cooperation on security and oil exploration.

The agreement highlighted improved relations after two wars over the last 15 years between the nations.

New hydrocarbon finds by Heritage Oil and Tullow Oil on the Ugandan side of Lake Albert, which lies on the border, also accelerated cooperation.

Oil exploration on the Congolese side is far behind that in Uganda, and estimates for reserves on the Congo side of the border are unknown.

Analysts said that the companies operating in the region were aiming is to establish operations on both sides of the lake and pump oil out by pipeline through Kenya.

EGYPT

Capital: Cairo
Monetary unit: Pound
Refining capacity: 726,250 b/cd
Oil production: 635,000 b/d
Oil reserves: 3.700 billion bbl
Gas reserves: 58.500 tcf

Overview

The Egyptian government was moving ahead in 2009 with plans to invest $7 billion into oil and natural gas exploration and oil field reconstruction in the 2008–2009 fiscal year.

In addition, the government was planning to invest $850 million by yearend 2009 to speed up construction of gas pipelines leading to Upper Egypt to provide energy for industrial and urban needs.

Exploration and Development

IFC aid

Kuwait Energy Co. reported in fourth quarter 2009 that it had secured $50 million from the IFC, part of the World Bank, for exploration and development in Egypt and Yemen. In Egypt, KEC has interests in:

- three onshore producing assets (Area A in the Gulf of Suez and the East Ras Qattara and Burg el Arab concessions, both in the Western Desert)
- one development asset (Abu Sennan in the Western Desert)
- an early exploration asset (Block 6 in Southern Egypt).

In mid-2009, KEC announced oil discoveries at Shukheir Northwest in Area A; and at Shahd Southeast and Rana Southeast in the East Ras Qattara, south of Alexandria.

It was planning to drill five exploration wells and five development wells before yearend.

In August, the company reported a new oil field, Al Zahraa, in East Ras Qattara.

Apache: Western Desert

Apache Corp. reported several oil and gas-condensate completions in mid-2009, including a new field discovery on its concessions in the Western Desert.

The Falcon-1x discovery on the Matruh concession tested 4,400 b/d of oil from the Cretaceous Alam El Buieb-3D formation. Falcon was expected to go on production by yearend.

In the Shushan concession 60 km (37 mi) west of Falcon, the Hydra-5x appraisal well tested 595,000 bcm/d (21 MMcfd) of gas and 3,744 b/d of condensate from Upper Safa.

Sondos field

Dana Gas Egypt said two wells were averaging 340,000 cu m/d (12 MMscfd) of dry gas in July 2009 from Sondos field, a February discovery on the West El Manzala concession in the Nile Delta.

The field's estimated reserves are 567,000 cu m/d (20 bcf). Gas is delivered to the company's South El Manzala plant.

The firm also raised production at El Basant field with completion of a development well in June.

North West Gemsa concession

Vegas Oil & Gas SA tested an average of 2,809 b/d of oil in mid-2009 from two pay zones in the Geyad-1X exploration well and completed the well as a potential producer.

It flowed 1,174 b/d and 37,500 cu m/d (1.32 MMscfd) on sustained tests of Miocene Kareem sandstones in the lower Kareem Rahmi formation zone.

The onshore well is on the North West Gemsa concession, 300 km (186 mi) southeast of Cairo.

Dana drilling program

Dana Petroleum was conducting a two-well exploration program in the West El Burullus offshore concession in the Nile Delta in 2009.

The first well, on the Papyrus prospect, was being readied for drilling in third quarter, targeting gas at a similar horizon to the 2008 discovery West El Burullus-1.

By yearend, the firm aimed to drill the second well on the Bamboo prospect, targeting deeper, high-pressure gas reservoirs.

The company also was performing workovers to offset a decline in the producing C2 well in East Zeit field in the Gulf of Suez.

Abu Qir concession

In first quarter 2009, Abu Qir Petroleum Co. hit pay with the NAQ PII-2 well in the offshore Abu Qir concession. The well tested a cumulative production of 52,400 cu m/d (1.85 MMcf/d) of natural gas and 850 b/d of condensate.

Additional surveys, interpretation, and well-drilling were being planned.

The firm was created in early 2009 by Egyptian General Petroleum Corp. and Edison SpA.

West Nile Delta

RWE DEA hit gas pay in early 2009 with the Ruby-3 exploration well in the West Nile Delta.

The Ruby-3 (Ji 50-2) exploration well, within the offshore West Mediterranean Deepwater concession, reached TD 1,957 m (6,421 ft), tapping gas-bearing sandstones in the Pliocene section.

Lease Sales

International 2009 Bid Round No. 1

The Egyptian General Petroleum Corp. invited bids in July 2009 for eight offshore exploration blocks under its International 2009 Bid Round No. 1.

Blocks 1–8, in the Gulf of Suez basin, have all undergone previous degrees of exploration.

Bids were to be submitted by November 1, 2009.

E. El Burullus Block

In May 2009, Enel and Total SA were awarded the E. El Burullus exploration block in the Nile Delta. Total will serve as operator.

The block is about 70 km (43 mi) offshore in water depths of 100–1,600 m (328–5,249 ft).

The initial 4-year exploration phase will include 3D seismic and drilling.

Drilling and Production

Al Basant

Dana Gas announced the start of production in April 2009 from its Al Basant gas-condensate discovery on the West Manzala concession in the Nile Delta.

It expected to reach estimated target production of 1.27 million cu m/d (45 MMscfd) by yearend.

Dana estimated Al Basant gas reserves at more than 3.5 bcm (123 bcf).

East El Wastani

Dana Gas reported the start of production in March 2009 from El Wastani East-2 sidetrack well EWE-2, the company's first highly deviated, horizontal well in Egypt.

It is in the East El Wastani development lease in the Nile Delta.

Initial gas production was 127,400 cu m/d (4.5 MMscfd) of gas, which was expected to increase.

Dana planned to continue efforts to develop and produce the remaining three discoveries: Salma, Azhar, and Sondos (Haggag) in 2009–2010.

Azhar-1 well

Dana Gas said that its Azhar-1 well may add as much as 2.8 bcm (100 bcf; 16 MMboe) to the company's reserves in Egypt.

In early 2009, the well hit 18 m (59 ft) of net pay in the Qawasim formation, and initial production tests yielded 427,600 cu m/d (15.1 MMcfd) of gas and 444 boe/d.

The well is in the West El Manzala concession, some 10 km (6.2 mi) from the firm's El Wastani gas processing facilities. The well reached TD 3,150 m (10,335 ft) in the upper Sidi Salim formation.

North Shadwan Block

Beach Petroleum NL conducted a successful test in late 2009 of its NS377-3 well on North Shadwan Block in the Gulf of Suez, off the Sinai Peninsula.

Initial flow was 1,400 b/d of oil, indicating a likely production rate of 1,500–2,000 b/d. The firm expects to bring the well on stream in mid-2010.

NS377-3 is the first development well of four planned for the NS377 and NS385 fields in the permit.

The shallow-water North Shadwan concession contains three existing undeveloped oil discoveries and several prospective exploration targets.

Combined ultimate recovery from these fields is expected to be 10–15 million bbl of oil.

Gas Sale Protest

In August 2009, the Egyptian-Israeli consortium Egyptian Mediterranean Gas (EMG) finalized terms of an agreement to supply natural gas to Israel's Dorad Energy, revising a 2007 agreement.

However, critics in Egypt said that the gas was being sold at prices considerably lower than those found on the international market. Others said that Egypt should not sell anything to Israel.

Planning for a new LNG receiving terminal in Israel was under way in 2009. Details are given in the Israel section of the Middle East chapter of this book.

LNG

Gas for Damietta LNG train

State-owned Egyptian Natural Gas Holding Co. (EGAS), clarifying earlier comments, reported in November 2009 that BP PLC and Eni SpA were working to find the volumes of gas needed for the Damietta LNG facility.

The two firms had only found 56.6 million cu m (2 tcf) of the 113.2 million bcm (4 tcf) needed to start up the second train at the plant.

Petrochemicals

"Gas City" concept

In mid-2009, Dana Gas was describing to Egyptian officials the plans of its JV to build a "gas city" in the country, with the potential, said Dana, to attract $20 billion in investments over 20–25 years.

A "gas city" is an integrated, self-contained industrial complex built to use natural gas as both fuel and feedstock for petrochemical production and heavy manufacturing.

Dana Gas and Crescent Petroleum formed a JV—Gas Cities Ltd.—to introduce the concept and to promote public-private sector partnerships in selected countries.

Industry, utilities and communication would be linked with commerce, education, residential and business facilities to create an integrated, self-sustaining unit, said Dana.

In Egypt, the JV would build the nucleus of the city and then rely on other entities to invest in various aspects of the complex.

The JV said it plans to build at least four gas cities in the Middle East and North Africa. The first, announced in 2008, would be in Kurdistan, at Chemchemal.

Alternative Energy

Power from wind

By 2020, Egypt said, it plans to generate 20% of its electricity from renewable sources and simultaneously reduce greenhouse gas emissions by 30%. This includes generation of 7,200 MW from wind energy projects by 2020.

The largest windfarm in the Middle East and Africa is in Zaafarana on the Gulf of Suez. It supplies 425 MW to Egypt's national grid.

The gulf has at least a dozen sites with attractive wind speeds, and in June, the government allocated 311,000 acres of land on the Red Sea coast specifically for new wind farms. Three were being planned with international partners, ranging in size from 200 to 300 MW.

The government also was seeking bids from private firms interested in designing, financing, building, owning and operating a 250-MW wind farm on the gulf between Zaafarana and Hurghada.

Also, in June, the World Bank announced that it would grant $28 million to Egypt for wind farm development and an associated power transmission system for offshore wind farms in the Gulf of Suez.

Power from sun

The strategy of Egypt's electric power sector envisions a growing role for solar energy.

Development of the country's first solar thermal project was under way in 2009, under the direction of Spain's Iberdrola SA.

The hybrid 140-MW facility in Kuraymat, about 95 km (59 mi) south of Cairo, will incorporate heat gathered with solar collectors with a combined-cycle, natural-gas-fueled power plant. Commissioning is anticipated in late 2010.

EQUATORIAL GUINEA

Capital: Malabo
Monetary unit: CFA franc
Refining capacity: 0
Oil production: 320,000 b/d
Oil reserves: 1.100 million bbl
Gas reserves: 1.300 tcf

Plans for Bioko

In early 2009, Germany's E.ON Ruhrgas AG and Unión Fenosa SA of Spain signed agreements to support the plans of the government to convert Bioko, the main island and home of the capital, into a major gas export hub.

The country wants to exploit its own deposits and also tap those of neighboring Nigeria.

It already exports up to 3.7 mtpy of LNG from a facility opened by Marathon Oil in 2007. But it wants to double or triple that amount by raising domestic production and making agreements to export gas from Nigeria and Cameroon.

E.ON has taken a 25% stake in a consortium set up by state-owned gas company Sonagas to build the needed infrastructure. Other participants include Galp Energia, Gasol PLC, and Gas Natural of Spain (which has a 95% stake in Union Fenosa).

Observers noted that uncertainties abound. For example, Nigeria and Cameroon have been reluctant to partner with Equatorial Guinea, and violence emanating from the Nigerian delta region could hamper efforts to transform Bioko.

The country also must deal with international allegations of human rights abuses and corruption

The government has applied to join the Extractive Industries Transparency Initiative, a global program that would subject its public accounts to much greater scrutiny. It must qualify for admission by March 2010.

Block EG-07

Saudi Trading & Research Co. (STARC) Ltd. and Glencore Exploration Ltd. signed a production-sharing agreement in mid-2009 with state oil firm, GEPetrol, for offshore Block EG-07.

The block (formerly Blocks D-14 and E-14) lies west of Blocks I and Q in Equatorial Guinea's segment of the Douala basin.

The energy ministry anticipates that additional PSCs will be signed before yearend 2010.

Aseng gas-condensate field

In July 2009, Noble Energy Inc. said it will proceed with bringing its $1.3-billion Aseng oil and gas-condensate project onstream by mid-2012.

Formerly known as Benita, Aseng is on Block I in the Douala basin. It will be the first oil development on Equatorial Guinea's side of the Douala basin and Noble's first operated development in the country.

Initial production from Aseng will be 50,000 b/d. It expects to recover 100–120 million bbl of hydrocarbon liquids over the project's life.

Noble also estimated that Aseng will produce 12.7–15.6 bcm (450–550 bcf) of gas as part of a gas monetization project once a pressure-maintenance phase is completed.

Another Douala find

Noble Energy Inc. said in early 2009 that its Carmen exploration well near the southeast corner of Block O in the Douala basin off Equatorial Guinea is an oil and gas discovery.

The well—drilled to 3,520 m (11,550 ft) in 46 m (150 ft) of water—cut 8 m (26 ft) of net oil pay along with 4 m (13 ft) of net gas pay Carmen is the first oil discovery on Block O.

Alba field: Marathon report

Marathon Oil Co. estimated in late 2009 that its net gas production from offshore Alba field in 2009 would be 12 million cu m/d (425 MMcfd) of gas and 42,000 b/d of condensate and NGLs.

Those figures compare to net production for calendar year 2008 of 10.4 million cu m/d (366 MMcfd), 29,400 b/d of condensate, and 11,100 b/d of NGL.

Ceiba and Okume Complex

Tullow Oil reported that for the first half of 2009, both the Ceiba and Okume Complex fields performed above expectations, with combined gross production averaging 105,000 b/d of oil.

Development drilling was continuing on the Okume Complex.

Gross combined production from the Ceiba and Okume complex was expected to average about 100,000 b/d of oil for 2009.

Bioko Island: Second LNG train?

Marathon said in late 2009 that it had held discussions with the government regarding a potential second LNG production train on Bioko Island, with a capacity of 4.4 million tpy.

An *OGJ* source said shipments from the existing Bioko LNG plant totaled 1.67 bcm (58.9 bcf) of gas in 2007, 4.19 bcm (148 bcf) in 2008, and 3.62 bcm (128 bcf) in 2009 through the end of October.

GABON

Capital: Libreville
Monetary unit: CFA franc
Refining capacity: 24,000 b/cd
Oil production: 230,000 b/d
Oil reserves: 2.000 billion bbl
Gas reserves: 1.000 tcf

Omoueyi permit

Maurel & Prom reported in early 2009 that its OMOC-1 well on the Omoueyi permit in southern Gabon on test stabilized at 1,000 b/d of oil from the Cretaceous Gas de Base formation, confirming the permit's potential and reinforcing exploration efforts.

The well targeted the Kissenda and Base sandstones to test a 6-m (20-ft) interval in the lower Kissenda formation and a 24-m (79-ft) interval in the upper Kissenda.

OMOC-1, which reached 1,020 m (3,346 ft), is south of the Onal production permit.

North Gabonese basin

Italy's Eni acquired rights in early 2009 to six exploration licenses totaling more than 8,000 sq km (3,089 sq mi) in Gabon.

The D3 and D4 blocks are in shallow water in the North Gabonese basin, and the E2, F3, F4, and F7 blocks are in the country's onshore basin. The licenses have 4-year initial terms.

2010 licensing round

Gabon announced in late 2009 that it will launch an oil licensing round for more than 40 blocks from two of its deepwater basins in May 2010. The round will close in November 2010.

CGGVeritas, hired to advise the government on the round, said it has seen some large structures below the salt where there is large potential for some of the blocks.

It was unclear whether the death in June of Pres. Omar Bongo Ondimba, who ruled for 41 years, would affect the actions of potential investors in exploration.

Gabon replaced its oil minister as part of a reorganization following elections made necessary by the death of Mr. Ondimba.

Ebouri field, Etame Marin Block

VAALCO Energy completed the EEBOM-3H well in early 2009 in North Ebouri field, its second horizontal development well there. First oil production began in April.

Also, a new production record of 24,993 b/d of oil from the Etame Marin Block was set soon thereafter. VAALCO expects production from the block will average 25,000 b/d of oil.

The company also plans to drill the Southeast Etame prospect; no timetable was given.

Perenco: Gabon position

In late 2009, Perenco acquired the Gabon subsidiary of Marathon Oil Corp. The effective date of the transaction is January 1, 2010.

The subsidiary holds a 56.25% working interest in three offshore production fields: Tchatamba Marin, Tchatamba South, and Tchatamba West.

The acquisition will add about 15,000 bo/d gross production Perenco's operated production in Gabon.

Tchatamba Marin, 24 km (15 mi) offshore and 160 km (100 mi) southeast of Port Gentil, began production in 1998. Tchatamba South and West fields came on stream in 1999 and 2000, respectively.

GHANA

Capital: Accra

Monetary unit: Cedi

Refining capacity: 45,000 b/cd

Oil production: 6,000 b/d

Oil reserves: 15.000 million bbl

Gas reserves: 800.000 bcf

Exploration and Development

Jubilee snapshot

Substantial work continued in 2009 by Tullow Oil PLC (operator), Kosmos Energy LLC, and their partners to appraise more completely and to further develop the massive Jubilee oil field.

Western Ghana Deep Water

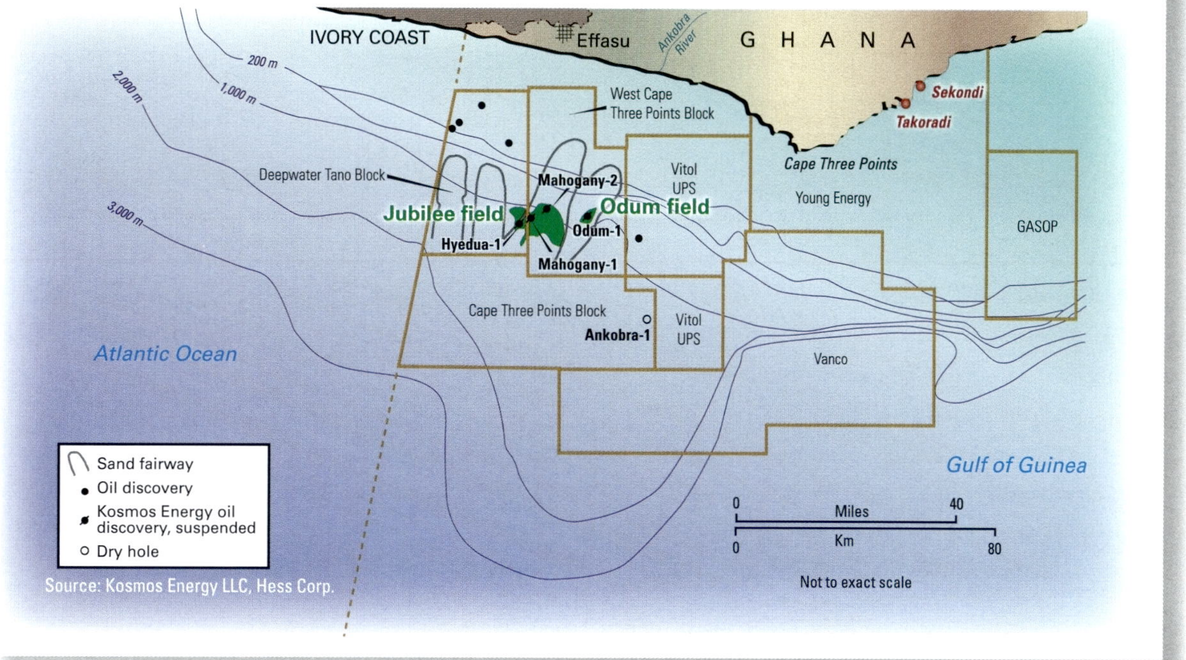

Jubilee field, Ghana

Well	Block	Water depth, m	Spud date	TD, m	Gross (net) reservoir, m	Production tests
Mahogany-1	West Cape Three Points	4,330	2007	3,826	270 (95)	
Hyedua-1	Deepwater Tano	1,530	2007	4,002	202 (41)	
Mahogany-2	West Cape Three Points	1,080	2008	3,443	193 (50)	5,200 bo/d, 5.5 MMcfd
Hyedua-2	Deepwater Tano	1,246	2008	3,663	(55)	16,750 bo/d, 21 MMcfd
Mahogany-3	West Cape Three Points	1,236	2008	4,028, deepened to 4,345	(33)	
Development well	West Cape Three Points		Q1 2009	TBD		
Adjacent to Jubilee:						
Odum-1	West Cape Three Points	955	Feb. 2008	3,386	60 (22)	
Ankobra-1	Cape Three Points	1,732	Sept. 2008	3,963		
Tweneboa-1	Deepwater Tano		Q1 2009	TBD		
Teak-1			2009	TBD		

Based on operator reports.

Tullow said in January 2009 that it planned to invest a total of €2.3 billion to develop Jubilee field, which straddles two offshore license blocks:

- West Cape Three Points Block covers 1,761 sq km (680 sq mi) in water 50–1,800 m (164–5,905 ft) deep. Operator Kosmos Energy is joined by interest partners Anadarko WTCP Co., Tullow Ghana Ltd., (an affiliate of Tullow Oil PLC); E.O. Group; Sabre Oil and Gas Ltd.; and Ghana National Petroleum Corp. (GNPC).
- Deepwater Tano Block covers 1,106 sq km (427 sq mi) in water 200–2,060 m (656–6,758 ft) deep. Interest partners with operator Tullow are Kosmos Energy; Anadarko; Sabre Oil and Gas; and GNPC.

Jubilee schedule

In July 2009, Ghana formally approved Phase 1 development of the Jubilee oil field.

In late 2009, Tullow Oil reported that development was on track to deliver first oil through an FPSO vessel late in 2010. The field holds an estimated 1.2 billion bbl of oil.

In an ongoing effort to appraise the eastern and southern extent of Jubilee, drilling of the final Mahogany well (Mahogany-Deep 2) is to begin in early 2010.

Stakes in Jubilee

As of mid-2009, shares in the field were: Tullow 34.7%; Kosmos Energy 23.49%; Anadarko Petroleum Corp. 23.49%; Sabre Oil & Gas Ltd. 2.81%; E.O. Group Ltd. 1.75%; and GNPC 13.75%.

In October 2009, ExxonMobil Corp. signed a binding agreement to buy a stake in Jubilee field from Kosmos Energy. The *Financial Times* said that the transaction was worth $4 billion.

The action would mark ExxonMobil's first entry into Ghana.

But also in October, GNPC said that it intended to buy Kosmos and to develop its stake in Jubilee field with a foreign oil company. Some observers believe that CNOOC heads the list of possible partners.

On December 1, 2009, GNPC said that its talks with Kosmos had reached a "critical stage."

Tano basin

Early-2009 evaluations suggested that the Tano basin may hold more oil than expected.

- Kosmos announced that the Mahogany-3 well—drilled about 5 km (3 mi) southeast of Mahogany-1—penetrated 16 m (52 ft) of oil-bearing sands equivalent to those in other wells.
- Tullow and partners reported in early 2009 that the deepwater Hyedua-2 appraisal well encountered 55 m (180 ft) of oil-bearing pay, producing 16,750 b/d of oil and 21 MMcfd on test.

> **Sierra Leone-to-Ghana: New "oil coast"**
>
> In September 2009, Anadarko Petroleum Corp. and partners Woodside Petroleum Ltd., Repsol-YPF SA, and Tullow Oil PLC, confirmed that they had found a new "oil coast" from Sierra Leone to Ghana.
>
> Bolstering their announcement was a report of positive results from the Venus B-1 well off the coast of Sierra Leone. Anadarko characterized Jubilee and Venus as "bookends," spanning 1,100km (700 mi) across two of the most highly prospective basins in the world.
>
> Woodside and Repsol each hold 25% of the Venus prospect. Anadarko owns 40% and Tullow 10%.

Odum find

Kosmos Energy said in late 2009 that its Odum-2 appraisal well confirmed its 2008 Odum discovery and proved the potential of the Campanian play on the West Cape Three Points Block.

The company had drilled nine consecutive successful exploration and appraisal wells offshore Ghana.

West Cape Three Points Block

Kosmos Energy reported in October 2009 that the Mahogany-4 appraisal well in West Cape Three Points Block penetrated net hydrocarbon-bearing sandstone pay of at least 43 m (141 ft).

The company said that Mahogany-3 and -4 have extended the potentially productive area on the block.

Tweneboa

Tullow Oil completed the Tweneboa-1 exploration well in the Deep Water Tano Block, west of Jubilee field, in March 2009.

It was drilled to 3,938 m (12,920 ft), encountering highly pressured oil-bearing sands.

A light hydrocarbon accumulation, with P50 oil and gas potential of 250 MMboe, was to be appraised with the Tweneboa-2 well in September.

Pipelines/Transportation

Takoradi Port oil service terminal

The Ghana Ports and Harbor Authority was poised to start building an oil service terminal at the Takoradi Port in September 2009, in preparation for the start of oil production in 2010.

About $50 million is needed for phase one of the project, which includes pipelay from offshore oil fields to the port. The project also includes provision of space for 10–15 supply vessels.

Refining and Products

Accra refinery

A South African firm, New Alpha Refinery (Ghana) Ltd., signed an MOU in mid-2009 with the government of Ghana to develop an oil refinery in the city of Accra.

The $6-billion refinery is expected to be the largest in West Africa, with an initial capacity of 200,000 b/d in 2015, and plans for expansion to 400,000 b/d.

Pumpuni refinery

Barclays Gedi Group Inc. was evaluating plans in early 2009 to build a $1.7-billion, 100,000-b/d refinery in Pumpuni, on the coast, southwest of Takoradi.

Company News

Tullow: Africa focus

Tullow Oil said in March 2009 that it planned to accelerate development of its two big oil discoveries in Uganda and Ghana, striving to produce oil from its Ghana fields by late 2010.

The company said that projects in Suriname and other countries will no longer be priorities.

The great majority of its £600-million 2009 capital expenditure was to be devoted to Africa, especially to completing Phase 1 of the offshore Jubilee field.

In January, the company had said it was targeting about 4 billion bbl of oil and gas resources in the Gulf of Guinea off Ghana and Ivory Coast.

Eni back in Ghana

Eni Ghana Exploration & Production Ltd. re-entered Ghana in 2009, where it had been active until the 1970s.

The firm will operate the Offshore Cape Three Points (OCTP) and Offshore Cape Three Points South (OCTPS) exploration licenses after acquiring a major interest from the Ghana unit of Vitol Group.

GNPC also has an interest in both blocks, which lie in the Tano-Cape Three Points basin.

The Sankofa-1 well was drilled in mid-2009 on the OCTP block, encountering high-quality sands. Sankofa field is a significant hydrocarbon discovery, 35 km (22 mi) east of Jubilee field.

IVORY COAST

Capital: Abidjan
Monetary unit: CFA franc
Refining capacity: 63,990 b/cd
Oil production: 30,000 b/d
Oil reserves: 100.000 million bbl
Gas reserves: 1.000 tcf

Tullow project report

Tullow Oil said in mid-2009 that it identified a significant "Jubilee-type" fan system in offshore Block CI-105, based on study of recently reprocessed 3D seismic data.

The company expected to drill this prospect, South Grand Lahou, by yearend.

Elsewhere, interpretation was nearing completion of 3D seismic across Block CI-102 and CI-103.

Baobab development

Canadian Natural Resources Ltd. (CN) was proceeding in 2009 with Phase II of its offshore Baobab development.

Three wells had been re-drilled as of mid-year, and a fourth re-drilling was in progress.

CN also was drilling further development wells on its Olowi field, aiming to reach plateau oil production of 20,000 b/d during 2010.

KENYA

Capital: Nairobi
Monetary unit: Kenyan shilling
Refining capacity: 90,000 b/cd
Oil production: 0
Oil reserves: 0
Gas reserves: 0

Missing oil imports

Kenyan authorities began a probe in early 2009 into the disappearance of $100 million in oil imports as part of a corruption scandal that threatened to disrupt fuel supplies to east Africa.

Kenya's energy ministry said that the oil was allegedly sold without the authorization of the banks and trading houses that financed it.

The case renewed concern about graft in Kenya, because Uganda, Rwanda, Burundi, eastern Congo, and south Sudan require reliable transport of commodities through the country.

Authorities suspected that documents were manipulated by some employees of the Kenya Pipeline Co.

Licensing trends

The energy ministry said in early 2009 that new applications for exploration licenses in Kenya had dropped almost to zero, from a record level of 14 issued in 2008.

Block 9: Anza basin

China's CNOOC planned to spud the Bogal-1-1 exploratory well in late 2009 on Block 9 in the north-central Anza basin, said Africa Oil Corp.

The well was to tap Jurassic limestone and Cretaceous sandstones at 5,000 m (16,400 ft).

Numerous other prospects had been identified.

Port project: Lamu

The government was in initial talks with China in late 2009 about a $3.5-billion port and transport-corridor project that could provide a new export route for Chinese oil.

The project would build a port in the coastal tourist area of Lamu, plus road and rail links to Kenya's borders with Ethiopia and southern Sudan.

Under the plan, CNOOC could prospect for oil in northern Kenya, and also would have exploration rights for a second block in the Lamu basin.

2,000 MW of green energy

The government of Kenya said in mid-2009 that it was making plans to invest $7 billion to advance "green energy" ventures in the country by 2013.

Kenya Electricity Generating Co., the nation's largest power utility, expects that about 2,000 MW of "green" generating capacity will be added by 2013.

Sources will include: 800 MW of wind power; 600 MW from clean coal; 500 MW of geothermal energy; 30–50 MW from biomass; and 30 MW from hydroelectric projects.

LIBYA

Capital: Tripoli
Monetary unit: Dinar
Refining capacity: 378,000 b/cd
Oil production: 1.545 million b/d
Oil reserves: 44.270 billion bbl
Gas reserves: 54.362 tcf

Overview

Libya has the largest oil reserves in Africa (44 billion bbl) and is the eighth-largest oil producer among the members of OPEC. Libya is aiming to boost oil output to 3 million b/d by 2016 or 2017, compared to 1.7 million b/d in 2007.

State-owned National Oil Company aims to increase production by encouraging the acquisition of additional seismic data and the drilling of at least 50 wildcat wells per year.

Libya is the second-largest oil supplier to Europe. NOC has offices in Europe: Umm Al-Jawaby Oil Service Co. Ltd. (London), and Mediterranean Oil Services GmbH (Düsseldorf).

Since the United States lifted economic sanctions in 2004, NOC has reinstituted lease sales, which are attracting international players and national oil companies.

New Supreme Council for energy

In late 2009, Libya replaced the Supreme Council of Oil and Gas Affairs with the new Supreme Council of Energy Affairs (SCEA).

The new body will oversee production and development targets, contract negotiations and the development of alternative energy and will have regulatory as well as executive functions.

But observers noted that its chair is not as reform-minded as the former head of the previous body, and that the SCEA will be linked to the National Security Council.

IHS Global Insight concluded that "the days of the reformers are over, and...Libya will pursue a more statist economic policy in the years ahead."

Nationalization threat

Col. Muammar Gadhafi said in early 2009 that Libya might nationalize its oil industry to control production in the face of plummeting oil prices.

He said that oil prices needed to increase to $100/bbl, and suggested that Libya might reject OPEC's quotas by withholding production from the global market to help boost prices.

According to state-run media, there was increasing pressure on the nation's legislative and executive bodies to vote for nationalization of oil companies.

Several analysts judged the nationalization threat highly unlikely because Libya was seeking foreign investment to improve its oil and gas industry.

Exploration and Development

$10 billion for oil field development

Libya said in third quarter 2009 that it plans to invest 12.1 billion dinars ($9.92 billion) in the development of 24 wells in fields that it calls "technically, financially, and economically proven."

The investment will be undertaken by NOC, its subsidiaries, and current foreign partners.

Fields of interest include Waha-Jalou and Nafoora-Oujlaa-Khleej.

Arous Al-Bahar prospect

In late 2009, the Libyan subsidiary of Hess Corp. re-entered a 2008 offshore discovery well east of Tripoli and found gas and condensate.

A1-54/01 on the Arous Al-Bahar prospect is in the nonproducing extension of the Sirte basin, 56 km (35 mi) offshore.

It encountered hydrocarbons in several intervals and flowed 765,000 cu m/d (27 MMscfd) of gas and 533 b/d of condensate on drillstem test.

Crude Diplomacy

Late 2009 saw a diplomatic flare-up in the UK after the release of a notorious Libyan prisoner from a Scottish prison in August.

Britain's leaders were pressured to explain any role they had in the release of Abdel Baset al-Megrahi—the only person convicted in the 1988 bombing of Pan Am Flight 103 over Lockerbie, Scotland.

Scottish justice officials said that al-Megrahi was released on medical compassion grounds.

However, Libyan leader Col. Muammar Gadhafi said publicly that senior British officials and members of the royal family influenced the decision to release him.

British leaders denied that claim as well as implications that there was a linkage of oil trade to the release. But opposition politicians said the matter demanded close examination.

Col. Gadhafi's son said that al-Megrahi's release was a constant point of discussion during trade talks, stretching back to the government of former Prime Minister Tony Blair.

Ghadames basin

Arabian Gulf Oil Co. (AGOC), a subsidiary of NOC, discovered oil in mid-2009 with a wildcat well in the Ghadames basin that reached TD 3,200 m (10,500 ft).

B1-NC4, drilled 190 km (118 mi) south of Tripoli, found oil and gas from both the Lower Acacus formation and the Mamouniyat formation.

Area 18

In May 2009, operators drilled two wells on Gulf of Sirte blocks.

A group led by Petrobras International Braspetro BV spudded the A1-18/01 well in Area 18, about 120 km (75 mi) east-northeast of Tripoli.

A sidetrack well was planned to target proven oil plays in Eocene and Cretaceous carbonates, as well as new gas plays in Jurassic and Triassic clastics.

Exploration Area off Libya

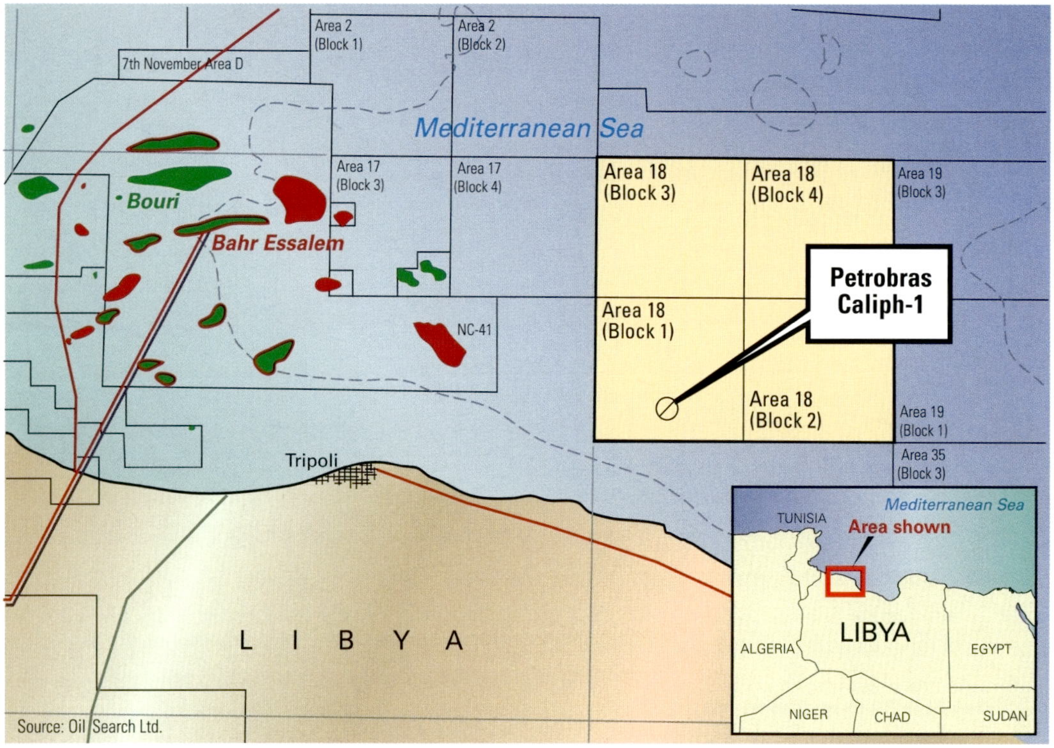

Area 20

The Libyan affiliate of ExxonMobil Corp. spudded the first deepwater exploration well off Libya in July 2009. A1-20/3 is in Contract Area 20 in the Sirte basin, northeast of Misrata.

Area 65

Sipex, the international arm of Algeria's Sonatrach gauged an oil and gas discovery in the Libyan portion of the Ghadames basin in early 2009. The find was in onshore Area 65.

Well A1–65/02 flowed 1,344 b/d of oil and 53,200 cu m/d (1.88 MMcfd) of gas from the Ordovician Memouniat formation. TD was 9,033 ft.

It is the first discovery on the block, 225 km (140 mi) southwest of Tripoli.

Block NC202

Spain's Repsol made its first offshore oil discovery in Libya in early 2009.

The well, drilled to 4,820 m (15,814 ft) in 50 m (164 ft) of water, was the first drilled in block NC202, about 15 km (9.3 mi) offshore, in the Sirte basin.

Initial production tests flowed 1,264 b/d of oil.

Blocks C17, C137

Total SA said in early 2009 that it will convert its petroleum agreements with NOC to exploration and production sharing agreements for Blocks C17 and C137.

The acreage is respectively located in the Sirte basin and the offshore Sabratha basin, 100 km (62 mi) from the Libyan coast.

Ghadames: Tenth find

Verenex Energy Inc. made its tenth discovery in the Ghadames basin in early 2009.

Well H1-47/02 tapped Ordovician Memouniat and Silurian Acacus formations. It gauged 1,315 b/d of oil and 459,000 cu m/d (16.2 MMcfd) of gas on test.

Alternative Energy

Smart Energy City: Sabratha

Libya was moving to develop by 2013 a Smart Energy City in mid-2009 as part of a $5-billion economic zone being set up for Libyan firms involved in various energy-related endeavors.

Construction had begun at Sabratha, 70 km (43 mi) west of Tripoli, in a JV between the Islamic investment bank Gulf Finance House BSC and Libya's State Fund for Economic and Social Development.

MADAGASCAR

Capital: Antananarivo
Monetary unit: Malagasy franc
Refining capacity: 0
Oil production: 0
Oil reserves: 0
Gas reserves: 0

Overview

Block 3113

In January 2009, Sino Union Petroleum & Chemical International Ltd. (Sunpec) told shareholders that it expected to see oil production "very soon" from Block 3113, its primary onshore interest.

In addition to the eight wells it already has drilled on the island, Sunpec said it planned to drill three more on the block in the first quarter.

Tsimiroro Block

In March 2009, Madagascar Oil said it produced the island's first oil in 60 years from an onshore steam injection pilot project at Tsimiroro, one of two heavy oil projects the firm is developing.

The firm also raised its estimate of Tsimiroro reserves by 30% to 1.7 billion bbl.

The company also said that 12 of 17 exploration wells drilled on the Tsimiroro Block yielded discoveries indicating 0.9–1.4 billion bbl of oil in place.

The private firm reported that its other project at Bemolanga, also in northwest Madagascar, is one of the biggest undeveloped bitumen reserves in the world, with an estimated 9.8 billion bbl of oil reserves.

On the brink of 'catastrophe'

International charities warned in April 2009 that Madagascar was on the brink of "catastrophe" and facing an unprecedented humanitarian crisis in the aftermath of a March coup that led to chronic shortages of food. Also, recent cyclones had caused extensive damage.

In addition, civil unrest led to destruction of an oil depot, in reaction to government closure of a radio station supportive of an opposition political party.

Exploration areas of interest: offshore NE Madagascar

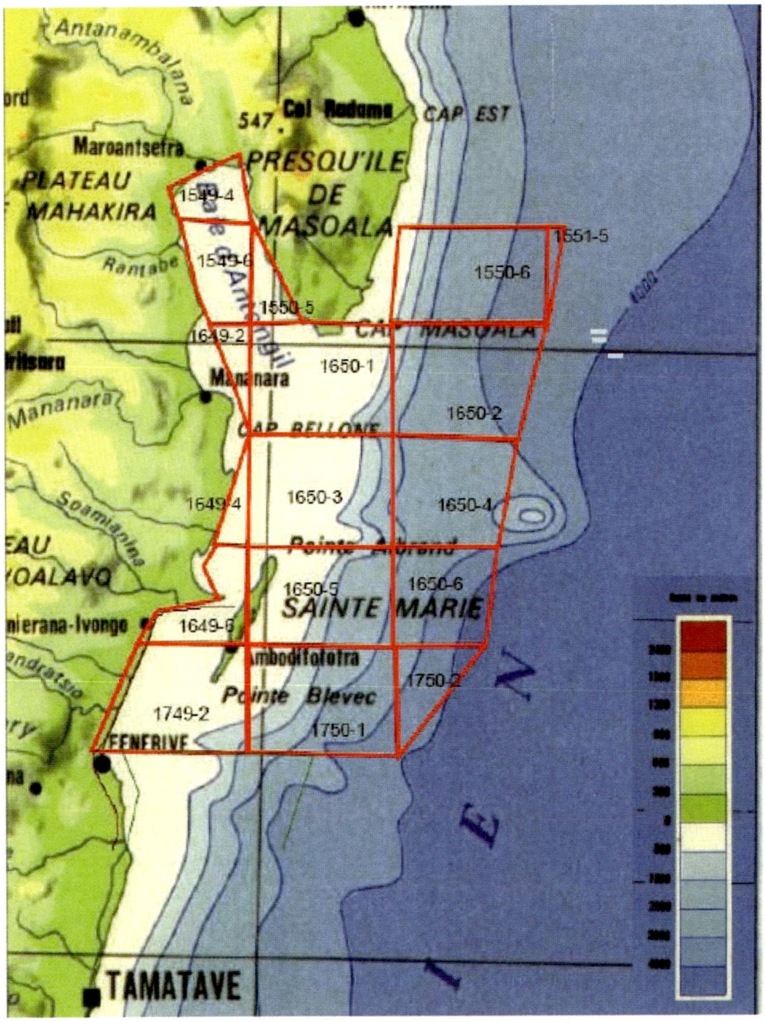

Exploration and Development

Ampasindava Block

Sterling Energy PLC said in mid-2009 that it would proceed with plans to drill its first exploration well off northwest Madagascar in 2010 as soon as the political situation on the island stabilized.

The company is targeting the giant Sifaka prospect in its Ampasindava Block, where water depths range from 200–3,000 m (656–9,842 ft).

Interest in Seychelles

In early 2009, two companies signed separate production agreements covering a total of 38,000 sq km (14,672 sq mi) in the 115-island archipelago nation of the Seychelles, northeast of Madagascar.

Exploration drilling offshore could resume in 2010 after a hiatus of 21 years.

The government planned another offshore licensing round later in the year.

The islands are seen as part of an extensive source rock system with strong hydrocarbon potential, extending from Kenya to Madagascar and also encompassing Mauritius and the Comoros Islands.

Mozambique Channel Permits

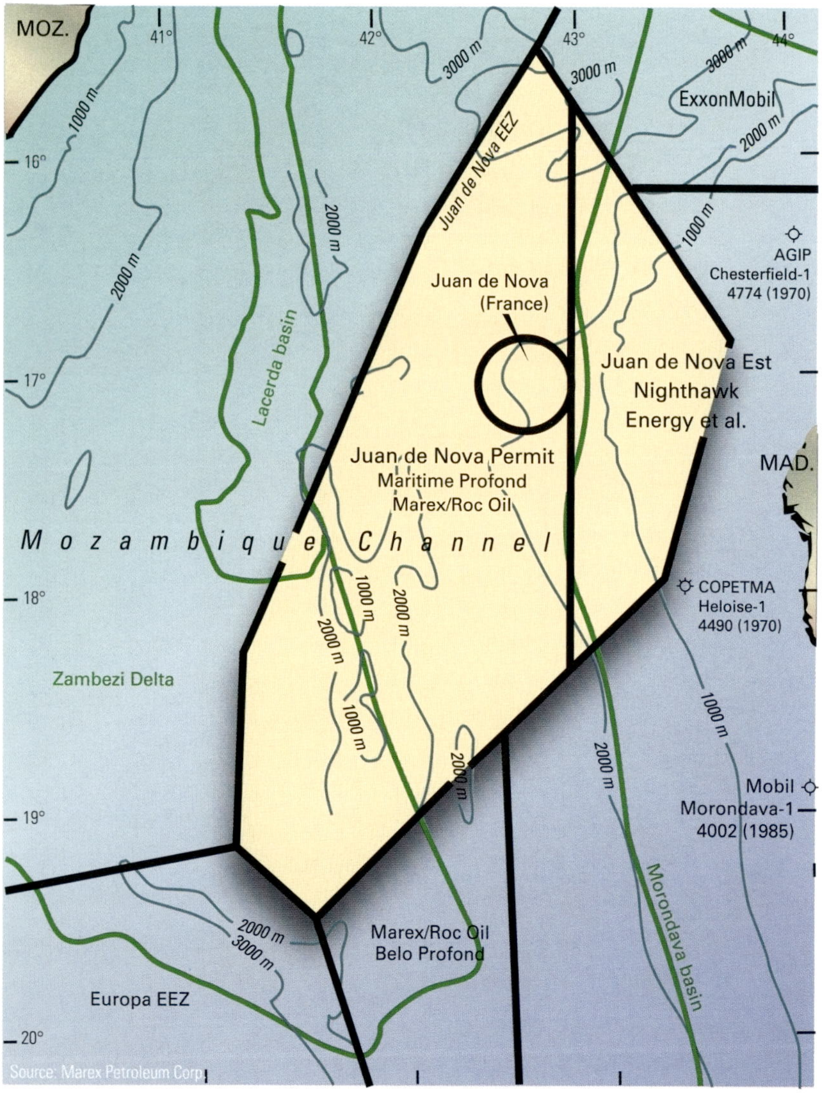

Source: Marex Petroleum Corp.

A drilling location was chosen in late 2008 on the flank of the structure, and in early 2009, the company completed an environmental assessment.

Mozambique Channel

The French government awarded two 5-year exploration permits in early 2009 for exploration offshore Juan de Nova, a tiny island possession in the Mozambique Channel between Madagascar and Mozambique.

One permit went to Marex Petroleum Corp. and Roc Oil Co. Ltd., the other—Juan de Nova Est—to three small UK firms. Drilling of two exploratory wells is expected by 2011.

Some analysts see possible large fields in the channel, between Madagascar's heavy oil fields on the east and Mozambique's Panda and Temane gas fields on the west.

MALI

Capital: Bamako
Monetary unit: CFA franc
Refining capacity: 0
Oil production: 0
Oil reserves: 0
Gas reserves: 0

Blocks 7 and 11

In late 2009, Heritage Oil was preparing to acquire significant 2D seismic over two blocks in the eastern part of the country, as part of a 2008 farm-in commitment with a subsidiary of Centric Energy Corp.

The seismic program plus drilling of a first exploration well will give Heritage a 75% interest in Blocks 7 and 11. The blocks cover more than 72,000 sq km (27,800 sq mi) in the Gao Graben, a Mesozoic rift basin with up to 4,572 m (15,000 ft) of Cretaceous sediments.

Gao Graben is considered geologically similar to the Albert Graben in Uganda.

MAURITANIA

Capital: Nouakchott
Monetary unit: Ouguiya
Refining capacity: 0
Oil production: 0
Oil reserves: 100.000 million bbl
Gas reserves: 1.000 tcf

Pipeline proposal

In early 2009, Mauritania was considering an offer from Baraka Petroleum Ltd. to assist in developing a proposed Mauritanian oil and gas pipeline system to accelerate exploitation of oil and gas reserves.

Baraka said such a system would make it possible for Mauritania to develop the still-to-be-delineated Taoudeni reserves.

One pipeline would run from the Taoudeni region in the east to the town of Zouerate in the northwest. Another link would extend south to Nema and from there, west to Nouakschott on the coast.

Nema also could become a hub for a southern link to neighboring Mali, said Baraka.

MOROCCO

Capital: Rabat
Monetary unit: Dirham
Refining capacity: 154,901 b/cd
Oil production: 300 b/d
Oil reserves: 0.752 million bbl
Gas reserves: 53.000 bcf

Overview

Energy planning

In early 2009, Morocco's energy ministry told an international conference that the country needs investment of more than $10 billion in the energy sector 2008–2015 to develop renewable energy resources and reduce its dependence on energy imports that now provide 97% of energy needs.

Wind and solar energy are seen as promising resources, and the government has embarked on a $3.2-billion investment project for 2009–2014 in the renewable energy sector. It wants renewable resources to provide 10% of energy needs by 2012, compared to the current 4%.

Three new windfarms (total generation capacity: 1,440 MW) were under construction in early 2009.

It also aims to have 1.7 million sq m (17.2 MM sq ft) of solar heating panels in use by 2025, compared to 100,000 sq m (1.08 MM sq ft) in 2009.

The investment project also calls for creating an industrial park in Oujda by 2010, to support renewable energy companies and facilitate private sector investment.

Morocco also is considering development of nuclear power for 2020–2025.

Exploration and Development

Rharb basin

Circle Oil PLC in mid-2009 identified the CGD-10 exploration well on the Sebou permit in the Rharb basin as a potential gas producer and was proceeding with development.

The well tested 110,450 cu m/d (3.9 MMscfd) from the Tertiary Lower Guebbas sequence, and testing of a promising secondary gas target as anticipated.

Five of six planned wells had been drilled on the permit northeast of Rabat, with the sixth scheduled.

Tanger-Larache area

Repsol hit pay in early 2009 with the Anchois-1 well in the Tanger-Larache area, 40 km (25 mi) from the coast. The well struck two gas columns totaling 90 m (295 ft).

Exploration of the Moroccan coast is part of a wider Repsol plan to assess gas-bearing sands off the African coast and the Canary Islands.

Essaouira Offshore

Canamens Energy Ltd. was initiating a round of offshore exploration in early 2009, following agreements made with Morocco's Office National Hydrocarbures et des Mines.

The company was preparing to reprocess and acquire new 2D data in the Essaouira Shallow Offshore [in water less than 500 m (1,640 ft) deep] and to then decide whether to seek an exploration permit.

In the Essaouira Deep Offshore area—water depths greater than 500 m (1,640 ft)—Canamens plans to reprocess and acquire new 2D data with an option to extend into a 3D and drilling commitment.

Sidi Moussa, Foum Draa concessions

In mid-2009, Morocco's ONHYM signed agreements to participate in an exploration consortium with a subsidiary of Ireland's Island Oil & Gas and Serica Energy PLC.

The focus is on the contiguous Sidi Moussa and Foum Draa offshore concessions, which cover about 12,700 sq km (4,903 sq mi) in the Agadir basin, 100 km (62 mi) southwest of Agadir.

Extensive seismic had been acquired previously, and some 40 prospects and leads had been identified, from the coast out to water depths of 2,000 m (6,562 ft).

The geological setting, on the Atlantic Margin, is said to be analogous to and on trend with oil-producing basins in West Africa.

Drilling and Production

Oil shale technology test

In mid-2009, San Leon Energy PLC was given 3 years by ONHYM to test its In Situ Vapor Extraction (IVE) technology to extract oil from shale on the Tarfaya exploration project.

Site tests were to begin by late 2009, with first process tests by yearend 2010.

San Leon is working with Mountain West Energy LLC to use IVE as an alternative to open pit mining.

The company has produced 62 l./tonne (16.4 gal/tonne) of oil using IVE under laboratory conditions and estimates that it could produce 500,000 b/d from oil shale in the Tarfaya region.

Oil shale reserves in Morocco are estimated at 53 billion bbl in place.

Alternative Energy

$9-billion solar project

The government of Morocco announced in late 2009 a $9-billion, public/private solar energy project that will produce 2,000 MW through five facilities around the nation.

The power generated is expected to account for 38% of the country's installed power capacity by 2020.

Hybrid solar-gas power plant

Work was proceeding in 2009 on construction of a 470-MW hybrid solar power plant in Ain-Beni-Mathar, scheduled to begin operation in 2010.

Builder Abener Energia SA said that the solar component of the plant will use more than 180,000 sq m (1.9 MM sq ft) of parabolic-trough collectors to generate 20 MW.

The other 450 MW will come from a combined-cycle section comprised of two gas turbines and a steam turbine. The latter will use waste heat from the gas turbines, plus output from the solar collectors.

Abener said that the facility will be the world's largest integrated solar combined cycle power plant.

MOZAMBIQUE

Capital: Maputo
Monetary unit: Metical
Refining capacity: 0
Oil production: 0
Oil reserves: 0
Gas reserves: 4.500 tcf

Buzi Block

In early 2009, state-owned Empresa Nacional de Hidrocarbonetos de Mocambique sold to Indonesia's PT Kalila Energy Ltd. a 75% stake in the Buzi Block for $30 million.

The partners will explore for gas in the central Sofala province through 2017 and will drill two exploration and two appraisal wells.

Buzi holds estimated reserves of 283–482 million cu m (10–17 bcf) of gas.

Refinery: Nacala

Ayr Logistics Ltd. said in late 2009 that it was resuming plans to construct a $5 billion, 300,000-b/d refinery in the district of Nacala-a-Velha, in Nampula province. The project had run into financing difficulties mid-year.

The government approved the project in 2007, and construction began in early 2008.

The planned start-up date is 2015.

Refinery: Maputo

OilMoz LDA said that it secured $50 million from financiers in early 2009 to study the feasibility of building an $8-billion, 350,000-b/d oil refinery in Maputo. A unit of Royal Dutch Shell PLC will assist.

The studies were to be completed by yearend 2009 for review by the government.

A preliminary schedule calls for construction to start in early 2010 and to be completed in 2013.

Mozambique's only oil refinery closed in 1985, leaving the country dependent on imports.

NIGERIA

Capital: Abuja
Monetary unit: Naira
Refining capacity: 505,000 b/cd
Oil production: 1.810 million b/d
Oil reserves: 37.200 billion bbl
Gas reserves: 185.280 tcf

Overview

Oil

Nigerian officials reported that the hydrocarbon industry accounted for 95% of the nation's foreign exchange income, 85% of revenues collected by the government, and 30% of GDP in 2008.

Nigeria is the largest oil exporter in Africa, shipping an estimated 1.88 million b/d at the end of 2008.

However, against a backdrop of inefficiency and a domestic insurgency, oil output slipped to an average of 2.1 million b/d in 2008, the lowest since 2002.

In August 2009, the central bank said that production had fallen to about 1 million b/d of oil.

The continued unrest, combined with uncertainty about a pending oil-sector reform law, led industry experts to conclude that Nigeria will struggle to pump even half of its target of 4 million b/d in 2010.

Refining

Nigeria has four crude oil refineries: one in Kaduna, two at Port Harcourt, and one in Warri.

The four have a total nameplate capacity of 445,000 b/d, but have never operated at full capacity, hampered by inefficient management, poor maintenance, sabotage, theft and inadequate power supplies.

As a result, Nigeria imports 85% of its domestic demand for fuel and petroleum products.

Fuel subsidies cost the country about $4.4 billion in 2008.

Natural gas

The country has several LNG projects on tap for monetizing primarily associated gas.

The administration is aiming to achieve better balance between domestic gas use and export.

Gas-fired electric power generation is planned to expand to nearly 15 GW by 2012, from none in 2007. This will likely raise gas use for power to more than 6 bcfd by 2011.

Gas demand for domestic industrial use also is expected to exceed 6 bcfd by 2013.

For export (by both pipeline and LNG), gas demand is expected to reach more than 11 bcfd by 2014 from only 4 bcfd in 2007.

To meet this level, NLNG Train 7 will come on line in 2012, and both Brass LNG Train 1 and OKLNG Trains 1–4 are to start up by 2013. Combined peak production is expected to reach 11 bcfd in 2014.

Combined natural gas domestic and export demand are expected to peak at more than 16 bcfd by 2013.

Oil sector reform

In 2009, Nigeria's national assembly was pondering a draft Petroleum Industry Bill described by the *Financial Times* as

the most comprehensive overhaul of the country's oil sector in five decades.

Supporters said it would address funding shortfalls, domestic gas shortages, and crippling fuel subsidies and would subject every upstream company to the same system of rents, royalties, and taxes.

There was concern, however, whether sufficient political will existed to enact it before 2011 elections.

Against that backdrop:

- ExxonMobil Corp., Chevron Corp., and Royal Dutch Shell were negotiating for the renewal of expiring leases on some of their Nigerian blocks—potentially paying huge "signature bonuses."
- In mid-year, China's CNOOC offered to buy 49% stakes in 23 oil blocks, including some of those up for renewal. The deal was said to be worth as much as $50 billion.
- Nigeria was said to be considering sale to the Chinese of some of its own majority holdings in the JVs with western groups (60% of the JVs operated by Chevron and ExxonMobil, 55% of Shell's).
- Local oil companies were positioning themselves to obtain undeveloped assets if proposed reforms require foreign groups to give up some of them.

Complicating the picture further, separate proposals being considered by the government would:

- divert 10% of JV profits to communities in the Niger Delta to placate the MEND insurgency there (see box).
- sell a further 9% of the state's share in the JVs to Nigerians who want to invest in the oil sector.

Impact of militants' actions

In its July 2009 *Oil Market Report*, the International Energy Agency estimated June oil production in Nigeria at 1.72 million b/d from its capacity of 2.55 million b/d.

A 300,000 b/d loss, due to a surge in violence largely by the MEND group, added to shut-ins totaling 500,000–600,000 b/d since the insurrection began in 2005, said IEA.

The threat affects 600 fields and 6,000 km (3,728 mi) of pipeline.

On June 25, NNPC said that 1.4 million b/d of production was offline. And in October, the central bank said that unrest in the delta had cost Nigeria $1 billion/month in lost revenues.

Islamist unrest

In addition to disruption attributed to MEND, Nigeria also battled a radical Islamist sect in 2009, called Al-Sunna wal Jamma, or Boko Haram.

In July, 4,000 people were displaced by violence that spread to Yobe, Kano, Borno and Plateau states.

Police said that the sect seeks to impose Taliban-style rule across Nigeria.

Islamic law was implemented in 12 northern states after Nigeria returned to civilian rule in 1999.

Exploration and Development

Discoveries reported

Discoveries made in Nigeria in 2009 included these:

- Afren PLC and Oriental Energy Resources Ltd: Ebok-4, -5; oil; offshore; OML 67
- Total SA: Etisong-1; oil; shallow offshore; OML 102
- Addax Petroleum Corp.: Njaba-2; oil; onshore; OML 124; Imo State

Ebok field

In early 2009, consultants estimated that 52–106 million bbl of oil is recoverable from Ebok field on OML 67. The field is 50 km (31 mi) offshore southeastern Nigeria.

Partners Afren PLC and Oriental Energy Resources Ltd. were studying an early production system that could deliver 15,000–25,000 b/d in early 2010 from several horizontal wells.

Petroleum Industry Bill: Key provisions

- Publish all licenses, leases, contracts, and payments to the government.
- Create new regulatory agencies.
- Simplify the industry's structure.
- Transform NNPC into an entity similar to Petrobras or Statoil, allowing it to raise money in financial markets rather than rely on the government for funds.
- Make NNPC's JVs with Royal Dutch Shell PLC, Total SA, Chevron Corp., and ExxonMobil Corp. independent companies, led by new management.

Violence in Niger Delta

Fighting continued in 2009 between the government and a rebel group called the Movement for the Emancipation of the Niger Delta (MEND), which emerged in late 2005.

The group seeks local control of oil production and wants federal reparations for pollution caused by the oil industry. Their tactics stress hit-and-run attacks on oil-industry facilities.

The conflict has been fanned by an interplay of criminality, kidnapping, ethnic rivalries, unhappiness with state and national governments, and industrial-scale theft of crude oil.

In the second half of 2009, a government amnesty program calmed the conflict, and in October, MEND extended a ceasefire to allow peace talks to begin.

Drilling and Production

Akpo field

In early 2009, the Nigerian subsidiary of Total SA began production from Akpo deepwater gas condensate field on OML 130, about 200 km (124 mi) offshore. Output is through an FPSO vessel.

It was expected to reach plateau levels in mid-year: 175,000 b/d of condensate and 9.06 million cu m/d (320 MMscfd) of gas from 44 wells.

Gas will go to both the domestic market and to the Bonny Island LNG plant.

Estimated proved and probable reserves are more than 28.3 bcm (1 tcf) of gas and 620 million bbl of condensate.

Oyo field

An affiliate of Eni SpA reported the start of production from Oyo oil field in late 2009.

Two subsea wells in 400 m (1,312 ft) of water about 75 km (47 mi) offshore showed initial production capacity of 25,000 b/d through an FPSO vessel.

Aje field

Providence Resources declared the offshore Aje field in OML113 a commercial discovery in early 2009. Chevron, technical advisor, was preparing a field development plan.

Ogoniland revival?

Oil production stopped in Ogoniland in 1993 when Royal Dutch Shell PLC was forced to cease operations amid massive human-rights and environmental protests.

(See The Netherlands section of the Europe chapter of this book for details about a 2009 legal settlement related to that situation.)

But a revival of hydrocarbons operations is now being discussed. In mid-2009, Wood Mackenzie said that Ogoniland formations might hold several trillion cubic feet of gas.

International Activities

Gazprom in Nigeria

Gazprom increased its presence in Nigeria in 2009, raising concern among European governments and groups dependent on Russian gas.

In June, the company announced a $2.5-billion JV—NiGaz Energy Co. Ltd.—with NNPC to explore and develop natural gas resources.

And in October 2009, it formed an alliance with independent Oando PLC to develop oil and gas assets and infrastructure in the West African sub-region and the Gulf of Guinea.

Nigeria's government has designed a "gas master plan" aimed at using more of the fuel to meet electricity shortages, so gas is expected to a focus of the alliance.

Gazprom also has expressed willingness to participate in a major project to build a $10-billion gas pipeline from Nigeria, across the Sahara, to Europe.

Gas Processing

Port Harcourt LPG plant

Ibeto Energy Development Company Ltd. was seeking U.S.-based engineering and construction firms in early 2009 to bid to develop a proposed $1-billion plant to produce, store and distribute LPG.

The facility would be sited in Bundu Ama, Port Harcourt.

The company was already accepting bids from European firms.

LNG

LNG snapshot

LNG is one of the fastest growing sectors of Nigeria.

The Nigerian LNG project (NLNG) was being upgraded in 2009 to improve gas utilization and reduce flaring. Train 6 was commissioned in mid-2009, and Train 7 was under construction.

Planned projects include the $8.5-billion Brass LNG facility on Brass Island in Bayelsa State. A final investment decision for the two-train, 10-mtpy plant, is expected in early 2010.

The four-train Olokola LNG facility (20 mtpy) is to be commissioned by mid-2010, with start-up of two trains.

The Bonny Island Seven Plus project trains are scheduled to follow in 2011–2012.

Funding for NLNG

NNPC said in August 2009 that it had secured full financing for its share of the expansion of the Nigerian Liquefied Natural Gas (NLNG) project.

The project will provide 8.8 million cu m/d (311 MMscfd) of natural gas for Trains 4 and 5.

Partners in the JV with NNPC are Royal Dutch Shell PLC, Total LNG Nigeria, and Eni SpA.

Start-up of Train 6 in 2009 brought total production capacity of the complex to about 22 mtpy.

Pipelines/Transportation

Trans-Sahara Gas Pipeline

The proposed $10-billion Trans-Sahara Gas Pipeline would transport as much as 30 bcm/year (1.06 tcf/year) of gas from Nigeria to Europe, passing through Algeria and Niger.

Details are given in the Algeria section of this chapter.

Gas Master Plan

NNPC short-listed 15 companies (12 foreign, 3 domestic) in early 2009 as potential investors in developing the country's gas sector under its Gas Master Plan.

That plan will offer three franchises around the gas-producing regions of the Southeast Delta, the Southwest Delta and the South-Central Delta.

Selected operators will set up a gas-gathering network, build gas processing plants and liquid storage facilities, and develop dry-gas transport infrastructure.

Refining and Products

Greenfield refinery

Nigerian officials said the government was in discussion with India's ONGC Videsh Ltd. in late 2009 regarding possible construction of a 180,000-b/d greenfield refinery in Nigeria. Location was not specified.

Lekki FTZ refinery

In early 2009, Oando PLC completed a feasibility study for the first phase of a 240,000-b/d greenfield refinery in the Lekki Free Trade Zone in Lagos.

That phase will build a finished-products reception terminal that the firm says will be the largest of its kind in sub-Saharan Africa. Bids for FEED work were being prepared.

Expansion to 360,000 b/d is envisioned for the second phase.

Lagos refinery

Nigeria's Capital Oil & Gas Industries Ltd. said it plans to build a refinery in Lagos with the state government.

The partners were carrying out an environmental assessment in early 2009 for the facility, expected to have a capacity of 100,000–200,000 b/d. Start of construction was anticipated by yearend.

Gasoline imports

In mid-2009, fuel marketers agreed to resume gasoline imports following a decision by the government to start making hundreds of millions of dollars in payment for fuel subsidies.

The dispute over the late payments caused Nigeria's worst fuel shortages in years and brought the government into conflict with major business operators.

Gas flaring

In early 2009, Nigerian officials said the country was considering slipping by three years, to yearend 2011, its target for ending gas flaring. The new date was requested by NNPC.

Elimination of flaring is a joint industry/NNPC initiative. However, NNPC has struggled to provide its share of funding needed for JV projects to build pipelines and gathering systems to stop the flaring.

Operators were skeptical, suggesting instead that 2013 would be feasible.

The World Bank in 2009 estimated that Nigeria flares 40% of the gas it produces, losing, by some estimates, $2.5 billion/year.

Company News

Oando goals

Oando PLC said in August 2009 that it was aiming to raise $1.3 billion in 2009 to fund its quest to become an "African major."

Besides expanding production, the firm aims to acquire dormant fields that western oil groups may have to relinquish under the nation's pending Petroleum Industry Bill.

Sinopec-Addax deal

In August 2009, China's Sinopec announced a $7.2-billion takeover of Addax Petroleum Corp., the largest independent producer in Nigeria.

Ogoni-Shell settlement

In June 2009, Royal Dutch Shell agreed to pay $15.5 million to the Ogoni people of the Niger Delta.

Details are given in The Netherlands section of the Europe chapter of this book.

SOUTH AFRICA

Capital: Cape Town
Monetary unit: Rand
Refining capacity: 485,297 b/cd
Oil production: 14,000 b/d
Oil reserves: 15.000 million bbl
Gas reserves: 0

Shale, sandstone evaluation

Falcon Oil & Gas Ltd. announced plans in late 2009 to evaluate the natural gas content of fractured Permian shales and sandstones in the Karoo basin, 193 km (120 mi) northeast of Cape Town.

The one-year appraisal will encompass study of data for 4,525 sq km (11,718 sq mi; 7.5 million acres), including review of the South African Petroleum Data Base.

No well-drilling is required.

Nine wells drilled in the area in the late 1960s and early 1970s encountered gas shows.

Clean-fuel specs

Sasol officials said in mid-2009 that South African refiners will need to invest $4.95 billion by 2014 to comply with clean-fuel specifications.

The firm warned of a clash between the high costs of improving fuel standards, a growing focus on cleaner fuels technology, and concerns about hampering economic growth.

South Africa's fuel meets Euro 2 emission standards, but EU countries have progressed to Euro 5 standards, which lower allowable emissions of nitrogen oxides and particulate matter.

Refiners have requested a 5-year delay to have sufficient time to meet Euro 4 standards.

New power tariffs for renewables

South Africa aims to generate 10,000 GW-hrs of power from renewable resources by 2013.

The National Energy Regulator of South Africa announced in early 2009 new tariffs for the feed-in to the grid of power generated from renewable sources.

The new tariffs cover wind, small hydroelectric, landfill gas and concentrating solar-collector technologies and guarantee the purchase price of electrical power they supply.

Also, the term of the revised tariffs was extended to 20 years from 15.

Sasol: Capital expenditures

In early 2009, Sasol Technology (Pty.) Ltd. said it would cut its capital expenditures for 2010–2012 by 40%, despite posting a sharp rise in profits.

The firm is the leading producer of liquid fuels from coal and gas, and its technology is generally most attractive when oil prices are high. The company said its planning assumes oil prices of $40–45/bbl for 2009 and 2010.

Sasol: Clean fuels research

In mid-2009, Sasol Technology opened its new Fuels Application Center in Cape Town, which will be used to test the impact of the company's range of fuels on vehicle exhaust emissions and performance.

Project Mafutha CTL plant

In mid-2009, sponsors of the $16-billion Project Mafutha coal-to-liquids (CTL) plant in Limpopo Province decided to extend until yearend the ongoing pre-feasibility study.

The greenfield CTL facility would produce 80,000 b/d of synthetic fuel using Sasol's proprietary Fischer-Tropsch processes and advanced refining methods to convert the F-T products to fuels.

Coal suppliers were vying for a role in the project, which will use about 25 mtpy of coal as a feedstock. Mafutha and its associated coal mine will be in northwestern Limpopo, in the Waterberg coal region.

A decision to proceed to a detailed feasibility phase likely will be made in March 2010.

SUDAN

Capital: Khartoum
Monetary unit: Pound
Refining capacity: 121,700 b/cd
Oil production: 500,000 b/d
Oil reserves: 5.000 billion bbl
Gas reserves: 3.000 tcf

Overview

Court ruling on oil fields

In mid-2009, the Permanent Court of Arbitration in The Hague readjusted the borders of Sudan's oil-rich Abyei region, effectively awarding control of key oil facilities there to the Khartoum-based northern government.

The oil fields had been appended to the Abyei region in 2005, after a peace agreement signed between the northern government and the Sudan People's Liberation Movement—a decision that Khartoum disputed.

In 2008, the two parties decided to refer the matter to the PCA after violent clashes in the region left 100 dead and sent thousands of others fleeing.

Altogether, the court ruling reduced the land mass of the Abyei region by 8,099 sq km (3,127 sq mi) by redrawing its northern, eastern, and western borders.

Khartoum will be able to keep the key Heglig and Bamboo oil fields, part of a block operated by the Greater Nile Petroleum Operating Co., a consortium led by CNPC.

The court ruling effectively awarded the Diffra oil field to Abyei.

Warrant for al-Bashir

On March 4, 2009, the International Criminal Court in The Hague issued a warrant for the arrest of Sudanese President Omar al-Bashir on charges of war crimes and crimes against humanity in Darfur.

The court said that there was insufficient evidence to support charges of genocide.

He is the first sitting head of state that the court has ordered arrested.

Pres. al-Bashir defied the ruling, saying that the ICC could "eat" any arrest warrant it issued.

African and Arab nations expressed concern that the court action will destabilize the region, bring even more conflict in Darfur, and threaten the peace deal that ended the Sudanese civil war.

Darfur violence

Violence, coupled with the threat of hunger and disease, continued to besiege Darfur in 2009 even as the intense fighting of 2003 and 2004 became more sporadic.

According to *Associated Press* reports, government attacks became more unpredictable, while inter-ethnic conflicts flared and banditry became endemic.

At yearend 2008, at least 2.7 million people had been displaced—nearly half of Darfur's population.

In January and February 2009, fighting between rebels and the government around Muhajeriya in south Darfur killed at least 17 people and displaced an additional 50,000.

The government aggravated insecurity by obstructing a UN-African Union peacekeeping mission, which, in early 2009, was still 6,000 troops short of its target of 19,000.

Oil-income reporting

In third quarter 2009, the UK group Global Witness reported that they found discrepancies in the reporting of Sudan's oil revenues. Such discrepancies could mean that the northern government was underpaying the south by hundreds of millions of dollars.

The report is titled, *Fuelling Mistrust: The need for transparency in Sudan's oil industry.*

Under the 2005 peace accord, both sides agreed to share the country's oil wealth, with the south receiving half the state revenues from the oil drilled from its territory.

Exploration and Development

Block 8

White Nile Petroleum Operating Co. discovered large volumes of dry gas in mid-2009 in Block 8, southeast of Khartoum, according to press reports.

The firm is a 50-50 combine of Sudan National Petroleum Corp. and Malaysia's Petronas, which said the finds came at the Hosan and Tawakul wells.

Potential was estimated at 453–566 bcm (16–20 tcf) but the finds are not yet proved commercial.

Sudan's only other gas discovery came in 2005 on Block 6 in the Heglig area.

Refining and Products

China use of Sudanese crude

China's CNPC announced that its Dalian Petrochemical unit processed 40,000 tonnes of high-acid oil imported from Sudan in March 2009.

Details are given in the China chapter of this book.

Khartoum refinery

In November 2009, CNPC signed three oil and gas cooperation agreements with the government of Sudan.

One agreement included an MOU regarding the second-phase expansion of Khartoum refinery.

Warap refinery

In October 2009, the government of southern Sudan approved plans to build a $2-billion, 50,000-b/d refinery in Akon, Warap state. It will process crude from Block 5A in the Unity field.

Construction will take 36 months, and could include a new pipeline.

The government also was planning a refinery for the Dar Blend oil fields in the Upper Nile region.

Help for Uganda

There were reports in October 2009 that the northern Sudanese government offered to help Uganda as it begins oil production from the Albertine rift.

That formation stretches from Southern Sudan through the Lake Albert Valley to southwest Uganda.

One account said that aid could include building a refinery.

TANZANIA

Capital: Dodoma
Monetary unit: Shilling
Refining capacity: 14,900 b/cd
Oil production: 0
Oil reserves: 0
Gas reserves: 230.000 bcf

Block 7

In late 2009, Dominion Petroleum Ltd. finished interpreting about 4,300 km (2,672 mi) of 2D seismic acquired in 2008 over offshore Block 7.

The results suggest the presence of numerous prospects and leads in mid-water depths. The largest prospect has potential recoverable hydrocarbons estimated at 1 billion bbl.

Dominion was seeking a farm-in partner and was formulating plans to acquire at least 500 sq km (193 sq mi) of 3D seismic over the acreage in 2010.

Delta Rufiji Block

Maurel & Prom Co. SA was making plans to spud an exploration well in August 2009 on the Delta Rufiji block.

The well is to probe the same formation in which the company discovered gas in the Mafia Deep well on the Bigwa Rufiji permit.

TUNISIA

Capital: Tunis
Monetary unit: Dinar
Refining capacity: 34,000 b/cd
Oil production: 82,000 b/d
Oil reserves: 425.00 million bbl
Gas reserves: 2.300 tcf

Hasdrubal offshore facilities

Gas began flowing in August 2009 through the new Hasdrubal facilities offshore Tunisia, according to operator BG Group. Commercial production was said to be imminent.

The net output will lift the company's Tunisian oil and gas production from the current 32,000 boe/d to 56,000 boe/d as Hasdrubal wells are brought onstream.

BG said it expects to become Tunisia's largest producer of gas, LPG, and liquids.

Desertec solar-power consortium

Desertec Industrial Initiative (DII) GmbH—a sustainable-energy initiative of 12 European companies—proposed plans in 2009 for a massive solar power project in the Sahara Desert.

Initial plans did not specify whether part of the project might be in Tunisia.

Details are given in the Germany section of the Europe chapter of this book.

UGANDA

Capital: Kampala
Monetary unit: Ugandan shilling
Refining capacity: 0
Oil production: 0
Oil reserves: 0
Gas reserves: 0

Overview

Development snapshot

Since the first confirmation of oil in Uganda in late 2006, exploration companies have identified hundreds of millions of barrels of oil in the Albertine Graben region.

That region includes some 23,000 sq km (8,880 sq mi) along Uganda's border with the Democratic Republic of the Congo (DRC), formerly Zaire.

Tullow Oil PLC, the most dominant of four companies with exploration licenses, has said it has found 800 million bbl as of 2009.

In March 2009, IOCs announced plans to boost production from fields around Lake Albert. In particular, Tullow said it found more oil in the lakes region.

The firm expects that production could start in 2010, rising to 100,000–150,000 b/d by 2019.

Because exploration has so far has covered only one-third of the Albertine Graben, experts say that Uganda's extractable deposits could exceed 2 billion bbl.

Uganda currently uses only about 11,000 b/d of oil.

Pres. Yoweri Museveni announced in October 2009 that a refinery would soon be built in Uganda to process the nation's oil, but details were still under discussion.

Policy change

Ugandan officials stated in April 2009 their intention to revise oil and gas policy by yearend.

The pending Resource Management Law and Revenue Management Law would establish a new institutional framework, separating policy from regulation and commercial activities. It would also create a state-owned Uganda Oil Co.

Exploration and Development

Block 5

In third quarter 2009, Tower Resources PLC reevaluated data for its Iti-1 wildcat on Block 5 and found that it has significant potential to contain oil just above basement.

Based on the findings, the company plans to drill another well on the Iti structure in early 2010.

Four finds on Block 2

Tullow Oil PLC announced oil finds in mid-2009 with three exploration wells—Ngara-1, Kigogole-3, and Nsoga-1—in the Butiaba region (Victoria Nile Delta play) of Block 2. It suspended drilling and will return to all three as future oil-producers.

Data from Ngara-1 will help assess two adjacent traps and guide planning of the next Block 2 drilling campaign in early 2010.

Earlier in the year, Tullow found oil with its Karuka-2 exploration well on the Vundu prospect, also on Block 2.

Uganda-Congo relations

In March 2009, Uganda and Congo (former Zaire) agreed to upgrade diplomatic relations to ambassadorial level. Analysts expected the move to end years of strained relations and open the way to better-coordinated commercial development of hydrocarbon reserves around Lake Albert.

Tullow Oil, already active on the Uganda side of Lake Albert, was negotiating in early 2009 with Congo (former Zaire) to re-acquire exploration rights to two blocks on that side of the lake.

Uganda's Lake Albert Rift Basin

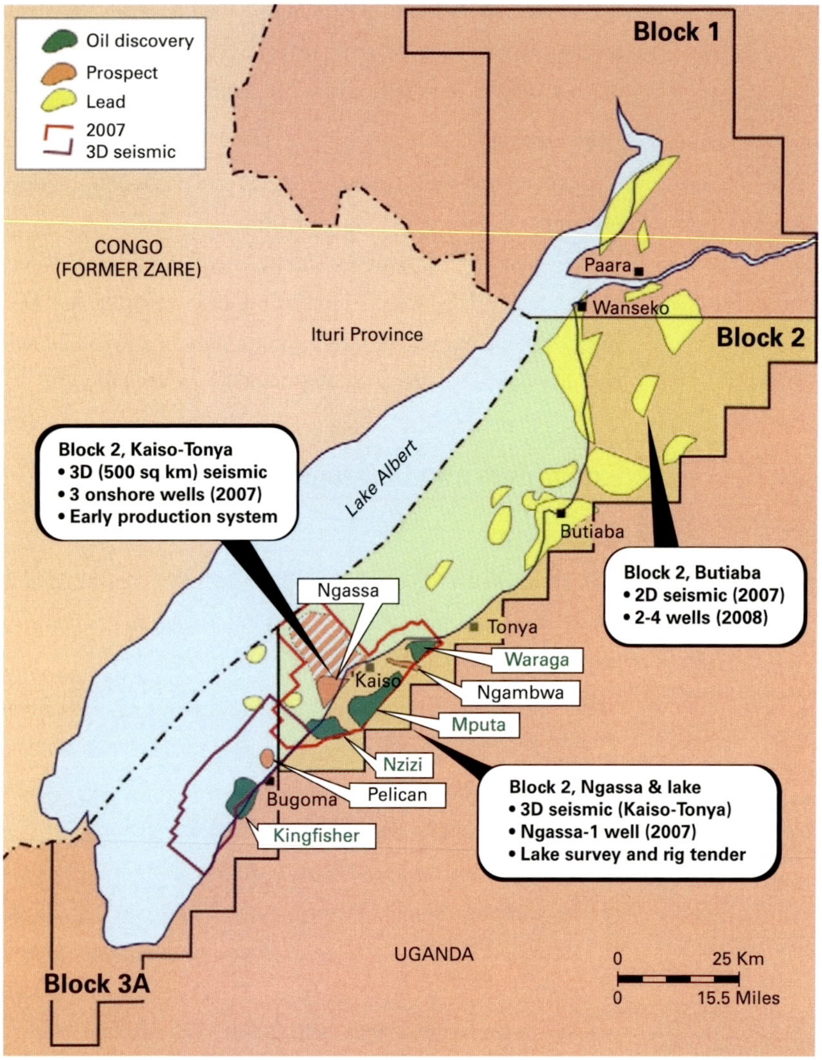

Source: Tullow Oil PLC

Ugandan reserves: 2 billion bbl

Ugandan officials said that the country's confirmed oil reserves reached 2 billion bbl in June 2009.

The corresponding figure in 2006 was just 300 million bbl.

Some analysts, however, believe that accurate assessment of reserves will require considerable further exploration over the next few years.

Chinese interest

In early 2009, it was reported that China's three main oil companies—CNPC, Sinopec, and CNOOC—were looking to acquire stakes in the Uganda blocks being worked by Tullow and partner Heritage Oil on the Ugandan side of the Lake Albert rift basin (Blocks 1, 2 and 3A).

All three Chinese firms were said to be interested in bidding for a stake in the Uganda development in a deal that could be worth as much as $500 million.

Some analysts expect Korea and Japan also will consider a role in Uganda.

Giraffe discovery

Heritage Oil said in early 2009 that it found enough recoverable oil with its Giraffe discovery on Block 1 to exceed the commercial threshold for development.

Giraffe is connected to the Buffalo discovery of late 2008, and the Buffalo-Giraffe complex is estimated to hold more than 400 million bbl of recoverable oil.

Heritage suspended Giraffe as a future production well.

Pipelines/Transportation

Eldoret-to-Kampala pipeline

In 2006, a unit of Libya's state-owned Tamoil Holdings won the tender to build the 360-km (224-mi) pipeline extension from Eldoret, Kenya, to Kampala, Uganda.

Tamoil said initial work would start in July, with completion scheduled for 2011 (4 years later than first planned). However, work had not started as of September.

Tamoil said it also plans to extend the line from Kampala to the Albert rift in Uganda to transport oil from there to Mombasa for global export.

Tullow Oil and Heritage Oil were said to favor construction of a new oil export line across Kenya to Mombasa. Heritage said in early 2009 that Asian and Middle East backers had offered to build such a line.

Concerned about delays, the Ugandan government was considering in late 2009 whether to build an alternative line to Tanzania's port of Dar es Salaam.

UGANDA DEVELOPMENT LAYOUT

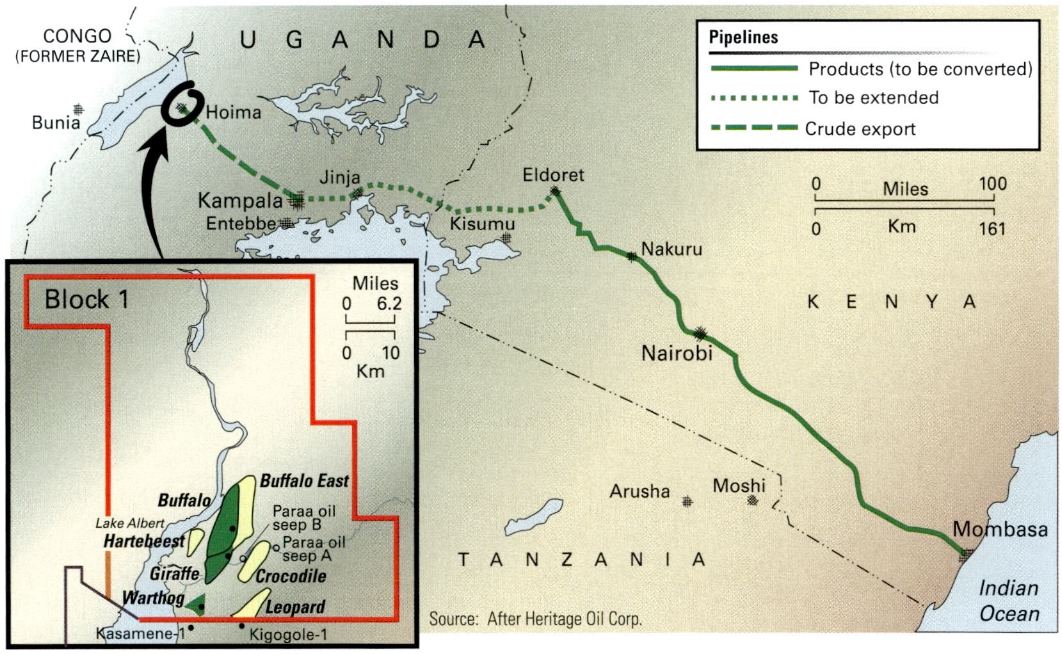

Source: After Heritage Oil Corp.

Refining and Products

Refinery plans

Uganda's stated policy is to refine domestically all oil produced in the country, with possible export of some excess crude, implying stronger support for construction of a refinery rather than pipelines.

Officials have expressed interest in a refinery with a capacity of at least 100,000 b/d.

Eni SpA has proposed constructing such a refinery, connected to the proposed export pipeline to Mombasa. However, officials with Heritage Oil in mid-2009 expressed the view that national oil production may not be sufficient to warrant a refinery of that size.

Assistance from Iran?

Following a visit to Tehran in mid-2009, Uganda's Pres. Yoweri Museveni said that he had signed an agreement with Iran to "consider cooperation in building an oil refinery in Uganda."

Iran also expressed readiness to invest in "the entire value chain" of Uganda's petroleum industry.

Company News

Jockeying for assets

In December 2009, Tullow Oil was pondering whether to thwart the sale by its partner, Heritage Oil, of the latter's Ugandan assets to Eni, the Italian energy group, for $1.35 billion.

Heritage decided in November 2009 to sell its 50% stakes in Blocks 1 and 3a along Lake Albert to Eni.

Because Tullow owns the other 50% of both those blocks, it can preempt the deal, but such action would require Ugandan government approval.

It was unclear at yearend whether Uganda would let Tullow preempt the deal, since Eni had shown Ugandan officials a full development plan for the Lake Albert region.

Other action in Africa

Several other African nations saw exploration and development action in 2009.

Benin

Capital: Porto-Novo

TGS–NOPEC Geophysical Co. began shooting 3,500 line-km of 2D seismic in the Gulf of Guinea in early 2009. The program will help to define the extent of the petroleum system in ultradeep water.

Liberia

Capital: Monrovia

- Anadarko Petroleum Corp. began shooting 4,700 sq km of 3D seismic in early 2009 on Blocks 15, 16, and 17 in the Atlantic, aiming to illuminate deepwater Cretaceous fan prospects.
- The government closed its third petroleum bidding round in November, and was poised to solicit bids in December for an ultra-deepwater licensing round (Blocks LB-1 through LB-5).

Namibia

Capital: Windhoek

A four-company international group made what appears to be a giant gas-condensate discovery in mid-2009 off northernmost Namibia in the Namibe basin. The energy ministry said the find could contain up to 14 tcf.

Niger

Capital: Niamey

A subsidiary of CNPC began preconstruction planning in March 2009 for the first refinery in Niger.

The 20,000-b/d Zinder facility, about 950 km (590 mi) east of Niamey, is to start up in mid-2011.

Rwanda

Capital: Kigali

In early 2009, ContourGlobal entered into an agreement with the government to build and operate a large gas-extraction unit and power-generation facility at Lake Kivu. The deep lake contains an estimated 55 bcm (1.94 tcf) of methane reserves in solution at a depth of 350 m (1,148 ft).

Sao Tome & Principe

Capital: Sao Tome

Sinopec JDZ Block 2 Ltd. planned to start drilling the deepwater Bomu-1 prospect on Block 2 in the Nigeria-Sao Tome Joint Development Zone by the end of August 2009. Addax Petroleum Corp. let a contract for the drilling of four wells in the same zone starting in the fourth quarter.

Senegal

Capital: Dakar

- First Australian Resources Ltd executed an agreement with Shell Exploration Co. in early 2009 to explore the offshore Sangomar, Rufisque, and Sangomar blocks.
- National Iranian Oil Refining and Distribution Co. made an agreement in mid-2009 with Senegal's national oil company to increase the capacity of an oil refinery to 64,000 b/d . NIORDC also will supply crude to the plant for one year.

Seychelles

Capital: Victoria

- Black Marlin Energy, East African Exploration, and Avana Petroleum started a second seismic survey in the Seychelles Islands in late 2009. A 2007 survey identified deep granitic features linked to those that underpin the Bombay High field.
- Seychelles Petroleum Co. completed an offshore oil-slick mapping and interpretation project in mid-2009 to prepare for a licensing round by early 2010. The government reportedly will offer 70,000 sq km (27,027 sq mi) of offshore acreage.

Venus Well Offshore Sierra Leone

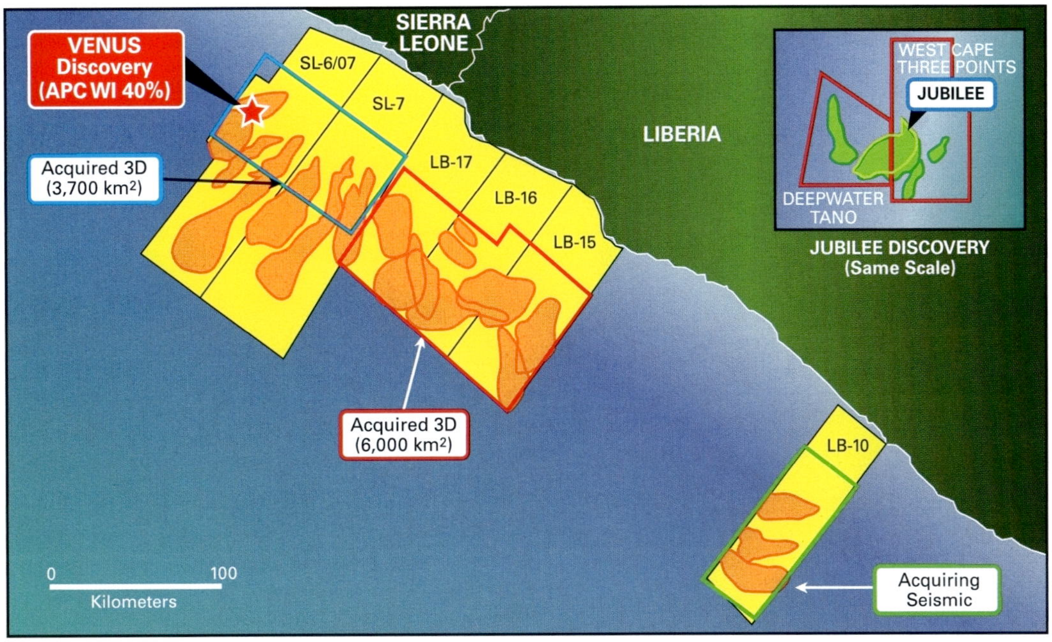

Sierra Leone

Capital: Freetown

Anadarko hit pay in Block SL 6/07 with its Venus exploration well in mid-2009.

The first deepwater test in the Sierra Leone-Liberian basin, Venus was drilled to TD of 18,500 ft (5,639 m) in about 5,900 ft (1,798 m) of water.

Tullow Oil PLC and partners said that the Venus discovery established a whole new active hydrocarbons system that spanned at least 1,000 km (621 mi) to the coast of Ghana.

Somalia

Capital: Mogadishu

In November 2009, Somali pirates seized a Greek-owned crude carrier 1,609 km (1,000 mi) off the Somali coast. The *Maran Centaurus*, carrying $20 million worth of crude, was the eleventh vessel seized in a surge of attacks in October and November.

Somaliland

Capital: Hargeisa

In February 2009, the government offered its first-ever petroleum exploration licenses, including eight blocks totaling more than 89,624 sq km (34,604 sq mi) of both offshore and onshore areas. Bids were due December 15, and concessions are to be awarded in March 2010.

Zambia

Capital: Lusaka

In mid-2009, the government launched its debut oil licensing round, for 23 blocks in the northwest, west, and eastern provinces.

Former Soviet Union

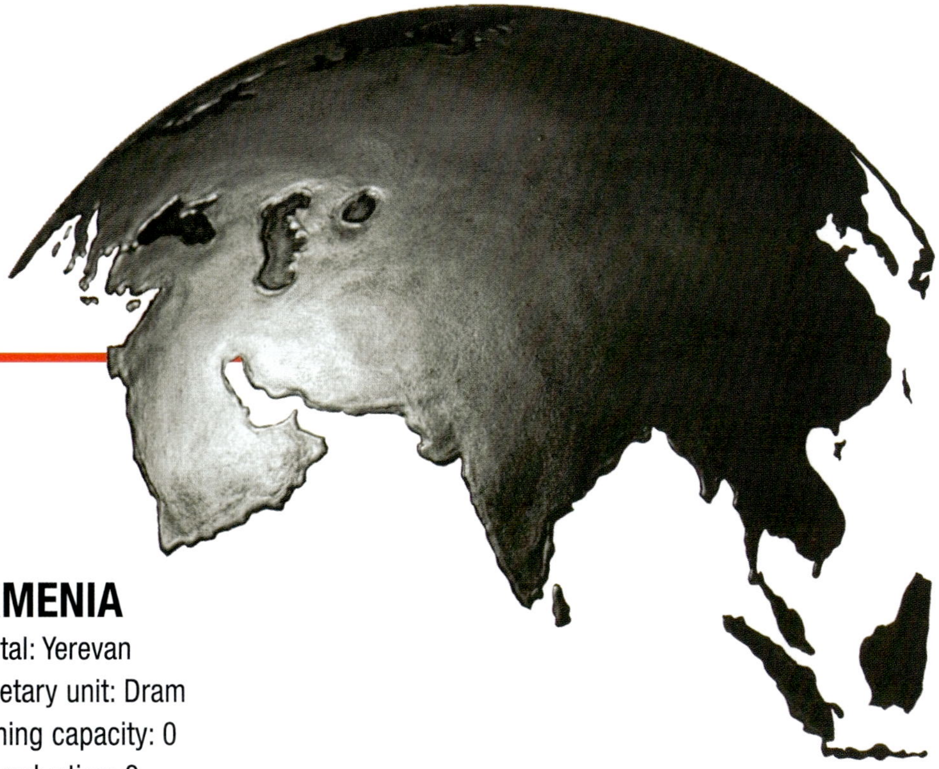

ARMENIA

Capital: Yerevan
Monetary unit: Dram
Refining capacity: 0
Oil production: 0
Oil reserves: 0
Gas reserves: 0

Fuel pipeline project with Iran

A joint Armenian-Iranian company was planning in mid-2009 for the construction of a pipeline for delivering gasoline and diesel fuel to Armenia from Iran.

Iranian officials also said they were interested in working with Armenia to build an oil refinery, a hydropower plant on the Aras River, and a railway line, in addition to the fuel pipeline.

Accord with Turkey

In October 2009, Armenia and Turkey met in a historic session and agreed to open their border and establish diplomatic relations after a century-long bad relationship.

It was expected that the accord would be ratified by the parliaments of both nations and that the border—closed for the past 16 years—could be reopened within months.

The diplomatic accord could reduce tensions in the Caucasus region and facilitate its growing role as a corridor for energy supplies bound for the West.

Major countries expressed support for the accord. The foreign ministers of the United States, Russia, France and the European Union were on hand in Zurich to witness the signing.

However, some 10,000 people rallied in Yerevan to oppose the signing, and protests also arose in Lebanon and France.

Discussions will continue on various issues, including the Armenian genocide in 1915 and occupation of Nagorno-Karabakh by Armenian troops in 1993.

New nuclear power plant

In early 2009, the Armenian government awarded a contract to WorleyParsons Ltd. to build a new nuclear power plant in Metsamor.

The new plant, of at least 1,000-MW capacity, would replace an aging reactor that currently generates about 40% of the total power produced in Armenia.

WESTERN FSU
0 100 200 300 Miles
0 200 400 Km
Barents Sea
Kara Sea
Eniseisk Gulf
NOVAYA ZEMLYA
Kolguyev Island
Kola Peninsula
White Sea
FINLAND
RUSSIA
FORMER SOVIET UNION
KAZAKHSTAN
UZBEKISTAN
TURKMENISTAN
BELARUS
UKRAINE
ROMANIA
BULGARIA
MOLDOVA
ESTONIA
LATVIA
LITHUANIA
GEORGIA
ARMENIA
AZERBAIJAN
TURKEY
IRAN
Black Sea
Caspian Sea
Aral Sea
Aegean Sea
Gulf of Finland
Helsinki
St. Petersburg (Leningrad)
Moscow
Arkhangel
Petrozavosk
Vyborg
Primorsk
Tallin
Tartu
Riga
Ventspils
Kaunas
Vilnius
Minsk
Grodno
Brest
Mozyr
Polotsk
Pskov
Novgorod
Kirishi
Valdai
Torzhok
Cherepovets
Yaroslavl
Vladimir
Nizhniy Novgorod
Kazan
Kirov
Perm
Okhansk
Izhevsk
Neftekamsk
Nizhnekamsk
Ufa
Salavat
Ishimbay
Krasnovolskiy
Magnitogorsk
Chelyabinsk
Sverdlovsk
Kurgan
Tyumen
Omsk
Atbasar
Karaganda
Orenburg
Orsk
Samara (Kuibyshev)
Syzran
Ul'yanovsk
Saratov
Volgograd
Astrakhan
Atyrau
Kulsary
Beineu
Aral'sk
Dzhusaly
Kungrad
Ryazan
Kaluga
Smolensk
Orel
Yelets
Michurinsk
Kursk
Voronezh
Kharkov
Lisichansk
Lugansk
Zaporozhye
Rostov
Kremenchug
Vinnitsa
Lvov
Dolina
Brody
Uzhgorod
Kishinev
Odessa
Kherson
Simferopol
Novorossiysk
Tuapse
Sochi
Krasnodar
Tikhoretsk
Stavropol
Grozny
Batumi
Tbilisi
Baku
Yerevan
Erzurum
Samsun
Istanbul
Ankara
Bucharest
Constanta
Burgas
Sofia
Aktau (Shevchenko)
Krasnovodsk
Ashgabat (Ashkhabad)
Tabriz
Astara
Rasht
Neka
Bandar Shah
Tehran
Rey
Naryan-Mar
Usinsk
Pechora
Ukhta
Vorkuta
Salekhard
Novyy Port
Nadym
Urengoi
Surgut
Nizhnevartovsk
Serov
Nizhnyaya Tura
Punga
Dudinka
Norilsk
Shtockmanovskoye
Tulomskaya
Severo-Kildinskoye
Murmanskoye
Admiralteyskaya
Leningradskoye
Rusanovskoye
Yamburg
Samotlor
Orenburgskoye
Karachaganak
Tengiz
Kashagan
Shah Deniz
Blue Stream
SCP
2 Lines
3 Lines
4 Lines
6 Lines

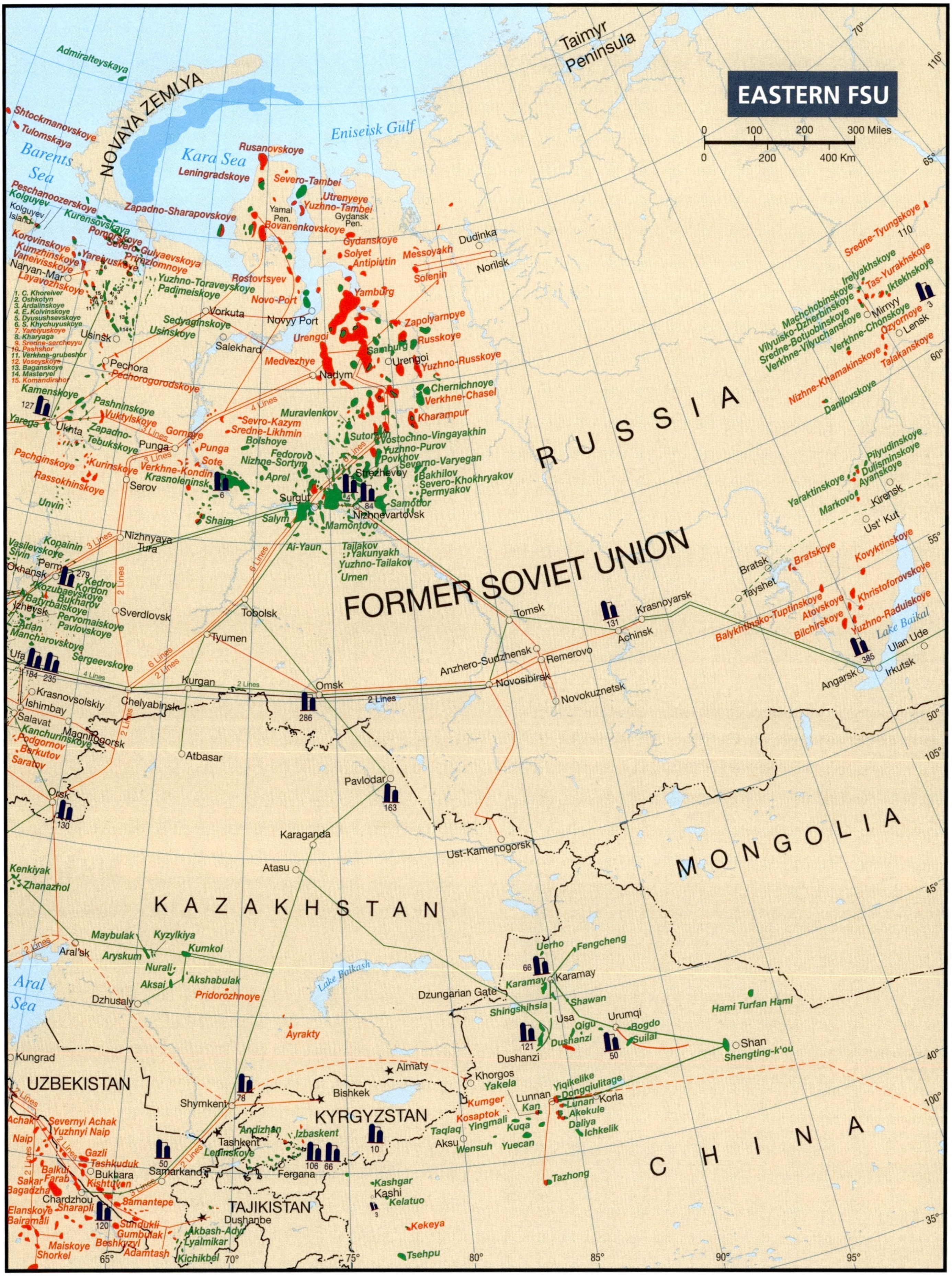
EASTERN FSU
0 100 200 300 Miles
0 200 400 Km
RUSSIA
FORMER SOVIET UNION
KAZAKHSTAN
MONGOLIA
CHINA
UZBEKISTAN
KYRGYZSTAN
TAJIKISTAN
NOVAYA ZEMLYA
Kara Sea
Barents Sea
Eniseisk Gulf
Taimyr Peninsula
Aral Sea
Lake Baikal
Lake Balkash
Yamal Pen.
Gydansk Pen.
Admiralteyskaya
Shtockmanovskoye
Tulomskaya
Rusanovskoye
Leningradskoye
Severo-Tambei
Utrenyeye
Yuzhno-Tambei
Bovanenkovskoye
Zapadno-Sharapovskoye
Peschanoozerskoye
Kolguyev
Kurensovskaya
Gydanskoye
Solyet
Antipiutin
Messoyakh
Solenin
Dudinka
Norilsk
Yamburg
Zapolyarnoye
Russkoye
Urengoi
Medvezhye
Nadym
Yuzhno-Russkoye
Chernichnoye
Verkhne-Chasel
Kharampur
Muravlenkov
Vorkuta
Novyy Port
Salekhard
Usinsk
Pechora
Ukhta
Surgut
Nizhnevartovsk
Samotlor
Mamontovo
Salym
Shaim
Serov
Nizhnyaya Tura
Perm
Sverdlovsk
Tobolsk
Tyumen
Kurgan
Omsk
Chelyabinsk
Ufa
Tomsk
Novosibirsk
Kemerovo
Novokuznetsk
Anzhero-Sudzhensk
Krasnoyarsk
Achinsk
Bratsk
Tayshet
Angarsk
Irkutsk
Ulan Ude
Mirnyy
Lensk
Kirensk
Ust' Kut
Pavlodar
Karaganda
Atasu
Atbasar
Ust-Kamenogorsk
Aral'sk
Dzhusaly
Kungrad
Shymkent
Tashkent
Samarkand
Bukhara
Chardzhou
Fergana
Bishkek
Almaty
Dushanbe
Karamay
Urumqi
Usa
Dushanzi
Shan
Khorgos
Lunnan
Korla
Aksu
Kashgar
Kashi
Tazhong
Kumkol
Kenkiyak
Zhanazhol
Hami Turfan Hami
Shengting-k'ou
Dzungarian Gate
Orsk
Magnitogorsk
Salavat
Ishimbay
Krasnovsolskiy
Naryan-Mar
Shtockmanovskoye
Sredne-Tyungskoye
Talakanskoye
Danilovskoye
Yaraktinskoye
Markovo
Kovyktinskoye
Bratskoye
Kristoforovskoye
Yuzhno-Raduiskoye
Balykhtinsko-Tuptinskoye
Bilchirskoye
Atovskoye
Achak
Naip
Gazli
Shurtan
Kokdumalak
Samantepe
Kumger
Kosaptok
Yakela
Tsehpu
Kekeya
Kelatuo

Construction is scheduled to start by early 2011. The goal is to start up the new plant by 2016 and then decommission the old one.

AZERBAIJAN

Capital: Baku
Monetary unit: Manat
Refining capacity: 398,978 b/cd
Oil production: 1.000 million b/d
Oil reserves: 7.000 billion bbl
Gas reserves: 30.000 tcf

Overview

Since gaining independence from Russia in 1991, Azerbaijan has taken a more pro-Western, foreign-investment-friendly approach than most other nations in the region. This has contributed to major expansion of the country's hydrocarbon production.

Most oil comes from the Azeri-Chirag-Guneshli structure, developed by Azerbaijan International Operating Co. The bulk of Azeri oil is exported through two major pipelines:

- Baku-Tbilisi-Ceyhan (BTC), with a terminal at the Turkish port on the Mediterranean
- Baku-Novorossiysk, with a terminal at the Russian port on the Black Sea.

Regarding natural gas, the country's production almost doubled from 1999–2009, with most of the gas coming from three fields: Azeri-Chirag-Guneshli, Bakhar, and Shah Deniz.

The latter is considered one of the world's largest gas discoveries in the last few decades.

Since 2007, Azerbaijan has been a net gas exporter, sending gas mainly to Turkey, but with small volumes also going to Georgia and Iran.

The major gas pipeline is the South Caucasus, known also as Baku-Tbilisi-Erzurum. It runs parallel to the BTC for most of its route, with a connection to the Turkish gas network.

Gazprom gas deal

Under a contract signed in mid-2009, Gazprom will import 500 million cu m of gas per year from Azerbaijan, through state-operated SOCAR, starting in January 2010.

Staff at *Oil & Gas* Journal saw the deal as an attempt by Gazprom to exert greater control over EU energy supplies.

Azerbaijan has up to 1.5 tcm (53 tcf) of proved gas reserves, including the Shah Deniz field in the Caspian Sea, with recoverable reserves of more than 1.3 tcm (50 tcf).

The EU has urged Azerbaijan to commit Shah Deniz gas to the planned Nabucco pipeline, which would bring Caspian gas directly to Europe across the Caucasus and Turkey.

Analysts said Azerbaijan would avoid pledging large volumes of gas to Gazprom that could undermine its independence from Russia.

Exploration and Development

Shafag, Asiman structures

BP and SOCAR signed an MOU in mid-2009 to jointly explore and develop the Shafag and Asiman structures in the Azerbaijan sector of the Caspian Sea.

The structures are on an unexplored block 125 km (78 mi) southeast of Baku, covering 1,100 sq km (425 sq mi). Water depth is about 700 m (2,297 ft) and reservoir depth is estimated at 7,000 m (22,966 ft).

Drilling and Production

Shah Deniz output

Oil & Gas Journal reported in mid-2009 that Azerbaijan expects to reach production of 9 bcm/year (318 bcf/year) by 2010 from the first stage of development of the massive Shah Deniz gas field.

Shah Deniz 2—expected to come online in 2014—might produce 10–15 bcm/year (353–530 bcf/year), according to state officials.

Caspian Sea Chirag project

KBR Inc. won a FEED contract from Azerbaijan International Oil Co. in early 2009 for the Chirag oil project in the Azerbaijan section of the Caspian Sea.

Focus of the work will be construction of a large drilling platform that will tie in to existing Azeri-Chirag-Gunashli (ACG) oil field development platforms and pipeline systems.

This is the next stage in the Chirag project, including incremental expansion of existing ACG field development.

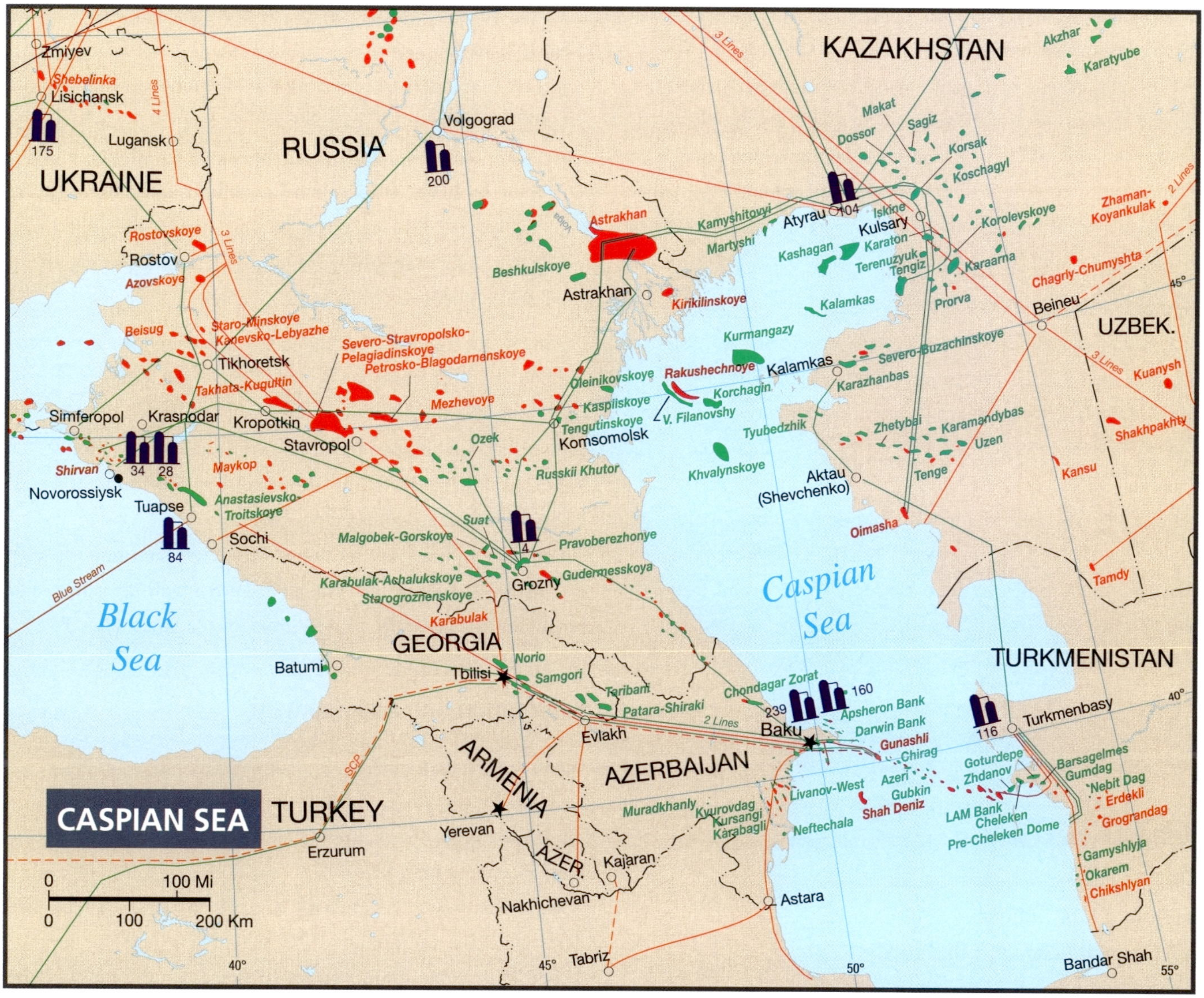

GEORGIA

Capital: Tbilisi

Monetary unit: Lari

Refining capacity: 0

Oil production: 1,000 b/d

Oil reserves: 35.000 million bbl

Gas reserves: 300 bcf

Aftermath of 2008 conflict

Russia invaded Georgia in August 2008 after Georgia's government tried to retake the breakaway region of South Ossetia by force. The conflict disrupted transit of oil and gas across the region, with Russian warplanes said to have targeted pipelines and railways.

Georgia and the United States, concerned about Russian aggression, agreed to step up physical security of energy transit across the trans-Caucasus to European markets as part of the U.S.-Georgia Charter on Strategic Partnership signed January 9, 2009.

The two nations said they "intend to…develop a new Southern Corridor to help Georgia and the rest of Europe diversify their supplies of natural gas by securing imports from Azerbaijan and Central Asia."

Rosneft, Abkhazia accord

In May 2009, Russia's Rosneft Oil Co. signed a 5-year exploration and development agreement with the Ministry of Economy of the Republic of Abkhazia.

The deal calls for upstream and downstream co-operation in oil and gas projects, which may extend to development of advanced technologies for exploration and for secondary recovery from discoveries offshore.

The government of Georgia quickly criticized Rosneft for signing the deal with the breakaway republic, calling the document a violation of Georgia's laws and territorial integrity.

In response to Georgia's criticism, Rosneft acknowledged that it is an arm of Russian state policy and said it intends to work in Abkhazia as in a sovereign state.

Deal with two Indian firms

In mid-January 2009, a 25-year agreement was signed between two Indian companies—Jindal Petroleum Pvt. Ltd. and Ensearch Petroleum Ltd.—and the International Oil and Gas Agency of Georgia, regarding exploration and production of oil and gas.

In 2008, the two firms won the international tender to explore in the regions of Kvemo Kartli, Guriya, Ajaria, and Kakheti. Jindal will get three blocks to develop and Ensearch will get two.

KAZAKHSTAN

Capital: Astana
Monetary unit: Tenge
Refining capacity: 345,093 b/cd
Oil production: 1.330 million b/d
Oil reserves: 30.000 billion bbl
Gas reserves: 85.000 tcf

Overview

Kazakhstan's oil and gas production has more than tripled over the last decade.

Most of the oil comes from Tengiz and Karachaganak fields. Kashagan field, discovered in 2000, is the largest oil field outside the Middle East and the fifth largest in the world in terms of reserves.

Since early 2008, the Kazakh government has had the lead role in the international consortium developing Kashagan, which is scheduled to come on stream in 2013, and as of January 2009, a new joint operating company was established for Kashagan, called the North Caspian Operating Co. BV.

Kazakhstan exports most of its oil through three major pipelines:

- The Atyrau-Samara pipeline provides a northbound link to the Russian distribution system.
- The Caspian Pipeline Consortium (CPC) connects Kazakhstan's oil fields in the Caspian Sea with Novorossiysk.
- The Kazakhstan-China Pipeline connects oil fields in Atasu in northwest Kazakhstan with Alashankou in China's northwestern Xinjiang region.

Almost all Kazakh natural gas is associated with oil production, with Karachaganak and Tengiz the largest sources. Gas from Kazakhstan—as well as from Turkmenistan and Uzbekistan—is shipped through the Central Asia-Center (CAC) Pipeline, which starts in Turkmenistan and runs through Uzbekistan and Kazakhstan to Russia.

Kazakhstan has three refineries with a total refining capacity of 345,000 b/d. Atyrau in the west, Pavlodar in the north, and Shymkent in the south all were built in the Soviet era.

Exploration and Development

Stake in Karachaganak

In December 2009, Kazakhstan said it was in talks with a foreign oil group to secure a stake for the government in the giant Karachaganak field.

Karachaganak is operated by a group made up of BG, Eni, ChevronTexaco and Lukoil.

The talks came amid tension between Kazakhstan and the group regarding tax issues and the consortium's decision to postpone an initiative to double oil production at Karachaganak until oil prices and development costs dropped.

The field holds an estimated 2.4 billion bbl of oil and 15 tcf of gas reserves.

Akkulka exploration contract

Tethys Petroleum Ltd. won an extension in late 2009 of the Akkulka exploration contract until March 10, 2011, which will allow further testing of deep and shallow prospects.

AKD01, the company's first deep exploration well on the block, went to TD of m (11,200 ft) and found two zones interpreted to contain hydrocarbons in two zones.

Kashagan development

Agip KCO let a $2.6 billion contract in May 2009 to Saipem SPA and Aker Solutions for a range of work on Phase 1 of Kashagan field in the north Caspian Sea.

Phase 1, due online in late 2012, is expected to lead to production of 300,000 b/d. It will be under the responsibility of Eni SPA.

Shell will manage the production operations after the start-up of Phase 1, with KazMunaiGaz assuming greater responsibility over time.

Production is expected to increase to 450,000 b/d during Phase 2. Under that phase, Shell will be responsible for offshore development, Eni for onshore plant, and ExxonMobil for drilling.

The field, 80 km (50 mi) southeast of Atyrau, is expected to reach plateau production of 1.5 million b/d by the end of the next decade. Total development cost for Kashagan is estimated at $136 billion.

Drilling and Production

Khvalynskoye gas field

KazMunaiGaz (KMG) estimated in October 2009 that Phase 1 development of Khvalynskoye natural gas field in the Caspian Sea will cost $5 billion.

In that same month, Total SA and GDF Suez Group reached a heads of agreement decision with KMG to take a combined 25% interest in Khvalynskoye.

Located in 25 m (82 ft) of water on the border between Kazakhstan and Russia, Khvalynskoye is a gas-condensate field to be developed by Russia's Lukoil, operator. Gas from the field will go to Russia.

Total and GDF Suez will invest $1 billion in the project. Details are given in the France section of the Europe chapter of this book.

Khvalynskoye is jointly owned with Lukoil, which said it had not concluded a production-sharing contract with KMG and would keep its 50% stake.

Tengizchevroil: Higher output

The Tengizchevroil (TCO) consortium, led by Chevron Corp., said in October 2009 that it planned to raise its oil production by yearend to 22.5 million tonnes (165 million bbl) from 21.5 million tonnes (158 million bbl).

Tengizchevroil is owned jointly by Chevron, ExxonMobil Corp., KazMunaiGas, and LukArco.

KMG-SOCAR agreements

In October 2009, KazMunaiGaz and Azerbaijan's SOCAR were reported to have reached several agreements to ensure that the rising volumes of Kazakh oil will reach world markets.

The agreements envisage shipments of Kazakh crude by tanker across the Caspian Sea and then by rail to Georgia's Black Sea ports of Batumi and Kulevi, as well as through the existing Baku-Supsa pipeline.

The so-called TransCaspian project also will include oil-loading terminals on the Kazakh coast of the Caspian, tankers and vessels, offloading terminals on the Azerbaijan coast of the Caspian, and equipment to link into the Baku-Tbilisi-Ceyhan line.

International Activities

Funds from French bank

In late 2009, France's BNP Paribas Bank said it would extend a €10.2-million loan to the Development Bank of Kazakhstan, largely earmarked for energy projects.

According to the Kazakh bank's press service, BNP Paribas opened a €200 million credit line for this.

Some of the funds will finance imports of Finnish equipment to build a gas turbine power plant at the Akshabulak oil field.

Kazakhstan, France energy accords

In late 2009, Kazakhstan and France signed a number of long-term agreements involving oil and gas.

Under one draft agreement, a consortium of French companies headed by Spie Capag (Vinci Group) will participate in siting a major export pipeline between the offshore Kashagan oil field and Baku that will enable oil to be carried to Europe.

The pipeline agreement will bypass Russia, diversifying Europe's sources of oil supply. At the same time, it will diversify Kazakhstan's markets.

China initiatives

In early 2009, China's CNPC bought a 49% stake in Kazakhstan's MangistauMunaiGaz from KazMunaiGaz for about $10 billion.

MMG has the rights to operate 36 oil fields, of which 14 are under development.

In the same time frame, state-controlled Development Bank of Kazakhstan received a $5-billion loan from the China Export-Import Bank, to be used to import equipment from China to develop the country's infrastructure. CNPC also provided $5 billion in assistance to KazMunaiGaz.

Pipelines/Transportation

Kazakh land-based export pipelines

In early 2009, the Oxford Institute for Energy Studies evaluated various land pipeline projects designed to export Kazakh oil and gas.

That evaluation, presented in full in two January 2009 issues of *Oil & Gas Journal*, included the following projects:

- The Central Asia-Center gas pipeline network
- The Caspian Littoral gas pipeline
- The Trans-Asian gas pipeline network
- The Kazakhstan-China Pipeline
- Trans-Caspian Gas Pipeline

Details and maps of these pipeline projects are presented in the Pipelines section of the Key Stats and Tables chapter toward the back of this book.

Turkmenistan-China gas line

In December 2009, leaders of Turkmenistan, Kazakhstan, Uzbekistan, and China met near the Turkmen-Uzbek border to inaugurate the new Turkmenistan-China natural gas pipeline.

Details are given in the Turkmenistan section of this chapter.

Kazakhstan-China oil pipeline

PetroChina, the flagship of state-owned CNPC, said that it expected the 20-mtpy crude oil pipeline being laid from Kazakhstan to China to start operation in 2010.

The pipeline, to span 3,000 km (1,864-mi) from the Caspian Sea to Xinjiang, China, has been under construction since the late 1990s.

The final stretch, planned for completion in third quarter 2009, will connect Kumkol and Kenkiyak, the end points of the first two stages of the pipeline.

The first stretch was completed in 2003, running from CNPC's oilfields in the Aktobe region to the oil hub of Atyrau on the Caspian Sea coast and on to Kenkiyak.

The second stretch, finished in 2005, runs from Kumkol through Atasu in northwestern Kazakhstan to the Alataw Pass at the Chinese border. From Alataw, the line carries oil to the Karamay refinery and the Dushanzi Petrochemical Corp. in the Xinjiang Autonomous Region.

Another gas line to Russia

Kazakhstan approved plans in May 2009 for construction of an additional gas pipeline to Russia that will bolster the control of Gazprom over central Asian gas exports.

The 1,600-km (994-mi) pipeline will carry 20 bcm/year (706 bcf/year) of gas from Turkmenistan and Kazakhstan north to Russia.

The EU has urged Kazakhstan and Turkmenistan to join the planned Nabucco pipeline project to bring Caspian and central Asian gas to the west without crossing Russia.

However, both countries refused to commit to Nabucco at a summit in Prague in May 2009.

At that summit, Azerbaijan, Turkey and Greece signed an agreement to open the so-called "southern gas corridor," reducing European dependence on Russian gas.

Refining and Products

Two refinery projects delayed

In early 2009, KazMunaiGas said it would not proceed with construction of refineries in Ceyhan and Batumi. The firm cited concerns about economic feasibility and the global financial crisis.

The company said that, instead, it would enhance production from three other refineries and increase national energy output.

The proposed 428,000-b/d Ceyhan refinery, which was to be built near the Ceyhan port in Turkey, was to begin construction in early 2009 and start operations by 2012. It would have become a center of crude delivery with three key pipeline routes passing through this region.

The Batumi refinery was to be located on the Georgian Black Sea coast. The refinery was conceived in early 2007 and would have processed 100,000–150,000 b/d.

Company News

KNOC-Sumbe deal

In December 2009, Korea National Oil Corp. acquired a controlling interest in Kazakhstan's Sumbe for about $335 million.

KNOC will buy an 85% stake while an unnamed Kazakh partner will take the remainder, subject to Kazakhstan government approval.

KNOC said it would continue its hunt for overseas oil assets in 2010 to expand its production capacity by six times, to 300,000 b/d, by 2012.

Sumbe has two oilfields in western Kazakhstan, with one under test production and the other still under exploration. The Kazakh oil company is expected to produce up to 20,000 b/d by 2014.

Eni-KazMunaiGas deal

In late 2009, Eni SpA and KazMunaiGas signed a preliminary agreement that could lead to as much as $50 billion in investments in Kazakhstan's oil and gas sectors.

Eni said it expected to make agreements with future investors by mid-2011.

The four-part agreement includes the following elements:

- Construction of a gas sweetening plant in Kazakhstan.
- Upgrading Kazakhstan's aging Pavlodar refinery
- Construction of a shipyard on the Caspian
- Joint Eni-KazMunaiGas exploration of the Isatay and Shagala oil blocks in the Caspian.

CIC-KazMunaiGas deal

China Investment Corp., the country's sovereign wealth fund, bought an 11% stake in late 2009 in the exploration and production arm of KazMunaiGas.

China has been increasing investment in central Asia in recent years, particularly in Kazakhstan, with which it shares a common border.

KYRGYZSTAN

Capital: Bishkek
Monetary unit: Som
Refining capacity: 10,000 b/cd
Oil production: 1,000 b/d
Oil reserves: 40.000 million bbl
Gas reserves: 200.000 bcf

Soh license

Swiss firm Manas Petroleum Corp. finished in third quarter 2009 the second of two wells that fulfill its 2009 commitments in its priority exploration areas in Kyrgyzstan.

Huday Nazar SPC-1 on the Soh license went to a TD of 2,213 m (7,260 ft) and penetrated a thick Jurassic coal and sandstone sequence in the deeper section. Evaluation was ongoing.

Assessment of shallower hydrocarbon anomalies in the Paleogene section suggests confirmation of the geological model but are not currently commercial.

Manas was evaluating data to select a future drilling location to gain commercial feasibility.

Previous evaluation showed that a potentially significant Jurassic oil-gas-condensate play is clearly present in the Soh permit in addition to the oil-dominated Paleogene play.

LATVIA

Capital: Riga
Monetary unit: Lats
Refining capacity: 0
Oil production: 0
Oil reserves: 0
Gas reserves: 0

Belarus oil cut-off

In July 2009, Belarus shut off the pipeline that carries over 10 mtpy of oil to Latvia from Russia.

Officials in Belarus said the shut-off was due to the poor condition of the pipe, claiming that they found more than a thousand defects and insisting that there had been leaks in recent years.

Daugavpils refinery

Latvia's Euro Oil Refinery said in September 2009 that it will start construction of an oil refinery in Laucesa, in Daugavpils district, near the Russian border, according to local media reports.

It will take about 3 years to build the €2-billion (US$2.86-billion) plant.

The plant will process 7.5 mtpy of oil products, including gasoline, diesel, and aviation fuel.

LITHUANIA

Capital: Vilnius
Monetary unit: Litas
Refining capacity: 190,000 b/cd
Oil production: 2,000 b/d
Oil reserves: 12.000 million bbl
Gas reserves: 0

Visaginas nuclear plant

In May 2009, the Ministry of Environment approved plans to build the $9 billion Visaginas nuclear plant near the site of the ageing Ignalina nuclear facility.

The new 3,400-MW plant will end the "energy island" status of the Baltic States and reduce their dependence on power generated in Russia.

Visaginas—a JV among Lithuania, Estonia and Latvia—will not come on line until 2018 at the earliest. Likely contenders to build it are Vattenfall AB, RWE AG, E.ON AG, and Electricite de France SA (EPA:

The first of the two Russian reactors at the troubled 35-year-old Ignalina plant was shut down in 2004, and the second unit was to be decommissioned by yearend 2009. Promise of closure of Ignalina was a condition of Lithuania's entry into the European Union in 2004.

A facility to store the nuclear fuel from Ignalina is not expected to be ready for use until mid-2011.

To help meet power needs until Visaginas is built, work began in early 2009 on an undersea cable between Lithuania and Sweden, to connect the Baltic States to the Scandinavian power grid by 2016.

Another link is planned with Poland.

Gazprom enters U.S. gas trade

In September 2009, Gazprom Marketing & Trading USA Inc. entered the U.S. market to begin trading and marketing of natural gas there.

Details are given in the United States section of the North America chapter of this book.

RUSSIA

Capital: Moscow
Monetary unit: Ruble
Refining capacity: 5.429 million b/cd
Oil production: 9.915 million b/d
Oil reserves: 60.000 billion bbl
Gas reserves: 1,680.000 tcf

Overview

Economic 'perfect storm'

The Russian oil industry faced a "perfect storm" of difficulties in early 2009, with simultaneous, sudden low oil prices, high taxes and tariffs, the devaluation of the ruble, and the global financial crisis.

The country also faces a shortage of skilled oil-industry workers as a result of structural, economic, and demographic changes after the breakup of the Soviet Union.

Russia, EU warning pact

Russia's Energy Ministry and the EU Energy Commission agreed in December 2009 to create an "early warning" mechanism regarding potential future disruptions of Russian energy supplies to the EU.

It covers gas, oil, and electricity.

Russian officials also said they would sign a new agreement with Ukraine for bilateral cooperation and also indicated Russia would take part in upgrading Ukraine's pipelines.

Energy targets

In October 2009, Russian officials described a new national energy strategy to guide energy development and use. They said it reflects the prevailing global economic scenario.

Targets for 2030 include:

- Oil production: 3.89–3.93 billion bbl
- Natural gas production: 880–940 bcm (31.1–33.2 tcf)
- Energy development investment: about $1.9 trillion
- Oil exports: 2.4 billion bbl/year
- Gas exports: 349–369 bcm (12.3–13.0 tcf)
- Portion of LNG production exported: 14–15%

In 2008, Russia produced 3.59 billion bbl of oil, 665 bcm (23.5 tcf) of natural gas and 1.02 trillion kWh of electricity. Exports in 2008 amounted to 1.8 billion bbl of oil and 174.3 bcm (6.15 tcf) of gas.

In 2009, crude production is expected to be 3.54 billion bbl, with gas production of 620–644 bcm (21.9–22.7 tcf).

Yamal gas projects

Russian officials and chief executives from 10 global energy companies met in September 2009 on the Yamal Peninsula, inside the Arctic Circle. Prime Minister Vladimir Putin urged the firms to help Gazprom develop new gas fields in the region.

A spokesman said that a plan for the exploration and development of the Yamal region, with untapped reserves of up to 12 tcm (424 tcf) of proven reserves of natural gas, could be ready by early 2010.

Gazprom had cut its investment program for 2009 by 17% (to $25.4 billion), and energy initiatives on the Yamal Peninsula were affected.

Start-up of Bovanenkovskoye, a gas field estimated to hold 4.9 tcm (173 tcf), was put off until 2012.

Gas demand

Inside Russia, gas demand fell sharply in 2009 as industrial output declined 14% in the first seven months, compared with 2008.

Russia is expected to recover more slowly from the global economic crisis than is Europe, with officials saying the country will not reach pre-crisis levels of growth until at least 2012.

Gazprom says it will trim investments in new fields until European demand improves. The firm's gas production is expected to be 450–510 bcm (15.9–18 tcf) in 2009, compared to 550 bcm (19.4 tcf) in 2008.

Gazprom said that its priority projects—including the Nord Stream pipeline and development of the Shtokman field—remain on schedule.

RUSSIAN NATURAL GAS CONSUMPTION

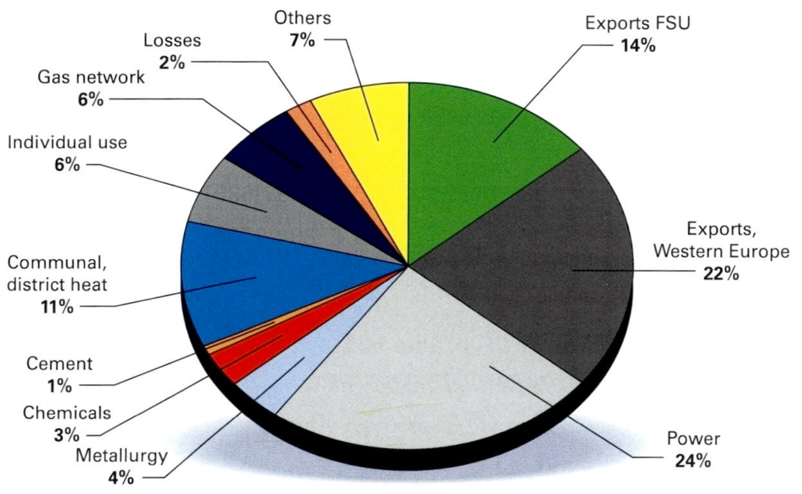

Source: Center for Energy Economics, http://www.beg.utexas.edu.energyecon/documents/RUSSIAN_NATURAL_GAS_FINAL.pdf

Exploration and Development

Kamchatka Peninsula

In late 2009, PetroKamchatka PLC was drilling its first exploration well on the Tigil block on the Kamchatka Peninsula.

The firm had secured seven onshore exploration licenses and identified numerous prospects and leads on the Tigil and Icha blocks.

Lagansky Block

Lundin Petroleum AB, Stockholm spudded the Petrovskaya-1 exploratory well in October 2009 on the Lagansky Block in the northern Caspian Sea off Russia.

The well is adjacent to the northwest of the company's giant Morskaya oil discovery in shallow water. The area is just shoreward from Korchagin field, which Lukoil aims to place on production in early 2010.

Petrovskaya-1 is to go to 2,100 m (6,890 ft), targeting Lower Cretaceous sandstones. Gross unrisked prospective resource is estimated at 300 MMboe.

Verkhnechonskoye field

In August 2009, an affiliate of TNK-BP let a contract for FEED work on a production facility at giant Verkhnechonskoye oil field in Eastern Siberia.

The contract covers a single facility to handle 140,000 b/d of crude oil to be tied into a new 85-km (53-mi) link to the Eastern Siberian Pacific Ocean (ESPO) pipeline.

The remote field, 1,100 km (984 mi) north of the regional capital of Irkutsk, holds reserves of about 1.4 million bbl.

Development cost is estimated at $4–5 billion, according to TNK-BP.

Early production of 20,000–30,000 b/d was anticipated by yearend 2009.

Yuri Korchagin field

Lukoil was poised at yearend 2009 to bring its Yuri Korchagin oil field on stream in the Caspian Sea following installation of offshore facilities, including a fixed offshore ice-resistant drilling platform.

It will be the first Lukoil North Caspian field in the Russian sector to be developed.

The field is 180 km (112 mi) from Astrakhan and 240 km (149 mi) from Makhachkala.

Able to reach 7,400 m (24,278 ft), the complex will drill 26 production wells.

The 3P field reserves are estimated at 570 MMboe, with a maximum production rate of oil and gas condensate at 2.3 mtpy and 1.2 bcm/year (42.4 bcf/year) of gas.

Barents Sea shelf

An assessment published in mid-2009 by the U.S. Geological Survey estimated that more than 76 billion boe may be technically recoverable on the Barents Sea shelf, off Russia and northern Norway.

Hydrates potential: Lake Baikal

In information publicized in mid-2009, Russian scientists said that the unexplored floor of Lake Baikal—the world's deepest at 1,680 m (5,512 ft)—holds significant deposits of clathrate hydrate, crystals containing natural gas.

The deposits are said to contain the equivalent of more than 1 tcm (35 tcf) of natural gas.

At present, mineral extraction is banned in the Lake Baikal region, which is a nature reserve.

Shtokman field

The massive Shtokman offshore gas and condensate field will become the main source of LNG exported by Russia to markets in the Atlantic Basin.

The field lies about 600 km (373 mi) northeast of Murmansk in the central Barents Sea shelf.

Proven reserves include 3.8 tcm (134 tcf) of gas and 37 million tons (271 million bbl) of gas condensate.

The stakeholders in the Shtokman Development Co. are Gazprom, Total SA, and Statoil ASA.

Details are given in the Norway section of the Europe chapter of this book.

Total-Novatek deal

In mid-2009, Russia approved an exploration JV deal between Total SA and Novatek, Russia's largest independent gas producer. The JV will develop the Termokarstovoye gas field in the Yamal-Nenets region of the Siberian Arctic.

Details are given in the France section of the Europe chapter of this book.

Arctic cooperation

In mid-2009, Gazprom and Statoil agreed to cooperate on exploration, development, and production operations in the Arctic areas of Russia and Norway.

Details are given in the Norway section of the Europe chapter.

Shtokman: Phase 1

Shtokman Development AG was poised in May 2009 to issue tenders for developing the giant Shtokman gas field (see box). Bids were to be returned by yearend.

The company plans to make an investment decision on Phase 1 in early 2010, which will produce 68 million cu m/d (2.4 bcfd) of gas from 20 wells. The gas will flow to a planned 7.5-mtpy LNG plant at Teriberka in the Murmansk Region.

Another portion will enter the North Stream pipeline from Teriberka to Vyborg.

Gazprom expects gas delivery to start in 2013, with LNG exports beginning in 2014.

Growing Arctic presence

According to an Associated Press report, an envoy to Pres. Dmitry Medvedev said in early 2009 that the country was making plans to modernize its icebreaker fleet, build a new research vessel, and station more researchers in the Arctic.

The Arctic could hold as much as 25% of the world's undiscovered oil and gas.

Russia also was reported to be preparing to resubmit its claim to the UN that an underwater mountain range crossing the polar region is part of its continental shelf.

Policy to stimulate E&P

In early 2009, the Ministry of Natural Resources and Environment noted a four-fold increase since 2004 in the exploration expenses of private Russian oil and gas companies.

Coupled with the sharp drop in crude oil prices and expansion of the global financial crisis, this reduced the attractiveness of exploration. However, the ministry also blamed an unfavorable fiscal system and the absence of stimulus to encourage exploration activities.

As a result, the ministry said it was developing economic measures to stimulate activities in the industry. Proposed initiatives include:

- Tax credits or deductions for exploration expenses against the mineral extraction tax (MET).
- Allowing regions to collect a share of auction revenues for areas explored at their own expense.

Drilling and Production

Samotlor field

In early 2009, TNK-BP said that western Siberia's super-giant Samotlor field should continue to produce until 2099. Discovered in 1965 in the Khanty-Mansi Autonomous area, Samotlor has produced 19.2 billion bbl of oil since development began in 1969.

The latest study found remaining recoverable volumes of more than 7 billion bbl of oil and 3.53 tcf of gas. TNK-BP said that Samotlor is one of the world's five largest oil fields.

Samotlor has 13,400 oil wells and 4,500 injection wells. in April 2009, it was producing 582,750 b/d.

TNK-BP said it aims to keep production at or slightly above this level at least until 2012. It plans to invest $1 billion/year in the field through 2011.

Urna and Ust-Tegus fields

TNK-BP reported it in February 2009 that it had started commercial production at Urna and Ust-Tegus fields in the eastern Uvat area of the Tyumen region in Siberia.

The fields hold an estimated 300 million tonnes (2.20 billion bbl) of oil in place, including 100 million tonnes (733 million bbl) of reserves.

Plans call for the production of 1.5 million tonnes (11 million bbl) of crude from the fields in 2009.

TNK-BP said it had invested $925 million to date in the Uvat project.

Russian Oil, Gas Exploration Expenses

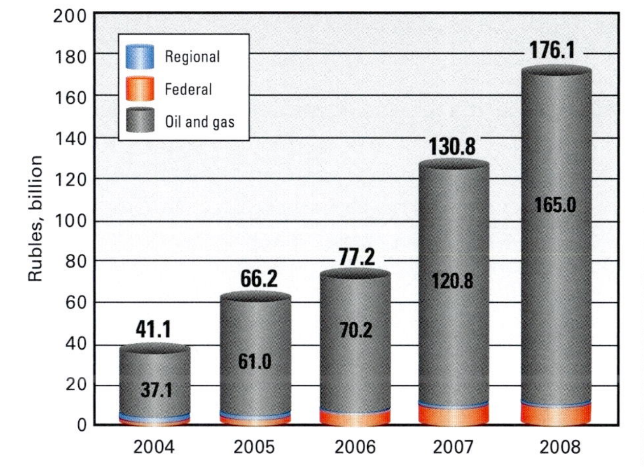

Russian Oil Production

	1,000 b/d
2009	9,915*
2008	9,750
2007	9,830
2006	9,498
2005	9,190
2004	8,887
2003	8,216
2002	7,405
2001	6,781
2000	6,325

*Estimate.

Oil Production at Supergiant Samotlor Field

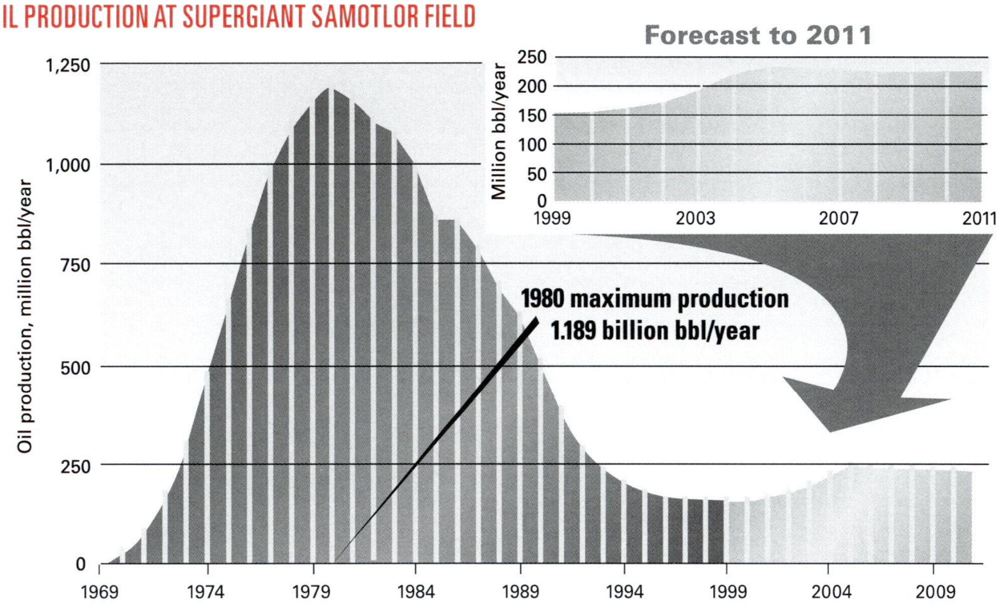

Source: TNK-BP

Lunskoye field

In January 2009, Sakhalin Energy reported production of first gas from the Lunskoye-A (Lun-A) platform in the Lunskoye field off northeastern Sakhalin Island.

Gas flows to an onshore processing facility and then to the Sakhalin-2 LNG plant in the south of the island. Lun-A was the third of three Sakhalin-2 platforms to be brought onstream.

First gas from Lunskoye field was used in March to finish commissioning the LNG plant.

Oil Production Record

In its November 2009 *Short-Term Energy Outlook*, the Energy Information Administration of the U.S. Dept. of Energy reported that Russia produced more than 10 million b/d of oil during August 2009, a new post-Soviet-era record.

International Activities

Gas accord with Turkmenistan

In late 2009, Russia and Turkmenistan signed amendments to an existing long-term gas purchase and sales contract, and Gazprom also settled an 8-month gas trade dispute.

Details are given in the Turkmenistan section of this chapter.

Venezuela project

In September 2009, a consortium of Russian companies signed an agreement with Venezuela's PDVSA to develop Block Junín 6 in the latter country's Orinoco Belt.

Details are given in the Venezuela section of the Latin America chapter of this book.

Accord with Nicaragua

The Nicaraguan government signed an agreement in August 2009 allowing exploration of land and offshore areas (Caribbean and Pacific) by a Russian consortium.

Details are given in the Nicaragua section of the Latin America chapter of this book.

Gas deal with Azerbaijan

Under a contract signed in mid-2009, Gazprom will import 0.5 bcm/year (17.7 bcf/year) of gas from Azerbaijan, through state-operated SOCAR, starting in January 2010.

Details are given in the Azerbaijan section of this chapter.

Sakhalin Island Project

Sakhalin-1

Gas delivery

In October 2009, the destination of gas produced from the Sakhalin-1 project was a subject of active discussion, with shareholders urging export of the gas to China.

However, Gazprom officials said that industrial development plans for the Far East and eastern Siberia would require that Sakhalin-1 gas be directed to domestic markets.

Sakhalin-2

First LNG plant

In February 2009, Russia inaugurated the $22-billion Sakhalin-2 project, one of the largest integrated oil and gas projects in the world. LNG production—the first for the project—began in March.

Natural gas will be transported from three offshore platforms to the Prigorodnoye liquefaction plant for the production of 9.6 mtpy of LNG, about 65% of which will go to Japan.

The project expected to send 50 LNG tankers in 2009 to Japan, South Korea and potentially to North America. Sakhalin LNG will be the first Russian gas supplied to these regions.

Officials said Sakhalin-2 could supply 5% of global LNG when it achieves full capacity in 2010.

First LNG export cargo

Sakhalin-2 sent its first LNG cargo in April 2009 to Japan, through the all-weather Prigorodnoye port.

The first train at the 9.6-mtpy liquefaction complex had begun production, and the second was slated to come on stream late in the year. Ramp-up to full production is expected in 2010.

Shareholders in Sakhalin Energy are Gazprom, Royal Dutch Shell PLC, Mitsui & Co. Ltd., and Mitsubishi Corp.

Lunskoye platform

In May 2009, Royal Dutch Shell was completing the last major offshore gas well at the Lunskoye platform, preparing the way for full LNG production capacity of Sakhalin-2.

Completion of the final gas well will facilitate start-up of its second LNG unit.

Sakhalin-3

Status report

The Sakhalin-3 project includes the Veninsky Block—being developed by Rosneft and Sinopec—and the Kirinsky, East Odoptinsky, and Ayashsky blocks, which in early 2009 had not been assigned.

In October, however, Gazprom did say that it plans to start production at Kirinsky in 2011 or 2012, not in 2014 as first planned. Kirinsky has gas reserves estimated at 75–100 bcm (2.65–3.53 tcf).

The development schedule will depend in part on progress on the SKV gas pipeline (see below).

According to Japanese press reports, Gazprom said that it likely will partner with Mitsui & Co. and Mitsubishi Corp. on the Sakhalin-3 project. Royal Dutch Shell, Mitsui and Mitsubishi already hold stakes in the Sakhalin-2 project.

Earlier reports said that Gazprom also invited Shell to participate in both Sakhalin-3 and -4.

Sakhalin-3 is expected to come on stream in 2017–2020, with peak production of 28.6 bcm/year (1.0 tcf/year).

Sakhalin pipelines

SKV gas pipeline

A pipeline already connects gas fields off Sakhalin Island to Khabarovsk on the Russian mainland.

However, Gazprom started construction in mid-2009 on a project to extend the line southwest from Khabarovsk to Vladivostok on the Pacific coast.

The $11-billion Sakhalin-Khabarovsk-Vladivostok (SKV) line is expected to be operational by 2012.

The government was seeking help from Japan to build the Khabarovsk-to-Vladivostok segment as well as an LNG export terminal at Vladivostok. Japan is interested in obtaining gas from Sakhalin-1.

Planned initial capacity of the SKV line is 7 bcm/year (247 bcf/year).

Gas input to the pipeline likely will come from the Sakhalin-1gas fields, operated by ExxonMobil. Those offshore fields—Oduptu, Arkutu, and Chayvoo—hold proven reserves of 485 bcm (17.1 tcf).

SKV: Extension to S. Korea

Gazprom and Korea Gas Corp. (Kogas) signed an MOU in mid-2009 to consider supplying Russian natural gas to South Korea by extending the SKV pipeline.

Kogas already gets 1.6 mtpy of LNG from the Sakhalin-2 project, and under terms of a 2008 agreement with Gazprom, will import 10 bcm/year (353 bcf/year) of Russian gas during 2015–2045.

Both overland and subsea routes for the 150-km (93-mi) extension were under study.

Gazprom in Nigeria

Gazprom increased its presence in Nigeria in 2009, raising some concern among European groups and governments dependent on Russian gas.

Details are given in the Nigeria section of the Africa chapter of this book.

Agreements with Namibia

Mid-2009 saw the first visit ever of a Russian president, Dmitry Medvedev, to Namibia.

The two nations signed several agreements related to petroleum products, fisheries and trade.

Taweelah-Fujairah gas pipeline

Gazprom's construction subsidiary started work in early 2009 on the $418-million Taweelah-Fujairah natural gas pipeline (TFP) project in the UAE.

Details are given in the UAE section of the Middle East chapter of this book.

OPEC relationship

Russia proposed in March 2009 sending a permanent envoy to the Secretariat of OPEC in an effort to coordinate policies, but again declined membership in the group.

Russian officials said that they want to exchange data and ideas. But they said that official membership would depend on formalizing "certain agreements" with the cartel.

Agreements with Gas Natural

Gazprom signed an MOU in early 2009 with Spain's Gas Natural SDG SA for collaboration on LNG marketing and for cooperation in areas such as CO_2 emissions trading, natural gas trading and power generation.

They also entered into an agreement on supply of gas by Russia to markets in which Gas Natural operates.

Gazprom said that volumes of gas supply would rise once Russia's Shtokman gas and condensate field development project is completed in 2014.

China deal: Oil for loans

In February 2009, Russia agreed to a $25-billion oil-for-loans deal with China.

Russia is to supply China with 30 million tonnes (220 million bbl) of crude from east Siberia over 20 years in return for loans to state-controlled oil-sector firms Transneft and Rosneft.

The Transneft loan supports construction of the ESPO oil pipeline (see below).

Pact with Serbia

In early 2009, Gazprom said that it will invest $1–2.5 billion in Serbia's oil and gas sector as part of an energy agreement.

The agreement resulted from Gazprom's purchase of a 51% stake in Serbia's largest oil and gas company, Naftna Industrija Sribje (NIS). Gazprom also will:

- enhance the technology and operations of refineries in Serbia
- build the 400-km (249-mi) Serbian section on the South Stream pipeline
- construct an underground storage depot at Banatski Dvor.

LNG

Yamal LNG Project

In mid-2009, Gazprom announced that Total SA and Royal Dutch Shell PLC were on the short list of candidates to participate in the Yamal LNG project.

That project would develop gas fields (such as Tambeiskaya) on the Yamal Peninsula and in offshore waters of the Kara Sea. Those fields hold an estimated 310–360 bcm (10.9–12.7 tcf) of natural gas.

A regional development plan envisions gas production beginning in 2030, and Gazprom is considering building an LNG plant to enable efficient shipment to market.

Yamal Peninsula

The Yamal Peninsula has 11 gas and 15 oil and condensate fields with an estimated 16 tcm (565 tcf) of explored gas reserves and about 22 tcm (777 tcf) of predicted gas reserves.

Condensate and oil reserves have been estimated at 230.7 million tons (1.69 trillion bbl) and 291.8 million tons (2.14 trillion bbl), respectively.

The Bovanenkovskoye field with 4.9 tcm (173 tcf) of gas is the most significant field on the peninsula. Gas production from the field is expected to start in 2011.

LNG to Lithuania?

In September 2009, Gazprom said it was prepared to supply LNG to Lithuania if a regasification terminal could be built at the Baltic port of Klaipeda.

The U.S. Trade and Development Agency said it would provide funds to study the feasibility of such a terminal at three potential sites. Completion of the study is expected in 2010.

Pipelines/Transportation

Nord Stream Pipeline

In December 2009, Russia and Germany granted permits that will allow the Nord Stream Pipeline to be routed through the two countries. The other needed permits were granted by Denmark in October and by Finland and Sweden in November.

Details are given in the Finland section of the Europe chapter of this book.

EUROPEAN DEPENDENCE ON RUSSIAN GAS

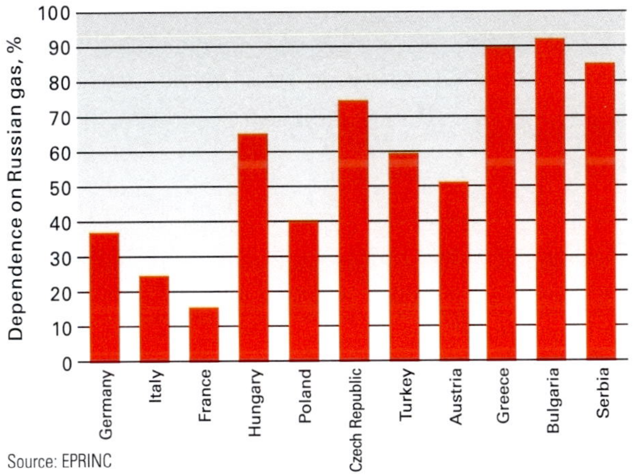

Source: EPRINC

ESPO Pipeline

Progress was made on several fronts on the massive Transneft project to construct the 4,720-km (2,933-mi) East Siberia-Pacific Ocean (ESPO) oil pipeline. When completed, the line will have the capacity to carry 80 mtpy (586 million bbl/year).

Stage 1

In December, Transneft completed the first 2,757-km (1,713-mi) segment from Taishet in the Irkutsk region to Skovorodino near the Chinese border. The $12.1-billion line can carry 30 mtpy (220 million bbl/year) of oil.

Also completed in December was a $2-billion export terminal at Kozmino, on Russia's Pacific coast.

Beginning in November, pipeline oil was loaded at Skovorodino onto railcars and carried to Kozmino. The rail line will be phased out when Stage 2 of the pipeline from Skovorodino to Kozmino is completed.

In December, Prime Minister Vladimir Putin inaugurated ESPO and shipment of its first oil to China.

Spur to China

About half of the oil flowing from Taishet to Skovorodino is earmarked for China.

Transneft said that construction of the 67-km (42-mi) spur from Skovorodino to Mohe at the Chinese border began in April, would be completed by September, and would begin operation by early 2011.

After completion of the spur, China will build a 960-km (597-mi) link from Mohe to Daquing.

Stage 2

Construction was to begin by late 2009 on the $10-billion second stage of the ESPO pipeline, spanning 1,963 km (1,220 mi) from Skovorodino to Kozmino.

When operational in 2015, it will deliver an additional 50 mtpy (367 million bbl/year) of oil to Kozmino.

Sources of oil for ESPO

Oil for ESPO shipment will come from fields in Tomsk Oblast and Khanty-Mansi Autonomous Okrug. The main source will be the Vankor field, from which production began in August 2009.

The line also will link to the existing Omstk-Irkutsk pipeline in Taishet to source crude from fields in East Siberia. One of those fields is East Talakan, which holds 124 million tonnes (909 million bbl).

Russian authorities said that oil for the ESPO line is to be branded Vsto, and will be light and medium-sour, superior to Urals export blend but inferior to Siberian Light.

Gas line to parallel ESPO?

In early 2009, Russian officials said that a natural gas pipeline might be constructed parallel to the ESPO oil pipeline. However, the plan was described as being in its very early stages of development.

Major FSU gas export basins

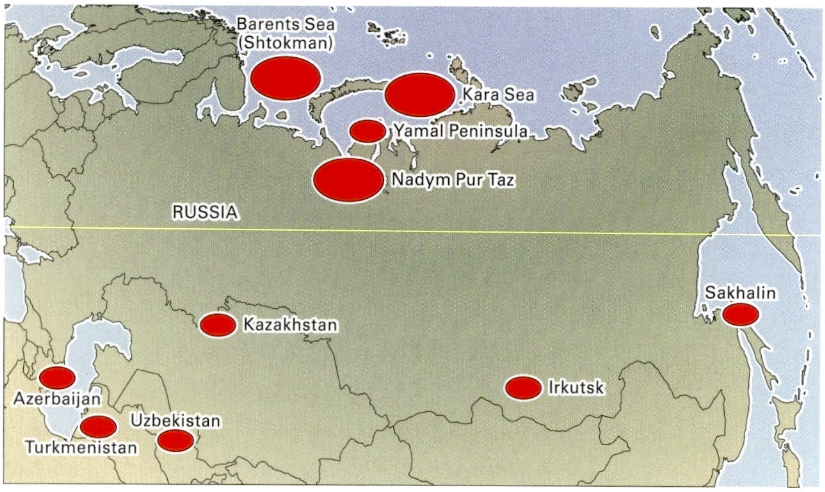

Kaliningrad gas-line segment

In third quarter 2009, Gazprom finished building the second stretch of the 139-km (86-mi) Minsk-Vilnius-Kaunas-Kaliningrad gas pipeline.

The line will help supply 2.5 bcm (88.3 bcf) of gas in 2010 to the Kaliningrad enclave, between Lithuania and Poland.

Gas storage in The Netherlands

In August 2009, Gazprom agreed to make use of a gas storage facility in The Netherlands that is being developed by a consortium led by the Abu Dhabi National Energy Co. (TAQA).

Details are given in the Netherlands section of the Europe chapter of this book.

Oil exports

In a mid-2009 evaluation, Energy Security Analysis Inc. predicted an increase over the next decade of more than 2 million b/d in light sweet crude exports from Russia.

Much of the growth in new oil production, from such areas as East Siberia and Timan-Pechora, is expected to be light and sweet, while output of medium sour will decline.

Expected changes to infrastructure will enable more light sweet to reach the market without commingling with medium sour crude.

Iran pipeline talks

Gazprom was talking in mid-2009 with Iran and Pakistan about possibly building a large Pakistan section of the long-awaited Iran-Pakistan-India (IPI) gas pipeline.

The project has been delayed because of disagreements that led India to pull out.

Details about the IPI are given in the Iran section of the Middle East chapter of this book.

Gazprom: Key role in Europe

An analysis published in early 2009 by the Center for Energy Economics at the University of Texas-Austin concluded that Gazprom's role in supplying European gas users will continue to be vital.

Researchers said that increased European LNG imports and planned pipeline projects are unlikely to provide timely gas supplies, and that continued cooperation with Gazprom will be mutually beneficial.

They characterized the EU as Gazprom's largest and best-paying customer.

Refining and Products

Refinery modernization

Aiming to boost refinery yields to more than 90% of installed capacity, Lukoil was laying plans in third quarter 2009 for a 10-year, $20-billion upgrade of six refineries: two overseas, four in Russia.

In the same time frame, the company said it also will make process improvements to comply fully with Euro-5 environmental standards.

The six refineries are in Volvograd, Perm, Nizhny Novgorod, and Ukhta in Russia; the Burgas refinery in Bulgaria; and the Odessa refinery in Ukraine.

Russian crude oil exports in 2020

Gazprom natural gas exports

	Gazprom exports; 2005-07, average	Projected Gazprom exports, 2020
	Billion cu m/year	
EU-27	142	[1]124–170
FSU	93	93
Non-EU Europe	20	20
Sakhalin II LNG, pipeline to China	0	39
Shtokman LNG	0	11
Total	**255**	**[2]287-332**

[1]Center for Energy Economics calculations. [2]"Concept of Long-Term Socio-Economic Development of the Russian Federation," Ministry for Economic Development of the Russian Federation, August 2008 (in Russian).

Projected crude balance and exports*

	Supply	Russia/CIS demand	Russia surplus	Transit volumes	Total exports
	Million b/d				
2008	9.8	5.4	4.1	1.0	5.1
2010	9.9	5.3	4.2	1.0	5.3
2015	10.1	4.9	4.8	1.3	6.1
2020	10.1	4.6	5.2	1.4	6.6

*In reported Russian data the amount from oil exported is consistently about 350,000 b/d less than the volume implied in the supply-demand balance. ESAI has made a downward adjustment to the Russia surplus in this table to reflect the statistical difference and more accurately predict actual exports.

Actual, projected crude exports to Europe[1]

Crude type Crude quality (API gravity, sulfur content)	Urals 31.8°, 1.35%	CPC blend 43.3°, 0.6%	Other light sweet
	1,000 b/d		
2008	3,864	634	130
[2]2020	3,320	1,340	420

[1]Includes estimated transit volumes from Kazakhstan and Azerbaijan. Figures exclude deliveries to CIS refineries. [2]Estimated.

Actual, projected crude exports to Asia

Crude type	Light sweet (ESPO blend and Sakhalin)	Other (includes medium sour and offshore output of undetermined quality)
	1,000 b/d	
2008	212	182
2020*	1,335	200

*Estimated.

European gas market, Gazprom's share

	Total imports, 2006	Total imports needed, 2020	EU-27 demand, 2020	EU-27 production, 2020	EU-27 projected shortfall, 2020
	Billion cu m/year				
EU-27	[1]396	[2]490	[3]688	[3]198	[3]151
Gazprom's share	[2]51%	[2]25-35%			

[1]European Commission, "Energy. Statistical Yearbook 2006." [2]CEE calculations. [3]"Natural Gas Demand and Supply. Long Term Outlook to 2030," EuroGas, 2007.

Vlissingen refinery deal

Total SA, which already owned 55% of the Vlissingen refinery in the Netherlands, acquired the remaining 45% share in the facility from Dow Chemical in mid-2009. No price was given.

Total then entered an agreement to sell the stake to Lukoil.

It said the sale to Lukoil was contingent on acceptance of the deal by the European Commission's competition regulators.

Lukoil supplies around 30% of the crude oil for Total's West European refineries.

Refinery to process ESPO crude

Rosneft Oil Co. announced in early 2009 its plans to develop a $14-billion petrochemical refinery in the Primorye region off the Pacific Coast. Construction was to start in late 2009.

Crude for the project will be supplied from the ESPO pipeline (see above).

The refinery will be built in two phases, with the first phase scheduled for commissioning in 2012, with the capacity to process 47 MMbbl/year (20 mtpy).

Petrochemicals

Tobolsk project

A consortium of Russian financiers and Italian engineering firms began construction in third quarter 2009 of a major petrochemical plant in Tobolsk.

Work will be conducted in two phases, with commissioning of the first phase (including a 500,000-tpy polypropylene unit) scheduled for mid-2012.

In the second phase, a 500,000-tpy polyethylene plant is to be built by late 2014.

Primorsky plant

In late 2009, Rosneft was reported to be talking with China's Sinopec about a potential jointly funded refining and petrochemical complex in the Primorsky region.

The $22-million complex, to be built by 2014, would process 20 mtpy (147 million bbl/year) of crude oil: 15 mtpy (110 million bbl/year) for refining and 5 mtpy (37 million bbl/year) for petrochemicals production.

The two firms have worked together since 2007 to develop oil and gas off Sakhalin Island.

Russia produces only about 2% of the world's petrochemical output, but is said to be planning to add capacity in the Far East region, aiming for export to Japan, South Korea and Vietnam.

Company News

Yukos ruling

In November 2009, the owners of the Yukos oil group claimed victory in a ruling that will let them seek up to $100 billion in damages from the Russian government for the state's takeover of the company.

An international tribunal in The Hague concluded that Moscow must abide by the terms of the Energy Charter Treaty, even though the Russian parliament has yet to ratify it.

Group Menatep, the Yukos holding company, can proceed with its plan to sue Moscow in the tribunal.

The company claims that Yukos was expropriated through a series of bankruptcies and asset sales.

Yukos was broken up in 2004 and sold for $33 billion in disputed back taxes.

Gazprom-SeverEnergia

Italy's energy companies Eni SpA and Enel SpA said in mid-2009 that they will assign to Gazprom a 51% stake in SeverEnergia, a Russian holding company with exploration and production licenses.

Details are given in the Italy section of the Europe chapter of this book.

Stake in MOL

Russian oil producer Surgutneftegaz bought a 21.2% stake in Hungary's key oil and gas group, MOL, in March 2009.

Selling the €1.4-billion stake was Austria's OMV AG, which abandoned a two-year takeover bid for the Hungarian group.

Observers saw the move as a sign of Russia's intent to strengthen its position in the EU energy market.

Gazprom's struggles

The dramatic fall in gas prices from mid-2008 to mid-2009 depressed revenues and disrupted the investment plans of Gazprom.

In June 2008, the firm had a market capitalization of $330 billion and was the world's third-largest company. As of April 2009, it was worth less than $90 billion and ranked thirty-seventh.

The company said in June 2009 that it expects an 11% drop in exports for the year to 142.1 bcm (5.0 tcf) and a 38% drop in export revenues to $40 billion.

It does not expect a return to 2008 production levels until after 2012.

The company outlined a $26.3-billion spending plan for 2009, but said it would have to set priorities for its projects in the turbulent market.

Top production ventures in its portfolio are Zapolyarnoye, Bovanenkovskoye, Urengoiskoye, Shtokman, and Severo-Kamennomisskoye fields.

GAZPROM'S KEY 2008 STATISTICS

Gas reserves: 33.1 trillion cu m
Condensate reserves: 1.3 billion tonnes
Oil reserves: 1.6 billion tonnes
Gas production: 549.7 billion cu m
Condensate production: 10.9 million tonnes
Oil production: 32 million tonnes

Source: Gazprom

GAZPROM'S GAS REVENUE: DOMESTIC AND FOREIGN SALES

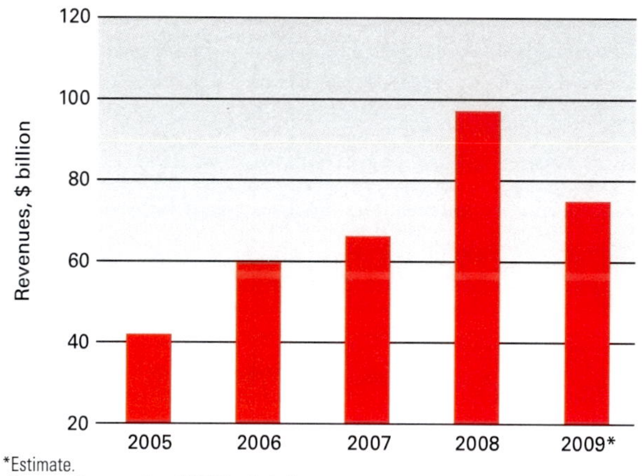

*Estimate.
Sources: Gazprom data, EPRINC calculations

GAZPROM PROPOSED PIPELINE PROJECTS

Planned pipeline	Estimated cost, $billion	Design capacity, billion cu m	Planned completion
Nord Stream	11.64	55	2011
South Stream	27.6–35	31–47	2013
Nabucco	11.5	31	2013–14

TAJIKISTAN

Capital: Dushanbe
Monetary unit: Ruble
Refining capacity: 0
Oil production: 0
Oil reserves: 12.000 million bbl
Gas reserves: 200.000 bcf

Khoja-Sartez field

In early 2009, Tethys Petroleum Ltd. began delivering gas to Kulob, Tajikistan, from Khoja-Sartez gas-condensate field under a 1-year contract.

Maximum initial volume was 2.3 MMcfd, through a 12-km (7.5 mi) pipeline.

Work to upgrade the gas delivery system was being considered.

The firm also was working to develop gas production from three existing gas wells in Komsomolsk for potential delivery to Dushanbe.

Bokhtar PSC area

Tethys Petroleum Ltd. assumed operation in early 2009 of the East Olimtoi-09 exploration well on the Bokhtar production sharing contract area.

Projected to reach 3,800 m (12,467 ft), the well will test a structure in the southeastern part of the PSC, just north of the Panj River. It will target oil and gas in the Lower Paleogene Bukhara and Alay formations that are productive nearby.

TURKMENISTAN

Capital: Ashgabat
Monetary unit: Manat
Refining capacity: 236,970 b/cd
Oil production: 220,000 b/d
Oil reserves: 600.000 million bbl
Gas reserves: 265.000 tcf

Overview

Turkmenistan is reported to have some of the world's largest gas reserves and is the second-largest gas producer, after Russia, among the Former Soviet Union countries.

According to government officials, the country produces 70–80 bcm/year (2.47–2.82 tcf/year) of gas and some 10 mtpy (73.3 million bbl/year) of oil.

Turkmenistan's gas export pipelines are controlled by Gazprom, but the government has said it is ready to bypass Russia and begin selling some of its natural gas to Europe.

Gurbanguly Berdymukhammedov, who became president in late 2006, strengthened links to Russia and Iran, and in 2009, was negotiating deals with Pakistan and India to deliver gas through the Turkmenistan-Afghanistan-Pakistan-India (TAPI) or Trans-Afghanistan pipeline.

Previous deals were made to sell 30 bcm/year (1.06 tcf/year) of Turkmen gas to China (2007); and to allow Germany's RWE AG to explore for and develop gas resources in Turkmenistan (2008).

Also in April 2008, Pres. Berdymukhammedov agreed to supply the EU with 10 bcm/year (353 bcf/year) of gas.

Several proposals for new multi-nation pipeline systems have been made in recent years, including Turkmenistan-China, Turkmenistan-Afghanistan-Pakistan, as well as Caspian and trans-Caspian routes.

Gas accord with Russia

In late 2009, Russia and Turkmenistan signed amendments to a long-term gas purchase and sales contract established in April 2003.

Under the new terms, the two countries will resume the purchase and supply of Turkmen gas starting in January 2010 at volumes of up to 30 bcm/year (1.06 tcf/year). Gas price will be governed by European gas market conditions.

The two nations also agreed to jointly implement projects related to construction of the Pre-Caspian gas pipeline and the East-West gas trunkline.

They also discussed opportunities for joint development of hydrocarbons on the Caspian Sea shelf.

Gas dispute with Russia

In late December 2009, Gazprom settled a months-long gas trade dispute with Turkmenistan.

The dispute began in April after a pipeline explosion halted Turkmen gas exports to Russia, choking off the country's main source of foreign revenue. Turkmenistan accused Gazprom of causing the blast, which Gazprom denied.

Facing a drop in Russian and European gas demand, Gazprom ordered Turkmenistan to renegotiate a 2008 contract to supply Russia with 50 bcm (1.77 tcf) of gas in 2009.

Turkmenistan refused, turning instead to China and Iran with offers of extra gas supplies.

It also stepped up talks with the EU to join the planned Nabucco pipeline project, which will bring Caspian and central Asian gas to Europe through the Caucasus and Turkey.

Under the late-2009 settlement, Russia agreed to buy up to 30 bcm/year (1.06 tcf/year) of gas from Turkmenistan starting in 2010.

Also, Gazprom said it would link untapped gas reserves in east Turkmenistan (such as the South Iolotan field) with a new pipeline running along the Caspian coast from central Asia to Russia.

New energy minister

Pres. Berdymukhamedov, following his dismissal of several key officials, appointed Oraznur Nurmiradov as Minister of Oil and Gas, Industry, and Mineral Resources in October 2009.

Those dismissed in the shake-up included the mineral resources minister, the head of the state gas company, and the chair of the state oil company.

Berdymukhamedov criticized the "poor construction quality" of Turkmenistan's energy installations, delays in providing gas consumers, and the absence of "needed measures to increase oil production."

Pipelines/Transportation

Turkmenistan-China gas line

In December 2009, leaders of Turkmenistan, Kazakhstan, Uzbekistan, and China met near the Turkmen-Uzbek border to inaugurate the new Turkmenistan-China natural gas pipeline.

The 1,800-km (1,118-mi) line begins at the Turkmen Bagtyyarlyk gas fields near the Amu Darya River, enters

Uzbekistan at Olot, then travels across southern Kazakhstan to Alashankou, in China's northwest Xinjiang Uyghur Autonomous Region.

Turkmenistan finished building its 188-km (117-mi) segment in November 2009.

At Alashankou, the line connects with the existing West-East Pipeline that carries gas to major Chinese cities such as Shanghai, Guangzhou, and Hong Kong.

The entire 7,000-km line also was dedicated in December 2009.

In addition to carrying gas from Turkmenistan and Uzbekistan, the $7.3-billion line also is to be supplied from Kazakhstan's Karachaganak, Tengiz and Kashagan gas fields.

China's CNPC previously signed a 30-year agreement for the supply of 30 bcm/year (1.06 bcf/year) of gas through the new line.

Turkmenistan gas to Iran

Turkmenistan was making plans to open a new pipeline in late 2009 to carry natural gas to Iran.

Turkmenistan usually sells most of its 78-bcm/year gas production to Gazprom, but halted supplies after a pipeline explosion in April disabled its main export link with Russia.

Pre-Caspian and East-West pipelines

The 1,700-km (1,056-mi) Pre-Caspian pipeline is intended to move gas from fields in the Caspian Sea and other locations in Kazakhstan and Turkmenistan to the Russian Federation.

In July 2008, Russia's Gazprom agreed to invest in the Turkmen part of the project. The capacity of the Turkmen section of the pipeline will reach up to 30 bcm/year (1.06 tcf/year).

The East-West pipeline, to serve as the main link with the Turkmen transmission system, will also provide gas supplies to Turkmenistan's population centers.

UKRAINE

Capital: Kiev
Monetary unit: Hryvnia
Refining capacity: 879,759 b/cd
Oil production: 75,000 b/d
Oil reserves: 395.000 million bbl
Gas reserves: 39.000 tcf

Overview

Gas supply issues with Russia

As 2008 came to a close, Russia and Ukraine were arguing about the size of gas debts owed by Ukraine. That dispute led, in January 2009, to degradation of gas supply—from Russia, transported through Ukraine—to 18 European countries.

It was reported that 90 million cu m/day of natural gas, destined for Ukraine, were completely cut off on January 1, while transit deliveries continued to the EU at a rate of 300 million cu m/d.

Lawsuits, meetings, and shuttle diplomacy ensued, leading to an international gas conference on January 17 in Moscow. After much negotiation, gas supplies were fully restored on January 21.

Various reports concluded that the gas crisis cost Gazprom more than $1.1 billion in revenue for the unsupplied gas. Other analyses said that Ukraine incurred losses from temporary shutdown of steel and chemical plants, and also lost $100 million in transit fees because of the cut-off.

In October, disagreement about the amount of gas Ukraine would import from Russia in 2010 raised concerns again about gas supply. Ukraine said it intends to imports less gas in 2010 because of the impact of economic recession on its industries. Gazprom insist that Ukraine fulfill its contractual obligations.

On December 15, Russia's energy minister said that he expected no problems with Ukraine over gas supplies at the start of 2010.

Oil tariff issues with Russia

Action by Ukraine and Russia in late December 2009 averted a potential disruption in the supply of oil to Ukraine over negotiation of an oil transit contract for 2010.

Officials in Kiev said Transneft, Russia's state oil pipeline company, had agreed to Ukraine's demand for a 22% increase in transit fees.

When Russian leaders objected to that demand, the EU triggered an early warning mechanism, advising oil consumers to brace for possible shortages. But oil traders said that there was enough capacity in other Russian oil pipelines to compensate for any drop in exports across Ukraine.

Oil shipments across Ukraine are far less important for Europe than gas shipments

IMF assistance sought

In mid-December 2009, Ukraine made an urgent appeal to the IMF for about $2 billion (€1.4 billion) in emergency loans to help the country meet its external obligations.

These include payments to Russia, which supplies natural gas to the European Union via Ukraine.

In October, the IMF suspended a $3.8-billion disbursement over the government's failure to implement key economic reforms.

Delay in instituting reforms demanded by the IMF was blamed in part on the impact of campaigning for presidential elections scheduled for January 2010.

Wheat-for-oil deal with Libya

Libya agreed in May 2009 to a deal under which the African country would be allowed to grow wheat for its own use on a designated 1,000-sq-km (386-sq-mi) site in Ukraine.

In return, Libya promised to provide 600,000 tonnes (4.4 million bbl) of oil to Ukraine's Kremenchug refinery.

Exploration and Development

Lelyaki field

The Kashtan Petroleum Joint Venture was making plans to further redevelop the Lelyaki oil field in Ukraine in 2009. The most recent work involved Well 307.

It was drilled to 1,960 m (6,431 ft), encountering oil-saturated porous zones in the lowermost K zone and also in the P1+P2 dolomite zone. Completed in the K zone, it was averaging 284 b/d of oil in early 2009.

Wells 304A and 308 were to be drilled next, and four other suspended wells were to be reentered and sidetracked in 2009.

Drilling and Production

Krasnapolianskoye field

In fourth quarter 2009, Transeuro Energy Corp. signed a contract and announced plans to sell natural gas in early 2010 from two wells in Ukraine's Krasnapolianskoye field. The firm estimates initial sales volume at 1.2 MMcfd.

The gas is to be converted to compressed natural gas for vehicle use.

Reserves are estimated by the government at 14 bcf, while Transeuro's third-party auditors assigned possible reserves of 2.7 bcf.

Coalbed methane tax exemption

Ukraine's president in mid-2009 signed into law legislation that exempts from taxation the production and use of coalbed methane for 10 years, starting January 1, 2010.

The law authorizes Ukraine's National Electricity Regulatory Commission to limit the price of methane if its production is funded from the state budget.

The government plans to grant state support and guarantees to the producers and suppliers of energy from coalbed methane.

Pipelines/Transportation

EC loans for gas sector

The European Commission in July 2009 approved loans worth up to $1.7 billion to help Ukraine buy, store, and move gas. The loans would come from the European Bank for Reconstruction and Development, the World Bank, and the European Investment Bank.

The EC said it expects an investment of $3.4 billion will be required between 2009 and 2015. The improvements are expected to increase system efficiency by 25%.

In return, Ukraine agreed to operate its pipeline network with increased transparency.

The Ukraine gas pipeline system spans 13,500-km (8,389 mi), is more than 40 years old, and is one of the most complex in the world.

EC Pres. Jose Manuel Barroso said the agreement "should provide the stability needed to significantly reduce the risk of a further gas crisis between Ukraine and Russia and therefore provide the security of supply that member states and our consumers expect."

The EC-funded project angered Russian officials who had expressed interest in participating in the modernization project.

Refining and Products

Kremenchug refinery upgrade

Ukrainian holding company Ukrtatnafta awarded a contract to a French firm in early 2009 for a first-of-its-kind upgrading gasoline production capabilities at the 360,000-b/d Kremenchug refinery—the largest in Ukraine.

The project will add processing units to enable the production of Euro-5 gasoline in 2011.

UZBEKISTAN

Capital: Tashkent
Monetary unit: Soum
Refining capacity: 222,271 b/cd
Oil production: 85,000 b/d
Oil reserves: 594.000 million bbl
Gas reserves: 65.000 tcf

North Urtabulak field

Tethys Petroleum Ltd. was working in early 2009 to boost oil production in North Urtabulak field in the Afghan-Tajik basin, after acquiring Rosehill Energy Ltd.

The latter holds a production enhancement contract for the field.

Tethys was considering further development drilling and attention to nearby fields.

Partnership with Sasol

Sasol Ltd. formed a JV mid-2009 with Petronas and Uzbekistan state oil and gas company Uzbekneftegaz and launched a study of the feasibility of building a 1.3-mtpy GTL plant in Uzbekistan.

The firms gave no start-up and no details on which gas fields could provide feed needed to produce diesel, kerosene, naphtha, and LPG.

The $2.5-billion facility will use Sasol's proprietary Slurry Phase Distillate process and will have a capacity of 36,000–40,000 b/d. Project cost is $2.5 billion.

Turkmenistan-to-China gas line

In December 2009, leaders of Turkmenistan, Kazakhstan, Uzbekistan, and China inaugurated the new Turkmenistan-China natural gas pipeline.

Details are given in the Turkmenistan section of this chapter.

China

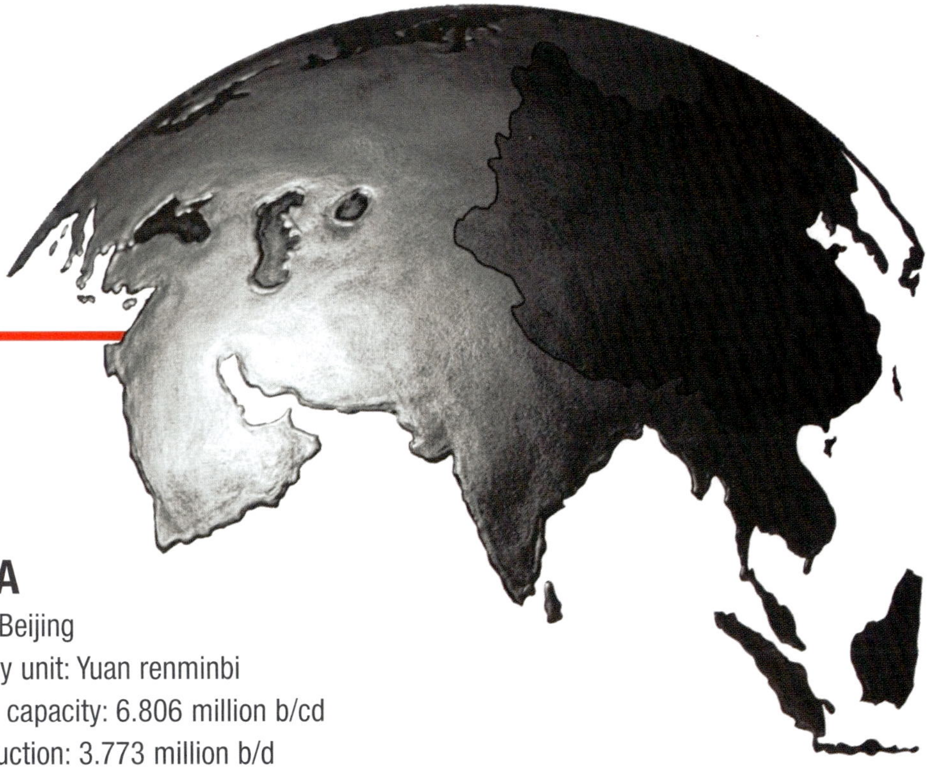

CHINA

Capital: Beijing
Monetary unit: Yuan renminbi
Refining capacity: 6.806 million b/cd
Oil production: 3.773 million b/d
Oil reserves: 20.350 billion bbl
Gas reserves: 107.000 tcf

Overview

General

China has rather limited oil and gas reserves, but demand for energy has been surging since the early 1990s. Securing supplies from abroad has become a major component of foreign policy.

Oil and gas resources are controlled by three state firms: China National Petroleum Corp. (CNPC), China Petroleum & Chemical Corp. (Sinopec), and China National Offshore Oil Corp. (CNOOC). The government maintains a majority stake in all three.

The National Development and Reform Commission (NDRC) is China's main energy planner.

CNPC and Sinopec operate almost all of China's refineries and the domestic pipeline network, while CNOOC has the most expertise with international transactions.

The country's first private sector company, the Great Wall Petroleum Group, was founded in 2005.

In 2009, notable energy-related developments in China included:

- a continued global campaign to acquire petroleum resources
- start of a second phase of expansion of strategic petroleum reserves
- continued strong oil production from Bohai Bay fields
- continued work on the major ESPO pipeline
- expansion of refining capacity
- expanded development of solar energy use as well as petrochemical facilities as parts of a major economic stimulus program.

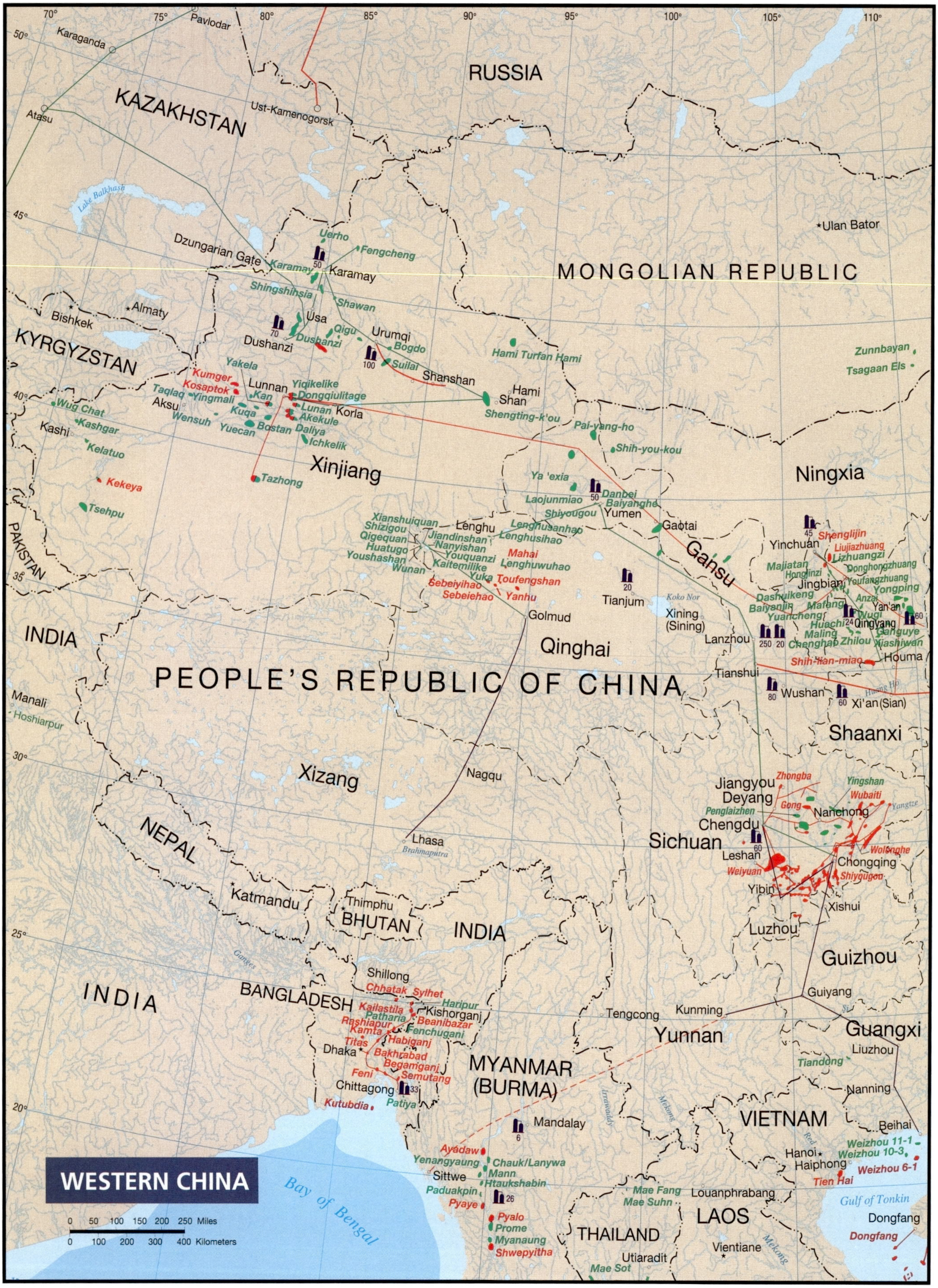
WESTERN CHINA
0 50 100 150 200 250 Miles
0 100 200 300 400 Kilometers
RUSSIA
KAZAKHSTAN
MONGOLIAN REPUBLIC
KYRGYZSTAN
PAKISTAN
INDIA
NEPAL
BHUTAN
BANGLADESH
MYANMAR (BURMA)
THAILAND
LAOS
VIETNAM
PEOPLE'S REPUBLIC OF CHINA
Xinjiang
Xizang
Qinghai
Gansu
Ningxia
Shaanxi
Sichuan
Guizhou
Yunnan
Guangxi
Pavlodar
Karaganda
Atasu
Ust-Kamenogorsk
Lake Balkhash
Dzungarian Gate
Ulan Bator
Almaty
Bishkek
Karamay
Uerho
Fengcheng
Shingshihsia
Shawan
Usa
Dushanzi
Qigu
Urumqi
Bogdo
Suilai
Shanshan
Hami Turfan Hami
Hami
Shan
Shengting-k'ou
Yakela
Kumger
Kosaptok
Lunnan
Yiqikelike
Dongqiulitage
Taqlaq
Yingmali
Kan
Aksu
Wensuh
Kuqa
Lunan
Akekule
Korla
Bostan
Yuecan
Daliya
Ichkelik
Tazhong
Wug Chat
Kashgar
Kashi
Kelatuo
Kekeya
Tsehpu
Pai-yang-ho
Shih-you-kou
Ya 'exia
Laojunmiao
Danbei
Baiyanghe
Shiyougou
Yumen
Gaotai
Xianshuiquan
Lenghu
Shizigou
Qigequan
Jiandinshan
Lenghusanhao
Lenghusihao
Nanyishan
Huatugo
Youquanzi
Youshashan
Wunan
Kaitemilike
Mahai
Lenghuwuhao
Yuka
Toufengshan
Sebeiyihao
Sebeiehao
Yanhu
Golmud
Tianjum
Koko Nor
Xining (Sining)
Lanzhou
Tianshui
Wushan
Xi'an (Sian)
Yinchuan
Shenglijin
Liujiazhuang
Lizhuangzi
Majiatan
Hongjinzi
Donghongzhuang
Youfangzhuang
Jingbian
Dashuikeng
Anzai
Yongping
Baiyanjin
Mafang
Yan'an
Yuancheng
Wuqi
Huachi
Qingyang
Maling
Ganguye
Chenghao
Zhilou
Xiashiwan
Shih-tan-miao
Houma
Huang Ho
Zunnbayan
Tsagaan Els
Manali
Hoshiarpur
Nagqu
Lhasa
Brahmaputra
Katmandu
Thimphu
Ganges
Jiangyou
Zhongba
Yingshan
Deyang
Wubaiti
Penglaizhen
Gong
Nanchong
Yangtze
Chengdu
Wolonghe
Leshan
Chongqing
Weiyuan
Shiyugou
Yibin
Xishui
Luzhou
Guiyang
Kunming
Tengcong
Liuzhou
Tiandong
Nanning
Beihai
Weizhou 11-1
Weizhou 10-3
Weizhou 6-1
Hanoi
Haiphong
Tien Hai
Gulf of Tonkin
Dongfang
Louanphrabang
Vientiane
Mekong
Mae Fang
Mae Suhn
Utiaradit
Mae Sot
Shillong
Chhatak
Sylhet
Haripur
Kailastila
Kishorganj
Patharia
Beanibazar
Rashiapur
Fenchugani
Kamta
Habiganj
Titas
Dhaka
Bakhrabad
Begamganj
Feni
Semutang
Chittagong
Patiya
Kutubdia
Mandalay
Irrawaddy
Ayadaw
Yenangyaung
Chauk/Lanywa
Sittwe
Mann
Htaukshabin
Paduakpin
Pyaye
Pyalo
Prome
Myanaung
Shwepyitha
Bay of Bengal

EASTERN CHINA
RUSSIA
MONGOLIAN REPUBLIC
PEOPLE'S REPUBLIC OF CHINA
NORTH KOREA
SOUTH KOREA
JAPAN
VIETNAM
LAOS
THAILAND
PHILIPPINES
TAIWAN
OKINAWA
Sea of Japan
Yellow Sea
East China Sea
South China Sea
Gulf of Tonkin
Bohai Gulf
Korean Channel
Osumi Strait
Formosa Strait
Heilongjiang
Jilin
Liaoning
Nei Mongol
Ningxia
Gansu
Shanxi
Hebei
Shandong
Beijing
Tianjin
Henan
Shaanxi
Jiangsu
Shanghai
Anhui
Hubei
Sichuan
Zhejiang
Hunan
Jiangxi
Guizhou
Fujian
Yunnan
Guangxi
Guangdong
HAINAN DAO
Komsomolsk
Khabarovsk
Vladivostok
Ulan Bator
Qing 'an
Daqing
Harbin
Qianguo
Huadian
Dongfeng
Fusong
Chongjin
Aershan
Zunnbayan
Tsagaan Els
Saihantala
Chifeng
Tieling
Fushun
Shenyang
Liaoyang
Jinzhou
Panshan
Anshan
Yingkou
Dandong
Pyongyang
Huai'an
Cangzhou
Hohot
Chin-hsi
Qilu
Changli
Jin Xian
Luda
Lushun (Dalian)
Renqiu
Huanghua
Onsan
Seoul
Shih-you-kou
Danbei
Yumen
Baiyanghe
Gaotai
Shandan
Yinchuan
Shenglijin
Liujiazhuang
Lizhuangzi
Donghongzhuang
Majiatan
Hongjinzi
Youfangzhuang
Dashuikeng
Baiyanjin
Mafang
Anzai
Wugi
Yuancheng
Huachi
Maling
Chenghao
Zhilou
Ganguye
Xiashiwan
Qingyang
Yulin
Changbei
Suide
Zichang
Jingbian
Yongping
Quinghuazhi
Yan'an
Shijiazhuang
Jinan
Yanshan
Kenli
Zibo
Zhaoyuan
Gaoqing
Weifang
Qingdao
Weicheng
Anyang
Wenmingzhai
Wennan
Wenzhong
Pucheng
Wenliu
Huzhuangji
Baimiao
Machang
Tianjum
Ulan
Koko Nor
Xining (Sining)
Lanzhou
Tianshui
Pingliang
Shih-lian-miao
Xianyang
Huang He
Luoyang
Zhengzhou
Wushan
Xi'an (Sian)
Fangcheng
Shuanghe
Lianyungang
Baoying
Liuzhuang
Anhui
Huangque
Fumin
Caozhuang
Lianmengzhuang
Zhenwu
Hefei
Nanjing
Yangtze
Wuhu
Hangzhou
Zhenhai
Ning-Bo
Donghai
Bo Yuan Ting
Ping Hu
Zhongba
Jiao
Jiangyou
Bajiaochang
Penglaizhen
Chengdu
Yingshan
Gong
Wubaiti
Nanchong
Suining
Wusheng
Dazu
Wolonghe
Weiyuan
Leshan
Chongqing
Shiyougou
Garze
Batang
Ceshui
Yingshan
Tuoshi
Qianjiang
Yichang
Wuhan
Mianyang
Anqing
Jiujuand
Yueyang
Jinxi
Hemgyang
Anshun
Dushan
Quanzhou
Kunming
Si
Liuzhou
Bose
Tiandong
Mashan
Nanning
Fuzhou
Quanzhou
Taipei
Chuhuangkeng
Chiting
Jinshuei
Cingcaohu
Paoshan
Yongheshan
Tiejhanshan
Sinying
Paonanghsi
Jinmen
Taoyuan
Pingtung
Kaohsiung
Guangzhou
Hong Kong
Lufeng 13-1
Huizhou 26-1
Xijiang 24-3
Xijiang 30-2
Lufeng 15-1
Huizhou 21-1
Liuhua 4-1
Liuhua 11-1
Lianjiang
Maoming
Beihai
Zhanjiang
Weizhou 11-1
Weizhou 10-3
Weizhou 6-1
Hanoi
Tien Hai
Red
Mekong
Irrawaddy
Haikou
Jinfeng
Wenchang 8-3
Wenchang 9-2
Wenchang 14-3
Wenchang 13-1
Wenchang 19-1
Wenchang 13-2
Dongfang
Yinggehai
Yacheng
Yacheng 13-1
Mae Fang
Mae Suhn
Vientiane
Ulsan
Pusan
Yosu
Hiroshima
Sasebo
Kashima
Nagasaki
Oita
0 100 200 300 Miles
0 200 400 Km
100° 105° 110 115° 120° 125 130° 135°
50° 45° 40° 35° 30° 25° 20° 15°

PetroChina: Output lower in 2009

PetroChina Co. Ltd., flagship of CNPC and the largest oil-producer in China, said in April that it planned to cut crude oil output by about 4% in 2009, due chiefly to predicted low oil prices and in order to reduce high development costs.

The company expected that both production and oil processing volume would drop from about 2.5 million b/d in 2008 to about 2.4 million in 2009.

Observers also noted that the Nanpu oil field, which has approximately 7.5 billion bbl of reserves and was discovered by CNPC in May 2007, is still in the initial survey and drilling stage.

CNOOC: 2009 plans

CNOOC said in early 2009 that it planned total capital expenditures of $6.76 billion for the year, up 19% over 2008. Of that total, $4.38 billion is for development, $1.11 billion for exploration, and $1.12 billion for production.

The company expects ten new projects to come on stream, including eight offshore China.

In addition, OML130 offshore Nigeria and the Tangguh LNG project onshore Indonesia were expected to start production by yearend. CNOOC has interests in both.

Sinopec refinery plans

Sinopec said in November 2009 that it expects its crude oil refining capacity to increase between 12 mtpy and 15 mtpy in the next three years.

In 2010, the company plans to process 205 million metric tons of crude (4.4 million b/d) and buy more than 1 million b/d of oil from Saudi Aramco.

Saudi Basic Industries Corp. will supply 210,000 b/d to Sinopec's Tianjin Refinery. That plant was under expansion in late 2009.

Strategic petroleum reserves

In a late-2009 analysis, FACTS Global Energy (Singapore) estimated that China would double its strategic petroleum reserve by 2011–2012.

The sites for Phase 1 SPR expansion are Zhenhai, Zhoushan, Huangdao, and Dalian. Construction was completed in late 2008. The tanks—all above ground and in coastal areas—were full as of September 2009, and can support 13–14 days of oil consumption or 25–26 days of net oil imports (based on 2008 levels).

The price paid for Phasel SPR crude purchases averaged $58/bbl.

The NDRC plans eight strategic stockpiling bases under Phase 2, with a total capacity of 169 million bbl. Some of the Phase 2 facilities will be in underground water-sealed caverns in inland provinces.

Phase 2 will include three categories:

- National SPR
- Commercial petroleum storage by state oil companies
- Commercial storage by local governments or companies

China SPR Phase I Sites

	Capacity, million bbl	Capacity, million cu m	Completion date	Investment, million yuan
Zhenhai (Sinopec)	32.7	5.2	August 2006	3,700
Zhoushan (Sinochem)	31.4	5.0	—	3,800
1st Phase	7.5	1.2	March 2007	—
2nd Phase	9.4	1.5	November 2007	—
3rd Phase	14.5	2.3	December 2007	—
Huangdao (Sinopec)	20.1	3.2	—	2,600
1st Phase	7.5	1.2	June 2007	—
2nd Phase	12.6	2.0	December 2007	—
Dalian (PetroChina)	18.9	3.0	December 2008	2,510
Total	**103.1**	**16.4**		**12,610**

China SPR Likely Phase 2 Sites

Company	Site	Capacity, million bbl	Capacity, million cu m	Status	Completion date
Sinopec	Pinghu, Baishawan, Zhejiang	12.6	2.0	Completed	Early 2009
PetroChina	Shanshan, Xinjiang	50.3	8.0	Building	—
	1st Phase	6.3	1.0	Completed	Dec. 24, 2008
	2nd Phase	44.0	7.0	Building	End-2009
PetroChina	Tieling, Liaoning	7.3	1.2	Building	End-2009
PetroChina	Jinzhou	3.8	0.6	Building	2010
Sinopec	Caofeidian	32.7	5.2	Land leveling	2011–12
Sinopec	Tianjin	40.2	6.4	Preliminary work	2011–12
PetroChina	Lanzhou	12.6	2.0	Completed environmental appraisal	2011–12
Sinopec	Zhanjiang, Guangdong	44.0	7.0	Planning	—

As of September 2009, one Phase 2 facility was complete, three were being built, and the other sites were being prepared for construction.

By early 2010, about 69 million bbl of Phase-2 storage were expected to be ready, with full capacity reached by yearend 2011 or early 2012.

China also had at least 300 million bbl of commercial crude storage capacity at the start of 2009. Sinopec owns about 50%, and CNPC-PetroChina about 40%; CNOOC, SinoChem, and others own the rest.

In mid-May 2009, the government announced plans to build strategic reserves for refined products, aiming to build storage for up to 10 million tonnes of refined products by 2011.

Securing oil supplies: Loans

Wood Mackenzie reported on China's use of its financial reserves to secure oil from a variety of producing countries in 2009 to meet growing domestic demand.

Of note was a $30-billion loan from state-owned China Development Bank (CDB) to the country's largest oil and gas producer, CNPC to support the latter's ambitious global strategy.

CHINA'S OIL SUPPLY-DEMAND BALANCE 2008

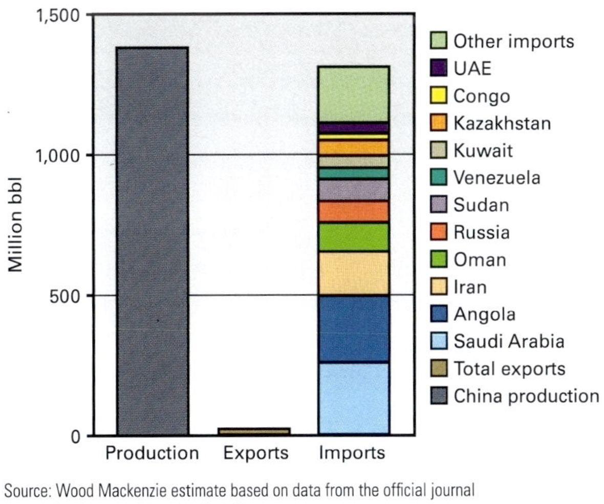

Source: Wood Mackenzie estimate based on data from the official journal "China Oil, Gas, and Petrochemicals"

CHINA'S NOCS: A 2008 SNAPSHOT

	Oil production	Oil reserves	Total assets	Total revenue	Profit
	Million bbl		Million $		
CNOOC	154	1,578.3	30,297.7	18,128.1	6,385.6
CNPC	796.3	14,231.6	264,532.4	183,184.7	13,188.4
Sinopec	296.8	2,841.0	110,277.7	216,201.5	3,770.2

Source: Company annual reports

And from February to August 2009, China extended loans totaling $46 billion in exchange for secure supplies of crude oil from other nations.

The deals included the following:

- CNPC entered a $5-billion financing deal with Kazakhstan's KazMunaiGaz.
- CDB committed to loan $25 billion to two Russian firms—$15 billion to OAO Rosneft and $10 billion to OAO Transneft—in return for supplies from East Siberian oil fields over 20 years.
- CDB and Sinopec signed an MOU with Brazil's Petrobras for reported loans of $10 billion to support investment in upstream and refining by Petrobras. Oil exports from Brazil to China are expected to reach 100,000 b/d in 2010.
- In return for $4 billion in financing, Venezuela's PDVSA agreed with CNPC to increase oil sales to China to 80,000–200,000 b/d.
- Sinopec and PDVSA signed an MOU to study joint field development opportunities in Venezuela.
- Angola confirmed getting another $1-billion loan from China (for a total of $5 billion in oil-backed loans since 2002).

Securing oil supplies: M&A activity

Wood Mackenzie also pointed to mergers, acquisitions and other initiatives in 2009 by China's major oil firms in pursuit of overseas resource opportunities with leading IOCs.

SOLAR ENERGY USE

China unveiled in mid-2009 the "Golden Sun" initiative, including proposed government subsidies of 50% during 2010–2013 for solar power projects with a minimum capacity of 500 MW.

Analysts estimated the cost of the initiative at $1.6 billion to $2.1 billion, with funds to come from the $30-billion "green technology" stimulus plan announced in July.

Aiming to boost domestic use of solar energy, the NDRC in mid-2009 raised its official targets for total installed solar-electric capacity, to 2GW by 2009, and to 20 GW by 2020.

M&A ACTIVITY: CHINESE COMPANY EXPENDITURE

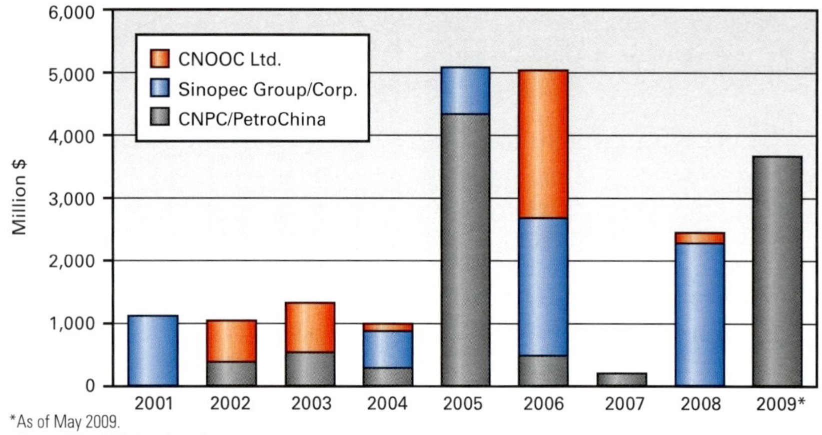

*As of May 2009.
Source: Wood Mackenzie estimates

China car sales

China's auto sales rose 90% (to 858,300 units) in August 2009 compared to August 2008.

Analysts said total vehicle sales could hit a record 12 million for the year, likely making China to the top global vehicle market.

The government in January instituted tax breaks on small cars and subsidies for vehicle purchases in rural areas as part of an economic stimulus initiative.

Ethnic riots in Xinjiang

Ethnic clashes between Han Chinese and Muslim Uighers in and near Urumqi, capital of the western province of Xinjiang, killed a reported 150 people in June and July 2009.

The rioting did not affect operations at Tahe oil field (owned by Sinopec) or at oil and gas fields in the Tarim basin (southern Xinjiang) and in Kelamayi (northwest of Urumqi) or the Dushanzi refinery, all owned by PetroChina.

The resource-rich Xinjiang region holds 20% of China's oil reserves. Also, the province's long borders with oil-producing Central Asian countries make it strategically important to Beijing.

A London-based risk analysis firm said in mid-2009 that Chinese projects around the world—including oil and gas activities—could become targets of Al-Qaeda, as a response to the deaths of the Uighers.

Areas said to be at risk include Yemen, Algeria, Niger, Chad, Kurdistan, Nigeria, and Gabon.

CNPC:

- was in talks with Petro-Canada and Suncor Energy Inc. in early 2009 to buy two oil fields in Libya and Syria for $5 billion (a deal delayed by the pending Suncor/Petro-Canada merger).
- was in talks buy a 75% stake in YPF, the Argentinean unit of Repsol.
- attempted (unsuccessfully) to buy Verenex Energy (Canada) to obtain interest in Block 47 in Libya.
- paid $3.3 billion for a 50% stake in Kazakh oil and gas producer MangistauMunaiGaz.

In other action:

- Sinopec bought Canada's Addax Petroleum for $7.2 billion—the Chinese firm's largest-ever acquisition—expanding Sinopec's upstream interests in West Africa and Iraq.
- Sinopec and CNOOC agreed to a $1.3-billion deal with Marathon for a stake in Block 32 in Angola. That country is second only to Saudi Arabia as an oil supplier to China.
- Press reports suggested the likely purchase of Talisman's assets in Trinidad and Tobago by a CNOOC-Sinopec JV for $323 million.
- PetroChina said it also was evaluating deals in Africa, Asia and South America in 2009.
- CNOOC was reportedly considering extending an offer for Kosmos, a small exploration company with a stake in the Jubilee field offshore Ghana, which holds an estimated 1.2 billion bbl of oil.

Exploration and Development

Pearl River Mouth basin

In September 2009, Newfield Exploration Co. (Houston) reported plans to pursue commercial development on Block 16/05 in the Pearl River Mouth basin.

The company said the block holds more than 30 million bbl of recoverable oil.

Newfield expects to start production from the block in 2012.

CNOOC has the right to participate in any development with a 51% interest.

Bohai Bay

CNOOC announced two oil strikes in Bohai Bay in 2009:

- in August, with the Qinhuangdao (QHD) 35-4 well.
- in March, with the Bozhong (BZ) 2-1 and QHD 29-2 wells.

The QHD well is located southeast of Qinnan Sag. The BZ well is about 6 km (4 mi) southeast of BZ 3-2 oil field, in the west of Bozhong Sag.

Liwan field

The third appraisal well on the Liwan 3-1 field in Block 29/26 in the South China Sea tested natural gas at an equipment-restricted rate of 52 MMcf/d, said Husky Energy.

The Liwan 3-1-4 well was drilled to a total vertical depth of 3,366 m (11,043 ft) below sea level, in water 1,450 m (4,757 ft) deep.

First gas production is targeted for 2013. Husky said future deliverability could exceed 150 MMcf/d.

Liaodong Bay

Discovery well JZ 20-2N-1 in the Liaodong Bay proved oil in mid-2009, according to CNOOC. The well penetrated oil pay zones with a total thickness of 33.4 m (110 ft). It flowed at an average rate of 1,900 b/d of oil and 3.4 MMcf/d of gas.

CNOOC was continuing to appraise the discovery's reserve scale and commercial viability.

Guizhou Province

A regional subsidiary of Sinopec agreed with Zunyi City in June 2009 to jointly develop oil and gas resources in southwestern Guizhou province.

The program will evaluate oil and gas accumulations near the city, aiming to start exploitation of the Chishui gas field by 2012.

Current Sinopec output of natural gas output in the Chishui region is 12 MMcm (424 MMcf) per year.

Xinjiang Province: Coalbed methane

TerraWest Energy Corp. was evaluating coalbed methane (CBM) prospects in early 2009 in the Liuhuanggou area near Urumqi in northwest China.

Wells drilled to 1,500 m (4,921 ft) on a PetroChina lease penetrated Jurassic J2X coals and intersected multiple coal seams.

China's second west-east gas pipeline now under construction runs through the area, and the Chinese government gives CBM producers preferential pipeline access.

Reserves

Oil resource assessment

The Oil and Gas Strategic Research Center at the Ministry of Land Resources estimated the country's total prospective oil resources (in 115 land and offshore basins) at about 2.3 billion b/d.

The five-year study by 17 institutes was led by the center.

Land resources account for about 2 billion b/d, and offshore oil accounts for about 325.7 million b/d.

Exploitable oil resources was estimated to be about 454.3 million b/d, including about 392.1 million b/d on land and about 62.1 million b/d offshore.

China's oil resources are mainly in the east, west and offshore. Prospective, geological and exploitable resources in those areas accounted for 82%, 79%, and 84%, respectively, of the nation's total.

Drilling and Production

Bohai Bay: Luda 27-2 field

CNOOC in late 2009 brought onstream the Luda (LD) 27-2 field in eastern Bohai Bay. The field is producing 11,000 b/d of oil from 11 wells.

It is in about 25 m (82 ft) of water, 104 km (65 mi) northwest of Qinhuangdao City.

At peak, production should reach 13,000 b/d.

LD 27-2 is adjacent to the LD 32-2 oil field—also a CNOOC project—which was expected to come onstream by yearend 2009.

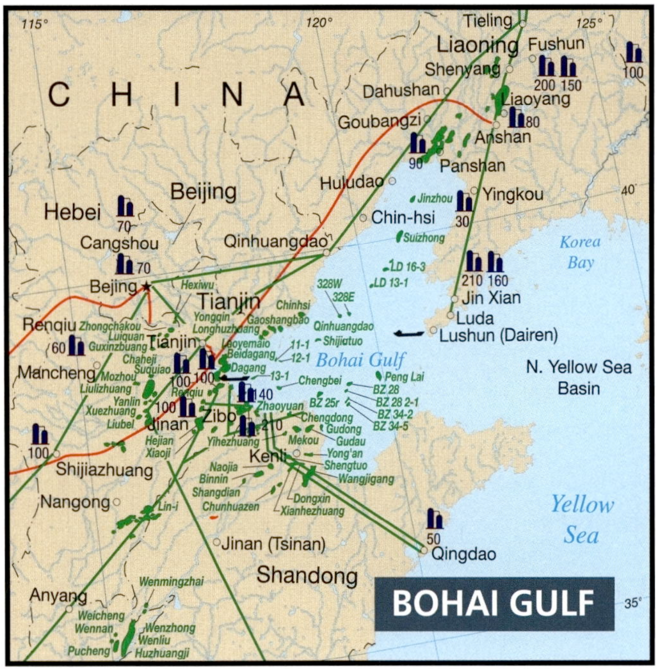

Bohai Bay: Qinhuangdao 33-1 field

Production began in mid-2009 from the Qinhuangdao (QHD) 33-1 field in Bohai Bay, said CNOOC, producing approximately 2,000 b/d of oil from two wells. Oil moves to the adjacent QHD 32-6 field by subsea pipeline for processing.

Bohai Bay: Bozhong 28-2S field

CNOOC reported the Bozhong (BZ) 28-2S oil field in the Bohai bay onstream in early 2009.

Production, which began with 4,000 b/d of oil from four wells, is expected to ramp up an average 25,000 b/d of oil by 2011.

The field is in water that averages about 20 m (65.6 ft) deep.

BZ 28-2S is being jointly developed with the adjacent BZ29-4, BZ28-2SN, and BZ34-1N fields.

Bohai Bay: Peng Lai (PL) 19-3 Block]

Operations began in early 2009 on Production Platform B, part of the Phase 2 project on Peng Lai (PL) 19-3 Block.

ConocoPhillips operates PL 19-3—China's largest offshore oil field—in partnership with CNOOC. The project is 235 km (146 mi) southwest of Tanggu.

Platforms D and E were expected to come online by yearend.

Bohai Bay: Zhao Dong Block

Roc Oil (Bohai) Co. (Sydney) announced plans in April 2009 to start a 12-well drilling program in the C and D oil fields at its Zhao Dong Block, which is already is producing 21,500 b/d of oil.

The block lies in shallow water, close to shore in Bohai Bay, southeast of Beijing.

The company started producing in March from the C-51 and C-52 wells in the extended reach area of the C and D fields. Well D-43 began producing in April.

The upcoming program aims to maintain production at the 2008 average of about 18,050 b/d.

China Sea: Yinggehai basin

Production began in third quarter 2009 from the Ledong (LD) 22-1 field. Initial production was about 30,000 cf/d of natural gas from five wells, said CNOOC.

LD 22-1, in the Yinggehai basin in the China Sea, is jointly developed with the LD 15-1 field.

Gas from the two fields will be processed at the Dongfang 1-1 gas terminal and then piped to customers in Hainan province.

Peak combined production from LD 22-1/15-1 is expected to be about 150 MMcf/d.

LD 15-1 was expected to begin production by yearend.

Tahe oil field

Sinopec said in early 2009 that it plans to invest $1.46 billion in the northwestern Xinjiang Uigher Autonomous Region to expand output from the Tahe oil field.

In 2009, it is expected to produce about 141,000 b/d of crude and 1.21 bcm of natural gas.

By 2020, Sinopec aims to produce 321,000 b/d of oil and 5 billion to 10 billion cu m/year of natural gas.

International Activities

JV deals for Canada oil sands

A unit of PetroChina entered a series of joint venture agreements in mid-2009 with Canada's Athabasca Oil Sands Corp. regarding oil sands development.

More details are presented in the Canada section of the North American chapter of this book.

Interest in Nigeria blocks

The *Financial Times* reported in September that CNOOC was talking with Nigeria about buying substantial interests in more than 20 blocks that could contain as much as 6 billion bbl of oil—about one-sixth of proven oil reserves in Nigeria.

Observers noted that western oil companies—including Shell, Chevron, Total, and ExxonMobil—also are interested in the blocks.

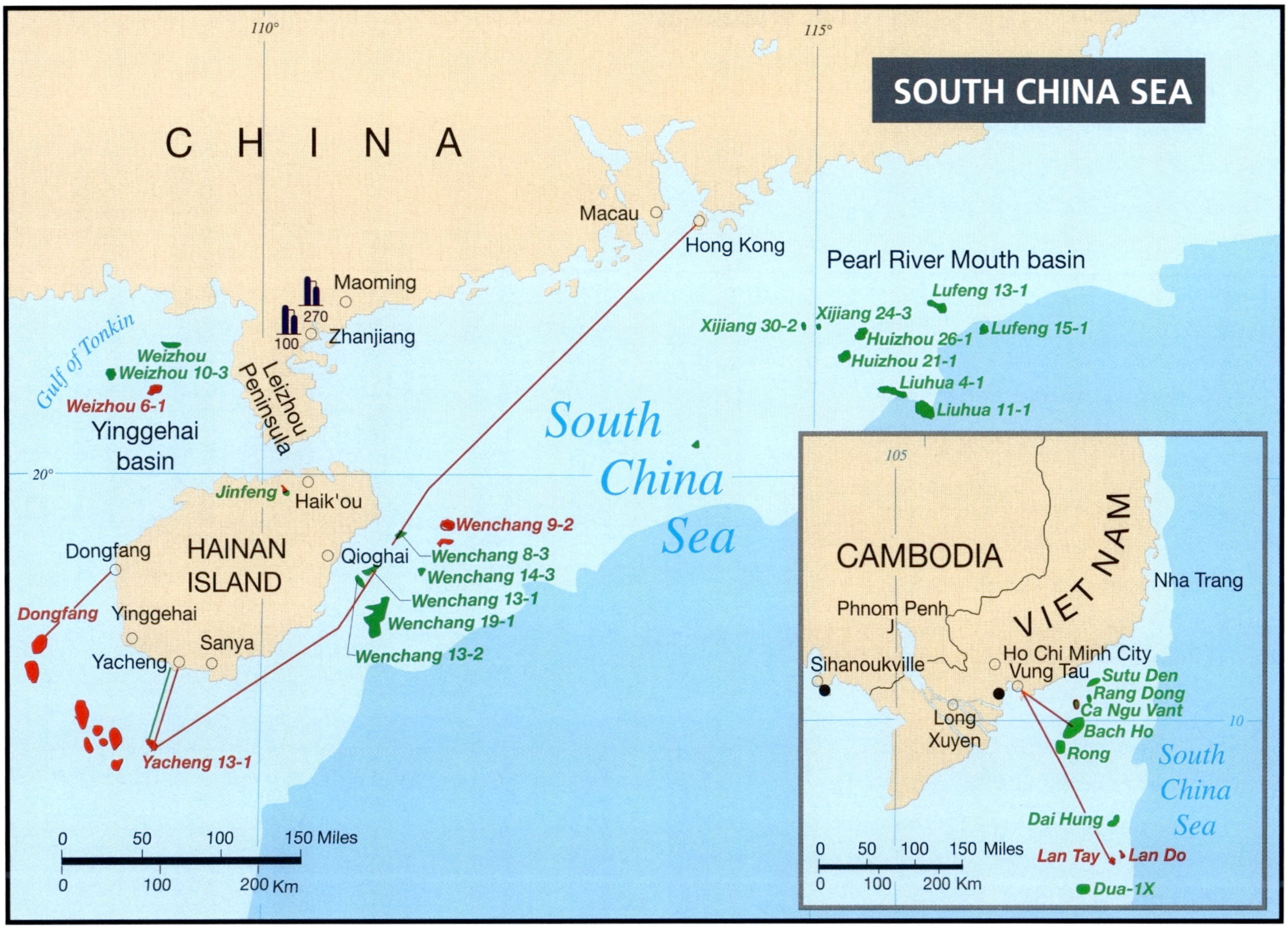

Gasoline to Iran

Chinese state oil companies began supplying gasoline to Iran in September 2009 and now provide as much as one-third of its imports. Iran usually imports 120,000 b/d.

The Chinese action threatened U.S.-led efforts to stop such imports as an element of pressure on Iran to curtail its nuclear program.

Other oil companies based in Asia and Europe also sell gasoline to Iran, but in 2008–2009, BP, Reliance (India) and other companies stopped such sales.

Iran is a major oil producer but its refineries are not in good condition. Also, fuel demand is high because the country also provides generous purchase subsidies.

Iran-China South Pars cooperation

In mid-2009, China was poised to begin a cooperative project with Iran's NIOC on the Phase 11 development of the South Pars gas field in the Persian Gulf.

Details are presented in the Iran section of the Middle East chapter of this book.

Venezuela oil production

The Venezuelan government reported in September that the country had signed a $16-billion investment deal with China for a three-year program to boost oil production in the Orinoco belt.

More details are given in the Venezuela section of the Latin America chapter of this book.

Deal for Australian LNG

In August 2009, in a deal with ExxonMobil Corp., PetroChina agreed to buy $41billion worth of natural gas produced by the Gorgon LNG project off the coast of Western Australia.

Announcement of the deal in Beijing—for purchase of 2.25 mtpy of LNG for 20 years—was said to be timed to defuse rising tensions between the two countries. Terms were reportedly agreed to early in 2009.

Oil deal with Ecuador

Ecuador signed a deal in August 2009 to supply 69 million bbl of oil over two years (2.88 million bbl/month) to China in return for a $1-billion advance payment.

The oil will be produced by Petroecuador and delivered to PetroChina.

China also offered a $1-billion loan, according to comments attributed to Ecuador's foreign ministry.

Petroecuador said revenues will be used for investment in the oil industry and for public works.

Ecuador's liquidity was of concern due to a crash in the commodities market and to its controversial default on $3.2 billion in foreign debt that it argued was illegitimate.

Refining Sudanese oil

CNPC announced that its Dalian Petrochemical processed 40,000 tonnes of high-acid oil imported from Sudan in March 2009.

Aiming to trim production cost, Dalian officials said they successfully blended Sudan's low-quality, acidic crude with normal crude to "pave the way for upcoming large-scale processing."

In a related a development, PetroChina was building a 10-mtpy refinery in southern Guangxi Zhuang Autonomous Region, designed specifically for processing Sudanese high-acid crude.

East China Sea pact

Japan and China announced in early 2009 that they will proceed with joint exploration for natural gas in the disputed East China Sea.

But both agreed that further talks are needed regarding development of the Kashi and Kusunoki gas fields and on China's unilateral development activities in the area.

Gas Processing

China's LPG output

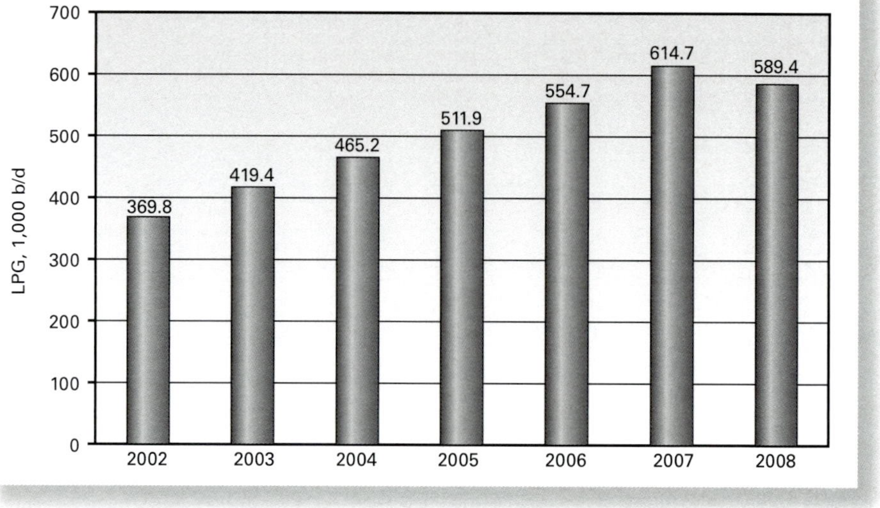

China's LPG imports

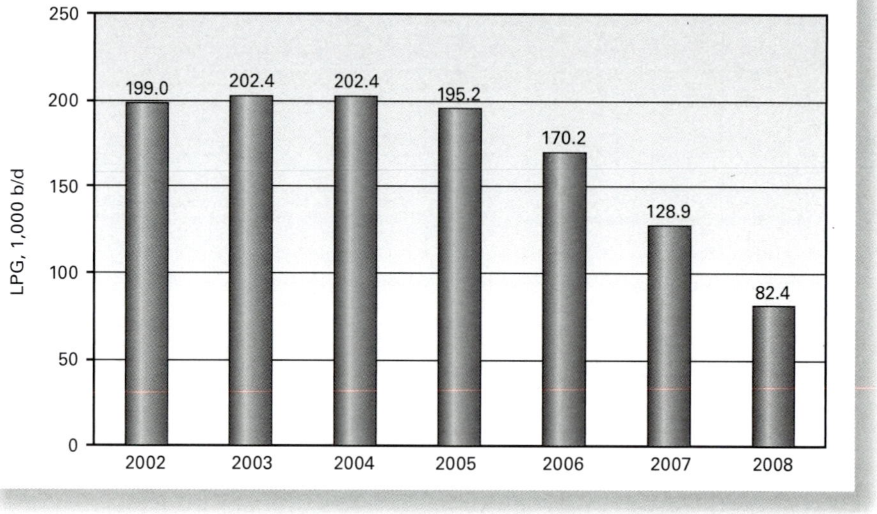

China's apparent use of LPG

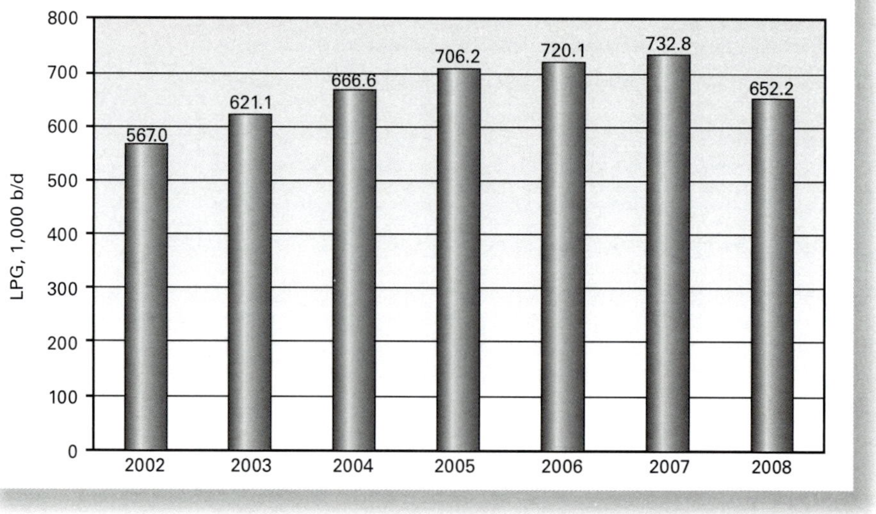

LPG: 2008 data

Chinese refinery production of LPG fell for the first time in history in 2008.

The country's imports of LPG also continued a decline that year that began in 2005.

And China's LPG consumption fell in 2008 for the first time since 2000.

These were the major findings of a June 2009 report by FACTS Global Energy (Singapore).

For 2009, the firm forecast that China's LPG output will recover as a result of expected higher refinery crude runs when a couple of local refineries come online. Imports, however, will continue their slide, while exports will "increase moderately."

LPG consumption is expected to recover only slightly in 2009 as markets continue to suffer from the global economic slump and natural gas makes inroads into traditional LPG demand.

About 65% of the LPG used in China is consumed as fuel by the residential sector. Some 25% goes for the industrial sector, 2% for vehicle use, and 8% for other uses.

LNG

Gorgon-Jansz LNG to PetroChina

ExxonMobil Corp. finalized an arrangement in August 2009 to supply PetroChina with 2.25 mtpy of LNG for 20 years, in a deal estimated at $50 billion (Aus.).

The transaction completes the sale by ExxonMobil of its share of LNG from the Gorgon-Jansz project in Australia.

Earlier in the month, ExxonMobil finalized a deal to supply Petronet (India) with 2 mtpy.

More details are given in the Australia section of the Asia-Pacific chapter of this book.

PetroChina: Jiangsu Province

In June 2009, the Ministry of Commerce approved a plan by PetroChina to build a 6.5-mtpy LNG project in Jiangsu Province.

The receiving terminal will be at the Xitaiyangsha Artificial Island on the Yellow Sea coast in Rudong County.

The facility will have processing capacity of 6.5 mtpy and an annual average natural gas supply capacity of 8.7 bcm.

The project is expected to become operational by mid-2011.

PetroChina: Shenzhen Province

Local media reported in early 2009 that PetroChina would team with CLP Holdings Ltd. and Shenzhen Gas Corp. on a proposed LNG terminal in Dachan Bay, in western Shenzhen Province.

Source gas for the plant would come from Xinjiang Province through the West-East pipeline, and from LNG shipments from overseas oilfields.

A pipeline would carry regasified LNG to CLP's Black Point power plant in Tuen Mun.

CNOOC: Fujian

CNOOC received the first LNG commercial cargo at its Fujian terminal in May 2009, from Indonesia's Bontang liquefaction plant.

A second cargo from BP's Tangguh LNG facility in Papua Barat was received in July. Fujian is to receive 2.6 mtpy from Tangguh.

Fujian is China's second terminal to begin operation.

CNOOC: Shanghai

In October 2009, the Shanghai LNG terminal received its first cargo, from Bintulu, Malaysia.

Owner CNOOC in 2008 signed a 25-year agreement to buy 2 mtpy from Qatargas Operating Co. Ltd., with supplies to start by yearend 2009 from the 7.8-mtpy Qatargas 2 project.

The terminal also has a 25-year supply contract with Malaysia for 3.03 mtpy.

CNOOC: Ningbo

Construction began in October on the Ningbo LNG terminal in Zhejiang Province, south of Shanghai. It will be China's fifth terminal. When completed in 2012, Phase 1 will have 3 mtpy of installed regasification capacity.

CNOOC: Zhejiang Province

In June 2009, CNOOC got approval from the NDRC to build an LNG project in cooperation with officials of eastern Zhejiang Province.

Total investment for Phase I was estimated to be just over $1 billion.

CNOOC: Guangdong Province

CNOOC signed an agreement in May 2009 with the city of Zhuhai (southern Guangdong Province) to build a $2.9-billion LNG terminal.

The project would also include facilities to manufacture offshore oil-field components.

Initial investment is estimated at $730 billion, with construction expected to begin by yearend 2009.

Pipelines/Transportation

Spur: Turkmenistan-to-China gas line

Officials said in October 2009 that Turkmenistan had completed its 188-km (117-mi) section of a 7,000-km (4,350-mi) natural gas pipeline from Turkmenistan, through Uzbekistan and Kazakhstan, to China.

The whole line was slated to start up by yearend 2009.

The pipeline starts from Turkmen gas fields near the Amu Darya River. It enters Uzbekistan at Olot, then flows on to southern Kazakhstan, and on to Alashankou in China, where it will connect to the West-East Gas Pipeline.

Gas lines from, through Myanmar

CNPC continued its construction of a $1.95-billion, 870-km (540-mi) pipeline to transport natural gas found in Blocks A-1 and A-3 off the coast of Myanmar to Yunan Province in mainland China.

The pipeline allows bypass of the Malacca and Lombok/Makassar Straits, through which more than 60% of China's imported oil moves.

CNPC has offered a stake in the pipeline to the consortium developing the gas field. That consortium is led by Korea's Daewoo International Corp. and also includes ONGC Videsh Ltd. (OVL), Korea Gas Corp., GAIL (India), and Myanmar's MOGE.

China will purchase gas from Myanmar under a 30-year contract. First gas is expected in early 2013.

CNPC also was poised to start construction in late 2009 of dual oil/gas pipelines, spanning nearly 4,000 km (2,485 mi) from Myanmar's western Arakan state to China's Yunnan Province.

In return for an estimated $29 billion paid over 30 years, China would receive oil shipped from the Middle East and natural gas from offshore Myanmar.

Russia-China gas pipeline

Russia's Gazprom announced in June 2009 an indefinite delay in plans to build a gas pipeline to China because of disagreement over price and concerns about adequate supplies from Siberian fields.

Details are given in the Russia section of the Former Soviet Union chapter of this book.

Kazakh-China oil pipeline

In July 2009, Chinese officials announced successful testing and initial operation of the Kenkiyak-to-Kumkol section (in Kazakhstan) of a major oil pipeline.

The line was developed by CNPC and the Kazakh oil company, KazMunaiGas.

The line runs chiefly east-west across Kazakhstan, from Atyrau on the northern edge of the Caspian Sea, to Alashankou in Xinjiang Province, near the Kazakh-Chinese border.

The pipeline will give China access to more oil from the Central Asian countries to its west, including the Kenkiyak field, which CNPC manages.

The first section of the line, in western Kazakhstan from Atyrau to Kenkiyak, was completed in 2003.

A second section, in eastern Kazakhstan from Atasu to Alashankou, was completed in 2005, as well as an extension from Kumkol to Atasu.

Henan Province projects

Officials in central Henan Province announced plans in April to invest about $1.5 billion in 2009 in oil and gas pipeline projects.

Other major investments also are planned for coal and power-generation initiatives.

The province expects to invest $1.55 billion to turn Zhengzhou into a major product-oil distribution and reserve center. Also, construction of 19 oil and gas pipeline sections was expected to be under way or planned by yearend, and construction of a 214,000-b/d refinery in Shanqiu was to be accelerated.

Oil pipeline in Chad

CNPC started work in mid-2009 on a 300-km (186-mi) oil pipeline in Chad. The line is slated for completion by 2013.

Details are given in the Chad section of the Africa chapter of this book.

ESPO pipeline section

CNPC began in May 2009 to build the 992-km (616-mi) section of the Eastern Siberia-Pacific Ocean pipeline in the Mohe region of China's Heilongjiang Province. Construction is scheduled for completion by October 2010.

The section will connect a 67-km (42-mi) spur line—running from Skovorodino in Russia to the Chinese-Russian border—with the city of Daqing. In April, Russia's Transneft started to build the spur.

CNPC also announced plans to build an underwater section of the ESPO pipeline across River Amur, starting in September 2009, aiming for completion in mid-2010.

Oil terminal at Kozmino Bay

Russia's Transneft announced plans to begin loading oil tankers at a terminal on Kozmino Bay (near the far-eastern Russian city of Nakhodka) in December 2009.

The terminal, designed to load 1 million b/d of crude oil, represents a mechanism for sending Russian crude east to Asian markets rather than west to Europe. It will be a terminus of the ESPO pipeline.

About 300,000 b/d will be loaded in 2009, of which about 100,000 b/d will go to China. Some 140,000 b/d will be set aside for spot or contract deals.

The ESPO pipeline will transport 300,000 b/d of crude from Russia to Daqing in China, in a 20-year supply agreement signed in February 2009.

PetroChina said it plans to boost oil storage capacity at Daqing by 200,000 b/d by 2010 to handle the Russian crude.

Oil storage terminals: Dalian

Brightoil Petroleum Holdings Ltd. signed an investment agreement in mid-2009 for construction of a large oil-storage terminal in the Changxing Island Port Area, near the city of Dalian.

Construction of the 8-million-cu-m (50.3-million-bbl) terminal and associated port facilities and pipelines is expected to take four years.

The terminal will store diesel, aviation fuel and crude oil.

Brightoil Petroleum also is building a 2.2-million-cu-m (13.8-million-bbl) oil-storage facility and port facilities at Waidiao Island, Zhoushan City, in Zhejiang province.

Kenyan port project

Kenya's government was in early talks with Beijing in late 2009 about a $3.5-billion port and transport-corridor project that could provide a new export route for Chinese oil.

Details are given in the Kenya section of the Africa chapter of this book.

Refining and Products

JV with Kuwait

Kuwait Petroleum Corp. (KPC) and Sinopec signed an MOU in October 2009 to establish a $9-billion, 300,000-b/d refinery and petrochemical complex on Donghai Island in Zhanjiang City, on the coast in Guangdong Province.

ESPO oil pipeline: Snapshot

The 4,700-km (2,920-mi) Eastern Siberia-Pacific Ocean (ESPO) pipeline is designed to carry 30 million tons/year of crude oil from fields in Tomsk Oblast and Khanty-Mansi Autonomous Okrug to Kozmino Bay, for export to China, Japan and Korea.

It would open up new export markets for Russia and reduce its dependence on European demand.

The ESPO line will also be linked to the existing Omsk-Irkutsk pipeline in Taishet to bring crude oil from East Siberian fields.

Oil is scheduled to begin flowing through the pipeline to China by early 2011.

The pipeline begins in Taishet in East Siberia, passes through Kazachinskoe, Skovorodino and Perevoznaya in Russia, and will eventually be routed to a Pacific port at Kozmino Bay.

A 67-km (42-mi) spur will connect Skovorodino to the Chinese-Russian border, at which point a 960-km (597-mi) section would run through Mohe to Daqing in China's northeast Heilongjiang Province.

After the spur is built, Transneft will develop a 1,963-km (1,220-mi) section from Skovorodino to Kozmino Bay in far eastern Russia, near the city of Nakhodka.

The Taishet-Skovorodino link was slated for commissioning by yearend 2009.

Construction of the Skovorodino-Kozmino Bay section was scheduled to begin in late 2009, and to be commissioned by 2013–2014.

Construction is scheduled for completion in 2013.

The refinery was initially planned for Nansha in Guangzhou Province, but the new site was chosen due to environmental concerns.

The facility, which will process primarily Kuwaiti crude, also will produce 1 mtpy of ethylene.

Sinopec will hold a 50% stake in the venture, which is scheduled to start operations in 2013.

Maoming: Expansion

Maoming Petrochemical Co., a unit of Sinopec, got approval in August 2009 from the NDRC to expand its refining capacity from 270,000 b/d to 510,000 b/d.

This will make Maoming Petrochemical the third Chinese refinery with capacity of more than 400,000 b/d. Others are the PetroChina Dalian Petrochemical Co. (Dalian), and the Sinopec Zhenhai Refining & Chemical Co. (Ningbo).

Tianjin complex

Construction of the Tianjin Petrochemical Integrated Project—a 50:50 JV of Sinopec and Saudi Basic Industries (SABIC)—was expected to be completed in the third quarter.

It will produce about 3.2 mtpy of petrochemical products, including 1 mtpy of ethylene.

The facility will produce ethylene and its downstream products, and possibly polycarbonate, using SABIC's technology and raw materials from the Tianjin Project.

Dushanzi refinery expansion

In mid-2009, expansion of PetroChina's major Dushanzi refinery in Xinjiang Autonomous Region was close to completion, targeting onstream operation by yearend.

The project aims to boost crude refining from about 850,000 bbl/year to 1.4 million bbl/year, and ethylene capacity from 220,000 tpy to 1.2 million tpy.

The crude to be processed will come from Kazakhstan.

PetroChina's investment in the project was estimated at $4 billion.

Gasoline, diesel price hikes

In late June 2009, Chinese officials unexpectedly raised gasoline prices by nearly 9% and diesel prices by nearly 10%, to the highest level ever. The increases were the largest since an 18% boost in June 2008.

Motorists paid just over $3/gal for gasoline.

Oil traders wondered if the increases would stifle demand that had only recently begun to return to growth after months of decline.

In June 2009, the price for benchmark light crude was about $71/bbl, compared to February lows of about $30/bbl.

Inner Mongolia

A subsidiary of PetroChina began a project in mid-2009 to expand by 100,000-b/d the capacity of its refinery in Hohhot, the capital of Inner Mongolia.

The $1.14-billion project will focus on improvements needed to process crude oil from CNPC's Chongqing oil field, as well as on a 150,000-mtpy polypropylene production unit.

Gasoline and diesel will be the refinery's main products.

Total output is designed to reach 4.64 mtpy, including byproducts polypropylene, propylene, benzene, and liquefied gas. Upgraded operation is to start in 2012.

Guangdong Province: Huizhou expansion

CNOOC revised the start date for a Phase II expansion of its 240,000-b/d Huizhou Oil Refinery (Daya Bay, Guangdong Province) to May 2010, somewhat later than originally planned.

The Phase I expansion was completed in April 2009.

No reason was given for the delay, but some observers pointed to a weakening in China's oil demand.

Refined-products depot: Guangzhou

A subsidiary of PetroChina signed an agreement with local development officials in mid-2009, under which PetroChina will build a two major related facilities in the Nansha district of Guangzhou (capital city of Guangdong Province).

The first is a refined-products terminal [765,000 cu m (4.8-million-bbl)]; the second is a 95,000-cu-m (600,000-bbl) refined-products port at Xiaohu Island.

The facility will be the largest refined-products depot ever built by PetroChina in southern China.

Construction was slated to start by yearend and take about two years.

Building refined-products reserves

China announced in May 2009 an initiative to build state reserves of refined oil products such as gasoline and diesel, diversifying strategic stockpiles away from unrefined crude oil.

The announcement was part of industry-specific plans for a $586-billion (Rmb4-trillion), three-year stimulus package.

The State Council said it aimed to lift China's crude oil refining capacity to 405 mtpy (8 million b/d) in 2011, up 18% from 2007.

Tieshan Port project

In May 2009, Sinopec agreed to invest up to $300 million to relocate and upgrade the oil refinery in the city of Beihai.

The project is part of a larger effort to enhance the role of the Tieshan Port Industry Zone as a competitive base for international petrochemical trade.

New coastal-city refineries

By 2012, China expects to build nine large oil refineries in its coastal region, according to plans released in March 2004 by the National Energy Administration.

The focus will be on the southeast coast. Plants in Ningbo, Shanghai and Nanjing—each with a capacity of about 640,000 b/d—would receive imports through Shanghai.

New refineries in Maoming, Guangzhou, Huizhou, Quanzhou, Tianjin and Caohuidian would each have a capacity of about 428,000 b/d).

Oriental Refinery: Tianjin

Construction of the $3.2-billion JV Sino-Russian Oriental Refinery project in the city of Tianjin is slated to start in 2010 and take two years.

Described as the largest joint project between China and Russia, the initiative will build a 214,000-b/d refinery as well as auxiliary facilities in the Binhai Chemical Zone in Tianjin Municipality.

Guangdong Province: Jieyang City

CNPC said in early 2009 that it expects to build a 214,000-b/d refinery in Jieyang City, Guangdong Province. The plan was approved by the national and provincial officials.

Media reported that the project would likely refine heavy oil imported from Venezuela.

In February, CNPC announced a research program for sweetening heavy crude, aiming for use in refineries able to process such Venezuelan oil.

In May 2008, CNPC signed an agreement with Venezuela to jointly exploit the heavy oil in the Orinoco Belt and to set up a large JV oil refinery in China to process that heavy oil.

Guangxi: Qinzhou

In early 2009, CNPC was considering postponing the commissioning date of a new 10-mtpy oil refining project at Qinzhou.

Located in Guangxi Autonomous Region, it would supply products such as gasoline and diesel to Guangxi and Guizhou Provinces, and part of Sichuan.

In response to weakened market conditions, CNPC said it would likely delay completion beyond the original yearend-2009 target date.

Henan Province: Shangqiu

CNPC said in early 2009 that it plans to build a 210,000-b/d oil refinery in Shangqiu City, in Henan Province. Two sites were under study: Yucheng County and Minquan County.

Anhui Province: Anqing

Sinopec and the government of Anhui Province signed an agreement in early 2009 to build a $1.5-billion integrated refinery and chemical project in the city of Anqing.

Construction is expected to begin at yearend 2009 on the 8-mtpy facility.

PetroChina: UK role?

The *Financial Times* reported in June 2009 that PetroChina was contemplating an investment in the Ineos refinery at Grangemouth, Scotland—its first move into the refining sector in Europe.

Details are given in the United Kingdom section of the Europe chapter of this book.

Refinery in Chad

CNPC signed agreements in August 2009 with the government of Chad for construction of a JV refinery in that country.

Details are given in the Chad section of the Africa chapter of this book.

Coal-to-Gas Plant

A subsidiary of CNOOC signed an agreement in early 2009 with the government of Datong city and the Datong Coal Mine Group Co. to build a $4.38-billion plant to convert coal to natural gas in Shanxi.

This represents the largest investment ever in coal-based natural gas in China.

The 4-bcm facility also will produce synthesis gas, gasoline, diesel, hydrogen, oxygen and other chemical products.

CNOOC also will build a pipeline to transport natural gas from northern Shanxi to several regions around the Bohai Sea. The pipeline is expected to be completed by 2013.

Petrochemicals

Stimulus plan

In March 2009, China unveiled a National Stimulation Plan for the Petrochemical Industry. The initiative will invest about $73 billion to develop the country's oil and petrochemical industry, aiming to boost the value of the sector by about $256 billion by 2011.

Most of the funds will support projects to produce products such as refined oil, ethylene, xylene, purified terephthalic acid, fertilizers and olefins. About 20 projects were identified for support.

Plans for new plants

China reported plans in mid-2009 to construct several new domestic petrochemical plants. According to its 2009–2011 industry blueprint, the State Council plans to raise ethylene production capacity by 51.2% to 15.5 mtpy.

Chinese refineries currently produce about 47.6% of the 21 mtpy of ethylene used in the country.

The new plants are in addition to current Chinese efforts to boost existing output capacity.

- PetroChina plans to complete a 1-mtpy addition to its 220,000-tpy ethylene plant in Dushanzi, Xinjiang Province, and to double the annual capacity of its ethylene plant in Daqing, Heilongjiang Province, to 1.2 mtpy.
- Sinopec was building a 1-mtpy ethylene plant in Ningbo, and started up in November a plant of similar size in Tianjin.

Observers said the action is aimed largely at enabling the country to compete with Middle Eastern ethylene producers, who are expected to more than double their production capacity of three major types of polyethylene (to 14.81 mtpy) in 2012 compared to 2007.

The Chinese and Middle East plans raised concern in Southeast Asia that production will outstrip demand and depress prices.

New ethylene plant: Shaanxi

In April 2009, Shaanxi Yanchang Petroleum (Group) Co. Ltd. announced plans to cooperate with CNPC to build a 1-mtpy ethylene project in Yan'an City in Shaanxi Province. Total investment is expected to be $3.36 billion.

Joint Sino-Russian plant

Sinopec was reported to be in discussion in late 2009 with Russia's Rosneft about a potential partnership to establish a jointly funded refining and petrochemical complex, to be located in the Primorsky region of eastern Russia.

Details are given in the Russia section of the Former Soviet Union chapter of this book.

Environmental Issues

Emissions targets: Prep for Copenhagen

In late November 2009, China said it intended to cut its greenhouse gas emissions per unit of economic output (its "carbon intensity") by 40–45% by 2020.

The announcement was seen as critical to reaching a global agreement to tackle climate change at the international Conference of the Parties (COP) 15, December 7–18, 2009 in Copenhagen.

In August, NDRC officials said that the country's carbon emissions will peak by 2050, the first time such a time frame was specified.

China is the world's largest emitter of carbon dioxide.

More details are given in the Climate Change Issues section of the Key Stats and Tables chapter of this book.

Company News

CIC stake in KazMunaiGas unit

In September 2009, China Investment Corp., the nation's sovereign wealth fund, bought an 11% stake in JSC KazMunaiGas Exploration Production.

CIC said it paid $939 million in the deal with the Kazakhstan-based oil and gas company.

Green Dragon/ConocoPhillips: CBM JV

Green Dragon Gas Ltd., a specialist in Chinese coal bed methane development, established a JV with ConocoPhillips in mid-2009.

The latter will make an initial payment of $20 million to a subsidiary of Green Dragon that holds CBM licenses in four Chinese provinces.

ConocoPhillips also will fund up to $30 million of capital expenditure and can elect to continue with a second phase.

Interest in Repsol YPF

Two large Chinese oil groups were reported to be discussing possible asset purchases and JVs with Spain's Repsol YPF in mid-2009.

One issue was possible sale to CNPC of Repsol's major stake in YPF. That Argentinean company accounts for about 65% of the Spanish group's oil production. Value of the deal was estimated at $17 billion. CNOOC also was said to be interested in a minority stake in YPF.

Takeover of Addax

Sinopec agreed in mid-2009 to a $7.2-billion takeover of Canada's Addax Petroleum Corp., which has interests in Africa (Nigeria, Gabon) and Iraqi Kurdistan (the Taq Taq oil field and refinery).

The deal was described as China's largest-ever outbound investment in the oil and gas sector.

PetroChina stake in SPC

PetroChina agreed in mid-2009 to pay $1.02 billion to Keppel Corp. for a minority stake (just over 45%) in Singapore Petroleum Co.

SPC's interests include oil and gas exploration in Australia, China, Indonesia, Vietnam, and Cambodia.

Angola oil field

In mid-2009, CNOOC and Sinopec agreed to buy a 20% stake in a subsea oil field on Block 32 off Angola for $1.3 billion from Marathon Oil Corp.

Details are given in the Angola section of the Africa chapter of this book.

Asia Pacific

AFGHANISTAN

Capital: Kabul
Monetary unit: Afghani
Refining capacity: 0
Oil production: 0
Oil reserves: 0
Gas reserves: 1.750 tcf

Three blocks offered

In November 2009, the Ministry of Mines received one bid in the First Afghan Hydrocarbon Bidding Round. Orient Petroleum International Inc. bid for rights to explore the Kashkari oil block.

The bid is to be opened in February 2010.

In March, the ministry had offered three blocks in the lightly explored Afghan-Tajik and Am Darya basins. The other two were Jangalikalan (the biggest) and Juma-Bashikurd.

All three contains oil and-or gas discoveries, most of which have never been produced.

Estimates of recoverable reserves are 19 bcm (671 bcf) of gas, 34 bcm (1.2 tcf) of gas, and 64.4 million bbl of oil, respectively. Possible reserves for Kashkari were estimated at 143.9 million bbl.

Based on its 2006 assessment of Afghanistan, the U.S. Geological Survey concluded that:

- much of petroleum resource potential and all the known oil and gas reserves are in the north
- most of the undiscovered oil is in the Afghan-Tajik basin to the east
- most of the undiscovered gas is in the Amu Darya basin that extends into Turkmenistan.

Election

Voting to choose a president was held on August 20, 2009. After weeks of vote-counting and investigation of irregularities, the Independent Election Commission declared Hamid Karzai the winner on November 2.

Evidence of widespread voting fraud left doubts about the legitimacy of the win by Karzai, who is to serve another 5-year term.

A key rival withdrew from a run-off contest that was to be held November 7, saying that the second vote also would lack credibility.

Afghanistan License Blocks

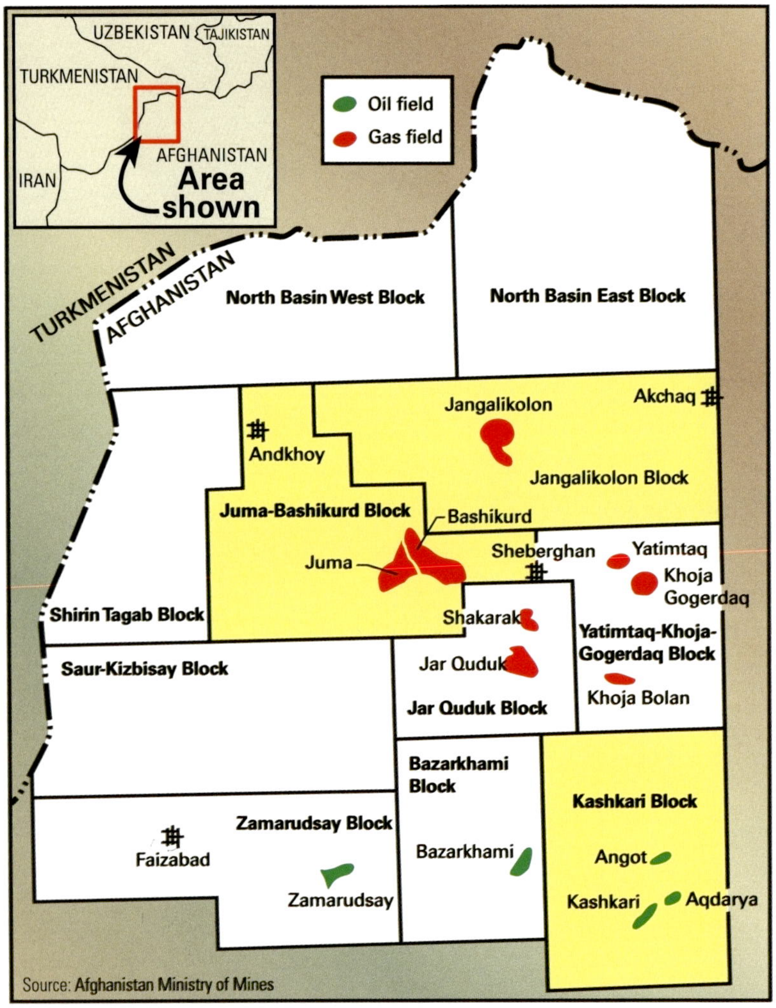

Source: Afghanistan Ministry of Mines

AUSTRALIA

Capital: Canberra
Monetary unit: Dollar
Refining capacity: 724,736 b/cd
Oil production: 465,000 b/d
Oil reserves: 3.318 billion bbl
Gas reserves: 110.000 tcf

Exploration and Development

Two finds in WA-374-P

A group led by Chevron Australia made another natural gas find in late 2009 in its exploration permit WA-374-P offshore Western Australia. The Satyr-1 wildcat intersected 130 m (427 ft) of net gas pay.

The group's Achilles-1 well in the same permit intersected a 100-m (62-ft) net gas pay zone in the Triassic-age Mungaroo sandstone reservoir target.

Both finds are expected to provide additional gas supplies to support the massive Gorgon domestic gas and LNG project.

Gas extension: Tasmania

In late 2009, Origin Energy Ltd. confirmed an extension to the Trefoil-1 gas discovery in the Bass basin permit T/18P off northern Tasmania with success at Trefoil-2 appraisal.

Several gas-bearing sands were intersected within the Eastern View Coal Measures.

Trefoil is close to the company's producing Yolla gas-condensate field.

Beetaloo subbasin: Shenandoah-1

Falcon Oil & Gas Ltd. said in late 2009 that it plans to reenter, deepen, and test the Shenandoah-1 exploration well in the Beetaloo subbasin of Australia's McArthur basin.

The well is southern part of Exploration Permit 98.

Consulting engineers released best estimates of 193 billion standard tank barrels (stb) of unrisked, undiscovered OIP and 10.9 tcm (385 tscf) of unrisked, undiscovered gas in place attributable to Falcon's interest. The prospective resource figures are 19 billion bbl and 1.8 tcm (64 tcf).

Eromanga basin: Chiton-1

Beach Petroleum NL and JV partner, Drillsearch Energy Ltd., made an oil discovery in mid-2009 in permit PEL91 in the South Australian part of the Eromanga basin.

The Chiton-1 wildcat found a 5.5-m (18-ft) oil interval in the primary Mesozoic-age Namur sandstone target. It is the first oil discovery in the permit.

A second well is planned in PEL91 about 8 km (5 mi) north of Chiton, targeting Namur sandstone.

Browse basin: Concerto-1

In mid-2009, Shell Australia found gas close to its earlier Prelude discovery on permit WA-371-P in the Browse basin off northwest Western Australia.

The discovery, in Concerto-1, was said to boost confidence in the company's plans to introduce a floating LNG development scheme to the region.

Prelude, 16 km (10 mi) away, was the first well in the series; Concerto-1 was number 10.

Development plans for Prelude could easily include the Concerto find, said Shell, and both could tie in to Crux field.

Prelude reserves are 56.6–85 bcm (2–3 tcf) of gas and 120 million bbl of condensate.

Cooper-Eromanga basin: Butlers-1

Beach Petroleum Ltd. reported encouraging oil flow of 2,600-b/d in mid-2009 its Butlers-1 discovery well in PEL 92. It is one of the most westerly producing wells in the Cooper-Eromanga basin.

APPEA: More oil discoveries needed

Australia's oil and condensate production is declining, and the country urgently needs new discoveries, according to the Australian Petroleum Production & Exploration Association Ltd.

The country was virtually 100% self-sufficient in 2000, but that figure dropped to just over 55% in 2009. Without new discoveries, it will drop to 32% self-sufficiency by 2017, said APPEA.

This would translate into a deficit in petroleum trade balance of $13.2 billion (Aus.) in 2009 and $28 billion (Aus.) in 2017.

Australian basins

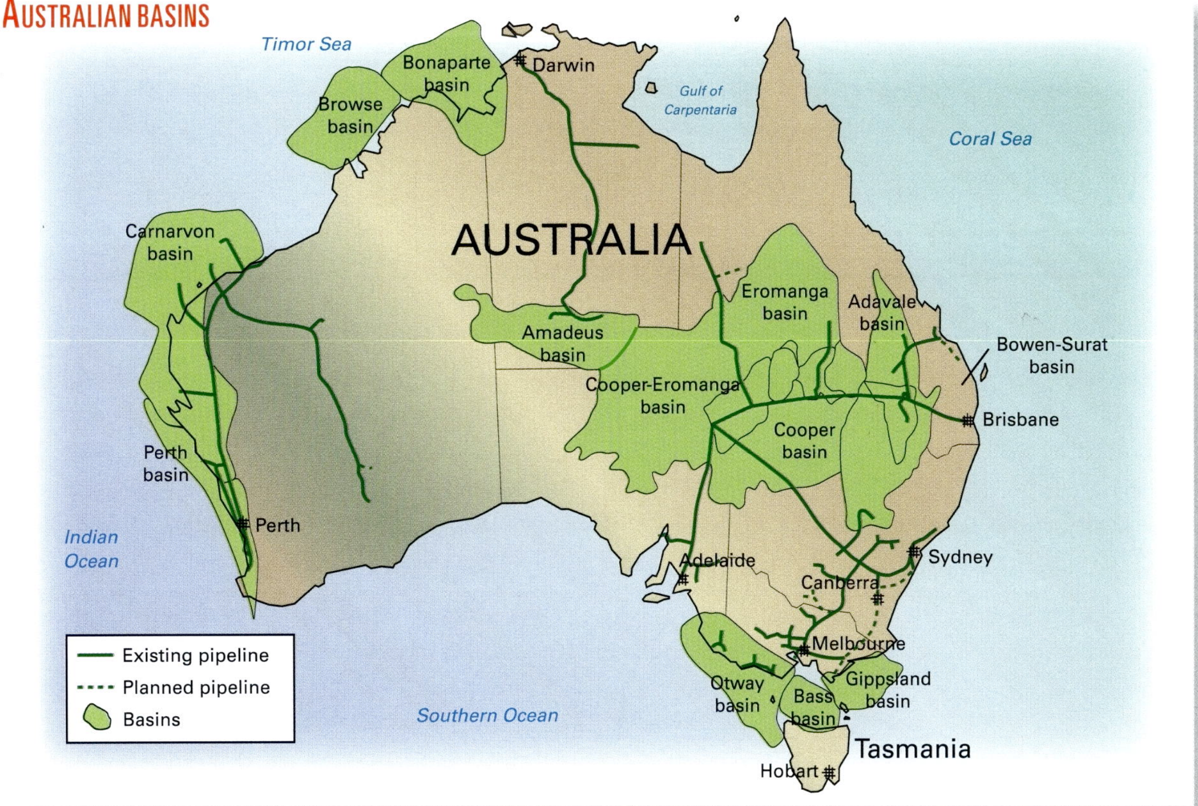

Oliver field

In mid-2009, Stuart Petroleum Ltd. upgraded in the estimated volume of hydrocarbons in the Oliver structure, below the firm's Oliver field. The field is in permit AC/P33, 700 km (435 mi) west of Darwin.

The gross mean potential for the field was raised by 165% over original estimates, to more than 22.7 bcm (800 bcf) of gas in-situ and more than 30 million bbl of condensate in place.

The mean average OIP was raised by 25% to 53 million bbl.

Estimated possible reserves are 42.5 bcm (1.5 tcf) of gas in place, 56 million bbl of condensate in place, and an oil leg of 100 million bbl in place.

The Oliver-2 appraisal well is to be drilled by early 2010.

Browse basin: Poseidon-1

The ConocoPhillips-Karoon Gas Australia Ltd. JV declared its Poseidon-1 wildcat a significant gas discovery in early 2009.

The find is in permit WA-315-P about 480 km (298 mi) north of Broome off Western Australian.

It is north northwest of the Inpex group's Ichthys gas-condensate field.

The well penetrated the Plover Formation reservoir containing gas sands.

The JV estimated proved reserves to be 85 bcm (3 tcf) of gas.

Victoria basin research

In early 2009, Australia's national science agency—Commonwealth Scientific and Industrial Research Organization—and the Dept. of Primary Industries for the state of Victoria (VicDPI) formed a 3-year research partnership.

The group will focus on research "into Victoria's basins to unearth new oil and gas reserves," sharing facilities, expertise and results with industry.

Ichthys field development

In the first half of 2009, Total SA and Inpex launched FEED work for the development of the Ichthys field in the Browse basin offshore Australia.

Two separate FEED contracts were awarded: one for offshore facilities development and a gas pipeline to Darwin; and one for the onshore LNG plant.

Ichthys contains estimated proved and probable reserves of around 530 million bbl of condensate and 363 bcm (12.8 tcf) of natural gas, according to Total.

Block JPDA 06-103

A two-well exploration program by Oilex Ltd. and Japan Energy Corp. was expected to begin by late 2009 on Block JPDA 06-103 in the Timor Sea, between Timor Leste and northwestern Australia.

Details are given in the Timor-Leste section of this chapter.

Carnarvon basin: Martell-1

Early in the year, Woodside Petroleum Ltd. found gas with the first well of its 2009 gas exploration drilling campaign in the Greater Pluto region off Western Australia.

The Martell-1 wildcat was drilled on permit WA-404-P in the Carnarvon basin, 100 km (62 mi) northwest of Pluto field.

The company said that the gas resource is needed to secure its proposed second LNG train at the Pluto project.

Reindeer field

Global economic conditions had cast doubt on development of Reindeer gas field off Western Australia and the associated Devil Creek processing plant, 45 km (28 mi) south of Dampier.

However, in early 2009, Santos Ltd. and a unit of Apache Corp. signed CITIC Pacific as the project's foundation gas buyer and the project was revived. Beginning in 2011, Santos will supply 1.98 bcm (69.75 bcf) of gas over 7 years to a CITIC mining project.

Located on permit WA-209-P, Reindeer has gas reserves of 11–17.3 bcm (390–610 bcf).

A 105-km (65-mi) subsea pipeline will carry the gas to the Devil Creek plant.

Lease Sales

Offshore permits awarded

In April 2009, Australia awarded 13 new offshore petroleum exploration permits in Commonwealth waters, related to the first round of the 2008 Acreage Release.

Forty-three bids were received for 17 areas included in the first closing round of the 2008 release.

The new permits are offshore Western Australia and the Territory of the Ashmore and Cartier Islands.

Five are in the Bonaparte basin, four in the Carnarvon basin, and four in Browse basin.

In fourth quarter 2009, Australia awarded 10 offshore exploration permits in Carnarvon, Roebuck, and Bonaparte basins.

Drilling and Production

Longtom gas field

In late 2009, Nexus Energy Ltd. brought on stream its wholly owned Longtom natural gas field in Bass Strait. Gas flows subsea to the Santos Ltd. onshore facilities in Gippsland, in eastern Victoria.

The field has 2P reserves of 9.4 bcm (333 bcf) of gas and 4 million bbl of condensate.

Bass Strait: Kipper field

Esso Australia secured in early 2009 the use of a semisub to be used in development drilling in 2010 on its Kipper gas-condensate field in Bass Strait. The drilling project could take up to 6 months.

Kipper, 45 km (28 mi) offshore in 100 m (62 ft) of water, has a confirmed resource of 17.6 bcm (620 bcf) of gas and 30 million bbl of condensate.

First gas production of is targeted for first half 2011.

CBM: India

In January 2009, India's ONGC Ltd. entered into an MOU with Arrow Energy Ltd. to cooperate in the coalbed methane (CBM) sector.

The two will jointly participate in the exploration, field development, and production of CBM from existing blocks in India owned by ONGC, as well as from certain areas of Australia.

ONGC estimates that India has 2.6–3.3 tcm (92–117 tcf) of CBM resources. The company has licenses for CBM blocks in the states of Gujarat, Jharkhand, Madhya Pradesh and West Bengal.

Gas Processing

Devil Creek plant

In late 2009, JV partners Apache Energy Ltd. and Santos Ltd. broke ground for the Devil Creek natural gas processing plant, in the Pilbara region of Western Australia, about 50 km (31 mi) south of Karratha.

The plant will supply as much as 5.92 million cu m/d (209 MMcfd) of gas into the Dampier-Bunbury trunk line. It will also produce as much as 500 b/d of condensate.

LNG

Status report: Major projects

A range of LNG projects were moving forward in Australia in 2009 (see table). Presented here are some highlights regarding their progress.

Browse Basin: Shell

- Shell found gas with its Concerto-1 well, 16 km (10 mi) from its earlier Prelude discovery. Both are on the WA-371-P permit.
- Shell awarded a FEED contract for a floating LNG facility at Prelude. The company says production from the $5-billion, 600,000-tonne facility would not start before 2016.

Browse Basin: Inpex/Total SA

- The principals awarded two FEED contracts for the project: one for work on offshore facilities, the other for the LNG plant in Darwin.

Pluto

- Woodside Petroleum Ltd. made a gas discovery with its Martell-1 wildcat well on permit WA-404-P in the Carnarvon basin. The find is 100 km (62 mi) northwest of Pluto field.
- Woodside announced in November 2009 that the foundation Pluto project was 82% complete and on schedule to deliver first LNG in early 2011.
- Petronet committed to buy 1.5 mtpy for 20 years of Pluto LNG output. A similar deal with PetroChina for 3 mtpy was in discussion but on uncertain ground at yearend 2009.

Wheatstone

- Chevron awarded the FEED contract for Wheatstone to Bechtel Group Inc.
- After a multi-well appraisal program, Chevron said that Wheatstone and nearby Iago fields hold enough gas to support a two-train export LNG development onshore as well as a domestic gas component.
- Tokyo Electric Power Co. (TEPCO) agreed to buy 4.1 mtpy of Wheatstone LNG for 20 years, as well as part interests in the project's gas fields and proposed gas processing facilities.
- Apache Energy Ltd. and Kuwait's Kufpec agreed to sell gas from their Julimar and Brunello fields, to supply the Wheatstone LNG project.

Greater Gorgon

- Chevron awarded the FEED contract for the project to Kellogg Joint Venture Group.
- Both state and federal Australian environmental regulatory agencies approved construction of the Gorgon project. The state's approval imposed operational conditions for the Barrow Island site.
- Partner ExxonMobil agreed to sell some of its LNG production to two firms. Petronet LNG of India will buy 1.5 mtpy for 20 years; and PetroChina will buy 2.25 mtpy for 20 years.
- The partners received conditional interim authorization from the Australian Competition and Consumer Commission to market gas in Western Australia.
- Federal and state Australian agencies offered five production licenses to the partnership for operation in the Io-Jansz and Gorgon fields that are the backbone of the project.
- In September, Chevron corporate officials formally approved an investment of $37 billion [$42 billion (Aus.)] in the Gorgon project.
- Also in September, the partnership made its final investment decision to proceed with the project.
- In early 2009, BP PLC said it will sell gas from its Io field to Shell as part of the Gorgon-Jansz project, and then repurchase it after it is processed at the Barrow Island LNG plant. Separate pipelines will link Jansz-Io and Gorgon fields to the proposed three-train plant on Barrow.

Gorgon, other Western Australia LNG

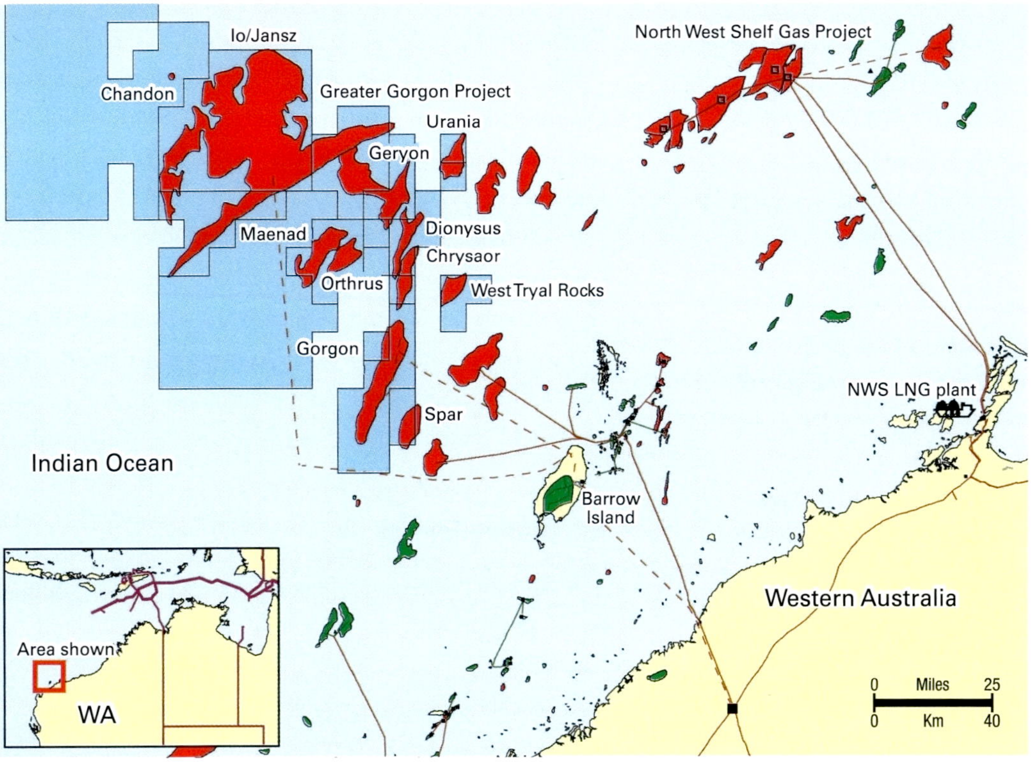

Source: Wood Mackenzie PathFinder.

Bonaparte Basin

- GDF Suez agreed to purchase a 60% stake in three fields in the Bonaparte basin owned by Santos Ltd. Full payment will be made when a final investment decision (still years away) is made. GDF also agreed to support construction of a floating LNG facility.

Gladstone: Fisherman's Landing

- Golar LNG signed a heads of agreement deal with Toyota Tsusho Corp. to sell that firm 1.5 mtpy of LNG from the Fisherman's Landing project for 12 years.
- The Queensland government approved the principals' environmental impact statement for the proposed LNG facility, which enables movement to the development approvals stage.

North West Shelf LNG section

In April 2009, the North West Shelf LNG venture—operated by Woodside Petroleum Ltd. on the Burrup Peninsula, in Western Australia—marked the 20th anniversary of the start of LNG production.

Gladstone: Origin/Conoco

- Australia Pacific LNG (APLNG)—the JV of Origin Energy Ltd. and ConocoPhillips Co—secured a site at Laird Point on Curtis Island from the state government for the JV's $29-billion LNG plant.
- The JV expected to award a design and engineering contract by yearend for development of a 400-km (249-mi) pipeline to bring coalbed methane to the LNG facility.

Gladstone: BG Group/QGC

- China's CNOOC Ltd. signed an agreement to buy 3.6 mtpy of LNG for 20 years from the plant. It also said it will (a) become an equity investor in the first LNG train, (b) buy an interest in the Surat basin gas reserves feeding the plant, and (c) establish a company with BG to build two LNG vessels in China.

Gladstone: Shell/Arrow

- Shell announced details of the proposed LNG facility, including the proposed production of 15–16 mtpy from four trains, rivaling the size of the Origin/Conoco facility. It said output would primarily be exported to Asia.
- Shell and Arrow formally agreed to develop Arrow's estimated 2 bcm (70 tcf) of coalbed methane resources in Queensland as feed for the LNG plant.

Gladstone: Santos/Petronas

- Santos Ltd. awarded FEED contracts to Bechtel Group, Foster Wheeler AG and Fluor Corp. for assessment and planning of various upstream and gas-transport facilities for its LNG facility.
- Petronas (JV partner with Santos) signed a supply agreement to buy 2 mtpy of LNG from the project for 20 years.

Gladstone: General

- In mid-2009, ConocoPhillips said that it might participate in reshuffling the multiple LNG projects planned for Gladstone. It sees merit in optimizing costs for development of CBM fields, constructing pipelines, and building LNG production facilities.

Pipelines/Transportation

QSN gas line

Natural gas flowed in May 2009 through the newly opened Queensland, South Australia, and New South Wales (QSN) pipeline. Epic Energy built the 180-km (112-mi) high-pressure line.

It will transport gas from the Wallumbilla gas hub in eastern Queensland through Epic's Southwest Queensland line to connect with the Moomba-Adelaide and Moomba-Sydney lines.

A second, 113-km (70-mi) line, commissioned by AGL Energy Ltd., will transport gas from a point about 8.5 km (5 mi) east of Miles in central-eastern Queensland to the Wallumbilla hub.

Alternative Energy

A range of projects and policy initiatives were being planned or in development in Australia in 2009, including those listed here.

Government actions/policies

- The Renewable Energy Target (RET) plan issued April 2009 by the Council of Australian Governments (federal and state) commits Australia to obtaining 20% of its energy needs from renewable sources by 2020.

MAJOR LNG PROJECTS IN AUSTRALIA: 2009

Project Name	Principals	Location	LNG Cap. (mpty)	First LNG	Comments
Browse Basin	Shell	Offshore NW Western Australia	3.6	2012	• Gas from Prelude, Concerto fields • Floating LNG facility proposed
Browse Basin	Inpex, Total SA	Offshore NW Western Australia	8	2015	• Gas from Ichthys field • Floating FPSO vessel • LNG plant in Darwin
Pluto	Woodside Petroleum	Northern Carnarvon Basin	4.3	2011	• Gas from offshore Pluto, Xena fields • LNG plant on Burrup Peninsula near Dampier
Wheatstone	Chevron	Offshore Western Australia	8.6	2016	• Gas from offshore Wheatstone and Iago gas fields • LNG plant in Ashburton North, near Onslow (Pilbara region)
Greater Gorgon	Chevron, Shell, ExxonMobil	Offshore Western Australia	15	2014	• Gas from Gorgon, Jansz, Io fields • LNG plant at Barrow Island • Geosequestration of extracted CO_2
Bonaparte Basin	GDF, Santos	Bonaparte Gulf (Timor Sea)	2	TBD	• Gas from Petrel, Tern and Frigate fields • Floating LNG facility proposed
Fisherman's Landing	Golar LNG, Perth Co. LNG, Arrow Energy	Gladstone	3	2012	• Convert coalbed methane to LNG • Gas from Surat basin
Gladstone	Origin Energy, ConocoPhillips	Gladstone: Laird Point, Curtis Island	14–16	2014	• Convert coalbed methane to LNG • Gas from Bowen and Surat basins
Gladstone	BG Group, Queensland Gas Co.	Gladstone: Curtis Island	7.4	2014	• Convert coalbed methane to LNG • Gas from Surat basin
Gladstone	Shell, Arrow Energy	Gladstone: Curtis Island	15–16	2014	• Convert coalbed methane to LNG • Gas from central Queensland
Gladstone	Santos, Petronas	Gladstone: Curtis Island	3–4	2014	• Convert coalbed methane to LNG • Gas from Bowen and Surat basins

- The RET plan also has set a target of generating 45,000 GW-hr of electricity in 2020 from renewable resources.
- Australia became a member of the International Renewable Energy Agency and attended the first meeting of the agency in June 2009.
- In June, the federal government was forced to suspend a program that provided rebates on solar power systems for households. It was extremely popular and exhausted allocated funding.

Solar energy

- The government earmarked about $1 billion for the Solar Flagships project, a long-term plan to develop solar power stations across the country. A call for tenders was to be issued by yearend, with winning applications announced by mid-2010.
- In May, the government said it would co-invest with industry to build the world's largest solar power station, with a capacity of about 1,000 MW.
- A 200-MW "solar tower" was being planned for Buronga, New South Wales. Solar-heated air rising through a large column would drive multiple turbines to generate electricity.
- A solar thermal plant planned for Whyalla in South Australia would use parabolic solar concentrators to produce superheated steam, routed through turbines to generate power.

Environmental Issues

ETS

The federal government said that it will make a third attempt, in February 2010, to gain legislative approval for its Emissions Trading Scheme (ETS). It was defeated in August and December 2009.

The scheme would set a limit on carbon pollution and allow companies to trade permits to pollute within that cap.

The opposition says the plan will raise costs because polluters will pass on the cost of polluting to consumers. The government contends that most families would not see higher costs.

IGCC-CCS power plant

Mitsubishi Heavy Industries Ltd. was negotiating in mid-2009 with an affiliate of the Queensland government about construction of an innovative power station.

The 530-MW Zero-Gen facility is called the world's first coal-fired power station that incorporates integrated gasification combined cycle (IGCC) technology and carbon capture and storage (CCS).

Operation by 2015 is targeted.

CCS institute

April 2009 saw the inauguration of the Australian Global Carbon Capture and Storage Institute, based in Canberra. The federal government will provide annual funding of up to $100 million.

Founding members and collaborating participants include 16 national governments and more than 40 major companies from the energy, technology, finance and research sectors.

Company News

Woodside, Santos: Funds for LNG projects

Woodside Petroleum Ltd. said in late 2009 that it plans to raise $2.5 billion (Aus.) in new equity to boost its deteriorating balance sheet and fund future LNG projects in Western Australia.

Santos Ltd. announced in mid-year its plans to raise up to $3 billion (Aus.) to help fund its LNG project in Papua New Guinea.

Total-Santos deal

In mid-2009, GDF Suez bought 60% of the interests of Santos Ltd. in the Petrel, Tern, and Frigate gas fields in Australia's Bonaparte Gulf.

Details are given in the France section of the Europe chapter of this book.

Santos: CBM assets

In mid-2009, Santos Ltd. increased its coalbed methane (CBM) assets with a corporate acquisition and an acreage buy in the Gunnedah basin of northern New South Wales.

The company acquired 19.9% interest in Eastern Star Gas Ltd. It also bought the 35% interest of Gastar Exploration Ltd in several Gunnedah basin permits and production areas operated by ESG.

BG-Pure Energy

In March 2009, BG Group won control of Pure Energy Resources after a months-long battle that also involved Arrow Energy and Arrow's JV partner, Royal Dutch Shell.

Pure holds coalbed methane (CBM) resources that BP wants to bolster its position in the CBM sector.

Bangladesh Offshore Bidding Tracts: 2008

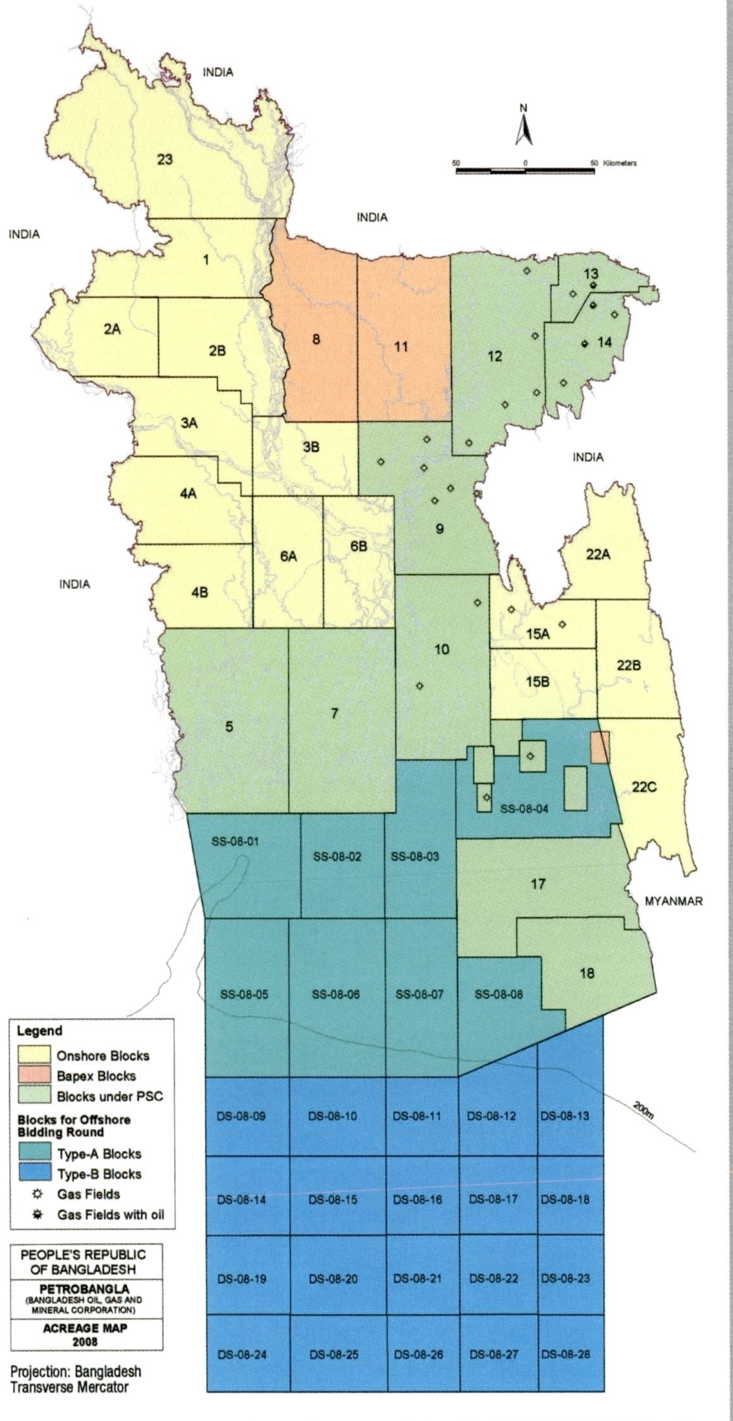

BANGLADESH

Capital: Dhaka
Monetary unit: Taka
Refining capacity: 33,000 b/cd
Oil production: 6,000 b/d
Oil reserves: 28.000 million bbl
Gas reserves: 6.900 tcf

Activity in 2009

Several foreign firms were exploring and producing gas in Bangladesh in 2009, including Tullow Oil PLC, Chevron Corp., Halliburton, Cairn Energy PLC, Total SA, and Santos Ltd.

Block 7 survey

In early 2009, U.S.-based Chevron Bangladesh was poised to launch a 3-month 2D seismic survey in Block 7 to delineate potential gas reserves and initiate subsequent exploration.

The company estimates that the block may hold 14.2 bcm (500 bcf) of gas.

The onshore and offshore survey could include up to 20 km of the Bay of Bengal within Bangladesh territory.

Focus: Deepwater gas

In mid-2009, Petrobangla told *Offshore* magazine that the country needs near-term investment of $8 billion for appraisal and development drilling to convert existing and probable and possible gas reserves into proved reserves of about 57 bcm (20 tcf).

The $8 billion would come from both IOCs and the government.

The expected focus will be on 20 deepwater blocks [2,000-m (6,562-ft) water depth] in the Bay of Bengal, each covering about 3,300 sq km (1,274 sq mi).

In 2008, ConocoPhillips Co. bid for eight deepwater blocks and Tullow Oil PLC for one, even though no seismic data were offered.

Petrobangla has approved Chevron's exploration plans, but it also wants to fast-track the evaluation of the other eight blocks (as well as development of ancillary facilities) and the seismic survey of the remaining eleven.

Assistance may come from Thailand. That country's PTTEP has an MOU with Petrobangla for work on "mutually beneficial energy projects," and also has stakes in Blocks 17 and 18.

Meeting gas demand

Gas production in 2009 was averaging 52.1 bcmd (1.84 tcfd) in Bangladesh, while domestic demand was 57.9 bcmd (2.046 tcfd). In 2020, demand is projected to reach 130.3 bcmd (4.6 bcfd).

The gas delivery system comprises 1,995 km (1,240 mi) of pipelines that can carry 56.6 million cu m/d (2 bcf/d). Plans for expansion include construction by 2011 of an 850-km (528-mi) pipeline. A 540-km (335-mi) gas pipeline also is planned.

Bangora field

In early 2009, Tullow Oil PLC appointed state-run Bangladesh Petroleum Exploration and Production Company (BAPEX) to produce gas at Bangora field, about 100 km (63 mi) southeast of Dhaka.

It marked the first time that BAPEX had been contracted by an international oil explorer.

Tullow plans to produce an additional 850,000 cu m/d (30 MMcfd) of gas from the field.

Located in Block 9, the field had been producing about 3.53 million cu m/d (100 MMcfd) gas from three wells, but technical problems prevented production at Well 3, on which BAPEX will focus.

Nuclear plant

In 2009, the government initiated plans to build a 1,000-MW nuclear power plant.

The $1.3-billion plant is to be built in Rooppur in the Pabna District, about 200 km (124 mi) from Dhaka, with a target to complete construction by 2019.

Companies from several nations—including Russia, China, and South Korea—were contending to secure a construction contract.

A technical team from Rosatom, Russia's nuclear energy corporation, visited Bangladesh in April.

BRUNEI

Capital: Bandar Seri Bagawan
Monetary unit: Dollar
Refining capacity: 8,600 b/cd
Oil production: 148,000 b/d
Oil reserves: 1.100 billion bbl
Gas reserves: 13.800 tcf

Block L

Nations Southeast Asia Ltd., operator of onshore Block L, shot 350 sq km (135 sq mi) of 3D seismic in the southwest part of the block in mid-2009, near the productive Seria oil field.

The firm plans to drill two exploration wells on the block, starting in first quarter 2010.

Brunei Shell Petroleum Co. announced a large dry, sweet gas discovery just offshore at Bubut, less than 1 km (0.6 mi) from the Block L boundary.

Block M, Belait field

The Borneo subsidiary of Australia's Tap Oil Ltd. identified in third quarter 2009 a contingent resource of 8–64 million boe in Brunei's undeveloped Belait onshore oil and gas field.

The field is 65 km (40 mi) southwest of Bandar Seri Begawan, on Block M.

Seismic work was completed and plans were being made to drill at least two wells in mid-2010.

A total of 18 wells have been drilled on Belait since its discovery in 1913.

Maritime boundary with Malaysia

In March 2009, Brunei reached agreement with Malaysia regarding their long-disputed maritime boundary.

It was part of a negotiation that set up a "commercial arrangement area" under which oil and gas revenues in the formerly disputed area will be shared between the two nations.

One promising part of the newly defined area is Block L, on which seismic evaluation was launched in late 2008.

Gas-to-methanol plant

Work continued in 2009 on the first natural-gas-based petrochemical plant in the Sultanate of Brunei.

Scheduled to come online in early 2010, the $400-million complex will convert methane gas to methanol at the Sungai Liang Industrial Park.

The plant is being designed by Mitsubishi Heavy Industries Ltd. to produce up to 2,500 tpd (850,000 tpy) of methanol as well as downstream products such as polymers, adhesives, and resins.

LNG assurances to Japan

In November 2009, Brunei reaffirmed that the sultanate will provide stable, long-term supplies of LNG to Japan. The confirmation followed personal intervention by the Japanese prime minister with his counterpart from Brunei.

Japan relies heavily on imported oil and gas, bringing into the country about 60 mtpy of LNG from several countries, including Indonesia, Malaysia, and Australia.

Recent difficulties with Indonesia led Japan to redouble efforts to diversify its sources of energy.

INDIA

Capital: New Delhi
Monetary unit: Rupee
Refining capacity: 2.836 million b/cd
Oil production: 664,000 b/d
Oil reserves: 5.625 billion bbl
Gas reserves: 37.960 tcf

Overview

Major oil and gas companies

Several companies (state-owned unless otherwise stated) dominate India's energy sector.

- Oil and Natural Gas Corp. (ONGC; Dehradun) is the largest and most active operator in India.
- Oil India Ltd. (OIL; Duliajan) is India's second largest domestic exploration and production company.
- Gujarat State Petroleum Corp. (GSPC) operates about a dozen oil and gas fields in India.
- GAIL (India) Ltd. (New Delhi)—formerly Gas Authority of India Ltd.—is the country's principal gas transmission and marketing company. It is also involved in exploration, production, and processing.
- Cairn India Ltd. (Gurgaon) was established as a separate entity in early 2007 by Cairn Energy PLC (Edinburgh, Scotland). Cairn Energy, a major participant in India's hydrocarbon sector since the late 1990s, holds a major interest in Cairn India Ltd.

Oil

Most of India's oil reserves are located in the Mumbai High, Upper Assam, Cambay, Krishna-Godavari, and Cauvery basins. The offshore Mumbai High field is by far India's largest producing field.

There are substantial undeveloped reserves in the offshore Bay of Bengal and in Rajasthan state.

Rising oil consumption and fairly stable production levels make India increasingly dependent on imports. EIA estimated that in 2008, the country produced an average of 650,000 b/d of crude oil, but consumed nearly 3 million b/d.

Refining

The Jamnagar refinery owned by Reliance Petroleum is India's largest. Reliance brought online a second facility at that site in early 2009, making it the largest refining complex in the world, with a total capacity of more than 1 million b/d.

Indian Oil Corp., Ltd., which operates 10 refineries, has announced plans to double its refining capacity by 2012.

Natural gas

EIA estimates that India produced 31.1 bcm (1.1 tcf) of natural gas in 2007, up slightly from 2006. The bulk of the production comes from the western offshore regions, especially the Mumbai High complex. ONGC and OIL account for the bulk of natural gas production.

The onshore fields in Assam, Andhra Pradesh, and Gujarat states are also significant sources of natural gas. The Bay of Bengal has also become an important source of natural gas for the country.

In 2007, India consumed roughly 4.2 bcm (1.5 tcf) of natural gas, some 2.8 bcm (100 bcf) more than in 2006.

Gas demand is expected to grow considerably, largely driven by demand in the power sector. The power and fertilizer sectors account for nearly three-quarters of natural gas consumption in India.

By 2030, EIA expects that gas use in India will be a major component of a doubling of natural gas demand in Asia.

India has been a net importer of natural gas since 2004. Net imports reached 10 bcm (353 bcf) in 2007.

Strategic Petroleum Reserves

In late 2009, India announced plans to increase its strategic petroleum reserves to 8.5 million tons (62.3 million bbl) by 2012. Current reserves are about 3.3 million tons (24 million bbl), held by Hindustan Petroleum Corp. Ltd., Bharat Petroleum Corp. Ltd., and Indian Oil Corp. Ltd.

Outlook for India's consumption

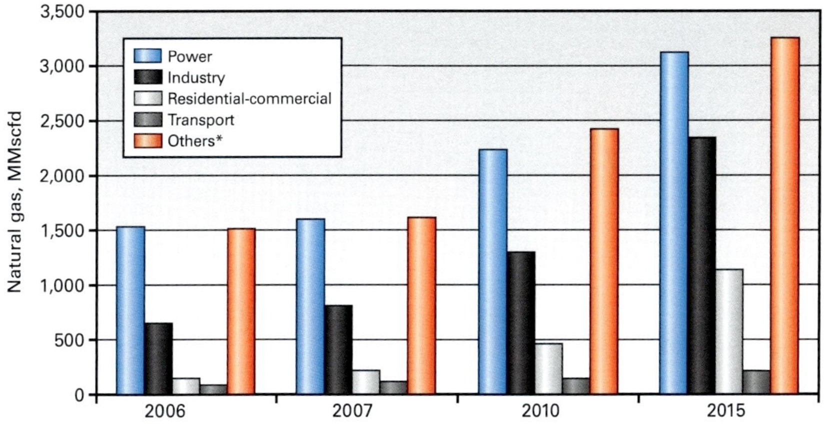

*Includes fertilizer use, LPG/C2-C3 shrinkage, and nonspecified others; excludes distribution losses.

Parliamentary elections

The Congress party swept parliamentary elections in May 2009, an outcome seen as a clear mandate to modernize the nation's economy. Manmohan Singh will continue as premier.

The win appears to give the Congress-led United Progressive Alliance the freedom to push ahead with economic reforms, yet also to pursue efforts to expand the economy.

Observers expect that the government will move ahead with privatization, greater private-sector participation in infrastructure development, and education reforms.

The additional reserves will be stored in underground rock or salt caverns at Padur in Tamil Nadu, Mangalore in Karnataka, and Vishakhapatnam in Andhra Pradesh.

Exploration and Development

Cambay: Block CB-ONN-2003/1

Reliance Industries Ltd. reported a Miocene oil discovery in the Cambay basin southwest of Ahmedabad that the company said "is expected to open future potential within the block."

Well CB10A-A1 in Part A of two-part Block CB-ONN-2003/1 flowed at 500 b/d of oil from the Babaguru formation.

Part A to the west covers 570 sq km (220 sq mi); Part B to the east covers 65 sq km (25 sq mi).

Reliance named the discovery Dhirubhai-43.

Cambay: Tarapur 1 field

GSPC and GeoGlobal Resources Inc. began development in mid-2009 of Tarapur 1 oil field in Cambay basin. Plans called for drilling 17 wells by yearend.

Just over 1.75 million standard tank barrels (stb) of proved developed non-producing oil reserves are attributed to the first six wells. Production in June from the first three averaged 466 b/d and 11,900 cu m/d (420,000 cfd).

Deen Dayal West field

The government approved the field development plan in late 2009 for the Deen Dayal West gas field in Block KG-OSN-2001/3.

The plan includes 4 existing wells (KG No. 8, KG No. 15, KG No. 17, and KG No. 28), and 11 new development wells (three slant wells and eight multilateral).

First gas production is slated for late 2011.

CBM: Raniganj Block

ONGC Ltd. said in early 2009 that it would begin the second (pilot) phase of coal bed methane (CBM) development by yearend on the Raniganj Block in West Bengal. The first phase of the three-stage exploration process was completed in late 2008.

ONGC also was in the advanced stages of developing the Jharia CBM project in Jharkhand.

The company estimates its CBM reserves in Bokaro and South Karanpura at 18 bcm (636 bcf).

CBM: India and Australia

In January 2009, ONGC Ltd. signed an MOU with Australia's Arrow Energy Ltd. to cooperate in the coalbed methane (CBM) sector.

At issue is potential cooperation on existing ONGC blocks from previous CBM license rounds in India and possible collaboration on Australian acreage.

ONGC has licenses for CBM blocks in Gujarat, Jharkhand, Madhya Pradesh and West Bengal states.

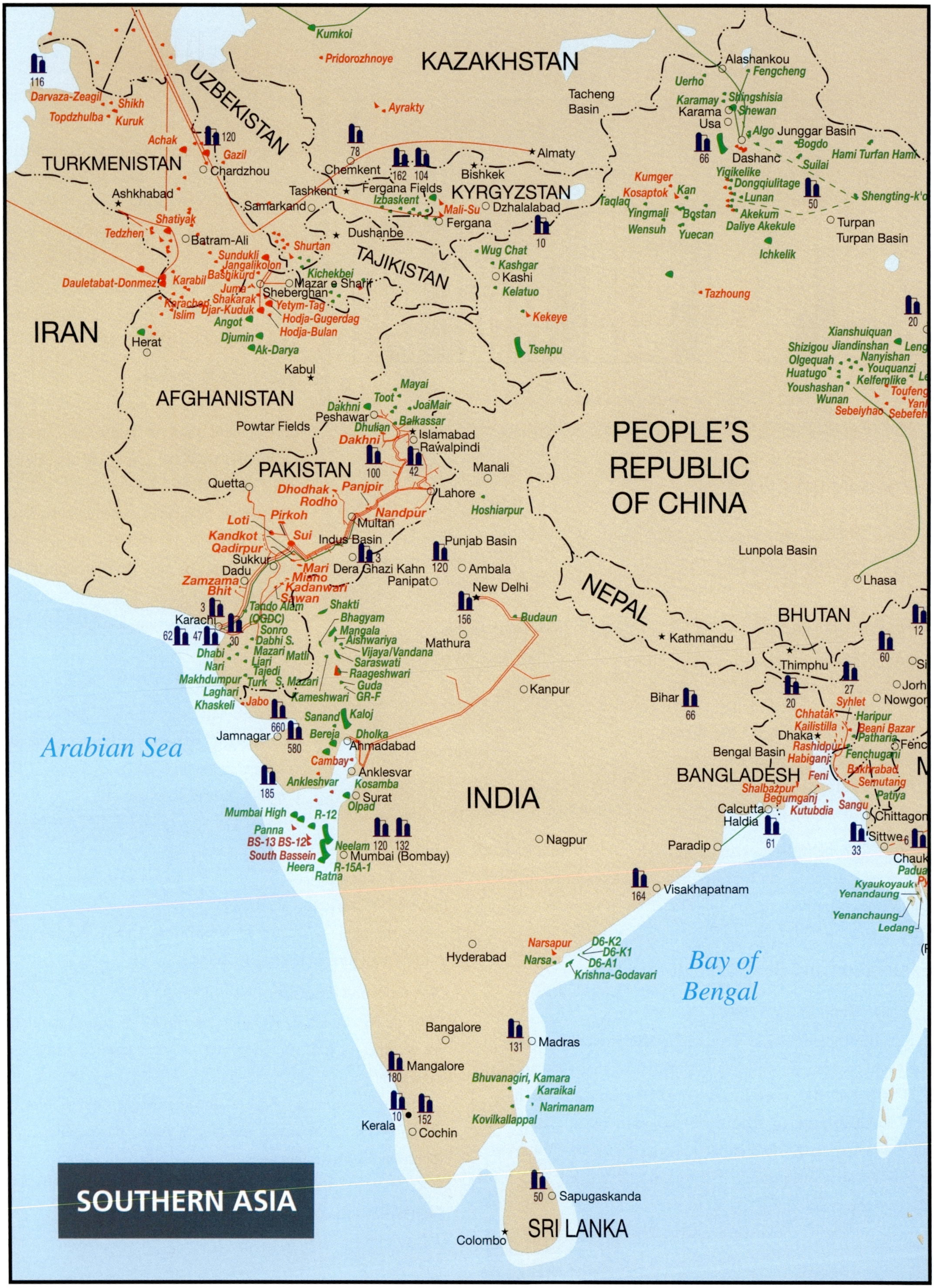
SOUTHERN ASIA
KAZAKHSTAN
UZBEKISTAN
TURKMENISTAN
KYRGYZSTAN
TAJIKISTAN
IRAN
AFGHANISTAN
PAKISTAN
PEOPLE'S REPUBLIC OF CHINA
NEPAL
BHUTAN
BANGLADESH
INDIA
SRI LANKA
Arabian Sea
Bay of Bengal
Kumkoi
Pridorozhnoye
Ayrakty
Darvaza-Zeagil
Shikh
Topdzhulba
Kuruk
Achak
Gazil
Chardzhou
Ashkhabad
Shatiyak
Tedzhen
Batram-Ali
Samarkand
Tashkent
Chemkent
Fergana Fields
Izbaskent
Mali-Su
Dzhalalabad
Fergana
Bishkek
Almaty
Dushanbe
Shurtan
Sundukli
Jangalikolon
Bashikurd
Karabil
Dauletabat-Donmez
Juma
Shakarak
Karachop
Islim
Djar-Kuduk
Yetym-Tag
Hodja-Gugerdag
Hodja-Bulan
Kichekbei
Mazar e Sharif
Sheberghan
Angot
Djumin
Ak-Darya
Herat
Kabul
Alashankou
Fengcheng
Uerho
Karamay
Shingshisia
Karama
Shewan
Usa
Tacheng Basin
Algo
Junggar Basin
Dashanc
Bogdo
Hami Turfan Hami
Suilai
Yigikelike
Dongqiulitage
Lunan
Akekum
Daliye Akekule
Shengting-k'o
Turpan
Turpan Basin
Kumger
Kosaptok
Kan
Taqlaq
Yingmali
Bostan
Wensuh
Yuecan
Ichkelik
Wug Chat
Kashgar
Kashi
Kelatuo
Tazhoung
Kekeye
Tsehpu
Xianshuiquan
Shizigou
Jiandinshan
Olgequah
Nanyishan
Huatugo
Youquanzi
Youshashan
Kelfemlike
Wunan
Toufeng
Sebeiyhao
Sebefeh
Mayai
Dakhni
Toot
JoaMair
Peshawar
Balkassar
Dhulian
Powtar Fields
Islamabad
Rawalpindi
Manali
Quetta
Dhodhak
Panjpir
Rodho
Lahore
Hoshiarpur
Loti
Pirkoh
Nandpur
Multan
Kandkot
Sui
Qadirpur
Indus Basin
Punjab Basin
Sukkur
Dadu
Mari
Dera Ghazi Kahn
Ambala
Zamzama
Miano
Panipat
Bhit
Kadanwari
Sawan
New Delhi
Lunpola Basin
Lhasa
Karachi
Tando Alam
(OGDC)
Shakti
Budaun
Bhagyam
Mangala
Aishwariya
Sonro
Dabhi S.
Mazari
Liari
Tajedi
Matli
Dhabi
Nari
Makhdumpur
Turk
Laghari
Khaskeli
Jabo
S. Mazari
Vijaya/Vandana
Saraswati
Raageshwari
Guda
Kameshwari
GR-F
Mathura
Kathmandu
Thimphu
Kanpur
Bihar
Sanand
Kaloj
Bereja
Dholka
Jamnagar
Ahmadabad
Cambay
Anklesvar
Ankleshvar
Kosamba
Surat
Olpad
Mumbai High
R-12
Panna
BS-13
BS-12
Neelam
South Bassein
Heera
Ratna
R-15A-1
Mumbai (Bombay)
Nagpur
Sylhet
Haripur
Chhatak
Kailistilla
Beani Bazar
Dhaka
Pathania
Rashidpur
Habiganj
Fenchuganj
Bengal Basin
Bakhrabad
Feni
Semutang
Shalbazpur
Begumganj
Patiya
Kutubdia
Sangu
Chittagong
Calcutta
Haldia
Sittwe
Paradip
Visakhapatnam
Kyaukoyauk
Yenandaung
Yenanchaung
Ledang
Narsapur
Narsa
D6-K2
D6-K1
D6-A1
Krishna-Godavari
Hyderabad
Bangalore
Madras
Mangalore
Bhuvanagiri, Kamara
Karaikai
Narimanam
Kovilkallappal
Kerala
Cochin
Sapugaskanda
Colombo

MONGOLIA
Ulan Bator
PEOPLE'S REPUBLIC OF CHINA
N. KOREA
S. KOREA
Pyongyang
Seoul
Unggi
Yellow Sea
South China Sea
Andaman Sea
TAIWAN
PHILIPPINES
HAINAN
VIETNAM
LAOS
THAILAND
CAMBODIA
MYANMAR (BURMA)
MALAYSIA
SABAH
SARAWAK
BHUTAN
Junggar Basin
Bogdo
Hami Turfan Hami
Suilai
Shengting-k'ou
Turpan
Turpan Basin
Pai-yang-ho
Ya'exia
Shih-you-kou
Danbei
Baiyanghe
Laojunmiao
Shiyougou
Lenghu
Xianshuiquan
Shizigou
Jiandinshan
Lenghusihao
Nanyishan
Olgequah
Youquanzi
Huatugo
Kelfemlike
Lenghuwuhao
Youshashan
Wunan
Toufengshan
Yanhu
Sebeiyhao
Sebefehao
Qinghai Hu Basin
Lanzou
Erlian Basin
Eerdous Basin
Shenglinjin
Liujiazhuang
Majiatan
Lizhuangzi
Hongjinzi
Donghongchuang
Dashuikeng
Zichang
Baiyanjin
Wuji
Yongping
Maling
Quinghazhi
Danbei
Gangua
Jiuqan (Juixi) Basin
Lhasa
Lhasa Basin
Basin
Chinkanshou
Changchun
Kongchuling
Tangluku
N. China Basin
Fahniu
Shuguang
Bohai Basin
Chin-hsi
Huanxing
Yuoy
Jinzhou
GanSuizhong 36-1
Beijing
Tiamin
Layemajo
Chinshsi
Gaoshangbao
Pohi Bay
Takang
Bozhong 28-1
Bozhong 34-2/4E
Yanlin
Shengil
Dawangzhuang
Qingdao
Lin-4
Zhongyuan
Subei Basin
Wecheng
Wenmingzhai
Wennan
Wenzhong
Pucheng
Wenliu
Fumim
Liuzhuang
Lianmengzhuang
Machang
Ahhui
Huanggue
Nanyang Basin
Wangil
Sha-Yen
Zhangdian
Xiamen
Welgang
Shuanghe
Dongzhuang
Qianjiang
Shanghai
Donghai
Bo Yuan Ting
E. China Basin
Ping Hu
Hangzhou
Guanghau
Wuhan
Wanchang
Zhangang
Xijiayuan
Tuoshi
Huayuan
Gaochang
Xingqou
Quingjiang Basin
Jianghan Basin
Zhongba
Zhongtaishan
Jinhuazhen
Leiyinpu
Guanan
Wubaiti
Gihua
Wolongha
Penglaizhen
Longnushi
Wanshunchang
Daxinxi
Bose Basin
Weiyuan
Shiyoug
Shilongxnou
Dian-Qian-Gui Basin
Fuzhow
Quanzhou
Chuwei
Taipei
Jinmen
Paoshan
Shoal Basin
Yungoshan
Sanshui Basin
Kaohsiung
Santou
Guiyang
Wanglongchang
Taihehums
Shiwan-Dashan Basin
Guangzhou
Hong Kong
Tiandong
Macao
K'unming
Maoming
Lufeng 13-1
Lufeng 15-1
Hizhou 26-1
Beihai
Xijiang 30-2
Liuhua 11-1
Weizhou 10-3
Weizhou 11-1
Fuizhou 21-1
Liuhua 4-1
Zhujiuang Basin
Weishou 6-1
Tien Hai
Haik'ou
Wenchang 9-2
Jinfeng
Wenchang 8-3
Wenchang 14-3
Wenchang 13-1
Dongfang
Wenchang 19-1
Wenchang 13-2
Yacheng 13-1
Digboi
Thimphu
Sibsagar
Jorhat
Nowgong
Myitkyina
Sylhet
Haripur
Chhatak
Kailistilla
Beani Bazar
Dhaka
Patharia
Fenchugani
Rashidpur
Habiganj
Bakhrabad
Feni
Semutang
Patiya
Kutubdia
Sangu
Chittagong
Ayadaw
Thanbayakan
Sittwe
Chauk/Lanywa
Chauk
Yenangyaung
Mann
Mae Fang
Paduakpin
Htaukshabin
Mae Suiti
Kyaukoyauk
Pyato
Pyaye
Yenandaung
Prome
Fang
Vietiane
Myanaung
Yenanchaung
Shwepyitha
Ledang
Yangon (Rangoon)
Phu Horm
Nam Phong
Mae Sot
Sirikit
KDN-1
Pyaye
Chonnabot
Khon Kaen
Syriam
Pyato
Shwepyitha
Yadana
Martaban
Bung Ya
Sawng
Neung
Bangkok
Phnom Penh
Nha Trang
Yetagun
Sriracha
Ho Chi Minh City
Plamuk
Kapong
Surat
Platong
Nang Nuan
Platong S.
Pladang
Tantawan
Chaba
Dara
Trat
Erawan
Sabun
Baanpot
Jakrawan
Pailin
Punan
Arthit
Songkhla
Bongkot
Bua Ban
Pilong
17-B-1
Damar
Jemeh
Bintang
Sepat
Noring
Inas
Bedong
Laho
Sutu Den
Rang Dong
Bach Ho
Rong
Dai Hung
Dua-1X
Bursa
Natuna
Guloc
Malampaya
West Linapacan
Camago
Matinlok
Nido
Cadlao
Manila
Limay
Batangas
S.W. Emerald
Barton
Tembungo
S. Furious
Erb W.
St. Joseph
Kinarut
Erb S.
Samarang
Glayze
Jintan
Bijan
Brunei
M1
M3
B12
M4
B11
F6
Lutong
F28
E11
E8
Bunya Nibung
Tarakan
50
70
20
17
100
30
172
170
42
50
100
110
100
70
29
270
565
817
74
140
310
730
10
70
60
146
134
160
206
24
100
50
50
56
104
240
200
70
540
300
270
154
270
12
60
27
20
33
6
25
26
62
156
174
193
140
180
93
9
45

CBM: Snapshot

In 2007, Great Eastern Energy Corp. Ltd. began the first coalbed methane operation in India, in West Bengal.

The Directorate of Hydrocarbons said that early studies indicate reserves of nearly 453 bcm (16 tcf) of CBM in the country.

Also as of early 2009, more than two dozen CBM blocks had been awarded, with 10 new blocks on offer in the states of Tamil Nadu, Chhattisgarh, Assam, Maharashtra, and Jharkhand.

Other Indian firms involved in CBM development in India include ONGC Ltd., Reliance Industries Ltd., and Essar Group. ONGC Ltd. is also working with an Australian company to develop CBM resources in that country.

RJ-ON-90/1: Commercial declaration

The government approved a declaration of commerciality in early 2009 for three 2007 gas and oil discoveries by Cairn Energy PLC in Rajasthan.

The declaration covers the Kameshwari West 2, 3, and 6 discoveries and an 822-sq-km (317-sq-mi) development area that was part of the northern appraisal area on the RJ-ON-90/1 license.

Lease Sales

NELP VIII results

The Indian government announced provisional results in October 2009 from its latest (eighth) exploration licensing round, NELP-VIII. The round was launched in April.

On offer were 70 exploration blocks, covering a sedimentary area of 164,000 sq km (63,321 sq m). Acreage included 24 deep-water and 28 shallow-water blocks in the western, eastern, and Andaman Sea sectors.

Notable provisional winners were Cairn Energy India; ONGC Ltd., BG E&P India; GSPC; Oil India Ltd.; a subsidiary of Bengal Energy Ltd.; National Thermal Power Corp.; and BHP Billiton.

Reserves

Krishna-Godavari basin

In mid-2009, Hardy Oil & Gas PLC upgraded the resource potential in its D3 and D9 offshore exploration blocks in the Krishna-Godavari basin off India.

Consulting engineers estimated a prospective resource of 10.8 tcf of gas and 143 million bbl of oil for the two blocks.

Drilling and Production

K-G basin

Production began in April 2009 from fields in the Krishna Godavari (K-G) basin, according to major developer Reliance Industries Ltd. (RIL).

Initial flow was from the Dhirubhai 1 and 3 discoveries in D6 Block.

The basin, discovered in 2000 off India's east coast, is expected to provide 40% of India's domestic hydrocarbon output when it reaches full production of 80 million standard cu m/d (2.8 bcfd).

It was brought into production just 6.5 years after discovery.

Gas from KG-D6 block will go to several companies operating 11 different power plants.

RIL said the expected supply of 18 million standard cu m/d (636 MMscfd) of gas to the power sector would largely eliminate their gas supply deficit and would result in generation of 4,000 MW of power.

PY-1 gas field

Production began in October 2009 from the PY-1 gas field offshore India, according to Hindustan Oil Exploration Co. (HOEC). Gas production is estimated to reach 1.4 million cu m/d (49 MMcfd).

Gas will be sold to GAIL for use in the local power generation market.

Mangala field

Cairn India Ltd. started oil production in third quarter 2009 at its Mangala field in Rajasthan state, which is part of the larger complex of Mangala, Bhagyam, and Aishwariya (MBA) fields.

The Mangala find represents the largest of 25 discoveries made by Cairn in the Barmer basin on Block RJ-ON-90/1.

Initial production was 30,000 b/d, expected to rise to 130,000 b/d by first-half 2010, and then to a peak of 175,000 b/d by yearend 2011.

Cairn said that the MBA fields hold nearly 1 billion bbl of recoverable oil recoverable, including proved plus probable gross reserves.

International Activities

MMS collaboration

The U.S. Minerals Management Service (MMS) signed an MOU in mid-2009 with Indian energy agencies regarding collaboration on energy development topics of mutual interest.

Details are in the United States section of the North America chapter of this book.

GAIL overseas initiatives

GAIL (India) Ltd. announced in early 2009 that it plans to spend $2.1 billion through 2012 on projects to expand overseas operations, aiming to tap possible markets in Russia, China, and Egypt.

Projects being evaluated include:

- a methanol plant in Russia, with ITERA Oil and Gas Co.
- gas distribution networks in selected Chinese cities, with China Gas Holdings Ltd.
- gas distribution networks in Egypt, with two subsidiaries of Royal Dutch Shell.

Venezuela: Heavy crude

A consortium of four leading Indian oil companies were making plans in early 2009 to bid for a minority stake in one of seven heavy-crude blocks in Venezuela's Orinoco Petroleum Belt in the state of Carabobo.

The consortium, led by ONGC Ltd., includes Reliance Industries Ltd., Indian Oil Corp. Ltd., and Oil India Ltd.

Potential investors will bid for a 40% share in the blocks; PDVSA will retain 60%.

The seven blocks on offer contain total estimated reserves of 272 billion bbl, including proven reserves of 80 billion bbl of heavy crude.

The Indian consortium plans to ship upgraded crude to RIL's refineries in Jamnagar and to IOCL's refinery being set up in Paradip.

China: Shwe gas project

ONGC Videsh Ltd.—a subsidiary of ONGC Ltd.—and GAIL India Ltd. were participating in 2009 in a project that aims to produce gas from the Shwe field offshore Myanmar and deliver it by pipeline to China's Yunnan Province.

The two were partners with Daewoo International Corp., Korea Gas Corp., and Myanmar Oil & Gas Enterprise in the somewhat controversial enterprise.

Details are given in the Myanmar section of this chapter.

LNG

Fuel for NTPC

In late 2009, National Thermal Power Corp. Ltd., India's largest power generation company, agreed to buy 1.2 million tons of LNG from Indian Oil Corp. Ltd.; GAIL India Ltd., and Bharat Petroleum Corp. Ltd., who will first purchase the gas from developers of Australia's offshore Gorgon gas fields.

The gas will fuel NTPC's Kayamkulam power station in Kerala.

Ennore

Work began in 2009 on the Ennore LNG terminal near Chennai in Tamil Nadu State. It will function as a storage and distribution hub for gas imports from Iran. An initial capacity of 2.5 mtpy is planned.

Maharashtra

Ratnagiri Gas and Power Pvt Ltd.—formerly Dabhol Power Co.—sought comments in late 2009 from companies that might be interested in leasing its LNG regasification terminal in Maharashtra, 330 km (205 mi) from Mumbai.

Formal bids will be solicited in 2010 for the terminal, which is part of a 1,950-MW power project.

Commissioning is slated for early 2010. Initial capacity will be 1.1 mtpy, growing to 2.7 mtpy by yearend 2010.

Iran project

In December 2009, ONGC Ltd. and a group of partners agreed to invest $10 billion in an LNG project and the development of two large gas fields in Iran.

Details are given in the Iran section of the Middle East chapter of this book.

Dahej expansion

Petronet LNG Ltd. said in late 2009 that it will invest $200 million to expand the Dahej LNG import terminal to a capacity of 15 mtpy by yearend 2012.Current nominal capacity is 10 mtpy.

The firm said that construction will take about 30 months from the date contracts are awarded.

July 2009 saw completion of an earlier project that doubled plant capacity from 5 mtpy to 10 mtpy.

Kochi Island

In May 2009, Petronet LNG Ltd. announced that the LNG regas terminal Kochi was on track to be fully operational in 2012.

The company already has a 25-year agreement with Ras Laffan LNG Co. Ltd. for the supply of 2.5 mtpy of LNG to be processed at Kochi.

Pipelines/Transportation

Iran-Pakistan-India pipeline

Significant progress was made in 2009 on the $8-billion Iran-Pakistan-India (IPI) gas pipeline project, also referred to as the Peace Pipeline.

Details are given in the Iran section of the Middle East chapter of this book.

Gas from West Asia

In 2009, India was reportedly considering development of a $3-billion deepwater multi-pipeline system to bring 8 bcm (282 bcf) of natural gas to the country from Iran, Iraq or Qatar.

A 2014 date has been proposed for the start of construction.

South Asia Gas Enterprise Pvt. Ltd (SAGE), a private Indian firm set up as a special JV, has been working with others to evaluate technical and economic aspects of the project since 2006.

The line is seen as a possible alternative to the Iran-Pakistan-India (IPI) pipeline.

CNG CORRIDOR

GAIL said in early 2009 that it would earmark about $1 billion through 2012 to set up a "compressed natural gas corridor," covering about 230 cities in India.

The project would develop retail CNG outlets to fuel vehicles along national and state highways, and also supply piped cooking gas for household use.

Six cities were under consideration for participation in the first phase of the project: Kakinada, Sonepat, Kota, Dewas, Mathura, and Meerut.

K-G basin to Madhya Pradesh

In early 2009, GAIL (India) Ltd. applied for approval to build a $1.6-billion gas pipeline from fields in the Krishna-Godavari (K-G) Basin in Andhra Pradesh to the state of Madhya Pradesh, to serve consumers in central and northern India.

The 1,100km (684-mi) line from Vijayawada to Vijaipur would carry 9.1–14.6 bcm/year (321–516 bcf/year).

GAIL: Seven projects by 2010

GAIL (India) Ltd. said in early 2009 that it plans to commission seven gas pipeline projects by yearend 2010. These include:

- The 610-km (379-mi) Dahej-Vijaipur II (DVPL-II) project
- The 500-km (311-mi) Vijaipur-Dadri project
- The Vijaipur-Dadri-Dadri-Bawana (VDBPL) project
- The 610-km (379-mi) Dadri-Bawana-Nangal project
- The Chainsa-Jhajjar-Bamnouli pipeline project
- The Chainsa-Jhajjar spur line project
- The 310-km Chainsa-Gurgaon-Jhajjar-Hissar project

GAIL has said that it wants to develop a pan-India gas pipeline grid by 2009–2010.

Paradip-Haldia oil line

In early 2009, IOC commissioned a 330-km (205-mi) crude oil pipeline in eastern India, connecting the port city of Paradip to Haldia, West Bengal.

The addition of this 235,000-b/d line brought the firm's total pipeline network to more than 10,000 km (6,214 mi), with cumulative throughput capacity of 1.5 million b/d of crude oil.

Refining and Products

Paradip refinery

In November 2009, Indian Oil Corp. Ltd. began building the 15-mtpy Paradip refinery complex in Orissa. Commissioning is scheduled for mid-2012. The plant is designed to convert the heaviest and toughest of crude.

Crude imports are expected to come from West Asia, and half of the refined products will be exported to the Asia Pacific region.

Plans to build a petrochemical complex at Orissa have been deferred.

Guru Gobind Singh refinery

Hindustan Petroleum Corp. Ltd. said in early 2009 that it plans to commission the Guru Gobind Singh refinery in Bhatinda, Punjab, by first quarter 2011.

The refinery project is a JV between HPCL and Mittal Energy Investment Pvt Ltd.

The plant will have the capability to produce 192,000 b/d of petroleum products conforming to Euro-4 emissions standards.

Shell refineries offered

Indian conglomerate Essar Group bid in August 2009 for three European refineries offered by Royal Dutch Shell as part of the latter's downstream restructuring plans.

Details are given in The Netherlands section of the Europe chapter of this book.

Jamnagar: Commissioning

In early 2009, Reliance Industries Ltd. was in the final stages of commissioning its 580,000-b/d refinery at Jamnagar. The new facility is 100% export-oriented.

Coupled with RIL's neighboring 660,000-b/d refinery, the new plant forms the world's largest refining complex, with a total capacity of 1.24 million b/d.

Jamnagar: Chevron stake

In early 2009, Reliance Industries Ltd. said that it will buy back Chevron Corp. the latter's 5% stake in a 580,000-b/d refinery in Jamnagar, at the original price of $300 million.

Chevron cited "attractive investment alternatives and continuing low global demand for refined products" as reasons for its decision to sell.

Jamnagar: Sulfur recovery

In September 2009, commercial operation began at a unique sulfur-processing facility at Reliance Petroleum's Jamnagar export refinery.

According to designer Black & Veatch, the unit is the first to pair "two existing and proven technologies in a way that guarantees removal of 99.9% of sulfur from a refinery's acid gases."

The new tail-gas treating unit doubles sulfur processing capacity at the complex to 4,050 tpd.

Total processing capacity at the refinery is 1.2 million b/d of a wide variety of crudes.

Vizag refinery

In September 2009, Total SA suspended its participation in the consortium considering construction of the $10-billion, 14-mtpy Vizag refinery, throwing into doubt the future of the project.

The refinery had been proposed by a consortium led by Hindustan Petroleum Co. Ltd.

In November 2008, Mittal Energy Investment Pvt Ltd. withdrew.

Petrochemicals

RIL bid for LyondellBasell

In late 2009, it was reported that Reliance Industries Ltd. offered to buy a controlling stake in LyondellBasell, the bankrupt Dutch-based U.S. petrochemical firm, for about $10 billion.

Details are given in The Netherlands section of the Europe chapter of this book.

Two plants: Assam

Brahmaputra Cracker & Polymer Ltd. awarded contracts in late 2009 to Lummus Technology for basic engineering of two new chemical plants—one producing 220,000 tpy of ethylene, the other 60,000 tpy of polypropylene. Both will be built in Lepetkata, Assam.

Alternative Energy

Solar energy

A range of projects and policy initiatives were being planned or in development in India in 2009, including these:

- Expected release of the government's $19-billon Solar Power Mission, to spur generation of 20,000 MW of solar power by 2020.
- Construction of a $10-billion, 3,000-MW solar power complex in Gurjarat State.
- Transformation of selected centers into "solar cities" (including Hamirpur, Shimla, and Nagpur).
- Planning for public-private projects to build more than 30 photovoltaic or solar-thermal power projects in Gujarat State, generating more than 700 MW by 2013.
- Plans to build photovoltaics plants in the states of Rajasthan, Haryana, Gujarat, West Bengal, Bihar, and Karnataka by Astonfield Renewable Resources Ltd.

Wind power

Generation of electric power from the wind also was a focus of attention in India in 2009. Projects in development included:

- A 100-MW project by NHPC Ltd. in Kukroo, Madhya Pradesh State
- Two 500-MW projects by power company NTPC Ltd. in Maharashtra and Karnataka states.

Environmental Issues

Voluntary carbon cut

India announced in December 2009 that it would reduce its carbon emissions intensity by 20–25% of 2005 levels by 2020.

But officials said that the reduction would be made voluntarily, not as part of any legally binding international agreement to combat global warming.

Measures to reach the goal include mandating fuel efficiency standards for all vehicles by December 2011, establishing an energy-efficient building code, and ensuring that half of all new coal-fired power generation uses clean-coal technology.

GHG emissions

According to five independent studies released by Indian officials in September 2009, the country's greenhouse gas (GHG) emissions are expected to at least quadruple over the next 20 years as the economy expands.

However, they said that India's per capita GHG output would be below that of developed nations

INDONESIA

Capital: Jakarta
Monetary unit: Rupiah
Refining capacity: 1.012 million b/cd
Oil production: 850,000 b/d
Oil reserves: 3.990 billion bbl
Gas reserves: 106.000 tcf

Overview

Role in regional oil, gas picture

Indonesia will see a 17.5% decline in oil production by 2018, with the country's crude volumes falling steadily to 800,000 b/d, according to Business Monitor International (BMI)

The finding as in the firm's early-2009 *Indonesia Oil and Gas Report*.

BMI forecasts that Indonesia will account for 10.91% of regional oil supply by 2013, while accounting for 4.32% of Asia-Pacific regional oil demand.

Earthquake

On September 30, 2009, a magnitude-7.6 earthquake struck Indonesia, off Sumatra. The UN said the death toll was at least 1,100 with tens of thousands made homeless.

Padang, a coastal city of 900,000 and capital of West Sumatra Province, suffered the most damage

Regarding natural gas, Asia-Pacific production should reach 483 bcm (17.1 tcf) in 2013, with consumption that year reaching 595 bcm (21.0 tcf).

Indonesia will account for 18% of regional gas supply by 2012, and for 7.47% of regional consumption.

Domestic gas market: A priority

In early 2009, Indonesian officials reaffirmed the need for the country to prioritize its domestic gas market over exports.

The government urged gas producers in the country to maintain production or produce more gas to meet rising demand from domestic and overseas markets.

Exploration and Development

First commercial CBM test well

In September 2009, Indonesia's first commercially drilled coalbed methane test well went to a TD of 594 m (1,950 ft in South Sumatra, cutting a combined 27 m (90 ft) of coal in five seams.

CBM-SE-02 well is in the Sekayu PSC block.

Operator PT Medco Energi Internasional Tbk said that ten more wells were planned by yearend.

Also participating were PT Ephindo, Batavia Energy Inc., and CBM Asia Development Corp.

CBM: Sanga Sanga Block

Another consortium operating in late 2009 was led by Virginia Indonesia Co. LLC.

VICO Indonesia—a JV of BP PLC and Eni SPA—signed a PSC with Indonesia to explore and develop CBM resources on the Sanga-Sanga Block in East Kalimantan.

The PSC covers 1,700 sq km (6,564 sq mi) in the Kutai basin. Preliminary estimates indicate a CBM resource potential of at least 113 bcm (4 tcf), subject to further appraisal.

In what has been called the first integrated CBM-to-LNG operation, gas produced will be piped to the Bontang LNG plant for liquefaction.

CBM: Kutei basin

In mid-2009, CBM Asia Development Corp. applied through its partner PT Ephindo for a second CBM PSC in the Kutei basin of East Kalimantan.

The application covers 139,000 acres next to the firm's Kutai West PSC.

By November 2011, CBM Asia must conduct a $5.6-million exploration and appraisal program to determine CBM commercial feasibility on Kutai West and submit a development plan.

CBM: PSCs granted

The government granted 15 CBM PSCs between May 2008 and August 2009 that represent $96 million in exploration commitments through 2012.

Indonesia's CBM resource is estimated at 453 tcf in place, 183 tcf attributed to the South Sumatra basin. An estimate earlier this decade put the resource at 337 tcf

Bengara I PSC

Salamander Energy struck pay in mid-2009 with the South Sebuku-1 (SS-1) exploration well in the Bengara I PSC, encountering gas-bearing sandstones in the Tabul and Meliat formations.

Drill stem tests on two zones flowed an aggregate rate of 309,700 cu m/d (10.9 MMcfd).

The discovery is 15 km (9 mi) from the South Sembakung gas/condensate field.

An appraisal well in Bengara I is planned for 2010.

Bengara II PSC

In early 2009, the government gave GeoPetro Resources Co. another year to submit its work program and 2009 budget to appraise and evaluate commerciality of an oil discovery on Bengara-II Block.

That block is in Tarakan basin, onshore East Kalimantan.

The oil discovery was made in 2008 on the Seberaba prospects.

Muara Bakau Block

Eni SpA announced its Jangkrik discovery in early 2009 on the offshore Muara Bakau Block in the Kutei basin. The well is in 400 m (1,312 ft) of water in the Makassar Strait off East Kalimantan.

Potential fast-track development was under study.

Natuna project

PT Pertamina was poised in mid-2009 to choose its technical partners to develop natural gas from the Natuna D-Alpha Block, 1,100 km (684 mi) north of Jakarta in Riau Islands Province.

Natuna is estimated to hold about 1.3 tcm (46 tcf) of recoverable gas reserves and 2.3 tcm (80 tcf) of potential reserves. However, the gas contains about 70% CO_2, making it difficult and expensive to recover and process. Pertamina hopes to begin production in 2017.

Development cost has been estimated at $40 billion.

Contenders for partnership with Pertamina are Royal Dutch Shell, Chevron Corp., ExxonMobil Corp., Statoil, Eni, Total SA, CNPC, and Petronas.

Pertamina has been granted a 40% share of the block by the government, after resolution of a dispute with ExxonMobil about the latter's rate of progress.

The other 60% share in the block is to be split among two or more partners.

Lease Sales

Indonesia is struggling to produce more oil from aging fields. First-quarter 2009 production averaged 946,000 b/d, down 14,000 b/d from the government's target of 960,000 b/d.

Of 16 oil and gas blocks offered between December 2008 and April 2009, only five won developers.

In a bid to raise production, the country offered and awarded exploration rights to a range of oil and gas blocks in 2009.

May awards

- South Block "A," North Sumatra (PT Renco, PT Prosys)
- East Pamai, onshore Central Sumatra (PT Nana Yamano)
- West Belida onshore South Sumatra (Orchard Energi, Bayu)
- Terumbu onshore and offshore Madura (Australia Worldwide Exploration Ltd.)
- Southeast Madura (PT Bama Bumi Santosa, PT Toba)
- Pasir, onshore East Kalimantan (PT Archipelago)
- S. Sesulu, offshore East Kalimantan (Hess Corp.)
- Kofiau, offshore West Papua (Biak Petroleum, Niko Resources)
- Offshore West Papua (Marathon, Komodo Energy, Kumawa Energy)

- Cendrawasih, offshore West Papua (Esso, ExxonMobil, and Biak Petroleum)
- Northern Papua, onshore and offshore North Papua (Sarmi Papua, Asia Oil).

September awards

- Andaman III block off North Sumatra (Talisman Energy Inc.)
- West Glagah Kambuna, off North Sumatra (PT Pertamina and Petronas)
- Halmahera Kofiau, off South Halmahera; East Bula, off Seram; and West Papua IV, off Papua (a consortium of Niko Resources and Black Gold Energy)

The energy ministry said that a total investment of $91.5 million had been committed for the projects in the first three years, and that each company would be expected to drill three exploration wells.

July offerings

Most of these 24 blocks are in eastern Indonesia. Seventeen were offered through regular tenders and 7 were offered directly.

The blocks offered under regular tender were: Tomini Bay I-V; Gorontalo Tomini I-II; North Bone; Kolaka Lasusua; Kabena; Jampea; Buton III; Menui Asera; Morowali; Sula I-II; and Bird's Head.

Offered directly were: Kubu; North East Ogan Komering; Offshore West Java; Blora; North Makassar; East Simenggaris; and Digul.

Drilling and Production

Cepu oil delayed

The Cepu Block, straddling the border of East Java and Central Java, it is said to hold proved reserves of 600 million bbl of oil and 48.1 bcm (1.7 tcf) of natural gas. The block's main field is Banyu Urip.

Very modest test production began in December 2008, but the tortuous process of ramping up to expected levels continued well into 2009.

After a March halt because of problems with pipeline permits, production re-started in August at about 1,500 b/d as operators aimed to reach 15,000 b/d by the end of that month.

But BPMigas officials told the government that production was likely to be no higher than 20,000 b/d through late 2011. They said that the tender for EPC work—on facilities needed to reach peak production levels—was still being developed at midyear.

The government had anticipated 20,000 b/d of output beginning in late 2009, rising to an eventual peak of 150,000–165,000 b/d.

Oyong Phase-2

Gas production began in September 2009 from the offshore Oyong Phase-2 project, according to operator Santos Group.

Oyong began oil production in September 2007 and has produced over 4 MMbbl to date.

Phase 2 involved modification of offshore facilities and construction of a 60-km (37 mi) pipeline to a new onshore gas processing facility in East Java.

Gross gas production rates of 1.4–2.1 bcmd (50–60 MMcfd) are expected at plateau.

Kambuna field

Serica Energy PLC reported the start of gas sales in August 2009 from Kambuna field, offshore Sumatra. First gas and condensate flowed from the Kambuna-4 well to onshore receiving facilities at Pangkalan Brandan.

First commercial gas entered the pipeline system for transport to the Belawan Power Plant.

Ramp-up to a plateau of 1.13 bcmd (40 MMscfd) of dry gas and 4,000 b/d of condensate is expected.

Pertamina production plans

PT Pertamina said in early 2009 that it planned to spend 11 trillion rupiah ($1 billion) to upgrade assets and boost production in Limau oil fields in South Sumatra, and Tambun gas fields in Bekasi, West Java.

Of the total upstream investment, 6 trillion rupiah was to go to PT Pertamina EP.

The company was aiming to boost oil production by 7% in 2009, to 171,250 b/d, from an estimated 159,000 b/d in 2008.

LNG

First Tangguh cargo

In July 2009, the Indonesia unit of BP PLC reported that the first cargo of LNG was lifted from the Tangguh project and shipped to POSCO's LNG re-gasification terminal in Gwangyang, South Korea.

Tangguh comprises six gas fields in the Wiriagar, Berau, and Muturi PSCs in the Bintuni area of Papua.

Gas is piped to two 3.8-mtpy onshore liquefaction trains.

Decline in 2009 output

Indonesian officials said that LNG output likely would fall in 2009, largely due to a decline in the supply of natural gas to the Bontang LNG facility. Gas fields facing output declines were not identified

An official with Bontang expected 307 cargoes in 2009, down from 317 in 2008.

At the Arun LNG plant, officials said that they expect no change from the 42 cargoes of 2008.

One cargo contains 125,000 cu m (4.4 MMcf).

Indonesian LNG Production

	Total	Bontang	Arun
	Number of cargoes		
2009*	349	307	42
2008	359	317	42
2007	372	320	52
2006	394	335	59

*Estimated.

Central Sulawesi plant

In November 2009, Japan Bank for International Cooperation threatened to withdraw its commitment to finance construction of an LNG plant in Central Sulawesi if Jakarta insists that the LNG produced there is for domestic consumption only.

JBIC had agreed to finance the project, hoping that the gas could be exported to Japan. But in June 2009, Indonesian officials said that gas from the Senoro and Matindok fields intended to supply the Donggi plant had to be sold domestically.

In May, concern arose that the two fields might have lower reserves than earlier estimated.

Shipments to Fujian

Two Indonesian LNG facilities sent cargoes to the CNOOC's Fujian LNG regasification plant in China in 2009: one from the Bontang liquefaction plant (May), the other from the Tangguh facility (July).

More details are given in the China chapter of this book.

Floating LNG plant?

In mid-2009, Indonesia urged Japan's Inpex Corp. to set aside its plans for a 4.5-mtpy, $19.6 billion floating LNG processing facility and to instead consider building an onshore facility in Saumlaki, in the Tanimbar Islands.

Officials want to see a facility in place by 2014 to process gas from the offshore Masela Block.

Saumlaki is about 150 km (93 mi) from the block, and the government said that it appears possible to link to Masela with a flexible subsea pipeline that could operate even in the deep water [2,500 m (8,200 ft)] of the Timor Trench between Masela and Saumlaki.

Indonesian officials put the cost of an onshore terminal at about one-fourth that of a floating terminal.

Tanker fleet: Sea change

The Indonesian National Shipowners' Association (INSA) said in early 2009 that its members need $6.5 billion to buy 127 ships to replace foreign-flagged tankers currently carrying oil and gas in Indonesian waters.

The group pointed to changes in law, effective January 2010, that will forbid the operation of foreign ships in Indonesian waters.

Refining and Products

Pertamina projects

Pertamina announced in mid-2009 its plan to build a $4-billion, 300,000-b/d refinery in the coastal city of Bojonegara, Banten Province, West Java.

JV partners will include Petrofield (M) Sdn Bhd, National Iranian Oil Refining and Distribution Co. (NIORDC), and STX Corp.

Also planned by 2014 are:

- A new 200,000-b/d facility at Tuban in East Java
- A second new-build refinery at Bojonegara
- Expansion of the capacity of the 125,000-b/d refinery at Balongan in West Java.

Batam Island

Setdco Group and a partner requested authority in mid-2009 to build a 300,000-b/d refinery at Tanjung Sauh on Batam Island near Singapore. Crude for the plant will come from the Middle East.

Also in mid-2009, Indonesian officials said that Gulf Petroleum Ltd. had initiated an application for approval to build a refinery in Batam.

CHINA
RUSSIA
NORTH KOREA
SOUTH KOREA
JAPAN
Sea of Japan
Pacific Ocean
Vladivostok
Nakhodka
Peter the Great Bay
Chongjin
42
Pohang
Ulsan
817
Pusan
10
Korean Channel
Tsushima Channel
Wakkanai
Soya
Katsu
Menas
Koetoi
Toyotomi
Kitatoyotomi
Teshio Fields
HOKKAIDO
Iskikari Bay
Ishikari
Atsuta
Barato
Tokachi Fields
Sapporo
Azuma
Fureoi
Kinausu
Tomakomai
133 180
Muroran
Oshamambe
Mori
Hakodate
Tsugaru Strait
Aomori
Nichi-Hachimori
Higashi-Hachimori
Sawame
Iwako
Sakake
Akita Fields
Surukawa
Kurokawa
Funakawa
Asakawa
Akita
Morioka
Yabase
Yuri
Kamekawa
Shonai
Innai
Kisagata
Okuoguni
Yokuka
Yamagata Fields
Chokaizan
Fukura
Sakata
Nogouchi
Mogaami
Shonai
Yamagata
Amarume
145
Sendai
Sendai Bay
Iwafune
Aga
Niitsu
Shibata
Niigata
Niigata-Plain
Higashi
Okawatsu
Mitsuke
Echigo Fields
Higashiyama
Nishiyama
Nagaoka
Kashiwazaki
Sekihara
Toyama Bay
Kubiki
Ojiya-Sanjo
Ubase
Gozu
Kubiki
Nadachi
Menji
Maki
Iwaki
Iwaki
Asakawa
Odagiri
Toyama
57
Kanazawa
HONSHU
Kashima
181
Choshi
Fukui
Wakasa Bay
Tokyo
Kawasaki
Chiba
Yokoshiba
Shirasato
Mobara
Yokohama
Goi
Negishi
Sodegaura
Chiba Fields
110 296 65
228 175 209 192
340
Kyoto
Nagoya
Shimizu
Shimizu
Yaizu
Sagara
Chita
152
Yokkaichi
Ise Bay
147 193
Himeyi
Hyogo
Osaka
Sakai
115
250
195
76
Mizushima
Yamaguchi
127 111
Hiroshima
Takamatsu
160
Wakayama
Marifu
114
Iwakuni
Sakaide
Kainan
Owase
Shimonoseki
Onoda
Ube
Tokuyama
Kikuma
50
Yahata
Kudamatsu
Matsuyama
Shimotsu
Ehime
120
Kii Strait
Fukuoka
Sasebo
Oita
160
Bungo Strait
SHIKOKU
Tosa Bay
Kashima
Nagasaki
Kumamoto
KYUSHU
Miyazaki
Okinawa
9000100
Off Map
Osumi Strait
JAPAN
0 50 100 Mi.
0 100 Km.

Balikpapan upgrade

In March 2009, Pertamina signed an MOU with Star Petro Energy (ETA Group) and Itochu Corp. to upgrade the 260,000-b/d refinery at Balikpapan in East Kalimantan.

The aim is to increase total capacity to 280,000 b/d, switch from sweet crude to cheaper sour crude, and add a 50,000 b/d cracking unit to process heavy residue into gasoline and petrochemical products.

A preliminary cost estimate for the project is $1.7 billion.

Gasoline import trend

Due the lack of refining capacity, a Pertamina official said in September 2009 that the firm expects gasoline imports to more than double from existing levels by 2017.

Because the nation's refineries can only produce 68.5 million bbl/year, gasoline imports would have to rise to 124.2 million bbl in 2017, from 55.3 million bbl in 2009.

Annual domestic gasoline consumption is expected to climb to 192.7 million bbl in 2017, from the forecast of 123.8 million bbl in 2009.

Petrochemicals

Polypropylene at Balongan

In early 2009, a consortium of international banks committed to provide $225 million to support a project by Pertamina to build a polypropylene unit at its refinery in Balongan, West Java.

The unit would have a production capacity of 200,000 tpy.

CTL projects

The Dept. of Energy and Mineral Resources said in early 2009 that South Africa's Sasol Ltd. plans to invest $10 billion by 2015 to develop coal liquefaction plants in eastern Kalimantan and Sumatra.

The facilities, with combined capacity of 1.1 MMb/d, would produce gasoline and other oil products.

Sasol was said to be conducting feasibility studies for the proposed ventures, which would lead to operational plants by 2015.

Initial output would be 80,000 b/d, growing in the future to 1.1 million b/d.

Company News

Pertamina spending

PT Pertamina, aiming for an 11% boost in oil production, said in late 2009 that it expects to increase capital expenditure to $4.15 billion in 2010—a 77% increase over 2009.

The Cepu block is expected to contribute significantly to the 2010 targeted oil production increase, with 20,000-b/d output planned at initial startup early in the year.

The firm also plans several refinery construction and upgrade projects (see above).

Tax debt dispute

Indonesia continued to argue with ExxonMobil Corp. and Kodeco Energy Co. Ltd. in mid-2009 about the companies' alleged tax debts, totaling $63 million.

The dispute hinges on how the tax is calculated and the amount owed.

JAPAN

Capital: Tokyo

Monetary unit: Yen

Refining capacity: 4.624 million b/cd

Oil production: 15.300 b/d

Oil reserves: 44.115 million bbl

Gas reserves: 738 bcf

International Activities

Iraq deal: Gharaf field

In December 2009, a consortium led by Japan Petroleum Exploration Co. won the rights to develop Gharaf oil field in Iraq.

Since 2005, Japex had been collaborating with the Iraqi Oil Ministry on technical issues, conducting geological research, and training hundreds of Iraqi oil and gas engineers.

It is estimated that investment of $7 billion will be needed to develop the field. Gharaf field is expected to produce 230,000 b/d.

Investment in Venezuela

Venezuela and Japan agreed in April 2009 to cooperate on 12 hydrocarbon projects.

Japan is expected to invest $33.5 billion in Venezuela's hydrocarbon sector through 2014, including projects for petrochemical production, oil refining, LNG, and oil exploration.

Details are given in the Venezuela section of the Latin America chapter of this book.

East China Sea pact

Japan and China announced in early 2009 that they will proceed with joint exploration for natural gas in the disputed East China Sea.

Details are given in the China chapter of this book.

Storing oil for ADNOC

In late 2009, Nippon Oil Corp. finalized an agreement with ADNOC for the UAE firm to store oil in Japan, in an effort to bolster the latter's national reserves.

ADNOC will store 3.9 million bbl of crude for 3 years at a reserve base in southern Japan.

Japanese officials said that the agreement also would give Japan a preferential right to purchase the stored oil reserves in times of emergency.

Abu Dhabi can sell its oil in other East Asian countries, such as South Korea and Taiwan, by using the Japanese base as a distribution terminal.

Abu Dhabi currently supplies nearly 40% of Japan's oil imports.

Saudi oil storage

Saudi Aramco was considering a proposal from Japan in late 2009 for joint use of oil storage facilities on Okinawa.

Details are given in the Saudi Arabia section of the Middle East chapter of this book.

LNG

Indonesia LNG plants

In November 2009, the Japan Bank for International Cooperation threatened to withdraw its commitment to finance the construction of the $1.6-billion Donggi Senoro LNG plant in Central Sulawesi, Indonesia.

Also at mid-year, the Indonesian government was urging Inpex Corp. to consider building an onshore LNG terminal in Saumlaki in the Tanimbar Islands, rather than a proposed floating LNG-processing facility.

Details are given in the Indonesia section of this chapter.

Russian gas pipeline

Gazprom was working in mid-2009 on the $11-billion Sakhalin-Khabarovsk-Vladivostok (SKV) gas pipeline. The line could carry gas from Sakhalin Island to Japan, and Russia was seeking help from Japan to build one segment of the line.

Details are given in the Russia section of the Former Soviet Union chapter of this book.

New terminal at Hitachi

In early 2009, Tokyo Gas Co. chose the coastal city of Hitachi as the site for its fourth LNG facility, the firm's first in 20 years. No details on size or gas supply were given.

Tokyo Gas already operates two LNG receiving facilities near Yokohama area and one in Sodegaura.

The new terminal is scheduled to start up by 2018.

Floating LNG platforms

In 2009, several Japanese firms—Inpex, JOGMEC, IHI Corp., JGC Corp., Chiyoda—were promoting development of technology to allow production of LNG at sea, on floating platforms.

They suggest that the LNG FPSO concept can lead to more stable procurement by visiting smaller fields usually ignored due to low profitability.

Proponents are said to have identified more than 130 undeveloped Asia-Pacific gas fields, with combined estimated reserves of 2.3 tcm (80 tcf), that could benefit from the new technology.

Refining and Products

Excess capacity

Analysis by Hawaii's East-West Center estimated in mid-2009 that Japan had excess refining capacity of 780,000 b/d for that year, and that this excess would grow to 1.4 million b/d in 2015.

The center pointed to the existence of massive global refining surpluses.

Researchers concluded that forming strategic alliances that come with crude supplies or product export outlets could help prolong the lives of Japan's refineries.

In particular, countries such as Vietnam, Indonesia, and perhaps China that have oil but insufficient refining capacity were seen as likely candidates for alliances with Japanese firms.

Japan Petroleum Demand: 1970-2020

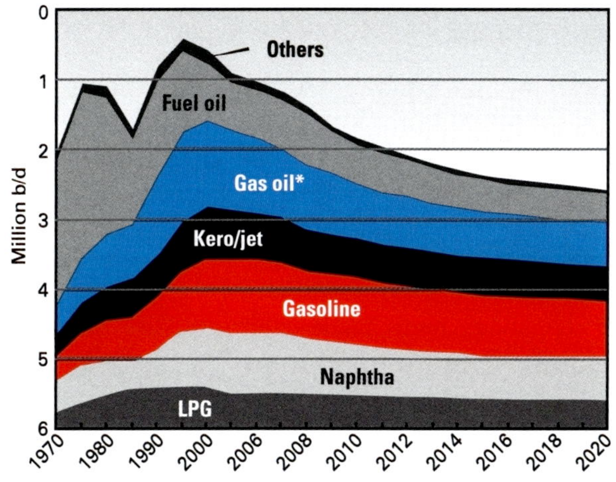

*Gas oil includes bunker gas oil. Fuel oil includes direct use of crude and bunker fuel.

Advanced automobiles

In April 2009, Honda's gasoline-electric hybrid Insight automobile outsold all other full-size passenger vehicles in Japan. It was the first time a hybrid model topped national sales rankings.

And in August, Nissan Motor Co. Ltd. unveiled its all-electric car, the Leaf, combining an 80-kW electric motor and a rechargeable lithium-ion battery.

It is to be sold in Japan and the United States in late 2010 and worldwide in 2012.

Japanese Refiners[1]

Nippon Oil Group[2]
1,317,000 b/d

Cosmo Oil
635,000 b/d

Japan Energy Group[3]
475,000 b/d

Idemitsu Kosan
640,000 b/d

Tonen General Group[4]
836,000 b/d

Show Shell Group[5]
707,000 b/d

Others:
Nansei Sekiyu, 100,000 b/d
Taiyo Oil, 120,000 b/d
Teikoku Oil, 5,000 b/d

[1]As of 2009; 4,835,000 b/d.
[2]Includes Nippon Oil and Kyushu Oil.
[3]Includes Japan Energy and Kashima Oil.
[4]Includes Tonen General and Kyokuto Petroleum.
[5]Includes Showa Yokkaichi, Seibu Oil, Toa Oil, and Fuji Oil.

Election results

On August 30, 2009, in a landslide election win, the Democratic Party of Japan (DPJ) ended a half-century of almost unbroken rule by the Liberal Democratic Party (LDP).

Observers expect the DPJ, led by Prime Minister Yukio Hatoyama, to break a deadlock in parliament and usher in a government that has promised to focus spending on consumers, cut wasteful outlays and reduce the power of bureaucrats.

Japan: Forecast Product Balances

	Output	Imports	Exports	Demand
	1,000 b/d			
2008				
LPG	145	433	3	507
Naphtha	376	420	1	780
Gasoline[1]	977	10	13	985
Kero/jet	638	7	65	589
Diesel	1,136	6	220	919
Fuel oil	538	91	56	526
Other	126	3	17	103
Total	**3,936**	**970**	**375**	**4,409**
[2]2009				
LPG	142	350	—	492
Naphtha	361	380	—	741
Gasoline[1]	991	2	10	983
Kero/jet	607	10	50	567
Diesel	1,029	5	150	884
Fuel oil	463	28	60	431
Other	110	3	10	103
Total	**3,703**	**777**	**280**	**4,200**
[2]2010				
LPG	140	350	—	490
Naphtha	365	340	—	705
Gasoline[1]	981	—	15	966
Kero/jet	608	10	50	568
Diesel	879	5	120	764
Fuel oil	443	20	50	413
Other	106	3	10	99
Total	**3,522**	**728**	**245**	**4,005**
[2]2015				
LPG	142	280	—	422
Naphtha	333	280	—	613
Gasoline[1]	893	—	15	878
Kero/jet	572	10	50	532
Diesel	757	6	100	663
Fuel oil	370	23	50	343
Other	91	3	10	84
Total	**3,158**	**602**	**225**	**3,535**

[1]Contains ethanol: 2008, 2,000 b/d; 2009, 5,000 b/d; 2010, 25,000 b/d; and 2015, 62,000 b/d. [2]2009-15 are forecasts.

Biggest nuclear plant restart

In August 2009, Tokyo Electric Power Co. restarted the Kashiwazaki-Kariwa nuclear power plant, just over two years after its shutdown for extended inspection and repair following a July 2007 earthquake.

The plant's seven reactors produce 8,212 MW.

Alternative Energy

Solar industry

Japan's share of the global solar power equipment market fell in 2007 to about 25%, from more than 50% in 2004. But in mid-2009, the picture was brightening.

Japan restarted subsidies to households that install solar panels; tight credit was hampering rival firms in China and Taiwan; and Japanese electronics companies outlined aggressive plans to expand.

Also, as part of its 2009 fiscal stimulus measures, Japan brought back subsidies to support the capital cost of installing solar panels on residential roofs.

And the industry expects institution of a "feed-in tariff" in 2010 that would require utilities to buy excess solar electricity from householders.

Environmental Issues

Climate-change policy

Following elections won by the DPJ in August 2009, observers expected that the country's climate-change policy was likely to become more ambitious and more closely aligned with Europe.

And in September, new Prime Minister Yukio Hatoyama pledged to proceed with plans for an aggressive cut in greenhouse gas emissions in spite of opposition from business groups.

His party had promised to set a target of cutting greenhouse gas emissions 25% from 1990 levels by 2020.

Company News

Mitsubishi Chemical

In April 2009, Mitsubishi Chemical Holdings Corp. said it plans to exit the polystyrene and PVC businesses. The firm's operations in those sectors generated annual sales of about $1 billion, accounting for more than 3% of the firm's annual revenues.

Driving the decision to leave the sector were the volatility of the commodity resins market and aggressive competition from manufacturers in the Middle East, China and other parts of Asia.

MALAYSIA

Capital: Kuala Lumpur
Monetary unit: Ringgit
Refining capacity: 514,832 b/cd
Oil production: 730,000 b/d
Oil reserves: 4.000 billion bbl
Gas reserves: 83.000 tcf

Exploration and Development

Finds off Sabah, Sarawak

Murphy Oil Corp. found oil in mid-2009 at one exploratory well and gas at another, both offshore.

A well on the Siakap North prospect in 1,311 m (4,300 ft) of water on Block K off Sabah, 10 km (6 mi) from Kikeh field, found oil-bearing pay sands of similar age and quality to those at Kikeh.

A well in 27 m (89 ft) of water on the East Patricia prospect in Block SK 309 off Sarawak cut 70 m (230 ft) of net natural gas pay. It is 37 km (23 mi) off the Bintulu onshore gas receiving facility under construction for the Sarawak gas development.

East Belumut oil field

Newfield Exploration Co. was accelerating Phase II development drilling at the offshore East Belumut oil field in PM 323.

It planned to drill three development wells in late 2009 and three in early 2010 from the existing production platform, aiming to boost 2010 oil production by 1 million bbl net.

Drilling and Production

Block B-17: Malaysia-Thailand JDA

Carigali-PTTEPI Operating Co. Sdn Bhd (CPOC) was poised to start natural gas production from Block B-17 in the Malaysia-Thailand Joint Development Area (JDA) in October 2009.

Thailand's PTT Exploration & Production (PTTEP), a stakeholder in CPOC, confirmed that the start-up was a year behind schedule.

First gas will come from Muda fields, with initial production of 4.9 million cu m/d (140 MMcfd), ramping up to 7.6 million cu m/d (270 MMcfd).

CPOC is an equal-partnership JV of PTTEP and Malaysia's Petronas.

$2.1 billion for EOR

Petronas Carigali signed a PSC in mid-2009 with ExxonMobil Exploration and Production Inc. (EMEPMI) for further development of seven offshore oil fields.

The two firms will spend a minimum of $2.1 billion on major EOR activities, rejuvenation of facilities and further development, and drilling activities in the fields.

Northern Fields development

The Malaysia unit of Talisman Energy Inc. reported the start of oil production in early 2009 from the Northern Fields development offshore Malaysia and Vietnam. The development is in the PM-3 Commercial Arrangement Area.

Production is expected to reach 40,000–50,000 boe/d (gross sales) by early 2010 with the commissioning of dry gas facilities in mid-2010.

Talisman expects to drill 16 development wells in the fields in 2009 and 13 more in 2010.

The project includes more than 70 reservoirs and 50 wells.

LNG

Supply deal with Santos

Australia's Santos signed a foundation LNG supply agreement in mid-2009 with its JV partner Petronas for the proposed Gladstone LNG (GLNG) project in Australia.

Petronas agreed to buy 2 mtpy of LNG from the project for 20 years, starting in 2014.

More details are given in the Australia section of this chapter.

Floating LNG system

A subsidiary of Petronas announced a JV in mid-2009 with shipping company MISC Berhad and Mustang Engineering Ltd. to develop an integrated floating LNG system for use offshore Malaysia.

The project team aims to start gas production by late 2013 from a 2-mtpy facility. They estimate the cost of a floating platform at one-third that of an onshore plant.

Refining and Products

Kedah Project

The Malaysian government announced in early 2009 that it will proceed with its $10-billion refinery project in Kedah Province at the Sungai Limau hydrocarbon hub.

The project will construct a 350,000-b/d refinery—the largest in Malaysia—and two major oil pipelines.

Construction of the Kedah Project was expected to start in late 2009 and to be completed in 2014.

Two Chinese firms and a private Malaysian firm are said to be funding the project.

Saudi Aramco is expected to be the major crude oil supplier, and China's CNPC has agreed to buy 200,000 b/d of the refinery's output for 20 years.

Perak State

In early 2009, Qatar-based Gulf Petroleum Ltd. re-confirmed plans to invest $5 billion in an integrated oil, gas and petrochemical complex in Manung, in Perak State. It will include a refinery with a production capacity of 100,000–150,000 b/d.

Maritime boundary with Brunei

In March 2009, Malaysia reached agreement with Brunei regarding a disputed maritime boundary. Details are given in the Brunei section of this chapter.

MALAYSIA-INDONESIA
MYANMAR (BURMA)
THAILAND
LAOS
CAMBODIA
VIETNAM
MALAYSIA
MALAYA
SABAH
SARAWAK
BRUNEI
SINGAPORE
INDONESIA
SUMATRA
JAVA
KALIMANTAN (BORNEO)
HAINAN DAO
ANDAMAN ISLANDS
NICOBAR ISLANDS
SIMEULUE
NIAS
SIBERUT
SIPORA
PAGAI UTARA
PAGAI SELATAN
ENGGANO
BANGKA
BELITUNG
KEPULAUAN LINGGA
NATUNA ISLAND (INDONESIA)
MADURA
BALI
SUMBAWA
PALAWAN
Bay of Bengal
Gulf of Martaban
Andaman Sea
Indian Ocean
Strait of Malacca
Gulf of Thailand
Gulf of Tonkin
South China Sea
Sulu Sea
Java Sea
Flores
Strait of Makassar
Sunda Strait
Noakhali
Chittagong
Patuakhali
Patiya
Shahbazpur
Thanbayakan
Mandalay
Sittwe
Ayadaw
Yenangyaung
Chauk/Lanywa
Mann
Htaukshabin
Paduakpin
Kyaukoyauk
Yenandaung
Pyaye
Ledang
Pyato
Prome
Yenanchaung
Myanaung
Shwepyitha
Yangon (Rangoon)
Mae Sot
Martaban
Yadana
Yetagun
Mae Fang
Mae Suhn
Chiang Mai
Nan
Vientiane
Uttaradit
Loei
Mae Sot
KDN-1
Sirikit
Pru Krathiam
Bung Ya
Phu Horm
Nam Phong
Chonnabot
Khon Kaen
Sisaket
Sawng Neung
Saraburi
Kanchanaburi
Bangkok
Sri Racha
Rayong
Phnom Penh
Sihanoukville
Nang Nuan
Plamuk
Kapong
Pakarang
Pladang
Tantawan
Surat
Chaba
Platong
Platong S.
Dara
Trat
Erawan
Baanpot
Pailin
Sabun
Jakrawan
Gomin
Punan
Khanom
Si Thammarot
Arthit
Bongkot
17-B-1
Songkhla
Bua Ban
Pilong
Lawit
Damar
Bintang
Jemeh
Sepat
Noring
Bedong
Inas
Larut
Bujang
Tangga
Bubu
Lerek
Tujoh
Dulang
Tabu
Barat
Tapis
Kakap
Besar
Duyong
Belumut
Angsi
Sotong
N. Belut
W. Belut
Malong
Mamung
Kerisi
Belanak
Terebuk
Sepat
Sotong
Sembilang
Tembang
Belida
Kijing
Keong
Udang
Buntal
Kepiting
Malong
Kerteh
Kuala Lumpur
Port Dickson
Pinang
Hanoi
Hai Phong
Tien Hai
Weizhou 11-1
Weizhou 10-3
Weizhou 6-1
Moaming
Macau
Hong Kong
Lufeng 13-1
Xijiang 24-3
Lufeng 15-1
Xijiang 30-2
Huizhou 26-1
Huizhou 21-1
Liuhua 4-1
Liuhua 11-1
Haik'ou
Jinfeng
Wenchang 9-2
Yacheng 13-1
Hue
Da Nang
Nha Trang
Ho Chi Minh City
Vung Tau
Long Xuyen
Sutu Den
Rang Dong
Bach Ho
Rong Dong
Dai Hung
Dua-1X
Bursa
Natuna
Subic Bay
Calauit
Galoc
Octon
West Linapacan
Malampaya
Camago
Destacado
Nido
Matinloc
Cadlao
Sampaguita
Kudat
S.W. Emerald
Tembungo
Erb W.
Kinarut
Barton
S. Furious
St. Joseph
Kota Kinabalu
Sandakan
Kikeh
Samarang
Kecil
Iron Duke
Glayzer
Labuan
Kecil
Jintan
Bijan
Seria
F14
Magpie
M1
M3
M4
F6
F28
F27
B11
B12
B23
E8
D12
J4
C2
W. Bayan
Temana
Patricia
Bandar Seri Begawan
Kuala Belait
Belait
Miri
Bintulu
Tawau
Mamburungan
Bunyu Nibung
Tarakan
Kuching
Batuputih
Sangkulirang
Kerindingon
Panyilatan
Serang
Attaka
Semberah
Badak
Tunu
Sanga-Sanga
Pamaguan
Handil
Bekapi
Balikpapan
Louise
Mutiara
Sepinggan
Sanggur
Bangkerak
Wailawi
Kerendan
Kambatin
Tanjung
Tapian Timur
Warukin Tengah
Warukin Selatan
Palangkaraya
Bandjarmasin
Pontianak
Palu
Kutaradja
Julu Rayeu
Perlak
Serang Jaya
Pangkalan Brandan
Belawan
Medan
Meulabah
Sibolga
Dumai
Pinang
Bangko
Buaya
Menggala
Mandar
Duri
Libo
Minik
Langgak
Mlias
Topaz
Mellbur
Zamrud
Lirik Fields
Pekanbaru
Batam
Singapore
N. Palai
S. Palai
Andan
Sago
Lirik
Molek
Binio
Bukittinggi
Jambi
Setiti
Bedjubang
Tempino
Limau Fields
Sumpal
Suban
Burong
Bentayan
Dayung
Keban
Sambar
Sungei Gerong
Tanjung Laban
Benakat
T. Djimar
Muraraenim Fields
Ogan Timur
Kampong Minyak
Taham
T. Tiga
Pagardewa
Palembang
Lang
Raja
Bukeras
Telukbetung
Intan
Indri
Winduri
Sundari
Krisna
Karmila
Cinta
Nora
AV
AA
KN
Arjuna
Rama
HZE
Cilegon
MB
E
Arimbi
Jakarta
Sidong
F
Cirebon
Jatibarang
Randegan
Bandung
Rembang
Ledok
Ngiobo
Petak
Camar
Semarang
Candi
Banyubang
Cepu
Kawengan
Kendangan
Maleo
Gresik
Waron
Cilacap
Jogjakarta
Wonokromo
Singaradja
Raba
33
25
6
26
62
174
193
156
270
104
140
5
40
85
125
114
93
126
605
462
290
9
45
260
48
109
125
348
3
0 50 100 150 200 250 Miles
0 100 200 300 400 Km
20°
15°
10°
5°
0°
5°
10°
95°
100°
105°
110°
115°
120°

Pingtung
270
Pacific Ocean
LUZON
PHILIPPINES
Quezon City
Manila
Limay
180
Tabango
93
MINDORO
Saddle Rock
Malajon
Linapacan
LEYTE
PANAY
Tacloban
Bacolod
Cebu
NEGROS
BOHOL
MINDANAO
Davao
Philippine Sea
Sea
Celebes Sea
Manado
Soasiu
Molucca Sea
Sorong
1-X TBM
Kasim
Klamono
Matoa
Salawati
Walio
Sele
Wiriagar
Teruk Berau C
Teruk Berau A
Manokwani
Mogo
Wasian
Steenkool
Wiriager Deep
Ubadari
Wos
Vorwata
Roabiba
Ofaweri
Fakfak
Ceram Sea
BURU
CERAM
Bula
Ambon
Kendari
Ujung Pandang
Banda Sea
Sea
ARU
TANIMBAR
FLORES
Dili
Ossulari
Matai
Aliambata
TIMOR
SUMBA
Kupang
Laminaria
Kitan
JPDA 06-103
Roti
Bayu
Jabiru
Challis
Cassini
Puffin
Timor Sea
Darwin
Bonaparte Gulf
AUSTRALIA
Arafura Sea
Gulf of Carpentaria
IRIAN JAYA
Hollandia
Bismarck Sea
Wewak
PAPUA NEW GUINEA
Madang
Elvala
Ketu
Elevala
Juha
Hides
Mt. Hagen
Goroka
S.E. Mananda
S.E. Hedinia
Agogo
Iagifu
2XGobe
4XGobe
S.E. Gobe
Lae
Bwata
Barikewa
Kuru
Omati
Puri
Lake Murray
Kikori
Baimuru
Uramu
Solomon Sea
Gulf of Papua
Pasca
Pandora-B1X
Pandora-1X
Port Moresby
33
Coral Sea
120°
125°
130°
135°
140°
145°
Samarang Kecil
Champion
Pelican
Glayzer
Iron Duke
SABAH
Petrel
Magpie
Osprey
Egret
Bandar Seri Begawan
9
Fairley
Fairley Baram
Ampa
S.W. Ampa
Baram
Tali
W. Lutong
Seria
BRUNEI
Kuala Belait
Rasau
SARAWAK

MONGOLIA

Capital: Ulaan Baatar
Monetary unit: Tugrik
Refining capacity: 0
Oil production: 3,216 b/d
Oil reserves: 0
Gas reserves: 0

Blocks IV and V

Capcorp, a unit of Petro Matad Ltd., was awarded two new production sharing contracts in mid-2009 by the Petroleum Authority of Mongolia (PAM).

The PSCs are for two petroleum blocks in the south-central part of Mongolia, some 500 km (310 mi) southwest of the capital, Ulaanbaatar.

Bogd Block IV (Block IV) and Ongi Block V (Block V) adjoin each other and jointly cover an area of about 71,000 sq km (27,400 sq mi).

Petro Matad said it will undertake work on both blocks through 2011.

Available data indicate the presence of sedimentary deposits up to 6,000 m (19,685 ft) thick, of sufficient organic richness and depth to be productive for the generation of hydrocarbons.

Darkhan-Uul refinery

In February 2009, Toyo Engineering Corp. and partner Mongolsekiu LLC completed a feasibility study regarding construction of a $1.2-billion oil refinery with a capacity of 44,000 b/d in Darkhan-Uul Province near the Russian border.

The team was seeking a certificate of guarantee from the government in order to apply for a loan from the Japanese government and the Japan Bank for International Cooperation.

Construction was slated to begin in mid-2009, with refinery operations expected to start in mid-2012.

The refinery is expected to meet Mongolia's needs for petroleum, diesel and aircraft fuel.

Crude for the project will be imported from Kazakhstan.

Coal gasification projects

In early 2009, Shenhua Group Corp. began construction of a coal gasification project next to its coal-to-oil project in Yijinhuoluo Banner of Ordos City, Inner Mongolia. The project was chosen as one of two demonstration projects by the National Development and Reform Commission.

The $2.92-billion project is expected to be completed by yearend 2010.

Phase I will lead to production of 2 bcm/year (70.6 bcf/year) of synthetic gas (syngas) by 2012. Products made from the syngas will go to Beijing, Tianjin and Tangshan regions.

China Datang Corp., a major Chinese power producer, also was preparing to build two large coal gasification projects, reported to be worth more than $5.8 billion. Construction was expected to begin by yearend 2009.

In late 2009, Clean Global Energy Ltd. (CGE) signed a $400-million deal to form a four-way JV company for an underground coal gasification project in Mongolia. Two partners are Chinese firms; the other is British.

A deal for purchase of 5–12 million cu m/d (177–424 MMcfd) of syngas has already been established.

MYANMAR

Capital: Naypyitaw
Monetary unit: Kyat
Refining capacity: 57,000 b/cd
Oil production: 20,000 b/d
Oil reserves: 50.000 million bbl
Gas reserves: 10.000 tcf

Shwe gas project

Daewoo International Corp. was proceeding in late 2009 with the Shwe gas field project offshore Myanmar. The firm awarded a $1.4-billion contract for construction of major components of offshore and onshore infrastructure, starting by early 2010.

Daewoo is developing two gas blocks in the Bay of Bengal, about 70 km (44 mi) off the coast. Blocks A-1 and A-3 have been estimated to hold a combined 218 bcm (7.7 tcf) of gas.

A subsea pipeline will carry gas from the production platform to Ramree Island.

Daewoo already has a 30-year purchase and supply deal with CNPC to supply 14.2 million cu m/d (500 MMcfd) of natural gas from the Shwe project, once it is operational in 2013.

Partners with Daewoo are Korea Gas Corp., GAIL India Ltd., ONGC Videsh Ltd., and Myanmar Oil & Gas Enterprise (MOGE).

Pipeline to China for Shwe gas

China's CNPC has begun building a $1.95-billion, 870-km (540-mi) onshore line from Ramree Island to Yunnan Province to carry gas produced from the Shwe field.

The pipeline will bypass the Malacca and Lombok/ Makassar Straits.

The Shwe Gas Movement—a group of Myanmar exiles in Bangladesh, India and Thailand—has called on China to stop building the gas pipelines through Myanmar. They insist that Myanmar's ruling junta should provide energy resources to its people rather than export them.

NEW ZEALAND

Capital: Wellington
Monetary unit: Dollar
Refining capacity: 107,000 b/cd
Oil production: 52,000 b/d
Oil reserves: 60.000 million bbl
Gas reserves: 1.200 tcf

Exploration and Development

Ministry: Deepwater focus

The Energy and Resources Ministry said in late 2009 that the government's plan for maximizing the country's petroleum potential will place particular focus on exploration of deep-water basins.

The petroleum sector accounts for around $3 billion/year of New Zealand's export revenue. The ministry said that this could increase to $30 billion/year by 2025 through development of unexplored basins.

Maari-Manaia: 100 million bbl

Estimated ultimate recovery of more than 100 million bbl of oil was confirmed in late 2009 for the OMV New Zealand-operated Maari field and adjacent Manaia field in the offshore portion of Taranaki basin, according to two of the fields' partners.

The new recovery figure is twice the volume initially estimated for the main Moki sands reservoir at Maari. New Zealand's next-largest producing area is the Tui fields at an original 50 million bbl.

OMV plans to place on production in early 2010 the 8,000-m (26,250-ft) extended reach Manaia-1 well—New Zealand's longest penetration. It tapped a Mangahewa reservoir on the Manaia structure.

Production from the M2A zone well in the Moki sands, above the main reservoir, was expected to start by yearend 2009.

AWE exploration plays

Australian Worldwide Exploration Ltd. was preparing to drill several play types in its mid-2009 exploration in the Taranaki basin.

AWE said that its first well, Hoki-1, will be a Cretaceous North Cape formation play in PEP 38401, about 100 km (62 mi) west of New Plymouth.

Two other wells will be extensions to the Tui oil field pool in the Kapuni Group F10 sands: Tui Northeast, with an estimated 10 million bbl recoverable, and Tui Southwest, with 5 million bbl.

Other play types of interest are the Moki sands (off D'Urville Island), and the Bahamas Pleistocene biogenic gas play, more than 100 km (62 mi) west of Taranaki, largely in PEP 38483.

PEP 38748

An internal assessment in late 2009 by TAG Oil Ltd. indicates a best-estimate resource of 19.1 million stock tank barrels (stb) of original oil in place on the Winchester exploration permit PEP 38748 in the onshore Taranaki basin.

TAG Oil said that the resources, assigned to six drill-ready prospects, were determined from 3D seismic data, well log data, core pyrolysis, and analog field data.

The oil is believed to be in the Miocene Mount Messenger formation.

Reinga basin: Taranaki similarities

CGGVeritas completed a 2D seismic survey in early 2009 of the unexplored Reinga basin, initially finding geology similar to the Taranaki basin, according to the Energy and Resources Ministry.

A thorough interpretation of the data by New Zealand's GNS Science will precede an expected block offering in late 2009.

2009 budget actions

In early 2009, the government announced that its 2009 budget would make provision for a 5-year continuation of a tax exemption for offshore oil and gas exploration.

The current exemption from tax on the profits of non-resident operators for offshore rigs and seismic vessels was introduced in 2004 and was scheduled to expire on December 31, 2009.

The government also announced that the budget would allocate $20 million over three years for acquisition of seismic data, in a program to be run by Crown Minerals, part of the Energy and Resources Ministry.

Lease Sales

PEPs 51311, 51311

New Zealand Oil & Gas Ltd. (NZOG) was awarded new Petroleum Exploration Permit (PEP) 51311, in early 2009. West and south of the Kupe development offshore Taranaki, it covers about 3,000 sq km (1,158 sq mi).

NZOG believes that the permit's proximity of Kupe provides a favorable chance for oil or wet gas.

The initial permit term is five years; a decision on drilling an exploration well is to be made at the end of the second year.

PEP 51558

Late in 2009, Australian Worldwide Exploration was awarded offshore PEP 51558 (named Parihaka) in the Taranaki basin in late 2009.

AWE and partners must acquire 200 km (124 mi) of 2D seismic data in the first two years and are required to drill an exploration well within three years.

Drilling and Production

Maari: Additional reserves

Maari, New Zealand's largest oil field, was originally estimated to hold in situ reserves of 68 million boe. It came on stream in February 2009, and by October, had produced more than 3 million bbl of oil.

In late 2009, the OMV New Zealand-led JV operating the field (Taranaki basin offshore) confirmed the presence of additional producible reserves in a separate reservoir, designated M2A, above the main Moki formation.

It lies in the main Maari mining license PMP 38160.

Kupe gas project

The New Zealand unit of Origin Energy Ltd. began commissioning the offshore Kupe field in December 2009, moving initial output by pipeline to a gas processing plant at Hawera.

Full start-up was expected in early 2010.

In full operation, Kupe is expected to provide 10–15% per year of the country's gas demand for 15–20 years, and more than 50% of the country's demand for LPG.

Over its lifetime, it is expected to provide 6.6 bcm (233 bcf) of gas, 1.1 million tonnes of LPG, and 14.7 million bbl of light crude.

Tui: Reserves search

The Tui oilfields are about 50 km (31 mi) offshore in the Taranaki basin. New Zealand's first stand-alone offshore oil development, the fields began producing in 2007.

Production totaled 9.12 MMbbl from July 2008 to June 2009, according to NZOG.

However, output is expected to 5.1 MMbbl by June 2010 as water production increases.

For that reason, partners were seeking in late 2009 to replenish reserves by exploring two near-field prospects.

South Island CBM

In early 2009, L&M Petroleum Ltd. was preparing to spud coalbed methane wells in the western Southland basin on South Island.

Bogle-1 is to further probe Beaumont formation coals encountered in the Goodwin-1 and Wairaki-1 wells, and also will test the deeper, gassier Morley coals.

Next to be drilled will be Mount Linton-1 and Belmont-1.

Four more wells were being considered in 2009, to confirm the extent of Beaumont coals in the Takitimu North and Longwoods coal trends.

Refining and Products

Marsden Point expansion

New Zealand Refining Co. Ltd. began a 35,000-b/d expansion of its 107,000-b/d Marsden Point refinery, modifying a 90,000-b/d distillation tower.

The project was scheduled for completion by yearend.

The Marsden Point facility is New Zealand's only refinery.

PAKISTAN

Capital: Islamabad
Monetary unit: Rupee
Refining capacity: 286,306 b/cd
Oil production: 63,000 b/d
Oil reserves: 313.000 million bbl
Gas reserves: 29.671 tcf

Finds in Karak, Sanghar

Pakistan's Oil & Gas Development Co. Ltd. (OGDCL) operates 40 oil and gas fields and is the country's largest oil and gas producer, with 35 oil and gas processing plants.

In mid-2009, the company discovered oil and natural gas with its Nashpa No. 1 exploratory well, drilled in Karak district in North-West Frontier Province.

The well, drilled to 4,384 m (14,383 ft) flowed 3,000 b/d of oil and 274,700 cu m/d (9.7 MMcfd) of natural gas on test.

OGDC also discovered oil and gas with its Baloch No. 1 exploratory well on Sinjhoro Block in the Sanghar district of Sindh province.

Drilled to 3,463 m (11,361 ft), the well flowed 780 b/d of oil and 99,120 cu m/d (3.5 MMcfd) of gas.

Sukhpur Block license

In a JV with Pakistan Petroleum Ltd. (PPL) and Royal Dutch Shell PLC, Italy's Eni SPA, as operator, won an exploration license in late 2009 for onshore Sukhpur Block in the Sindh province.

The area is north of Karachi, near Eni-operated producing areas of Bhit and Badhra.

News reports said that Eni was likely to double its investment in Pakistan to $3 billion in 2010. The company also is exploring offshore.

The Italian company has been in Pakistan since 2000 and was the first international company involved in exploration and production of gas.

Tal Block production

The Pakistan unit of Hungary's MOL began gas production in late 2009 from the Tal Block in North West Frontier Province, 30 km (19 mi) from the tense border with Afghanistan.

It had announced a discovery earlier in the year from its Maramzai-1 exploratory well there.

The company expected production to reach 7.1 million cu m/d (250 MMcfd) within weeks, following inauguration of a central gas processing facility.

MOL was planning further work in the Margalla region, near Islamabad.

Tenders for LPG plants

OGDCL was preparing in mid-2009 to invite tenders for construction of several LPG installations.

Proposed installation sites are reported to be Qadirpur, Dhodhak, Dahkni, Loti, Pirkoh, Hundi/Sari, Missa Keswal, and Sadqal.

Earlier in the year, OGDCL called for tenders for LPG and NGL plants downstream of the facilities at the Bobi Oil Complex in the Sanghar district, and at the Kunnar gas-processing plant.

LPG plays an important role in Pakistan's energy mix, used in domestic, commercial, industrial and vehicle applications. More than 20 million households are reliant on LPG.

The government is encouraging the use of LPG in areas where natural gas is not available is to reduce deforestation.

Mashal LNG Project

In 2009, the economic downturn created financing difficulties for Pakistan's leading proposed LNG terminal project, at Port Qasim, Karachi, according to FACTS Global Energy Group.

The government approved the 3.5-mtpy Mashal LNG Project of Sui Southern Gas Corp. at the port in early 2007, choosing a development consortium led by 4Gas BV.

Two other projects proposed for Port Qasim have not advanced as far as Mashal LNG.

GAS IMPORTS

In a mid-2009 analysis, FACTS Global Energy Group concluded that it will be several years before increased gas imports reach Pakistan.

Analysts said that LNG will not reach Pakistan before 2015, and that pipeline gas from Iran is not expected before 2014.

Details are in their report, *Iran-Pakistan Pipeline Deal: Future Role of Gas and LNG in Pakistan's Power Sector.*

IPI/Peace Pipeline

In mid-2009, Iran and Pakistan were moving ahead on a long-awaited gas pipeline between the two countries.

The $7.6-billion project—called both the Iran-Pakistan-India (IPI) Pipeline and the Peace Pipeline—has been delayed by disagreements over transit fees that have put India's participation in doubt.

Construction of the line could begin in 2010, and building the 1,000-km (621-mi) section within Pakistan is estimated to take up to 4 years.

More details are given in the Iran section of the Middle East chapter of this book.

New refineries

In late 2009, government officials said that three refineries in the planning stages would all be operational by yearend 2012. They are:

- 250,000-b/d Khalifa Coastal refinery, in Hub, in Balochistan Province
- 115,000-b/d Bosicor Oil Pakistan Ltd. facility, in Hub, Balochistan Province
- 100,000-b/d Trans-Asia Refinery Ltd. facility, at Port Qasim in Karachi.

The country has seven existing refineries:

- 100,000-b/d Pak-Arab refinery
- 62,050-b/d National refinery
- 47,110-b/d Pakistan Refinery Ltd. facility
- 42,000-b/d Attock refinery
- 30,000-b/d Bosicor Pakistan facility
- 2,500-b/d Dhodak Refinery Ltd. facility
- 2,646-b/d Enar Petrotech Services Ltd. facility.

PAPUA NEW GUINEA

Capital: Port Moresby
Monetary unit: Kina
Refining capacity: 32,500 b/cd
Oil production: 37,000 b/d
Oil reserves: 88.000 million bbl
Gas reserves: 8.000 tcf

Exploration and Development

Subthrust play: Coral Sea

Blue Energy Ltd. was pursuing a campaign in 2009 to evaluate a potential new play type (subthrust) and a large fairway in an unexplored basin in the Coral Sea, off Papua New Guinea.

The area of focus is within the area of several permits held by Blue Energy: PPLs 271, 272, 273, 274, and 330. They span deepwater offshore to transition to land on the southern peninsula of PNG.

The deepwater rocks are continental crust, and the crust appears to be thickening due to thrusting.

For that reason, the company postulates a possible subthrust structural play below the metamorphic and oceanic crust.

PAPUA NEW GUINEA SUBTHRUST PLAY AREA

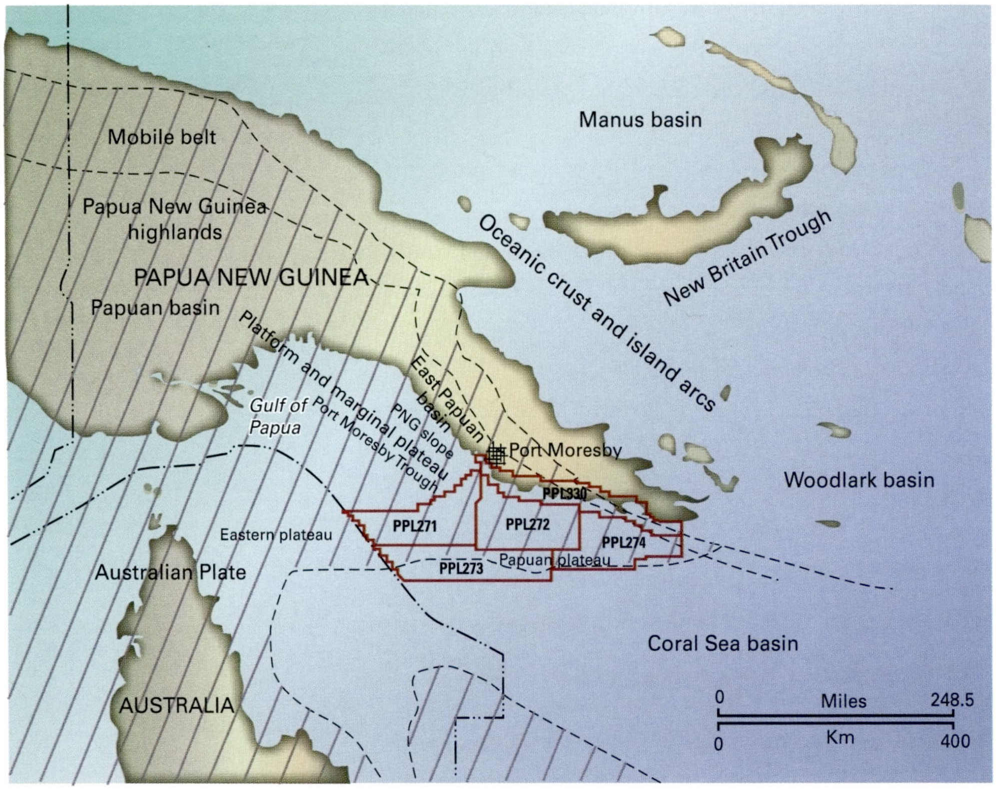

PPL 257: Seismic program

Eaglewood Energy Inc. awarded a contract in late 2009 for a 370-line km (230-mi) seismic acquisition in offshore license PPL 257. Focus will be on the Buna and Kumasi structures.

Previous seismic data established the presence of several significant structures in PPL 257 with multi-tcf gas potential, Eaglewood Energy said. The data also show seismic anomalies that have been previously interpreted as direct hydrocarbon indicators on the crest of structures.

Panakawa oil prospect

New Guinea Energy Ltd. was working to fast-track the drilling in 2009 of its Panakawa oil prospect on the PPL 267 onshore permit, after completion early in the year of a seismic survey.

It was planning to start a drilling program by yearend.

The Panakawa prospect is very close to the Fly River, which should expedite delivery of drilling equipment.

Nippon Oil farm-in

In early 2009, the exploration subsidiary of Nippon Oil Corp. acquired stakes in four exploration licenses from Oil Search Ltd. (OSL) for potential natural gas and oil fields in PNG.

The unit aims to begin a joint exploration program with OSL in 2009, and believes that production at some of the fields could begin in 2010.

The four licenses are: PPL 219 and PPL 239 (both onshore); and PPL 234 and PPL 244 (offshore).

Nippon's purchase was its second investment in PNG since December 2008.

Elk/Antelope resources

InterOil Corp. reported in early 2009 that the independently estimated contingent resource for its Elk/Antelope discovery is 3.4 tcf of gas and 59 million bbl of condensate.

Antelope-1, on PPL 238 in the eastern Papuan basin, flowed 382 MMscfd and 5,000 b/d in early March.

Drilling and Production

Record gas flow: Antelope-2

In late 2009, InterOil Corp. reported that it set a new record at its onshore Antelope field, logging a 20-million cu m/d (705-MMcfd) flow rate from the Antelope-2 appraisal gas well.

The gas flow was accompanied by 11,200 b/d of condensate.

Antelope-2 was drilled to evaluate the southern extent of the Antelope reef reservoir.

LNG

PNG LNG Project

In late December, Esso Highlands Ltd., subsidiary of Exxon Mobil Corp., approved EPC contracts for major elements of the Papua New Guinea LNG (PNG LNG) Project.

The $17-billion project in the Southern Highlands is slated to become operational in 2014.

It will commercialize undeveloped petroleum resources in the Angore, Hides and Juha fields, as well as associated gas resources in the Agogo, Gobe, Kutubu and Moran fields that are currently operational.

Gas will be piped to a processing facility northwest of Port Moresby.

In December 2009, Sinopec entered a long-term agreement with Exxon Mobil to buy about 20 mtpy of LNG from the project over a period of 20 years.

FEED work and the design phase for the project were completed in November.

After being transported to the coast through a 306-km (190-mi) pipeline, gas will move through a 402-km (250-mi) subsea pipeline to a gas-processing terminal before being shipped to customers in Asia.

LNG plant in Port Moresby

Also in late December 2009, InterOil Corp announced that the PNG national government signed the company's project agreement for construction of a $7-billion, two-train LNG plant in Port Moresby.

It would be built near InterOil's Napa Napa refinery and have an initial capacity of 9 mtpy.

Developers will be InterOil subsidiary Liquid Niugini Gas Ltd. and two JV partners.

Plans call for production of 8 mtpy of LNG by early 2015; capacity could be expanded to 10 mtpy.

Small-scale LNG

Japan's Hitachi Ltd. and Toyo Engineering Corp. began a feasibility study in late 2009 into possible development of small- and medium-scale natural gas fields in Papua New Guinea.

The firms see the project as a potential entry into the global LNG market, producing gas from numerous small, undeveloped gas fields in Indonesia and Australia as well as PNG.

They would aim to develop less expensive LNG plants than are typically constructed.

Company News

Talisman buys Rift

In mid-2009, Rift Oil PLC, which has found more than 56 bcm (2tcf) of natural gas in Papua New Guinea, was acquired for £114.8m cash by Talisman Energy Inc.

Santos sells PNG interest

Santos Ltd. executed a conditional agreement in mid-2009 to sell its interest in retention lease PRL 5 in the PNG forelands to Thailand's P3 Global Energy Co. Ltd.

PRL 5 contains Ketu and Elevala gas-condensate fields.

PHILIPPINES

Capital: Manila
Monetary unit: Peso
Refining capacity: 273,000 b/cd
Oil production: 30,000 b/d
Oil reserves: 138.500 million bbl
Gas reserves: 3.480 tcf

Exploration and Development

SC 60 drilling

Shell Philippines Exploration BV and its partners {Kuwait Foreign Petroleum (Kufpec) and South China Resources Ltd. (SCR)} were preparing to start drilling activities at their Service Contract 60 (SC 60) in Northeast Palawan in mid-2009.

Rigorous study of 3D seismic data uncovered a number of leads, several of which were elevated to "prospect" status. The team has a commitment to drill one well by February 2010.

SC 52 potential

E.F. Durkee & Associates, Inc. reported in April 2009 their analysis of a prospect in the Philippines—one of the smallest licenses in the region.

They concluded that the Cagayan basin, Northern Luzon, was worth reevaluation, and in particular, the area encompassed by SC 52.

In mid-2005, indicators suggested that SC 52 was a highly prospective block for gas. Promising formations on the block included Nassiping dome and the Baggao-San Jose prospect.

The latter is a Miocene-aged reef and possibly the largest onshore reef prospect yet identified in the Philippines.

Analysts said that Baggao reef has characteristics similar to those of the Malampaya reef, which potentially holds 56.6–113.3 bcm (2–4 tcf) of gas.

Sandakan basin

In early 2009, ExxonMobil Corp. was planning to invest as much as $100 million exploring for oil and gas in southwestern Philippine waters, according to press reports.

In partnership with a private firm, ExxonMobil was preparing to drill two deepwater exploration wells in 2009.

SC 56 covers more than 8,600 sq km (3,320 sq mi) in as much as 3,000 m (9,842 ft) of water in the Sulu Sea, northeast of Borneo Island in the Sandakan basin.

Reserves

Yakal, Tindalo reserves

In early 2009, Nido Petroleum Ltd. and JV partner Kairiki Energy Ltd. reported the reserves associated with the recent Yakal and Tindalo oil discoveries in offshore SC 54.

Those figures are 10.7 MMbbl and 24.5 MMbbl, respectively.

In December, the JV made the final investment decision to develop Tindalo field, which is in 100 m (328 ft) of water on Block A of SC 54, off Northwest Palawan Island.

Production is to begin by mid-2010 and reach a plateau of 7,000–15,000 b/d.

Most likely recoverable volume is 5.1 million bbl, with a range of 1.5–9.1 million bbl.

Refining and Products

Limay refinery project

Petron Corp. was finalizing construction plans in third quarter 2009 for a $1.5-billion project to enhance operation of the 180,000-b/d Limay refinery, in Bataan Province.

The planned expansion will allow full conversion of residual products into LPG, diesel, gasoline and propylene. Products from the reworked facility will meet Euro-4 fuel quality standards, which have been proposed for use in the country by 2012.

NORTHERN LUZON EXPLORATION AREA

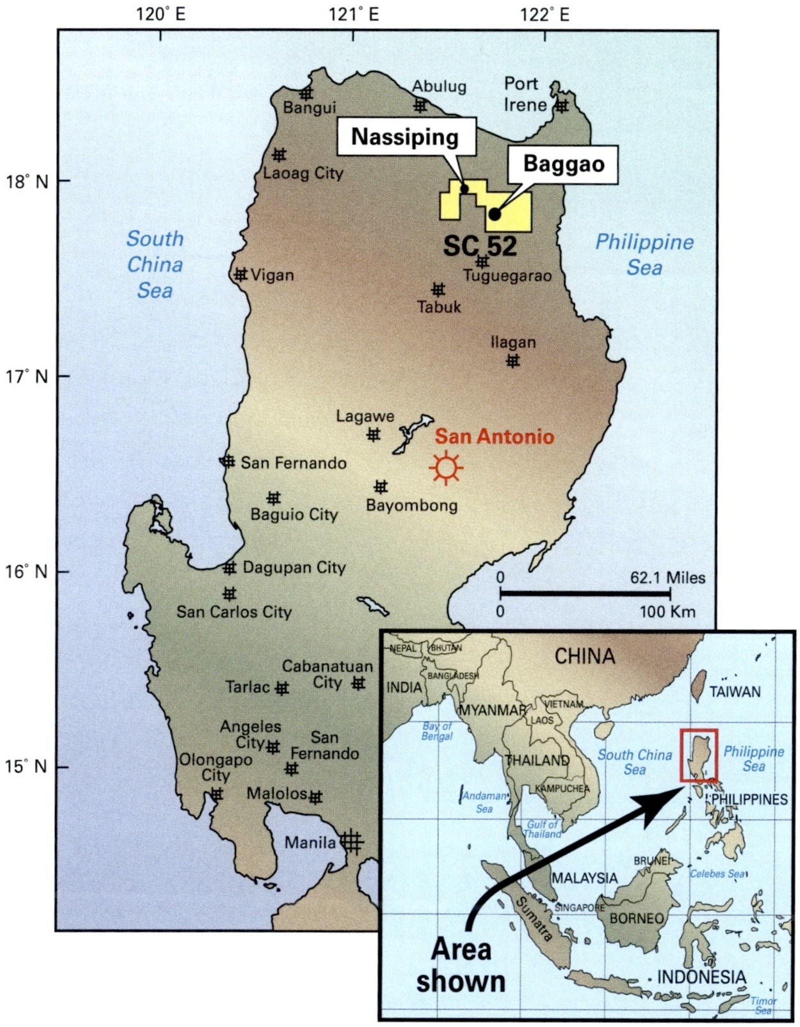

Seafarer Ban

In early 2009, the government banned Filipino seafarers from the waters off Somalia in the Gulf of Aden, threatening a serious disruption of world shipping.

Pirates had attacked scores of ships in that region over several months, and the government said 105 of about 320 crew held hostage off Somalia were Filipinos.

The Philippines provides at least 10% of the world's 1.5 million ocean-going seafarers.

Labor officials directed ships with Filipino crew to stay at least 200 nautical miles off Somalia's east coast, and, in the Gulf of Aden, within a transit corridor patrolled by international naval forces.

Alternative Energy

Renewables energy push

In late 2009, the government's Dept. of Energy awarded 87 contracts worth a combined $2.2 billion in the renewable energy sector to 18 companies. The contracts entail the development of biomass, geothermal, hydropower, ocean, solar and wind energy resources, with a total power generation potential of 4,042 megawatts (MW).

The department aims to increase renewable energy generation capacity from 4,500 MW to 9,000 MW during 2003–2013, and to have renewable sources provide about 40% of primary energy requirements.

SINGAPORE

Capital: Singapore
Monetary unit: Dollar
Refining capacity: 1.357 million b/cd
Oil production: 8,553 b/d
Oil reserves: 0
Gas reserves: 0

Natural gas market

In a 2008 analysis, *Oil & Gas Journal* looked at emerging Asia-Pacific LNG markets.

Singapore relies on gas imports, having no domestic gas fields. Imports come by pipeline from Malaysia, from Indonesia's offshore West Natuna fields, and from Indonesia's Sumatra fields.

In 2000, the gas share of power generation was about 18.5%. It rose sharply to about 74% in 2007, and is expected to reach 83% by 2015.

OGJ concluded that a small LNG receiving terminal could diversify Singapore's gas supply and create price competition for pipeline imports.

Singapore's demand

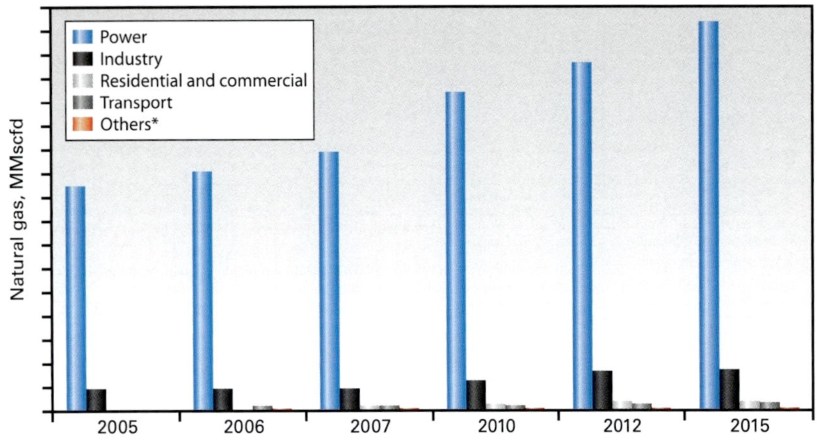

*Includes oil refineries and nonspecified others; excludes distribution losses.

Singapore gas prices

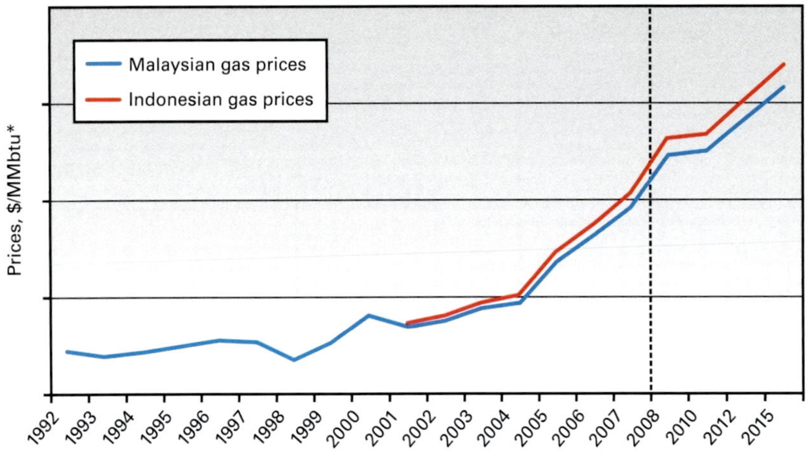

*1992-2007, annual average; 2008-15, estimated annual average.

Oil storage caverns

Hyundai Engineering and Construction Co. won a $594-million construction contract in early 2009 to develop caverns and associated facilities for the first phase of a rock-cavern oil-storage project near the town of Jurong.

Construction was to begin by yearend.

The first phase of work will see the construction of five caverns with a total capacity of 1.48 million cu m (52.3 MMcf), a 7-km (4.3-mi) tunnel, and galleries. The storage areas will be designed to hold gas oil, crude oil, condensate and naphtha. The first two caverns are expected to be ready by 2013.

A second phase add 1.3 million cu m (46 MMcf), primarily for stock storage.

The cavern facility is an initiative of state-owned JTC Corp. to augment the storage capacity of Jurong Island. Facilities on the manmade island refine about 1.3 MMb/d of crude oil.

PetroChina stake in SPC

PetroChina agreed in mid-2009 to pay $1.02 billion to Keppel Corp. for a stake of just over 45% in Singapore Petroleum Co. SPC's interests include oil and gas exploration in Australia, China, Indonesia, Vietnam, and Cambodia.

The agreement marks the first direct acquisition of an overseas stake directly by PetroChina, rather than through parent company, CNPC.

Global Insight commented that the purchase may have been made in this "commercial" manner to deflect potential criticism that it was an action by the Chinese government.

SOUTH KOREA

Capital: Seoul
Monetary unit: Won
Refining capacity: 2.702 million b/cd
Oil production: 30,400 b/d
Oil reserves: 0
Gas reserves: 1.766 tcf

LNG

LNG from Yemen

In October 2009, Total SA announced that the 6.7-mtpy Yemen LNG plant produced its first LNG from the first of two planned trains. The second train was under construction.

The firm expected to export its first cargo by yearend to South Korea.

More details are given in the Yemen section of the Middle East chapter of this book.

Kogas-Kitimat agreement

Canada's Kitimat LNG Inc. signed an MOU in mid-2009 to supply Korea Gas Corp. (Kogas) with as much as 40% of Kitimat LNG's production.

More details are given in the Canada section of the North America chapter of this book.

Kogas terminals

Kogas operates three LNG import terminals. The oldest, at Inchon, started up in 1996. At that facility, 18 tanks can store nearly 2.5 million cu m (88.3 MMcf) of LNG.

According to data for 2007, South Korea ranked second in total LNG imports, behind Japan.

Kogas was scheduled to receive the first cargoes from Russia's Sakhalin-2 project in mid-2009.

Pipelines/Transportation

Myanmar-to-China pipeline

Daewoo International Corp. was proceeding in late 2009 with the controversial Shwe gas field project offshore Myanmar. The project includes construction by CNPC of an 870-km (540-mi) onshore line to transport the gas to China's Yunnan Province.

Details are given in the Myanmar section of this chapter.

Sakhalin gas to S. Korea?

Kogas and Russia's Gazprom signed an MOU in mid-2009 to consider supplying Russian natural gas to South Korea by extending the Sakhalin-Khabarovsk-Vladivostok (SKV) pipeline.

Details are given in the Russia section of the Former Soviet Union chapter of this book.

Refining and Products

Onsan expansion

S-Oil Corp. started a $1.12-billion expansion at its Onsan refinery in Ulsan in mid-2009.

To be completed in 2011, the initiative will add a 900,000-tpd paraxylene plant and a benzene plant with an output potential of 280,000 tpd.

The expansion is part of a strategy to help S-Oil meet future demand arising from Asia and China.

To ensure a supply of naphtha for its petrochemical plants, S-Oil also plans to increase its crude-refining capacity from the present 580,000 b/d output to 630,000 b/d.

Ulsan expansion

In early 2009, Horizon Terminals Ltd. [(subsidiary of Emirates National Oil Co. (ENOC)] and Taeyoung Industry Corp. completed a $76-million expansion project at Horizon Taeyoung Korea Terminals (HTKT) (Ulsan) to augment storage capacity.

Storage capacity was raised from 99,100 cu m (3.5 MMcf) to 231,000 cu m (8.2 MMcf), making it the second-largest storage facility at Ulsan.

Horizon Terminals owns a 50% stake in HTKT, a JV established with Taeyoung.

Petrochemicals

Sector outlook

The government's Financial Supervisory Service announced a decision to downsize the construction and shipbuilding sector in early 2009, and industry experts were expecting that the petrochemical sector could be next.

Weakening demand and resultant slumping prices for products had hurt the sector, and investments in new production facilities or capacity expansion were likely to be reduced or dropped.

The Samsung Economic Research Institute (SERI) projected that domestic demand for petrochemicals in Korea was likely to fall by 1.6% from 2008, to about 9.35 million tons in 2009. Compounding the problem: an expected slump in demand from downstream industries such appliances and automobiles.

SERI predicted marginal growth of 5.3% in South Korea's exports of petrochemicals to 11.2 million tons in 2009 versus 2008. From 2007 to 2008, such exports declined by 36.6%.

South Korea's petrochemical sector depends entirely on imported crude from the Persian Gulf, mainly Saudi Arabia.

BTX plant

In late 2009, Hyundai Oilbank Co. Ltd. and Japan's Cosmo Oil Co. Ltd. formed a 50:50 JV to develop a benzene, toluene and xylene (BTX) plant in South Korea.

Cosmo will invest $600 million; Hyundai Oilbank will contribute land and related production facilities.

The JV will acquire an existing 380,000-tpy paraxylene plant from Hyundai Oilbank, at Hyundai Oilbank's Daesan refinery complex.

While operating the plant, the JV also will build the new 800,000-tpy BTX plant at the same site. Construction is to be completed in 2013.

Company News

KNOC-Harvest Energy Trust

In December 2009, Korea National Oil Corp. was poised to buy Canadian firm Harvest Energy Trust for $1.8 billion (Can.) plus the assumption of $2.3 billion (Can.) in debt.

Details are given in the Canada section of the North America chapter of this book.

North Korea: Confrontation

In May 2009, North Korea rattled diplomats as well as energy markets by:

- exploding a below-ground a nuclear device with an estimated yield of 20 kilotons
- test-firing three short-range missiles
- restarting a plant that reprocesses spent nuclear fuel into weapons-grade plutonium.

In June, the Big Five nuclear powers, along with Japan and South Korea, agreed on new UN sanctions against the country. North Korea said that the sanctions cannot halt its nuclear missile program.

TAIWAN

Capital: Taipei
Monetary unit: New Taiwan dollar
Refining capacity: 1.310 million b/cd
Oil production: 800 b/d
Oil reserves: 2.380 million bbl
Gas reserves: 220.000 bcf

Taichung terminal

In July 2009, Taiwan's Chinese Petroleum Corp. (CPC) inaugurated the country's second LNG terminal at Taichung in the north, according to press reports. The start-up of the nearly $955-million terminal was more than a year behind schedule.

For 2009, the 3-mtpy plant will supply 1.68 million tonnes of vaporized LNG through a subsea pipeline to the Tatan power plant in northern Taiwan. The rest will go to industrial and residential markets.

Start-up of the Taichung terminal will allow CPC to import nearly 1.6 million tonnes of LNG from Qatar in 2009, rising to 2.5 million in 2010 and to 3 million in 2011.

Combined output from Taichung and the existing Yungan terminal in southern Taiwan will enable CPC to move as much as 10.5 mtpy of LNG.

Formosa refinery expansion

Formosa Petrochemical Corp., Taiwan's only private oil refiner, said in late 2009 that it plans to invest $803 million to boost its refining capacity to 593,000 b/d by yearend 2012.

Current capacity at the firm's complex in Mailiao is 540,000 b/d.

Longer-term, the company said its $8.67 billion "fifth-phase" expansion will aim to raise capacity to 764,000 b/d. Plans for that phase must still undergo environmental review.

CPC petrochem plans

In mid-2009, CPC announced that it will invest $1.4 billion in the petrochemical industry over 10 years to set up new plants and to upgrade facilities.

The upgrades will shift output slates to increase production of higher-value products, such as diesel.

Over 10 years, CPC intends to invest $26 billion to improve efficiency and build new facilities to feed domestic chemical and fuel markets and to compete with Formosa Petrochemical Corp.

CPC owns and operates three refineries (total capacity: 720,000 b/d) and three plants that process naphtha. It also has an overall production capacity of 1.08 mtpy of ethylene.

Offshore windfarms

The government passed the Renewable Energy Development Act in June 2009, under which it will provide incentives to promote domestic renewable power generation.

The government sees wind power as the fastest way to achieve its target of producing 15% of electricity needs (8,450 MW) from renewable sources by the year 2025.

Taiwan's west coast has more than 100 wind turbines, producing more than 420,000 MW-hr/year.

In late 2009, a subsidiary of Scotland's SeaEnergy PLC and Taiwan Generations Corp. (TGC) entered into an agreement to jointly develop offshore wind power projects. One major initial project will be the $1.23-billion, 600-MW Changhua facility in Taiwan Strait, off the west coast.

> **TYPHOON MORAKOT**
>
> Typhoon Morakot struck Taiwan on August 7, 2009. The center of the storm made landfall in Hualien County along the central-east coast and then passed over the central-northern part of the island.
>
> However, the worst effects were in the south, where as much as 80 inches of rain were reported in some areas, causing the worst flooding in 50 years. More than 500 people died.

Also, Taiwan Power Co. announced a $244-million plan to build 50 new wind turbines by 2015.

The firm currently has 106 wind turbines in operation and 56 under construction.

THAILAND

Capital: Bangkok
Monetary unit: Baht
Refining capacity: 584,100 b/cd
Oil production: 236,000 b/d
Oil reserves: 430.000 million bbl
Gas reserves: 12.079 tcf

Exploration and Development

Bo Rang B field

Carnarvon Petroleum Ltd. and JV partner Pan Orient Energy Corp. completed testing in late 2009 of the BR-3D1 appraisal well in Bo Rang B oil field declared the field commercial.

The first horizontal appraisal in the field, the well flowed at 700 b/d of oil from a fractured volcanic reservoir target at a depth of 446 m (1,529 ft).

The JV was planning three more horizontal appraisals in the Bo Rang B field by yearend.

Block L27/43, Khorat basin

Coastal Energy Co. spudded the Phu Kheng-1 exploration well on Block L27/43 in northeast Thailand's Khorat basin in mid-2009.

Projected depth was 2,070 m (6,791 ft) subsea to target Jurassic and Triassic gas shows.

The Phu Kheng formation has a mean gross prospective resource of 38.5 bcm (1.36 tcf).

NSE-I1 well

Pan Orient Energy Corp. said in mid-2009 that its NSE-I1 well discovered commercial hydrocarbons in a previously untested volcanic reservoir.

The well was pumping 75 b/d of oil and 200 b/d of water.

The well encountered the first volcanic target zone at 637 m (2,090 ft) TVD and the second at 750 m (2,460 ft) TVD.

Bua Ban field

In early 2009, the government approved a 20-year production license for Coastal Energy Co. surrounding the Bua Ban field in the shallow waters of the Gulf of Thailand.

Coastal estimated that the field contains 21.8 MMbbl of proven plus probable reserves.

The firm was aiming to develop Bua Ban field beginning in mid-2009, with first production expected in third quarter. Production is expected to average 9,500 b/d of oil by yearend.

Drilling and Production

Bongkot field

The consortium led by Thailand's PTT Exploration & Production PLC (PTTEP) was preparing plans in mid-2009 to ramp up production from its greater Bongkot field by 58% to 24.6 million cu m/d (870 MMcfd) of gas by 2012.

The incremental delivery of 9 million cu m/d (320 MMcfd) will come from the major sister field in the Gulf of Thailand, Bongkot South, which is slated to come on stream by mid-2013.

The main Bongkot field, 600 km (373 mi) south of Bangkok, has been in production since 1993.

Bualuang field

Salamander Energy completed its drilling program on the offshore Bualuang oil field in early 2009.

Two horizontal wells were completed in the crestal area of the field, and on test, flowed at more than 18,000 b/d of oil.

Both wells consisted of about 250-m (280-ft) horizontal sections in a target zone 2 to 3 m (7 to 10 ft) below the top reservoir.

Natuna D Alpha Block

Indonesia was negotiating in early 2009 with Thailand's PTT PLC for the purchase of undisclosed volumes of natural gas from the Natuna D Alpha Block.

Bua Ban Field, Gulf of Thailand

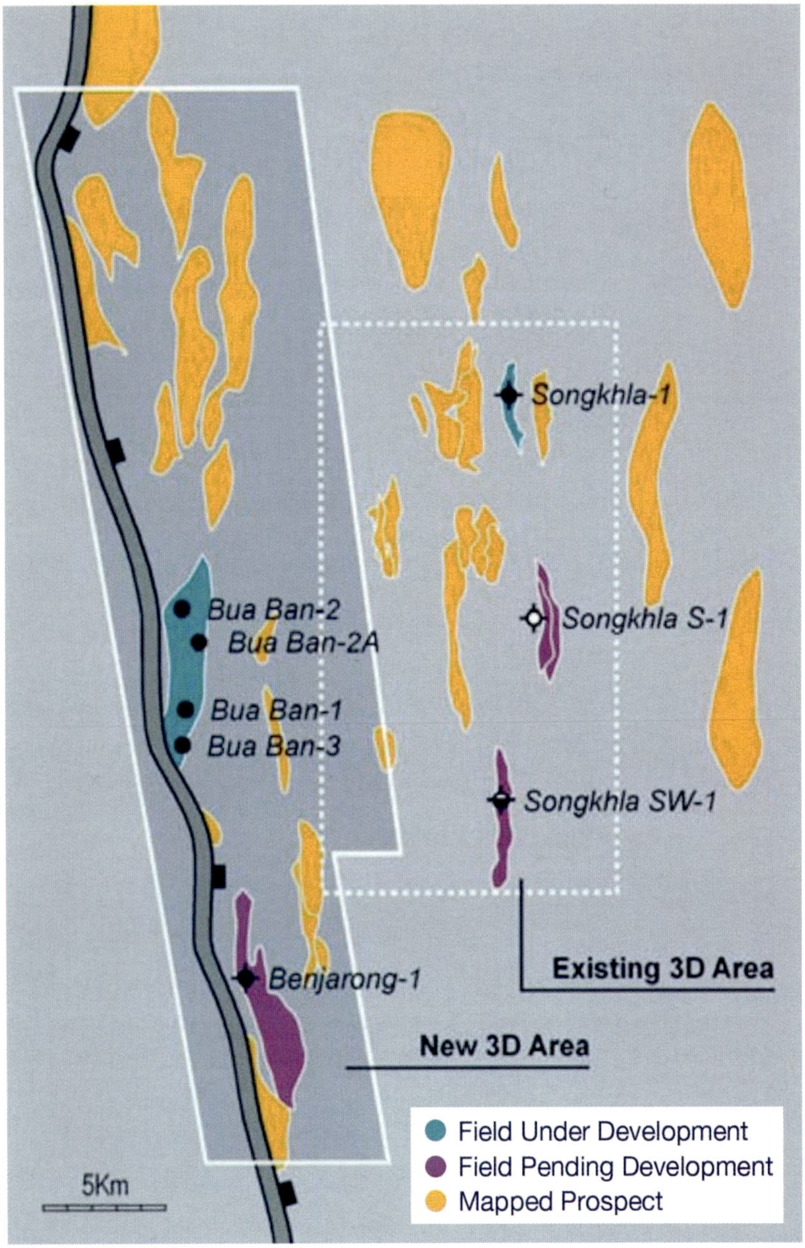

State-owned PT Pertamina had not yet decided on a new partner for developing the block.

Songkhla field

Coastal Energy Co. started production from Songkhla field in the Gulf of Thailand in early 2009. The field was producing 11,000 b/d from the Songkhla A-01 and A-03 wells.

Production from the Songkhla A-07 exploration well was scheduled for March.

Also, in late 2009, the A-08 development well struck oil, flowing more than 3,500 b/d of oil from the Lower Oligocene primary reservoir.

Petrochemicals

Project suspension reaffirmed

In December 2009, Thailand's Supreme Court largely upheld a September injunction that halted development on $12 billion worth of petrochemical and power projects.

The projects, at Map Ta Phut Industrial Park near Bangkok, had been blocked because of allegations that companies had not carried out out full environmental-impact surveys before licenses were issued.

The court allowed 11 of 76 projects to proceed on the grounds that they posed no threat to the environment, but blocked 65.

Alternative Energy

First nuclear power plant

As of mid-2009, state-owned Electricity Generating Authority of Thailand was one-third through a feasibility study to identify possible locations for Thailand's first nuclear power plant.

Potential sites are in the Chai Nat, Chon Buri, Nakhon Si Thammarat and Surat Thani provinces.

The program would aim to start up the first plant by 2020.

Company News

PTTEP: 2009 budget

PTTEP cut its 2009 capital expenditures in mid-2009 by 12% to cope with the global economic slowdown. As a result, said officials, only 35 wells would be drilled worldwide in 2009, not the 45 first planned.

TIMOR-LESTE

Capital: Dili
Monetary unit: U.S. dollar
Refining capacity: 0
Oil production: 100,000 b/d
Oil reserves: 553.8 million bbl
Gas reserves: 7.06 tcf

Block JPDA 06-103

A two-well exploration program by Oilex Ltd. and Japan Energy Corp. was expected to begin by late 2009 on Block JPDA 06-103 in the Timor Sea, between Timor-Leste and northwestern Australia.

The block is replete with large structures off the northeast flank of oil fields in the Flamingo Trough.

Interpretation of seismic identified more than 20 prospects and leads.

The prospect inventory indicates a potential mean estimated prospective resources in excess of 100 MMbbl of oil. Oilex identified 10 of the prospects as potential drilling targets.

The first two targets and their mean prospective resources are Lore at 195 million bbl and Lolotoe at 90 million bbl.

Oilex said that the area's prospectivity is implied by the discovery of relatively nearby fields such as Kitan (and Kuda Tasi and Jahal discoveries north of Kitan), Laminaria, Corallina, and Bayu-Undan.

Early estimates indicate 30–40 million bbl of oil recoverable from Kitan.

Montara project

Thai state company PTTEP Australasia awarded a $116-million (Aus.) contract in early 2009 to Clough Australia to install oil facilities for the Montara project in the Timor Sea.

The project includes development of the main Montara field along with the outlying Swift and Skua fields, all about 690 km (430 mi) west of Darwin in the Ashmore-Cartier region of the Timor Sea.

They lie in 80 m (263 ft) of water in the AC/L7 and AC/L8 production licenses.

PTTEP expects the Montara to start producing crude and gas in the fourth quarter.

FLAMINGO TROUGH OIL FIELDS, STRUCTURES

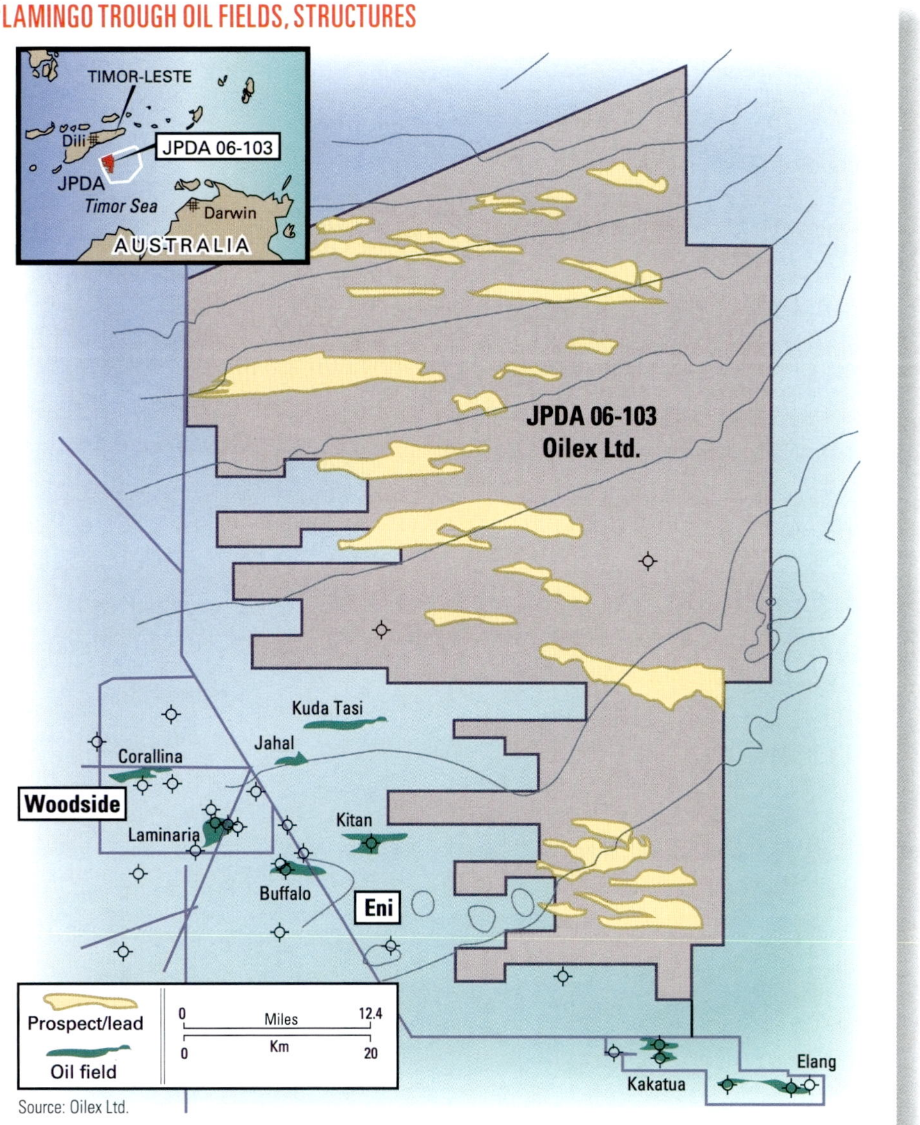

Source: Oilex Ltd.

VIETNAM

Capital: Hanoi
Monetary unit: Dong
Refining capacity: 140,000 b/cd
Oil production: 300,000 b/d
Oil reserves: 600.000 million bbl
Gas reserves: 6.800 tcf

Lac Da Nau well

Exploration well 15-01/05 LDN-1X on the Lac Da Nau prospect hit pay in late 2009, according to a subsidiary of PetroVietnam Exploration and Production Corp.

The company reported that the well, in 45 m (148 ft) of water, produced 4,200 b/d of oil on test.

This is the first well and the first discovery on block 15-01/05.

Nuclear power

In late 2009, Vietnam's national assembly approved a nuclear development plan. It includes construction of two power plants in the southern province of Ninh Thuan.

The first will be at Phuoc Dinh, in the Thuan Nam district. Construction activity is expected to begin by 2014. The second plant will be at Vinh Hai, in the Ninh Hai district.

Total investment for the two is estimated at $11 billion. Discussions were under way with several international energy firms.

Vietnam is proposing to develop nuclear power projects with a total generating capacity of 2,000 MW by 2020.

Te Giac Trang field

In late 2009, the Ministry of Industry and Trade approved the development plan for the Te Giac Trang field in the Cuu Long basin, offshore Vietnam, according to operator Hoang Long Joint Operating Co.

First oil from the field is targeted for mid-2011.

Seven exploration/appraisal wells have been drilled, with an average oil and gas flow of about 11,300 boe/d from each well.

Block 07/03: Red Emperor

Red Emperor well on offshore Block 07/03 proved oil and gas in mid-2009, said Pan Pacific Petroleum NL (PPP). It was drilled to a measured depth of 3,810 m (12,500 ft).

The well encountered about 90 m (295 ft) net pay within multiple stacked reservoir layers. Two zones flowed at a combined 3,265.4 b/d of oil and 229,000 cu m/d (8.1 MMcfd) of gas on test.

Block 16-1: TGD, VT

In April 2009, the government approved the applications of the Hoang Long Joint Operating Co. for the Te Giac Den (TGD) and Voi Trang (VT) appraisal areas in Block 16-1.

The partners were reprocessing seismic over the TGD area and aiming to drill an appraisal well in early 2010.

Blocks B, 48/95, 52/97

In mid-2009, Mitsui Oil Exploration Co., its partners, and PetroVietnam started the design of a $4-billion natural gas project on Blocks B, 48/95, and 52/97.

Output of up to 14.2 million cu m/d (500 MMcfd) is expected in the first 5 years, following start-up no earlier than 2014. It will supply domestic power plants.

Tanker fleet

Early 2009 saw the laying of the keel for a long-planned 105,000-tonne Aframax oil tanker at the Dung Quat Ship Building Co, in Quang Ngai Province.

The ship is scheduled to be completed by late 2010, one of three new ships that will be placed in the service of the country's new 140,000-b/d Dung Quat refinery.

In midyear, a second Aframax oil tanker was to be delivered from Dung Quat Shipping Industry Co.

In December 2008, PetroVietnam's shipping arm said it planned to invest as much as $3 billion through 2015 to upgrade its crude and oil product tanker fleet to meet demand from Vietnam's refineries.

Refinery review

In July 2009, the $3.1-billion, 140,000-b/d Dung Quat refinery—the country's first—produced its first run of commercial-grade gasoline.

The government aims to have at least three major refineries up and running by 2013.

Taiwan's Formosa Heavy Industries Corp. got approval from Vietnam in late 2009 to build a $12.5-billion oil refinery and petrochemical complex in the Vung Ang Economic Zone in Ha Tinh Province.

Other refineries in various stages of development are at:

- Nghi Son in Thanh Hoa Province
- Vung Ro in Phu Yen Province
- Can Tho in the Mekong Delta
- Nam Van Phong in Khanh Hoa Province
- Long Son Island in Ba Ria-Vung Tau Province.

Several of these projects also will include petrochemical plants.

Other Action in Asia-Pacific

Several other nations in the Africa-Pacific region saw exploration and development action in 2008 and early 2009.

Fiji

Capital: Suva

Savoy Energy Corp. and Masi Corp. Holdings Ltd. were forming a JV (Savoy-Masi Petroleum Corp. Ltd.) in mid-2009 to license properties for oil exploration, recompletion, and workover.

Republic of Palau

Capital: Ngerulmud

- In mid-2009, Fugro Airborne Surveys gathered aeromagnetic data in the offshore North block concession, seeking to map basement rock and lithology. The company will prepare a map of the basement surface and possible sedimentary packages over the concession area.
- Palau Pacific Exploration was developing plans in mid-2009 to drill a well off northern Palau to evaluate carbonate reef systems.
- Also in mid-year, the government was formulating a hydrocarbon law and drawing plans for an exploration license round.

Sri Lanka

Capital: Colombo

- A unit of Cairn India began a 3D seismic program in late 2009 in the offshore Mannar basin. The firm holds a license for Block SL 2007-01-001. Completion of data processing is expected in mid-2010.
- In May 2009, the insurgent Tamil Tigers admitted defeat in their 25-year war for a separate homeland, surrendering to government forces.

Key Stats

Section 1:
Future energy supply

IEA: Peak oil comment

The Economist in December 2009 attributed to the chief economist of the International Energy Agency (IEA) the statement that if no big new discoveries are made, "the output of conventional oil will peak in 2020 if oil demand grows on a business-as-usual basis."

The magazine called it "a new and striking claim," noting that the IEA had not previously committed itself to a firm prediction of when oil supplies might cease to grow.

In the IEA energy outlook released in November 2009, the agency said only that conventional oil is "projected to reach a plateau sometime before" 2030.

After examining historical production trends of 800 oilfields in 2008, the IEA concluded that the decline in annual output from fields that are past their prime could average 8.6% in 2030.

"Even if oil demand were to remain flat, the world would need to find more than 40 MMb/d of gross new capacity—equal to four new Saudi Arabia's—just to offset this decline," said the agency's chief economist.

Total SA: Peak oil comment

In February 2009, CEO Christophe de Margerie of Total SA told the *Financial Times* that the world will never be able to produce more than 89 million b/d of oil, citing high costs in areas such as Canada and political restrictions in countries such as Iran and Iraq.

He also said he had revised his forecast for 2015 oil production downward by at least 4 million b/d because of the early-2009 economic crisis and the collapse in oil prices.

Modeling largest oil fields

RSK (UK) Ltd., provider of valuation and advisory services for production and exploration properties, has developed modeling methods for use in forecasting future oil supply.

In mid-2009, they shared some of their findings regarding the contribution of giant oil fields to global oil supply, and called for a cooperative government/industry approach to modeling in order to better understand future global oil productive capacity.

Giant Fields' Shares of…

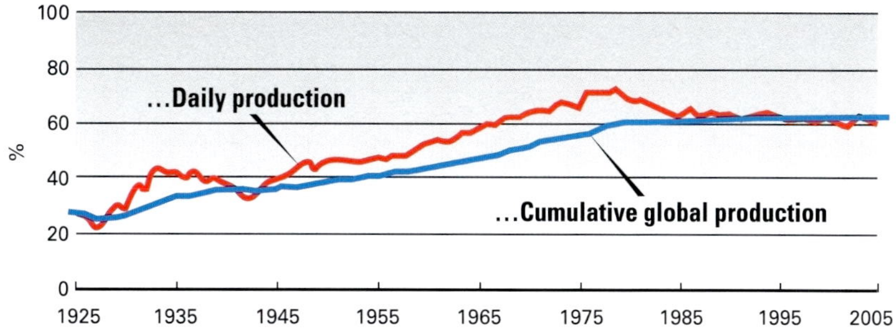

Source: Hook, M., Soderbergh, B., Jokobsson, K., Alklett, K., 2009, "The evolution of giant oil field production behavior," *Natural Resources Research*, Vol. 18, No. 1, 2009, pp. 39–56.

How Production Depends on Giant Fields*

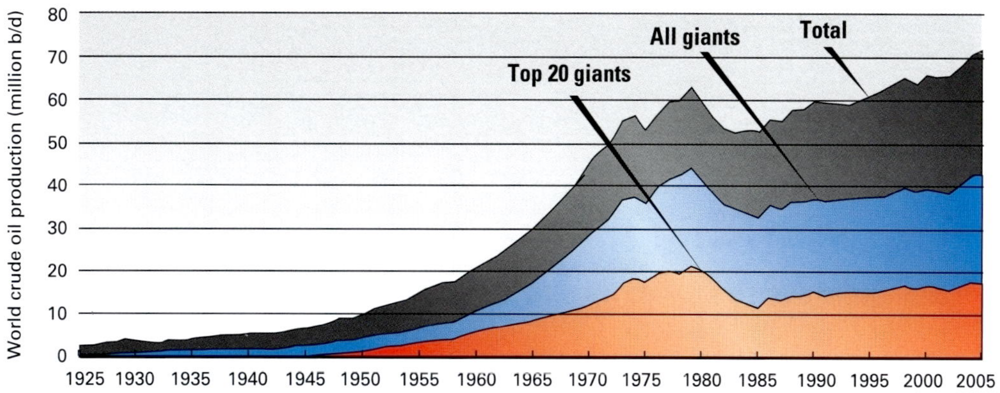

*Note correlation in last several decades between total production and production from the giants.
Sources: Hook, M., Soderbergh, B., Jokobsson, K., Alklett, K., 2009, "The evolution of giant oil field production behavior," *Natural Resources Research*, Vol. 18, No. 1, 2009, pp. 39–56. Robelius, F., "Giant Oil Fields—The Highway to Oil: Giant Oil Fields and their Importance for Future Oil Production," PhD dissertation, Dept. of Nuclear and Particle Physics, Uppsala University, SE-75121, Uppsala, Sweden, 2007.

Giant Field Discoveries vs. Present Reserves*

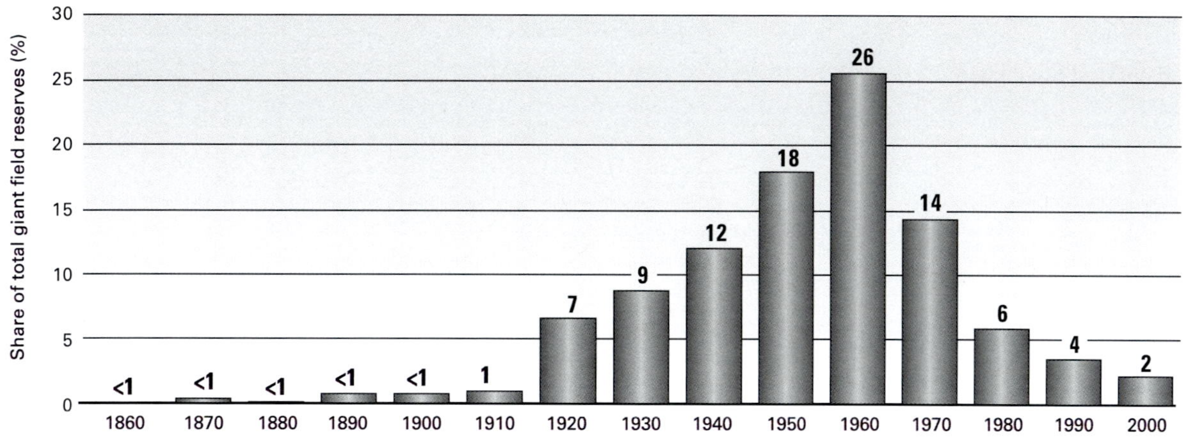

*Reserves discovered over time in giant fields as share of present day volumes in giant fields.

Oil shale resource base

In early 2009, analysts with INTEK Inc. and the U.S. Dept. of Energy published a comprehensive review of the potential of oil shale as a global energy resource.

Kerogen, comprising 10–60 gal/ton of shale, can be extracted and processed to yield high-quality jet fuel, diesel, naphtha, and other products.

Oil shale occurs in nearly 100 major deposits in 27 countries. Estimates indicate that in-place hydrocarbons contained in the shales may exceed 10 trillion bbl. The western United States has the largest amounts.

The two organizations also reported on several new approaches being used to develop oil shale resources. This included:

- The in situ conversion process: Royal Dutch Shell PLC
- The Electrofrac process: ExxonMobil Corp.
- Use of superheated air: PetroProbe (a unit of Earth Sciences Inc.)
- Use of radio frequency energy and critical fluids: Schlumberger.
- Use of fuel cell technology: Independent Energy Partners
- Use of retort-based pyrolysis in Brazil, China, Estonia, and Alberta

Resource volume

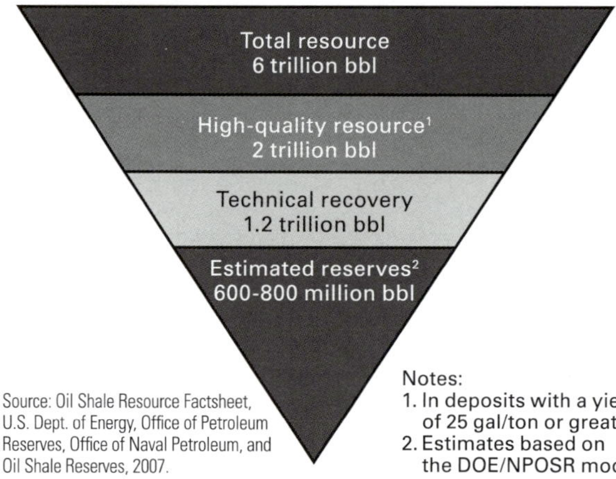

Source: Oil Shale Resource Factsheet, U.S. Dept. of Energy, Office of Petroleum Reserves, Office of Naval Petroleum, and Oil Shale Reserves, 2007.

Notes:
1. In deposits with a yield of 25 gal/ton or greater
2. Estimates based on the DOE/NPOSR model

Resource density comparison

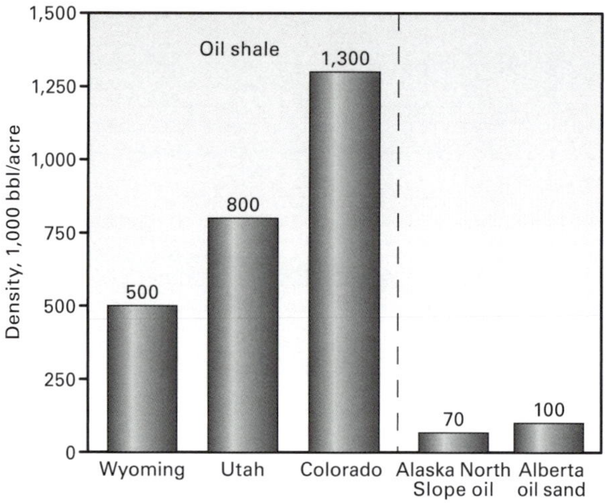

Source: Oil Shale Resource Factsheet, U.S. Dept. of Energy, Office of Petroleum Reserves, Office of Naval Petroleum, and Oil Shale Reserves, 2007.

Countries with oil shale deposits

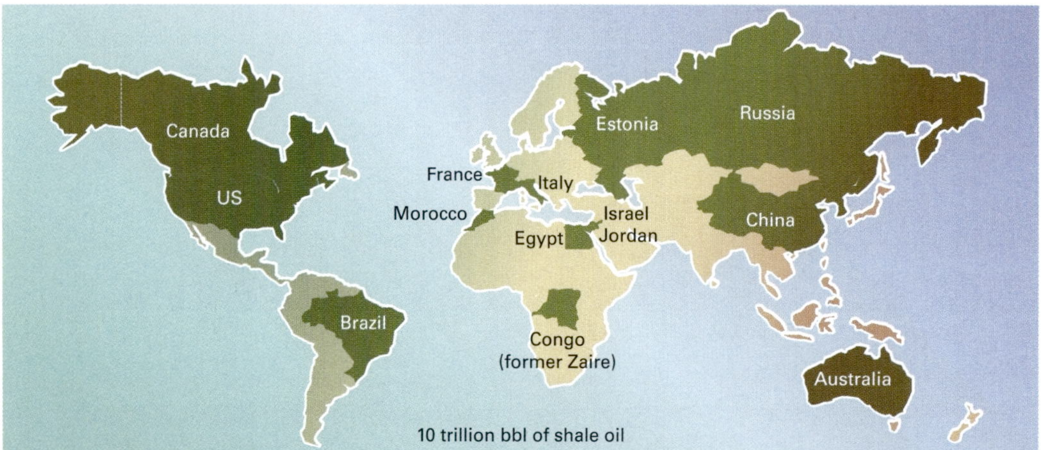

Source: Oil Shale Resource Factsheet, U.S. Dept. of Energy, Office of Petroleum Reserves, Office of Naval Petroleum and Oil Shale Reserves, 2007.

US OIL SHALE RESOURCES

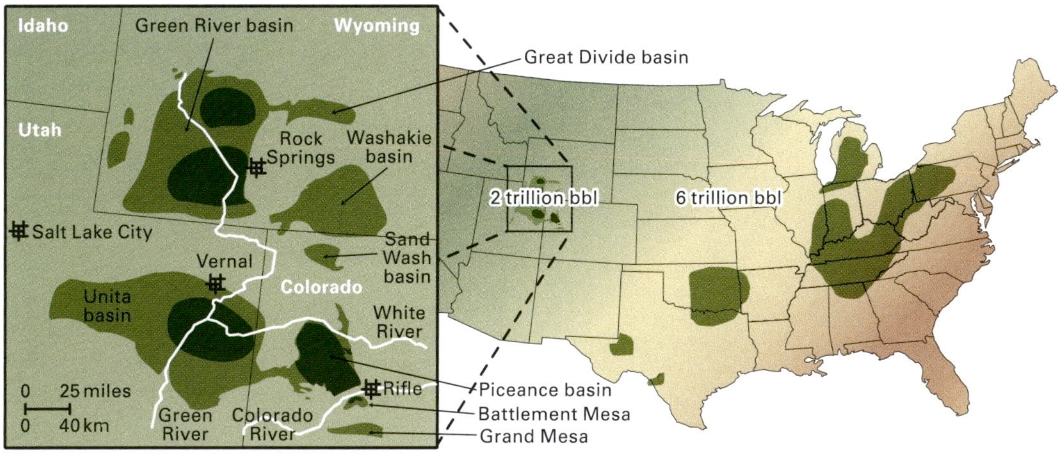

Source: Oil Shale Resource Factsheet, U.S. Dept. of Energy, Office of Petroleum Reserves, Office of Naval Petroleum and Oil Shale Reserves, 2007.

U.S. OIL SHALE RICHNESS

	Richness, gal/ton		
	5–10	10–25	25–100
Colorado, Wyoming, and Utah (Green River), billion bbl	4,000	2,800	1,200
Central and Eastern states, billion bbl	2,000	1,000	
Alaska, billion bbl	Large	200	250
Total, billion bbl	6,000+	4,000	2,000+

Source: Reference 2 and others

USGS: Arctic resources

In mid-2009, the U.S. Geological Survey (USGS) assessed undiscovered oil and gas resources north of the Arctic Circle. The Circum-Arctic Resource Appraisal (CARA) includes probabilistic estimates of the numbers, sizes, and aggregate volumes of yet-to-find oil, gas, and NGLs in undiscovered accumulations, both offshore and onshore.

The CARA study looked at the potential for new discoveries north of the Arctic Circle of conventional oil and/or gas in accumulations larger than 50 MMbbl of oil or 300 bcf (8.5 bcm) of natural gas. Unconventional resources such as coal bed methane, heavy oil and bitumen, and gas hydrates are not evaluated.

The study assessed 48 geologic "assessment units" (AUs) and assessed each using a quantitative probabilistic methodology.

Oil

The USGS estimates that 44–157 billion bbl of oil could be added to proved reserves from new discoveries north of the Arctic Circle, and that there is a 50% chance of finding 83 billion bbl of oil. Approximately 73% of this amount (almost 61 billion bbl), is expected to be found offshore.

Most of the resource is concentrated in just seven AUs:

- The Alaska Platform: 27.9 billion bbl of oil
- Canning-Mackenzie AU: 6.4 billion bbl
- North Barents basin: 5.3 billion bbl
- Northwest Greenland Rifted Margin AU: 4.9 billion bbl
- Northeast Greenland shelf, South Danmarkshavn basin: 4.4 billion bbl
- Northeast Greenland shelf, North Danmarkshavn salt basin: 3.3 billion bbl
- Yenisey-Khatanga AU, east of the northern West Siberian basin: 5.3 billion bbl (primarily onshore).

Gas

Like the discovered accumulations, undiscovered gas is much more abundant than oil in the arctic, and most resources are expected to be offshore.

The USGS estimates a high probability that more than 770 tcf (21.8 tcm) of undiscovered, conventional gas occurs north of the Arctic Circle, and that as much as 2,990 tcf (84.7 tcm) may be found.

The mean estimate is 1,547 tcf (43.8 tcm), of which almost 77% is expected to be found mainly in Russian territory under relatively shallow water in the continental shelves of the South Kara Sea and the east Barents Sea.

Two-thirds of the undiscovered gas is in just four AUs:

- South Kara Sea (607 tcf; 17.2 tcm)
- South Barents basin (184 tcf; 5.2 tcm)
- North Barents basin (117 tcf; 3.3 tcm)
- Alaska Platform (122 tcf; 3.5 tcm).

EIA: Arctic resources

In October 2009, the U.S. Energy Information Administration (EIA) released its study of natural gas resources in the arctic, including those in Canada, Denmark/Greenland, Finland, Iceland, Norway, Russia, Sweden, and the United States.

EIA concluded that several factors add uncertainty to the scale and pace of Arctic resource development, including:

- The prospect of recovering 5,000–16,000 tcf (142–453 tcm) of shale gas in more accessible geologic provinces
- The extreme risk and expense of Arctic operations
- Unresolved sovereignty claims
- Likely demands for environmental safeguards.

Nonetheless, growing European gas demand may be a strong incentive for Russian Arctic gas development, EIA concluded.

Desertec: Solar electricity

In July 2009, more than a dozen large companies met in Europe to discuss formation of a consortium called Desertec. The project would build a massive network of solar power stations in Africa and Arabia, capturing the sun's energy to make steam to drive turbines connected to electrical generators.

The power generated would flow to Europe through high-voltage subsea cables.

Further details are given in the Germany section of the Europe chapter of this book.

World wind energy

The *World Wind Energy Report 2009*, published in midyear, projected 25% market growth for the global wind sector in 2009 compared to 2008.

The report, by the World Wind Energy Association, stated that about 5,400 MW of new capacity was added through September, based on data from 11 of the world's top 15 wind-power-generation nations. It projected total global wind power generating capacity would reach 30,300 MW by yearend 2009.

The group expects that figure to reach 152,000 MW by 2009, and says that the two primary wind energy generators are likely to be China and the United States.

Liquid fuels from algae

Oil & Gas Journal noted in September 2009 that established investors, including one notable oil major, were putting money into early-stage companies developing algae-based biofuel.

ExxonMobil Corp. said it plans to invest $300 million in a multi-year collaboration with Synthetic Genomics Inc., founded by noted genomics expert J. Craig Venter. That investment aims to help find a biochemical method to alter the molecules of algal oil, leaving behind a hydrocarbon.

ExxonMobil is also said to be spending another $300 million internally on algae research.

Methane hydrates

Natural gas hydrates are a potentially significant energy source—more than 200,000 tcf (5,664 tcm) in the United States, according to estimates by the USGS. In comparison, the total amount of conventional U.S. natural gas is estimated to be about 1,700 tcf (48.1 tcm).

However, at a hearing held by a subcommittee of the House Energy and Natural Resources Committee in July 2009, experts said that much work is needed to determine if they can be economically produced.

Production is technically feasible with existing technology, but the process also could degrade seabed stability and release methane, a potent greenhouse gas, into the oceans and atmosphere.

Nuclear power plants

World Nuclear News reported that construction of 12 nuclear reactors began in 2009, but only two nuclear power plants began operations. The year also brought the closure of two nuclear power plants in Lithuania and France.

The report estimated that worldwide, 54 nuclear reactors were under construction, including one each in Russia and South Korea, as well as 10 in China.

In 2010, eight new reactors are expected to be commissioned, and two existing plants will go live after augmentation.

In 2009, global nuclear power generating capacity increased by 808 MW through capacity augmentation at existing plants.

As of yearend 2009, global nuclear power generating capacity was about 372,673 MW.

China, which operates 11 nuclear power plants, plans to build 100 new reactors by 2020.

India has also announced that it will ramp up its nuclear reactor fleet to generate 20,000 MW by 2020.

Section 2:
Exploration and reserves

Recent discoveries

Several major oil and gas discoveries have been made since 2006—from Tupi, Iara, and Guara fields off Brazil, to Jubilee offshore Ghana, to Tiber in the U.S. Gulf of Mexico.

Will their output be enough to avoid the supply constraints that many in the industry foresee (as early as 2014) as the world tries to recover from the recession that began in 2008?

Probably not, concluded the *Financial Times* in a September 2009 evaluation, noting that major fields take a long time to be developed fully.

The IEA's supply forecasters said that the speed of the economic recovery would be a major factor in keeping up with higher energy demand.

Recent oil and gas discoveries

Year	Field	Nation	Reserves (MMboe)
2006	Tupi	Brazil	5,007
2008	Iara	Brazil	2,956
2008	Guara	Brazil	1,550*
2007	Jubilee	Ghana	1,376
2009	Miran West	Iraq	1,100
2009	Perla (gas)	Venezuela	1,000
2007	Carioca	Brazil	867
2008	Tsentralnaya	Russia	632
2009	Tiber	U.S. GoM	750*
2006	Kaskida	U.S. GoM	430
2009	Venus	Sierra Leone	200*
2009	Venus	Sierra Leone	200*

Source: *Financial Times* *Estimate

If the economy returns to 4.5–5% growth rates, the world will need about 4 million b/d of oil more output to meet demand if it does not want to risk a price spike, such as the mid-2008 jolt.

Oil, gas reserves rise; oil output declines

New estimates for Australia, China, and Turkmenistan raised reported worldwide oil and gas reserves, according to an annual survey of proved reserves, published in December 2009 by *Oil & Gas Journal.*

Global oil reserves total 1.35 trillion bbl, up from 1.34 trillion in the previous (late-2008) survey.

Total gas reserves are 6.6 quadrillion cu ft (186.9 tcm), up almost 355 tcf (10 tcm) from a year ago.

OGJ compiles the estimates of proved reserves from an annual survey of official sources, including government agencies and ministries.

A decade of reserves changes

Jan. 1	World (1,000 bbl)	OPEC (1,000 bbl)	World gas, bcf
2010	1,354,182,395	951,277,000	6,609,346
2009	1,342,207,320	944,017,000	6,254,363
2008	1,331,698,077	927,482,000	6,185,693
2007	1,317,447,415	902,343,000	6,182,692
2006	1,293,344,534	901,659,000	6,101,158
2005	1,277,701,992	885,188,000	6,040,208
2004	1,265,811,583	869,521,000	6,068,302
2003	1,212,880,852	819,007,000	5,501,424
2002	1,031,100,681	818,842,000	5,451,332
2001	1,028,457,585	814,398,710	5,278,484

Source: OGJ Worldwide Production Reports

Because oil demand fell in response to the recession, *OGJ* estimated that worldwide oil production in 2009 would drop by 3% from 2008 to average 70.5 million b/d.

Russian oil production

	1,000 b/d
2009	9,915*
2008	9,750
2007	9,830
2006	9,498
2005	9,190
2004	8,887
2003	8,216
2002	7,405
2001	6,781
2000	6,325

*Estimate.

Russian production rebounded from its 2008 slump. *OGJ* estimated that oil production in Russia would be up almost 2% from 2008, averaging 9.9 million b/d.

Big boosts to production in Azerbaijan and Kazakhstan helped push up 2009 output for Eastern Europe and the former Soviet Union by 3%.

For details on individual countries, see Table 1: World Reserves and Production, in the Statistical Tables section of this chapter.

IHS CERA: Upstream cost trends

Upstream cost declines evident in the first half of 2009 appeared to be flattening, reported IHS Cambridge Energy Research Associates at yearend.

A Look at Upstream Cost Trends

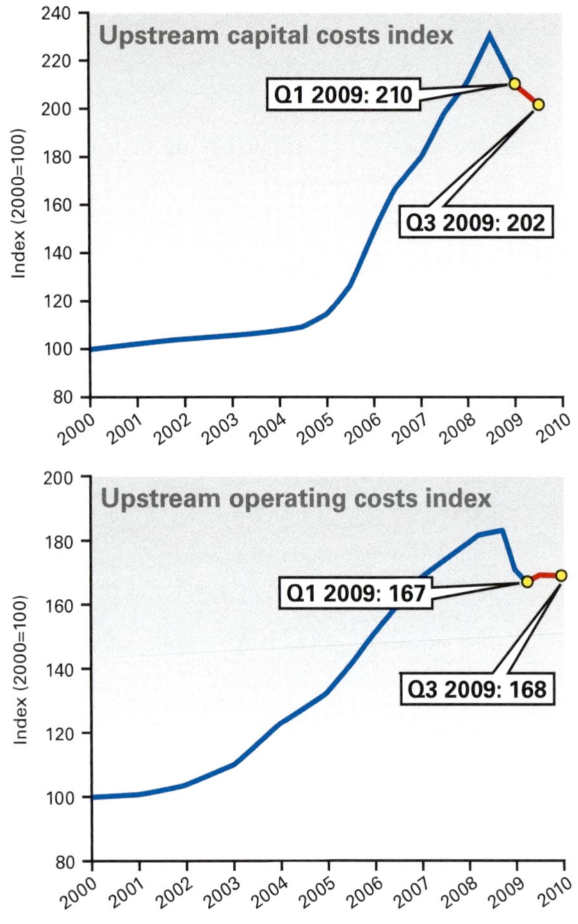

Source: IHS Cambridge Energy Research Associates

In the third quarter, the firm's capital-cost index fell more slowly than earlier in the year, while its operating-cost index rose slightly from first-quarter level.

The indexes assess changes since 2000, costs of which are assigned the value 100.

E&P in the Mediterranean, Black Sea

Offshore expenditure is set to rise sharply across the Mediterranean and Black Sea, according to the *Spends and Trends 2009* activity overview released by Scottish Enterprise in late 2009.

The group said that offshore investments across the region could top $41 billion from 2009–2013, up 45% compared to the $28.7 billion estimate for 2004–2008.

The survey foresees Egypt and Libya together attracting around 60% of Mediterranean offshore investments. It is important to note, says SE, that its data do not include:

- recent deepwater success in Libya's Sirte basin
- exploration successes in 2009 in the deepwater Israeli Levantine basin
- development drilling in the North Aegean Sea off Greece.

The report provides an overview of recent and planned E&P offshore programs for each country.

Mediterranean Expenditure Summary by Country

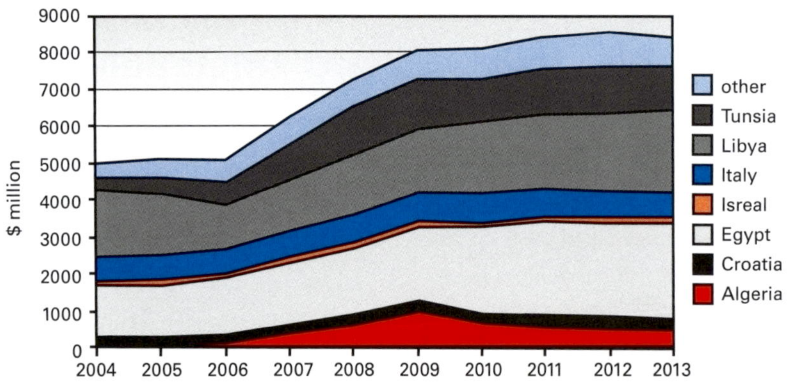

Offshore expenditure trends

A Douglas-Westwood report commissioned by Norway's INTSOK, issued in mid-2009, revealed that overall expenditure levels of 19 target countries and 26 key onshore and offshore market sectors is expected to total $2 trillion for the 2009–2013 period.

Within the offshore target markets, expenditure is expected to grow from $163 billion in 2009 to $222 billion in 2013. Key drivers include the strong growth in energy demand from the developing countries, increasing tightness in global energy supplies, and a resultant growth in oil prices.

Douglas-Westwood believes that operators are increasingly moving to prospects in deeper water.

In a November update, Douglas-Westwood projected that offshore operations and maintenance expenditure could total more than $330 billion from 2010–2014, with the largest share allocated to North America.

Unconventional gas: Huge potential

Global natural gas resources could be more than quadrupled if the world adopted U.S. technology and expertise to tap unconventional sources, according to a mid-2009 report by consultancy PFC Energy.

In Europe, said PFC, with significant unconventional shale gas and coalbed methane, this approach could help reduce dependence on gas from Russia. And globally, it could ease the transition from high-carbon coal to cleaner-burning gas for electricity production.

PFC says that global unconventional natural gas resources, based on 1997 geological estimates that could rise with new technologies, total 3,250 tcf (92 tcm). In contrast, global conventional natural gas reserves are estimated at 620 tcf (17.6 tcm), said PCF.

Mature fields: Snapshot

IPC Petroleum Consultants, Inc. offered this look at mature oil and gas fields, compiled in early 2009.

Oil

- Worldwide, more than 40,000 oil fields have been discovered. The top 100 account for 50% of current production and 65% of reserves.
- Of the known discovered reserves of these fields (estimated at 2.4 trillion bbl), 94% are in fewer than 1,500 major fields (those with more than 100 million bbl of ultimate reserves).
- The United States has more than 30,000 oil fields, of which about 300 are majors that hold 80% of total U.S. reserves.
- The North Sea has 265 oil fields, of which 90 are majors that account for 80% of total reserves.
- Saudi Arabia has 96 oil fields.
- The United States, Russia, and Saudi Arabia together hold 600 billion bbl of ultimate reserves—some one-fourth of the total oil reserves discovered globally to date.
- For these three countries, it is estimated that 80% of current oil production comes from mature fields, or fields that have passed their peak.
- More than 90% of the world's oil fields are classified as small and hold just 3% of reserves. A small oil field has less than 25 million bbl of reserves.

Natural gas

Tallying gas fields is more complex because gas can be associated with oil production, or come from gas-only conventional formations or unconventional formations (such as coalbeds and shales).

- It is estimated that there are about 28,000 nonassociated gas fields globally.
- There are about 365 giant nonassociated gas fields, compared with 370 giant oil fields.
- Giant gas fields account for roughly 75% of the total ultimate reserves discovered [13.8 quadrillion cu ft (390.8 tcm), or 2.3 trillion boe]. Giant oil fields account for a similar percentage.
- A giant oil or gas field is defined as one with reserves of more than 500 million boe [3.5 tcf (100 bcm) of gas].
- More than 90% of the world's gas fields are classified as small and hold 12% of total reserves. A small gas field has less than 0.25 tcf (7 tcm) of reserves.
- There are about 24,000 U.S. gas fields; about 300 are major, with reserves greater than 1 tcf (28.3 bcm).
- World wide, 2000–2008, an average of 214 oil fields and 58 gas fields were discovered annually. Average size: 28 million bbl and 1.7 tcf (48.1 bcm), respectively.

Evaluating mature-field reserves

IPC Petroleum Consultants, Inc. described in early 2009 the importance of mature oil and gas fields, as well as a new model it developed for estimating remaining reserves of such fields.

The firm described its model as an alternative to volumetric methods, material balance analysis, and decline analysis. Their model requires knowledge of only a field's peak production rate.

The firm claims that the model provides results with a 95% confidence level for individual fields produced without any operational limitations that ultimately repress their true peak rate.

Model results are presented in the figure titled, "Production, Reserves Potential of Mature Fields," and in the table titled, "Estimates of Ultimate Reserves for Selected Fields."

Production, reserves potential of mature fields

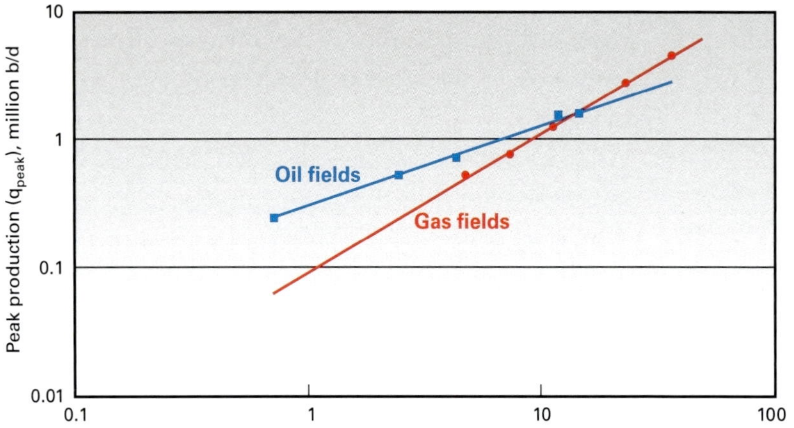

Estimates of ultimate reserves for selected fields

Field (country) *OIL*	Production began	Peak year	Peak output, 1,000 b/d	Ultimate reserves — Billion bbl of oil — Decline[1]	Model[2]	Difference, %
Abkatun (Mexico)	1979	2004	548	2.4	2.6	8.3
Brent (UK)	1975	1984	483	2.0	2.1	5.0
•Cusiana (Colombia)	1992	1998	250	0.71	0.75	5.6
•Forties (UK)	1975	1979	523	2.5	2.4	4.0
Furrial (Venezuela)	1986	1998	454	2.0	2.0	0.0
Gullfaks (Norway)	1982	1994	530	2.5	2.5	0.0
Marlim (Brazil)	1991	2002	602	2.7	3.0	11.1
Oseberg (Norway)	1988	1994	503	2.3	2.3	0.0
•Prudhoe Bay (U.S.)	1977	1988	1,500	12.0	13.0	8.3
•Romashkino (Russia)	1950	1972	1,645	16.0	15.0	6.2
•Statfjord (Norway/UK)	1979	1991	641	3.5	3.4	2.8
GAS			1,000 boe/d (bcfd)	Billion boe/d (tcf)		
•Groningen (Netherlands)	1963	1977	1,400 (8.6)	12 (73)	12 (73)	0.0
•Medvezhye (Russia)	1972	1983	1,200 (7.1)	11 (68)	10 (61)	9.0
•Orenberg (Russia)	1971	1979	770 (4.6)	7.5 (45)	6.7 (40)	10.6
•Shatlyk (Turkmenistan)	1973	1977	550 (3.3)	4.8 (29)	4.8 (29)	0.0
•Urengoy (Russia)	1981	1987	4,700 (28)	37 (222)	37 (222)	0.0
•Yamburg (Russia)	1985	1998	2,800 (17)	23 (138)	23 (138)	0.0

[1]Values established by decline analysis; [2]Estimates using Equations 2 & 3. • = Field used in Fig. 1 correlations.

Production costs in major oil producing regions

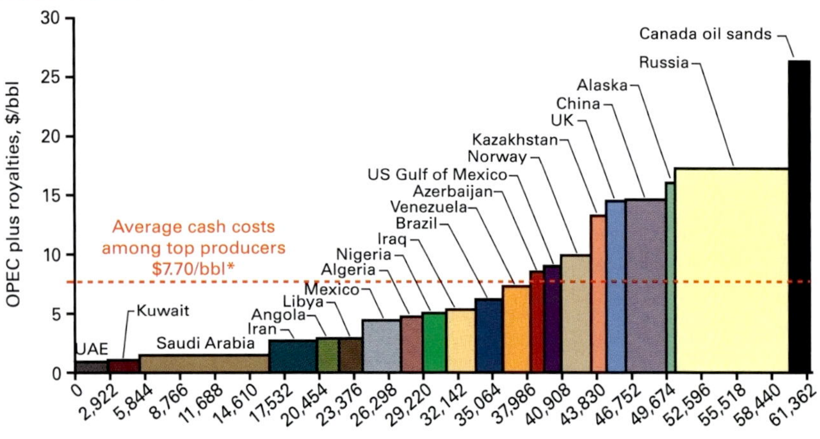

2009 production (1,000 b/d). Column width indicates share of group production.

*Average is $12.50/bbl excluding OPEC countries.

Source: Deutsche Bank

Oil production costs

An analysis by Deutsche Bank in early 2009 found that, in the short term, oil prices likely would have to fall to $20/bbl and below for nonmembers of OPEC to shut in a large amount of production.

However, noting the downtrend in investment at that time, the bank concluded that the downside risks to supply forecasts were increasing. This suggested that upside price risks, once demand recovers, would be considerable.

The bank also reviewed risks to current production in mature basins, as well as what oil price would be required for projects to deliver an economic return in today's growth markets of Angola, Brazil, the U.S. Gulf of Mexico, and Nigeria's deep waters.

Production at risk in mature provinces*

	Oil prices below $20/bbl	Oil prices below $30/bbl	Oil prices below $40/bbl
	1,000 b/d		
UK	471	132	70
Norway	228	47	20
Canada oil sands	1,610	460	—
Alaska	18	15	—
Russian export	1,033	—	—

*Excludes US Lower 48.

Source: Deutsche Bank

Offshore spending growth

Despite the global recession in mid-2009, Douglas-Westwood and Energyfiles (the latter now part of Datamonitor Group) said in mid-2009 that they expect offshore spending to grow strongly.

From 2004–2008, capital expenditures totaled $578 billion, and operating expenditures totaled $379 billion.

For 2009–2013, they expect capex of $807 billion and opex of $549 billion opex.

Details are in their *World Offshore Oil & Gas Production and Spend Forecast 2009–2013*.

Section 3:
Drilling and production

OGJ drilling report: Mid-2009

In its mid-2009 drilling analysis, *OGJ* reported that rig counts in the United States and Canada averaged about 37% lower in the first half of the year than they did in the first half of 2008.

Elsewhere in the world, comparing the same period showed declines of 10% in Africa, 7% in Europe, 5% in the Middle East, 4% in Asia-Pacific, and 2% in the Western Hemisphere outside the United States and Canada.

The international rig totals did not include figures for Iran, Iraq, Russia, or China onshore.

The *OGJ* forecast for 2009, highlighted the following:

- Operators will drill 36,788 wells in the United States, down from an estimated 52,097 wells drilled in 2008.
- All operators will drill 3,534 exploratory wells of all types, down from an estimated 5,474 last year.

Oil & Gas Journal well forecast for 2009

	First half 2009			Full year 2009			
State	Total wells	Exploratory wells	Field wells	Total wells	Exploratory wells	Field wells	Total footage (1,000)
Alabama	175	9	166	360	17	343	1,632
Alaska	60	8	52	125	13	112	848
Arkansas	415	71	344	880	147	733	6,186
California land	895	31	864	1,915	61	1,854	4,927
California offshore	3	—	3	7	—	7	44
Colorado	765	200	565	1,555	400	1,155	10,215
Florida	1	—	1	1	—	1	12
Illinois	130	39	91	295	84	211	726
Indiana	63	19	44	133	40	93	212
Kansas	1,829	132	1,697	3,914	274	3,640	13,656
Kentucky	414	14	400	882	26	856	2,487
Louisiana	1,169	116	1,053	2,534	281	2,253	24,475
North	835	88	747	1,885	228	1,657	17,679
South	122	6	116	235	11	224	2,367
Offshore	212	22	190	414	42	372	4,429
Michigan	177	28	149	365	55	310	626
Mississippi	105	10	95	200	18	182	1,866
Montana	75	10	65	145	18	127	980
Nebraska	17	4	13	36	8	28	184
Nevada	1	—	1	2	1	1	14
New Mexico - East	255	16	239	565	34	531	4,710
New Mexico - West	205	3	202	448	6	442	2,808
New York	20	1	19	43	1	42	171
North Dakota	244	50	194	520	98	422	6,544
Ohio	375	33	342	720	61	659	3,150
Oklahoma	1,072	47	1,025	2,214	93	2,121	17,048
Pennsylvania	2,700	273	2,427	5,724	595	5,129	20,360
South Dakota	4	—	4	8	1	7	32
Tennessee	46	13	33	98	28	70	223
Texas	4,045	357	3,688	8,208	708	7,500	73,062
Dist. 1	131	11	120	255	21	234	1,700
Dist. 2	230	34	196	470	67	403	4,560
Dist. 3	277	34	243	590	70	520	5,296
Dist. 4	340	31	309	625	55	570	6,506
Dist. 5	491	17	474	966	31	935	11,367
Dist. 6	560	67	493	1,135	136	999	12,813
Dist. 7-B	255	11	244	525	22	503	3,418
Dist. 7-C	320	16	304	650	30	620	5,029
Dist. 8	432	25	407	890	52	838	6,268
Dist. 8-A	312	30	282	655	63	592	3,689
Dist. 9	378	12	366	797	27	770	5,181
Dist. 10	297	63	234	610	124	486	6,805
Offshore	22	6	16	40	10	30	430
Utah	277	52	225	585	109	476	4,631
Virginia	240	25	215	495	53	442	1,342
Washington	1	1	—	1	1	—	15
West Virginia	865	104	761	1,790	220	1,570	8,302
Wyoming	955	39	916	2,020	83	1,937	10,779
U.S. total	**7,598**	**1,705**	**15,893**	**36,788**	**3,534**	**33,254**	**222,267**
Western Canada	**4,305**	**903**	**3,402**	**9,854**	**2,037**	**7,817**	**43,759**
Alberta	3,114	604	2,510	6,515	1,244	5,271	26,874
Saskatchewan	650	122	528	2,150	404	1,746	9,333
Brit. Columbia	469	168	301	1,014	367	647	6,862
Manitoba	72	9	63	175	22	153	690
NWT + Yukon Terr.	4	3	1	9	–	9	57
Eastern offshore	3	1	2	7	2	5	47
Eastern land	42	5	37	95	10	85	192

- The Baker Hughes Inc. count of active U.S. rotary rigs will average 1,235 rigs/week, down from 1,867 in 2008 and 1,768 in 2007.
- Operators will drill 9,854 wells in western Canada, down from an estimated 16,290 in 2008.

OGJ100: Production, reserves leaders for 2008

Leading *OGJ's* OGJ100 list of oil production leaders for 2008 was Saudi Aramco, which reported total output of 3.25 billion bbl. This compares to 3.1 billion bbl in 2007.

With regard to 2008 reserves, Aramco, NIOC, and Iraq National Oil Co. led the survey, just as they did in the previous year.

The survey, evaluating the 100 leading oil and gas producers based outside the United States, shows that most of these firms posted improved financial results from a year earlier. However, higher operating costs offset strong oil and gas prices for some of the companies.

OGJ100: Oil production and reserve leaders

Rank	Company	Production, million bbl
1	Saudi Arabian Oil Co.	3,248.5
2	National Iranian Oil Co.	1,423.5
3	Petroleos de Venezuela SA	1,180.8
4	Petroleos Mexicanos	1,019.1
5	Abu Dhabi National Oil Co.	894.3
6	BP PLC	876.4
7	PetroChina Co. Ltd.	870.7
8	Iraq National Oil Co.	863.2
9	Kuwait Petroleum Corp.	846.8
10	OAO Rosneft	776.3
11	Petroleo Brasileiro SA	722.7
12	Nigerian National Petroleum Corp.	708.1
13	OAO Lukoil	702.9
14	Sonangol	668.0
15	National Oil Corp. (Libya)	627.8
16	Royal Dutch Shell	617.9
17	Total SA	531.4
18	Sonatrach	481.8
19	OJSC Surgutneftegas	453.7
20	StatoilHydro	380.0
	Total	**17,893.8**

Rank	Company	Reserves, million bbl
1	Saudi Arabian Oil Co.	259,900.0
2	National Iranian Oil Co.	136,150.0
3	Iraq National Oil Co.	115,000.0
4	Kuwait Petroleum Corp.	101,500.0
5	Petroleos de Venezuela SA	99,377.0
6	Abu Dhabi National Oil Co.	92,200.0
7	National Oil Corp. (Libya)	43,660.0
8	Nigerian National Petroleum Corp.	36,220.0
9	OAO Rosneft	17,694.0
10	Qatar Petroleum Corp.	15,210.0
11	OAO Lukoil	14,458.0
12	Sonatrach	12,200.0
13	PetroChina Co. Ltd.	11,221.3
14	Petroleos Mexicanos	10,501.3
15	Petroleo Brasileiro SA	9,109.0
16	Sonangol	9,040.0
17	Total SA	5,695.0
18	BP PLC	5,665.0
19	Petroleum Development Oman LLC	5,500.0
20	Petronas	5,460.0
	Total	**1,005,760.6**

OGJ150: Sharp earnings decline in 2008

To qualify for OGJ150 evaluation (tables in this column and on next two pages), oil and gas producers must have their headquarters in the United States, be publicly traded, and hold oil or gas reserves in the United States.

The group posted a sharp decline in 2008 earnings despite an increase in revenues. High demand for equipment, employees, and services led to a surge in capital and exploration expenditures and operating costs.

Earnings and production through the first half of 2008 were propelled by high commodity prices and strong worldwide demand growth. But the second half saw oil and gas demand and prices thrown into rapid descent.

{Editor's note: The OGJ150 group, previously the OGJ200, contained only 141 companies in the 2009 survey (of year 2008 performance). In the previous year, there were 147 firms in the compilation.}

Worldwide demand for oil products slipped to average 85.4 million b/d in the fourth quarter of 2008, down from the year-earlier average of 87.8 million b/d, according to IEA.

Some key changes from 2008

How company appeared on last year's list	Why change?	How company appears on this year's list
Equitable Supply	Changed name to...	EQT Production
DTE Oil & Gas Inc.	Changed name to...	DTE Gas Resources
Energen Resources Corp.	Now listed as parent company...	Energen Corp.

The following companies sold their US producing properties, liquidated, or became private since the last survey: Bayou City Exploration Inc., LL&E Royalty Trust

Top 20 in total revenue

Rank	Company	Total revenue, $1,000
1	ExxonMobil Corp.	477,359,000
2	Chevron Corp.	273,005,000
3	ConocoPhillips	246,182,000
4	Marathon Oil Corp.	78,569,000
5	Hess Corp.	41,094,000
6	Murphy Oil Corp.	27,513,000
7	Occidental Petroleum Corp.	24,480,000
8	Anadarko Petroleum Corp.	15,723,000
9	Devon Energy Corp.	15,211,000
10	Apache Corp.	12,389,750
11	Chesapeake Energy Corp.	11,629,000
12	XTO Energy Inc.	7,695,000
13	EOG Resources Inc.	7,127,143
14	El Paso Corp.	5,363,000
15	Dominion Energy Inc.	4,312,000
16	Noble Energy Inc.	3,901,000
17	Questar Corp.	3,491,800
18	Williams Cos. Inc.	3,121,000
19	Plains Exploration & Production Co.	2,403,471
20	Pioneer Natural Resources Co.	2,338,287
	Total	**1,262,907,451**

20 Fastest-Growing Companies[1]

Rank by total assets	Company	Stockholders' equity 2008 ($1,000)	Stockholders' equity 2007 ($1,000)	Stockholders' equity Change, %	Net income 2008 ($1,000)	Net income 2007 ($1,000)	Net income Change, %	Long-term debt 2008 ($1,000)	Long-term debt 2007 ($1,000)
8	XTO Energy Inc.	17,347,000	7,941,000	118.4	1,912,000	1,691,000	13.1	11,959,000	6,320,000
87	GeoResources Inc.	140,995	68,031	107.3	13,522	3,069	340.6	40,000	96,000
61	Clayton Williams Energy Inc.	314,682	160,806	95.7	140,534	5,990	2,246.1	347,225	430,175
69	Arena Resources Inc.	482,295	257,811	87.1	83,617	34,442	142.8	—	—
41	Berry Petroleum Co.	827,544	459,974	79.9	133,529	129,928	2.8	1,131,800	445,000
37	Concho Resources Inc.	1,325,154	775,398	70.9	278,702	25,360	999.0	630,000	325,404
32	Cabot Oil & Gas Corp.	1,790,562	1,070,257	67.3	211,290	167,423	26.2	831,143	330,000
27	Southwestern Energy Co.	2,507,830	1,646,500	52.3	567,946	221,174	156.8	674,200	977,600
45	Continental Resources Inc.	948,708	623,132	52.2	320,950	28,580	1,023.0	376,400	165,000
97	Credo Petroleum Corp.[2]	62,211	41,140	51.2	5,993	5,760	4.0	—	—
23	Range Resources Corp.	2,457,833	1,728,022	42.2	346,158	230,569	50.1	—	—
50	Bill Barrett Corp.	1,087,798	773,511	40.6	107,647	26,754	302.4	—	—
31	Energen Resources Corp.	1,913,290	1,378,658	38.8	321,915	309,233	4.1	207,557	208,467
33	Encore Acquisition Co.	1,314,128	948,155	38.6	430,812	17,155	2,411.3	1,319,811	1,120,236
129	Texas Vanguard Oil Co.	9,990	7,210	38.6	2,780	1,354	105.3	—	—
53	Comstock Resources Inc.	1,062,085	771,644	37.6	251,962	68,901	265.7	210,000	680,000
117	Spindletop Oil & Gas Co.	13,036	9,515	37.0	3,521	1,808	94.7	—	—
47	CNX Gas Corp.	1,384,874	1,023,237	35.3	239,073	135,678	76.2	15,386	5,799
19	Questar Corp.	3,418,000	2,577,900	32.6	683,800	507,400	34.8	2,078,900	1,021,200
14	Noble Energy Inc.	6,309,000	4,808,807	31.2	1,350,000	943,870	43.0	2,241,000	1,851,000

[1]Companies were selected on the basis of growth in stockholder's equity. Only companies with positive net income for both 2007 and 2008 were considered. Companies were not considered if they had a decline in net income for 2008, were subsidiaries of another company, or became public within the last year. [2]Year ending Oct. 31.

Top 20 in Spending and U.S. Net Wells Drilled

Rank	Company	Capital, exploratory spending, $1,000
1	ExxonMobil Corp.	20,769,000
2	Chevron Corp.	19,666,000
3	ConocoPhillips	19,099,000
4	Devon Energy Corp.	9,375,000
5	Chesapeake Energy Corp.	9,177,000
6	Marathon Oil Corp.	7,146,000
7	Apache Corp.	5,972,846
8	Anadarko Petroleum Corp.	4,801,000
9	EOG Resources Inc.	4,718,860
10	Occidental Petroleum Corp.	4,664,000
11	Hess Corp.	4,438,000
12	XTO Energy Inc.	3,661,000
13	Petrohawk Energy Corp.	3,121,736
14	El Paso Corp.	2,757,000
15	Questar Corp.	2,485,700
16	Forest Oil Corp.	2,338,488
17	Murphy Oil Corp.	2,185,960
18	Newfield Exploration Co.	2,067,000
19	Noble Energy Inc.	1,971,000
20	Southwestern Energy Co.	1,755,888
	Total	**132,170,478**

Rank	Company	US net wells drilled
1	Chesapeake Energy Corp.	1,733.0
2	Anadarko Petroleum Corp.	1,590.8
3	EOG Resources Inc.	1,545.0
4	XTO Energy Inc.	1,247.0
5	Devon Energy Corp.	1,068.9
6	Williams Cos. Inc.	1,050.0
7	Noble Energy Inc.	929.7
8	Chevron Corp.	860.0
9	ConocoPhillips	849.0
10	Occidental Petroleum Corp.	772.4
11	Pioneer Natural Resources Co.	566.0
12	CNX Gas Corp.	552.0
13	Equitable Supply	533.2
14	Range Resources Corp.	490.2
15	El Paso Corp.	450.0
16	ExxonMobil Corp.	442.0
17	Forest Oil Corp.	437.0
18	Quicksilver Resources Inc.	419.7
19	Atlas America Inc.	419.0
20	Exco Resources Inc.	399.5
	Total	**16,354.4**

Top 20 in Liquids Reserves

Rank	Company	U.S. liquids reserves, million bbl
1	ConocoPhillips	1,928.0
2	ExxonMobil Corp.	1,644.0
3	Occidental Petroleum Corp.	1,547.0
4	Chevron Corp.	1,470.0
5	Anadarko Petroleum Corp.	692.0
6	Apache Corp.	514.8
7	Devon Energy Corp.	484.0
8	Pioneer Natural Resources Co.	448.9
9	XTO Energy Inc.	343.3
10	Hess Corp.	227.0
11	EOG Resources Inc.	205.8
12	Noble Energy Inc.	198.0
13	Whiting Petroleum Corp.	180.0
14	Denbury Resources Inc.	179.1
15	Marathon Oil Corp.	178.0
16	Encore Acquisition Co.	134.5
17	Berry Petroleum Co.	125.3
18	Chesapeake Energy Corp.	120.6
19	Newfield Exploration Co.	110.8
20	Continental Resources Inc.	106.2
	Total	**10,837.3**

Rank	Company	Worldwide liquids reserves, million bbl
1	ExxonMobil Corp.	7,576.0
2	Chevron Corp.	7,305.0
3	ConocoPhillips	5,817.0
4	Occidental Petroleum Corp.	2,212.0
5	Apache Corp.	1,081.1
6	Hess Corp.	970.0
7	Anadarko Petroleum Corp.	926.0
8	Devon Energy Corp.	781.0
9	Marathon Oil Corp.	636.0
10	Pioneer Natural Resources Co.	463.0
11	XTO Energy Inc.	343.3
12	Noble Energy Inc.	311.0
13	EOG Resources Inc.	225.0
14	Whiting Petroleum Corp.	180.0
15	Denbury Resources Inc.	179.1
16	Murphy Oil Corp.	173.6
17	Newfield Exploration Co.	140.3
18	Encore Acquisition Co.	134.5
19	Berry Petroleum Co.	125.3
20	Chesapeake Energy Corp.	120.6
	Total	**29,699.8**

Top 20 in Liquids Production

Rank	Company	U.S. liquids production, million bbl
1	ConocoPhillips	157.0
2	Chevron Corp.	154.0
3	ExxonMobil Corp.	104.0
4	Occidental Petroleum Corp.	96.0
5	Anadarko Petroleum Corp.	54.0
6	Devon Energy Corp.	41.0
7	Apache Corp.	35.1
8	XTO Energy Inc.	26.2
9	Marathon Oil Corp.	23.0
10	EOG Resources Inc.	20.0
11	Noble Energy Inc.	18.0
12	Pioneer Natural Resources Co.	15.1
13	Kinder Morgan CO2 Co. LP	15.0
14	Hess Corp.	15.0
15	Whiting Petroleum Corp.	12.4
16	Denbury Resources Inc.	11.5
17	Chesapeake Energy Corp.	11.2
18	Encore Acquisition Co.	10.1
19	Continental Resources Inc.	9.1
20	Cimarex Energy Co.	8.4
	Total	**836.0**

Rank	Company	Worldwide liquids production, million bbl
1	ExxonMobil Corp.	724.0
2	Chevron Corp.	604.0
3	ConocoPhillips	494.0
4	Occidental Petroleum Corp.	171.0
5	Hess Corp.	97.0
6	Apache Corp.	96.9
7	Devon Energy Corp.	81.0
8	Anadarko Petroleum Corp.	80.0
9	Marathon Oil Corp.	77.0
10	Murphy Oil Corp.	38.7
11	Noble Energy Inc.	32.0
12	XTO Energy Inc.	26.2
13	EOG Resources Inc.	22.5
14	Pioneer Natural Resources Co.	18.2
15	Kinder Morgan CO2 Co. LP	15.0
16	Whiting Petroleum Corp.	12.4
17	Denbury Resources Inc.	11.5
18	Chesapeake Energy Corp.	11.2
19	Newfield Exploration Co.	11.2
20	Encore Acquisition Co.	10.1
	Total	**2,633.9**

Top 20 in Gas Reserves

Rank	Company	U.S. gas reserves, bcf
1	XTO Energy Inc.	11,802.9
2	ExxonMobil Corp.	11,778.0
3	Chesapeake Energy Corp.	11,327.5
4	ConocoPhillips	10,920.0
5	Devon Energy Corp.	8,369.0
6	Anadarko Petroleum Corp.	8,150.0
7	EOG Resources Inc.	4,889.0
8	Williams Cos. Inc.	4,339.0
9	Ultra Petroleum	3,355.8
10	Occidental Petroleum Corp.	3,153.0
11	Chevron Corp.	3,150.0
12	Equitable Supply	3,097.3
13	Pioneer Natural Resources Co.	2,917.0
14	Apache Corp.	2,537.2
15	Range Resources Corp.	2,213.5
16	Southwestern Energy Co.	2,175.5
17	Newfield Exploration Co.	2,109.8
18	El Paso Corp.	2,091.0
19	Questar Corp.	2,028.5
20	Cabot Oil & Gas Corp.	1,886.0
	Total	**102,290.0**

Rank	Company	Worldwide gas reserves, bcf
1	ExxonMobil Corp.	31,402.0
2	ConocoPhillips	24,948.0
3	Chevron Corp.	23,075.0
4	XTO Energy Inc.	11,802.9
5	Chesapeake Energy Corp.	11,327.5
6	Devon Energy Corp.	9,885.0
7	Anadarko Petroleum Corp.	8,105.0
8	Apache Corp.	7,917.0
9	EOG Resources Inc.	7,339.2
10	Occidental Petroleum Corp.	4,601.0
11	Williams Cos. Inc.	4,339.0
12	Ultra Petroleum	3,355.8
13	Marathon Oil Corp.	3,351.0
14	Noble Energy Inc.	3,315.0
15	Equitable Supply	3,097.3
16	Pioneer Natural Resources Co.	2,979.8
17	Hess Corp.	2,773.0
18	Range Resources Corp.	2,213.5
19	Southwestern Energy Co.	2,175.5
20	El Paso Corp.	2,138.0
	Total	**170,140.5**

Top 20 in Gas Production

Rank	Company	U.S. gas production, bcf
1	ConocoPhillips	896.0
2	Chesapeake Energy Corp.	775.4
3	Anadarko Petroleum Corp.	750.0
4	Devon Energy Corp.	726.0
5	XTO Energy Inc.	697.4
6	ExxonMobil Corp.	555.0
7	Chevron Corp.	549.0
8	EOG Resources Inc.	436.0
9	Williams Cos. Inc.	406.0
10	Apache Corp.	248.8
11	El Paso Corp.	230.0
12	Occidental Petroleum Corp.	215.0
13	Southwestern Energy Co.	192.3
14	Newfield Exploration Co.	167.9
15	Marathon Oil Corp.	164.0
16	Pioneer Natural Resources Co.	154.3
17	Questar Corp.	151.9
18	Noble Energy Inc.	145.0
19	Ultra Petroleum	138.6
20	Exco Resources Inc.	131.2
	Total	**7,729.7**

Rank	Company	Worldwide gas production, bcf
1	ExxonMobil Corp.	2,468.0
2	ConocoPhillips	2,095.0
3	Chevron Corp.	1,876.0
4	Devon Energy Corp.	940.0
5	Chesapeake Energy Corp.	775.4
6	Anadarko Petroleum Corp.	750.0
7	XTO Energy Inc.	697.4
8	EOG Resources Inc.	603.5
9	Apache Corp.	592.0
10	Williams Cos. Inc.	406.0
11	Marathon Oil Corp.	359.0
12	Occidental Petroleum Corp.	307.0
13	Noble Energy Inc.	281.0
14	Hess Corp.	272.0
15	El Paso Corp.	233.0
16	Southwestern Energy Co.	192.3
17	Newfield Exploration Co.	167.9
18	Pioneer Natural Resources Co.	158.9
19	Questar Corp.	151.9
20	Forest Oil Corp.	141.4
	Total	**13,467.8**

Oil, gas supply trends

In a late-2009 analysis, IPC Petroleum Consultants Inc. noted that global crude oil production has increased by just 8 million b/d in the last 30 years and has remained flat at 71 million b/d since 2004.

As the world recession eases, said IPC, supply and demand of crude oil will inevitably lead again to the imbalances of the mid-2000s that brought on $145 oil prices in mid-2008.

Although worldwide gas production capacity would exceed demand over the next 25 years, there are tight spots of supply over the medium term, not due to a lack of reserves as in the case of oil, but rather to a need of massive investments to develop existing reserves.

IPC expects that in the near future, gas prices will inevitably move towards more symmetry with those of oil. In late 2009, gas was about one-half the price of oil.

Global gas supply by region

	2006	2015	2030
		Bcf/day	
Eurasia	82	92	103♦
N. America	73	76	73↓
Asia-Pacific	37	50	62♦
Middle East	31	46	96♦♦♦♦
W. Europe	29	27	21↓
Africa	19	28	43♦♦
L. America	13	18	27♦♦
World	**284**	**337**	**425**

♦ = Degree of output growth. ↓ = Output decline. Source: IEA08

Global crude oil production since 1900

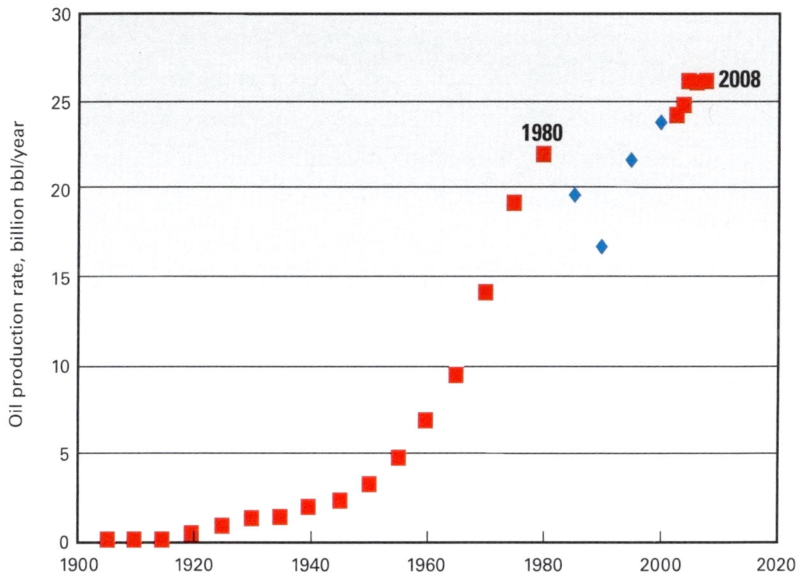

Global oil E&P trends

	Discoveries	Production
	Billion bbl/year	
1930s	19	1.6
1940s	26	2.7
1950s	35	5.4
1960s	55	11.0
1970s	39	20.0
1980s	21	21.0
1990s	15	23.0
2000s	10	26.0

Sources: IHS; EIA

Global gas E&P trends

	Discoveries	Production
	Tcf/year	
1930s	23	2
1940s	46	5
1950s	86	13
1960s	244	29
1970s	290	43
1980s	100	58
1990s	129	79
2000s	100	97

Sources: IHS; EIA

Global land-offshore oil production

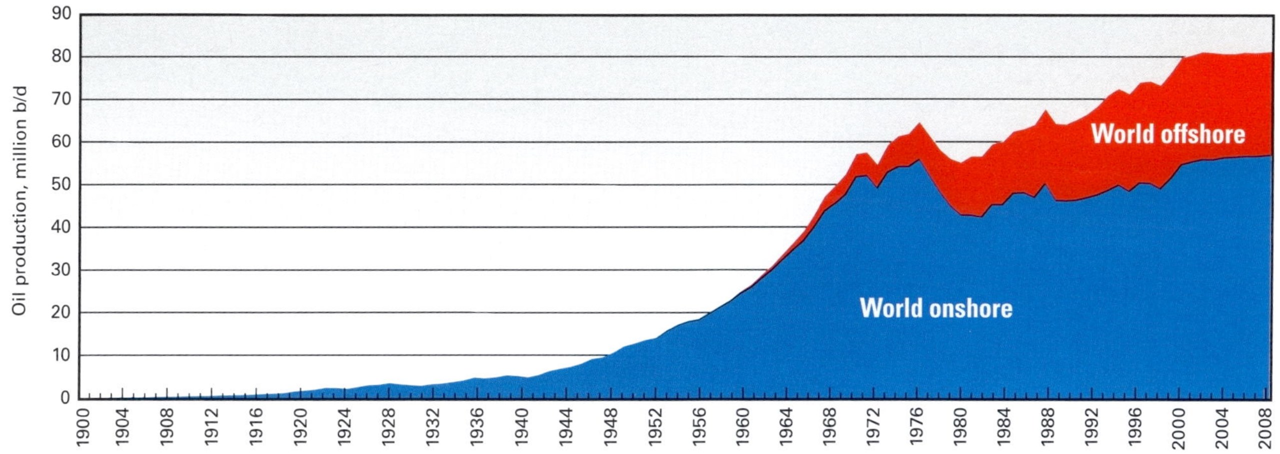

GLOBAL OIL* SUPPLY BY REGION

	2008	2030
	Million b/d	
North America	14.1	11.4↓
Europe	4.2	2.1↓
Eurasia	13.3	16.5♦♦
Asia-Pacific	8.1	5.1↓
Latin America	7.3	5.9↓
Middle East	25.8	43.2♦♦♦♦
Africa	10.8	13.8♦♦
Subtotal	84.0	98.0
Unconventional	2.0	8.0
Total	**86.0**	**106.0**

*Oil includes crude oil, NGLs, and refinery gains.
↓ = Output decline. ♦↓ = Degree of output growth. Source: IEA08

Drilling outlook stabilizes

OGJ reported that the fundamental outlook for the global drilling market gained strength in October 2009, driven by improving crude oil prices amid signs of an economic recovery across much of the world.

Schlumberger Ltd. officials were optimistic that oil producers might boost spending in 2010 if they gain confidence that oil prices will stay around $80/bbl.

The December 2009 contract for benchmark U.S. light, sweet crude settled at $81.19/bbl on the NYMEX on October 22, marking the first time in more than a year that oil settled higher than $80/bbl.

Middle East, Africa drilling expenditures

Drilling and workover expenditure across the Middle East and North Africa could increase by more than $10 billion to reach $27.9 billion by 2014, according to Douglas-Westwood.

The firm shared that information in its *Middle East & North Africa Oilfield Services Market Report 2010–2014.*

The report provides data for Algeria, Egypt, Iran, Iraq, Kuwait, Libya, Oman, Qatar, Saudi Arabia, Syria, UAE, and Yemen.

U.S. drilling-rig trend

In November 2009, *OGJ* presented a comprehensive review of worldwide drilling activity.

The U.S. working drilling rig count began to reverse its steep drop in October as expectations for crude and gas prices and demand became more bullish.

The Baker Hughes Inc. U.S. rig count for the week ending October 23 was 1,048, up 8 rigs from the previous week. Although this is down considerably from the 2,031 rigs working in September 2008, the count was up 172 rigs from the 2009 low point of 876 rigs in June.

U.S. WELLS SPUD BY OPERATOR (JAN.–SEPT. 2009)*

Rank	Operator	Wells spud	% of total	Footage drilled	% of total	Average footage
1	Chesapeake Operating Inc.	809	5.67	7,228,360	6.10	8,935
2	XTO Energy Inc.	644	4.51	6,735,247	5.60	10,458
3	EOG Resources Inc.	541	3.79	5,031,771	4.20	9,301
4	Kerr-McGee Oil & Gas Onshore LP	379	2.65	3,155,885	2.60	8,327
5	EnCana Oil & Gas (USA) Inc.	357	2.50	4,002,393	3.40	11,211
6	SEECO Inc.	347	2.43	1,431,175	1.20	4,124
7	Noble Energy Inc.	329	2.30	2,508,168	2.10	7,624
8	Devon Energy Production Co. LP	318	2.23	3,140,972	2.60	9,877
9	BP America Production Co.	265	1.86	2,374,969	2.00	8,962
10	ConocoPhillips Co.	234	1.64	1,760,624	1.50	7,524
11	COG Operating LLC	187	1.31	1,416,492	1.20	7,575
12	Burlington Resources Oil & Gas Co. LP	185	1.30	1,617,093	1.40	8,741
13	Williams Production RMT Co.	185	1.30	1,484,624	1.20	8,025
14	Questar Exploration & Production Co.	159	1.11	2,058,664	1.70	12,948
15	Energen Resources Corp.	134	0.94	632,519	0.50	4,720
16	Newfield Production Co.	128	0.90	836,045	0.70	6,532
17	ExxonMobil Development Co.	109	0.76	1,367,687	1.10	12,548
18	Atlas Resources Inc.	104	0.73	833,970	0.70	8,019
19	Bill Barrett Corp.	102	0.71	771,209	0.60	7,561
20	Ultra Resources Inc.	101	0.71	1,356,450	1.10	13,430
21	OK Arena Operating Co.	101	0.71	494,900	0.40	4,900
22	BP Exploration (Alaska) Inc.	99	0.69	871,773	0.70	8,806
23	Dominion Exploration & Production Inc.	98	0.69	461,792	0.40	4,712
24	Anadarko E&P Co. LP	98	0.69	1,142,077	1.00	11,654
25	Texas Keystone Inc.	96	0.67	537,720	0.50	5,601
26	Swepi LP	95	0.67	1,271,552	1.10	13,385
27	Marathon Oil Co.	93	0.65	1,067,766	0.90	11,481
28	Apache Corp.	93	0.65	845,620	0.70	9,093

U.S. WELLS SPUD BY OPERATOR (JAN.–SEPT. 2009)* ...CONT.

Rank	Operator	Wells spud	% of total	Footage drilled	% of total	Average footage
30	Quicksilver Resources Inc.	87	0.61	787,500	0.70	9,052
29	Equitable Production Co.	92	0.64	373,103	0.30	4,055
31	Newfield Exploration Mid-Continent Inc.	85	0.60	799,115	0.70	9,401
32	Chesapeake Appalachia LLC	84	0.59	572,349	0.50	6,814
33	HighMount Expl. & Prod. Texas LLC	84	0.59	670,475	0.60	7,982
34	Cabot Oil & Gas Production Corp.	83	0.58	797,500	0.70	9,608
35	Mewbourne Oil Co.	81	0.57	715,755	0.60	8,836
36	Range Production Co.	78	0.55	763,423	0.60	9,787
37	Chevron USA Inc.	74	0.52	487,481	0.40	6,588
38	EQT Production Co.	70	0.49	310,708	0.30	4,439
39	Petrohawk Operating Co.	69	0.48	921,405	0.80	13,354
40	Range Resources-Appalachia LLC	68	0.48	490,756	0.40	7,217
41	Laredo Petroleum Inc.	66	0.46	891,575	0.70	13,509
42	Range Resources-Pine Mountain Inc.	62	0.43	414,098	0.30	6,679
43	Murfin Drilling Co. Inc.	62	0.43	279,550	0.20	4,509
44	Samson Lone Star LLC	61	0.43	738,531	0.60	12,107
45	Wexpro Co.	61	0.43	775,396	0.60	12,711
46	Venoco Inc.	59	0.41	410,258	0.30	6,954
47	New Dominion LLC	59	0.41	283,500	0.20	4,805
48	Legend Natural Gas III LP	58	0.41	508,000	0.40	8,759
49	Endeavor Energy Resources LP	53	0.37	640,250	0.50	12,080
50	St. Mary Land & Exploration Co.	52	0.36	582,195	0.50	11,196
51	El Paso Exploration & Production Co. LP	49	0.34	612,952	0.50	12,509
52	ConocoPhillips Alaska Inc.	47	0.33	316,717	0.30	6,739
53	SandRidge Exploration & Production LLC	46	0.32	518,646	0.40	11,275
54	Henry Resources LLC	45	0.32	496,000	0.40	11,022
55	Fortuna Energy Inc.	44	0.31	292,471	0.20	6,647
56	OXY USA WTP LP	44	0.31	464,320	0.40	10,553
57	Whiting Oil & Gas Corp.	43	0.30	440,183	0.40	10,237
58	Hilcorp Energy Co.	41	0.29	483,551	0.40	11,794
59	Summit Petroleum LLC	38	0.27	412,200	0.30	10,847
60	Petro-Canada Resources (USA) Inc.	37	0.26	290,939	0.20	7,863
61	Williams Production Mid-Continent Co.	36	0.25	284,516	0.20	7,903
62	Broad Oak Energy Inc.	35	0.25	338,453	0.30	9,670
63	EXCO Production Co. LP	35	0.25	418,252	0.40	11,953
64	American Warrior Inc.	33	0.23	146,930	0.10	4,452
65	Texas Energy Operations LC	33	0.23	266,000	0.20	8,061
66	Hess Corp.	32	0.22	305,122	0.30	9,535
67	Shell Western E&P	32	0.22	443,737	0.40	13,867
68	Fasken Oil and Ranch Ltd.	32	0.22	355,400	0.30	11,106
69	SandRidge Energy Inc.	32	0.22	415,200	0.30	12,975
70	Kriebel Minerals Inc.	31	0.22	165,000	0.10	5,323
71	East Resources Inc.	31	0.22	167,707	0.10	5,410
72	Rosetta Resources Operating LP	31	0.22	288,804	0.20	9,316
73	Comstock Oil & Gas LP	30	0.21	349,772	0.30	11,659
74	Presco Western LLC	30	0.21	178,800	0.10	5,960
75	Unioil	29	0.20	224,171	0.20	7,730
76	Mariner Energy Inc.	28	0.20	291,722	0.20	10,419
77	U.S. Energy Development Corp.	28	0.20	79,134	0.10	2,826
78	Lewis Petro Properties Inc.	28	0.20	238,900	0.20	8,532
79	Petroleum Development Corp.	28	0.20	224,291	0.20	8,010
80	Continental Resources Inc.	27	0.19	283,106	0.20	10,485
81	Carrizo Oil & Gas Inc.	27	0.19	241,200	0.20	8,933
82	Southwestern Energy Production Co.	26	0.18	207,250	0.20	7,971
83	Williams Production Gulf Coast LP	26	0.18	209,289	0.20	8,050
84	Anadarko Petroleum Corp.	25	0.18	379,421	0.30	15,177
85	Pioneer Natural Resources USA Inc.	25	0.18	261,884	0.20	10,475
86	Forest Oil Corp.	25	0.18	311,655	0.30	12,466
87	Unit Petroleum Co.	24	0.17	261,750	0.20	10,906
88	KEBO Oil & Gas Inc.	24	0.17	175,475	0.10	7,311
89	Linn Operating Inc.	24	0.17	218,760	0.20	9,115
90	NFR Energy LLC	24	0.17	298,500	0.30	12,438
91	Stephens Production Co.	24	0.17	158,175	0.10	6,591
92	Chaparral Energy LLC	23	0.16	105,195	0.10	4,574
93	Aruba Petroleum Inc.	23	0.16	207,000	0.20	9,000
94	Clean Gas Inc.	23	0.16	105,100	0.10	4,570
95	VAL Energy Inc.	23	0.16	112,450	0.10	4,889
96	Cholla Petroleum Inc.	22	0.15	136,578	0.10	6,208
97	Cimarex Energy Co.	22	0.15	294,691	0.20	13,395
98	Daugherty Petroleum Inc.	22	0.15	100,957	0.10	4,589
99	Henry Petroleum LP	22	0.15	237,400	0.20	10,791
100	Key Oil Co.	22	0.15	115,490	0.10	5,250
Total for top 100 operators		**9,469**		**83,581,656**		**2,278**
Total for all operators		**14,277**		**119,319,265**		**2,812**

*The list only includes wells deeper than 2,500 ft.

Source: RigData

TOP 25 U.S. DRILLING CONTRACTORS—ONSHORE AND OFFSHORE (JAN.–SEPT. 2009)*

Rank	Company	Jan.	Feb.	Mar.	Apr.	May	June	July	Aug.	Sept.	Avg. monthly rig use	2009 fleet	2008 fleet
1	Nabors Drilling USA LP	198	176	146	126	110	98	99	103	116	130	332	334
2	Helmerich & Payne IDC	172	156	135	115	105	103	101	101	103	121	214	194
3	Patterson-UTI Drilling Co. LLC	146	107	77	59	57	59	69	74	85	81	334	331
4	Precision Drilling Oilfield Svcs. Corp.	99	81	76	61	56	49	51	58	64	66	162	27
5	Nomac Drilling LLC	76	72	75	68	67	71	76	76	82	74	98	64
6	Unit Drilling Co.	50	38	29	26	25	20	21	24	25	29	87	88
7	Ensign United States Drilling Inc.	48	32	28	22	22	21	22	24	23	27	60	56
8	Trinidad Drilling LP	32	26	31	30	29	36	36	37	36	33	56	46
9	Pioneer Drilling Co.	41	36	31	22	19	20	18	21	19	25	66	64
10	Union Drilling Inc.	33	31	25	23	22	24	21	20	24	25	54	57
11	Cactus Drilling Co. LLC	37	29	25	21	20	20	20	21	22	24	46	43
12	Bronco Drilling Co. Inc.	32	27	17	15	11	11	9	8	12	16	52	54
13	Goober Drilling LLC	27	22	18	17	14	13	13	14	14	17	38	37
14	Unit Texas Drilling LLC	23	17	16	11	9	9	15	14	14	14	46	35
15	Hercules Drilling Co. LLC	24	14	11	10	12	9	12	10	9	12	40	56
16	Rowan Drilling Co. Inc.	23	21	18	15	13	12	14	15	14	16	32	31
17	Nabors Well Services Co.	16	16	15	14	13	12	13	15	15	14	82	76
18	Nabors Offshore Corp.	18	14	15	10	12	9	7	3	3	10	44	44
19	Big Dog Drilling Co.	19	11	9	6	6	5	6	10	11	9	25	25
20	Key Energy Services Inc.	13	11	9	12	7	5	9	14	12	10	47	44
21	Scandrill Inc.	20	18	18	17	14	14	14	15	15	16	20	19
22	Capstar Drilling LP	16	11	8	5	5	7	11	12	14	10	27	27
23	Cyclone Drilling Inc.	15	8	5	4	4	4	4	6	7	6	33	27
24	Felderhoff Bros. Drilling LLC	17	12	10	10	11	10	10	10	10	11	21	23
25	Lariat Services Inc.	18	15	10	8	5	6	6	7	8	9	34	47
	Active rigs	**1,213**	**1,001**	**857**	**727**	**668**	**647**	**677**	**712**	**757**	**807**	**2,050**	**1,849**

*Some rigs that worked earlier in the year may no longer be marketed.

Source: RigData

US RIG COUNTS BY BASIN

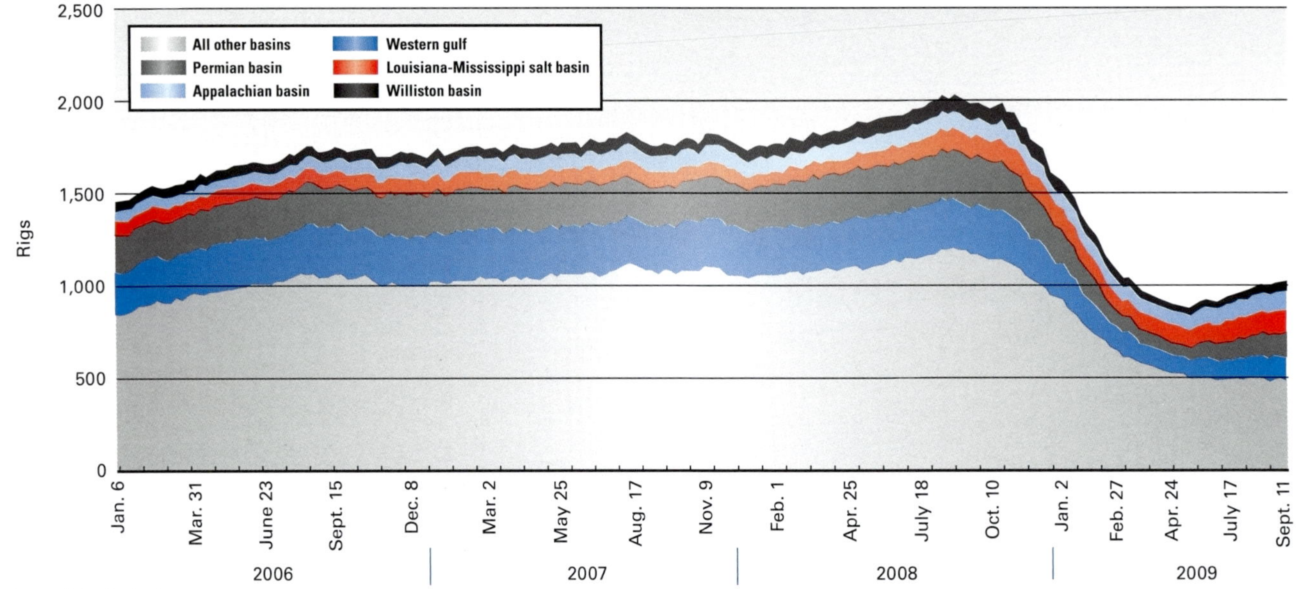

Source: Baker Hughes Inc.

Rig count trend

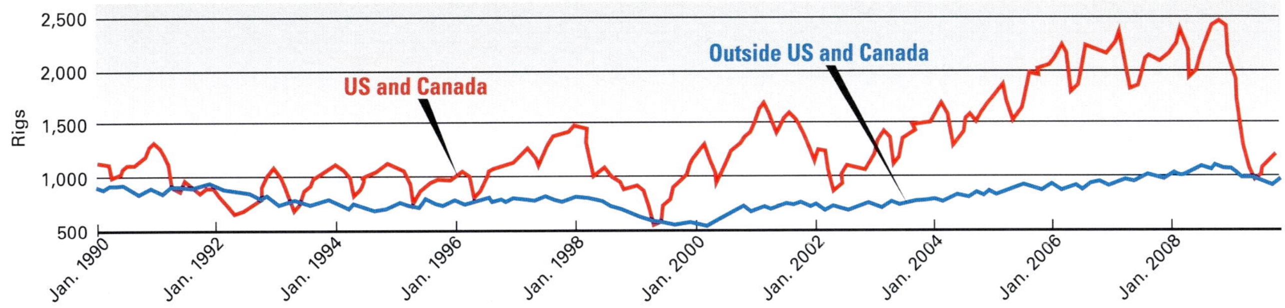

Source: Baker Hughes Inc.

Canada drilling data

As reported by *OGJ* in November 2009, the number of oil and gas wells expected to be drilled in Canada in 2009 was 8,787—almost a 50% decline from the 16,000 completed in 2008.

These data are from the Canadian Association of Oilwell Drilling Contractors.

The decline is partly attributed to a new resource royalty framework in Alberta that took effect in January 2009.

Most drilling in Canada occurs in Alberta, British Columbia, and Saskatchewan. Because of drilling rules, drilling there usually peaks in February and declines in the summer.

In 2009, the low point was 70 working rigs in May.

How U.S., Canada drilling compare

Average number of active rigs per week, year to date

	Jan. 1–June 27, 2008	Jan. 1–June 26, 2009
Alabama	4.1	3.2
Alaska	7.4	8.5
Arkansas	44.0	47.0
California	38.1	23.5
Land	36.9	22.5
Offshore	1.2	1.0
Colorado	114.0	59.5
Florida	0.1	0.5
Kansas	10.9	17.8
Kentucky	10.0	9.4
Louisiana	149.5	147.0
N. land	52.2	78.2
S. inland waters	20.0	6.4
S. land	25.5	16.5
Offshore	51.9	45.9
Michigan	0.8	0.0
Mississippi	11.3	11.1
Montana	11.3	2.2
Nebraska	0.1	0.2
Nevada	3.2	4.5
New Mexico	73.6	41.8
New York	6.3	2.0
North Dakota	58.7	49.2
Ohio	11.8	7.9
Oklahoma	203.1	109.1
Pennsylvania	20.1	28.4
South Dakota	1.7	0.2
Tennessee	4.8	2.4
Texas	886.8	469.2
Offshore	8.8	3.9
Inland waters	2.5	0.3
Dist. 1	23.9	10.5
Dist. 2	34.3	19.6
Dist. 3	62.2	36.5
Dist. 4	91.6	44.3
Dist. 5	182.8	106.4
Dist. 6	121.1	78.1
Dist. 7B	32.0	13.4
Dist. 7C	60.2	24.3
Dist. 8	128.3	54.4
Dist. 8A	22.5	15.3
Dist. 9	40.5	25.1
Dist. 10	76.2	37.0
Utah	40.3	20.2
Virginia	4.2	3.8
West Virginia	26.4	23.5
Wyoming	71.0	43.5
U.S. total	**1,817.0**	**1,139.0**
U.S. land	**1,732.0**	**1,079.0**
U.S. offshore	**62.4**	**53.4**
Canada total	**341.2**	**210.7**

Source: Baker Hughes Inc.

Western Canada Rig Count

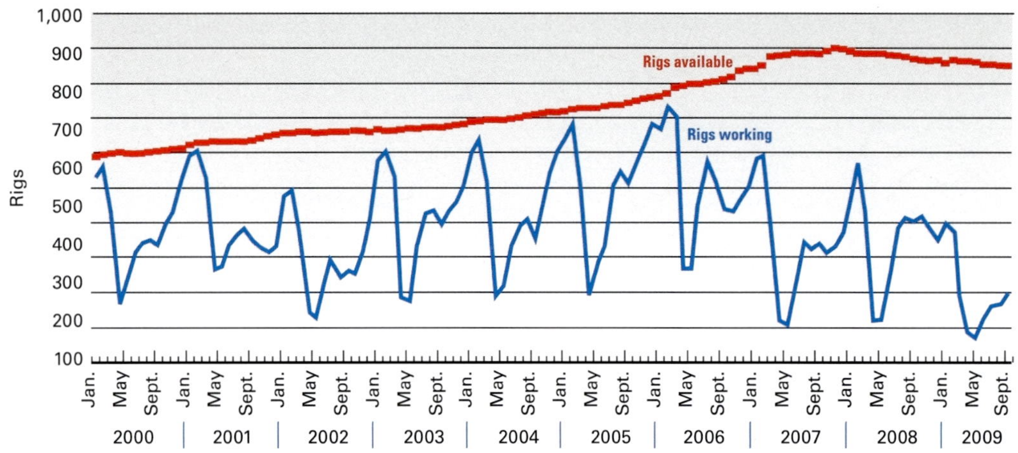

Source: Canadian Association of Oilwell Drilling Contractors

Drilling outside U.S., Canada

As reported by *OGJ* in November 2009, the working rig count outside of the United States and Canada in September 2009 was 986, up from 947 in August according to Baker Hughes.

Rig Count Outside US, Canada

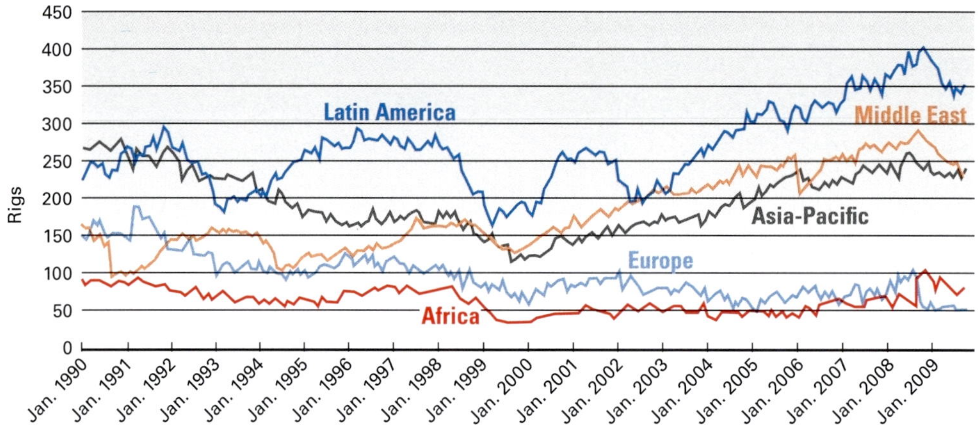

Source: Baker Hughes Inc.

World Rig Count

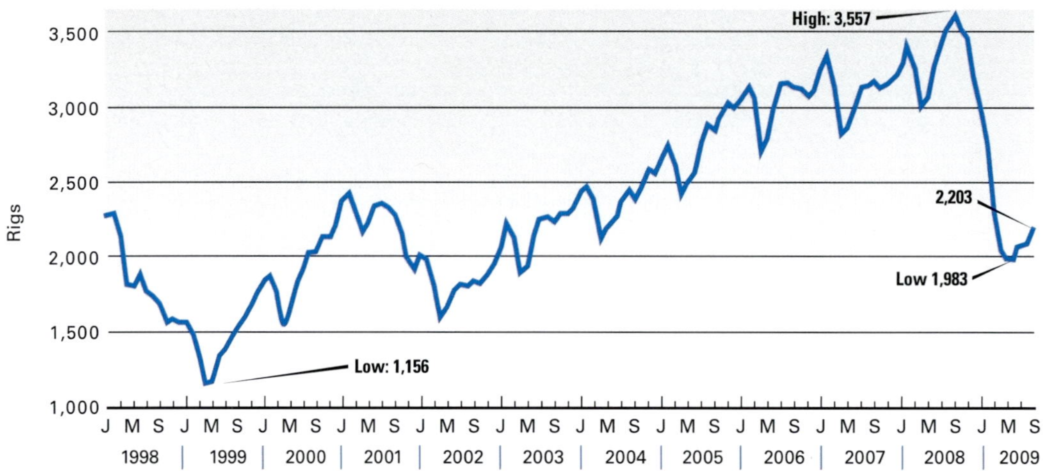

Source: Baker Hughes Inc.

World drilling data

As reported by *OGJ* in November 2009, the Baker Hughes world rig count fell from a high of 3,557 rigs in October 2008 to a low of 1,983 rigs in June 2009, then rose to 2,203 working rigs in September 2009.

Baker Hughes does not include land rigs drilling in China and Russian or activity in Iran, Iraq, Kuwait, Sudan, and elsewhere.

Worldwide rig count

2009	Latin America	Europe	Africa	Middle East	Far East	Canada	US	World
January	381	93	58	274	238	377	1,553	2,974
February	374	81	59	264	242	413	1,320	2,753
March	358	95	61	262	236	196	1,105	2,313
April	349	86	62	253	236	74	995	2,055
May	357	82	62	253	239	72	918	1,983
June	343	77	64	247	236	125	895	1,987
July	351	73	57	249	244	175	931	2,080

Source: Baker Hughes Inc.

Worldwide rig count

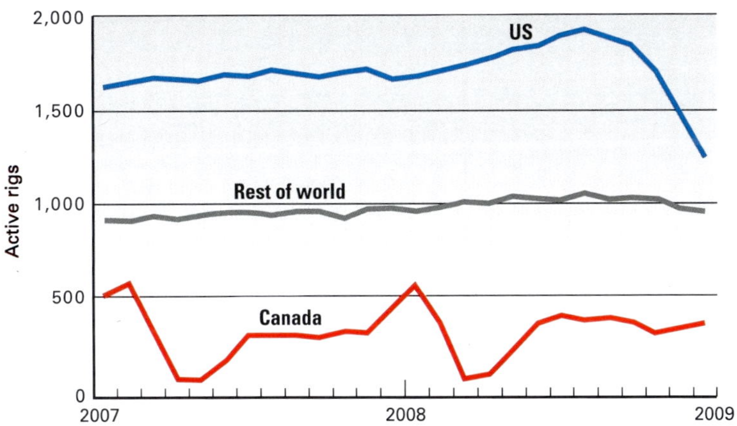

Source: Baker Hughes

International rig count

Avg. number of active rigs per month, year to date

	Jan. 2008–May 2008	Jan. 2009–May 2009
WESTERN HEMISPHERE		
Argentina	82	56
Bolivia	2	3
Brazil	46	61
Chile	1	2
Colombia	40	29
Ecuador	7	10
Mexico	99	129
Peru	7	5
Trinidad	6	1
Venezuela	81	67
Other	1	1
Total Western Hemisphere*	**373**	**365**
ASIA-PACIFIC		
Australia	26	20
Brunei	3	4
China-offshore	20	23
India	83	75
Indonesia	65	67
Japan	3	3
Malaysia	11	14
Myanmar(Burma)	8	3
New Zealand	5	4
Papua New Guinea	4	2
Philippines	1	3
Taiwan	0	0
Thailand	11	12
Vietnam	7	7
Other	3	0
Total Asia-Pacific	**249**	**238**
AFRICA		
Algeria	27	26
Angola	5	4
Congo	2	2
Gabon	2	1
Kenya	0	0
Libya	15	14
Nigeria	9	6
South Africa	0	0
Tunisia	4	3
Other	2	3
Total Africa	**67**	**60**
MIDDLE EAST		
Abu Dhabi	12	11
Dubai	1	1
Egypt	52	50
Iran	—	—
Iraq	—	—
Jordan	0	1
Kuwait	12	12
Oman	54	52
Pakistan	20	20
Qatar	11	9
Saudi Arabia	78	70
Syria	21	22
Yemen	14	11
Other	1	1
Total Middle East	**275**	**261**
EUROPE		
Croatia	0	0
Denmark	2	3
France	1	1
Germany	8	9
Hungary	3	3
Italy	5	3
Netherlands	3	3
Norway	18	22
Poland	2	2
Romania	20	10
Turkey	5	5
United Kingdom	20	21
Other	7	6
Total Europe	**94**	**87**
Total World	**1,058**	**1,012**

*Excludes US and Canada

Source: Baker Hughes Inc.

Emerging U.S. gas shale plays

Region	Shale play	Basin	Geologic age	Development stage
Midcontinent	Eagle Ford	Maverick	Cretaceous	Evaluation
Midcontinent	Woodford	Anadarko	Mississippian-Silurian	Evaluation
Midcontinent	Bend	Palo Duro	Pennsylvanian	Exploration
Southwest	Barnett	Permian	Mississippian	Exploration
East	Lower Huron	Appalachian	Devonian	Development
East	Utica	Appalachian	Ordovician	Exploration
Gulf Coast	Chattanooga, Floyd	Appalachian	Mississippian	Exploration
Gulf Coast	Conasauga	Appalachian Thrust Belt	Cambrian	Exploration
Gulf Coast	Tuscaloosa Marine	Gulf Coast	Cretaceous	Exploration
Rockies	Mancos, Manning Canyon	Uinta	Cretaceous	Exploration
Rockies	Pierre, Mancos, Niobrara	Raton	Cretaceous	Exploration
Rockies	Gothic, Hovenweep	Paradox	Pennsylvanian	Evaluation
Rockies	Baxter	Vermillion	Cretaceous	Evaluation
California	Monterey, McClure	San Joaquin	Miocene	Exploration

Emerging North American gas shales

Advanced Resources International Inc. (ARI) reported in October 2009 on drilling activity in the emerging gas shale areas of North America.

Shale reservoirs currently contribute more than 8 bcfd (227 MM cu m/d) to North American gas supplies, up from just 2 bcfd (57 MMcmd) in 2004. Leading the way are the Barnett [5 bcfed (142 MMcmed)], Fayetteville [1.3 bcfed (37 MMcmed)], and fast-rising Haynesville [1.1 bcfed (31 MMcmed)] formations.

{Editor's note: For maps of the six major U.S. gas shale areas and data on shale-gas composition, see the Gas Processing and Products section in this Key Stats chapter.}

Global shale potential

ARI also noted in October 2009 that early-stage exploration of promising gas shale resources was under way in a range of countries outside the United States.

- In Europe, promising gas shale targets exist in northern Poland, the Paris basin, southern Sweden, and in Russia's Dnieper-Donets basin.
- In Australia, companies are evaluating gas shales in the Cooper basin, the Southern Georgina basin, and the Amadeus basin.
- Promising gas shale potential also appears to exist in Algeria, Morocco, Colombia, and Venezuela.

Microbial EOR

Researchers at UAE University reported in mid-2009 on work aimed at characterizing the types of reservoirs suitable for use of microbial enhanced oil recovery (MEOR) technology.

They describe a decade of work involving 10 field cases with 221 producing wells in Malaysia, the United States, Argentina, and China.

MEOR field trials

Location	Reference	Oil field	No. of wells	No. of positive wells	Reservoir temp., °C.	Reservoir permeability, md
Malaysia	1	Bakor	3 producers	1	48	50–4,000
US	2	North Blowhorn Crreek	4 injectors and 3 producers	8	<55	0.7–141
Argentina	3	Piedras Coloradas	6 producers	6	76	5–10 and 120 in two formations
Argentina	4	Papagayos Vizcacheras	1 injector and 9 producers	9	92	1,000
China	5	Changqing	27 producers	Not reported	43–54	2–149
China	6	Shengli	3 producers	Not reported	66	89–263
China	7	Wenmingzhai	12 injectors and 29 producers	25	85	60–726
China	8	Dagang	5 injectors and 7 producers	7	73	259 and 468
China	9	Chaoyanggou	60 producers	43	55	203, average 10
China	9	Pubei	10 producers	8	50	103
China	9	Chaoyanggou	2 injectors and 10 producers	7	55	10
China	10	Fuyu	6 producers	4	30	240

Offshore Europe oil production

In a comprehensive look in mid-2009 at global oil production, Douglas-Westwood concluded that:

- Europe remains one of the world's largest offshore producers, with nearly 600 offshore fields developed.
- small companies are playing an important role in sustaining offshore Europe production.

The global oil and gas production context*

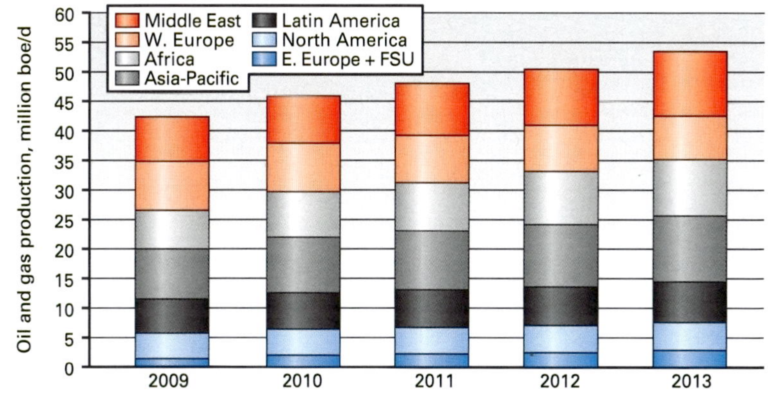

*Excludes natural gas liquids.

Source: The World Offshore Oil & Gas Production and Spend Forecast

FPS prospects off Western Europe, 2009–13

Country	FPSO	FPSS	TLP	Total
Ireland	1			1
Italy	1			1
Norway	2	1		3
UK	11	2	1	14
Total	**15**	**3**	**1**	**19**

Source: World Floating Production Report, 2009-13

Deepwater circular drilling rig

In mid-2009, The COSCO Shipyard Group's Qidong shipyard completed construction of the *Sevan Driller*, the world's first deepwater circular drilling rig.

The rig is designed it to drill to 12,192 m (40,000 ft) and in 3,810 m (12,500 ft) of water.

Due for delivery in third quarter 2009, the rig has a 6-yr fixed contract with Petrobras for work in the Santos basin off Brazil.

Offshore Europe oil and gas production*

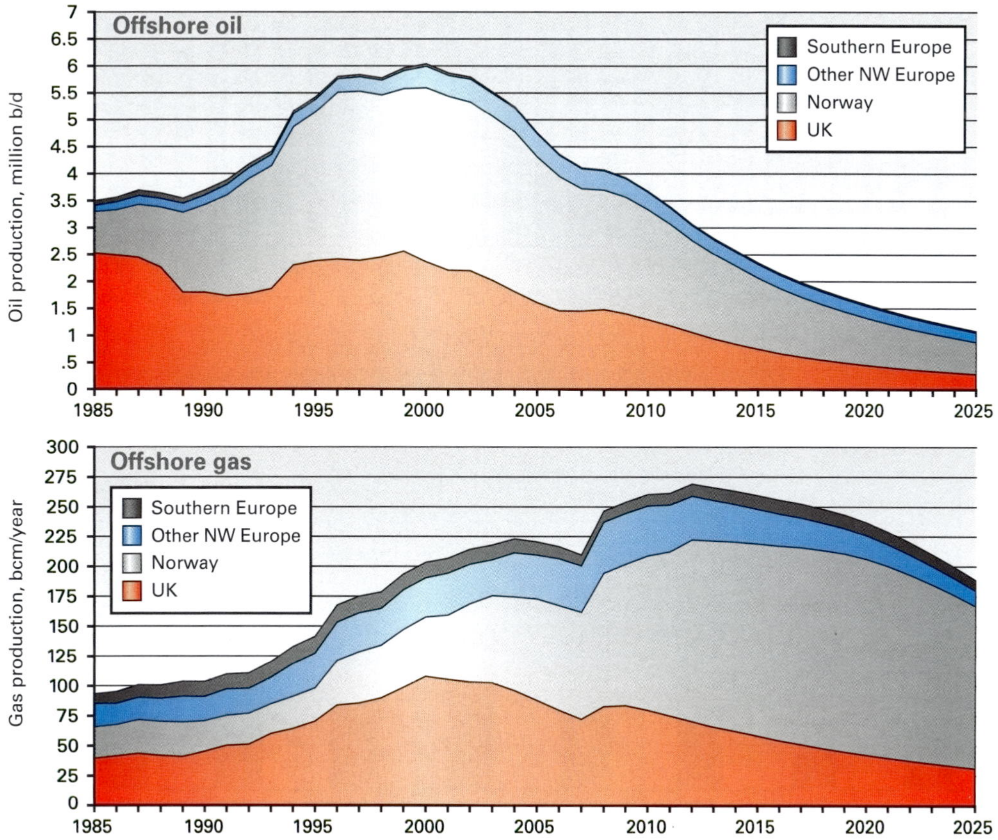

*Excludes natural gas liquids.

Source: Energyfiles Ltd.

Global floating production market

Infield Energy Analysts forecast that from 2009–2013, the total global floating production systems market will exceed $85 billion, an increase of $32 billion over the previous five years.

Over the period, 229 units are expected to be installed, with Asia (61) leading the way.

Petrobras is the most active player (expected $15-billion expenditure), followed by Total, Chevron, Shell, ExxonMobil, and BP.

FPSOs will command the majority of the spending ($48.6 billion).

International Maritime Associates Inc. presented historical data for the floating production systems market in early 2009, and also expressed concern about pressures it expected would push down demand in 2009.

Deepwater expenditure growth

Despite lower expected expenditures during 2009 and 2010 relative to 2008, a study by Douglas-Westwood in early 2009 forecast that deepwater expenditures will climb to $35 billion by 2013.

The firm's *World Deepwater Market Report 2009–13* expects cumulative deepwater project expenditures to total $162 billion in the 2009–2013 period.

Floating Production System Order Backlog

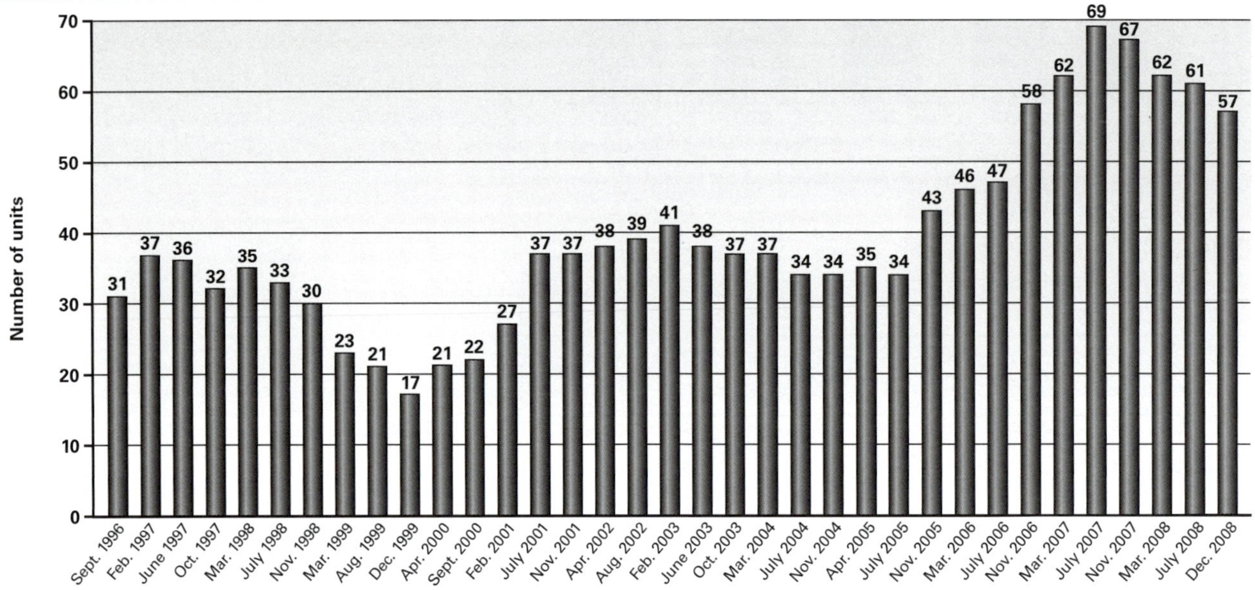

Note: Excludes storage-only units and jack up mobile offshore production units (MOPUs).

Worldwide Deepwater Market Expenditures

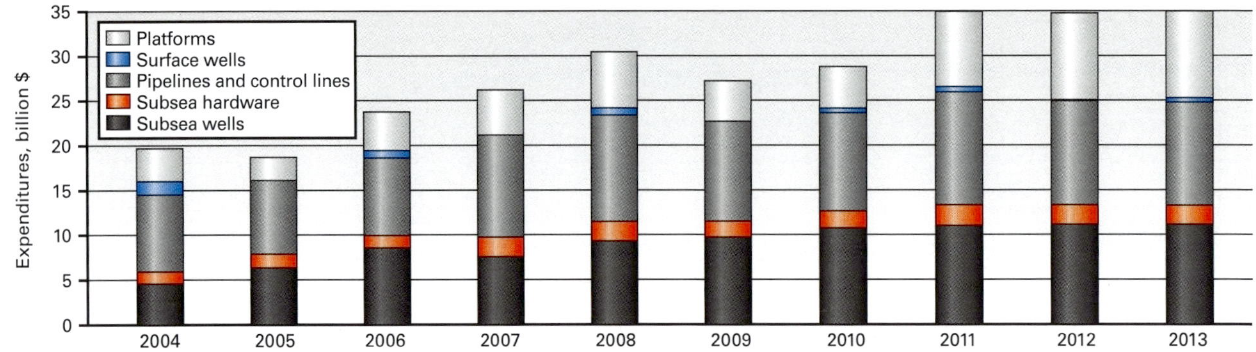

Source: Douglas-Westwood

Section 4:

Investment and markets

Where funds will go for US projects

	2009, million $	Change 2009–2008, %	2008, million $	Change 2008–2007, %	2007, million $
Exploration-production					
Drilling-exploration	174,621	–27.1	239,646	10.7	216,462
Production	33,178	–27.1	45,533	10.7	41,128
OCS lease bonus	1,080	–84.3	6,883	119.1	3,142
Subtotal	**208,879**	**–28.5**	**292,062**	**12.0**	**260,732**
Other					
Refining	10,140	–22.0	13,000	57.0	8,280
Petrochemicals	50	–95.0	1,000	19.0	840
Marketing	1,950	–35.0	3,000	20.0	2,500
Crude and products pipelines	5,164	16.5	4,431	146.8	1,796
Natural gas pipelines	10,374	63.6	6,343	45.2	4,367
Other transportation	840	–30.0	1,200	23.7	970
Mining, other energy	900	–25.0	1,200	20.0	1,000
Miscellaneous	3,750	–25.0	5,000	22.0	4,100
Subtotal	**33,168**	**–5.7**	**35,174**	**47.5**	**23,853**
Total	**242,047**	**–26.0**	**327,236**	**15.0**	**284,585**

Canadian spending plans

	2009, million $ (Can.)	Change 2009–2008, %	2008, million $ (Can.)	Change 2008–2007, %	2007, million $ (Can.)
Exploration-production					
Drilling-exploration	25,045	–8.0	27,223	20.0	22,686
Production	9,896	–8.0	10,757	20.0	8,964
Subtotal	**34,941**	**–8.0**	**37,980**	**20.0**	**31,650**
Oil sands*	15,000	–25.0	20,000	10.7	18,065
Other					
Refining	1,500	–25.0	2,000	–42.3	3,465
Petrochemicals	10	–95.0	200	–39.9	333
Marketing	400	–27.3	550	–18.8	677
Crude and products pipelines	485	46.5	331	292.6	84
Natural gas pipelines	28	–91.9	338	—	—
Other transportation	225	–25.0	300	4.5	287
Miscellaneous	550	–15.4	650	2.2	636
Subtotal	**3,197**	**–26.8**	**4,369**	**–20.3**	**5,482**
Total	**53,138**	**–14.8**	**62,349**	**13.0**	**55,197**

*In situ, mining, and upgrading.

Cuts in capital spending for 2009

In an April 2009 survey, *Oil & Gas Journal* reported that oil and gas producers in the United States and around the world were slashing 2009 capital budget plans, in reaction to the worldwide economic downturn, lower oil and gas prices, and tight credit markets.

Oil and gas industry capital spending in the United States was expected to fall 26% in 2009 to $242 billion.

In Canada, total spending was seen dropping 15% to $53 billion (Can.), but spending plans indicated an increase in Mexico. Outside North America, upstream oil and gas capital expenditures were projected to decline from 2008 levels.

Many small oil and gas producers slashed their original 2009 capital spending plans. The larger operators mostly stayed with their plans, but many of these spending outlooks call for a decline in 2009 outlays from last year's levels.

In a mid-year assessment, Ernst and Young found that the world's biggest oil companies had cut back investment far less than had been expected, and that national oil companies (NOCs) had kept investment largely steady compared with 2008.

In total, the firm found that in 2009 the largest NOCs plan to invest more than $275 billion, while big IOCs had cut their expenditures to $100 billion, from the $122 billion they invested in 2008.

In 2010, investment is expected to remain at levels similar to 2009, at around $325 billion, down from a five-year peak of $370 billion in 2008.

However, the firm said there would be a few major new capital projects implemented through 2010, such as Gorgon in Australia, Rumaila in Iraq, Blocks 17 and 31 in Angola, and the Kearl project in the Canadian oil sands sector.

Wood Mackenzie: Global upstream view

In an October 2009 report, Wood Mackenzie said it saw hints of recovery in global upstream spending.

The firm predicted that investment in the upstream sector will return to growth by 2011, reaching around $350 billion in 2012.

The report stressed that upstream spend will be restored on a modest upward trend in 2011.

Global reserves: Down in 2008

Although global investment by oil and gas companies in 2008 for exploration and development projects totaled $492 billion—a 21% increase from 2007—both oil and gas reserves fell.

That was the finding in the 2009 Global Upstream Performance Review (released in September) of 232 oil and gas companies by IHS Herold Inc. and Harrison Lovegrove & Co. Ltd.

High oil prices during most of 2008 helped industry to generate record cash flow of $590 billion, up 36% from 2007.

In 2008, world oil reserves declined nearly 3%, primarily due to a 5.2-billion bbl decline in revisions that stemmed from reduced commodity prices. Natural gas reserves grew by 3%, the same as the past 5 years. Gas production accelerated nearly 5% to 44.2 tcf (1.25 tcm).

Offshore spending trend

According to a mid-2009 analysis by Scottish Enterprise, total offshore expenditure showed a steady rise from $123 billion in 2004 to $148 billion in 2008, and forecasts indicate continued increase to $161 billion in 2013.

The cumulative 2009–2013 forecast total of $784 billion is 16% higher than the 2004–2008 total.

The firm expects a decline in offshore expenditure in more-mature areas (e.g., U.S Gulf of Mexico), but a rapid increase in West Africa, the Mediterranean, and the Caspian Sea.

ICE regulation

In mid-2009, the InterContinental Exchange (ICE) agreed to be regulated, after the U.S. Commodity Futures Trading Commission (CFTC) effectively ruled in July that one of ICE's natural gas contracts should no longer be exempt.

ICE declared itself a registered CFTC entity on July 27 and announced that it would begin submitting enhanced market statistics for its cash-settled Henry Hub gas swap market to the CFTC immediately.

Recession impact on independents

Analysts speaking at a June industry conference noted that independent oil and gas producers that rely heavily on cash flow were facing bigger financing challenges than major or national oil companies as the global recession depressed oil and gas demand and prices fell.

The possible consequences range from a new wave of acquisitions to slower development of natural gas from domestic shale formations.

One bright spot: Costs of steel, concrete, and other critical E&P components were dropping, after nearly doubling from 2000 to 2008, according to the U.S. EIA.

Worldwide oil, gas deals: 2008

The total transaction value for worldwide upstream corporate and asset deals plunged to $104 billion in 2008 compared with an annual average of nearly $160 billion during 2005–2007.

That was a key finding in the 2009 Global Upstream Mergers and Acquisitions Review, an annual assessment by IHS Herold Inc. and Harrison Lovegrove & Co. Ltd., released in March 2009.

The report analyzes more than 280 major upstream transactions announced in 2008.

The firms concluded that the upstream M&A market appeared ripe for considerable consolidation.

Though first-half 2008 activity was frenetic, the pace slowed dramatically in the second half, particularly in the United States and Canada.

WORLD UPSTREAM DEAL COUNT*

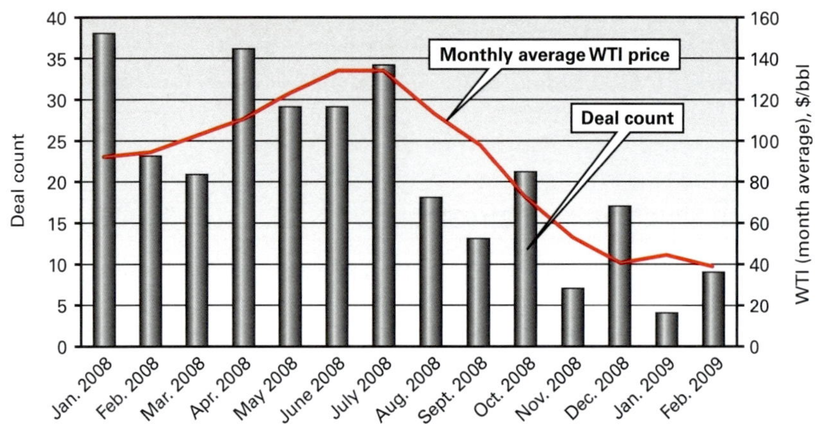

*Announced deals greater than $10 million. Sources: John S. Herold Inc. and Harrison Lovegrove & Co. Ltd.

Oil, gas tax changes

Friedman, Billings, Ramsey & Co. Inc. said it saw "ample signs" in early 2009 that U.S. state and federal government will take a bigger tax bite from both upstream and downstream oil and gas operations to cover pending financial shortfalls in 2009.

Regarding federal law, the firm said changes were likely in rules for inventory accounting, Section 199 deductions, surcharges on GoM deepwater production, royalty deductions, and tax treatment of intangible drilling costs.

FBR said that "2009 will kick off what we estimate to be a 2-year countdown to higher extraction costs."

IOC RISK, OPPORTUNITY IN TOP 20 BOE PROVED RESERVES COUNTRIES*

Country	Overall risk score (0-100)	Opportunity score (0-100)	Proved gas reserves, tcf	Proved oil reserves, billion bbl	Proved reserves, billion boe
Russia	70	45	1,576.8	79.43	342.22
Saudi Arabia	35	60	253.0	264.21	306.38
Iran	60	30	981.7	138.40	302.02
Qatar	20	85	904.1	27.44	178.11
Iraq	70	45	111.9	115.00	133.66
UAE	25	70	215.1	97.80	133.64
Venezuela	75	45	181.9	87.04	117.35
Kuwait	35	55	63.0	101.50	112.00
Nigeria	70	50	187.0	36.22	67.39
US	20	70	211.1	29.44	64.62
Kazakhstan	60	40	67.2	39.83	51.03
Libya	40	65	52.8	41.46	50.26
Algeria	55	70	159.4	12.27	38.84
Canada	25	65	57.5	27.66	37.26
China	50	50	66.5	15.49	26.58
Norway	25	55	104.6	8.17	25.60
Indonesia	40	40	105.9	4.37	22.03
Malaysia	30	60	87.4	5.36	19.92
Australia	20	55	88.6	4.16	18.93
Turkmenistan	55	40	94.2	0.60	16.30

*International oil companies' risk, opportunities in countries ranked in descending orderof 2007 proved oil plus gas boe reserves holdings. Source: BP Statistical Review, June 2008

Risk, opportunity analyses

UK-based David Wood & Associates (DWA) presented results of a risk/opportunity analysis tool for possible use by international oil companies in making decisions about resource development projects.

Details about the tool and related methodology were published in the February 9, 2009, issue of *Oil & Gas Journal.*

Attributes related to "country risk" included such issues as corruption, political instability, and inflexible fiscal terms.

Attributes related to opportunity analysis included such issues as degree of access to large reserves, finding/development costs, and ease of operation.

Using a numerical scale for both risk and opportunity yields an aggregate score that can be useful for comparing investment and operational options. Analytical results can then be used to create graphical representations to aid decision-making.

TOP 24 OIL AND GAS RESERVES NATIONS AS FUNCTION OF IOC RISK, OPPORTUNITY*

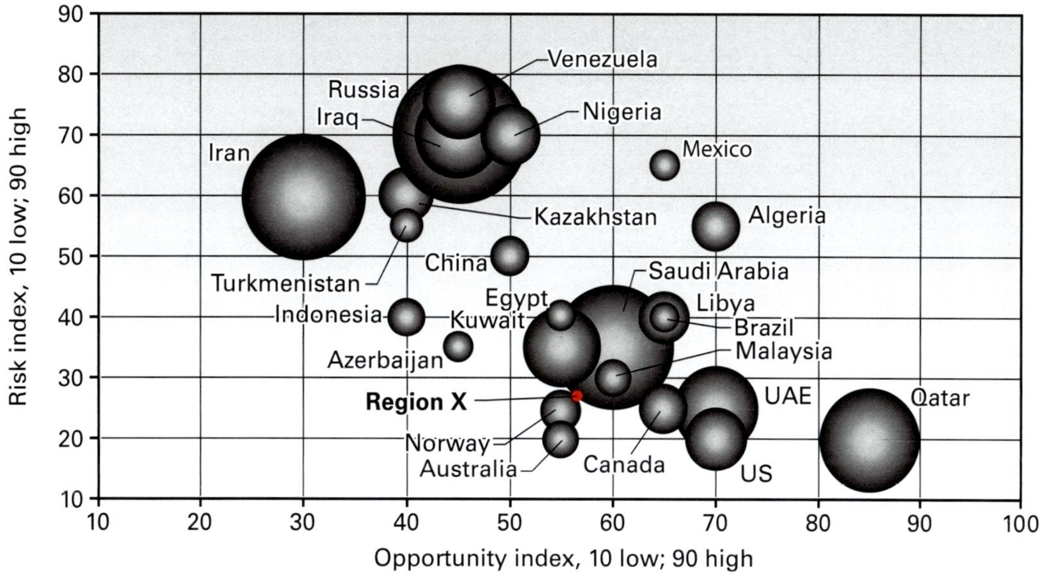

*Assessment conducted July 2008. Risk-opportunity indices evaluated for the top crude oil plus natural gas (boe) proved reserves-holding nations. Bubble diameters are proportional to proved crude oil plus proved gas reserve holdings as reported by BP Statistical Review (June 2008).
Sources: David Wood & Associates, BP Statistical Review (2008)

For interested analysts

A more detailed report and workbook on E&P Risk Management by David Wood & Associates are available for purchase through the *OGJ* Online Research Center. For additional information, contact: www.ogjresearch.com

Topic: Research Reports

Report Title: Fiscal design strategies and fiscal instability impact oil & gas risk and opportunity analysis and international oil company decision-making

E-mail: orcinfo@pennwell

Clean energy investments

In a mid-2009 analysis, the *Financial Times* concluded that investors in clean technology, battered during the financial crisis of 2007–2008, were rewarded during second quarter 2009.

New investment in clean energy—the bulk of the sector—plunged from $41 billion in fourth quarter 2007 to only $13.3 billion in first quarter 2009, according to consultant New Energy Finance.

But share prices rose 36% in the second quarter, according to a widely watched sector index.

Also, new investment in clean energy globally jumped to $36.2 billion in the second quarter, and was seen as likely continuing at similarly high levels.

Two factors behind the increases were concerns about rising oil prices and recent stimulus packages put in place by governments around the world.

Section 5:
Trade and tankers

Gas oil: Out to sea

A December 2009 analysis by Deutsch Bank AG estimated that nearly 100 million bbl of gas oil (diesel and heating oil) were being held in floating storage around the globe, much of it off Europe, symptomatic of the global downturn. Vessel-based storage began in early 2009.

Analysts expected that the overhang would continue to weigh on the gas oil complex and margins into 2010. This would further challenge refiners, particularly in the United States and Europe, who would be struggling to cope with the rapid increase in new refinery capacity in China and India.

There is no official reporting system for gas oil stored in vessels at sea. A number or organizations make their own estimates, including IEA, OPEC, and ICAP Shipping International Ltd.

Two factors contributing to the phenomenon were the impact of the recession on oil flows (crushing shipping rates), and the addition of a major number of new vessels to the global fleet.

Crude: VLCC storage

At the end of August 2009, Norway's Frontline Ltd., the largest independent oil tanker shipping group, estimated that 40–45 VLCCs—about 10% of the world fleet—were storing crude oil, down from a peak of 60 that were storing crude in April.

New ship orders plunge

As 2009 came to a close, the *Financial Times* looked at the state of the shipbuilding industry.

Their analysis pointed to an expected increase in insolvencies as one of the steepest collapses ever in order levels combines with reluctance of banks to finance ship construction.

Clarkson Research Services Ltd. reported that only 28.8 million deadweight tonnes (dwt) of ships were ordered January-November 2009. In 2007, the peak of the shipping boom, orders totaled 272 million dwt.

Half the orders placed in 2009 were in China, the world's number two shipbuilder. Chinese state companies were the only ship-owners worldwide ordering vessels in any numbers.

During 2009, three shipyards in Korea (the world leader) several small yards in China, one yard in Japan, three German yards, and a Norwegian and U.S. yard filed for insolvency. Denmark's only shipyard also is to be closed.

MAJOR TRADE MOVEMENTS 2008

Trade flows worldwide
(million tonnes)

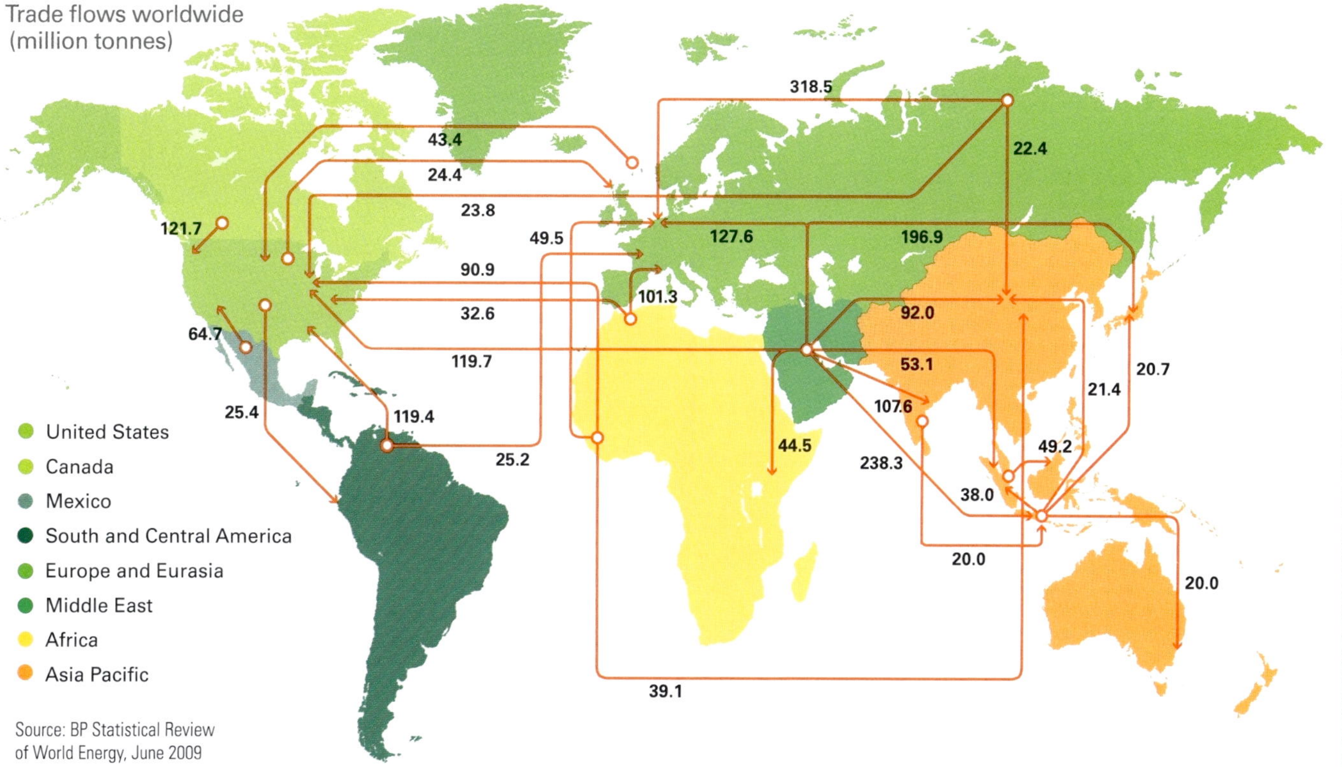

Source: BP Statistical Review
of World Energy, June 2009

MAJOR TRADE MOVEMENTS 2008

Trade flows worldwide
(million tonnes)

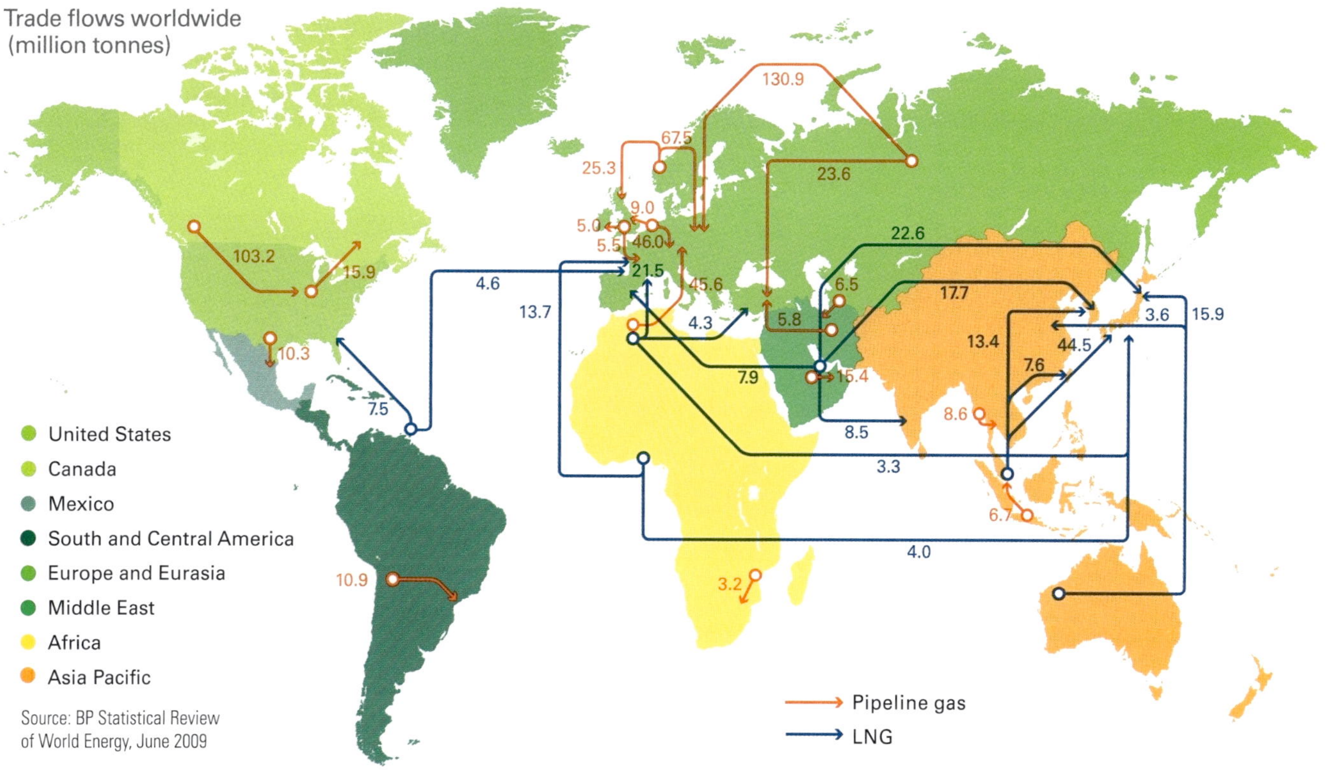

Source: BP Statistical Review
of World Energy, June 2009

Marine CNG

Recent offshore natural gas opportunities in U.S. waters often lack access to pipelines to support offshore drilling and transport gas to shore.

Analysts with Sutherland Assbill & Brennan LLP described in early 2009 the potential advantages of an emerging alternative method for moving stranded gas from production sites to market: compressed natural gas (CNG) vessels.

The technology and its application are still in the formative stage and working through commercial and regulatory hurdles, but they said it could become an important transportation mechanism.

Earnings: 2008

Clarkson Research Services Ltd. reported that tanker market earnings for April-September 2008 firmed 43.7%, accelerating strength seen in the previous six months.

Reasons included: large tankers being withdrawn for conversion to offshore structures or bulk carriers; and scrapping of some older vessels. The result kept vessel supply constrained even as high crude and product prices eroded demand.

VLCC Fleet Development

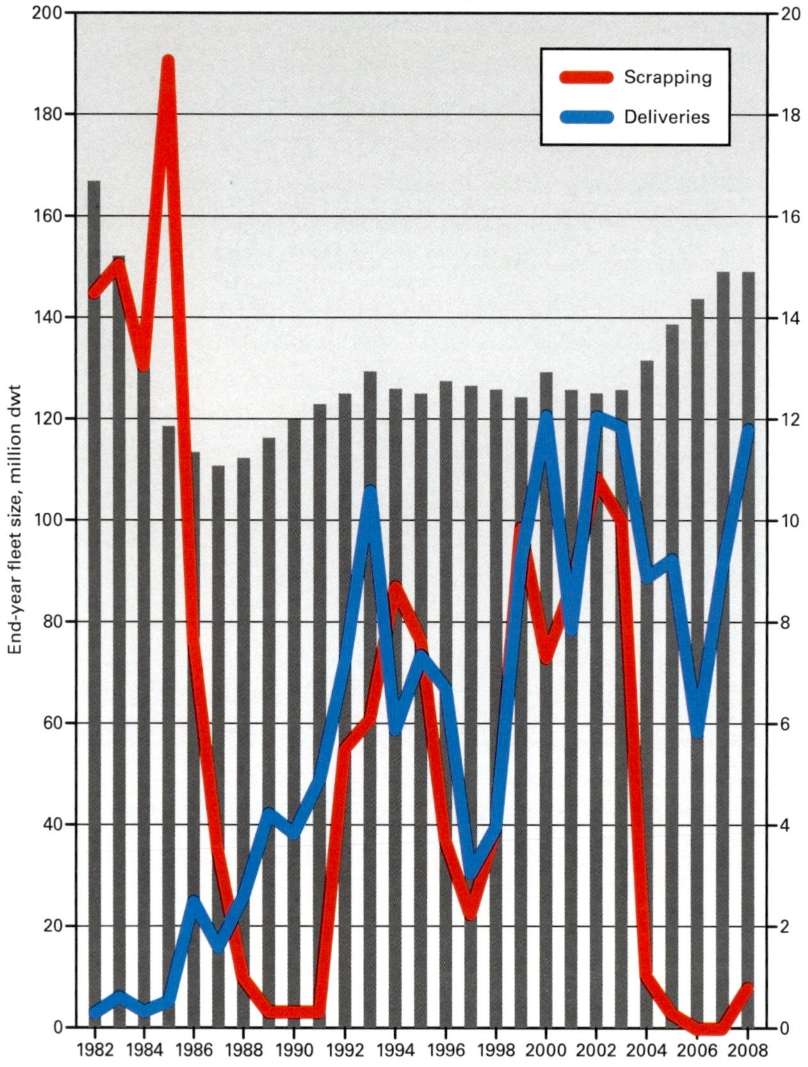

Source: Clarkson Research Services

Tanker Earnings

Vessel type	Sept. 2007–March 2008 ($/day)	March–Sept. 2008 ($/day)	% change
VLCC (modern)	85,492	112,256	31.3
VLCC (early '90s)	79,262	104,621	32.0
Suezmax (modern)	51,806	86,661	67.3
Aframax (modern)	35,439	58,535	65.2
Products (dirty)	29,202	42,596	45.9
Products (clean)	22,373	30,243	35.2
Weighted average	34,280	49,272	43.7

VLCC Market Summary

World-scale rates	End 2007	Sept. 2008	% change
AG-West	180.0	83	–54.2
AG-Far East	275.0	95	–65.5
Med-UK	205.0	125	–39.0
Revenue, $/day			
Average spot earnings, early '90s	186,368.0	47,318	–74.6
1-year time-charter rate, early '90s	52,500.0	62,500	19.0
Asset values, $ million			
Newbuild price, 300,000 dwt	146.0	160	9.3
5-year old, 300,000 dwt	135.0	165	22.2
Tonnage supply, million dwt			
Fleet	148.3	147.6	–0.5
Orderbook	56.4	81.4	44.5

Section 6:
LNG—Industry outlook

SUEZMAX EARNINGS, TIME CHARTER RATES*

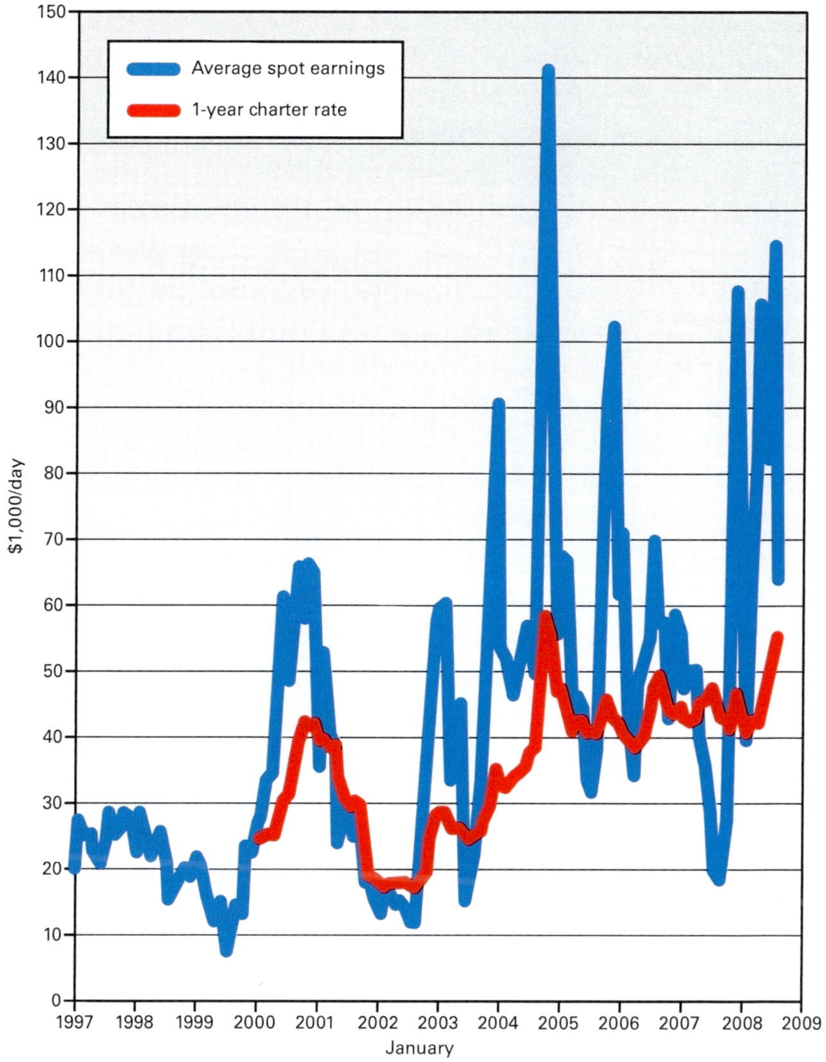

*Modern vessels only.

Source: Clarkson Research Services

Global production capacity: Rising

As of October 2009, nearly 30 mtpy of new LNG production capacity came on line. Key projects included the following:

- Startup of the first train of Yemen LNG in October.
- Sakhalin added two trains of 4.8 million tpy each earlier in the year.
- Qatargas 2 started up its 7.8-million tpy Train 4 and aimed to start up the 7.8-mtpy Train 5 by yearend.
- In Asia, debottlenecking at Malaysia Dua was to have added 1.5 mtpy at some point in the year. And Indonesia's 7.6-mtpy Tangguh formally began operations but has been plagued by outages and is currently shut down.
- Expected to start up very early in 2010 is Peru LNG's 4.4-mtpy plant.

Toyo Engineering, Hitachi plan

In October 2009, Toyo Engineering Corp. and Hitachi Ltd. said they will join forces to enter the global market for LNG plants, according Japan's *Nikkei Business Daily* (NBD).

Details about their plans to target small, undeveloped gas fields are given in the Papua New Guinea section of the Asia-Pacific chapter of this book.

North America: LNG exporter?

Speakers at the 24th World Gas Conference in October 2009 noted two major forces in the evolution of the LNG market over the past 3 years:

- decisive transformation to a global gas trade with robust demand growth
- development of North America's unconventional gas supply.

Some at the event went so far as to suggest that, in the not-too-distant future, North America could become self-sufficient in gas supply and become an LNG-exporting region.

FLNG optimism

In a mid-2009 review, Douglas-Westwood said that the floating liquefied natural gas (FLNG) business is underpinned by compelling long-term fundamentals and poised to be a major growth sector.

The firm's *World FLNG Market Report 2010–2016* forecasts expenditures of $23 billion through 2016.

The report predicts that Australasia and Africa will account for the largest proportions of that spending, (each with a 23% share, or $5.2 billion).

North America has the greatest number of FLNG project—mostly import terminals—but will account for only $1.6 billion (7%) of global capex.

Shifting trends

Prospects for global LNG demand remained "bleak," said Barclays Capital in a mid-2009 research report. The LNG market was proving itself "truly global," as supply-demand fundamentals for the first 6 months forced regional prices to converge, said the report.

Demand deteriorated sharply in Asia and spot prices for most of the Pacific basin collapsed to below those in Europe and the United States.

Major Asian economies that import LNG—chiefly Japan and Korea—continued to struggle, and emerging markets—mainly China—were only partially offsetting the slowing demand, says the report.

In the Atlantic Basin market, relative weakness in U.S. gas prices compared with European benchmarks was proving to be "a limiting factor for U.S. LNG imports."

Europe, however, continued to unload record LNG cargoes, its natural gas prices maintaining relative strength.

Given these global supply-demand trends, Barclays said it expected U.S. LNG imports to grow only modestly for the rest of the year, averaging 1.6 bcfd for all of 2009.

In May 2009 comments, FACTS Global Energy (Singapore) also concluded that the 2009 LNG outlook was grim.

Gas supply gluts, an economic malaise depressing energy demand, and plunging petroleum prices characterized global markets.

European LNG trends

Purvin & Gertz shared the pessimistic outlook of Barclays and FACTS Global Energy in a May commentary and also gave its perspective on trends in the European gas and LNG market.

The quantity of LNG supply imported into Europe as dedicated or fixed supply increased to 29–30 million tonnes in both 2007 and 2008, from an estimated 26.7 million tonnes in 2006.

The LNG share of total supply increased to 73% in 2007 from 63%, but dropped back to 65% in 2008.

The firm concluded that the single biggest challenge facing sellers of gas into Europe was likely to be the deteriorating economic situation.

It saw emerging evidence of declining gas demand in the industrial and power generation sectors of the main LNG importing countries even in late 2008, particularly in Italy, Spain, and the UK.

Existing European LNG Import Terminals

Country	Location	Operator	LNG capacity, million tpy
North Atlantic Europe			
France	Montoir-de-Bretagne	GdF	7.6
UK	Isle of Grain	Grain LNG (National Grid Transco)	10.0
UK	Teesside	Excelerate Energy	—
South Atlantic Europe			
Portugal	Sines	Galp Atlantico (Transgas)	4.0
Spain	Bilbao	Bahia de Bizkala Gas	5.3
Spain	Huelva	Enagas	8.7
Spain	Mugardos	Reganosa	2.7
Spain	Sagunto	Planta de Regasificacion de Sagunto	2.8
Mediterranean Europe			
France	Fos-sur-Mer	GdF	7.3
Greece	Revithoussa Island	DEPA	1.0
Italy	La Spezia	Snam Rete Gas	2.6
Spain	Barcelona	Enagas	10.5
Spain	Cartagena	Enagas	7.7
Turkey	Izmir	Egegaz LNG	4.3
Turkey	Marmara Ereglisi	Botas	4.3
Total capacity			**78.7**

European LNG Importers, 2006-08

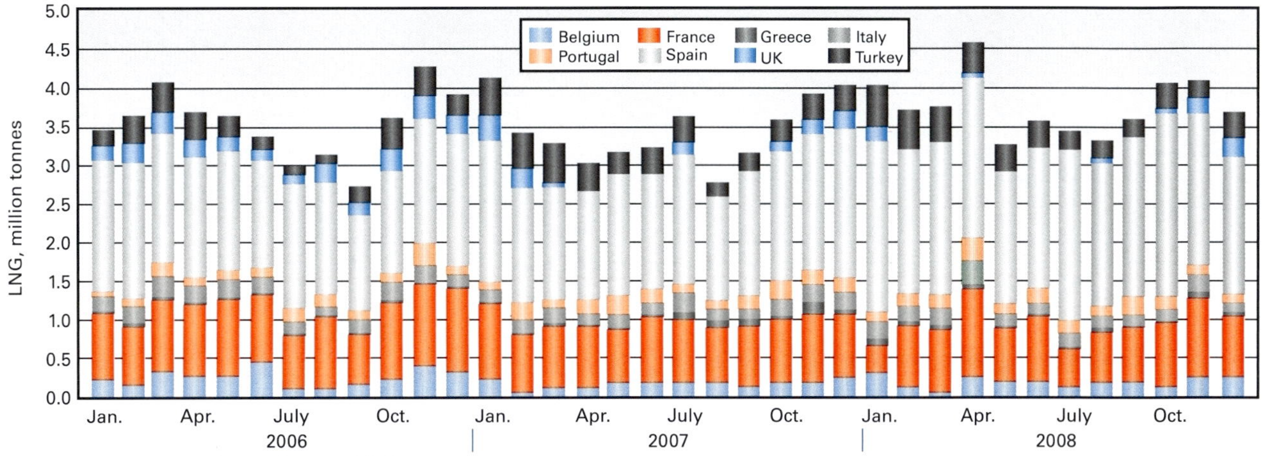

Source: Waterborne Energy Inc., European Waterborne LNG Report

European LNG Sources, 2006-08

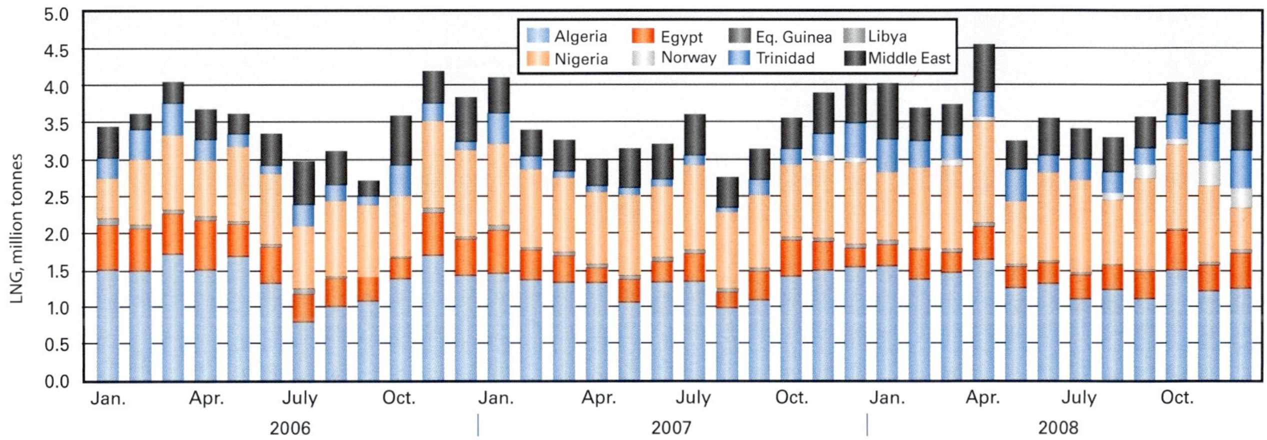

Source: Waterborne Energy Inc., European Waterborne LNG Report

Type of European LNG Imports, 2006-08

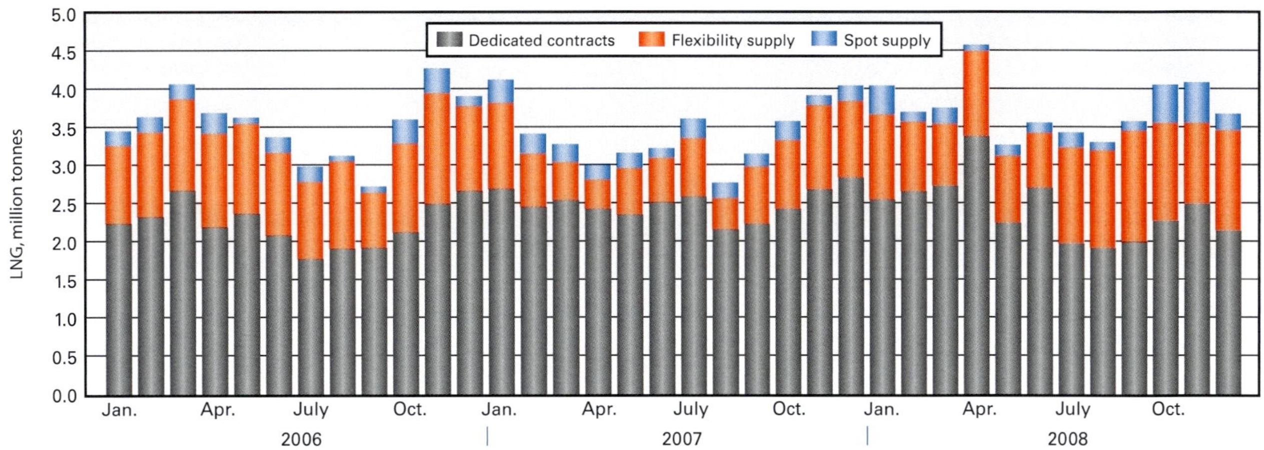

Source: Purvin & Gertz estimates based on Waterborne Energy Inc., European Waterborne LNG Report data

INTRA-ATLANTIC BASIN TRADE, 2006-08

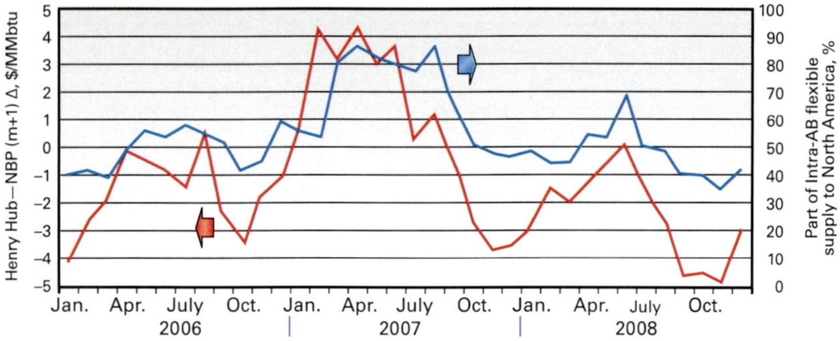

Source: Purvin & Gertz analysis

ATLANTIC BASIN TRADE

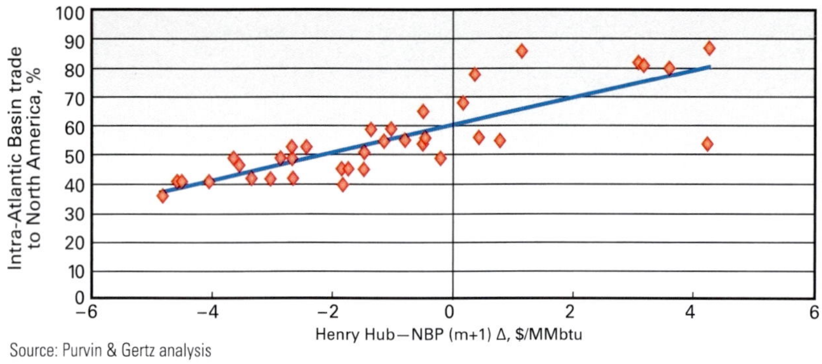

Source: Purvin & Gertz analysis

The firm's mid-2009 forecast was for the rate of gas demand growth to slow sharply in 2009 and be stronger in 2010.

But the firm saw many diverging trends within that forecast: Flat or declining industrial demand in France, Italy, and the UK; and declining or flat power generation demand in Italy, Spain, and the UK.

First FSRU

OGJ reported in early 2009 that Petrobras received the world's first floating storage and regasification unit vessel (FSRU), the *Golar Spirit*, following commissioning tests on the vessel's regasification trains.

It went into service at the LNG terminal of Petrobras, off Pecem, Ceara State.

Each of its three regas units can send out up to about 130 MMcfd, though two units normally operate while the third remains on standby.

EUROPEAN GAS, LNG SUPPLY, DEMAND OUTLOOK*

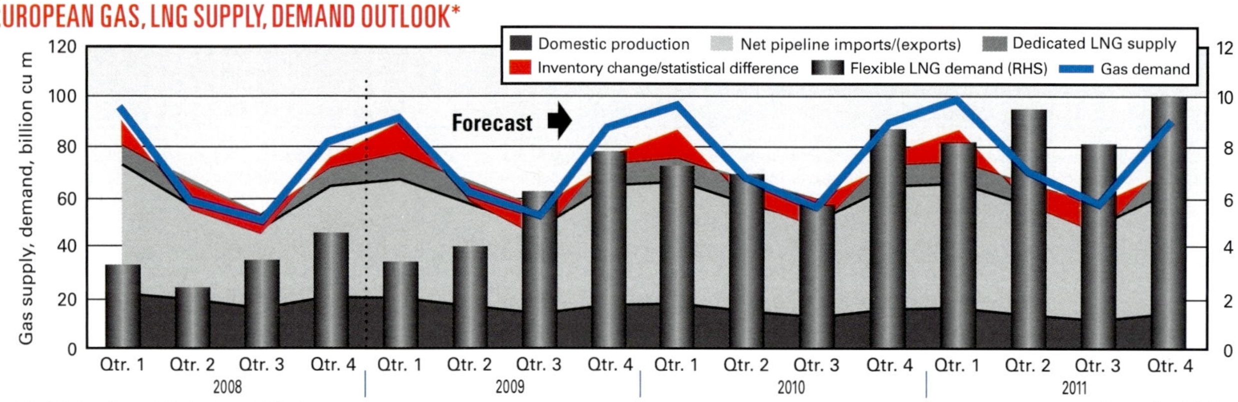

*For Belgium, France, Italy, Spain, and UK only.

Source: Purvin & Gertz

EUROPEAN LNG IMPORT TERMINALS UNDER CONSTRUCTION

Country	Project	Location	Operator	LNG capacity, million tpy	Start-up year
North Atlantic Europe					
Netherlands	Gate Terminal	Rotterdam	Gasunie/Vopak/RWE	7.8	2011
UK	Dragon LNG	Milford Haven	Dragon LNG (4Gas, BG, Petronas)	6.6	2009
UK	South Hook LNG	Milford Haven	ExxonMobil, Qatar Petroleum	7.8	2009
South Atlantic Europe					
Spain	Mugardos expansion	Mugardos	Reganosa	2.7	2013
Mediterranean Europe					
France	Fos Cavaou	Fos Cavaou	GdF/Total	7.3	2009
Italy	Isola di Porto Levante (offshore)	North Adriatic Sea	QP/ExxonMobil/Edison	6.3	2009
Italy	OLT Offshore Cross Gas	Livorno, Tuscany	Offshore LNG (E.On/Iride)	2.7	2011
Spain	Barcelona expansion	Barcelona	Enagas	2.0	2009–10
Total capacity				**43.2**	

Asia-Pacific LNG demand

In early 2009, the East-West Center and FACTS Global Energy (Singapore) examined likely trends in Asian LNG demand for the next several years.

They also noted that the Asian market not only drove incremental demand growth in recent years, but also accounted for about two-thirds of the global LNG market.

A key conclusion was that there appears to be only limited room for incremental LNG supplies to go to the Asia-Pacific market in the next few years. But this will likely lead to higher imports to markets such as the United States, which has a large liquid gas market as well as substantial storage capacity.

The two firms expect to see continued weakness in spot and short-term LNG prices for several years.

Global LNG: Supply on rise; project costs falling

OGJ summarized a range of views expressed at the February 2009 CERAWeek 2009 conference.

- Global LNG supply by 2013 will jump as nearly 100 mtpy of liquefaction projects currently near start-up or under construction come on line.
- In a global market that for the first time faces an abundance of LNG supply, vessels, and regasification, prices will move toward convergence.
- For new projects, debt financing will be available, even as the world's financial markets struggle to recover from collapse. Project developers, however, may have to accept more onerous terms and offer lower risk to participating banks.
- Project costs, which in late 2008 began to move lower, will continue to back away from the unsustainable levels of last year.
- Natural gas demand, contracting in early 2009 in the global recession, will inevitably recover, providing larger and newer markets for the widened global gas trade.

Asia-Pacific LNG import forecast

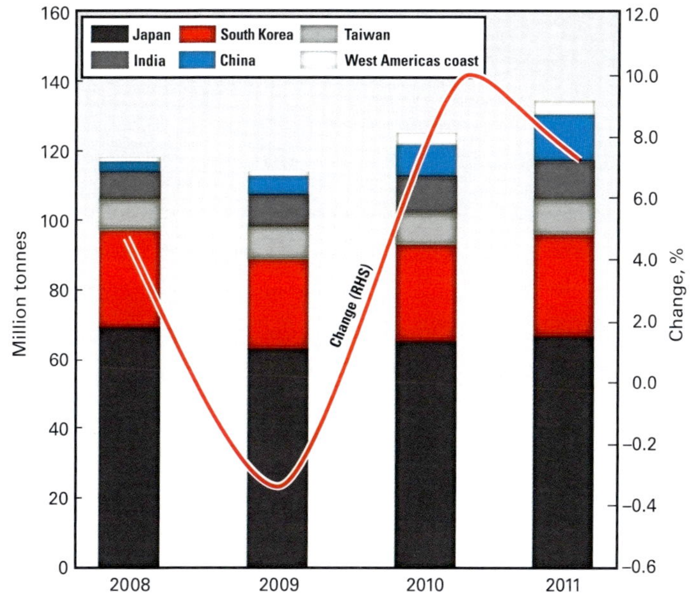

Asia-Pacific LNG import by country*

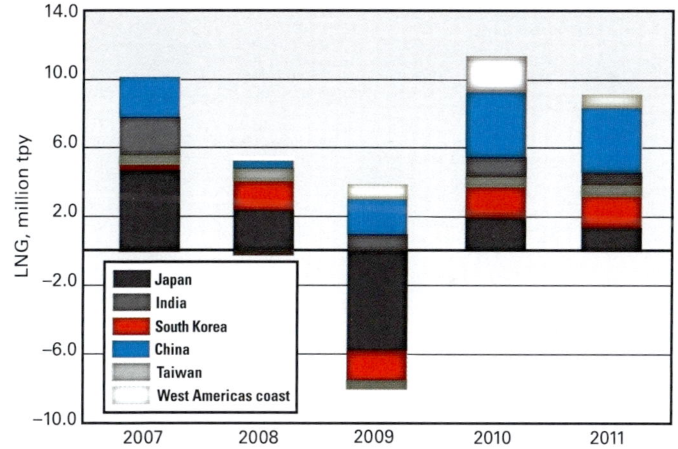

*Year-on-year change.

Asia-Pacific uncontracted LNG demand

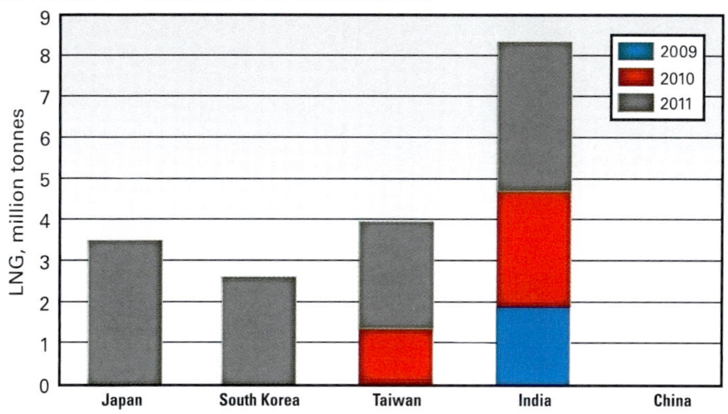

Section 7:
LNG—Liquefaction and receiving

Floating LNG concept: Shell

In July 2009, Shell Gas & Power Developments BV announced that it had signed an agreement with a consortium of Technip and Samsung for design, construction, and installation of several floating LNG (FLNG) facilities over a period of up to 15 years.

The companies also signed a FEED contract for Shell's 3.5-mtpy "FLNG solution."

Shell conceived its technology, said the announcement, to place gas liquefaction directly over offshore gas fields and thereby precluding long-distance pipelines and extensive onshore infrastructure.

Artist's rendering: Floating LNG facility for offshore gas liquefaction

Image courtesy of Shell Gas & Power Developments BV, The Hague, The Netherlands.

Regas units for FLNG facilities

Testing was under way in mid-2009 of LNG regasification packages for use on three floating LNG facilities. The packages are manufactured by Hamworthy PLC.

Brazil

Three propane regas skids were being installed on the *Golar Winter*, a floating storage and regasification unit (FSRU) chartered by Petrobras.

The vessel will be moored in Guanabara Bay, Rio de Janeiro, to send vaporized LNG into the local onshore grid. Each skid can deliver up to 7 MMcmd (247 MMcfd).

Boston

The Neptune terminal offshore Boston was the site for installation in June of three other Hamworthy systems, on the shuttle regas vessel (SRV) *Suez Neptune*.

A second SRV also is slated for operation at Neptune starting in mid-2010.

Each set of three regas skids on each SRV will have a capacity of 210 tonnes/hr of LNG.

Dubai

The third vessel receiving regas skids was the FSRU *Golar Freeze*, a converted LNG carrier.

That vessel will be operated by Dubai Supply Authority and Shell, at the Jebel Ali port.

Delivery of the regas skids was slated for September 2009. FSRU operation is to begin in 2010, delivering up to 14 MMcmd (494 MMcfd; 3 mtpy) of regasified LNG into the Dubai gas network.

Kitimat FLNG project

Bermuda-based Teekay Corp. announced in March 2009 an agreement with Merrill Lynch Commodities, Inc. to jointly develop a project to convert the *Arctic Spirit* into an FLNG plant.

The vessel would be moored at a pier near Kitimat, BC, Canada, a site on which construction of an LNG receiving terminal is to begin in 2010.

The converted vessel would have the capacity to liquefy 2.1–2.8 MMscmd (75–100 MMscfd) of pipeline quality gas, or, approximately 0.5 mtpy of LNG.

Section 8:
LNG—Carrier fleet

QatarGas: Q-Flex, Q-Max carriers

In early 2009, Qatargas Operating Co. Ltd. (QatarGas) chartered a Q-Flex and a Q-Max LNG carrier.

Both were delivered to Nakilat, the Qatari-listed shipping company from Daewoo Shipbuilding & Marine Engineering Co. Ltd. in South Korea.

The carriers will transport LNG produced by the Qatargas-3 project to the United States.

The *Al Sadd* Q-Flex carrier—one of the world's largest at 210,000 cu m (7.42 MMcf)—was delivered to Nakilat from Daewoo Shipbuilding & Marine Engineering Co. Ltd. in South Korea.

The Mekaines Q-Max carrier, with a capacity of 266,000 cu m (9.39 MMcf), was delivered to Nakilat from the Samsung Heavy Industries Co. Ltd. shipyard, also in South Korea.

QGTC: Q-Max carrier

In July 2009, Qatar Gas Transport Co. (QGTC) took delivery of the 263,000-cu m (9.29 MMcf) *Al Samriya* Q-Max carrier, which will transport LNG to Europe from the Qatargas-2 project of Qatar Liquefied Gas Co. Ltd.

It was built by Daewoo Shipbuilding & Marine Engineering Co,

The vessel joins five other recently delivered Q-Max ships—*Mozah*, *Al Ghuwairiya*, *Umm Slal*, *Lijmiliya*, and *Bu Samra*—all now on long-term charter to Qatargas Operating Co.

Q-Max ships are said to carry up to 80% more cargo but require 40% less energy per unit of cargo than conventional LNG carriers due to economies of scale and engine efficiency.

Section 9:
LNG—Risks and opportunities

Natural gas oversupply impact

Wood Mackenzie observed that the timing, duration, and extent of the oversupply evident in mid-2009 would vary among regions, with different implications for different sellers and buyers.

The firm said that several recent events had shifted market balance away from sellers, including:

- the global slide into economic recession
- the dramatic growth in gas availability from unconventional sources in North America
- and the largest wave of new LNG supply ever to hit the global market.

In combination, these factors contributed to a global gas glut, driving spot prices for gas down everywhere.

PACIFIC LNG SUPPLY-DEMAND*

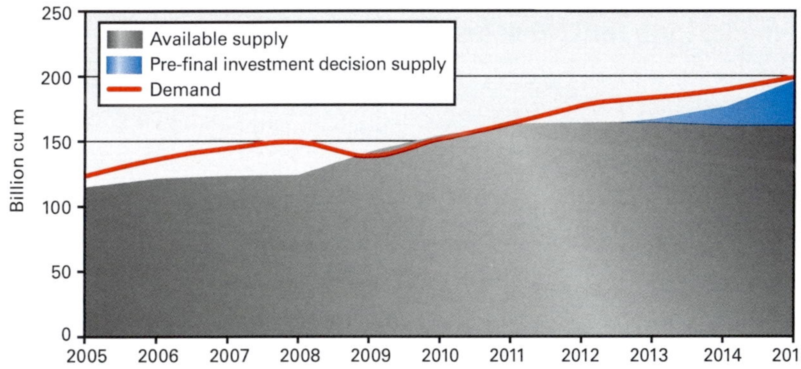

*Years run April-March.

Source: Wood Mackenzie Global Gas Model

ATLANTIC NATURAL GAS PRICING

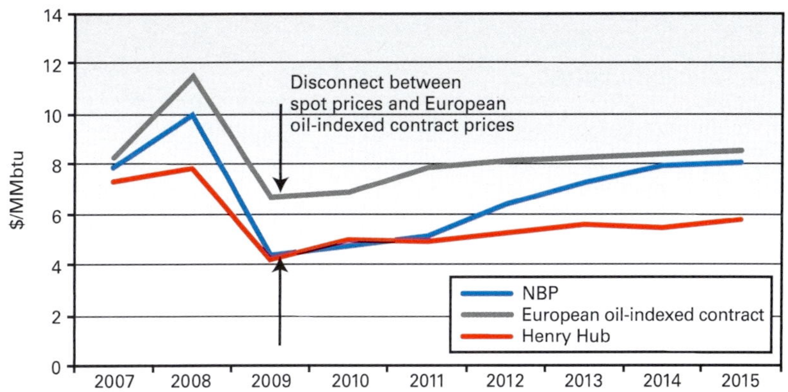

Source: Wood Mackenzie Global Gas Model

WoodMac concluded that this situation likely will get worse and the market will increasingly favor buyers.

Beyond the FPSO vessel

Offshore staff commented in early 2009 on emerging trends for using LNG-related technology to exploit remote offshore gas fields, moving beyond FPSO vessels.

Recent progress has been made on floating LNG (FLNG) units, which enable exploitation of gas reserves of 1 tcf (28.3 bcm) or less. Major issues include footprint, maintainability of liquefaction trains under rough sea conditions, and safety regulations.

Another area of attention is E-LNG technology, aimed at use of steam turbines as compressor drives in the liquefaction of natural gas at sea. Siemens AG is one of the key players. Advantages are said to be high availability and high reliability.

Another variation is "all-electric" E-LNG technology that uses waste heat recovery, hybrid schemes and variable-speed drive systems to reduce emissions.

New model form for LNG sales

In late 2009, the nonprofit Association of International Petroleum Negotiators published a new model form master LNG sale and purchase agreement.

The group's aim is to create a secondary market for LNG that will "facilitate trading and arbitrage of LNG cargoes."

AIPN's drafting committee (with more than 150 industry representatives) worked for 3 years on the effort, holding meetings and workshops in seven countries on five continents.

The group has 2,600 members in more than 80 countries, representing international oil and gas companies, governments, law firms, multilateral organizations, and academic institutions.

Coal vs. LNG: Emissions

A study by PACE Global Energy Services in early 2009 concluded that coal-fired power plants produce 161% more greenhouse gas emissions through their life cycles than plants fueled by LNG.

The study, done for advocacy group the Center for Liquefied Natural Gas, also found that two cleaner coal technologies—integrated gasification combined cycle (IGCC) and advanced ultra-supercritical coal (SCPC)—produce 70% more GHG emissions than LNG.

Section 10:
Refining and Products

Gloomy economics

At yearend 2009, *OGJ* reported that refining economics had been quite gloomy all year and that the outlook was not particularly good.

Refiners were shutting in capacity to shore up margins in light of dismal demand for transportation fuels and heating oil.

The removal in late 2009 of large volumes of heavy sour crude from the world oil market crushed light-sweet/heavy-sour spreads, thereby reducing upgrading margins.

- Europe's largest independent refiner, Petroplus, operated at only 61% of capacity during third quarter 2009. The firm's UK Teesside refinery was closed for most of the year and likely will be converted into a storage facility.
- The fate of Royal Dutch Shell PLC's Stanlow refinery was uncertain.
- Total SA's Dunkirk refinery was shut down as of yearend.
- In the United States, Sunoco Inc. shut down its refinery in Eagle Point, NJ, and sold its Tulsa refinery.
- Valero Energy Corp. announced that it intends to permanently shut its 210,000-b/d Delaware City, DE, refinery.

IEA: More hopeful

Late in the year, the International Energy Agency said that despite the bleak outlook, particularly in OECD countries, necessary changes appeared to be taking place that could signal the end of the worst for refiners.

Looking at utilization of upgrading units at U.S. refineries, IEA found that in recent months, catalytic cracking unit throughputs moved above their 2008 levels. The agency speculated that this might point to the reemergence of gasoline as the binding constraint on U.S. refining activity.

In contrast, weak hydrocracking and coking-unit throughputs demonstrated a response to weak distillate markets and the narrowing of heavy sour crude discounts. This suggests that refiners were adapting to current market circumstances.

Changes in crude flows

Big changes in global crude flows impacted U.S. refiners in 2009, said IEA.

- Imports of crude from Saudi Arabia had dropped at yearend to about 700,000 b/d from about 1.5 million b/d.
- Imports from Mexico fell to about 1 million b/d in August from a five-year average of 1.6 million b/d.
- Imports from Canada increased, and refiners had to adapt to new crudes that have offset some of the large declines in imports from Saudi Arabia and Mexico.

WORLDWIDE REFINING*

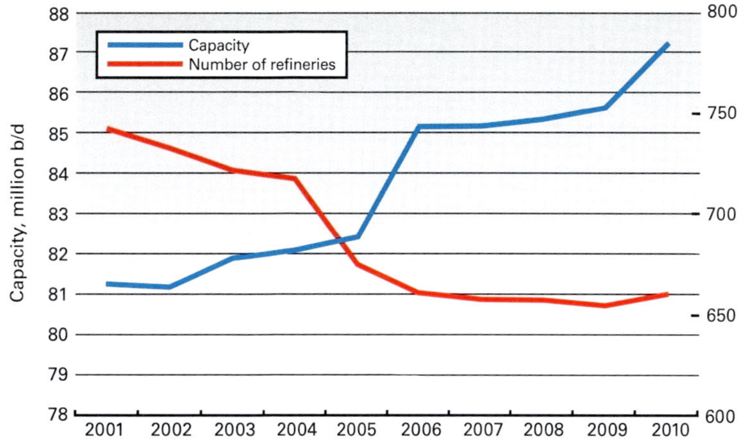

*As of Jan. 1 of each year.

Changes in fuel oil markets

Narrower fuel oil cracks in 2009 also caused changes in fuel oil markets, according to IEA.

The agency noted three important changes.

- Fuel oil yields in the OECD were up at in late 2009, suggesting that refineries were responding to shifting product cracks.
- Above-average fuel oil exports from the former Soviet Union and Saudi Arabia were seen as possibly helping restore typical fuel oil discounts.
- The start of long-term or even permanent refinery closures will define a "new normal" for the industry, removing uncompetitive capacity.

OGJ subscribers can download, free of charge, the text version of the *OGJ Worldwide Refining Report* 2009 tables from www.ogjonline.com. Scroll down to "Additional Information," click "OGJ Subscriber Surveys," then "Worldwide Refining."

OGJ: Global refining survey

In its survey at yearend 2009, *OGJ* found that global crude distillation capacity in 2009 shot ahead of capacity for 2008.

Worldwide refining capacity set a record for the eighth year in a row.

The survey showed a total capacity of 87.2 million b/cd in 2009 for 661 refineries, an increase of more than 1.6 million b/cd over 2008. The previous year's refinery survey (as of January 1, 2009) listed a global capacity of 85.6 million b/cd in 655 refineries.

The 2009 capacity growth far surpassed the total growth for the previous 3 years.

HOW THE WORLD'S LARGEST REFINERS RANK

Rank: Jan. 1, 2010	Rank: Jan. 1, 2009	Company	Crude capacity, b/cd[1]
1	1	ExxonMobil Corp.	5,797,000
2	2	Royal Dutch Shell PLC	4,630,239
3	3	Sinopec	3,971,000
4	4	BP PLC	3,328,390
5	5	ConocoPhillips	2,778,200
6	8	Valero Energy Corp.	2,704,660
7	6	Petroleos de Venezuela SA	2,678,000
8	9	China National Petroleum Corp.	2,615,000
9	7	Total SA	2,594,608
10	12	Chevron Corp.[2]	2,492,000
11	10	Saudi Aramco	2,433,000
12	11	Petroleo Brasileiro SA	1,997,000
13	13	Petroleos Mexicanos	1,703,000
14	14	National Iranian Oil Co.	1,451,000
15	15	Nippon Oil Co. Ltd.	1,317,000
16	16	Rosneft	1,293,000
17	17	OAO Lukoil	1,217,000
18	20	Marathon Oil Corp.	1,196,000
19	18	Repsol YPF SA	1,105,000
20	19	Kuwait National Petroleum Co.	1,085,000
21	21	Pertamina	993,000
22	22	Agip Petroli SPA	904,000
23	23	Sunoco Inc.	825,000
24	25	SK Corp.	817,000
25	24	Flint Hills Resources	816,525

[1]Includes partial interests in refineries not wholly owned by the company.
[2]Includes holdings in Caltex Oil and GS Caltex.

Six new plants (net) were added during 2009, reversing a 10-year contraction that saw more than 100 refineries closed (mostly smaller, less-efficient ones).

Companies with 200,000+ b/cd refining capacity in Asia, the U.S., Western Europe

Rank	Company	No. of refineries	Crude capacity, b/cd[1]
Asia[2]			
1	Sinopec	27	3,971,000
2	China National Petroleum Corp.	25	2,615,000
3	ExxonMobil Corp.	10	1,937,500
4	Royal Dutch Shell PLC	13	1,324,875
5	Nippon Oil Co. Ltd.	7	1,317,000
6	Reliance Petroleum Ltd.	2	1,240,000
7	Pertamina	8	1,011,825
8	SK Corp.	1	817,000
9	Indian Oil Co. Ltd.	10	787,290
10	Chinese Petroleum Corp.	3	770,000
11	GS Caltex Corp.	1	[3]730,000
12	Tonen/General Sekiyu Seisei KK	4	[4]628,250
13	Idemitsu Kosan Co. Ltd.	4	608,000
14	Hyundai Oil Refinery Co.	3	589,500
15	Cosmo Oil Co. Ltd.	4	565,250
16	S-Oil Corp.	1	[5]565,000
17	Formosa Petrochemical Co.	1	540,000
18	Chevron Corp.	6	523,667
19	BP PLC	4	345,864
20	Saudi Aramco	6	327,652
21	Hindustan Petroleum Corp. Ltd.	2	296,250
U.S.			
1	ConocoPhillips	13	2,226,200
2	Valero Energy Corp.	13	2,209,660
3	ExxonMobil Corp.	7	2,065,000
4	BP PLC	6	1,477,525
5	Marathon Oil Corp.	7	1,196,000
6	Royal Dutch Shell PLC	8	[6]959,350
7	Chevron Corp.	5	941,000
8	Petroleos de Venezuela SA	4	[7]849,400
9	Sunoco Inc.	4	825,000
10	Flint Hills Resources (Koch Industries)	3	816,525
11	Motiva Enterprises LLC[8]	3	732,000
12	Tesoro Corp.	7	658,000
13	Saudi Aramco	3	[9]370,000
14	LyondellBasell	1	268,000
15	Alon USA	3	241,000
16	Husky Energy Inc.	2	237,500
17	EnCana Corp.	2	226,000
Western Europe[10]			
1	Total SA	15	2,264,587
2	ExxonMobil Corp.	9	1,665,500
3	Royal Dutch Shell PLC	11	1,549,940
4	AgipPetroli SPA	10	876,117
5	BP PLC	8	868,954
6	Petroplus International NV	7	796,000
7	Repsol YPF SA	5	709,200
8	Turkish Petroleum Refineries Corp.	4	613,275
9	ConocoPhillips	4	610,125
10	Compania Espanola de Petroles SA (CEPSA)	3	427,000
11	Ineos Group Holdings Inc.	2	402,800
12	OMV AG	3	398,635
13	ERG Group	4	396,214
14	Preem Raffinaderi AB	2	316,000
15	Hellenic Petroleum SA	3	313,000
16	Neste Oil	6	[11]309,000
17	Statoil AS	3	304,210
18	Galp Energia SA	2	304,172
19	Saras SPA	1	300,000
20	Petroleos de Venezuela SA	8	294,550
21	Chevron Corp.	1	210,000

[1]Includes partial interest in refineries not wholly owned by the company. [2]Asia includes Australia, Bangladesh, Brunei, China (and Taiwan), India, Indonesia, Japan, Malaysia, Myanmar, New Zealand, North Korea, Pakistan, Papua New Guinea, the Philippines, Singapore, South Korea, Sri Lanka, and Thailand. [3]Includes Caltex's 50% stake. [4]Includes ExxonMobil Corp.'s 50% stake. [5]Includes Saudi Aramco's 35% stake. [6]Includes Shell's stakes in Motiva and its 50% stake in the Deer Park, Tex., refinery. [7]Consists of PDVSA's ownership of Citgo and its 50% stake in the ExxonMobil Chalmette, La., refinery. [8]50/50 joint venture between Shell and Saudi Aramco. [9]Consists of 50% stake in Motiva. [10]Western Europe includes Austria, Belgium, Denmark, Finland, France, Germany, Greece, Ireland, Italy, the Netherlands, Norway, Portugal, Spain, Sweden, Switzerland, Turkey, and the UK. [11]Includes 50% stake in AB Nynas refineries.

US PROCESSING CAPABILITY*

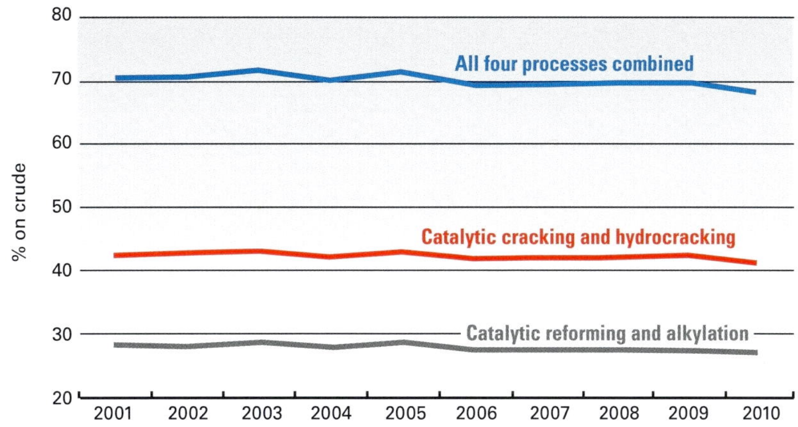

*As of Jan. 1 of each year.

EU PROCESSING CAPABILITY*

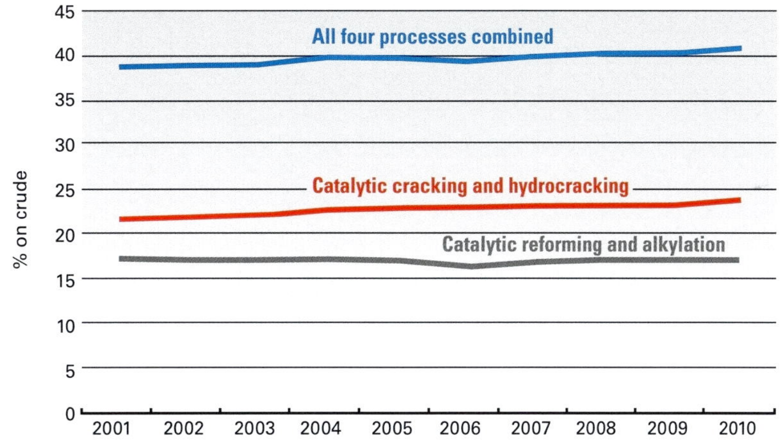

*As of Jan. 1 of each year.

ASIA PROCESSING CAPABILITY*

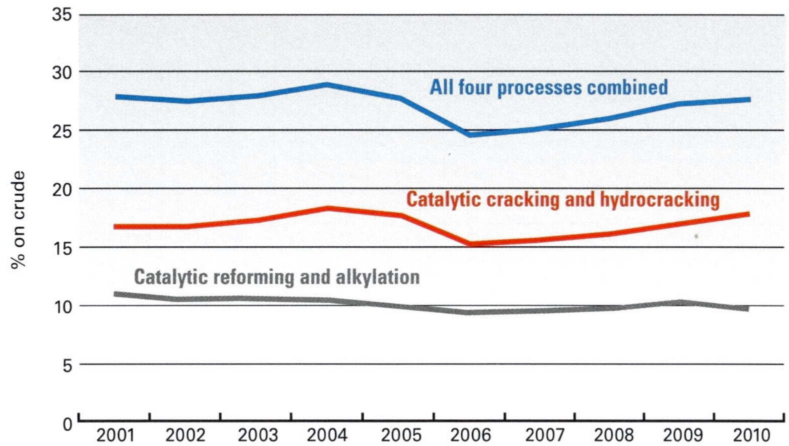

*As of Jan. 1 of each year.

Regional look at worldwide refining operations

Region	No. of refineries	Crude distillation (b/cd)	Vacuum distillation (b/cd)	Catalytic cracking (b/cd)	Catalytic reforming (b/cd)	Catalytic hydrocracking (b/cd)	Catalytic hydrotreating (b/cd)	Coke, tonnes/day
Africa	45	3,278,382	507,904	205,765	458,427	61,754	830,876	1,841
Asia	161	23,527,629	4,686,771	3,019,525	2,114,856	1,180,755	9,721,163	20,410
Eastern Europe	89	10,343,580	3,903,261	877,089	1,474,382	330,392	4,273,893	12,570
Middle East	44	7,245,365	1,982,275	357,550	654,397	596,891	2,038,063	3,300
North America	154	21,342,753	9,279,757	6,546,834	4,230,491	1,897,133	16,249,568	132,074
South America	66	6,571,153	2,846,885	1,309,507	402,095	132,400	1,904,061	24,640
Western Europe	102	14,914,517	5,817,689	2,255,960	2,186,185	1,179,706	9,977,420	12,709
Total	**661**	**87,223,380**	**29,024,542**	**14,572,230**	**11,520,833**	**5,379,031**	**44,995,044**	**207,544**

World's largest refineries

	Company	Location	Crude capacity, b/cd
1	Paraguana Refining Center	Cardon/Judibana, Falcon, Venezuela	940,000
2	SK Corp.	Ulsan, South Korea	817,000
3	GS Caltex Corp.	Yeosu, South Korea	730,000
4	Reliance Industries Ltd.	Jamnagar, India	660,000
5	ExxonMobil Refining & Supply Co.	Jurong/Pulau Ayer Chawan, Singapore	605,000
6	Reliance Industries Ltd.	Jamnagar, India	580,000
7	ExxonMobil Refining & Supply Co.	Baytown, Tex.	576,000
8	S-Oil Corp.	Onsan, South Korea	565,000
9	Saudi Arabian Oil Co. (Saudi Aramco)	Ras Tanura, Saudi Arabia	550,000
10	Formosa Petrochemical Co.	Mailiao, Taiwan	540,000
11	ExxonMobil Refining & Supply Co.	Baton Rouge, La.	504,500
12	Hovensa LLC	St. Croix, Virgin Islands	500,000
13	Kuwait National Petroleum Co.	Mina Al-Ahmadi, Kuwait	466,000
14	Shell Eastern Petroleum (Pte.) Ltd.	Pulau Bukom, Singapore	462,000
15	BP PLC	Texas City, Tex.	451,250
16	Citgo Petroleum Corp.	Lake Charles, La.	440,000
17	Marathon Petroleumn Co. LLC	Garyville, La.	436,000
18	Shell Nederland Raffinaderij BV	Pernis, Netherlands	404,000
19	Sinopec	Zhenhai, China	403,000
20	Saudi Arabian Oil Co. (Saudi Aramco)	Rabigh, Saudi Arabia	400,000
21	Saudi Aramco-Mobil	Yanbu, Saudi Arabia	400,000

IHS CERA downstream capital costs index

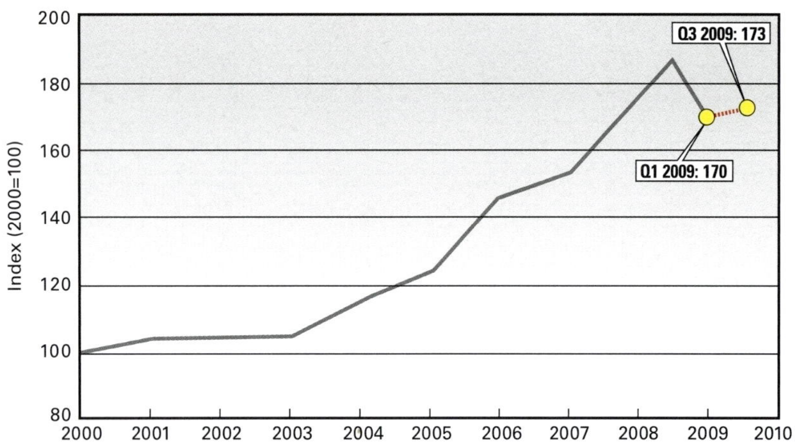

Source: IHS Cambridge Energy Research Associates

Capital costs trend

As of late 2009, costs of designing and building refineries and petrochemical plants ended a brief decline, according to IHS Cambridge Energy Research Associates.

The firm's downstream capital costs index, which assigns year-2000 costs a value of 100, rose to 172.5 for the 6 months ending in third quarter 2009, from 170 in the previous 6-month period.

The first-quarter index was down 9% from the 2008 peak.

Sulfur reduction

In its late-2009 evaluation of the sulfur sector, Purvin & Gertz, Inc. concluded that U.S. refiners had successfully managed a large number of product quality changes related to sulfur reduction while also processing higher-sulfur crude oils.

The firm sees more changes coming to U.S. refined-product sulfur qualities on the heels of gasoline and transportation diesel fuel improvements already implemented.

The largest impact will be in the marine sector, in which portions of traditional high-sulfur residual bunker will be replaced with distillate.

US Gulf Coast Sulfur

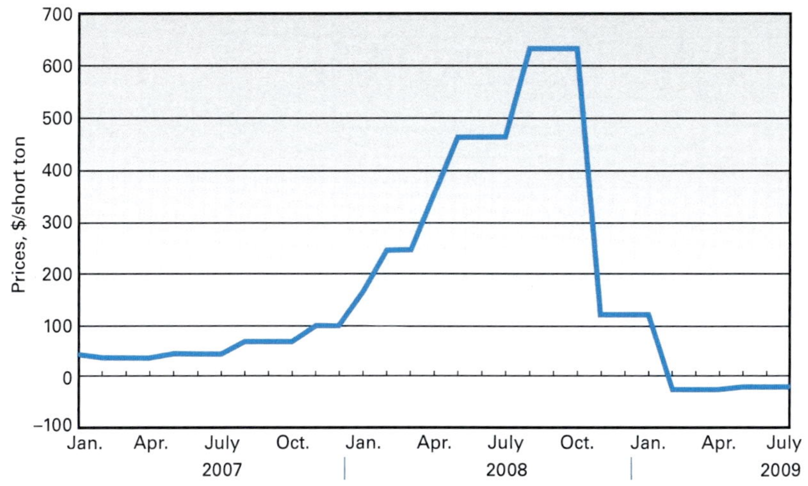

US Residual Fuel Market

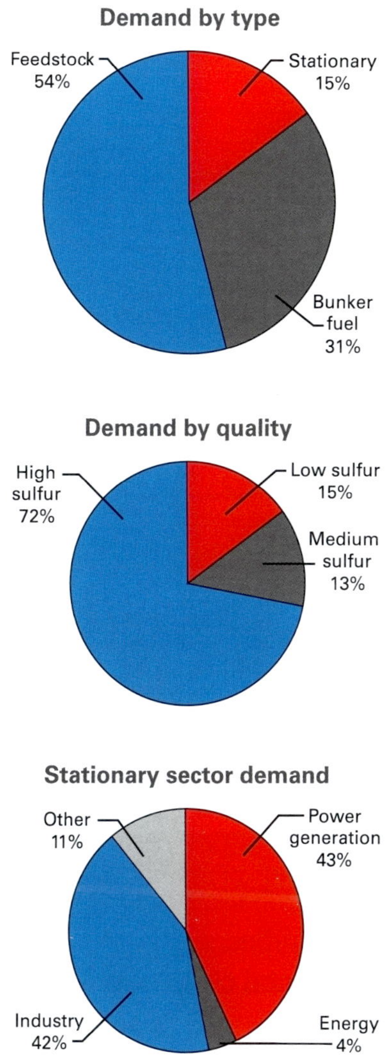

Global Crude Sulfur

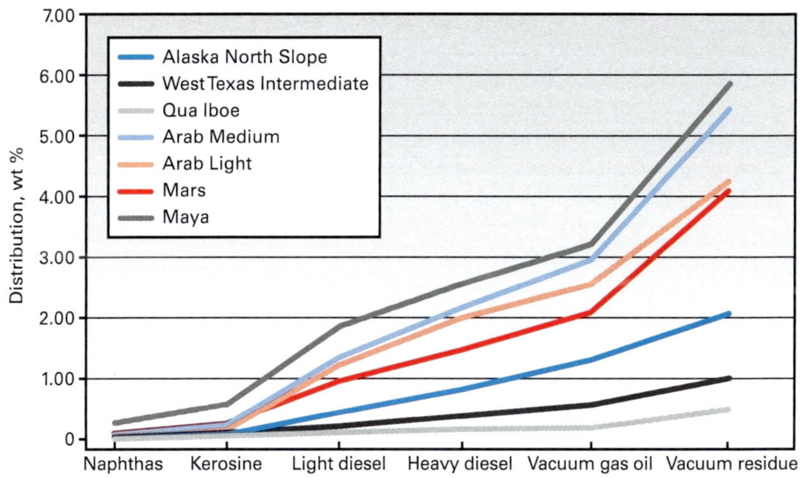

Diesel Sulfur

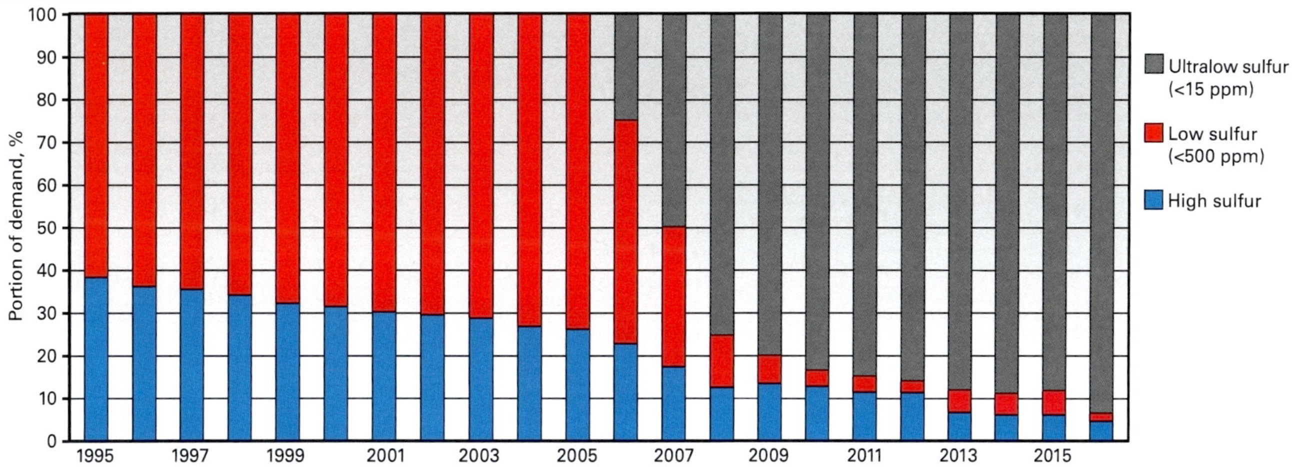

Sulfur in Crude

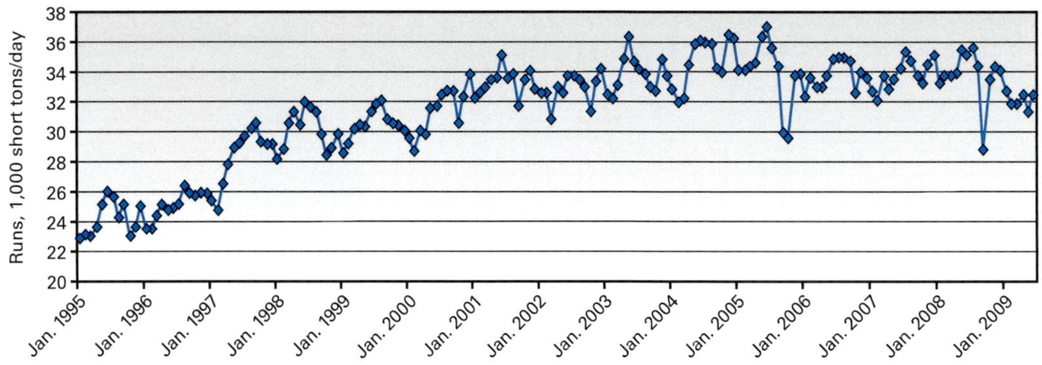

Sulfur in Marine Bunker Fuel

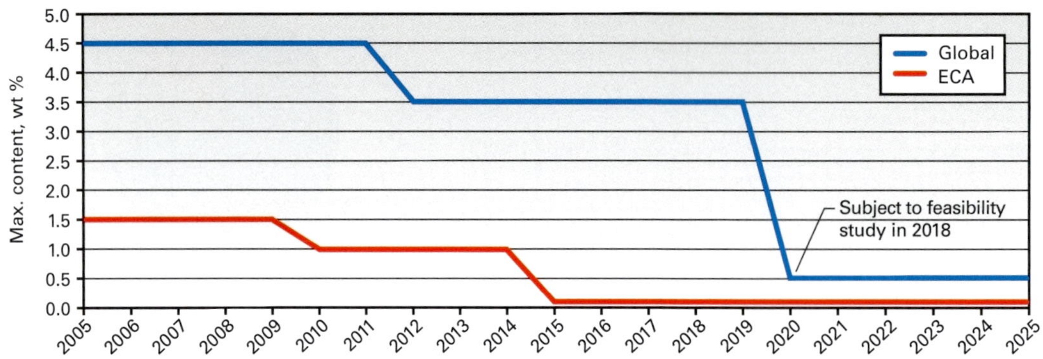

European refiners: Environmental challenges

In a mid-2009 look at refining in Europe, Purvin & Gertz noted that refiners there face new challenges driven by environmental concerns.

[The study defined Europe as 38 countries to the west of Belorussia, Moldova, and Ukraine, most of which (27) are members of the European Union.]

Main challenges include:

- Adapting to changing demand patterns, both in Europe and in its trading partners.
- Participating in European efforts to reduce greenhouse-gas emissions.
- Adjusting to the effects of future product-quality legislation.

European Refinery Ownership Profile

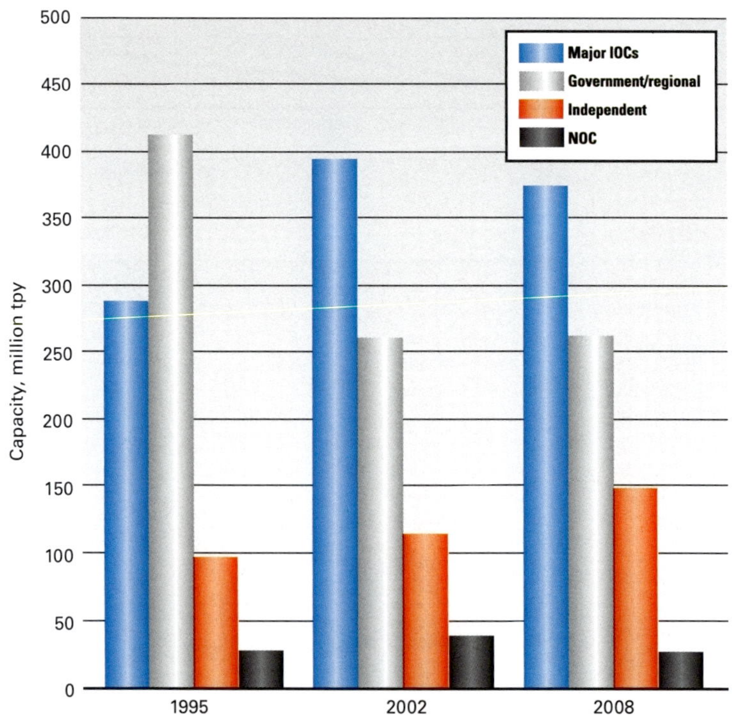

European Refinery Crude Distillation Profile

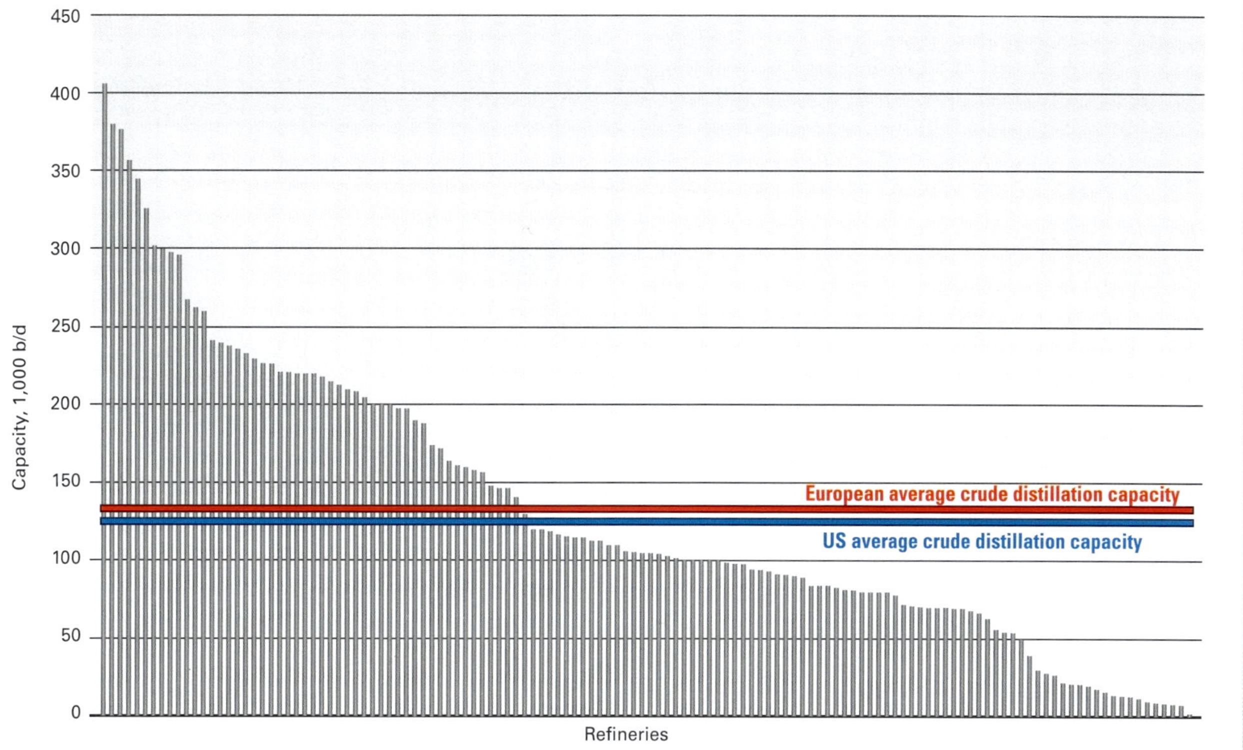

Europe Net Trade: 2007 and 2020

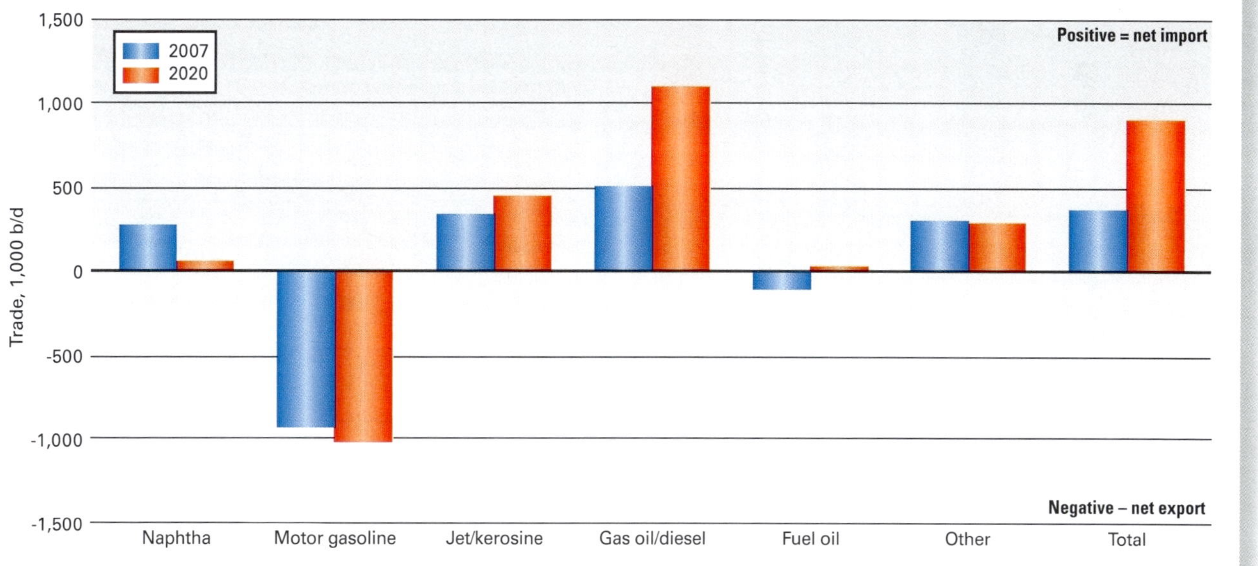

European demand, production

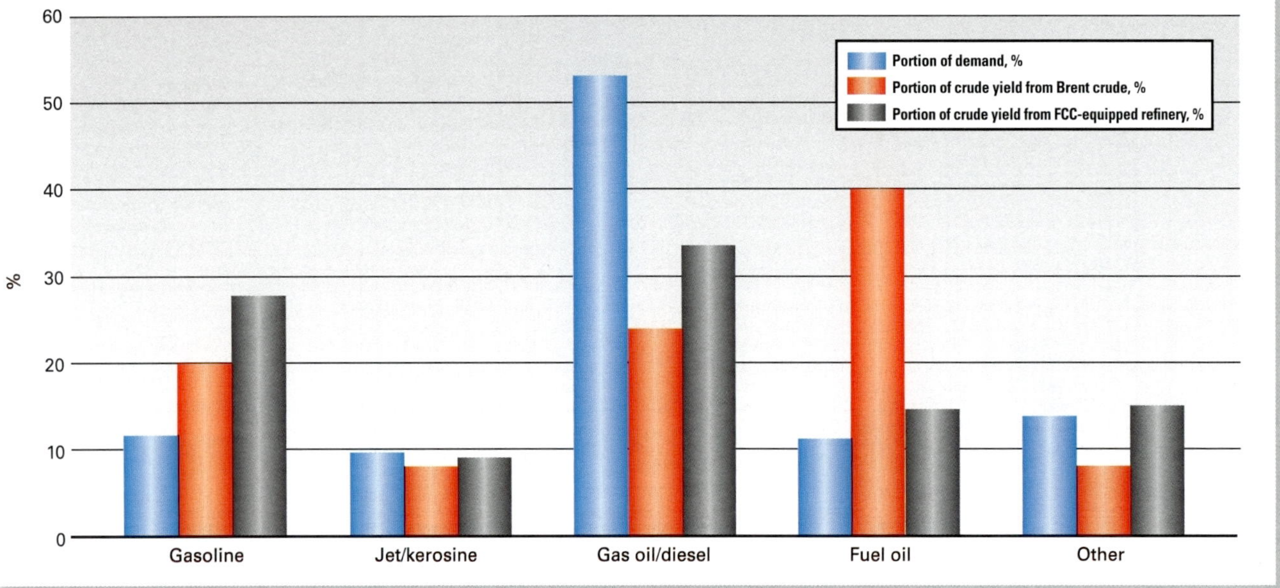

Europe demand change, 2007-20

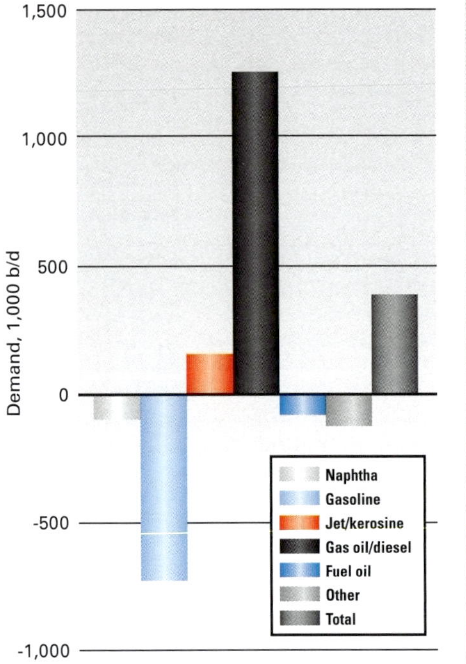

CO_2 price, impact on refining costs

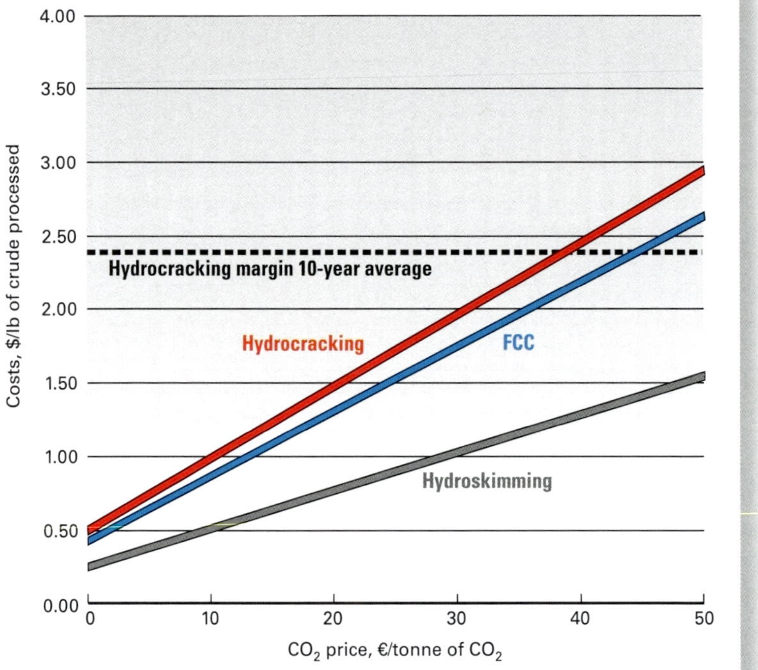

European challenges: *OGJ* view

An *OGJ* review of European refining in mid-2009 echoed the Purvin & Gertz view of a challenging year due to drastic changes in world oil prices.

With crude prices returning to the $60–70/bbl range at midyear, falling margins for fuel producers, high operating costs, falling demand, and a supply glut in new capacity were major problems.

A string of upgrades and conversion projects were postponed or canceled due to the global recession and the tightening of credit availability.

Worldwide Incremental Topping, Cracking Capacities

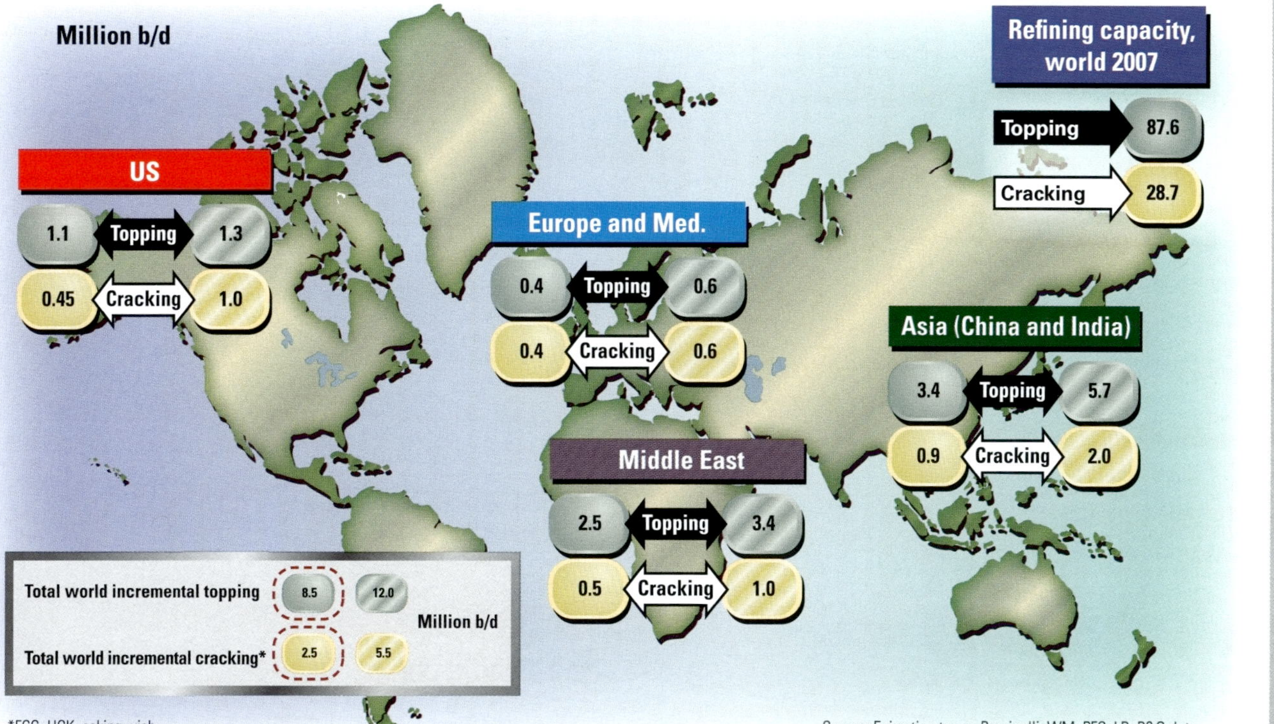

*FCC, HCK, coking, visb.

Source: Eni estimates on Parpinelli, WM, PFC, LB, P&G data

Note: Figures on left refer to projects expected to be realized before 2015. Figures on right refer to projects announced with probability to be realized after 2015 and before 2020.

Fuel Demand Trends in Europe, US

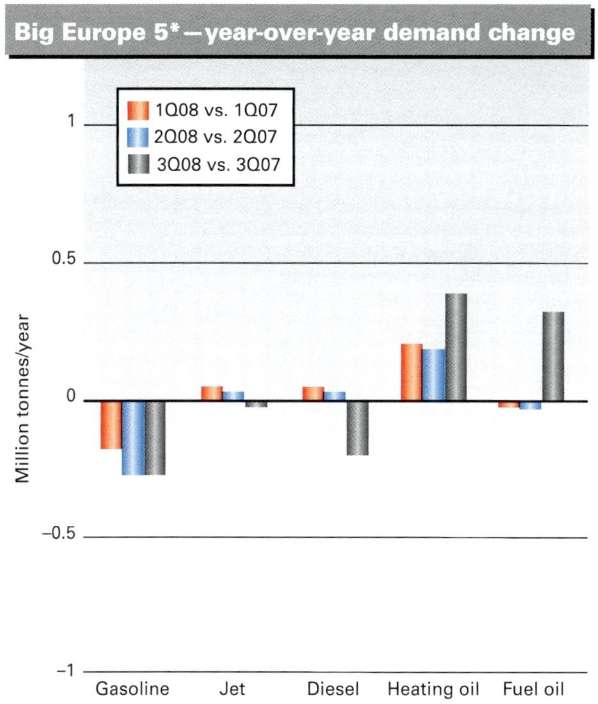

*France, Italy, Germany, Spain, and UK.

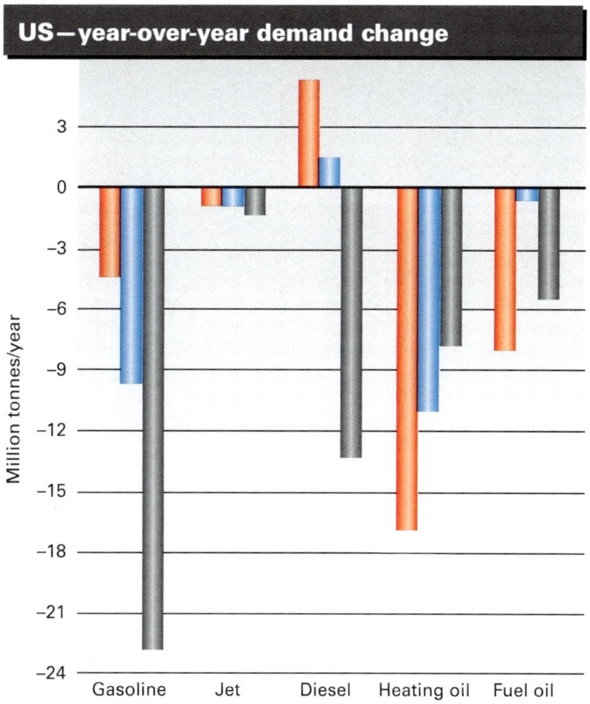

Source: Eni SPA after PFC

Gasoline plus cellulosic ethanol

Royal Dutch Shell PLC in June 2009 conducted a month-long demonstration, selling gasoline blended with 10% cellulosic ethanol at a Shell service station in Ottawa, ON. The ethanol was made from wheat straw by Iogen Energy Corp.

Shell also was working with Germany's Choren to launch a new 15,000-tpy biofuels plant that will convert biomass, such as woodchips, into synthetic fuel.

India Product Balance

2009	Ouput	Imports	Exports	Demand
LPG[1]	330	64	—	395
Naphtha[1]	406	100	195	311
Gasoline[2]	514	—	248	266
Kero/jet	363	25	98	290
Diesel[2]	1,423	20	373	1,069
Fuel oil	299	18	81	236
Other	295	1	—	296
Total	3,629	229	995	2,863
Dubai Crude	—	—	—	220

[1]Nonrefinery LPG: 75 Naphtha: 37. [2]Ethanol: 34 Biodiesel: 32.

2010	Output	Imports	Exports	Demand
LPG[1]	351	62	—	412
Naphtha[1]	435	85	186	334
Gasoline[2]	601	9	332	278
Kero/jet	403	30	154	279
Diesel[2]	1,593	15	508	1,099
Fuel oil	327	10	108	229
Other	319	—	11	308
Total	4,029	211	1,300	2,940
Dubai crude	0	—	—	0

[1]Nonrefinery LPG: 80 Naphtha: 37. [2]Ethanol: 46 Biodiesel: 68.

China Product Balance

2009	Ouput	Imports	Exports	Demand
LPG[1]	633	54	15	662
Naphtha[1]	672	33	25	665
Gasoline[2]	1,561	25	81	1,495
Kero/jet	286	99	95	290
Diesel[2]	2,850	30	53	2,842
Fuel oil	474	259	85	648
Other	817	75	45	847
Total	7,292	575	399	7,448
Dubai crude	220	—	—	220

[1]Nonrefinery LPG: — Naphtha: —. [2]Ethanol: 43 Biodiesel: 19.

2010	Output	Imports	Exports	Demand
LPG[1]	679	25	15	689
Naphtha[1]	755	128	20	863
Gasoline[2]	1,669	5	121	1,553
Kero/jet	301	93	90	305
Diesel[2]	3,007	6	48	2,965
Fuel oil	477	207	25	659
Other	839	74	40	873
Total	7,728	538	359	7,907
Dubai crude	223	—	—	223

[1]Nonrefinery LPG: Naphtha. [2]Ethanol: 67 Biodiesel: 32.

Asia-Pacific perspective

The East-West Center and FACTS Global Energy (Singapore) examined the effect of the global economic downturn in mid-2009 on the oil industry in the Asia-Pacific region.

Notably, slowing economies in the West reduced demand for exports from the Asia-Pacific, which in turn reduced manufacturing output, investment, and economic growth.

Asia-Pacific Demand, Distillation Additions

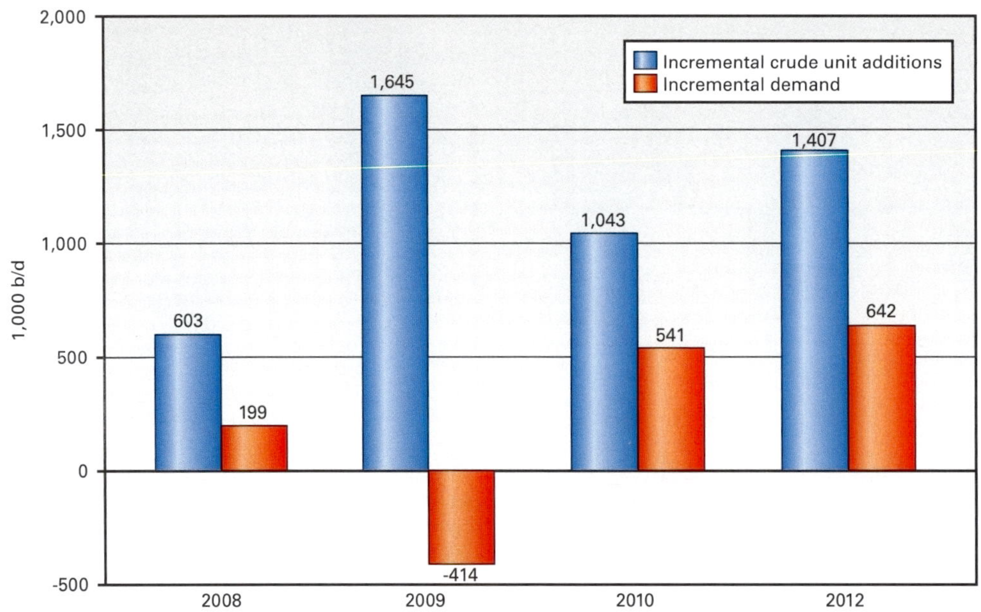

The researchers estimated that Asia-Pacific demand for oil products would decrease by 414,000 b/d in 2009. This could pose problems because the region was poised to add 2.7 million b/d of refining capacity over the 2009–2010 period.

It is expected that refiners will have to take actions such as running refineries at much lower throughputs, delaying construction plans, or even closing existing refineries.

On average, refinery utilization in 2009 and 2010 is expected to fall to levels last seen in 2002.

Asia-Pacific Refinery Utilization, Crack Margins

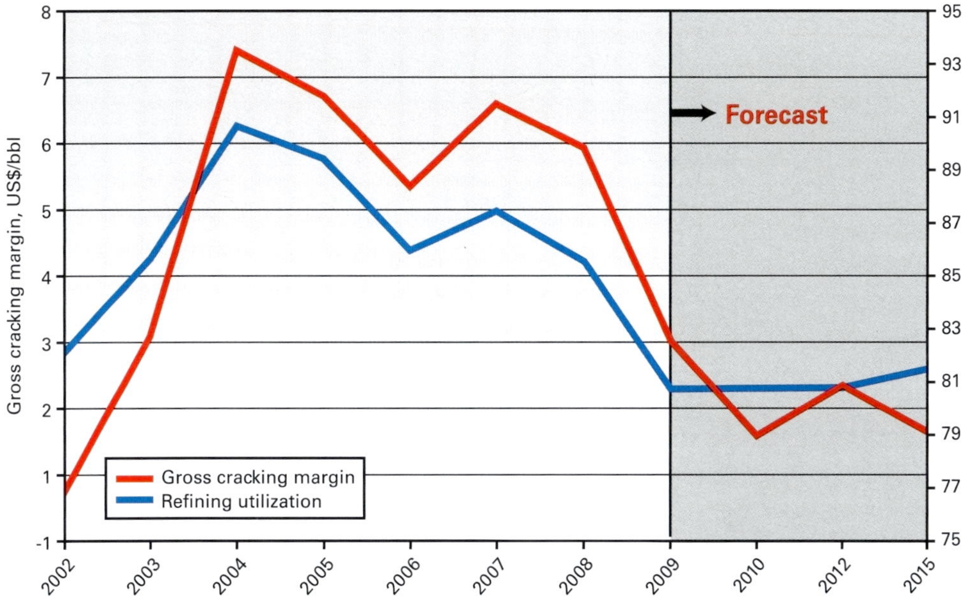

Refinery Utilization in Select Countries

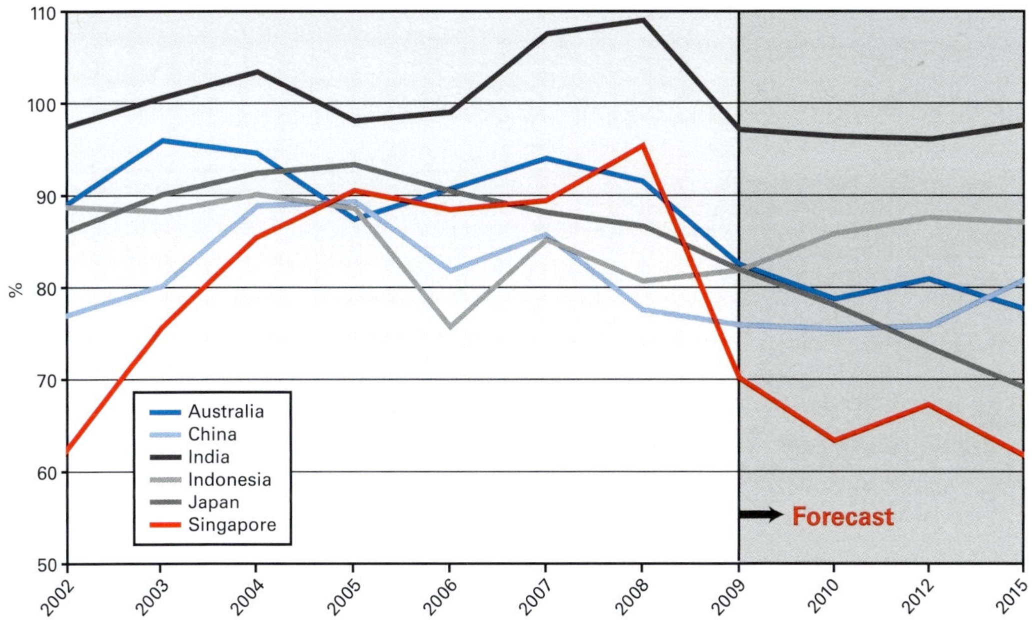

Section 11:

Gas processing and products

Processing U.S. shale gas

Researchers at Bryan Research and Engineering Inc. noted in early 2009 that gas processing requirements for U.S. gas shales can vary regionally because not all such gas has the same composition.

Fayetteville shale gas composition

Well	C_1	C_2	C_3	CO_2	N_2
Avg.	97.3	1.0	0	1.0	0.7

Fayetteville shale

Barnett shale gas composition*

Well	C_1	C_2	C_3	CO_2	N_2
1	80.3	8.1	2.3	1.4	7.9
2	81.2	11.8	5.2	0.3	1.5
3	91.8	4.4	0.4	2.3	1.1
4	93.7	2.6	0.0	2.7	1.0

*Normalized to the reported compounds. Adapted from Hill, Ronald J.; Jarvie, Daniel M.; Zumberge, John; Henry, Mitchell; and Pollastro, Richard M., "Oil and gas geochemistry and petroleum systems of the Fort Worth Basin," AAPG Bulletin, Vol. 91, No. 4 (April 2007), pp. 445-473.

Texas Barnett shale

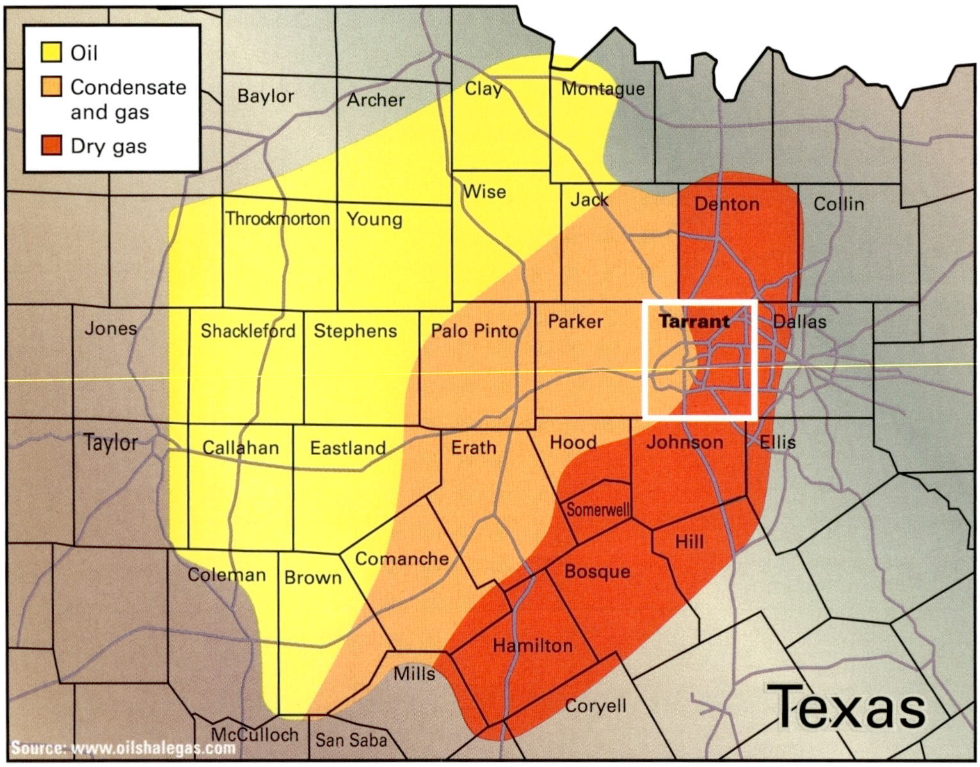

New Albany Shale

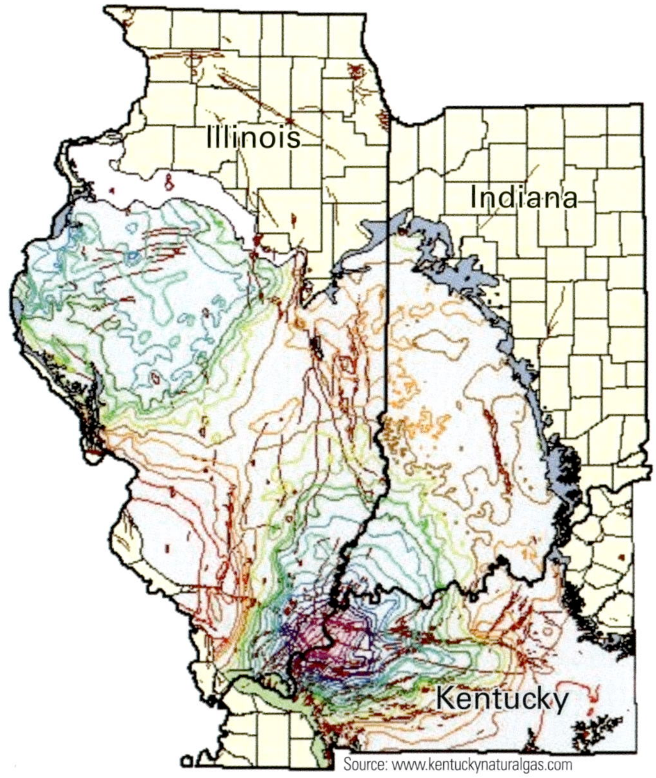

Source: www.kentuckynaturalgas.com

New Albany Shale Gas Composition*

Well	C_1	C_2	C_3	CO_2
1	87.7	1.7	2.5	8.1
2	88.0	0.8	0.8	10.4
3	91.0	1.0	0.6	7.4
4	92.8	1.0	0.6	5.6

*Compositions normalized to reported compounds; nitrogen content was not reported. Adapted from Martini, Anna M.; Walter, Lynn M.; and McIntosh, Jennifer C., "Identification of microbial and thermogenic gas components from Upper Devonian black shale cores, Illinois and Michigan basins," AAPG Bulletin, Vol. 92, No. 3 (March 2008), pp. 327–339.

Processors must be concerned about elevated ethane and nitrogen levels, the requirements of urban gas processing, and other special gas handling requirements.

BR&E examined six shale plays, their compositions, and processing needs.

Marcellus Shale Gas Composition

Well	C_1	C_2	C_3	CO_2	N_2
11	79.4	16.1	4.0	0.1	0.4
2	82.1	14.0	3.5	0.1	0.3
3	83.8	12.0	3.0	0.9	0.3
4	95.5	3.0	1.0	0.3	0.2

Marcellus Shale Fairway

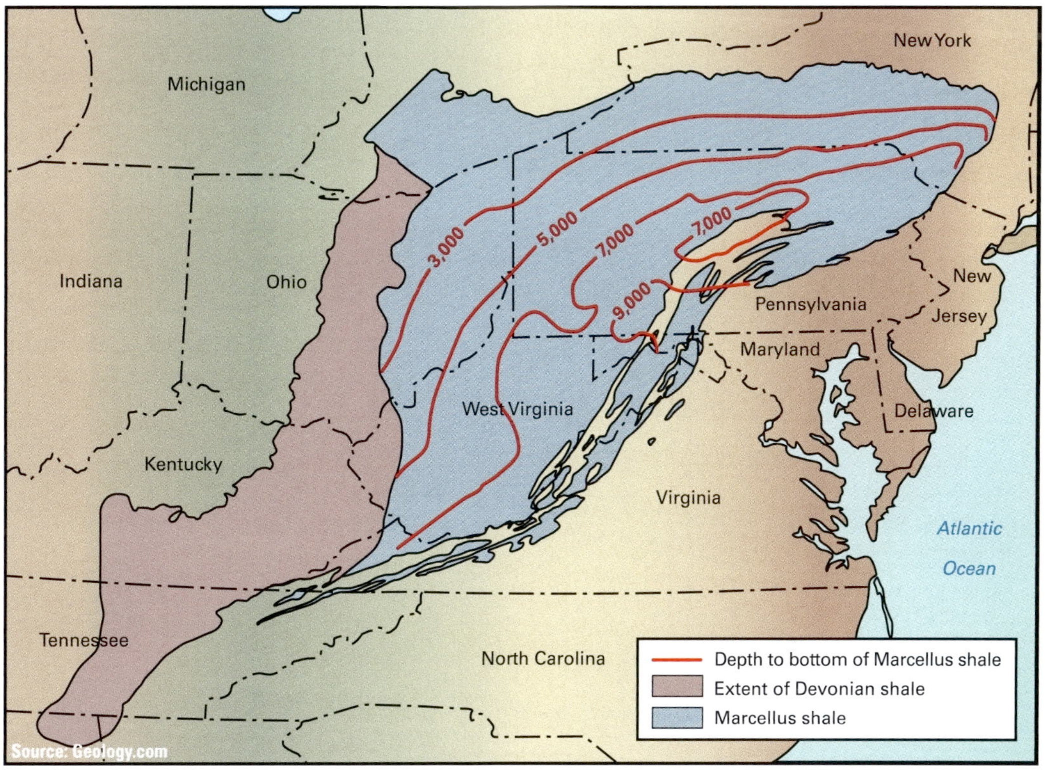

Haynesville area

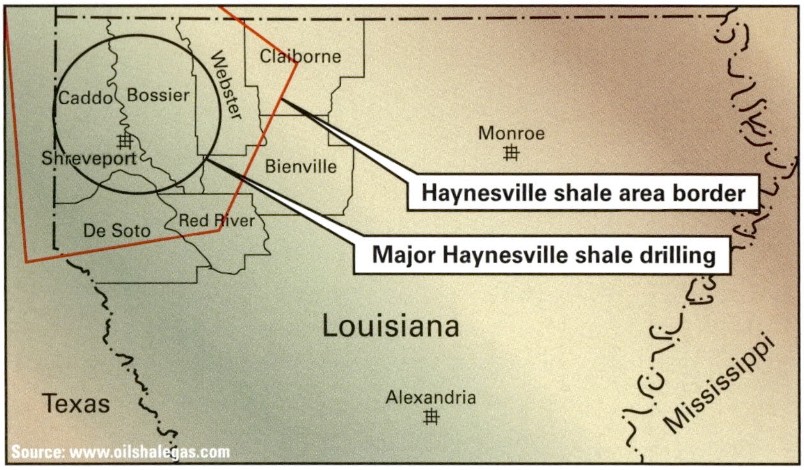

Haynesville shale gas composition

Well	C_1	C_2	C_3	CO_2	N_2
Avg.	95.0	0.1	0	4.8	0.1

Antrim shale

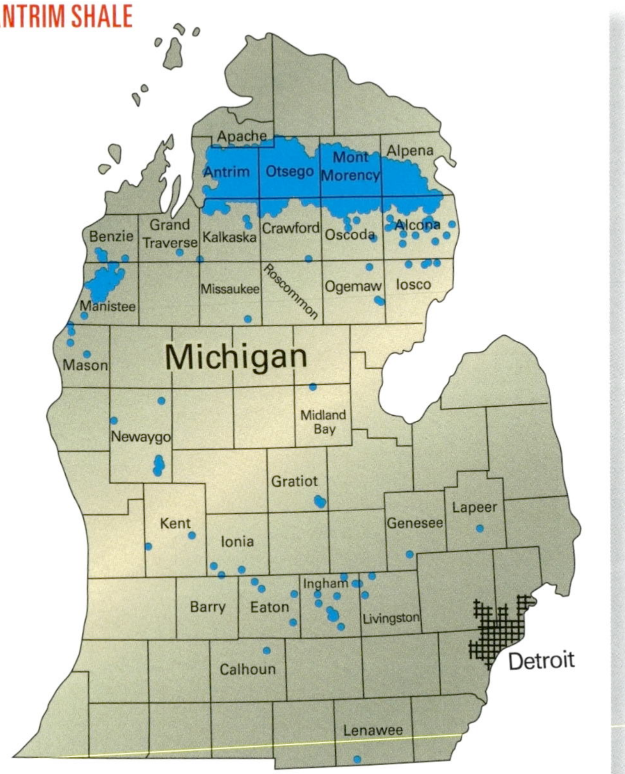

Antrim shale gas composition*

Well	C_1	C_2	C_3	CO_2	N_2
1	27.5	3.5	1.0	3.0	65.0
2	57.3	4.9	1.9	0	35.9
3	77.5	4.0	0.9	3.3	14.3
4	85.6	4.3	0.4	9.0	0.7

*Normalized to the reported compounds. Adapted from Martini, Anna M.; Walter, Lynn M.; Ku, Tim C.W.; Budai, Joyce M.; McIntosh, Jennifer C.; and Schoell, Martin, "Microbial production and modification of gases in sedimentary basins: A geochemical case study from a Devonian shale gas play, Michigan basin," AAPG Bulletin, Vol. 87, No. 8 (August 2003), pp. 1,355-375.

NGL/condensate plans: Middle East

An IEA report in July 2009 indicated that NGL and condensate projects slated for start-up in 2009–2011 in some OPEC member nations will add 2 million b/d to capacity at peak output.

Saudi Arabia, Qatar, Iran, and the UAE will provide 90% of the increase. Notable projects include:

- Saudi Arabia: nonassociated Hawiyah NGL recovery project (300,000 b/d by 2011).
- Saudi Arabia: Khursaniyah gas processing plant (290,000 b/d by yearend 2009).
- Saudi Arabia: Khurais oil field development (adding 70,000 b/d of condensate capacity in 2009).
- Qatar: six projects at North field (adding 545,000 b/d of gas liquids capacity by yearend 2010).
- Iran: completion of Phases 6–10 of South Pars project (adding 245,000 b/d of condensate and NGL capacity in 2009 and 2010).
- UAE: start-up of Habshan OGD3 processing (adding 340,000 b/d of capacity by yearend 2010).

International LPG market

In mid-2009, Purvin & Gertz reported on 2008 performance of the global LPG market; on the impact of the global economic recession, new leadership in the United States (the world's largest LPG producer), and the collapse of crude prices; and on the market outlook to 2012.

World LPG Supply

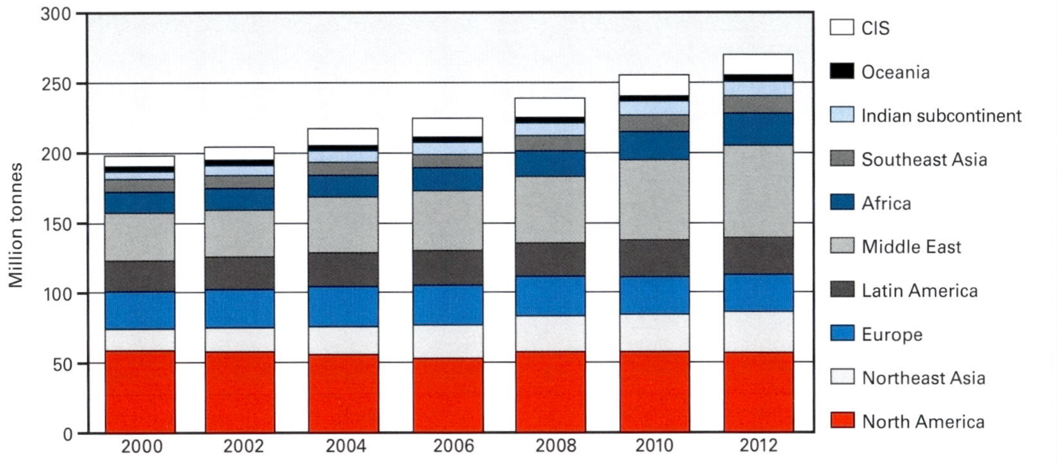

World LPG Demand

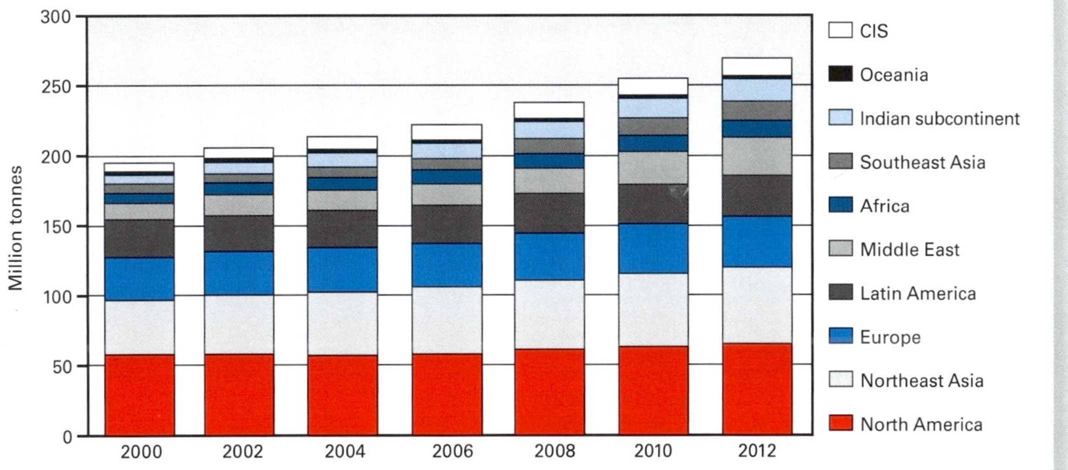

LPG Demand, Sectors

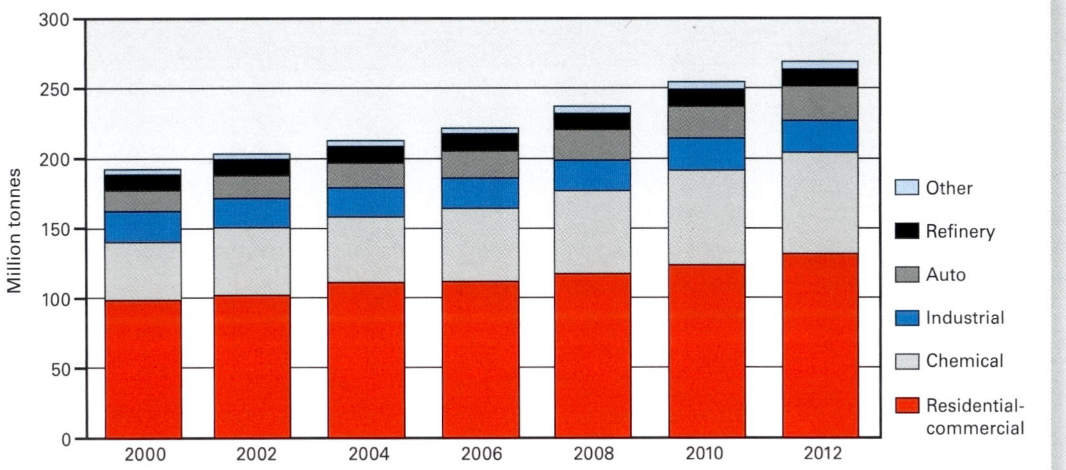

Regional LPG trade

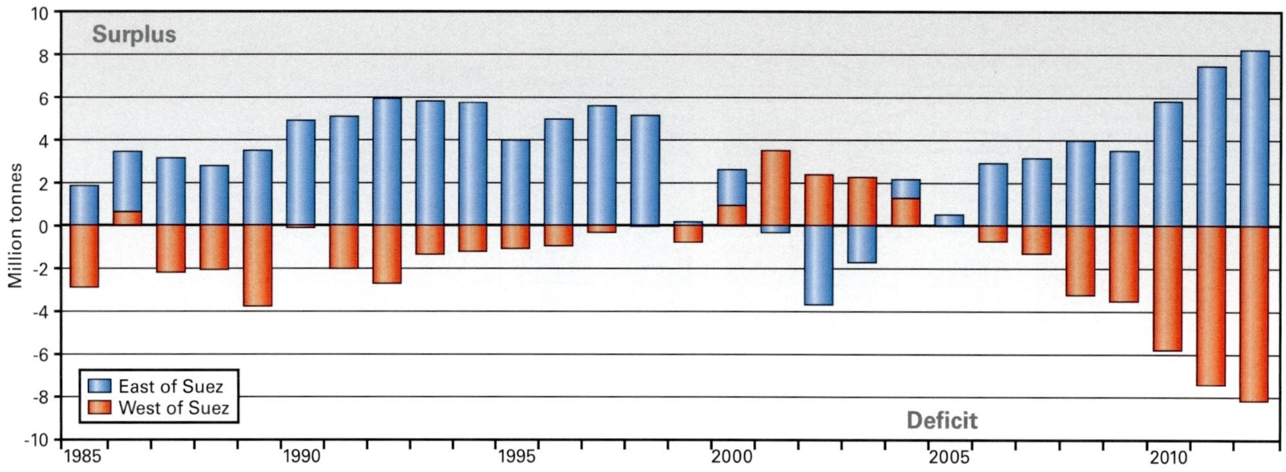

Global gas processing: 2008

In mid-2009, *OGJ* looked back at global gas processing activity in 2008.

That activity level remained flat despite the fact that natural gas production rose in every region of the world, spurred by rising gas prices. Global gas production growth was nearly 7%.

The Middle East led the way with more than 18% [1.8 tcf (51 bcm)] growth. Countries of the Western Hemisphere produced more than 1.6 tcf (45.3 bcm) of additional gas (4.8% growth) over 2007.

World's 20 top gas producing countries: 2008

Country	Production, bcf	Share, %
Russia/FSU	28,490.0	26.4
U.S.	21,453.0	19.9
Canada	5,602.2	5.2
Iran	3,515.0	3.3
Norway	3,503.9	3.3
Algeria	3,300.0	3.1
Netherlands	3,030.0	2.8
China	2,828.7	2.6
Indonesia	2,730.0	2.5
UK	2,594.1	2.4
Saudi Arabia	2,570.0	2.4
Mexico	2,532.4	2.3
Qatar	2,170.0	2.0
Malaysia	1,725.0	1.6
Egypt	1,610.0	1.5
UAE	1,565.0	1.5
Argentina	1,491.0	1.4
Pakistan	1,463.0	1.4
Trinidad	1,379.7	1.3
Australia	1,350.8	1.3
Top 20 total	**94,903.8**	**88.1**
Other	12,873.6	11.9
Total	**107,777.4**	**100.0**

Source: *OGJ*, Mar. 9, 2009, p. 64

Worldwide gas processing roundup

Region	[1]2007	[2]2008	Change	Change, %
U.S.				
Gas capacity, MMcfd	71,062.7	72,342.7	1,280.0	1.8
Gas throughput, MMcfd	45,501.7	45,897.2	395.5	0.9
NGL production, 1,000 gpd	76,227.6	75,757.6	–470.0	–0.6
Proved reserves, bcf	211,085.0	237,726.0	26,641.0	12.6
CANADA				
Gas capacity, MMcfd	53,010.0	53,010.0	—	—
Gas throughput, MMcfd	29,946.6	29,230.2	–716.4	–2.4
NGL production, 1,000 gpd	32,521.8	32,524.8	3.0	—
Proved reserves, bcf	58,200.0	57,906.0	–294.0	–0.5
WESTERN EUROPE				
Gas capacity, MMcfd	24,495.0	24,495.0	—	—
Gas throughput, MMcfd	9,501.2	9,501.2	—	—
NGL production, 1,000 gpd	9,689.6	9,689.6	—	—
Proved reserves, bcf	160,132.0	157,105.0	–3,027.0	–1.9
EASTERN EUROPE				
Gas capacity, MMcfd	2,796.0	2,796.0	—	—
Gas throughput, MMcfd	1,426.5	1,426.5	—	—
NGL production, 1,000 gpd	10,142.4	10,142.4	—	—
Proved reserves, bcf	2,026,709.0	2,005,781.0	–20,928.0	–1.0
LATIN AMERICA				
Gas capacity, MMcfd	18,448.2	18,448.2	—	—
Gas throughput, MMcfd	12,427.3	12,343.3	–84.0	–0.7
NGL production, 1,000 gpd	35,929.9	33,991.9	–1,938.0	–5.4
Proved reserves, bcf	273,140.0	277,198.0	4,058.0	1.5
MIDDLE EAST				
Gas capacity, MMcfd	39,264.2	39,264.2	—	—
Gas throughput, MMcfd	27,318.0	27,318.0	—	—
NGL production, 1,000 gpd	86,323.1	86,323.1	—	—
Proved reserves, bcf	2,548,900.0	2,591,653.0	42,753.0	1.7
AFRICA				
Gas capacity, MMcfd	17,254.2	17,407.2	153.0	0.9
Gas throughput, MMcfd	9,642.4	9,642.4	—	—
NGL production, 1,000 gpd	16,777.3	16,684.0	–93.3	–0.6
Proved reserves, bcf	489,630.0	494,078.0	4,448.0	0.9
ASIA-PACIFIC				
Gas capacity, MMcfd	24,960.2	24,960.2	—	—
Gas throughput, MMcfd	20,290.8	20,227.8	–63.0	–0.3
NGL production, 1,000 gpd	25,799.1	25,593.1	–206.0	–0.8
Proved reserves, bcf	415,393.0	430,412.0	15,019.0	3.6
TOTAL-excl. U.S.				
Gas capacity, MMcfd	180,227.8	180,380.8	153.0	0.1
Gas throughput, MMcfd	110,552.8	109,689.4	–863.4	–0.8
NGL production, 1,000 gpd	217,183.2	214,949.0	–2,234.3	–1.0
Proved reserves, bcf	5,972,104.0	6,014,133.0	42,029.0	0.7
TOTAL-incl. U.S.				
Gas capacity, MMcfd	251,290.5	252,723.5	1,433.0	0.6
Gas throughput, MMcfd	156,054.5	155,586.5	–468.0	–0.3
NGL production, 1,000 gpd	293,410.8	290,706.5	–2,704.3	–0.9
Proved reserves, bcf	6,183,189.0	6,251,859.0	68,670.0	1.1

[1]Proved gas reserve totals for 2006 are as of Jan.1, 2008 (*OGJ*, Dec. 24, 2008, p. 24). [2]Proved gas reserve totals for 2008 are as of Jan.1, 2009 (*OGJ*, Dec. 22, 2008, p. 22).

World's Major Gas Reserves

Country	Estimated proved reserves, tcf	Share, %
Russia	1,680.0	26.9
Iran	991.6	15.9
Qatar	891.9	14.3
Saudi Arabia	258.0	4.1
U.S.	237.7	3.8
Abu Dhabi	198.5	3.2
Nigeria	184.2	2.9
Venezuela	170.9	2.7
Algeria	159.0	2.5
Iraq	112.0	1.8
Kazakhstan	85.0	1.4
Turkmenistan	94.0	1.5
Indonesia	106.0	1.7
Malaysia	83.0	1.3
China	80.0	1.3
Norway	81.7	1.3
Uzbekistan	65.0	1.0
Egypt	58.5	0.9
Canada	57.9	0.9
Kuwait	62.9	1.0
Libya	54.4	0.9
Netherlands	50.0	0.8
Ukraine	39.0	0.6
Australia	30.0	0.5
Oman	30.0	0.5
Argentina	15.6	0.2
UK	12.1	0.2
Subtotal	**5,888.9**	**94.2**
Others	365.5	5.8
Total	**6,254.4**	**100.0**

Source: OGJ, Dec. 22, 2008, p. 22

World's Top 10 NGL Producing Countries: 2008

Country	Production, bcf	Share, %
U.S.	75,757.6	26.0
Saudi Arabia	37,044.0	12.7
Canada	32,524.8	11.2
Mexico	19,711.0	6.8
Kuwait	18,684.8	6.4
Iran	12,917.5	4.4
Australia	10,733.3	3.7
Russia/FSU	9,127.9	3.1
UAE	8,978.6	3.1
Venezuela	7,312.9	2.5
Top 10 subtotal	**232,792.4**	**80.0**
Rest of world	58,234.1	20.0
Total worldwide	**291,026.5**	**100.0**

US Gas Processing Trends

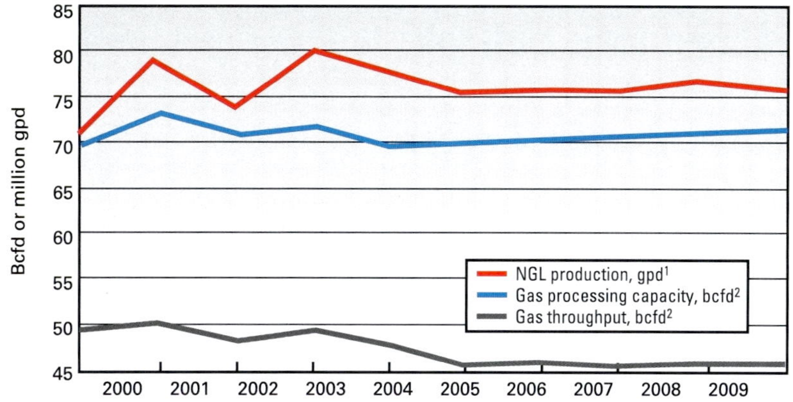

[1]Based on a 12-month average for previous year. [2]As of Jan. 1 of each year.

US LPG Differential, Price Trends: 2008

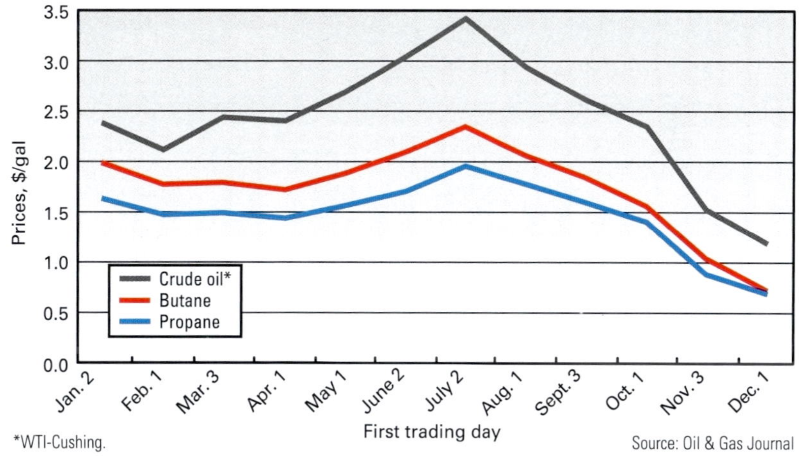

*WTI-Cushing.

Source: Oil & Gas Journal

Section 12:
Petrochemicals outlook

U.S. propane: Late 2008, early 2009

Petral Worldwide Inc. reported that the financial crisis and ensuing economic recession had substantial and overriding impact on U.S. propane supply-demand trends during winter 2008–2009.

The collapse of crude prices in fourth quarter 2008 prompted OPEC producers (notably Saudi Arabia) to make major cuts in crude production. In turn, those cuts reduced associated-gas production and gas plant propane-butane recovery volumes.

U.S. PROPANE INVENTORIES

Month	PADD 2	PADD 3	Other regions	Total
	Million bbl			
July 2008	21.3	20.5	5.78	47.5
August	24.6	22.6	7.05	54.2
September	27.8	24.5	6.88	59.2
October	29.9	22.7	6.91	59.5
November	32.6	21.6	6.56	60.7
December	31.3	18.4	5.66	55.4
January 2009	26.4	13.6	4.59	44.6
February*	21.0	9.2	3.94	34.2
March*	16.8	8.4	3.80	29.0

*Based on weekly survey estimates.

At the end of September 2008, the United States had a propane inventory deficit of more than 10 million bbl relative to 2006.

Despite this deficit and strong retail demand, inventory levels ended the winter heating season with a year-to-year gain of about 8 million bbl.

US ETHYLENE FEEDSTOCK DEMAND FOR PROPANE

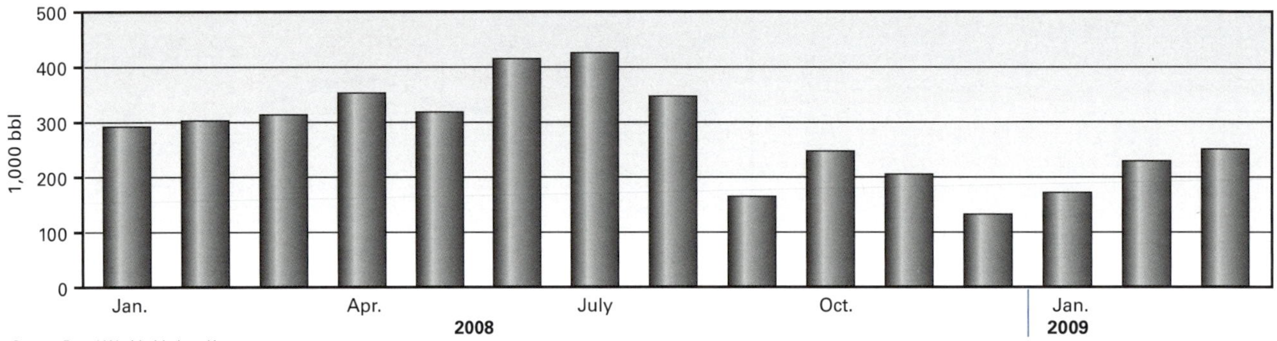

Source: Petral Worldwide Inc., Houston

US GAS-PLANT PROPANE PRODUCTION

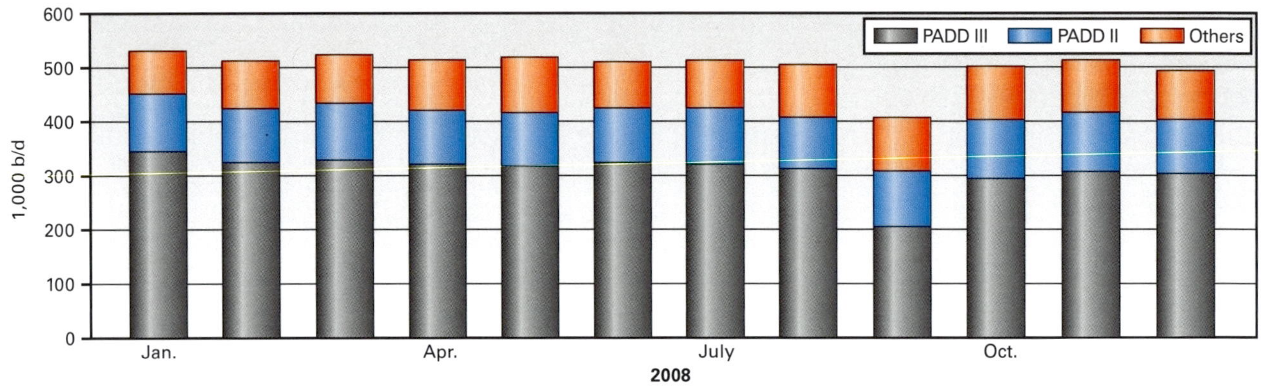

Source: US Energy Information Administration

Total US Propane Production

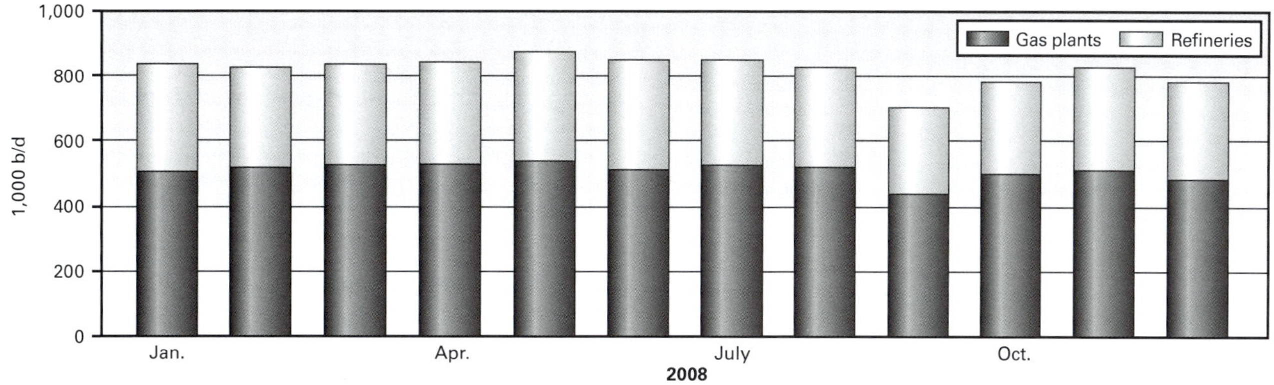

Source: US Energy Information Administration

US Propane Inventory

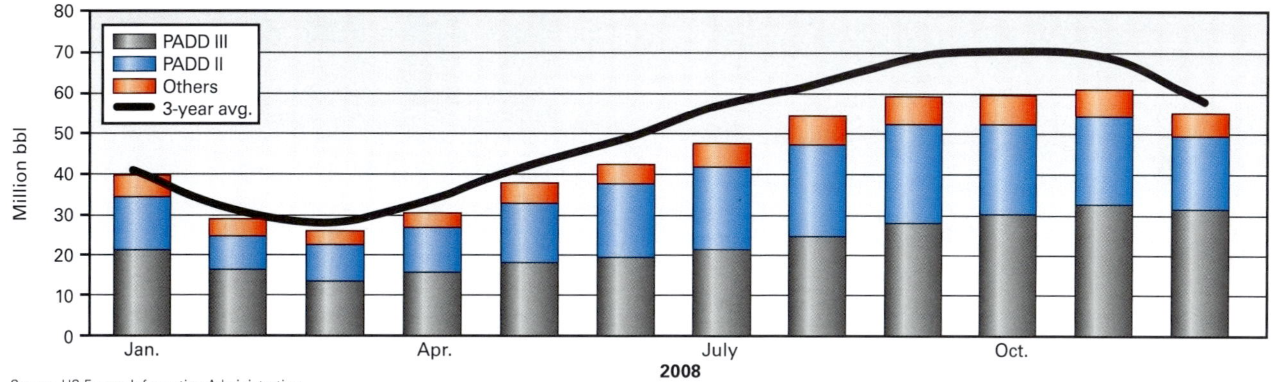

Source: US Energy Information Administration

Ethylene production: 2008

In its mid-2009 survey, *OGJ* reviewed the 2008 global ethylene market, noting that worldwide, producers added more than 7 million tons/yr (mtpy) of capacity.

Primarily due to the start-up of new plants in the Middle East, that growth represented record additions. Capacity grew by only 2 mtpy in 2007 and by 245,000 tpy in 2006.

Global ethylene production capacity as of January 1, 2009, was 126.7 mtpy, an increase from 119.6 mtpy of capacity reported in the prior year's survey. The additional capacity represents an increase of nearly 6%.

The latest survey lists five new world-scale ethylene production plants (combined capacity nearly 6 mtpy), and new trains in two existing plants. No plant shutdowns were reported.

According to *OGJ* construction data, almost 11 mtpy of new capacity was slated to come on stream in 2009.

Top 10 Ethylene Complexes*

	Company	Location	Capacity, tpy
1	Nova Chemicals Corp.	Joffre, Alta.	2,812,000
2	Formosa Petrochemical Corp.	Mailiao, Taiwan, China	2,550,000
3	Arabian Petrochemical Co.	Jubail, Saudi Arabia	2,250,000
4	ExxonMobil Chemical Co.	Baytown, Tex.	2,197,000
5	ChevronPhillips Chemical Co.	Sweeny, Tex.	1,844,000
6	Dow Chemical Co.	Terneuzen, Netherlands	1,800,000
7	Ineos Olefins & Polymers	Chocolate Bayou, Tex.	1,752,000
8	Equistar Chemicals LP	Channelview, Tex.	1,750,000
9	Yanbu Petrochemical Co.	Yanbu, Saudi Arabia	1,705,000
10	Equate Petrochemical Co.	Shuaiba, Kuwait	1,650,000

*As of Jan. 1, 2009.

Regional Capacity Breakdown

	Ethylene capacity, tpy		Change	
	Jan. 1, 2009	Jan. 1, 2008	tpy	%
Asia-Pacific	33,362,000	33,002,000	360,000	1.09
Eastern Europe	8,571,000	8,512,000	59,000	0.69
Middle East, Africa	19,312,000	12,342,000	6,970,000	56.47
North America	35,407,000	35,707,700	–300,700	–0.84
South America	5,083,500	5,083,500	–	–
Western Europe	24,918,000	24,438,000	–10,000	–0.04
Total capacity	**126,653,500**	**119,575,200**	**7,078,300**	**5.92**

National ethylene capacities

Country	Jan. 1, 2009 — Ethylene capacity, tpy	Jan. 1, 2008 — Ethylene capacity, tpy	Change, tpy
Algeria	133,000	133,000	—
Argentina	838,500	838,500	—
Australia	502,000	502,000	—
Austria	500,000	500,000	—
Azerbaijan	330,000	330,000	—
Belarus	193,000	193,000	—
Belgium	2,460,000	2,460,000	—
Brazil	3,500,000	3,500,000	—
Bulgaria	400,000	400,000	—
Canada	5,531,000	5,531,000	—
Chile	45,000	45,000	—
China	7,348,000	6,988,000	360,000
China, Taiwan	3,621,000	3,621,000	—
Colombia	100,000	100,000	—
Croatia	90,000	90,000	—
Czech Republic	544,000	485,000	59,000
Egypt	330,000	330,000	—
Finland	330,000	330,000	—
France	3,373,000	3,373,000	—
Germany	5,757,000	5,757,000	—
Greece	20,000	20,000	—
Hungary	660,000	660,000	—
India	2,515,000	2,515,000	—
Indonesia	520,000	520,000	—
Iran	4,734,000	1,214,000	3,520,000
Israel	200,000	200,000	—
Italy	2,170,000	2,170,000	—
Japan	7,265,000	7,265,000	—
Kazakhstan	130,000	130,000	—
Kuwait	1,650,000	800,000	850,000
Libya	350,000	350,000	—
Malaysia	1,649,000	1,649,000	—
Mexico	1,384,000	1,384,000	—
Netherlands	3,965,000	3,975,000	–10,000
Nigeria	300,000	300,000	—
North Korea	60,000	60,000	—
Norway	550,000	550,000	—
Poland	700,000	700,000	—
Portugal	330,000	330,000	—
Qatar	1,030,000	1,030,000	—
Romania	844,000	844,000	—
Russia	3,490,000	3,490,000	—
Saudi Arabia	9,400,000	6,800,000	2,600,000
Serbia and Montenegro	200,000	200,000	—
Singapore	1,980,000	1,980,000	—
Slovakia	220,000	220,000	—
South Africa	585,000	585,000	—
South Korea	5,630,000	5,630,000	—
Spain	1,430,000	1,430,000	—
Sweden	625,000	625,000	—
Switzerland	33,000	33,000	—
Thailand	2,272,000	2,272,000	—
Turkey	520,000	520,000	—
Ukraine	630,000	630,000	—
UAE	600,000	600,000	—
United Kingdom	2,855,000	2,855,000	—
United States	28,492,000	28,792,700	–300,700
Uzbekistan	140,000	140,000	—
Venezuela	600,000	600,000	—
Total	**126,653,500**	**119,575,200**	**7,078,300**

Mideast ethylene production capacity

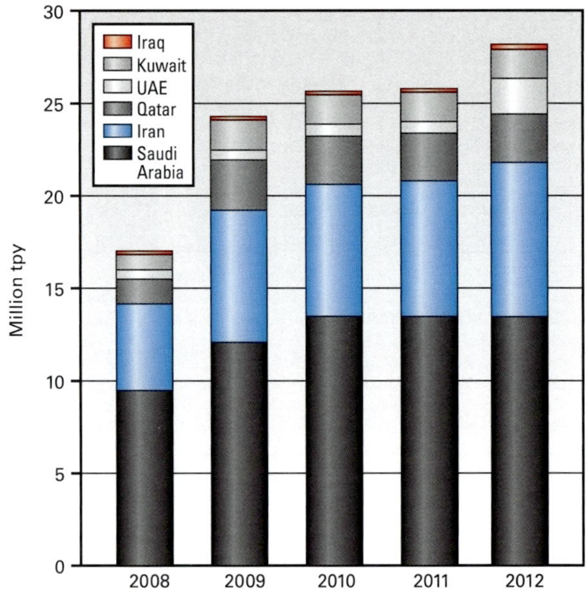

Source: FACTS Global Energy, OGJ, Mar. 23, 2009, p. 44

Ethylene closures

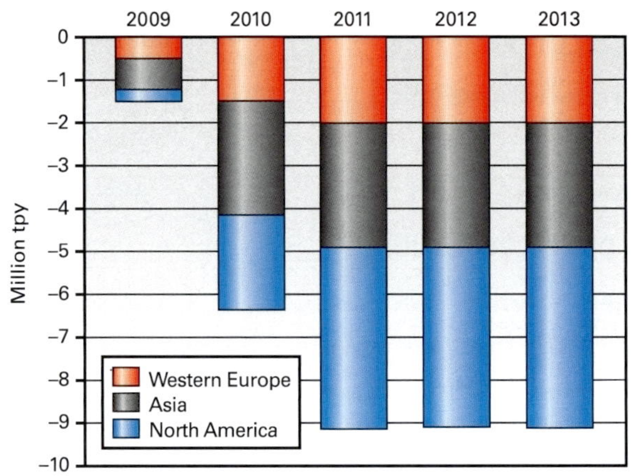

Source: CMAI, Houston

Top 10 ethylene producers[1]

	Company	Number of sites	Capacity, tpy — Of entire complexes	Capacity, tpy — With only company of partial interests
1	Dow Chemical Co.	15	14,005,000	10,710,750
2	Saudi Basic Industries Corp.	8	10,235,000	8,455,000
3	ExxonMobil Corp.	15	11,470,000	8,352,000
4	Royal Dutch Shell PLC	10	8,613,300	6,489,300
5	LyondellBasell[2]	8	6,200,000	5,450,000
6	Ineos	4	4,656,000	4,286,000
7	National Petrochemical Co.	7	4,734,000	4,234,000
8	Formosa Petrochemical Corp.	2	4,091,000	4,091,000
9	Sinopec	10	4,375,000	4,075,000
10	ChevronPhillips Chemical Co.[3]	5	4,307,000	4,052,000

[1]As of Jan. 1, 2009. [2]Includes subsidiary Equistar Chemicals LP. [3]Ownership: Chevron Corp. 50%, ConocoPhillips 50%.

Global Ethylene Capacity Additions

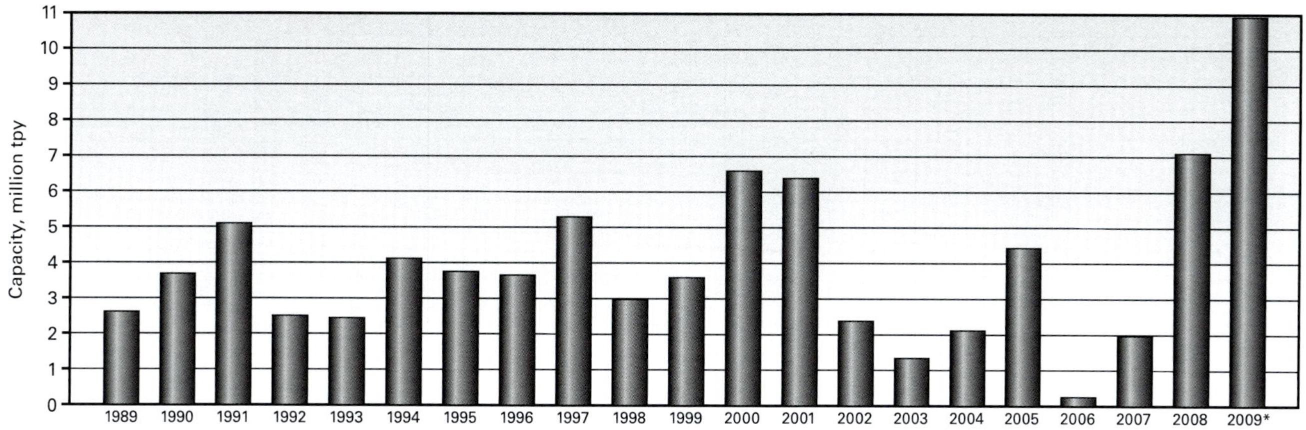

*Estimate.

Source: ChemSystems, White Plains, NY (1989-99 data). OGJ Energy Database (2000-09 data)

Ethylene Expansions, 2009–13

Location	Company	2009	2010	2011	2012	2013
		Ethylene capacity, tpy				
Arzew, Algeria	Total, Sonatrach				1,100,000	
Nanjing, China	BASF-YPC Co. Ltd.	750,000				
Heilongjian Province, China	Daqing Petroleum & Chemical Co.					600,000
Fujian Province, China	Fujian Petrochemical Co. Ltd.	800,000				
Fushun, China	PetroChina			800,000		
Ningbo, China	PetroChina					1,200,000
Sichuan, China	PetroChina				800,000	
Dushanzi, Zinjang, China	PetroChina Dushanzi Petrochemical Co.	1,000,000				
Maoming, Guangdong, China	Sinopec		320,000			
Quanzhou City, China	Sinopec					800,000
Ningbo, Zhejiang, China	Zenhai Refining & Chemical Co.		1,000,000			
Vadinar, Gujarat, India	Essar Gujarat Petrochemicals Ltd.					1,300,000
Dibrugarh, Assam, India	GAIL India Ltd.					220,000
Panipat, India	Indian Oil Co. Ltd.	800,000				
Dahej, India	Oil & Natural Gas Corp.		1,100,000			
Gachsaran, Iran	Gachsaran Olefin				1,000,000	
Ilam, Iran	Ilam Petrochemical Co.				153,000	
Assaluyeh, Iran	Kavyan Petrochemical Co.	2,000,000				
Assaluyeh, Iran	Morvarid Petrochemical Co.	500,000				
Sohar, Oman	Dow Chemical Co.			850,000		
Ras Laffan, Qatar	Qatar Chemical Co. Ltd., Chevron Corp., Total AS	1,300,000				
Ras Laffan, Qatar	Qatar Petroleum Co., ExxonMobil Corp.					1,600,000
Mesaieed, Qatar	Qatar Petroleum Co., Honam Petrochemical Co.				900,000	
Al-Jubail, Saudi Arabia	Eastern Petrochemical Co.	1,300,000				
Al-Jubail, Saudi Arabia	National Petrochemical Industrialization				1,008,000	
Rabigh, Saudi Arabia	Saudi Aramco, Sumitomo Chemical Co.	1,500,000				
Yanbu, Saudi Arabia	Saudi Basic Industries Corp.					1,300,000
Al-Jubail, Saudi Arabia	Saudi Polymers			1,200,000		
Al-Jubail, Saudi Arabia	Tasnee Petrochemicals					1,000,000
Singapore	ExxonMobil Chemical Corp.			1,000,000		
Singapore	Shell Eastern Petroleum Ltd.		800,000			
Kaoshiung Linyuan, Taiwan	Chinese Petroleum Corp.			1,000,000		
Lin Yuan, Taiwan	Chinese Petroleum Corp.					600,000
Map Ta Phut, Thailand	PTT Polyethylene Co. Ltd.	1,000,000				
Rayong, Thailand	Siam Cement PLC, Dow Chemical Co.		900,000			
Trinidad & Tobago	Westlake Chemical Corp.		570,000			
Ruwais, Abu Dhabi, UAE	Abu Dhabi Polymers Co. Ltd.				1,500,000	
El Tablazo, Venezuela	Polinter				1,000,000	
Jose, Anzoategui, Venezuela	Pequiven				1,050,000	
Total		**10,950,000**	**4,690,000**	**4,850,000**	**8,511,000**	**8,620,000**

Source: *Oil & Gas Journal* construction survey

Surplus ethylene capacity

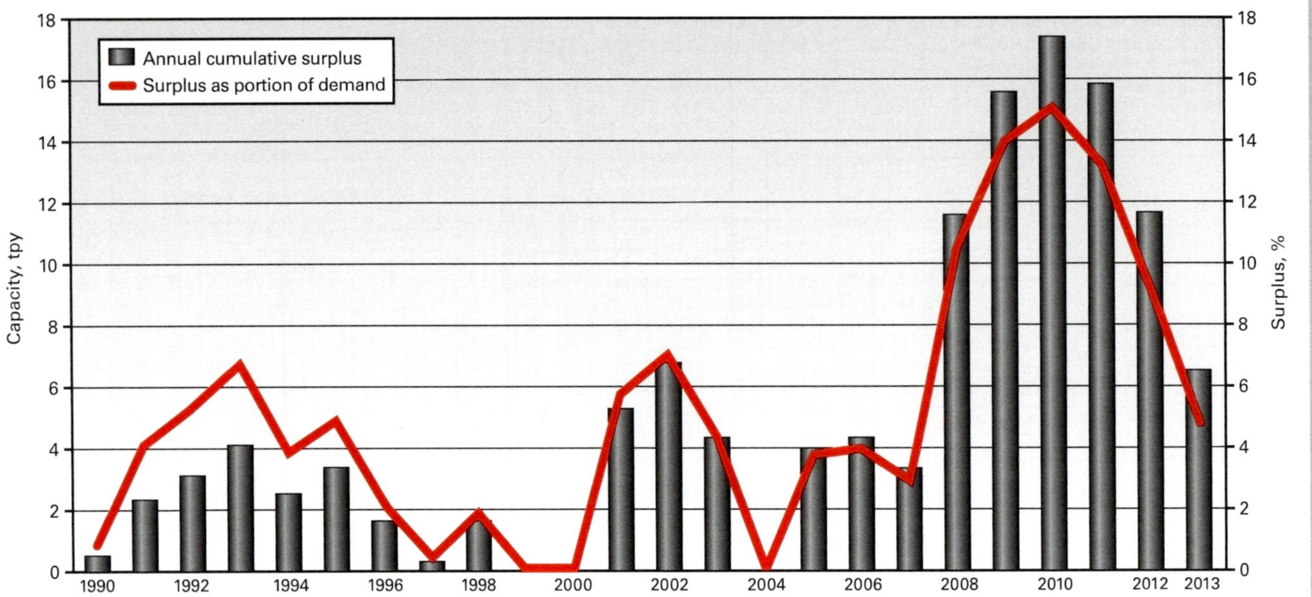

Source: CMAI, Houston

Mideast olefin production: Oversupply?

FACTS Global Energy (Honolulu) reported in early 2009 that, with an abundance of cheap feedstock, the Middle East was poised to become one of the world's largest exporters of petrochemicals.

Several new mega petrochemical projects were under construction in Saudi Arabia, Iran, Qatar, the UAE, and Kuwait. Upon their completion, Middle East ethylene production capacity will increase to about 28.1 mtpy in 2012, from 16.9 mtpy in 2008.

Mideast ethylene capacity additions

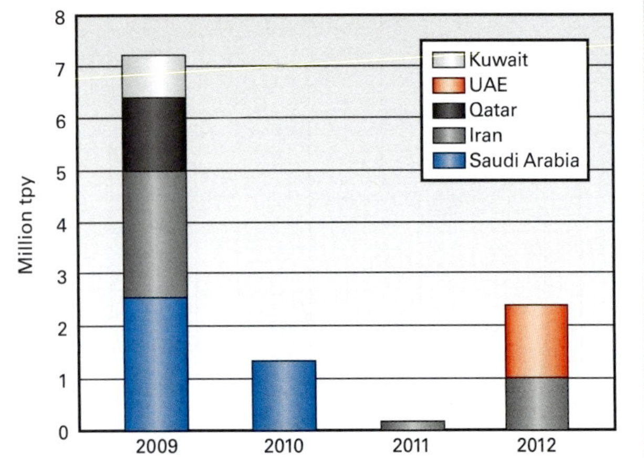

Propylene production capacity will increase to 7.0 mtpy in 2012, from 3.5 mtpy in 2008.

FGE noted that the global economic recession was depressing ethylene and polyethylene demand and that the new projects, especially those coming online in 2009, might face an oversupplied market.

Mideast ethylene capacity, 2008-12

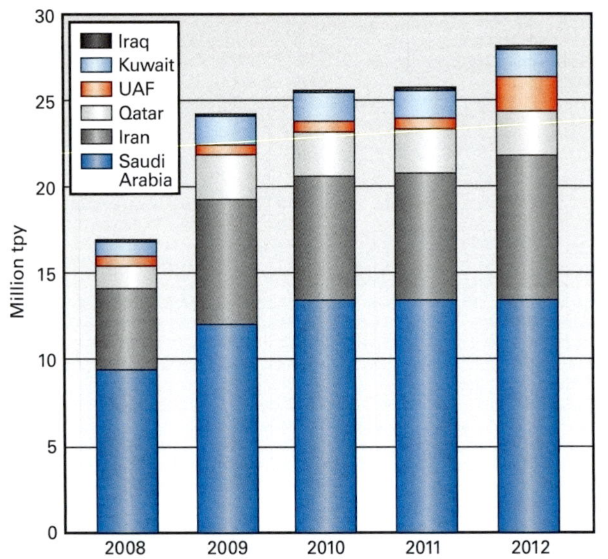

New Ethylene, Propylene Capacity: Saudi Arabia

Project	Start-up	Ethylene — Capacity, 1,000 tpy —	Propylene
PetroRabigh	Early 2009	1,300	900
Sharq III	Early 2009	1,300	—
Saudi Ethylene & Polyethylene Co.	Early 2009	—	460
Saudi Kayan	2009–10	1,350	550
Total		**3,950**	**1,910**

Iran's Ethylene Capacity, 2008

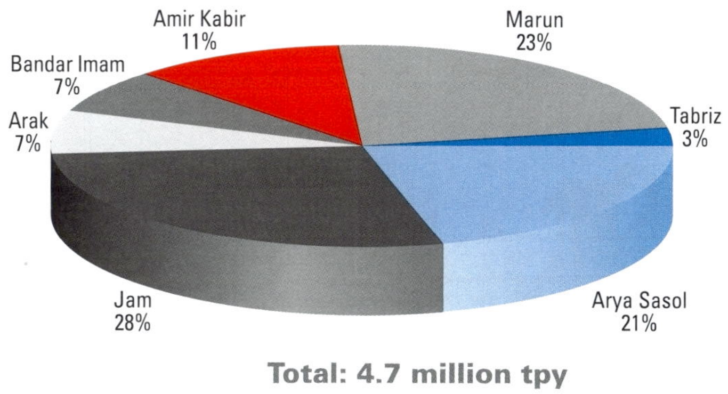

Mideast Propylene Capacity, 2008-12

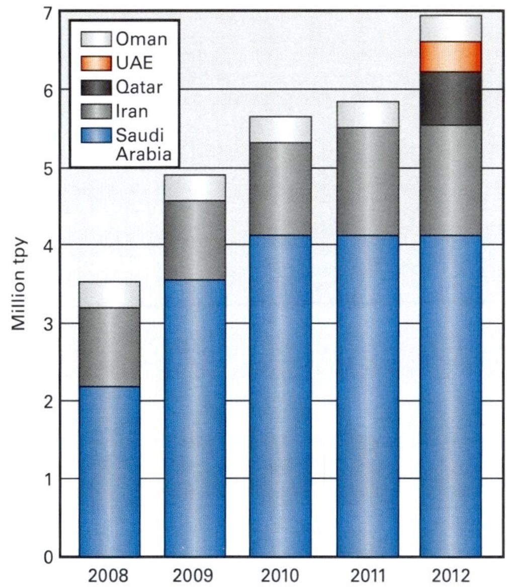

Iran's Propylene Capacity, 2008

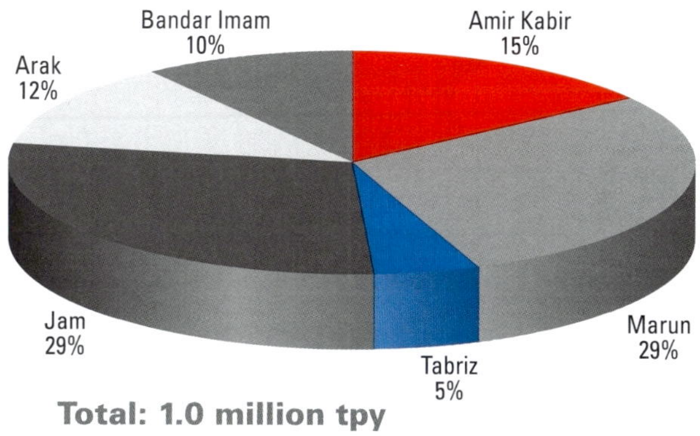

Mideast Propylene Capacity Additions

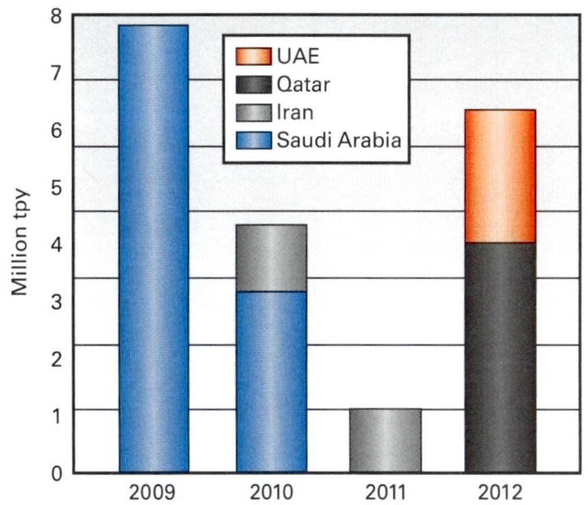

New Ethylene, Propylene Capacity: Iran

Project	Start-up	Ethylene — Capacity, 1,000 tpy —	Propylene
11th Olefin (Kavyan Petrochemical Co.)	2009	2,000	—
13th Olefin (Ilam Petrochemical Co.)	2011–12	153	120
Gachsaran Olefin	2012	1,000	—
Assaluyeh Olefin (Morvarid Petrochemical Co.)	2009	500	—
Fanavaran Petrochemical Co.)	2010–11	—	120
Total		**3,653**	**240**

Section 13:

Pipelines

U.S. pipeline sector: 2008

In September 2009, *Oil & Gas Journal* reported on the 2008 U.S. natural gas pipeline sector, noting that operators saw their profits reach new highs that year.

Profits rose by more than 7% compared with 2007, despite a 9% drop in revenues year-over-year. Net profits totaled $5.1 billion against revenues of $19.8 billion.

The resulting earnings as a portion of revenues—25.78%—was the highest figure yet recorded by *OGJ* for the industry.

Oil pipeline profit growth also outstripped revenue growth: 4.67% vs. 2.75%.

Additions to gas plants totaled nearly $12.2 billion, a roughly 89% increase from 2007 levels.

Operators also increased planned capacity expansions. Proposed mileage increased by more than 242%, and compression plans rose by more than 270% to total 664,755 hp.

Oil Pipeline Investment

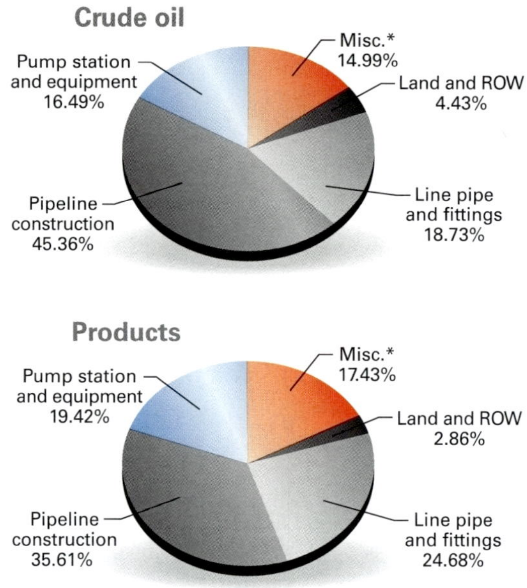

*Generally includes delivery systems, communications, office furniture and equipment, vehicles and other work equipment, and other property.
Source: US oil pipeline company annual reports (Form 6) to FERC for 2008

Top 10 U.S. Interstate Oil Pipeline Companies—2008

	Company	Mileage	Company	Trunkline traffic, million bbl-miles	Company	Income, $1,000
1	Magellan Pipeline Co. LP	8,658	Colonial Pipeline Co.	686,727	Kinder Morgan Operating LP "A"	734,163
2	Mid-America Pipeline Co.	7,833	Enbridge Energy LP	432,111	Whiting Oil & Gas Corp.	258,509
3	Plains Pipeline LP	7,575	Marathon Pipeline LLC	161,240	Enbridge Energy LP	247,973
4	ConocoPhillips Pipe Line Co.	6,694	Explorer Pipeline Co.	120,529	Shell Pipeline Co. LP	225,126
5	Sunoco Pipeline LP	5,904	Plains Pipeline LP	114,624	Colonial Pipeline Co.	193,144
6	Colonial Pipeline Co.	5,593	TE Products Pipeline Co. LP	108,217	NuStar Logistics LP	164,279
7	TE Products Pipeline Co. LP	4,676	Mid-America Pipeline Co. LLC	106,665	Mid-America Pipeline Co. LLC	159,119
8	ExxonMobil Pipeline Co.	4,557	Plantation Pipe Line Co.	105,577	Plains Pipeline LP	150,616
9	TEPPCO Crude Pipeline LP	3,967	ConocoPhillips Transportation Alaska Inc.	78,991	Marathon Pipeline LLC	142,638
10	BP Pipelines North America Inc.	3,776	ExxonMobil Pipeline Co.	73,491	Sunoco Pipeline LP	141,640
	Total	**59,233**	**Total**	**1,988,172**	**Total**	**$2,417,207**
	Part of all companies	**40.34%**		**57.84%**		**61.48%**
	Top 10 total—2007	59,448		2,058,820		$2,131,461

Source: US FERC Form 6: Annual Report of Oil Pipeline Companies, Dec. 31, 2008

Top 10 U.S. Interstate Gas Pipeline Companies—2008

	Company*	Transmission mileage	Company*	Volumes moved for fee, MMcf	Company*	Net income, $1,000
1	Northern Natural Gas Co.	15,018	Transcontinental Gas Pipe Line Corp.	2,577,642	Natural Gas Pipeline Co. of America	384,621
2	Tennessee Gas Pipeline Co.	14,463	ANR Pipeline Co.	1,982,666	Northern Natural Gas Co.	250,798
3	Columbia Gas Transmission Corp.	11,531	Natural Gas Pipeline Co. of America	1,925,335	Distrigas of Massachusetts LLC	248,473
4	El Paso Natural Gas Co.	10,237	Columbia Gulf Tranmission Co.	1,910,403	Transcontinental Gas Pipe Line Co.	235,164
5	Transcontinental Gas Pipe Line Corp.	9,972	Tennessee Gas Pipeline Co.	1,801,283	Texas Eastern Transmission Corp.	207,298
6	ANR Pipeline Co.	9,580	El Paso Natural Gas Co.	1,740,860	Dominion Transmission Inc.	203,114
7	Texas Eastern Transmission LP	9,202	Texas Eastern Transmission Corp.	1,432,742	Columbia Gas Transmission Corp.	193,276
8	Natural Gas Pipeline Co. of America	9,008	CenterPoint Energy Gas Transmission Co.	1,252,757	Gas Transmission Northwest Corp.	177,659
9	Southern Natural Gas Co.	7,593	Northern Natural Gas Co.	1,143,582	Panhandle Eastern Pipe Line Co.	174,456
10	Gulf South Pipeline Co. LP	6,466	Gulf South Pipeline Co. LP	1,048,042	Southern Natural Gas Co.	149,176
	Total 2008	**103,070**	**Total 2008**	**16,815,312**	**Total 2008**	**$2,224,035**
	Part of majors	**53.58%**		**43.99%**		**44.24%**
	Part of all companies	**52.05%**		**43.36%**		**43.57%**
	Total—2007 Top 10	102,736		16,387,585		$2,147,348

*All FERC-classified as "major."

Source: US FERC Form 2: annual reports for natural gas companies, Dec. 31, 2008

Major cost components—10 years

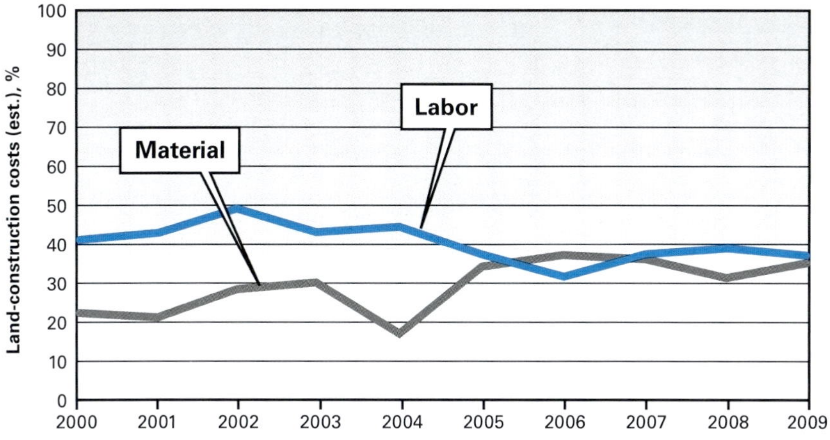

Source: US FERC

U.S. interstate pipeline mileage

Year	Gas[1,2] (Miles)	Oil (Miles)	Total[1] (Miles)
1999	180,489	155,904	336,393
2000	186,151	152,823	338,974
2001	180,961	154,877	335,838
2002	190,899	149,619	340,518
2003	188,178	139,901	328,079
2004	190,117	142,200	332,317
2005	188,847	131,334	320,181
2006	189,012	140,407	329,419
2007	192,189	147,235	339,424
2008	192,384	146,822	339,206

[1]FERC-defined major gas pipelines only; transmission mileage. See GAS COMPANIES table for definition of major and nonmajor companies and details of companies reporting mileage for 2008.
[2]Totals revised from initial publication.
Source: US FERC annual reports: Form 6, oil pipelines; Form 2, gas pipelines

Pipeline construction costs–estimated

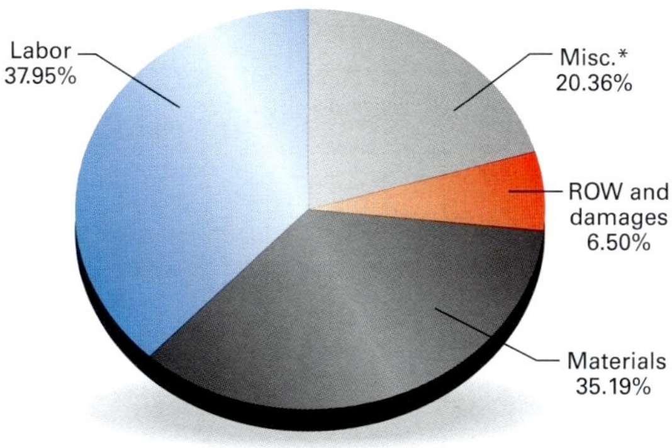

*Generally includes surveying, engineering, supervision, administration and overhead, interest, contingencies and allowances for funds used during construction (AFUDC), and regulatory filing fees.

Source: US FERC construction-permit filings, July 1, 2008, to June 30, 2009

Nord Stream Pipeline permits

In November 2009, Finland and Sweden granted permits that will allow the Nord Stream Pipeline to be routed through the Exclusive Economic Zones (EEZ) of the two countries.

The three other needed EEZ permits were granted by Denmark in October, and by Germany and Russia in December.

The Baltic Sea pipeline will carry natural gas 1,223 km (760 mi) from Russia to the European Union.

Construction of the two parallel lines that comprise Nord Stream is planned to start in 2010, with the first pipeline operational in 2011 and the second in 2012.

Gazprom holds 51% of the joint venture. German energy companies E.ON Ruhrgas AG and Wintershall AG each hold 20%, and Dutch company Nederlandse Gasunie NV holds the remaining 9%.

Nabucco agreement

In July 2009, the prime ministers of Turkey, Bulgaria, Romania, Hungary and Austria signed an agreement under which the five nations will allow the 3,300-km (2,050-mi), $10-billion Nabucco pipeline to pass through their territories.

Details on the project are given in the Turkey section of the Europe chapter of this book.

Spur to tap ITGI gas pipeline

Greek, Bulgarian and Italian companies agreed in July 2009 to build a 160-km (99-mi) pipeline that would help to end Bulgaria's reliance on Russian natural gas. A completion date of 2012 was set, but observers believe 2015 may be more likely.

The $167-million Greece-Bulgaria Interconnector (GBI) would provide Bulgaria access to the Italy-Turkey-Greece Interconnector (ITGI) pipeline carrying gas from Azerbaijan to Turkey, Greece and Italy.

Details are given in the Bulgaria section of the Europe chapter of this book.

Trans-Sahara Gas Pipeline

The planned $10-billion Trans-Sahara Gas Pipeline aims to transport as much as 30 bcm/yr (1.06 tcf/yr) of gas from Nigeria to Europe, passing through Algeria and Niger.

Although no date has been set for the start of work on the 4,100-km (2,548-mi) line, its first shipment of gas is scheduled for 2015.

Details are given in the Algeria section of the Africa chapter of this book.

Europe: Southern Corridor gas transit network

At a May 2009 energy summit in Prague, key gas suppliers and transit countries agreed to participate in establishing the Southern Corridor Network, referred to as the "Silk Road" of the oil and gas sector.

The Southern Corridor project was initiated by the EU in 2008 in a bid to free itself from the Russian energy monopoly. The project comprises three major pipelines and also a trans-Caspian line, which will link natural gas reserves in the Caspian basin, Iran, and Iraq to Europe.

The 3,300-km (2,050-mi) Nabucco pipeline is the critical link of the project

The deal would bind the EU and delegate nations—Azerbaijan, Egypt, Georgia, Iraq, Kazakhstan, Turkey, Turkmenistan, and Uzbekistan—in a single contract for energy transport.

GASENE PIPELINE

KEYSTONE PIPELINE PROJECT

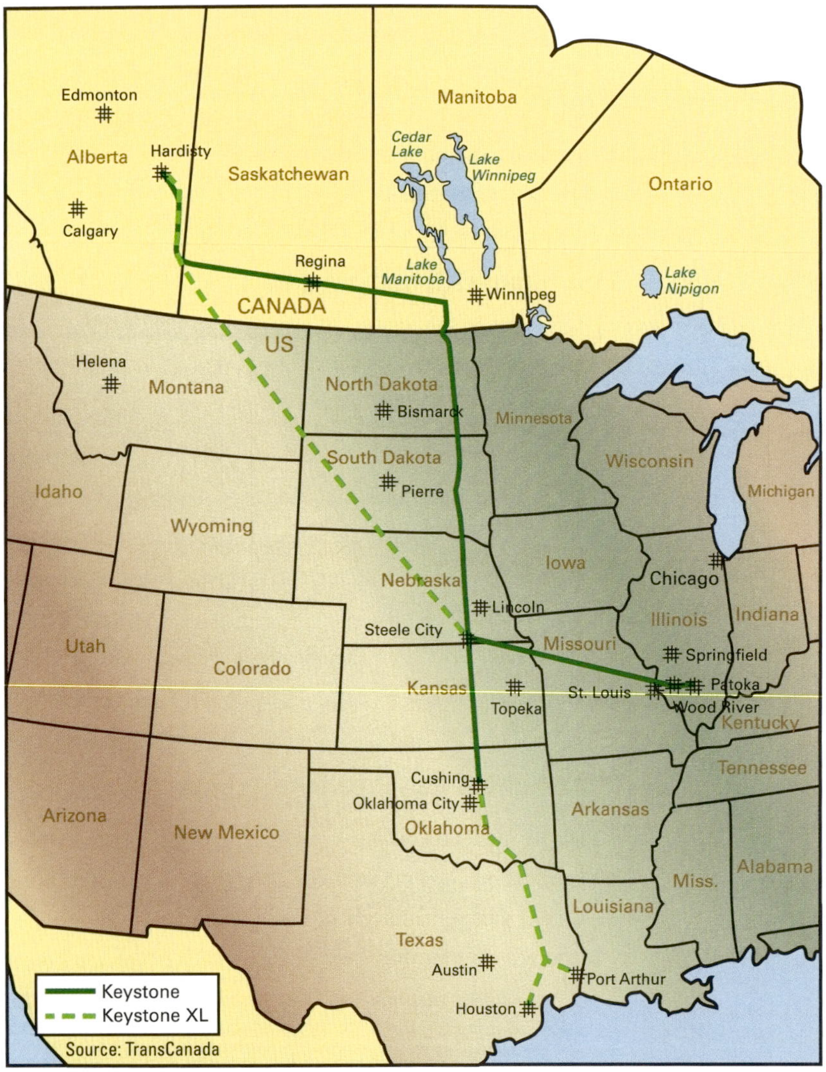

Global pipeline expansion plans

In early 2009, *OGJ* released its survey of global pipeline expansion plans for 2009 and beyond, based on data gathered from pipeline companies.

Planned pipeline construction to be completed in 2009 rose more slowly than in 2008, increasing by more than 13% from the previous year, driven by large natural gas transportation projects in both the United States and Asia-Pacific, but constrained by a drop in planned crude mileage.

The downturn in worldwide pipeline construction trends reflects the current economic unrest but runs counter to the forecasts of the U.S. EIA regarding energy consumption, which show continued growth, even if more slowly than predictions from 2008.

Trans-Asian, Kazakhstan-China pipelines

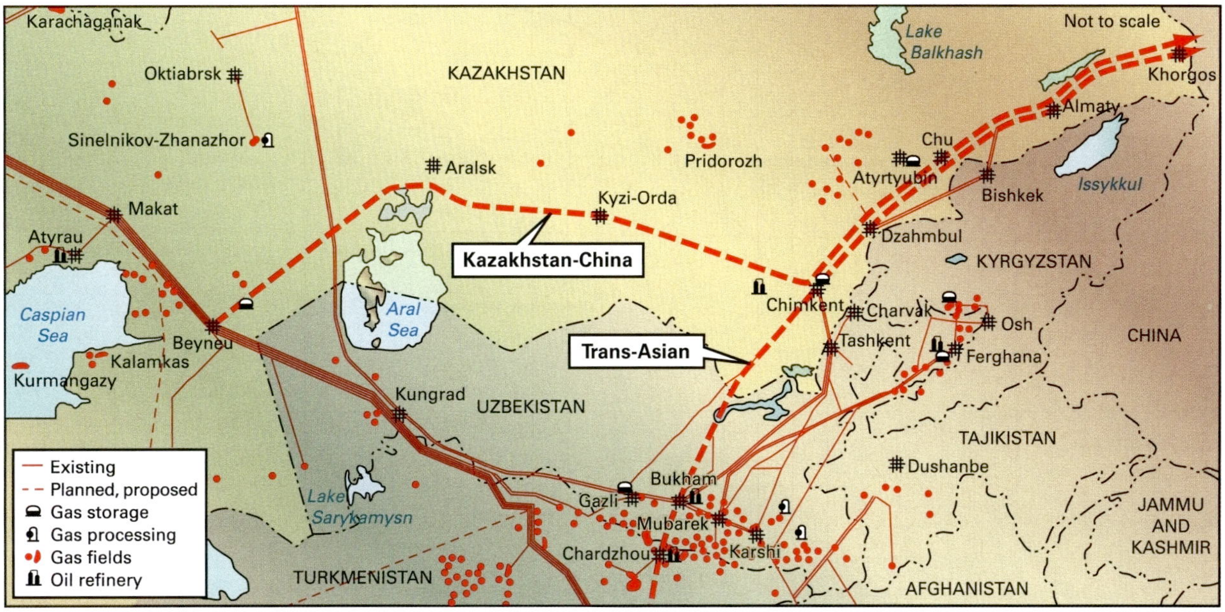

Pipeline construction in 2009[1]

Area	4–10 in.	12–20 in.	22–30 in.	32+ in.	Total
			Miles		
GAS PIPELINES					
U.S.	0	313	324	3,351	3,988
Canada	34	0	0	0	34
Latin America	208	841	0	195	1,244
Asia-Pacific[2]	311	366	0	1,135	1,812
Europe[3]	0	300	115	0	415
Middle East	0	0	56	1,439	1,495
Africa	50	72	130	0	252
Total gas	**603**	**1,892**	**625**	**6,120**	**9,240**
CRUDE PIPELINES					
U.S.	0	0	1,054	507	1,561
Canada	0	0	236	240	476
Latin America	0	0	146	0	146
Asia-Pacific[2]	0	72	888	0	960
Europe[3]	0	0	0	0	0
Middle East	0	152	0	40	192
Africa	0	0	0	0	0
Total crude	**0**	**224**	**2,324**	**787**	**3,335**
PRODUCT PIPELINES					
U.S.	60	1,424	0	0	1,484
Canada	0	28	0	0	28
Latin America	0	715	0	0	715
Asia-Pacific[2]	112	0	0	0	112
Europe[3]	0	0	0	0	0
Middle East	0	0	0	0	0
Africa	350	0	0	0	350
Total product	**522**	**2,167**	**0**	**0**	**2,689**
WORLD TOTALS					
Gas	603	1,892	625	6,120	9,240
Crude	0	224	2,324	787	3,335
Product	522	2,167	0	0	2,689
Total	**1,125**	**4,283**	**2,949**	**6,907**	**15,264**

[1]Projects planned to be completed in 2009. [2]Regions east of the Ural Mountains and south of the Caucasus Mountains, excluding the Middle East. [3]Regions west of the Ural Mountains and north of the Caucasus Mountains.

Pipeline construction in 2009 and beyond[1]

Area	4–10 in.	12–20 in.	22–30 in.	32+ in.	Total
			Miles		
GAS PIPELINES					
U.S.	25	325	1,427	7,526	9,303
Canada	0	0	253	380	633
Latin America	400	212	1,989	1,177	3,778
Asia-Pacific[2]	0	573	0	19,283	19,856
Europe[3]	0	124	859	7,963	8,946
Middle East	0	0	43	3,855	3,898
Africa	0	0	0	2,354	2,354
Total gas	**425**	**1,234**	**4,571**	**42,538**	**48,768**
CRUDE PIPELINES					
U.S.	57	157	0	2,394	2,608
Canada	0	0	866	1,934	2,800
Latin America	0	0	0	0	0
Asia-Pacific[2]	0	0	682	2,685	3,367
Europe[3]	0	83	0	515	598
Middle East	0	0	43	303	346
Africa	0	0	0	0	0
Total crude	**57**	**240**	**1,591**	**7,831**	**9,719**
PRODUCT PIPELINES					
U.S.	0	812	314	460	1,586
Canada	0	387	0	0	387
Latin America	0	352	0	0	352
Asia-Pacific[2]	0	0	0	0	0
Europe[3]	0	0	0	0	0
Middle East	0	1,341	0	0	1,341
Africa	0	434	0	0	434
Total product	**0**	**3,326**	**314**	**460**	**4,100**
WORLD TOTALS					
Gas	425	1,234	4,571	42,538	48,768
Crude	57	240	1,591	7,831	9,719
Product	0	3,326	314	460	4,100
Total	**482**	**4,800**	**6,476**	**50,829**	**62,587**

[1]Projects under way at the start of or set to begin in 2009 and be completed in 2010 or later. Includes some probable major projects whose installation will begin in 2010 or later. [2]Regions east of the Ural Mountains and south of the Caucasus Mountains, excluding the Middle East. [3]Regions west of the Ural Mountains and north of the Caucasus Mountains.

Forecast pipeline construction

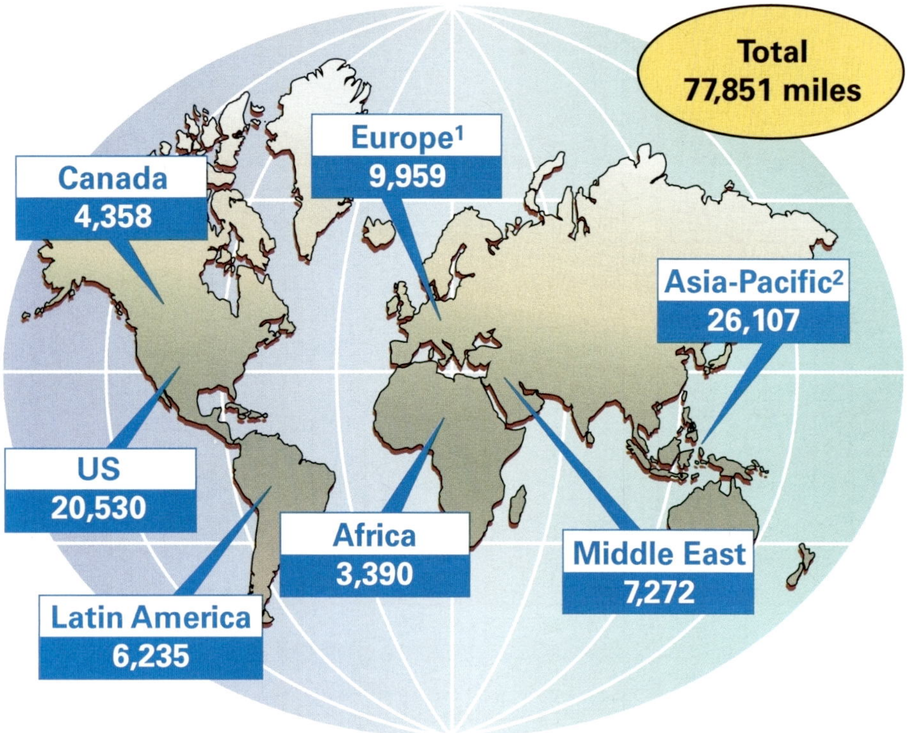

[1]Including Russia and former Soviet republics west of the Ural Mountains. [2]Including Russia and former Soviet republics east of the Ural Mountains.

Caspian region oil shipments, 2005–15

Route	2005	2007	2010	2015 Low	2015 High
	Million tons				
China	6.2	9.3	9.7	6.7	21.2
Iran	3.5	7.9	5.3	3.0	8.0
Russia	53.4	53.5	62.0	51.0	69.0
Georgia-Azerbaijan	17.4	33.9	75.1	74.3	107.1
Total	**80.5**	**104.6**	**152.1**	**135.0**	**205.3**
	%				
China	8	9	6	5	10
Iran	4	8	3	2	4
Russia	66	51	41	38	34
Georgia-Azerbaijan	22	32	49	55	52

Transport of new Caspian oil

Georgia's Ministry of Economic Development described in early 2009 pipeline projects aimed at bringing new Caspian oil production to international markets in a way that bypasses Russian transport routes.

Caspian region oil production will increase from 80.4 million tons in 2005 to 135–205.3 million tons in 2015. This oil will come from increases in output from Tengiz field in Kazakhstan, Kashagan field in the northern Caspian Sea just off the coast of Kazakhstan, and ACG field in the southern Caspian Sea just off the coast of Azerbaijan.

Limited expansion potential on the Russian-operated Caspian Pipeline Consortium (CPC) line, plus a desire by the region's shippers to avoid additional use of Russian lines, will lead to transport of most of this new production by the Baku-Tbilisi-Ceyhan (BTC) pipeline instead.

CASPIAN OIL DISTRIBUTION, 2010*

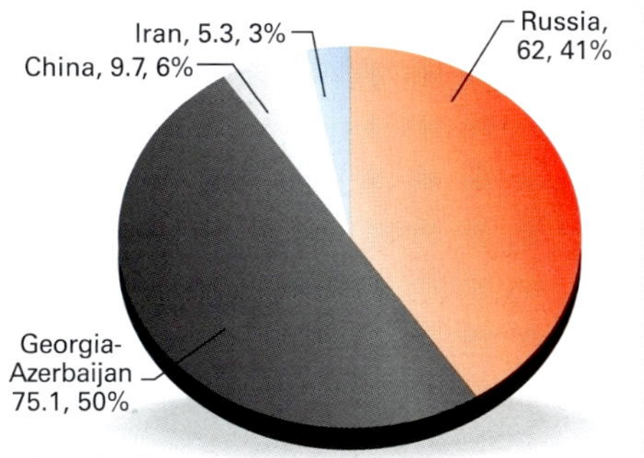

*Volume in million tons/year.

CASPIAN OIL DISTRIBUTION*

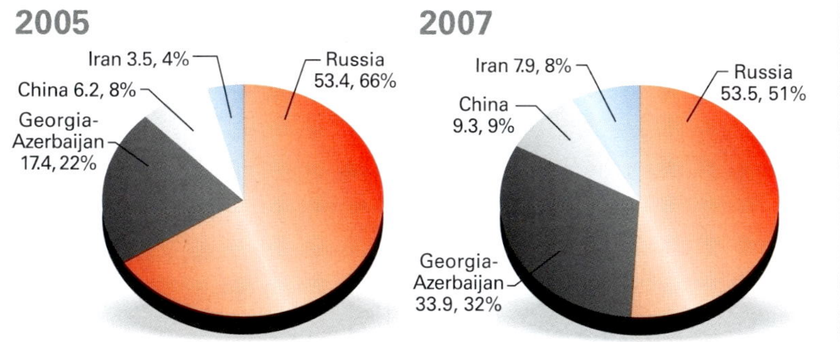

*Volumes in million tons/yr.

CASPIAN OIL DISTRIBUTION, 2015 WITH CPC*

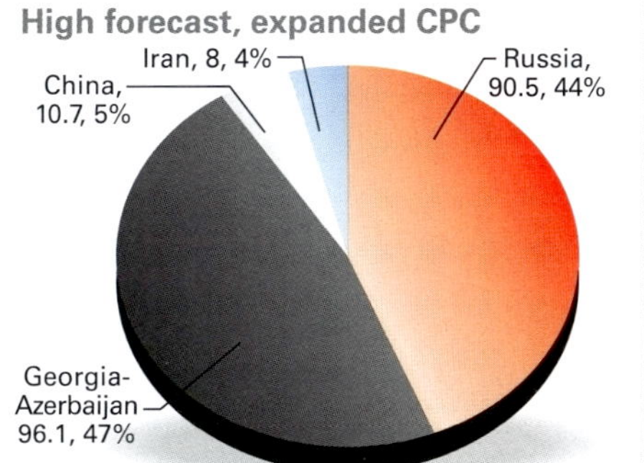

*Volume in million tons/year.

CASPIAN OIL DISTRIBUTION, 2015*

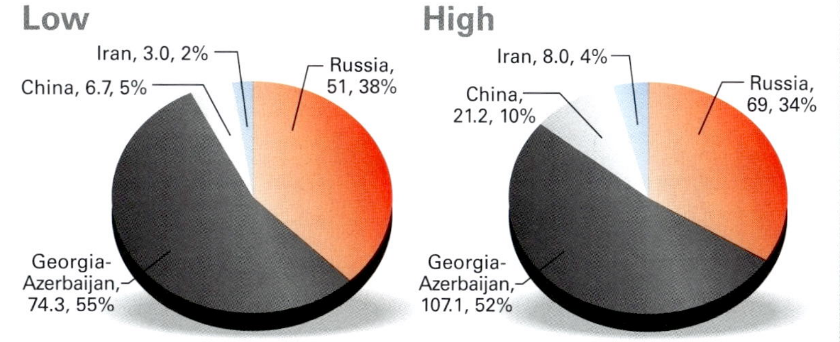

*Volumes in million tons/yr.

CASPIAN OIL TRANSPORT ROUTES

Source: Caucasus Transport Corridors for Oil and Oil Products—Draft, ECSSD World Bank, June 2008

Caspian oil pipelines

Pipeline	Route	Length, km	2006 estimated traffic	Current capacity	Low 2015 capacity	High 2015 capacity
			Million tons/year			
Atyrau-Samara	Uzen in southwestern Kazakhstan to Caspian port of Atyrau, linking to Russian pipeline system at Samara	695	15.5	20	25	25
Baku-Novorossiysk (AIOC Northern Route)	Baku via Chechnya to Novorossiysk, terminating at Black Sea oil terminal	1,400	4.4	6	6	6
Baku-Novorossiysk (Chechnya bypass, link to Makhachkala)	Baku via Dagestan to Tikhoretsk, terminating at Novorossiysk Black Sea oil terminal	283	—	—	—	—
Baku-Supsa (AIOC Western Route)	Baku (Azerbaijan) to Supsa (Georgia), terminating at Black Sea port	885	5.6	7	7	7
BTC	Baku via Tbilisi (Georgia) to Ceyhan (Turkey), terminating at Mediterranean Sea port	1,780	7.7	50	80	80
CPC	Tengiz (Kazakhstan) to Russian Black Sea oil port at Novorossiysk	1,580	32.3	30	32	67
Iran oil swap	Caspian tankers deliver to Iranian port at Neka, pipeline shipment to Tehran	335	5.2	12.5	18	20
Karachaganak-Atyrau	Karachaganak field to Atyrau, connecting to Atyrau-Samara and CPC	635	7	7	7	7
Kazakhstan-China (Atashu-Alashanko-Dushanzi section)	South Turgay fields to Chinese refineries	987	10	10	20	20
Kenkiyak-Orsk	Abtyubinsk fields (Kazakhstan) to Orsk refinery (Russia)	400	6.5	6.5	6.5	6.5
Kenkiyak-Atyrau	Aktobe fields to rest of system	448	12	12	12	12
Total		**9,428**	**106.2**	**161**	**213.5**	**250.5**

Sources: BP Statistical Review of World Energy, British Petroleum, 2007; World Energy Outlook 2006, International Energy Agency (IEA), OECD/IEA 2006; Aqtau Port Development, Master Planning and Feasibility Study, Traffic Forecasts, October 2007.

Section 14:
Climate Change Issues

Government and industry officials around the world in 2009 were exploring ways to address concerns about global climate change, as affected by hydrocarbon development, delivery and use.

Presented here are highlights of some of those initiatives and related discussions.

Before Copenhagen

Ahead of the UN's December 2009 Copenhagen conference, the EU committed to cutting emissions by 20% by 2020, compared with 1990 levels; Japan to 25% (if others take similar steps); and the United States provisionally to 3–4%.

For the first time, China also offered to trim greenhouse gas (GHG) output, pledging to reduce its "carbon intensity" (use of fossil fuels per unit of economic output) by 40–45%. India, Brazil and South Africa followed suit with their own voluntary targets.

Scientists say that deeper reductions are needed because GHG emissions have increased an average of 2–3% per year in the past decade.

Some U.S. experts predict that a big enough rise in temperatures could lead to serious damage from coastal flooding, droughts, species die-offs and other impacts of climate change.

Copenhagen aim: A new treaty

A legally binding international agreement—a treaty—requiring further emissions cuts by richer nations was the goal set in Bali, Indonesia, in 2007 when the UN set a two-year timetable leading to Copenhagen.

A new pact would succeed the first phase of the 1997 Kyoto Protocol, whose initiative to achieve relatively modest emissions cuts by 37 nations expires in 2012.

It was hoped a new regime would encompass the United States, which rejected Kyoto.

The next deadline in the quest for a treaty will be the 2010 UN climate conference in Mexico City.

Copenhagen Summit: No treaty, but...

The long-awaited 2009 United Nations Climate Change Conference (also called the Copenhagen Summit) was held December 7–18, 2009, in Copenhagen, Denmark, with more than 110 world leaders in attendance.

The event included the 15th Conference of the Parties (COP 15) to the UN Framework Convention on Climate Change, and the 5th Meeting of the Parties (COP/MOP 5) to the Kyoto Protocol.

Its goal was to develop a framework for what could become an agreement, containing:

- commitments on emissions cuts from rich countries
- curbs on emissions from the developing world
- financing pledges from rich nations to help poor nations reduce emissions and cope with the effects of climate change.

However, the conference ended by issuing only a nonbinding Copenhagen Accord (see below) rather than the desired framework for a treaty.

...a Copenhagen Accord

The key elements of the Copenhagen Accord, with no legal obligation, were:

- Nations agreed to cooperate in reducing emissions, "with a view" to scientists' warnings to keep temperatures from rising more than 2°C (3.6°F) above preindustrial levels.
- Developing nations will report every two years on their voluntary actions to reduce emissions. Those reports would be subject to "international consultations and analysis"—a concession to the United States by China, which had seen this as an intrusion on its sovereignty.
- Richer nations will finance a $10-billion/yr, three-year program to fund poorer nations' projects to deal with drought and other climate-change impacts, and to develop clean energy.
- Signers also set a "goal" of mobilizing $100 billion/yr by 2020 for adaptation and mitigation purposes.
- In a U.S. concession to China and other developing nations, wording was dropped that would have set a goal of reducing global emissions by 50% by 2050. Developing nations thought that such a goal would hamper efforts to raise their people from poverty.

EC grants for CCS

In December 2009, the European Commission confirmed that it will approve 280 million euros ($412 million) in funding for six carbon capture and storage (CCS) projects.

The projects are in Britain, Spain, Germany, Poland, The Netherlands, and Italy.

Soros: $1 billion for green technology

Billionaire George Soros announced in October 2009 that he plans to invest as much as $1 billion in clean-energy technologies.

He also will form a new climate-policy organization, the Climate Policy Initiative, to combat global warming and intends to provide funding for it of $10 million/year for the next decade.

Shell: Speed up CCS

Shell International Exploration & Production in late 2009 called for accelerated deployment of CCS technology, arguing that delay will increase the costs of retrofitting power plants and industrial facilities.

The company has extensive experience in pipeline transportation and use of CO_2 for enhanced oil recovery (EOR), and also has eight CCS projects under way, four of which are industrial scale.

One of the four will take CO_2 from the company's Pernis refinery and transport it by pipeline for storage in two depleted natural gas fields near Barendrecht in The Netherlands.

IEA includes climate-change scenario

In 2009, for the first time, the International Energy Agency included a climate-change policy scenario as part of its annual forecast.

The *2009 World Energy Outlook* noted that 60% of global CO_2 emissions come from energy, and that action to curb such emissions will be critical.

The scenario projects impacts of taking steps that would stabilize greenhouse gas concentrations at about 450 ppm CO_2 equivalent by 2030. Under this scenario, the most severe weather and rising sea level consequences are avoided by limiting the overall temperature increase to 2°C.

"Without a change in policy, the world is on a path for a rise in global temperature of up to 6°C, with catastrophic consequences for our climate," IEA said.

Reaching the goals of the 450 Scenario would require cumulative incremental investment of $10.5 trillion in low-carbon energy technologies and energy efficiency by 2030—largely offset by fuel cost savings—as well as starting to decrease fossil fuel demand after 2020.

In the same report, IEA warned that:

- the price of carbon credits will have to more than double from the $21/tonne seen in Europe in late 2009 if high-tech solutions to climate change are to be economically attractive.
- The world's use of fossil fuels—coal, oil, and natural gas—will have to peak by the early 2020s in order to bring about substantial reductions in atmospheric levels of CO_2

CCS test: Longannet

In October 2009, Scottish and Southern Energy PLC completed a successful 1-MW CCS pilot project at the Longannet coal-fired power plant in Fife, Scotland. An operational commercial-scale plant at the site is scheduled to start up in 2014.

EPA: GHG health-threat finding

In December 2009, the U.S. Environmental Protection Agency issued a finding based on its study showing that greenhouse gases threaten public health.

The "endangerment" finding is needed before the EPA can regulate carbon dioxide and five other greenhouse gases released from automobiles, power plants, and factories under the federal Clean Air Act.

Details are given in the United States section of the North America chapter of this book.

France: Carbon permits caution

In the run-up to the late-2009 UN climate-change conference in Copenhagen, the French finance ministry expressed concern about unregulated "dark pools" in the carbon trading market and called for creation of a single European regulator to prevent future abuses.

Details are given in the France section of the Europe chapter of this book.

Business: Clear CO_2 framework needed

A business coalition called Combat Climate Change said in November 2009 that a binding international agreement and related legislation on climate change must be in place before needed private-sector investment will be made to address the issue.

More than 60 large companies from around the world make up the group, including BP PLC, General Electric Co., Unilever, China National Offshore Oil Co., and Gazprom.

The group estimates that 80% of the money required to implement climate-change-mitigation measures will come from the private sector.

Climategate

In November 2009, e-mails written by some of the world's leading climate scientists at the University of East Anglia's Climate Research Unit in the UK were stolen by hackers and published on Web sites run by climate-change skeptics.

The skeptics claimed that the e-mails were evidence that scientists manipulated data in order to strengthen their argument that human activities were causing global warming.

Supporters of the scientists said that the documents simply describe typical give-and-take among researchers, the media, and others as various data, methods, and theories are being evaluated.

The International Energy Agency argues that important technology, such as CCS and widespread use of electric and hybrid cars, would be economic only if a high price for carbon penalizes those extracting and using dirtier energy sources, such as coal and gasoline.

IEA was optimistic about global action to keep atmospheric GHGs from exceeding 450 ppm—a level that scientists believe gives the world a 50% chance of keeping the increase in global temperatures within limits thought to be safe.

High cost of CCS

The addition of carbon capture and storage (CCS) solutions to fossil-fueled power plants can add up to 78% to the cost of electricity produced.

That was the finding in late 2009 from report by the Global Carbon Capture and Storage Institute about the world's current CCS projects. Other findings:

- Pipeline is the most cost-effective way ($3–4/tonne) of moving CO_2.
- Finding and characterizing a site for CO_2 storage can cost $15–150 million.
- The uncertainly of locating a suitable storage site raises storage cost by $3.50–7.50/tonne.

CO_2 capture cost

A study by INTEK Inc. released in late 2009 described the characteristics and costs of a range of technologies under development for capturing CO_2.

Estimates by the UN Intergovernmental Panel on Climate Change have estimated capture cost at $34–61/ton of CO_2 captured, depending on site and technology characteristics.

However, partnerships with enhanced oil and gas production projects, or the sale of pure CO_2 for industrial or commercial use, can yield values of up to $86/ton of CO_2 captured, transported, and sold.

Technologies studied include:

- Amine absorption, using a derivative of ammonia to treat flue gases.
- Hot and cold methanol
- Pressure swing absorption
- Potassium carbonate
- Membranes
- Combinations of amine and membranes.

Available industrial CO_2

Industrial source	CO_2 available, bcf/year
Fossil-fuel plants	21,763
Refineries	2,082
Cement plants	403
Hydrogen plants	343
Ammonia plants	126
Ethanol plants	108
All	**24, 825**

Average pipeline tariffs

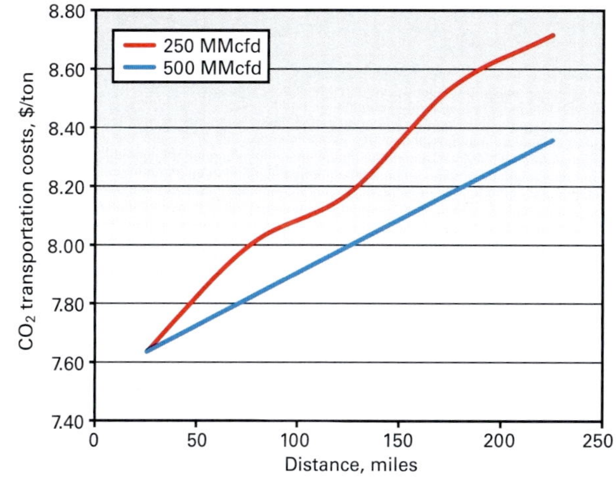

Source-specific CO_2 capture costs

Technology	$/ton
Fossil-fuel power plants	42–69
Refineries	39–61
Cement plants	39–61
Hydrogen plants	7–13
Ammonia plants	7–13
Ethanol plants	7–13
New integrated-gasification combined cycle plants	28–44

US annual carbon dioxide emissions

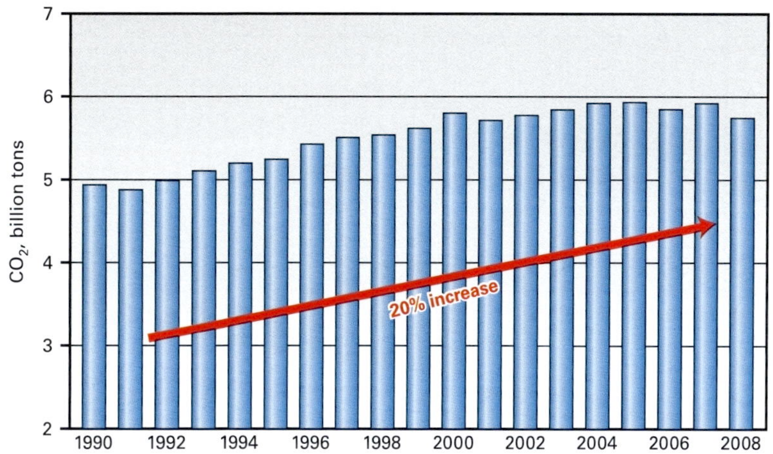

Source: US Energy Information Administration

Industrial CO_2 Sources

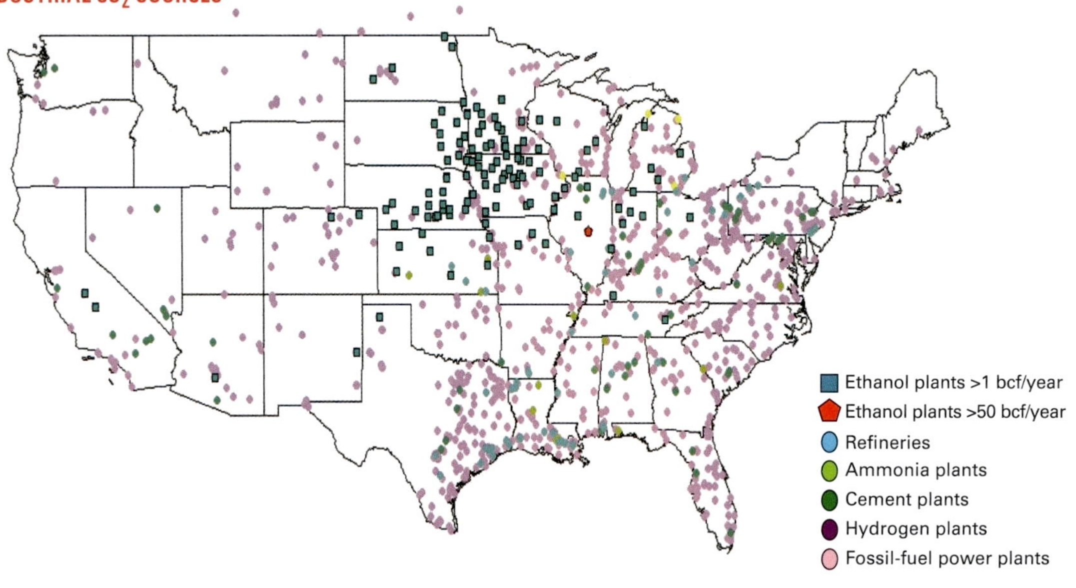

US CO_2 EOR Projects

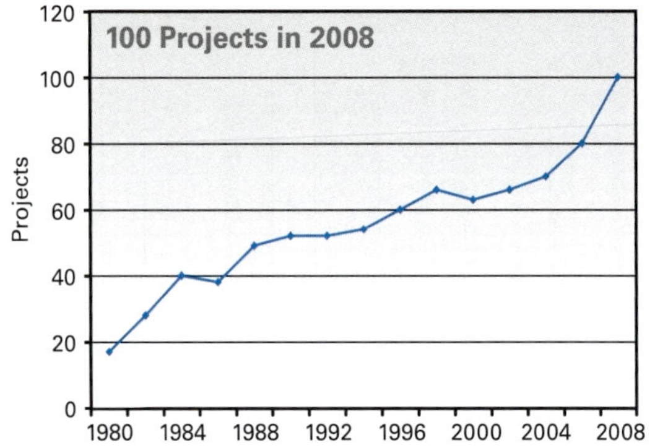

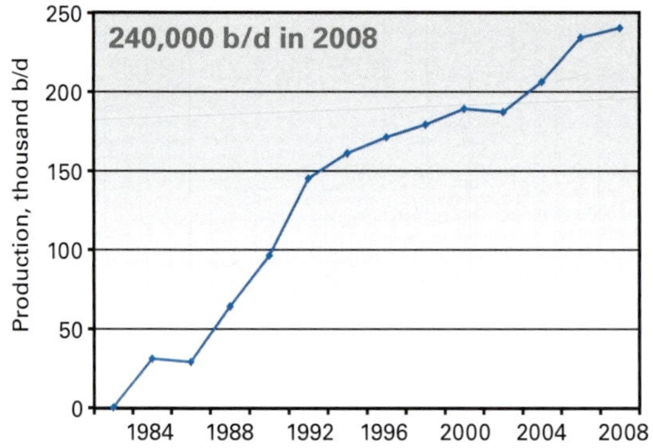

EOR CO_2 Sequestration

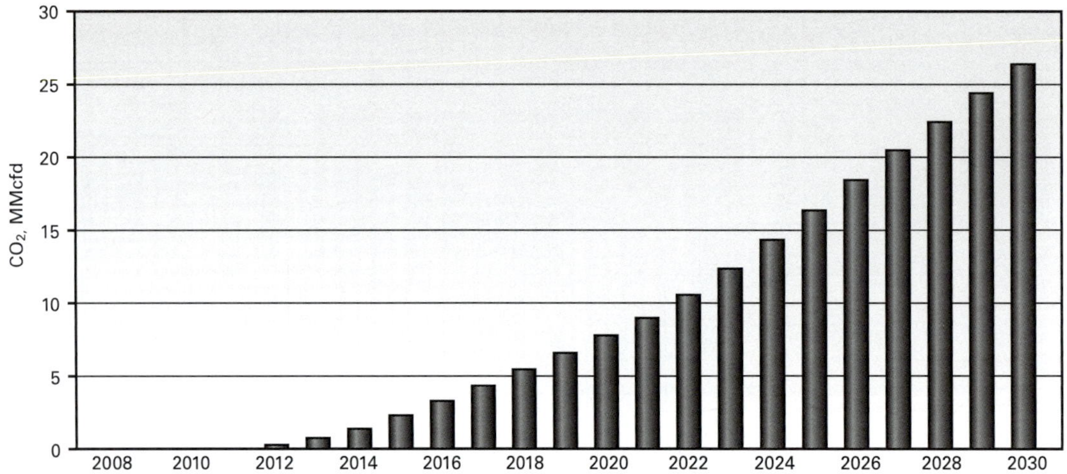

Deep Saline Aquifers

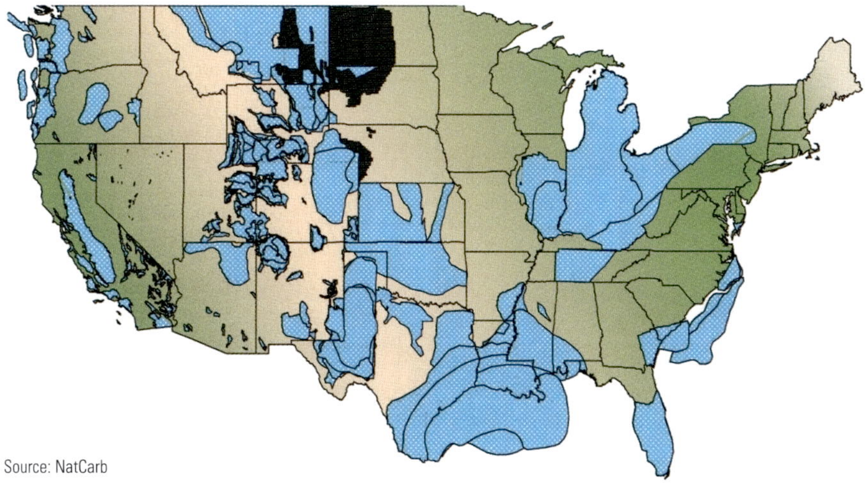

Source: NatCarb

Candidate US CO_2 EOR Fields

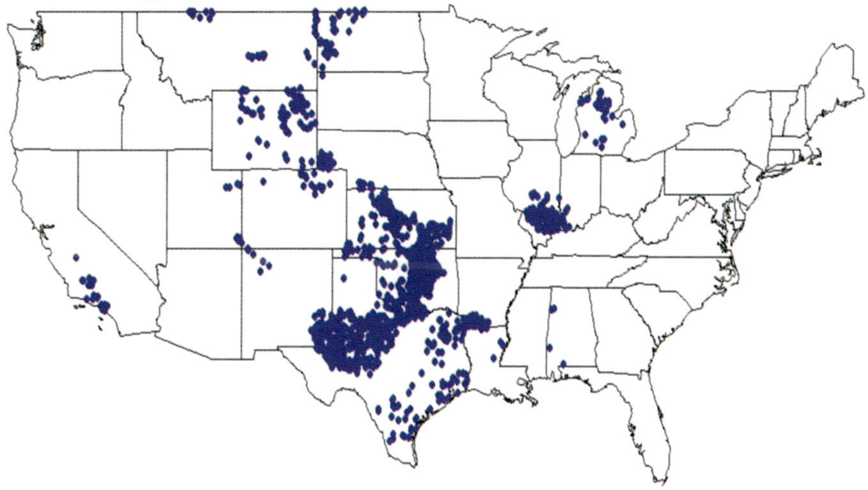

Unminable Coal Seams

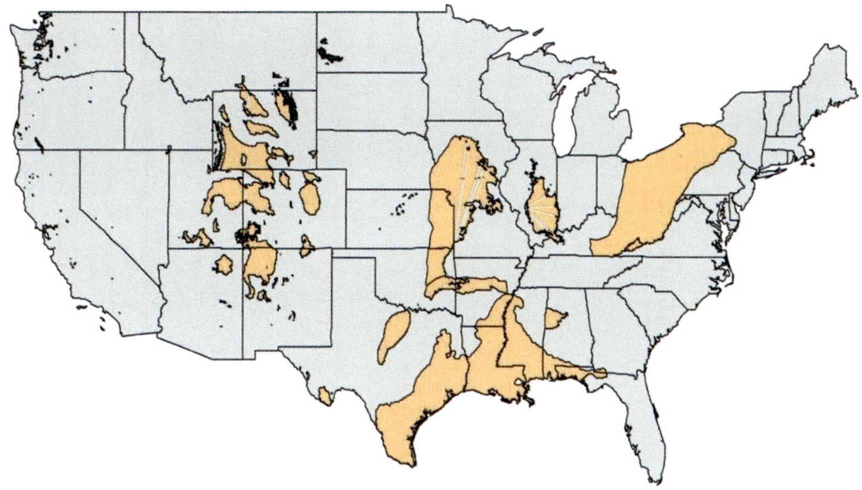

Sequestration, value options

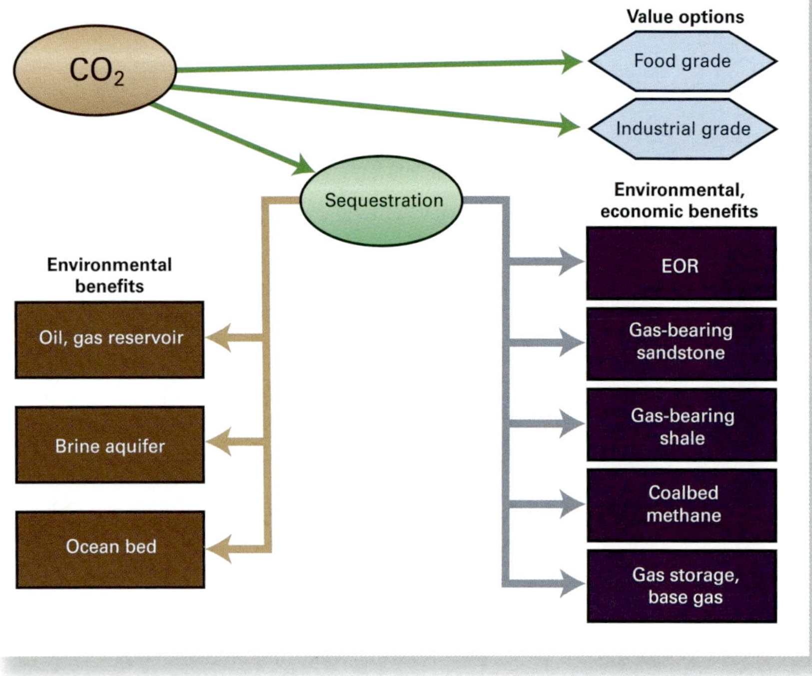

Costs, benefits of CO_2 capture and sequestration

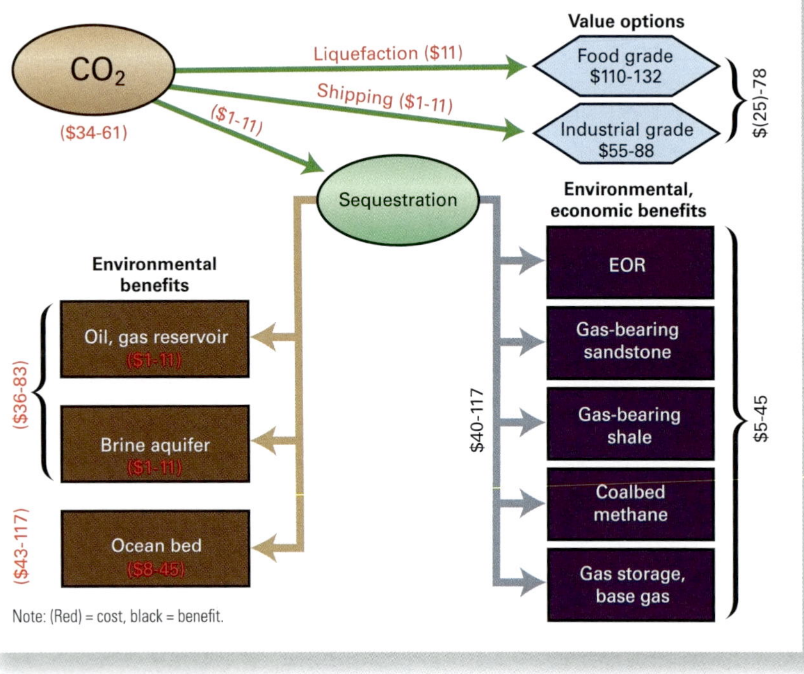

Border taxes and "carbon leakage"

Under discussion in late 2009 in the UK was the possible use of border tax adjustments to deal with "carbon leakage," the situation in which companies move to countries with lax carbon emissions rules.

Proponents say that such taxes would help to level the playing field by penalizing imports from countries lacking adequate controls on carbon.

The main alternative to such taxes is for countries to compensate domestic industries for any unfair competition. In the case of cap-and-trade systems, the compensation would take the form of a free allocation of carbon permits.

Arctic sea-ice thinning

Findings reported in the Catlin Arctic Survey, released in October 2009, warned that the Arctic sea will be ice-free in the summer within 20 years, shortening the global time frame on climate change by a decade.

They are based on study of floating sea-ice thickness collected during a 73-day trek made by researchers earlier in the year off the Canadian coast in the northern Beaufort Sea.

The research team covered 435 km (270 mi), making 1,500 drilled ice-floe measurements. The data, evaluated by the Polar Ocean Physics Group at the University of Cambridge, revealed sea ice in the region was significantly thinner than satellite images previously showed, cutting the period forecast for summer sea ice melt.

Such acceleration will mean storage of more heat in the Arctic Ocean that will be returned to the atmosphere in the autumn, changing wind patterns and affecting rainfall, food production, and human water supplies.

Late formation of sea ice will also mean an extra one to two months for shipping and possibly oil exploration in the north.

CO_2 transport guidelines

In September 2009, Det Norske Veritas (DNV) issued the first guidelines setting out criteria for the development, design, construction, testing, operation, and maintenance of steel pipelines transporting CO_2.

Developed as a joint industry project called CO2PIPETRANS, the guidelines address gaseous, liquid, and dense-phase operating conditions and apply to:

- new or converted offshore and onshore pipelines transporting fluids containing predominantly CO_2
- pipeline transportation of CO_2 captured from hydrocarbon streams and anthropogenic CO_2
- pipeline transportation of natural sources for EOR.

DNV also continued its work in 2009 on developing a standardized risk management approach for use in considering protection of human health and the environment associated with broader deployment of CCS technologies.

Impact of N_2O

A 2009 report of research sponsored by the International Council for Science (ICSU) concluded that, so far, the production of biofuels has aggravated rather than helped reduce global warming.

In particular, it points to the underestimated negative impact of nitrous oxide. N_2O is almost 300 times more potent than CO_2 in its potential, over the course of century, to warm the planet.

Researchers concluded that the amount of N_2O released by farming biofuel crops probably cancels out any advantage offered by reduced emissions of CO_2.

Rising sea levels

At a scientific conference in March 2009 in Copenhagen, a group of environmental experts announced that sea levels appear to be rising almost twice as rapidly as had been forecast in 2007 by the UN's Intergovernmental Panel on Climate Change.

The experts were from the University of Colorado-Boulder, the Centre for Australian Weather and Climate Research, the Potsdam Institute for Climate Impact Research, and the University of California-Irvine.

They said that new knowledge about the rate of polar ice-melting in the Antarctic and Greenland led to the new assessment of sea-level rise.

"Green" stimulus funding

In early 2009, governments around the world were crafting stimulus programs to deal with the unfolding global financial crisis and developing recession.

Economists estimated the total amount of stimulus funding at $2.8 trillion.

And many governments were earmarking major portions of stimulus money to projects aimed at moving away from fossil fuel use and toward low-carbon alternatives. Politicians and others were referring to this spending as a "green new deal."

EU: CO_2 permit prices

In early 2009, the price of CO_2 in the EU had fallen so low that is no longer provided an incentive to pursue low-carbon development. Experts pondered how long the low price would persist.

Permits to emit CO_2, issued under the EU's emissions trading scheme (ETS) were trading in early February at about €11.80 ($15.12, £10.42), according to analysts.

In mid-2008, the price was more than €30.

Economists say that a level of €35 to €45 is needed to spur "green" investment.

The Maldives: No fossil fuels

In March 2009, Pres. Mohamed Nasheed of the Maldives announced that his country would try to stop using fossil fuels—and thus eliminate most of its greenhouse-gas emissions—by 2020.

The tiny country is an archipelago of 26 atolls, 700 km (435 mi) southwest of Sri Lanka in the Indian Ocean.

The average elevation of the country is less than 2 m (7 ft) above sea level, so it could be the first nation to disappear if and when sea levels rise as a result of global climate change.

Table 1: World Reserves and Production

	ESTIMATED PROVED RESERVES				OIL PRODUCTION			
	Jan. 1, 2010		Jan. 1, 2009		Producing oil wells*, Dec. 31, 2008	Estimated 2009, 1,000 b/d	Change from 2008, %	Actual 2008, 1,000 b/d
COUNTRY	Oil, 1,000 bbl	Gas, bcf	Oil, 1,000 bbl	Gas, bcf				
Canada	175,214,000	61,950	178,092,000	57,906	62,519	2,530.0	-2.6	2,597.0
United States	19,121,000	244,656	21,317,000	237,726	512,560	5,337.0	7.8	4,950.0
North America	**194,335,000**	**306,606**	**199,409,000**	**295,632**	**575,079**	**7,867.0**	**4.2**	**7,547.0**
Argentina	2,520,300	14,070	2,616,000	15,600	15,799	610.0	0.1	609.3
Bolivia	465,000	26,500	465,000	26,500	361	40.0	0.0	40.0
Brazil	12,801,500	12,862	12,623,900	12,890	11,955	1,950.0	7.6	1,812.5
Chile	150,000	3,460	150,000	3,460	315	2.5	8.7	2.3
Colombia	1,355,000	3,955	1,355,000	3,739	7,600	660.0	13.8	580.0
Cuba	124,000	2,500	124,000	2,500	251	50.0	0.0	50.0
Ecuador	6,500,000	282	4,660,000	315	1,296	470.0	-5.6	498.0
Guatemala	83,070	0	83,070	0	20	14.0	7.7	13.0
Mexico	10,404,000	12,702	10,501,200	13,162	3,052	2,605.0	-6.9	2,798.8
Peru	447,382	11,800	415,769	11,842	5,035	106.0	30.1	81.5
Trinidad & Tobago	728,300	15,400	728,300	18,770	3,903	107.0	-5.1	112.7
Venezuela	99,377,000	175,970	99,377,000	170,920	15,669	2,170.0	-7.7	2,350.0
Other Latin America	88,089	4	88,471	5	1,280	16.8	-0.6	16.9
Latin America	**135,043,641**	**279,505**	**133,187,710**	**279,703**	**66,536**	**8,801.3**	**-1.8**	**8,965.0**
Austria	50,000	570	50,000	570	905	18.4	5.7	17.4
Denmark	1,060,000	2,165	1,060,000	2,165	232	265.0	-7.6	286.7
France	101,150	250	103,300	245	421	18.3	-6.6	19.6
Germany	276,000	6,200	276,000	6,200	1,141	55.5	-9.2	61.1
Greece	10,000	35	10,000	70	12	1.6	33.3	1.2
Ireland	0	350	0	350	0	0.0	0.0	0.0
Italy	423,678	2,466	406,500	3,325	205	80.0	-20.0	100.0
Netherlands	100,000	50,000	100,000	50,000	203	25.5	-25.0	34.0
Norway	6,680,000	81,680	6,680,000	81,680	801	2,025.0	-7.1	2,180.0
Spain	150,000	90	150,000	90	16	1.9	-24.0	2.5
Turkey	262,200	215	300,000	300	897	45.0	10.0	40.9
United Kingdom	3,084,480	10,312	3,410,000	12,110	1,317	1,305.0	-7.8	1,415.5
Western Europe	**12,197,508**	**154,333**	**12,545,800**	**157,105**	**6,150**	**3,841.2**	**-7.6**	**4,158.9**
Abu Dhabi	92,200,000	198,500	92,200,000	198,500	1,200	2,135.0	-12.3	2,435.0
Bahrain	124,560	3,250	124,560	3,250	496	29.5	-10.3	32.9
Dubai	4,000,000	4,000	4,000,000	4,000	200	92.0	-11.5	104.0
Iran	137,620,000	1,045,670	136,150,000	991,600	1,128	3,725.0	-4.7	3,910.0
Iraq	115,000,000	111,940	115,000,000	111,940	1,685	2,400.0	-1.0	2,424.0
Israel	1,940	1,075	1,940	1,075	6	0.0	0.0	0.0
Jordan	1,000	213	1,000	213	4	0.0	0.0	0.0
Kuwait	101,500,000	63,000	101,500,000	62,860	790	2,010.0	-13.0	2,310.0
Neutral Zone	5,000,000	1,000	5,000,000	1,000	578	540.0	-5.3	570.0
Oman	5,500,000	30,000	5,500,000	30,000	2,298	800.0	6.7	750.0
Qatar	25,410,000	899,325	15,210,000	891,945	421	765.0	-9.8	848.0
Ras al Khaimah	100,000	1,200	100,000	1,200	7	0.7	-12.5	0.8
Saudi Arabia	259,900,000	263,000	264,210,000	257,970	1,560	7,920.0	-11.0	8,900.0
Sharjah	1,500,000	10,700	1,500,000	10,700	49	43.0	-14.0	50.0
Syria	2,500,000	8,500	2,500,000	8,500	136	365.0	-6.4	390.0
Yemen	3,000,000	16,900	3,000,000	16,900	1,649	270.0	-4.6	283.0
Middle East	**753,357,500**	**2,658,273**	**745,997,500**	**2,591,653**	**12,207**	**21,095.2**	**-8.3**	**23,007.7**
Algeria	12,200,000	159,000	12,200,000	159,000	1,285	1,240.0	-9.6	1,372.0
Angola	9,500,000	9,600	9,040,000	9,530	1,064	1,790.0	-5.5	1,894.0
Cameroon	200,000	4,770	200,000	4,770	255	74.0	-11.7	83.8
Chad	1,500,000	0	1,500,000	0	521	145.0	0.0	145.0
Congo (former Zaire)	180,000	35	180,000	35	150	25.0	0.0	25.0
Congo (Brazzaville)	1,600,000	3,200	1,600,000	3,200	460	240.0	0.0	240.0
Egypt	3,700,000	58,500	3,700,000	58,500	1,491	635.0	-2.3	650.0
Equatorial Guinea	1,100,000	1,300	1,100,000	1,300	73	320.0	0.0	320.0
Gabon	2,000,000	1,000	2,000,000	1,000	395	230.0	9.5	210.0
Ivory Coast	100,000	1,000	100,000	1,000	9	30.0	0.0	30.0
Libya	44,270,000	54,362	43,660,000	54,380	1,543	1,545.0	-10.4	1,725.0

	ESTIMATED PROVED RESERVES				OIL PRODUCTION			
	Jan. 1, 2010		Jan. 1, 2009					
COUNTRY	Oil, 1,000 bbl	Gas, bcf	Oil, 1,000 bbl	Gas, bcf	Producing oil wells*, Dec. 31, 2008	Estimated 2009, 1,000 b/d	Change from 2008, %	Actual 2008, 1,000 b/d
Mauritania	100,000	1,000	100,000	1,000	0	0.0	0.0	0.0
Morocco	752	53	752	53	7	0.3	0.0	0.3
Mozambique	0	4,500	0	4,500	0	0.0	0.0	0.0
Nigeria	37,200,000	185,280	36,220,000	184,160	2,524	1,810.0	-7.2	1,950.0
South Africa	15,000	0	15,000	0	28	14.0	0.0	14.0
Sudan	5,000,000	3,000	5,000,000	3,000	9	500.0	2.0	490.0
Tanzania	0	230	0	230	0	0.0	0.0	0.0
Tunisia	425,000	2,300	425,000	2,300	237	82.0	-3.9	85.3
Other Africa	23,428	6,120	23,428	6,120	11	6.0	0.0	6.0
Africa	**119,114,180**	**495,250**	**117,064,180**	**494,078**	**10,062**	**8,686.3**	**-6.0**	**9,240.4**
Albania	199,140	30	199,140	30	1,945	9.8	-1.0	9.9
Azerbaijan	7,000,000	30,000	7,000,000	30,000	58	1,000.0	11.7	895.0
Belarus	198,000	100	198,000	100	0	32.0	-3.0	33.0
Bulgaria	15,000	200	15,000	200	281	1.0	0.0	1.0
Croatia	73,350	1,080	79,300	1,080	846	13.8	-6.8	14.8
Czech Republic	15,000	140	15,000	140	0	4.1	-6.8	4.4
Georgia	35,000	300	35,000	300	283	1.0	0.0	1.0
Hungary	26,566	286	20,180	286	875	13.7	-4.9	14.4
Kazakhstan	30,000,000	85,000	30,000,000	85,000	1,006	1,330.0	9.9	1,210.0
Kyrgyzstan	40,000	200	40,000	200	0	1.0	0.0	1.0
Lithuania	12,000	0	12,000	0	0	2.0	-33.3	3.0
Poland	96,375	5,820	96,375	5,820	512	12.0	-26.8	16.4
Romania	600,000	2,225	600,000	2,225	6,000	90.0	-3.2	93.0
Russia	60,000,000	1,680,000	60,000,000	1,680,000	105,339	9,915.0	1.7	9,750.0
Serbia	77,500	1,700	77,500	1,700	646	15.0	0.0	15.0
Slovakia	9,000	500	9,000	500	0	1.0	0.0	1.0
Tajikistan	12,000	200	12,000	200	0	0.0	0.0	0.0
Turkmenistan	600,000	265,000	600,000	94,000	2,460	220.0	0.0	220.0
Ukraine	395,000	39,000	395,000	39,000	2,494	75.0	-3.2	77.5
Uzbekistan	594,000	65,000	594,000	65,000	2,190	85.0	-5.6	90.0
Eastern Europe & FSU	**99,997,931**	**2,176,781**	**99,997,495**	**2,005,781**	**124,935**	**12,821.4**	**3.0**	**12,450.4**
Afghanistan	0	1,750	0	1,750	0	0.0	0.0	0.0
Australia	3,318,000	110,000	1,500,000	30,000	1,305	465.0	0.3	463.8
Bangladesh	28,000	6,900	28,000	5,000	40	6.0	-7.7	6.5
Brunei	1,100,000	13,800	1,100,000	13,800	779	148.0	-7.9	160.7
China	20,350,000	107,000	16,000,000	80,000	71,542	3,773.0	-1.0	3,810.0
China, Taiwan	2,380	220	2,380	220	71	0.8	0.0	0.8
India	5,624,640	37,960	5,624,640	37,960	3,686	664.0	-1.9	677.0
Indonesia	3,990,000	106,000	3,990,000	106,000	8,331	850.0	-0.9	857.3
Japan	44,115	738	44,115	738	145	15.3	-19.0	18.9
Malaysia	4,000,000	83,000	4,000,000	83,000	788	730.0	-3.5	756.7
Myanmar	50,000	10,000	50,000	10,000	450	20.0	5.3	19.0
New Zealand	60,000	1,200	60,000	1,200	72	52.0	-3.9	54.1
Pakistan	313,000	29,671	339,000	31,266	204	63.0	-5.1	66.4
Papua New Guinea	88,000	8,000	88,000	8,000	46	37.0	-9.8	41.0
Philippines	138,500	3,480	138,500	3,480	11	30.0	100.0	15.0
Thailand	430,000	12,079	441,000	11,198	1,178	236.0	3.5	228.1
Vietnam	600,000	6,800	600,000	6,800	34	300.0	8.2	277.3
Asia-Pacific	**40,136,635**	**538,598**	**34,005,635**	**430,412**	**88,682**	**7,390.1**	**-0.8**	**7,452.6**
Total World	**1,354,182,395**	**6,609,346**	**1,342,207,320**	**6,254,363**	**883,691**	**70,502.6**	**-3.2**	**72,822.0**
Total OPEC	**944,017,000**	**3,216,020**	**931,999,000**	**3,151,742**	**39,330**	**32,136.0**	**3.8**	**30,948.9**

Source: *Oil & Gas Journal* survey data as of Jan. 1, 2010.

NOTES: All reserves figures are reported as proved reserves recoverable with current technology and prices except Former Soviet Union, some eastern European countries, and Canada's gas reserves; these figures include proved plus some probable.

Other Latin America: Barbados, Belize, Suriname. Other Africa: Benin, Ethiopia, Ghana, Namibia, Rwanda, Somalia.

* Does not include shut-in, injection, or service wells.

Some totals might not compute exactly due to rounding and other factors.

Table 2: World Oil Consumption

	2006	2007	2008	2006	2007	2008	Change 2008 over 2007	2008 share of total
COUNTRY	million tonnes per year (mtpy)			thousand barrels per day (1,000 b/d)			%	
North America								
United States	943.8	942.3	884.5	20,687	20,680	19,419	-6.4	22.5
Canada	99.6	102.8	102.0	2,246	2,323	2,295	-1.0	2.6
Mexico	86.8	89.4	90.0	1,970	2,027	2,039	0.5	2.3
Total	1,130.2	1,134.5	1,076.6	24,904	25,030	23,753	-5.4	27.4
Latin America								
Argentina	20.6	23.2	24.2	432	487	508	4.1	0.6
Brazil	92.8	99.8	105.3	2,102	2,274	2,397	5.3	2.7
Chile	12.3	16.1	16.8	264	346	358	3.8	0.4
Colombia	10.8	10.7	10.7	240	234	234	0.3	0.3
Ecuador	8.2	8.9	9.3	182	196	204	4.0	0.2
Peru	6.8	7.1	7.9	147	154	172	10.8	0.2
Venezuela	27.4	30.0	32.5	607	665	719	8.1	0.8
Other Latin America	64.1	64.2	63.6	1,323	1,324	1,308	-1.3	1.6
Total	243.0	260.0	270.3	5,297	5,681	5,901	3.7	6.9
Europe/Eurasia								
Austria	14.2	13.4	13.4	294	278	278	-0.1	0.3
Azerbaijan	4.9	4.5	3.3	99	92	71	-25.9	0.1
Belarus	8.0	7.6	7.7	165	157	159	0.9	0.2
Belgium/Luxembourg	41.1	41.0	41.3	839	832	836	0.4	1.1
Bulgaria	5.2	5.3	5.4	116	117	120	2.1	0.1
Czech Republic	9.8	9.7	9.9	208	206	210	1.4	0.3
Denmark	9.3	9.3	8.9	197	196	189	-4.0	0.2
Finland	10.6	10.6	10.5	225	226	225	-1.2	0.3
France	93.0	91.3	92.2	1,956	1,921	1,930	0.7	2.3
Germany	123.6	112.5	118.3	2,624	2,393	2,505	4.9	3.0
Greece	22.2	21.7	21.4	454	445	438	-1.5	0.5
Hungary	7.8	7.7	7.7	169	169	169	-0.2	0.2
Iceland	1.0	1.0	0.9	20	21	19	-9.1	*
Republic of Ireland	9.3	9.4	9.0	195	198	190	-4.4	0.2
Italy	86.7	84.0	80.9	1,831	1,759	1,691	-3.9	2.1
Kazakhstan	11.0	11.6	10.9	227	244	229	-6.4	0.3
Lithuania	2.8	2.8	3.1	59	59	64	9.4	0.1
Netherlands	49.0	46.0	46.5	1,043	962	982	0.8	1.2
Norway	10.0	10.2	9.8	217	222	210	-4.4	0.2
Poland	23.3	24.2	24.9	516	535	545	2.6	0.6
Portugal	14.4	14.4	13.7	300	302	286	-5.1	0.3
Romania	10.3	10.3	10.6	219	223	227	2.2	0.3
Russian Federation	127.1	126.2	130.4	2,709	2,706	2,797	3.1	3.3
Slovakia	4.0	4.1	4.3	82	86	89	3.2	0.1
Spain	78.1	78.8	77.1	1,602	1,617	1,574	-2.5	2.0
Sweden	15.5	14.7	14.5	322	308	302	-1.5	0.4
Switzerland	12.6	11.3	12.1	269	243	258	6.4	0.3
Turkey	29.5	30.5	32.3	635	656	690	5.8	0.8
Turkmenistan	5.3	5.4	5.5	118	121	123	1.7	0.1
Ukraine	15.0	15.3	15.5	318	325	326	0.5	0.4
United Kingdom	82.3	79.2	78.7	1,785	1,714	1,704	-1.0	2.0
Uzbekistan	5.1	5.4	5.5	105	111	113	1.5	0.1
Other Europe/Eurasia	26.9	27.9	29.2	562	585	610	4.3	0.7
Total	968.5	947.6	955.5	20,463	20,031	20,158	0.6	24.3

COUNTRY	2006	2007	2008	2006	2007	2008	Change 2008 over 2007	2008 share of total
	million tonnes per year (mtpy)			thousand barrels per day (1,000 b/d)			%	
Middle East								
Iran	80.8	81.3	83.3	1,693	1,693	1,730	2.1	2.1
Kuwait	13.6	13.6	15.3	267	269	300	11.7	0.4
Qatar	3.3	4.0	4.6	75	90	104	16.5	0.1
Saudi Arabia	87.4	96.1	104.2	1,841	2,054	2,224	8.1	2.7
United Arab Emirates	19.5	20.7	22.9	402	425	467	10.6	0.6
Other Middle East	73.7	74.4	76.6	1,539	1,554	1,599	2.7	2.0
Total	278.3	290.1	306.9	5,817	6,084	6,423	5.5	7.8
Africa								
Algeria	11.5	12.9	14.0	260	288	311	8.3	0.4
Egypt	28.7	30.6	32.6	610	650	693	6.4	0.8
South Africa	25.3	25.8	26.3	537	549	558	1.7	0.7
Other Africa	60.9	60.7	62.3	1,290	1,288	1,319	2.4	1.6
Total	126.3	129.9	135.2	2,606	2,776	2,881	3.8	3.4
Asia-Pacific								
Australia	41.4	41.7	42.5	918	925	936	1.5	1.1
Bangladesh	4.5	4.5	4.6	93	94	95	1.9	0.1
China	346.1	362.8	375.7	7,382	7,742	7,999	3.3	9.6
China Hong Kong	14.9	16.0	14.5	305	324	294	-9.4	0.4
India	120.4	128.5	135.0	2,580	2,748	2,882	4.8	3.4
Indonesia	55.2	56.4	57.4	1,173	1,201	1,217	1.5	1.5
Japan	237.5	229.3	221.8	5,213	5,039	4,845	-3.5	5.6
Malaysia	21.0	22.0	21.8	459	481	475	-1.4	0.6
New Zealand	7.2	7.2	7.3	156	156	158	1.2	0.2
Pakistan	17.6	19.2	19.3	356	388	389	0.2	0.5
Philippines	13.3	14.0	13.4	284	300	288	-4.0	0.3
Singapore	44.0	47.4	49.9	853	916	958	5.0	1.3
South Korea	105.5	108.3	103.3	2,317	2,389	2,291	-4.9	2.6
Taiwan	51.7	52.5	50.1	1,097	1,123	1,074	-4.8	1.3
Thailand	39.3	38.2	36.7	834	823	797	-4.1	0.9
Other Asia-Pacific	28.1	29.4	30.0	601	629	641	1.8	0.8
Total	1,147.7	1,177.4	1,183.4	24,620	25,277	26,339	0.2	30.1
Total World	3,894.0	3,939.4	3,927.9	83,797	84,878	84,455	-0.6	100.0

Source: Adapted from *BP Statistical Review of World Energy 2009* (June 2009).

NOTES: "Oil Consumption" includes inland demand plus international aviation and marine bunkers and refinery fuel and loss. Consumption of fuel ethanol and biodiesel is also included.

Some totals might not compute exactly due to rounding and other factors.

Growth rates are adjusted for leap years.

Differences between these world consumption figures and world production statistics are accounted for by stock changes, consumption of non-petroleum additives and substitute fuels, and unavoidable disparities in the definition, measurement or conversion of oil supply and demand data.

Annual changes and shares of total are calculated using million tonnes per year figures.

* Less than 0.05%

Table 3: World Petroleum Product Consumption

REGION/PRODUCT	2003	2004	2005	2006	2007	2008	Change 2008 over 2007	2008 share of total
	thousand barrels per day (1,000 b/d)						%	
United States								
Light distillates	9,275	9,518	9,548	9,599	9,597	9,220	-3.9	47.5
Middle distillates	5,886	6,116	6,198	6,226	6,199	5,790	-6.6	29.8
Fuel oil	763	859	914	683	718	615	-14.4	3.2
Others	4,109	4,239	4,143	4,178	4,166	3,794	-8.9	19.5
Total	20,033	20,732	20,802	20,687	20,680	19,419	-6.1	100.0
North America (including U.S.)								
Light distillates	10,675	10,979	11,020	11,131	11,205	10,857	-3.1	45.7
Middle distillates	6,861	7,133	7,249	7,297	7,318	6,924	-5.4	29.2
Fuel oil	1,271	1,340	1,396	1,055	1,082	946	-12.5	4.0
Others	5,243	5,446	5,359	5,420	5,425	5,026	-7.4	21.2
Total	24,050	24,898	25,023	24,904	25,030	23,753	-5.1	100.0
South & Central America								
Light distillates	1,496	1,452	1,551	1,637	1,746	1,801	3.2	30.5
Middle distillates	1,655	1,776	1,833	1,873	2,040	2,126	4.2	36.0
Fuel oil	754	740	744	766	810	814	0.6	13.8
Others	925	971	1,002	1,021	1,086	1,160	6.8	19.7
Total	4,830	4,938	5,120	5,297	5,681	5,901	3.9	100.0
Europe								
Light distillates	3,923	3,875	3,764	3,615	3,491	3,365	-3.6	20.9
Middle distillates	7,137	7,347	7,595	7,853	7,763	7,937	2.2	49.3
Fuel oil	1,952	1,888	1,855	1,773	1,667	1,633	-2.0	10.1
Others	3,158	3,221	3,298	3,273	3,137	3,179	1.4	19.7
Total	16,170	16,331	16,513	16,514	16,058	16,114	0.3	100.0
Former Soviet Union								
Light distillates	937	918	934	993	1,053	1,096	4.1	27.1
Middle distillates	1,085	1,108	1,148	1,208	1,274	1,324	4.0	32.7
Fuel oil	643	578	558	615	511	455	-11.0	11.2
Others	1,080	1,157	1,132	1,133	1,136	1,170	3.0	28.9
Total	3,745	3,760	3,772	3,948	3,973	4,045	1.8	100.0
Middle East								
Light distillates	1,059	1,177	1,241	1,325	1,326	1,379	4.0	21.5
Middle distillates	1,723	1,767	1,849	1,950	2,013	2,124	5.5	33.1
Fuel oil	1,317	1,390	1,463	1,450	1,532	1,644	7.3	25.6
Others	1,039	1,036	1,064	1,092	1,213	1,276	5.2	19.9
Total	5,138	5,370	5,617	5,817	6,084	6,423	5.6	100.0
Africa								
Light distillates	605	632	637	617	634	666	5.1	23.1
Middle distillates	1,066	1,070	1,119	1,137	1,181	1,232	4.3	42.8
Fuel oil	380	390	437	417	416	429	3.1	14.9
Others	486	509	534	525	545	554	1.7	19.2
Total	2,537	2,600	2,728	2,696	2,776	2,881	3.8	100.0
China								
Light distillates	1,489	1,673	1,776	1,853	1,983	2,175	9.7	27.2
Middle distillates	1,945	2,285	2,528	2,723	2,879	3,178	10.4	39.7
Fuel oil	880	956	904	938	892	716	19.8	8.9
Others	1,489	1,858	1,777	1,867	1,988	1,930	-2.9	24.1
Total	5,803	6,772	6,984	7,382	7,742	7,999	3.3	100.0

REGION/PRODUCT	2003	2004	2005	2006	2007	2008	Change 2008 over 2007	2008 share of total
	thousand barrels per day (1,000 b/d)						%	
Japan								
Light distillates	1,788	1,793	1,822	1,813	1,790	1,675	-6.4	34.6
Middle distillates	1,904	1,856	1,876	1,758	1,619	1,500	-7.4	31.0
Fuel oil	755	660	683	664	664	712	7.2	14.7
Others	993	960	963	979	966	959	-0.7	19.8
Total	5,440	5,269	5,343	5,213	5,039	4,845	-3.8	100.0
Asia-Pacific (including China, Japan)								
Light distillates	6,283	6,644	6,781	6,826	7,147	7,196	0.7	28.4
Middle distillates	8,246	8,774	9,030	9,089	9,151	9,496	3.8	37.5
Fuel oil	3,547	3,558	3,552	3,632	3,614	3,403	-5.9	13.4
Others	4,526	4,923	4,921	5,073	5,364	5,244	-2.2	20.7
Total	22,601	23,899	24,283	24,620	25,277	25,339	0.2	100.0
World								
Light distillates	24,976	25,676	25,929	26,145	26,601	26,359	-0.9	31.2
Middle distillates	27,773	28,974	29,823	30,407	30,740	31,163	1.4	36.9
Fuel oil	9,865	9,884	10,004	9,708	9,631	9,324	-3.2	11.0
Others	16,457	17,263	17,308	17,537	17,905	17,609	-1.7	20.8
Total	79,071	81,796	83,065	83,797	84,878	84,455	-0.5	100.0
European Union								
Light distillates	3,671	3,619	3,520	3,367	3,255	3,139	-3.5	21.3
Middle distillates	6,592	6,795	6,996	7,221	7,109	7,234	1.8	49.0
Fuel oil	1,743	1,703	1,705	1,651	1,553	1,516	-2.4	10.3
Others	2,861	2,916	2,985	2,972	2,829	2,876	1.6	19.5
Total	14,867	15,032	15,205	15,211	14,746	14,765	0.1	100.0
OECD								
Light distillates	17,495	17,747	17,753	17,739	17,764	17,149	-3.5	36.3
Middle distillates	16,684	17,095	17,483	17,676	17,459	17,088	-2.1	36.1
Fuel oil	4,294	4,205	4,237	3,784	3,692	3,513	-4.8	7.4
Others	9,803	10,026	10,015	10,074	9,915	9,552	-3.7	20.2
Total	48,276	49,073	49,490	49,274	48,830	47,303	-3.1	100.0
Other EMEs								
Light distillates	6,544	7,011	7,241	7,412	7,785	8,115	4.2	24.5
Middle distillates	10,004	10,771	11,192	11,523	12,007	12,751	6.2	38.5
Fuel oil	4,928	5,102	5,209	5,309	5,428	5,356	-1.3	16.2
Others	5,574	6,079	6,161	6,330	6,855	6,886	0.5	20.8
Total	27,050	28,963	29,804	30,575	32,075	33,108	3.2	100.0

Source: Adapted from *BP Statistical Review of World Energy 2009* (June 2009).

NOTES: Other EMEs excludes Former Soviet Union.
Annual changes and shares of total are calculated using thousand barrels daily figures.
Some totals might not compute exactly due to rounding and other factors.
'Light distillates' consists of aviation and motor gasolines and light distillate feedstock (LDF).
'Middle distillates' consists of jet and heating kerosenes, and gas and diesel oils (including marine bunkers).
'Fuel oil' includes marine bunkers and crude oil used directly as fuel
'Others' consists of refinery gas, LPGs, solvents, petroleum coke, lubricants, bitumen, wax, other refined products, and refinery fuel and loss.

Table 4: World Oil Production

COUNTRY	1990	2000	2002	2003	2004	2005	2006	2007	Actual 2008	Est. 2009
	1,000 b/d									
North America										
Canada	1,508.0	2,035.0	2,212.9	2317.4	2,417.8	2,368.8	2,517.2	2,618.1	2,597.0	2,530.0
United States	7,220.0	5,822.0	5,746.0	5,681.6	5,418.7	5,178.4	5,101.7	5,064.5	4,950.0	5,337.0
Total	8,728.0	7,857.0	7,958.9	7,999.0	7,836.5	7,547.2	7,618.9	7,682.6	7,547.0	7,867.0
Latin America										
Argentina	473.0	750.7	738.1	707.0	738.3	645.4	640.5	623.5	609.3	610.0
Bolivia	19.2	27.6	30.9	32.2	35.0	41.6	44.8	44.0	40.0	40.0
Brazil	633.0	1,128.0	1,495.0	1,535.7	1,479.2	1,634.0	1,722.3	1,747.8	1,812.5	1,950.0
Chile	20.3	7.0	10.0	10.0	10.0	10.0	10.0	2.6	2.3	2.5
Colombia	445.0	687.3	578.5	541.3	527.6	526.1	527.5	531.0	580.0	660.0
Cuba	15.0	44.0	40.0	40.0	40.0	38.6	38.7	50.0	50.0	50.0
Ecuador	287.0	401.1	392.5	417.8	525.5	532.0	534.6	500.0	498.0	470.0
Guatemala	4.0	20.7	23.6	23.9	20.2	18.5	16.1	16.0	13.0	14.0
Mexico	2,633.0	3,012.0	3,177.0	3370.4	3,383.3	3,334.2	3,256.3	3,082.6	2,798.8	2,605.0
Peru	132.0	98.0	92.8	87.3	88.6	111.2	115.6	78.0	81.5	106.0
Trinidad & Tobago	151.0	119.0	130.6	136.5	122.9	144.5	142.8	120.0	112.7	107.0
Venezuela	2,118.0	3,028.0	2,285.0	2,006.7	2,208.3	2,705.8	2,561.7	2,390.0	2,350.0	2,170.0
Other Latin America	—	—	—	12.8	12.3	13.0	14.1	15.7	16.9	16.8
Total	6,935.7	9,323.4	9,007.6	8,921.6	9,191.1	9,754.9	9,625.0	9,201.2	8,965.0	8,801.3
Europe										
Albania	—	5.9	6.2	6.0	6.1	6.1	8.9	9.6	9.9	9.8
Austria	24.0	19.1	18.6	18.2	17.1	17.3	17.4	17.4	17.4	18.4
Bulgaria	5.0	1.0	1.0	1.0	1.0	1.0	1.0	1.0	1.0	1.0
Croatia	—	23.0	20.7	20.7	19.1	17.5	16.7	15.9	14.8	13.8
Czech Republic	3.0	3.2	4.8	5.8	5 .9	4.8	4.9	4.5	4.4	4.1
Denmark	118.0	363.0	370.8	367.7	389.1	377.3	342.2	311.7	286.7	265.0
France	62.0	28.4	26.4	24.5	22.9	21.7	21.4	19.6	19.6	18.3
Germany	78.0	62.5	74.1	76.0	70.3	71.5	70.3	68.3	61.1	55.5
Greece	15.0	5.1	3.3	2.4	2.3	1.8	1.8	1.5	1.2	1.6
Hungary	40.0	25.8	22.0	23.7	22.5	19.0	16.4	15.8	14.4	13.7
Italy	97.0	90.8	84.4	95.8	115.0	115.0	110.4	107.9	100.0	80.0
Lithuania	—	—	—	—	—	4.3	3.6	3.1	3.0	2.0
Netherlands	71.0	29.3	43.7	45.8	44.0	29.0	26.3	40.1	34.0	25.5
Norway	1,574.0	3,212.0	3,148.8	3,065.0	2,955.1	2,697.6	2,492.0	2,271.0	2,180.0	2,025.0
Poland	3.0	14.3	16.6	17.6	20.1	19.9	17.5	17.6	16.4	12.0
Romania	160.0	122.9	116.1	115.2	103.0	100.0	98.2	97.6	93.0	90.0
Serbia	—	—	—	15.5	15.0	15.0	15.0	15.0	15.0	15.0
Slovakia	—	—	—	1.0	1.0	1.0	0.3	0.1	1.0	1.0
Spain	14.3	4.6	6.5	6.6	5.3	3.3	2.9	2.9	2.5	1.9
Turkey	70.0	56.0	46.9	44.4	42.0	42.0	41.5	40.6	40.9	45.0
United Kingdom	1,860.0	2,513.7	2,305.8	2,103.2	1,888.3	1,678.0	1,518.2	1,524.1	1,415.5	1,305.0
Total	4,194.3	6,580.6	6,333.8	6,056.1	5,745.1	5,243.1	4,826.9	4,585.3	4,331.8	4,003.6
Middle East										
Bahrain	42.0	102.3	174.1	33.8	174.2	174.4	171.9	169.9	32.9	29.5
Iran	3,120.0	3,681.7	3,430.0	3,784.2	3,931.7	3,890.8	3,890.8	3,980.0	3,910.0	3,725.0
Iraq	2,083.0	2,566.7	2,014.2	1,328.3	2,002.5	1,810.0	1,900.0	2,090.0	2,424.0	2,400.0
Israel	—	—	—	0.1	0.1	0.1	0.1	—	—	—
Kuwait	1,080.0	1,765.0	1,600.0	1,870.0	2,050.0	2,130.0	2,210.0	2,160.0	2,310.0	2,010.0
Neutral Zone	315.0	630.0	538.0	610.0	600.0	580.0	580.0	560.0	570.0	540.0
Oman	658.0	933.3	904.2	828.3	764.2	757.5	740.0	710.0	750.0	800.0
Qatar	387.0	688.3	644.2	736.7	782.5	798.3	822.5	800.0	848.0	765.0
Saudi Arabia	6,215.0	7,995.0	6,850.0	8,480.0	8,750.0	9,060.0	8,930.0	8,200.0	8,900.0	7,920.0
Syria	385.0	522.5	508.3	527.8	503.3	460.0	422.5	389.2	390.0	365.0
United Arab Emirates	2,101.0	2,230.5	1,873.5	2,287.2	2,353.7	2,453.0	2,605.7	2,530.7	2,589.8	2,270.7
Yemen	179.0	354.0	350.0	350.0	350.0	412.8	354.3	320.3	283.0	270.0
Total	16,566	21,469	18,886.6	20,836.4	22,262.2	22,526.9	22,627.8	21,910.2	23,007.7	21,095.2

	1990	2000	2002	2003	2004	2005	2006	2007	Actual 2008	Est. 2009
COUNTRY	1,000 b/d									
Africa										
Algeria	797.0	808.0	850.0	1,110.8	1,206.7	1,351.7	1,347.5	1,360.0	1,372.0	1,240.0
Angola	480.0	740.0	894.2	879.7	986.2	1,238.1	1,410.0	1,698.2	1,894.0	1,790.0
Cameroon	164.0	85.0	69.0	65.0	74.0	82.5	87.4	85.2	83.8	74.0
Chad	—	—	—	—	175.4	180.0	150.0	145.0	145.0	145.0
Congo (Brazzaville)	161.0	265.0	254.1	240.3	240.0	240.0	240.0	240.0	240.0	240.0
Congo (Zaire)	28.0	25.0	22.7	25.5	20.0	20.0	20.0	25.0	25.0	25.0
Egypt	873.0	810.0	751.3	750.1	709.2	695.0	670.0	645.0	650.0	635.0
Equatorial Guinea	—	167.5	200.0	241.6	320.0	350.0	320.0	320.0	320.0	320.0
Gabon	289.0	325.0	248.0	240.4	235.0	234.2	235.0	230.0	210.0	230.0
Ivory Coast	2.1	12.0	10.0	25.8	30.0	30.0	30.0	30.0	30.0	30.0
Libya	1,369.0	1,410.0	1,315.8	1,427.5	1,546.7	1,640.0	1,708.3	1,710.0	1,725.0	1,545.0
Morocco	—	—	—	0.2	0.2	0.2	0.2	0.2	0.3	0.3
Nigeria	1,808.0	2,030.0	1,945.0	2,145.0	2,342.5	2,406.7	2,219.2	2,130.0	1,950.0	1,810.0
South Africa	—	25.9	20.1	16.5	26.2	24.0	20.0	14.4	14.0	14.0
Sudan	—	185.0	200.0	255.0	286.7	290.0	414.2	472.0	490.0	500.0
Tunisia	93.0	75.9	72.2	65.8	69.7	71.1	68.2	94.6	85.3	82.0
Other Africa	—	—	—	6.0	6.0	6.0	6.0	6.0	6.0	6.0
Total	6,048.4	6,971.6	6,858.6	7,495.2	8,274.5	8,859.9	8,946.0	9,205.6	9,240.4	8,686.3
Former Soviet Union										
Azerbaijan	—	275.0	300.0	306.3	300.0	443.3	642.5	830.0	895.0	1,000.0
Belarus	—	35.0	35.0	35.0	37.0	35.5	36.0	35.2	33.0	32.0
Georgia	—	2.0	2.0	2.8	3.0	2.0	2.0	1.0	1.0	1.0
Kazakhstan	—	675.0	818.3	888.0	970.0	994.2	1,064.2	1,100.0	1,210.0	1,330.0
Kyrgyzstan	—	1.0	2.0	2.0	2.0	1.0	1.0	1.0	1.0	1.0
Russia	—	6,325.0	7,404.6	8,240.0	8,886.7	9,190.0	9,498.3	9,830.0	9,750.0	9,915.0
Turkmenistan	—	140.0	180.0	210.0	200.0	195.0	165.0	198.0	220.0	220.0
Ukraine	—	73.0	85.0	80.0	84.0	98.0	95.0	80.0	77.5	75.0
USSR	11,500.0	—	—	—	—	—	—	—	—	—
Uzbekistan	—	155.0	149.0	144.0	150.0	113.0	108.0	114.0	90.0	85.0
Total	11,500.0	7,896.2	8,975.9	9,908.1	10,632.7	11,072.0	11,612.0	12,189.2	12,277.5	12,659.0
Asia-Pacific										
Australia	582.0	700.0	622.6	518.2	440.4	471.5	469.8	451.4	463.8	465.0
Bangladesh	0.9	3.3	5.0	3.7	4.0	4.0	3.6	4.0	6.5	6.0
Brunei	143.0	177.9	188.4	193.4	189.2	186.5	203.5	179.0	160.7	148.0
China	2,755.0	3,237.3	3,405.7	3,422.9	3,485.2	3,627.1	3,684.4	3,739.4	3,810.0	3,773.0
China, Taiwan	2.5	0.6	0.9	0.8	0.9	0.9	1.0	1.0	0.8	0.8
India	679.0	646.0	661.8	669.4	687.7	659.1	680.2	687.3	677.0	664.0
Indonesia	1,274.0	1,267.3	1,116.3	1,013.3	971.7	945.0	892.5	840.0	857.3	850.0
Japan	10.5	12.7	12.5	14.2	14.3	15.6	15.5	16.8	18.9	15.3
Malaysia	605.0	690.0	770.0	800.8	859.2	770.0	750.8	760.0	756.7	730.0
Myanmar	13.0	8.0	10.0	13.0	13.0	13.0	13.0	13.0	19.0	20.0
New Zealand	39.0	36.0	30.0	23.9	21.0	13.9	15.6	40.7	54.1	52.0
Pakistan	60.0	53.3	60.9	61.1	62.1	64.5	64.8	68.4	66.4	63.0
Papua New Guinea	—	69.7	47.0	48.2	46.0	46.0	49.0	46.7	41.0	37.0
Philippines	5.0	1.1	14.4	14.8	12.4	15.3	15.2	16.3	15.0	30.0
Thailand	41.0	111.0	128.7	159.7	154.1	186.0	186.9	212.2	228.1	236.0
Vietnam	40.0	304.0	330.7	333.2	340.0	340.0	344.6	310.0	277.3	300.0
Total	6,249.9	7,318.2	7,404.9	7,290.6	7,301.3	7,358.5	7,390.4	7,386.1	7,452.6	7,390.1
Total World	**60,317**	**67,234**	**65,426.3**	**68,507.0**	**71,243.4**	**72,362.1**	**72,647.0**	**72,160.0**	**72,822.0**	**70,503.5**

Table 4: World Oil Production, cont'd.

COUNTRY	1990	2000	2002	2003	2004	2005	2006	2007	Actual 2008	Est. 2009
	Million tonnes per year (mtpy)									
North America										
Canada	75.1	101.4	110.2	115.4	120.4	118.0	125.4	130.4	129.3	126.0
United States	359.6	289.5	286.1	282.9	269.8	257.9	254.1	252.2	246.5	265.8
Total	434.7	390.9	396.3	398.3	390.2	375.9	379.4	382.6	375.8	391.8
Latin America										
Argentina	23.6	37.4	36.8	35.2	36.8	32.1	31.9	31.1	30.3	30.4
Bolivia	1.0	1.4	1.5	1.6	1.7	2.1	2.2	2.2	2.0	2.0
Brazil	31.5	56.2	74.4	76.5	73.7	81.4	85.8	87.0	90.3	97.1
Chile	1.0	0.3	0.5	0.5	0.5	0.5	0.5	0.1	0.1	0.1
Colombia	22.2	34.2	28.8	27.0	26.3	26.2	26.3	26.4	28.9	32.9
Cuba	0.7	2.2	2.0	2.0	2.0	1.9	1.9	2.5	2.5	2.5
Ecuador	14.3	20.0	19.5	20.8	26.2	26.5	26.6	24.9	24.8	23.4
Guatemala	0.2	1.0	1.2	1.2	1.0	0.9	0.8	0.8	0.6	0.7
Mexico	131.1	150.0	158.2	167.8	168.5	166.0	162.2	153.5	139.4	129.7
Peru	6.6	4.9	4.6	4.3	4.4	5.5	5.8	3.9	4.1	5.3
Trinidad & Tobago	7.5	5.9	6.5	6.8	6.1	7.2	7.1	6.0	5.6	5.3
Venezuela	105.5	150.8	113.7	99.9	110.0	134.7	127.6	119.0	117.0	108.1
Other Latin America	—	—	—	0.6	0.6	0.6	0.7	0.8	0.8	0.8
Total	345.5	464.3	448.5	444.3	457.7	485.8	479.3	458.2	446.5	438.3
Europe										
Albania	—	0.3	0.3	0.3	0.3	0.3	0.4	0.5	0.5	0.5
Austria	1.2	1.0	0.9	0.9	0.9	0.9	0.9	0.9	0.9	0.9
Bulgaria	0.2	—	—	—	—	—	—	—	—	—
Croatia	—	1.1	1.0	1.0	1.0	0.9	0.8	0.8	0.7	0.7
Czech Republic	0.1	0.2	0.2	0.3	0.3	0.2	0.2	0.2	0.2	0.2
Denmark	5.9	18.1	18.5	18.3	19.4	18.8	17.0	15.5	14.3	13.2
France	3.1	1.4	1.3	1.2	1.1	1.1	1.1	1.0	1.0	0.9
Germany	3.9	3.1	3.7	3.8	3.5	3.6	3.5	3.4	3.0	2.8
Greece	0.7	0.3	0.2	0.1	0.1	0.1	0.1	0.1	0.1	0.1
Hungary	2.0	1.3	1.1	1.2	1.1	0.9	0.8	0.8	0.7	0.7
Italy	4.8	4.5	4.2	4.8	5.7	5.7	5.5	5.4	5.0	4.0
Lithuania	—	—	—	—	—	0.2	0.2	0.2	0.1	0.1
Netherlands	3.5	1.5	2.2	2.3	2.2	1.4	1.3	2.0	1.7	1.3
Norway	78.4	160.0	156.8	152.6	147.2	134.3	124.1	113.1	108.6	100.8
Poland	0.1	0.7	0.8	0.9	1.0	1.0	0.9	0.9	0.8	0.6
Romania	8.0	6.1	5.8	5.7	5.1	5.0	4.9	4.9	4.6	4.5
Serbia	—	—	—	0.8	0.7	0.7	0.7	0.7	0.7	0.7
Slovakia	—	—	—	—	—	—	—	—	—	—
Spain	0.7	0.2	0.3	0.3	0.3	0.2	0.1	0.1	0.1	0.1
Turkey	3.5	2.8	2.3	2.2	2.1	2.1	2.1	2.0	2.0	2.2
United Kingdom	92.6	125.2	114.8	104.7	94.0	83.6	75.6	75.9	70.5	65.0
Total	208.7	327.8	315.4	301.6	286.1	261.1	240.4	228.3	215.7	199.4
Middle East										
Bahrain	2.1	5.1	8.7	1.7	8.7	8.7	8.6	8.5	1.6	1.5
Iran	155.4	183.4	170.8	188.4	195.8	193.8	193.8	198.2	194.7	185.5
Iraq	103.7	127.8	100.3	66.1	99.7	90.1	94.6	104.1	120.7	119.5
Israel	—	—	—	—	—	—	—	—	—	—
Kuwait	53.8	87.9	79.7	93.1	102.1	106.1	110.1	107.6	115.0	100.1
Neutral Zone	15.7	31.4	26.8	30.4	29.9	28.9	28.9	27.9	28.4	26.9
Oman	32.8	46.5	45.0	41.2	38.1	37.7	36.9	35.4	37.4	39.8
Qatar	19.3	34.3	32.1	36.7	39.0	39.8	41.0	39.8	42.2	38.1
Saudi Arabia	309.5	398.2	341.1	422.3	435.7	451.2	444.7	408.4	443.2	394.4
Syria	19.2	26.0	25.3	26.3	25.1	22.9	21.0	19.4	19.4	18.2
United Arab Emirates	104.6	111.1	93.3	113.9	117.2	122.2	129.8	126.0	129.0	113.1
Yemen	8.9	17.6	17.4	17.4	17.4	20.6	17.6	16.0	14.1	13.4
Total	825.0	1,069.2	940.5	1,037.7	1,108.6	1,121.8	1,126.9	1,091.1	1,145.8	1,050.5

COUNTRY	1990	2000	2002	2003	2004	2005	2006	2007	Actual 2008	Est. 2009
	Million tonnes per year (mtpy)									
Africa										
Algeria	39.7	40.2	42.3	55.3	60.1	67.3	67.1	67.7	68.3	61.8
Angola	23.9	36.8	44.5	43.8	49.1	61.7	70.2	84.6	94.3	89.1
Cameroon	8.2	4.2	3.4	3.2	3.7	4.1	4.4	4.2	4.2	3.7
Chad	—	—	—	—	8.7	9.0	7.5	7.2	7.2	7.2
Congo (Brazzaville)	8.0	13.2	12.7	12.0	12.0	12.0	12.0	12.0	12.0	12.0
Congo (Zaire)	1.4	1.2	1.1	1.3	1.0	1.0	1.0	1.2	1.2	1.2
Egypt	43.5	40.3	37.4	37.4	35.3	34.6	33.4	32.1	32.4	31.6
Equatorial Guinea	—	8.3	10.0	12.0	15.9	17.4	15.9	15.9	15.9	15.9
Gabon	13.4	16.2	12.3	12.0	11.7	11.7	11.7	11.5	10.5	11.5
Ivory Coast	0.1	0.6	0.5	1.3	1.5	1.5	1.5	1.5	1.5	1.5
Libya	68.2	70.2	65.5	71.1	77.0	81.7	85.1	85.2	85.9	76.9
Morocco	—	—	—	—	—	—	—	—	—	—
Nigeria	90.0	101.0	96.9	106.8	116.6	119.9	110.5	106.1	97.1	90.1
South Africa	—	1.3	1.0	0.8	1.3	1.2	1.0	0.7	0.7	0.7
Sudan	—	9.2	10.0	12.7	14.3	14.4	20.6	23.5	24.4	24.9
Tunisia	4.6	3.8	3.6	3.3	3.5	3.5	3.4	4.7	4.2	4.1
Other Africa	—	—	—	0.3	0.3	0.3	0.3	0.3	0.3	0.3
Total	301.2	347.2	341.5	373.3	412.0	441.2	445.5	458.4	460.2	432.6
Former Soviet Union										
Azerbaijan	—	13.7	14.9	15.3	14.9	22.1	32.0	41.3	44.6	49.8
Belarus	—	1.7	1.7	1.7	1.8	1.8	1.8	1.8	1.6	1.6
Georgia	—	0.1	0.1	0.1	0.1	0.1	0.1	—	—	—
Kazakhstan	—	33.6	40.7	44.2	48.3	49.5	53.0	54.8	60.3	66.2
Kyrgyzstan	—	—	0.1	0.1	0.1	0.1	0.1	0.1	—	—
Russia	—	315.0	368.7	410.4	442.5	457.7	473.0	489.5	485.6	493.8
Turkmenistan	—	7.0	9.0	10.5	10.0	9.7	8.2	9.9	11.0	11.0
Ukraine	—	3.6	4.2	4.0	4.2	4.9	4.7	4.0	3.9	3.7
USSR	572.7	—	—	—	—	—	—	—	—	—
Uzbekistan	—	7.7	7.4	7.2	7.5	5.6	5.4	5.7	4.5	4.2
Total	572.7	393.2	447.0	493.4	529.5	551.4	587.3	607.0	611.4	630.4
Asia-Pacific										
Australia	29.0	34.9	31.0	25.8	21.9	23.5	23.4	22.5	23.1	23.2
Bangladesh	—	0.2	0.2	0.2	0.2	0.2	0.2	0.2	0.3	0.3
Brunei	7.1	8.9	9.4	9.6	9.4	9.3	10.1	8.9	8.0	7.4
China	137.2	161.2	169.6	170.5	173.5	180.6	183.5	186.2	189.7	187.9
China, Taiwan	0.1	—	—	—	—	—	—	—	—	—
India	33.8	32.2	33.0	33.3	34.2	32.8	33.9	34.2	33.7	33.1
Indonesia	63.4	63.1	55.6	50.5	48.4	47.1	44.4	41.8	42.7	42.3
Japan	0.5	0.6	0.6	0.7	0.7	0.8	0.8	0.8	0.9	0.8
Malaysia	30.1	34.4	38.3	39.9	42.8	38.3	37.4	37.8	37.7	36.4
Myanmar	0.6	0.4	0.5	0.6	0.6	0.6	0.6	0.6	0.9	1.0
New Zealand	1.9	1.8	1.5	1.2	1.0	0.7	0.8	2.0	2.7	2.6
Pakistan	3.0	2.7	3.0	3.0	3.1	3.2	3.2	3.4	3.3	3.1
Papua New Guinea	—	3.5	2.3	2.4	2.3	2.3	2.4	2.3	2.0	1.8
Philippines	0.2	—	0.7	0.7	0.6	0.8	0.8	0.8	0.7	1.5
Thailand	2.0	5.5	6.4	8.0	7.7	9.3	9.3	10.6	11.4	11.8
Vietnam	2.0	15.1	16.5	16.6	16.9	16.9	17.2	15.4	13.8	14.9
Total	310.9	364.4	368.7	363.1	363.6	366.5	368.0	367.8	371.1	368.0
Total World	**3,003.4**	**3,348.2**	**3,257.9**	**3,411.6**	**3,547.6**	**3,603.6**	**3,617.8**	**3,593.6**	**3,626.5**	**3,511.1**

Source: *Oil & Gas Journal* survey data as of Jan. 1, 2010.

NOTES: Some totals might not compute exactly due to rounding and other factors.
Other Latin America includes Barbados and Suriname. Other Africa includes Benin, Ethiopia, Ghana, Namibia, Rwanda, and Somalia.
Slovakia production is included in the Czech Republic total before 2000.
USSR includes total of all Former Soviet Union countries before 2000.

Table 5: Historical Oil Reserves

	January 1										
	1990	2000	2002	2003	2004	2005	2006	2007	2008	2009	2010
COUNTRY	billion bbl										
North America											
Canada	6.1	4.9	4.9	180.0	178.9	178.8	178.8	179.2	178.6	178.1	175.2
United States	25.9	21.0	22.0	22.4	22.7	21.9	21.4	21.6	21.0	21.3	19.1
Latin America											
Argentina	2.3	2.8	3.0	2.9	2.8	2.7	2.3	2.5	2.6	2.6	2.5
Brazil	2.8	7.4	8.5	8.3	8.5	10.6	11.2	11.8	12.2	12.6	12.8
Colombia	2.0	2.6	1.8	1.8	1.8	1.5	1.5	1.5	1.5	1.4	1.4
Ecuador	1.5	2.1	2.1	4.6	4.6	4.6	4.6	4.5	4.5	4.7	6.5
Mexico	56.4	28.4	26.9	12.6	15.7	14.6	12.9	12.3	11.7	10.5	10.4
Venezuela	58.5	72.6	77.7	77.8	77.8	77.2	79.7	80.0	87.0	99.4	99.4
Europe											
Norway	11.5	10.8	9.4	10.3	10.4	8.5	7.7	7.9	6.9	6.7	6.7
United Kingdom	4.2	5.2	4.9	4.7	4.7	4.5	4.0	3.9	3.6	3.4	3.1
Middle East											
Iran	92.8	89.7	89.7	89.7	125.8	125.8	132.5	136.3	138.4	136.2	137.6
Iraq	100.0	112.5	112.5	112.5	115.0	115.0	115.0	115.0	115.0	115.0	115.0
Kuwait	94.5	94.0	94.0	94.0	96.5	99.0	101.5	99.0	101.5	101.5	101.5
Neutral Zone	5.2	5.0	5.0	5.0	5.0	5.0	5.0	5.0	5.0	5.0	5.0
Oman	4.2	5.3	5.5	5.5	5.5	5.5	5.5	5.5	5.5	5.5	5.5
Qatar	4.5	3.7	15.2	15.2	15.2	15.2	15.2	15.2	15.2	15.2	25.4
Saudi Arabia	254.9	261.0	259.3	259.3	259.4	259.4	264.3	259.8	264.3	264.2	259.9
Syria	1.7	2.5	2.5	2.5	2.5	2.5	2.5	2.5	2.5	2.5	2.5
United Arab Emirates	96.2	96.2	97.8	97.8	97.8	97.8	97.8	97.8	97.8	97.8	97.8
Yemen	3.0	4.0	4.0	4.0	4.0	4.0	4.0	3.0	3.0	3.0	3.0
Africa											
Algeria	9.2	9.2	9.2	9.2	11.3	11.8	11.4	12.3	12.2	12.2	12.2
Angola	2.0	5.4	5.4	5.4	5.4	5.4	5.4	8.0	9.0	9.0	9.5
Egypt	4.5	2.9	2.9	3.7	3.7	3.7	3.7	3.7	3.7	3.7	3.7
Gabon	0.7	2.5	2.5	2.5	2.5	2.5	2.5	2.0	2.0	2.0	2.0
Libya	22.8	29.5	29.5	29.5	36.0	39.0	39.1	41.5	41.5	43.7	44.3
Nigeria	16.0	22.5	24.0	24.0	25.0	35.3	35.9	36.2	36.2	36.2	37.2
Former Soviet Union											
Azerbaijan	—	—	1.2	7.0	7.0	7.0	7.0	7.0	7.0	7.0	7.0
Kazakhstan	—	—	5.4	9.0	9.0	9.0	9.0	30.0	30.0	30.0	30.0
Russia	—	—	48.6	60.0	60.0	60.0	60.0	60.0	60.0	60.0	60.0
USSR	24.0	59.0	—	—	—	—	—	—	—	—	—
Asia-Pacific											
Australia	1.7	2.9	3.5	3.5	3.5	1.5	1.4	1.6	1.5	1.5	3.3
Brunei	1.4	1.4	1.4	1.4	1.4	1.4	1.4	1.1	1.1	1.1	1.1
China	24.0	24.0	24.0	18.3	18.3	18.3	18.3	16.0	16.0	16.0	20.4
India	7.5	4.8	4.8	5.4	5.4	5.4	5.8	5.6	5.6	5.6	5.6
Indonesia	8.2	4.9	5.0	5.0	4.7	4.7	4.3	4.3	4.4	4.0	4.0
Malaysia	3.0	3.9	3.0	3.0	3.0	3.0	3.0	3.0	4.0	4.0	4.0
Total World	1,001.6	1,016.0	1,031.6	1,212.9	1,265.8	1,277.7	1,292.5	1,309.3	1,331.7	1,322.6	1,334.6

Source: *Oil & Gas Journal* data as of Jan. 1, 2010.

NOTES: Data are for nations with major reserves. Total World includes other nations as well.
Some totals might not compute exactly due to rounding and other factors.
Canada's large increase in 2002–03 was due to including its oil sands.
USSR includes total of all Former Soviet Union countries before 2000.

Table 6: World Oil Imports and Exports

COUNTRY/REGION	2002	2003	2004	2005	2006	2007	2008	Change 2008 over 2007	2008 share of total
	thousand barrels per day (1,000 b/d)							%	
IMPORTS									
United States	11,357	12,254	12,898	13,525	13,612	13,632	9,756	-28.4	24.7
Europe	11,895	11,993	12,538	13,261	13,461	13,953	10,859	-22.2	27.5
Japan	5,070	5,314	5,203	5,225	5,201	5,032	4,069	-19.1	10.3
Rest of World	16,291	17,191	18,651	19,172	20,287	22,207	14,767	-33.5	37.4
Total World	44,613	46,752	49,290	51,182	52,561	54,824	39,451	-28.0	100.0
EXPORTS									
United States	904	921	991	1,129	1,317	1,439	139	-90.3	0.4
Canada	1,959	2,096	2,148	2,201	2,330	2,457	1,931	-21.4	4.9
Mexico	1,966	2,115	2,070	2,065	2,102	1,975	1,459	-26.1	3.7
South & Central America	2,965	2,942	3,233	3,528	3,681	3,570	2,479	-30.6	6.3
Europe	2,234	2,066	1,993	2,149	2,173	2,273	281	-87.6	0.7
Former Soviet Union	5,370	6,003	6,440	7,076	7,115	8,334	6,235	-25.2	15.8
Middle East	18,062	18,943	19,630	19,821	20,204	19,680	17,925	-8.9	45.4
North Africa	2,620	2,715	2,917	3,070	3,225	3,336	2,618	-21.5	6.6
West Africa	3,134	3,612	4,048	4,358	4,704	4,830	4,466	-7.5	11.3
East & Southern Africa	—	—	—	—	—	—	340	n/a	0.9
Asia-Pacific	3,848	3,978	4,189	4,243	4,312	5,274	1,320	-75.0	3.3
Unidentified	1,551	1,361	1,631	1,542	1,359	1,656	257	-84.5	0.7
Total World	44,613	46,752	49,290	51,182	52,561	54,824	39,451	-28.0	100.0

Source: Adapted from *BP Statistical Review of World Energy 2009* (June 2009).

NOTES: Some totals might not compute exactly due to rounding and other factors.
'Asia-Pacific' starting in 2008 includes Australasia, China, India, Japan, Singapore and Other Asia-Pacific.
'Rest of World' imports include Canada, Mexico, South and Central America, Former Soviet Union, Middle East, North Africa, West Africa, East and Southern Africa, Asia-Pacific (but not Japan), and Unidentified trade.
'Unidentified' exports include changes in the quantity of oil in transit, movements not otherwise shown, unidentified military use, and the like. Bunkers are not included as exports.
Intra-area movements (e.g., between nations in Europe) are excluded.
Annual changes and shares of total are calculated using thousand barrels per day figures.

Table 7: World Oil Trade Movements

	2008 IMPORTS TO													Total
	North America		Latin America		Europe	Africa	Asia-Pacific						Rest of World	
2008 EXPORTS FROM	United States	Canada	Mexico	South and Central America			Australasia	China	India	Japan	Singapore	Other Asia-Pacific		
	1,000 b/d													
United States	—	13.1	17.0	25.4	25.4	1.8	0.8	0.8	0.7	3.6	4.3	1.4	1.6	94.6
Canada	121.7	—	—	0.1	1.6	—	—	*	*	*	*	*	*	123.6
Mexico	64.7	1.4	—	4.3	7.7	—	—	—	1.9	—	0.1	—	*	98.1
S. & Central America	119.4	1.0	1.3	—	25.2	1.1	—	16.5	5.8	0.1	7.8	0.1	0.1	178.3
Europe	43.4	8.3	5.7	4.8	—	16.9	*	0.2	0.5	1.4	5.3	1.4	9.7	97.6
FSU	23.8	1.6	0.2	3.0	318.5	1.1	0.6	22.4	2.0	8.2	5.0	6.6	11.8	404.8
Middle East	119.7	6.3	0.6	5.8	127.6	44.5	5.4	92.0	107.6	196.9	53.1	238.3	3.1	1,000.7
N. Africa	32.6	8.9	0.2	5.1	101.3	1.0	0.3	4.2	4.3	0.4	0.1	3.1	—	161.5
W. Africa	90.9	5.2	*	15.0	49.5	4.5	—	39.1	16.6	1.1	0.1	6.7	0.1	228.8
E. & S. Africa	—	—	—	*	0.6	—	—	10.6	0.8	4.9	0.2	0.6	—	17.8
Australasia	1.8	*	*	—	*	—	—	0.9	0.2	2.6	3.5	6.6	*	15.6
China	0.8	*	—	4.0	0.3	0.4	0.2	—	0.1	1.3	2.1	9.2	0.3	18.8
India	0.3	—	—	1.6	3.3	—	—	0.2	—	1.4	6.8	20.0	0.8	34.4
Japan	—	—	0.5	0.1	1.2	—	2.6	4.9	0.6	—	4.6	2.4	0.1	17.1
Singapore	—	*	*	1.0	2.4	1.6	12.1	4.5	2.9	1.4	—	49.2	0.8	76.0
Other Asia-Pacific	5.3	*	0.6	1.5	2.5	0.5	20.0	21.4	5.8	20.7	38.0	—	0.2	116.5
Unidentified	12.3	2.5	—	—	14.9	—	1.7	—	—	0.2	—	—	—	31.7
Total	636.5	48.4	26.0	71.9	680.9	73.4	43.8	217.8	149.7	244.2	130.9	345.7	28.6	2,697.8
	million tonnes													
United States	—	270	352	529	508	37	16	16	14	75	89	29	34	1,967
Canada	2,459	—	—	3	34	—	—	*	*	1	*	*	*	2,498
Mexico	1,289	28	—	88	155	—	—	—	38	—	1	—	*	1,609
S. & Central America	2,418	21	26	—	509	23	—	334	116	1	163	1	2	3,616
Europe	900	168	118	100	—	352	1	5	11	29	110	28	202	2,023
FSU	492	33	4	61	6,428	22	12	453	40	165	105	133	236	8,184
Middle East	2,398	125	12	116	2,564	899	108	1,844	2,167	3,960	1,072	4,800	64	20,128
N. Africa	665	179	4	103	2,039	20	5	83	87	9	2	63	—	3,260
W. Africa	1,825	105	1	301	992	90	—	783	332	21	1	134	3	4,587
E. & S. Africa	—	—	—	*	12	—	—	213	16	99	5	12	—	357
Australasia	35	*	*	—	1	—	—	19	3	52	70	132	*	312
China	15	*	—	83	6	9	5	—	1	27	43	191	6	388
India	6	—	—	32	70	—	—	5	—	30	142	417	17	718
Japan	—	—	11	2	24	—	54	103	12	—	97	51	2	356
Singapore	—	*	1	22	50	33	250	93	61	29	—	1,026	17	1,583
Other Asia-Pacific	108	1	12	32	52	11	403	442	117	422	786	—	5	2,391
Unidentified	252	51	—	—	307	—	35	—	—	5	—	—	—	650
Total	12,872	981	541	1,473	13,751	1,495	889	4,393	3,016	4,925	2,585	7,017	588	54,626

Source: Adapted from *BP Statistical Review of World Energy 2009* (June 2009).

NOTES: Some totals might not compute exactly due to rounding and other factors.
'Unidentified' includes changes in the quantity of oil in transit, movements not otherwise shown, unidentified military use, and the like.
* Less than 0.05.

Table 8: World Oil Balance

	Avg. 2005	Avg. 2006	Avg. 2007	2008				Avg. 2008
				1Q08	2Q08	3Q08	4Q08	
SUPPLY, million b/d								
OECD								
United States	8.32	8.33	8.46	8.67	8.75	8.18	8.46	8.51
Canada	3.09	3.29	3.42	3.38	3.22	3.40	3.40	3.35
Mexico	3.78	3.71	3.50	3.29	3.19	3.15	3.12	3.19
North Sea	5.17	4.77	4.53	4.44	4.31	4.06	4.37	4.29
Other OECD	1.51	1.48	1.54	1.52	1.57	1.59	1.59	1.57
Total OECD	21.88	21.58	21.45	21.30	21.05	20.37	20.94	20.91
Non-OECD								
OPEC								
OPEC Crude								
Algeria	1.80	1.81	1.83	1.83	1.82	1.82	1.82	1.82
Angola	1.25	1.41	1.74	2.00	2.01	1.94	1.97	1.98
Ecuador	0.53	0.54	0.51	0.52	0.50	0.50	0.50	0.50
Iran	4.14	4.03	3.91	4.00	4.00	4.10	4.10	4.05
Iraq	1.88	2.00	2.09	2.30	2.42	2.43	2.35	2.38
Kuwait	2.53	2.54	2.46	2.58	2.60	2.62	2.54	2.59
Libya	1.63	1.68	1.70	1.79	1.75	1.70	1.70	1.74
Nigeria	2.63	2.44	2.35	2.22	2.11	2.18	2.15	2.17
Qatar	0.84	0.85	0.85	0.91	0.94	0.95	0.90	0.92
Saudi Arabia	9.55	9.15	8.72	9.20	9.32	9.57	8.96	9.26
United Arab Emirates	2.54	2.64	2.60	2.71	2.71	2.71	2.59	2.68
Venezuela	2.56	2.51	2.43	2.44	2.41	2.38	2.35	2.39
OPEC NGL	2.87	2.96	3.03	3.07	3.09	3.12	3.07	3.09
Refinery proc. gain	0.05	0.05	0.05	0.05	0.05	0.05	0.05	0.05
Total OPEC	34.95	34.74	34.39	35.72	35.84	36.18	35.16	35.72
Non-OPEC								
Former Soviet Union	11.76	12.15	12.60	12.59	12.60	12.42	12.46	12.52
China	3.79	3.86	3.91	3.94	4.00	3.97	3.98	3.97
Other Non-OECD	12.17	12.21	12.05	12.21	12.13	12.31	12.36	12.26
Total Non-OECD	**62.68**	**62.96**	**62.96**	**64.46**	**64.57**	**64.89**	**63.97**	**64.47**
Total World Supply	**84.56**	**84.54**	**84.40**	**85.76**	**85.62**	**85.26**	**84.91**	**85.38**

Table 8: World Oil Balance, cont'd

	Avg. 2005	Avg. 2006	Avg. 2007	2008				Avg. 2008
				1Q08	2Q08	3Q08	4Q08	
DEMAND, million b/d								
OECD								
OECD North America								
United States	20.80	20.69	20.68	20.04	19.76	18.90	19.30	19.50
U.S. Territories	0.34	0.34	0.33	0.27	0.29	0.30	0.23	0.27
Canada	2.34	2.25	2.31	2.31	2.19	2.25	2.26	2.25
Mexico	2.07	2.08	2.14	2.12	2.19	2.14	2.07	2.13
Total OECD N. America	25.55	25.36	25.46	24.74	24.43	23.59	23.86	24.15
OECD Asia-Pacific								
Australia & New Zealand	1.08	1.08	1.10	1.10	1.11	1.10	1.12	1.11
Japan	5.33	5.20	5.04	5.45	4.63	4.34	4.71	4.78
South Korea	2.19	2.18	2.24	2.35	2.11	2.10	2.14	2.17
Total OECD Asia-Pacific	8.60	8.46	8.38	8.90	7.85	7.54	7.97	8.06
OECD Europe								
France	1.99	1.98	1.97	2.01	1.95	1.95	2.04	1.99
Germany	2.65	2.69	2.47	2.49	2.43	2.71	2.65	2.57
Italy	1.75	1.74	1.69	1.66	1.64	1.64	1.62	1.64
United Kingdom	1.82	1.80	1.74	1.73	1.73	1.65	1.73	1.71
Other OECD Europe	7.44	7.45	7.47	7.43	7.32	7.57	7.38	7.42
Total OECD Europe	15.66	15.67	15.34	15.32	15.06	15.52	15.41	15.33
Total OECD	**49.80**	**49.49**	**49.16**	**48.96**	**47.36**	**46.64**	**47.25**	**47.55**
Non-OECD								
Former Soviet Union	4.16	4.27	4.27	4.23	4.22	4.47	4.48	4.35
Non-OECD Europe	0.75	0.77	0.79	0.79	0.80	0.80	0.80	0.80
China	6.70	7.26	7.58	7.94	8.07	7.78	7.54	7.83
Other Asia	8.66	8.90	9.17	9.64	9.74	9.06	8.83	9.31
Other Non-OECD	13.97	14.51	15.17	15.29	16.12	16.53	15.69	15.91
Total Non-OECD	**34.24**	**35.71**	**36.98**	**37.89**	**38.95**	**38.64**	**37.33**	**38.20**
Total World Demand	84.04	85.20	86.14	86.86	86.31	85.28	84.58	85.75
Supply/Demand Balance	0.52	- 0.66	- 1.74	- 1.10	- 0.69	- 0.02	0.33	- 0.37

Source: U.S. Energy Information Administration, www.eia.doe.gov/emeu/ipsr/balance.html, "December 2009 *International Petroleum Monthly*." Data as of January 11, 2010.

NOTES: Some totals might not compute exactly due to rounding and other factors.
OECD = Organization for Economic Cooperation and Development.

Table 9: Summary of Operating Refineries Worldwide

		Charge capacity, b/cd		
Country	Number of refineries	Crude	Catalytic cracking	Catalytic reforming
Albania	2	26,300	0	3,500
Algeria	4	450,000	0	88,900
Angola	1	39,000	0	1,900
Argentina	10	627,075	141,180	56,630
Aruba	1	235,000	0	0
Australia	7	724,736	228,200	165,403
Austria	1	208,600	26,250	32,725
Azerbaijan	2	398,978	57,750	24,466
Bahrain	1	253,650	35,100	13,500
Bangladesh	1	33,000	0	1,800
Belarus	2	493,323	44,000	92,000
Belgium	4	797,629	132,684	104,517
Bolivia	2	41,200	0	12,100
Brazil	13	1,908,278	505,287	24,386
Brunei	1	8,600	0	5,700
Bulgaria	1	115,240	23,300	4,060
Cameroon	1	37,000	0	6,500
Canada	18	2,039,401	503,500	368,503
Chile	3	226,800	50,540	26,460
China	54	6,806,000	588,000	178,000
China, Taiwan	4	1,310,000	217,900	115,000
Colombia	5	285,850	90,000	0
Congo, Republic of	1	21,000	0	2,000
Costa Rica	1	24,000	0	1,200
Croatia	3	250,317	51,002	49,368
Cuba	4	301,400	14,700	20,000
Czech Republic	3	183,000	0	27,470
Denmark	2	174,400	0	21,990
Dominican Republic	2	50,000	0	8,200
Ecuador	3	176,000	18,000	12,800
Egypt	9	726,250	0	62,240
El Salvador	1	22,000	0	3,000
Eritrea	1	14,564	0	1,465
Finland	2	258,500	53,600	46,700
France	13	1,983,668	390,805	290,676
Gabon	1	24,000	0	1,400
Germany	15	2,410,662	349,171	405,850
Ghana	1	45,000	14,000	65,000
Greece	4	423,000	75,550	49,200
Hungary	1	161,000	24,000	29,600
India	18	2,835,540	503,305	41,673
Indonesia	8	1,011,825	101,450	92,970
Iran	9	1,451,000	35,000	164,700
Iraq	9	637,500	0	82,000
Ireland	1	71,000	0	11,000
Israel	2	220,000	49,500	26,500
Italy	17	2,337,229	320,000	287,069
Ivory Coast	1	63,990	0	12,330
Jamaica	1	36,000	0	3,700
Japan	30	4,623,690	934,980	785,225
Jordan	1	90,400	4,000	10,900
Kazakhstan	3	345,093	38,356	51,586
Kenya	1	90,000	0	8,260
Kuwait	3	936,000	36,000	46,620
Kyrgyzstan	1	10,000	0	0
Liberia	1	15,000	0	2,000
Libya	5	378,000	0	20,250
Lithuania	1	190,000	43,200	45,900
Macedonia	1	50,000	0	10,860
Malaysia	6	514,832	42,700	86,070
Martinique	1	17,000	0	3,019

Table 9: Summary of Operating Refineries Worldwide, cont'd.

		Charge capacity, b/cd		
Country	Number of refineries	Crude	Catalytic cracking	Catalytic reforming
Mexico	6	1,540,000	380,500	279,300
Morocco	2	154,901	5,040	24,359
Myanmar	3	57,000	0	0
Netherlands	6	1,205,714	103,342	149,753
Netherlands Antilles	1	320,000	50,000	20,000
New Zealand	1	107,000	0	25,840
Nicaragua	1	20,000	0	3,000
Nigeria	4	505,000	82,700	70,070
North Korea	2	71,000	0	7,300
Norway	2	319,000	49,000	36,400
Oman	1	85,000	0	16,000
Pakistan	7	286,306	0	26,365
Papua New Guinea	1	32,500	0	0
Paraguay	1	7,500	0	0
Peru	6	192,950	32,000	2,100
Philippines	3	282,000	19,000	51,000
Poland	4	492,950	32,985	67,514
Portugal	2	304,172	40,500	50,182
Puerto Rico	1	73,000	0	21,000
Qatar	2	338,700	60,000	29,400
Romania	10	516,557	109,478	61,763
Russia	40	5,428,497	330,817	745,735
Saudi Arabia	7	2,080,000	103,600	193,160
Senegal	1	25,030	0	1,590
Serbia & Montenegro	2	214,826	18,950	18,822
Sierra Leone	1	10,000	0	0
Singapore	3	1,357,000	80,000	167,970
Slovakia	1	115,000	18,000	21,000
Slovenia	1	13,500	0	0
South Africa	4	485,297	104,025	77,143
South Korea	6	2,701,500	212,000	261,470
Spain	9	1,271,500	191,300	196,750
Sri Lanka	1	50,000	0	5,300
Sudan	3	121,700	0	1,900
Suriname	1	7,000	0	0
Sweden	5	437,000	29,700	70,660
Switzerland	2	132,000	20,400	28,000
Syria	2	239,865	0	31,242
Tanzania	1	14,900	0	2,500
Thailand	4	584,100	91,990	97,770
Trinidad & Tobago	1	168,000	24,000	18,000
Tunisia	1	34,000	0	3,300
Turkey	6	714,275	28,935	65,662
Turkmenistan	2	236,970	15,151	52,540
Ukraine	6	879,759	70,100	144,711
United Arab Emirates	5	773,250	34,350	25,875
United Kingdom	11	1,866,168	444,723	339,051
United States	130	17,763,499	5,662,834	3,582,688
Uruguay	1	50,000	12,000	12,000
Uzbekistan	3	222,271	0	23,487
Venezuela	5	1,282,100	231,800	49,500
Vietnam	1	140,000	0	0
Virgin Islands	1	500,000	140,000	105,000
Yemen	2	140,000	0	14,500
Zambia	1	23,750	0	5,320
Total	661	87,223,379	14,572,230	11,520,833

Source: *Oil & Gas Journal* survey data as of Jan. 1, 2010.

NOTES: Some totals might not compute exactly due to rounding and other factors.
b/cd = barrels per calendar-day.

Table 10: World Refining Capacity

	1980	1990	2000	2003	2004	2005	2006	2007	2008	2009	2010
Country						1,000 b/cd					
North America											
Canada	2,222	1,852	1,912	1,983	1,988	2,017	2,017	2,041	1,969	2,029	2,039
United States	17,720	16,244	16,541	16,623	16,698	16,775	17,126	17,273	17,447	17,380	17,764
Total	19,942	18,096	18,453	18,606	18,686	18,792	19,143	19,314	19,416	19,409	19,803
Latin America											
Argentina	676	689	662	639	639	625	625	625	626	626	627
Bolivia	74	58	48	63	63	47	47	41	41	41	41
Brazil	1,205	1,397	1,783	1,865	1,914	1,920	1,908	1,908	1,908	1,908	1,908
Chile	139	147	205	205	205	227	227	227	227	227	227
Colombia	193	227	286	286	286	286	286	286	286	286	286
Cuba	160	160	301	301	301	301	301	301	301	301	301
Ecuador	86	145	176	176	176	176	176	176	176	176	176
Guatemala	16	16	20	16	16	—	—	—	—	—	—
Jamaica	—	—	—	—	—	—	36	36	36	36	36
Mexico	1,394	1,514	1,525	1,684	1,684	1,684	1,684	1,540	1,540	1,540	1,540
Peru	170	172	182	191	191	193	193	193	193	193	193
Trinidad & Tobago	456	300	160	160	165	165	175	168	168	168	168
Venezuela	1,446	1,201	1,239	1,282	1,282	1,282	1,282	1,282	1,282	1,282	1,282
Other Caribbean	—	—	—	—	1,125	1,045	1,224	1,229	1,227	1,228	1,195
Other Latin America	—	—	—	—	—	—	131	131	131	131	131
Total	8,620	7,315	7,973	6,868	8,047	7,951	8,295	8,143	8,142	8,143	8,111
Europe											
Albania	—	—	26	26	26	26	26	26	26	26	26
Austria	280	204	209	209	209	209	209	209	209	209	209
Belgium	1,064	614	719	791	793	803	858	791	798	798	798
Bulgaria	—	—	115	115	115	115	115	115	115	115	115
Croatia	—	—	236	260	250	250	250	250	250	250	250
Czech Republic	455	455	186	198	198	198	198	198	183	183	183
Denmark	214	187	135	176	176	176	176	176	174	174	174
Finland	336	241	200	252	252	252	252	252	252	256	259
France	3,385	1,820	1,902	1,903	1,951	1,951	1,979	1,959	1,932	1,986	1,984
Germany	2,986	1,507	2,275	2,267	2,289	2,323	2,344	2,417	2,417	2,417	2,411
Greece	431	385	383	407	401	401	413	413	423	423	423
Hungary	290	220	232	161	161	161	161	161	161	161	161
Ireland	56	56	71	71	71	71	71	71	71	71	71
Italy	4,131	2,804	2,341	2,300	2,313	2,321	2,324	2,337	2,337	2,337	2,337
Netherlands	1,828	1,381	1,188	1,207	1,222	1,228	1,222	1,212	1,227	1,208	1,206
Norway	264	295	358	310	310	310	310	310	315	315	319
Poland	385	385	382	350	350	350	467	497	493	493	493
Portugal	378	313	304	304	304	304	304	304	304	304	304
Romania	608	617	499	501	501	517	517	517	517	517	517
Slovakia	—	—	115	115	115	115	115	115	115	115	115
Spain	1,455	1,293	1,316	1,322	1,272	1,272	1,272	1,272	1,227	1,272	1,272
Sweden	458	428	427	424	424	434	434	434	437	437	437
Switzerland	137	132	132	132	132	132	132	132	132	132	132
Turkey	356	725	691	719	714	714	714	714	714	714	714
United Kingdom	2,527	1,831	1,785	1,789	1,817	1,825	1,877	1,887	1,858	1,858	1,866
Other Europe	—	—	—	—	—	—	278	279	278	278	278
Total	4,933	4,409	16,227	16,309	16,366	16,458	17,018	17,048	16,965	17,049	17,054
Middle East											
Bahrain	250	243	249	249	249	249	255	249	262	262	254
Iran	921	530	1,474	1,474	1,474	1,474	1,451	1,451	1,451	1,451	1,451
Iraq	168	319	348	418	588	598	598	598	598	598	638
Israel	195	180	220	220	220	220	220	220	220	220	220
Kuwait	645	819	865	889	889	889	889	889	889	889	936
Oman	—	—	—	85	85	85	85	85	85	85	85
Qatar	11	62	58	200	200	200	200	200	200	200	339
Saudi Arabia	487	1,484	1,710	1,745	1,745	1,745	2,095	2,095	2,080	2,080	2,080

Table 10: World Refining Capacity, cont'd.

Country	1980	1990	2000	2003	2004	2005	2006	2007	2008	2009	2010
	1,000 b/cd										
Syria	223	244	242	240	240	240	240	240	240	240	240
United Arab Emirates	14	180	429	514	514	514	781	781	781	781	773
Yemen	175	162	120	130	130	130	130	140	140	140	140
Other Middle East	—	—	—	—	—	—	90	90	90	90	90
Total	3,163	4,447	5,928	6,164	6,334	6,344	7,034	7,038	7,036	7,036	7,246
Africa											
Algeria	122	465	503	450	450	450	450	450	450	450	450
Angola	36	32	39	39	39	39	39	39	39	39	39
Cameroon	—	43	35	42	42	42	42	42	37	37	37
Congo (Brazzavillef)	—	21	21	21	21	21	21	21	21	21	21
Congo (Zaire)	16	17	17	15	—	—	—	—	—	—	—
Egypt	234	489	578	726	726	726	726	726	726	726	726
Gabon	20	24	17	17	17	17	17	17	24	24	24
Ivory Coast	50	69	69	65	65	65	65	65	64	64	64
Libya	138	329	348	343	343	380	380	378	378	378	378
Madagascar	—	—	—	15	15	15	—	—	—	—	—
Morocco	72	155	157	155	155	155	155	155	155	155	155
Nigeria	160	433	439	439	439	439	439	439	505	505	505
South Africa	478	434	467	490	520	490	505	488	485	485	485
Sudan	26	25	82	122	122	122	122	122	122	122	122
Tanzania	—	—	—	—	—	—	15	15	15	15	15
Tunisia	34	34	34	34	34	34	34	34	34	34	34
Other Africa	—	—	—	—	—	—	221	222	223	223	223
Total	1,665	2,824	3,046	2,973	2,988	2,995	3,231	3,213	3,278	3,278	3,278
Former Soviet Union											
Azerbaijan	—	—	—	442	399	399	399	399	399	399	399
Belarus	—	—	—	493	493	493	493	493	493	493	493
Georgia	—	—	—	106	106	—	—	—	—	—	—
Kazakhstan	—	—	—	427	427	345	345	345	345	345	345
Kyrgyzstan	—	—	—	10	10	10	10	10	10	10	10
Lithuania	—	—	—	—	263	263	190	190	190	190	190
Russia	—	—	—	5,435	5,435	5,433	5,341	5,339	5,429	5,429	5,429
Turkmenistan	—	—	—	237	237	237	237	237	237	237	237
Ukraine	—	—	—	1,025	1,025	880	880	880	880	880	880
USSR	10,950	12,300	9,762	—	—	—	—	—	—	—	—
Uzbekistan	—	—	—	222	222	222	222	222	222	222	222
Total	10,950	12,300	9,762	8,397	8,617	8,282	8,117	8,115	8,205	8,205	8,205
Asia-Pacific											
Australia	725	675	812	848	755	755	702	705	707	696	725
Bangladesh	31	31	33	33	33	33	33	33	33	33	33
Brunei	—	—	—	9	9	9	9	9	9	9	9
China	1,600	2,200	4,347	4,528	4,528	4,650	6,246	6,246	6,246	6,446	6,806
India	557	1,080	1,858	2,135	2,135	2,255	2,255	2,256	2,256	2,256	2,836
Indonesia	528	714	993	993	993	993	993	993	993	993	1,012
Japan	5,509	4,198	4,998	4,767	4,703	4,707	4,672	4,677	4,651	4,691	4,624
Malaysia	—	—	—	516	516	545	545	545	515	515	515
Myanmar	26	32	32	57	57	57	57	57	57	57	57
New Zealand	74	95	98	106	104	104	104	104	107	107	107
Pakistan	98	121	143	234	269	269	269	269	267	276	286
Papua New Guinea	—	—	—	—	—	33	33	33	33	33	33
Philippines	253	284	401	420	333	333	333	333	282	282	273
Singapore	1,093	1,049	1,788	1,259	1,319	1,337	1,337	1,337	1,344	1,344	1,357
South Korea	601	867	2,540	2,560	2,544	2,577	2,577	2,577	2,577	2,607	2,702
Taiwan	425	570	770	920	920	1,220	1,220	1,290	1,290	1,290	1,310
Thailand	186	215	713	703	703	703	703	729	729	729	584
Other Asia-Pacific	—	—	—	—	—	—	121	121	124	121	261
Total	11,986	12,181	19,645	20,088	19,921	20,595	22,209	22,314	22,220	22,485	23,530
Total World	79,514	74,535	81,556	81,878	82,055	82,409	85,043	85,185	85,262	85,604	87,223

Table 10: World Refining Capacity, cont'd.

Country	1980	1990	2000	2003	2004	2005	2006	2007	2008	2009	2010
	million tonnes per year (mtpy)										
North America											
Canada	110.7	92.2	95.2	98.8	99.0	100.4	100.4	101.6	98.1	101.0	101.5
United States	882.5	809.0	823.7	827.8	831.5	835.3	852.8	860.0	868.7	865.4	884.4
Total	993.2	901.2	918.9	926.6	930.5	935.7	953.2	961.6	966.8	966.4	985.9
Latin America											
Argentina	33.7	34.3	33.0	31.8	31.8	31.1	31.1	31.1	31.2	31.2	31.2
Bolivia	3.7	2.9	2.4	3.1	3.1	2.3	2.4	2.0	2.1	2.0	2.1
Brazil	60.0	69.6	88.8	92.9	95.3	95.6	95.0	95.0	95.0	95.0	95.0
Chile	6.9	7.3	10.2	10.2	10.2	11.3	11.3	11.3	11.3	11.3	11.3
Colombia	9.6	11.3	14.2	14.2	14.2	14.2	14.2	14.2	14.2	14.2	14.2
Cuba	8.0	8.0	15.0	15.0	15.0	15.0	15.0	15.0	15.0	15.0	15.0
Ecuador	4.3	7.2	8.8	8.8	8.8	8.8	8.8	8.8	8.8	8.8	8.8
Guatemala	0.8	0.8	1.0	0.8	0.8	—	—	—	—	—	—
Jamaica	—	—	—	—	—	—	1.8	1.8	1.8	1.8	1.8
Mexico	69.4	75.4	75.9	83.9	83.9	83.9	83.9	76.7	76.7	76.7	76.7
Peru	8.5	8.6	9.1	9.5	9.5	9.6	9.6	9.6	9.6	9.6	9.6
Trinidad & Tobago	22.7	14.9	8.0	8.0	8.2	8.2	8.7	8.4	8.4	8.4	8.4
Venezuela	72.0	59.8	61.7	63.8	63.8	63.8	63.8	63.8	63.8	63.8	63.8
Other Caribbean	—	—	—	—	56.0	52.0	61.0	61.2	61.1	61.1	59.5
Other Latin America	—	—	—	—	—	—	6.5	6.5	6.5	6.5	6.5
Total	429.2	364.0	396.9	342.0	400.6	395.8	413.1	405.4	405.4	405.4	403.9
Europe											
Albania	—	—	1.3	1.3	1.3	1.3	1.3	1.3	1.3	1.3	1.3
Austria	13.9	10.2	10.4	10.4	10.4	10.4	10.4	10.4	10.4	10.4	10.4
Belgium	53.0	30.6	35.8	39.4	39.5	40.0	42.7	39.4	39.7	39.7	39.7
Bulgaria	—	—	5.7	5.7	5.7	5.7	5.7	5.7	5.7	5.7	5.7
Croatia	—	—	11.8	12.9	12.4	12.4	12.5	12.5	12.5	12.5	12.5
Czech Republic	22.7	22.7	9.3	9.9	9.9	9.9	9.9	9.9	9.1	9.1	9.1
Denmark	10.7	9.3	6.7	8.8	8.8	8.8	8.8	8.8	8.7	8.7	8.7
Finland	16.7	12.0	10.0	12.4	12.4	12.4	12.5	12.5	12.5	12.8	12.9
France	168.6	90.6	94.7	94.8	97.2	97.2	98.6	97.5	96.2	98.9	98.8
Germany	148.7	75.0	113.3	112.9	114.0	115.7	116.7	120.3	120.3	120.3	120.0
Greece	21.5	19.2	19.1	20.2	20.0	20.0	20.6	20.6	21.1	21.1	21.1
Hungary	14.4	11.0	11.6	8.0	8.0	8.0	8.0	8.0	8.0	8.0	8.0
Ireland	2.8	2.8	3.5	3.5	3.5	3.5	3.5	3.5	3.5	3.5	3.5
Italy	205.7	139.6	116.6	114.5	115.2	115.6	115.7	116.4	116.4	116.4	116.4
Netherlands	91.0	68.8	59.2	60.1	60.8	61.1	60.8	60.3	61.1	60.2	60.0
Norway	13.1	14.7	17.8	15.4	15.4	15.4	15.4	15.4	15.7	15.7	15.9
Poland	19.2	19.2	19.0	17.4	17.4	17.4	23.2	24.7	24.5	24.5	24.5
Portugal	18.8	15.6	15.1	15.1	15.1	15.1	15.1	15.1	15.1	15.1	15.1
Romania	30.3	30.7	24.9	24.9	24.9	25.7	25.7	25.7	25.7	25.7	25.7
Slovakia	—	—	5.7	5.7	5.7	5.7	5.7	5.7	5.7	5.7	5.7
Spain	72.5	64.4	65.5	65.8	63.3	63.3	63.3	63.3	63.6	63.3	63.3
Sweden	22.8	21.3	21.3	21.1	21.1	21.6	21.6	21.6	21.8	21.8	21.8
Switzerland	6.8	6.6	6.6	6.6	6.6	6.6	6.6	6.6	6.6	6.6	6.6
Turkey	17.7	36.1	34.4	35.8	35.6	35.6	35.6	35.6	35.6	35.6	35.6
United Kingdom	125.8	91.2	88.9	89.1	90.5	90.9	93.5	94.0	92.5	92.5	92.9
Other Europe	—	—	—	—	—	—	13.9	13.9	13.9	13.9	13.9
Total	245.6	761.0	808.2	811.7	814.7	819.3	847.3	848.8	847.2	848.9	849.1
Middle East											
Bahrain	12.4	12.1	12.4	12.4	12.4	12.4	12.7	12.4	13.0	13.0	12.6
Iran	45.9	26.4	73.4	73.4	73.4	73.4	72.3	72.3	72.2	72.2	72.2
Iraq	8.4	15.9	17.3	20.8	29.3	29.8	29.8	29.8	29.7	29.7	31.7
Israel	9.7	9.0	11.0	11.0	11.0	11.0	11.0	11.0	11.0	11.0	11.0
Kuwait	32.1	40.8	43.1	44.3	44.3	44.3	44.3	44.3	44.3	44.3	46.6
Oman	—	—	—	4.2	4.2	4.2	4.2	4.2	4.2	4.2	4.2
Qatar	0.5	3.1	2.9	10.0	10.0	10.0	10.0	10.0	10.0	10.0	16.9
Saudi Arabia	24.3	73.9	85.2	86.9	86.9	86.9	104.3	104.3	103.6	103.6	103.6
Syria	11.1	12.2	12.1	12.0	12.0	12.0	11.9	11.9	11.9	11.9	11.9
United Arab Emirates	0.7	9.0	21.4	25.6	25.6	25.6	38.9	38.9	38.9	38.9	38.5
Yemen	8.7	8.1	6.0	6.5	6.5	6.5	6.5	7.0	7.0	7.0	7.0
Other Middle East	—	—	—	—	—	—	4.5	4.5	4.5	4.5	4.5
Total	151.4	221.6	295.4	307.0	315.6	316.1	350.4	350.4	350.3	350.3	360.7

Table 10: World Refining Capacity, cont'd.

Country	1980	1990	2000	2003	2004	2005	2006	2007	2008	2009	2010
	million tonnes per year (mtpy)										
Africa											
Algeria	6.1	23.2	25.0	22.4	22.4	22.4	22.4	22.4	22.4	22.4	22.4
Angola	1.8	1.6	1.9	1.9	1.9	1.9	1.9	1.9	1.9	1.9	1.9
Cameroon	—	2.1	1.7	2.1	2.1	2.1	2.1	2.1	1.8	1.8	1.8
Congo (Brazzaville)	—	1.0	1.0	1.0	1.0	1.0	1.0	1.0	1.0	1.0	1.0
Congo (Zaire)	0.8	0.8	0.8	0.7	—	—	—	—	—	—	—
Egypt	11.7	24.4	28.8	36.2	36.2	36.2	36.2	36.2	36.2	36.2	36.2
Gabon	1.0	1.2	0.8	0.8	0.8	0.8	0.9	0.9	1.2	1.2	1.2
Ivory Coast	2.5	3.4	3.4	3.2	3.2	3.2	3.2	3.2	3.2	3.2	3.2
Libya	6.9	16.4	17.3	17.1	17.1	18.9	18.9	18.8	18.8	18.8	18.8
Madagascar	—	—	—	—	—	0.7	—	—	—	—	—
Morocco	3.6	7.7	7.8	7.7	7.7	7.7	7.7	7.7	7.7	7.7	7.7
Nigeria	8.0	21.6	21.9	21.9	21.9	21.9	21.8	21.8	25.1	25.1	25.1
South Africa	23.8	21.6	23.3	24.4	25.9	24.4	25.1	24.3	24.2	24.2	24.2
Sudan	1.3	1.2	4.1	6.1	6.1	6.1	6.1	6.1	6.1	6.1	6.1
Tanzania	—	—	—	—	—	—	0.7	0.7	0.7	0.7	0.7
Tunisia	1.7	1.7	1.7	1.7	1.7	1.7	1.7	1.7	1.7	1.7	1.7
Other Africa	—	—	—	—	—	—	11.0	11.1	11.1	11.1	11.1
Total	82.9	140.4	151.2	147.3	148.0	148.3	160.7	160.0	163.1	163.1	163.1
Former Soviet Union											
Azerbaijan	—	—	—	22.0	19.9	19.9	19.9	19.9	19.9	19.9	19.9
Belarus	—	—	—	24.6	24.6	24.6	24.6	24.6	24.6	24.6	24.6
Georgia	—	—	—	5.3	5.3	—	—	—	—	—	—
Kazakhstan	—	—	—	21.3	21.3	17.2	17.2	17.2	17.2	17.2	17.2
Kyrgyzstan	—	—	—	0.5	0.5	0.5	0.5	0.5	0.5	0.5	0.5
Lithuania	—	—	—	—	13.1	13.1	9.5	9.5	9.5	9.5	9.5
Russia	—	—	—	270.7	270.7	270.5	266.0	265.8	270.3	270.3	270.3
Turkmenistan	—	—	—	11.8	11.8	11.8	11.8	11.8	11.8	11.8	11.8
Ukraine	—	—	—	51.0	51.0	43.8	43.8	43.8	43.8	43.8	43.8
USSR	545.3	612.5	486.1	—	—	—	—	—	—	—	—
Uzbekistan	—	—	—	11.1	11.1	11.1	11.1	11.1	11.1	11.1	11.1
Total	545.3	612.5	486.1	418.2	429.3	412.5	404.4	404.2	408.7	408.7	408.7
Asia-Pacific											
Australia	36.1	33.6	40.4	42.2	37.6	37.6	34.9	35.1	35.2	34.7	36.1
Bangladesh	1.5	1.5	1.6	1.6	1.6	1.6	1.6	1.6	1.6	1.6	1.6
Brunei	—	—	—	0.4	0.4	0.4	0.4	0.4	0.4	0.4	0.4
China	79.7	109.6	216.5	225.5	225.5	231.5	311.0	311.0	311.0	320.9	338.9
India	27.7	53.8	92.5	106.3	106.3	112.3	112.3	112.3	112.3	112.3	141.2
Indonesia	26.3	35.6	49.5	49.5	49.5	49.5	49.4	49.4	49.4	49.4	50.4
Japan	274.3	209.1	248.9	237.4	234.2	234.4	232.6	232.9	231.6	233.6	230.2
Malaysia	—	—	—	25.7	25.7	27.1	27.1	27.1	25.6	25.6	25.6
Myanmar	1.3	1.6	1.6	2.8	2.8	2.8	2.8	2.8	2.8	2.8	2.8
New Zealand	3.7	4.7	4.9	5.3	5.2	5.2	5.2	5.2	5.3	5.3	5.3
Pakistan	4.9	6.0	7.1	11.7	13.4	13.4	13.4	13.4	13.3	13.7	14.3
Papua New Guinea	—	—	—	—	—	1.6	1.6	1.6	1.6	1.6	1.6
Philippines	12.6	14.1	20.0	20.9	16.6	16.6	16.6	16.6	14.0	14.0	13.6
Singapore	54.4	52.2	89.0	62.7	65.7	66.6	66.6	66.6	66.9	66.9	67.6
South Korea	29.9	43.2	126.5	127.5	126.7	128.3	128.3	128.3	128.3	129.8	134.5
Taiwan	21.2	28.4	38.3	45.8	45.8	60.8	60.8	64.2	64.2	64.2	65.2
Thailand	9.3	10.7	35.5	35.0	35.0	35.0	35.0	36.3	36.3	36.3	29.1
Other Asia-Pacific	—	—	—	—	—	—	6.0	6.0	6.2	6.0	13.0
Total	596.9	606.6	978.2	1,000.4	992.0	1,025.4	1,105.6	1,111.0	1,106.0	1,119.5	1,171.4
Total World	3,959.5	3,711.5	4,060.8	4,077.5	4,086.0	4,103.6	4,234.7	4,241.4	4,250.6	4,262.2	4,342.8

Source: *Oil & Gas Journal* survey data as of Jan. 1, 2010.

NOTES: Some totals might not compute exactly due to rounding and other factors.
b/cd = barrels per calendar-day.
Other Caribbean includes Aruba, Dominican Republic, Martinique, Netherlands Antilles, Puerto Rico, and Virgin Islands.
Other Latin America includes Costa Rica, El Salvador, Nicaragua, Paraguay, Suriname, and Uruguay.
Other Europe includes Macedonia, Serbia & Montenegro, and Slovenia.
Other Middle East is Jordan.
Other Africa includes Eritrea, Ghana, Kenya, Liberia, Senegal, Sierra Leone, and Zambia.
USSR includes total of all Former Soviet Union countries before 2002.
Brunei and Malaysia were totaled with Singapore before 2002.
Other Asia-Pacific includes North Korea, Sri Lanka and Vietnam.
Vietnam first added for 2010 compilation.

Table 11: World Refinery Throughput

	2003	2004	2005	2006	2007	2008	Change 2008 over 2007	2008 share of total
Region	thousand barrels per day (1,000 b/d)						%	
United States	15,304	15,475	15,220	15,242	15,156	14,645	-3.4	19.5
Canada	1,880	1,957	1,882	1,835	1,834	1,796	-2.1	2.4
Mexico	1,436	1,436	1,415	1,407	1,395	1,392	-0.2	1.9
South and Central America	5,040	5,430	5,402	5,356	5,470	5,479	0.2	7.3
Europe and Eurasia	19,991	20,454	20,845	20,902	20,705	20,661	-0.2	27.5
Middle East	5,566	5,805	6,007	6,276	6,366	6,403	0.6	8.5
Africa	2,301	2,304	2,477	2,367	2,470	2,495	1.0	3.3
Australasia	823	820	757	749	767	756	-1.4	1.0
China	4,823	5,382	5,916	6,155	6,563	6,851	4.4	9.1
Japan	4,118	4,038	4,136	4,026	3,995	3,946	-1.2	5.2
Other Asia-Pacific	9,263	9,916	10,035	10,392	10,711	10,754	0.4	14.3
Total World	70,545	73,018	74,093	74,707	75,432	75,179	-0.3	100.0

Source: Adapted from *BP Statistical Review of World Energy 2009* (June 2009).
Includes data from Energy Security Analysis, Inc. (ESAI).

NOTES: Input to primary distillation units only.
Some totals might not compute exactly due to rounding and other factors.
Annual changes and shares of total are calculated using thousand barrels per day figures.

Table 12: World Refining Margins

	1999	2000	2001	2002	2003	2004	2005	2006	2007	Dec. 2008	Dec. 2009	2008–2009 Change, %
Market	$/bbl											
U.S. Gulf Coast												
West Texas sour	2.15	5.54	2.91	3.81	3.83	5.67	12.05	9.69	9.53	5.58	3.49	-37.4
Composite U.S. gulf refinery	—	—	—	4.54	4.16	7.87	13.42	10.33	13.01	6.50	4.07	-37.4
Arabian light	3.14	7.27	5.02	5.45	5.51	8.81	13.22	10.58	15.39	7.19	4.02	-44.0
Bonny light	1.74	6.40	0.97	1.54	2.89	5.27	7.74	1.87	3.39	-1.08	-0.91	-16.2
U.S. PADD II Region												
Chicago (WTI)	2.47	4.60	2.52	4.30	3.18	3.31	7.90	7.58	6.87	6.18	1.94	-68.6
U.S. East Coast												
NY Harbor (Arabian medium)	2.50	6.38	4.15	3.38	3.45	5.90	9.14	4.48	14.29	6.70	4.32	-35.5
East Coast composite-RFG	—	—	2.88	4.33	4.15	6.53	10.06	6.42	12.35	3.50	2.34	-33.2
U.S. West Coast												
Los Angeles (ANS)	6.55	12.48	4.33	3.57	6.95	10.03	8.60	19.56	12.31	9.30	9.35	0.6
NW Europe												
Rotterdam (Brent)	2.19	4.71	0.88	0.76	2.21	0.52	0.65	-0.57	4.02	3.34	0.95	-71.7
Mediterranean												
Italy (Urals)	1.67	3.28	0.05	0.74	1.88	7.38	5.60	7.34	3.95	3.95	-2.04	-151.7
Far East												
Singapore (Dubai)	0.78	3.89	0.63	1.98	3.43	4.24	1.46	-1.39	4.65	-0.29	-0.73	152.7

Source: Jacobs Consultancy, Inc.; in *Oil & Gas Journal*, January 11, 2010.

NOTES: WTI = West Texas Intermediate.
ANS = Alaska North Slope.

Table 13: World Natural Gas Consumption

Country	2006	2007	2008	2006	2007	2008	Change 2008 over 2007	2008 share of total
	billion cubic meters (bcm)			bcfd			%	
North America								
United States	614.1	652.6	657.2	59.4	63.1	63.4	0.6	22.0
Canada	96.9	96.7	100.0	9.4	9.4	9.7	3.2	3.3
Mexico	60.9	63.1	67.2	5.9	6.1	6.5	6.2	2.2
Total	771.9	812.4	824.4	74.7	78.6	79.5	1.3	27.6
South and Central America								
Argentina	41.8	43.9	44.5	4.0	4.2	4.3	1.0	1.5
Brazil	20.6	22.0	25.2	2.0	2.1	2.4	14.2	0.8
Chile	7.8	4.6	2.6	0.8	0.4	0.3	-43.2	0.1
Colombia	7.0	7.4	8.2	0.7	0.7	0.8	9.3	0.3
Ecuador	0.7	0.5	0.6	0.1	0.1	0.1	7.4	**
Peru	1.8	2.7	3.4	0.2	0.3	0.3	26.6	0.1
Venezuela	31.5	32.2	32.4	3.0	3.1	3.1	0.3	1.1
Other S/C America	24.0	24.6	26.3	2.3	2.4	2.5	6.6	0.9
Total	135.2	137.9	143.0	13.1	13.3	13.8	3.5	4.7
Europe and Eurasia								
Austria	9.4	8.9	9.5	0.9	0.9	0.9	6.3	0.3
Azerbaijan	9.1	8.0	9.3	0.9	0.8	0.9	16.0	0.3
Belarus	19.0	18.8	19.2	1.8	1.8	1.9	1.9	0.6
Begium & Luxembourg	17.0	16.9	17.0	1.6	1.6	1.6	0.5	0.6
Bulgaria	3.2	3.3	3.3	0.3	0.3	0.3	-0.3	0.1
Czech Republic	9.3	8.4	8.7	0.9	0.8	0.8	3.1	0.3
Denmark	5.1	4.6	4.6	0.5	0.5	0.4	0.5	0.2
Finland	4.2	3.9	4.0	0.4	0.4	0.4	0.8	0.1
France	44.1	42.5	44.2	4.3	4.1	4.3	3.7	1.5
Germany	87.2	82.9	82.0	8.4	8.0	7.9	-1.3	2.7
Greece	3.2	4.0	4.2	0.3	0.4	0.4	4.2	0.1
Hungary	12.7	12.0	12.0	1.2	1.2	1.2	-0.1	0.4
Ireland	4.5	4.8	5.0	0.4	0.5	0.5	4.3	0.2
Italy	77.4	77.8	77.7	7.5	7.5	7.5	-0.4	2.6
Kazakhstan	20.3	19.5	20.6	2.0	1.9	2.0	5.1	0.7
Lithuania	3.2	3.6	3.2	0.3	0.3	0.3	-10.5	0.1
Netherlands	38.1	37.0	38.6	3.7	3.6	3.7	3.9	1.3
Norway	4.4	4.3	4.4	0.4	0.4	0.4	2.9	0.1
Poland	13.7	13.7	13.9	1.3	1.3	1.3	1.4	0.5
Portugal	4.1	4.2	4.6	0.4	0.4	0.4	8.3	0.2
Romania	18.1	16.1	14.5	1.8	1.6	1.4	-9.8	0.5
Russian Federation	419.2	425.7	420.2	40.8	41.2	40.5	-1.6	13.9
Slovakia	6.0	5.7	5.7	0.6	0.5	0.6	1.1	0.2
Spain	33.7	35.1	39.0	3.3	3.4	3.8	10.7	1.3
Sweden	0.9	1.0	1.0	0.1	0.1	0.1	-7.6	**
Switzerland	3.0	2.9	3.1	0.3	0.3	0.3	5.4	0.1
Turkey	30.5	35.1	36.0	3.0	3.4	3.5	2.5	1.2
Turkmenistan	18.4	21.3	19.0	1.8	2.1	1.8	-10.9	0.6
Ukraine	65.1	62.7	59.7	6.3	6.1	5.8	-5.0	2.0
United Kingdom	90.1	90.9	93.9	8.7	8.8	9.1	3.0	3.1
Uzbekistan	41.9	45.9	48.7	4.1	4.4	4.7	5.9	1.6
Other Europe/Eurasia	16.3	16.8	17.1	1.6	1.6	1.6	1.2	0.6
Total	1,132.4	1,138.3	1,143.9	109.6	110.0	110.4	0.2	37.8
Middle East								
Iran	108.7	113.0	117.6	10.5	10.9	11.3	3.8	3.9
Kuwait	12.5	12.1	12.8	1.2	1.2	1.2	5.5	0.4
Qatar	19.6	19.7	19.8	1.9	1.9	1.9	0.4	0.7
Saudi Arabia	73.5	74.4	78.1	7.1	7.2	7.5	4.7	2.6
United Arab Emirates	43.4	49.3	58.1	4.2	4.8	5.6	17.5	1.9
Other Middle East	33.8	34.8	40.7	3.3	3.4	3.9	16.7	1.3
Total	291.5	303.3	327.1	28.2	29.3	31.6	7.6	10.8
Africa								
Algeria	23.7	24.3	25.4	2.3	2.4	2.4	4.1	0.8

Table 13: World Natural Gas Consumption, cont'd.

Country	2006	2007	2008	2006	2007	2008	Change 2008 over 2007	2008 share of total
	billion cubic meters (bcm)			bcfd			%	
Egypt	36.5	38.4	40.9	3.5	3.7	3.9	6.3	1.3
Other Africa	23.7	26.5	28.6	2.3	2.6	2.8	7.6	0.9
Total	83.9	89.2	94.9	8.1	8.6	9.2	6.1	3.1
Asia-Pacific								
Australia	24.0	24.9	23.5	2.3	2.4	2.3	-5.7	0.8
Bangladesh	15.3	16.3	17.3	1.5	1.6	1.7	6.0	0.6
China	56.1	69.5	80.7	5.4	6.7	7.8	15.8	2.7
China Hong Kong SAR	2.9	2.2	2.6	0.3	0.2	0.3	15.4	0.1
India	37.3	40.1	41.4	3.6	3.9	4.0	3.0	1.4
Indonesia	35.9	34.0	38.0	3.5	3.3	3.7	11.5	1.3
Japan	83.7	90.2	93.7	8.1	8.7	9.0	3.6	3.1
Malaysia	29.0	28.5	30.7	2.8	2.8	3.0	7.2	1.0
New Zealand	3.7	4.1	3.8	0.4	0.4	0.4	-6.3	0.1
Pakistan	36.1	36.5	37.5	3.5	3.5	3.6	2.6	1.2
Philippines	2.9	3.4	3.5	0.3	0.3	0.3	1.1	0.1
Singapore	8.6	8.6	9.2	0.8	0.8	0.9	6.6	0.3
South Korea	35.6	38.5	39.7	3.4	3.7	3.8	2.8	1.3
Taiwan	11.1	11.8	12.8	1.1	1.1	1.2	8.4	0.4
Thailand	33.3	35.4	37.4	3.2	3.4	3.6	5.6	1.2
Other Asia-Pacific	12.3	13.0	13.6	1.2	1.3	1.3	4.3	0.4
Total	427.8	456.8	485.3	41.4	44.2	46.8	5.9	16.0
Total World	**2,842.7**	**2,938.0**	**3,018.7**	**275.0**	**284.3**	**291.3**	**2.5**	**100.0**

Source: Adapted from *BP Statistical Review of World Energy 2009* (June 2009). Includes data from Cedigaz.

NOTES: Annual changes and shares of total are calculated using million tonnes of oil equivalent figures.
As far as possible, the data above represent standard cubic meters (at 15°C and 1,013 mbar). Because they are derived directly from tonnes of oil equivalent using an average conversion factor, they do not necessarily equate with gas volumes expressed in specific national terms.
The difference between these world consumption figures and the world production figures in Table 14 is due to variations in stocks at storage facilities and liquefaction plants, together with unavoidable disparities in the definition, measurement, or conversion of gas supply and demand data.
Growth rates are adjusted for leap years.
** Less than 0.05%

Table 14: World Natural Gas Production

Country	2006	2007	2008	2006	2007	2008	Change 2008 over 2007	2008 share of total
	billion cubic meters (bcm)			bcfd			%	
North America								
United States	524.0	540.5	582.2	50.7	52.3	56.2	7.5	19.3
Canada	188.4	184.1	175.2	18.2	17.8	16.9	-5.1	5.7
Mexico	51.6	54.0	54.9	5.0	5.2	5.3	1.3	1.8
Total	764.0	778.7	812.3	73.9	75.3	78.4	4.1	26.7
South and Central America								
Argentina	46.1	44.8	44.1	4.5	4.3	4.3	-1.9	1.4
Bolivia	12.9	13.8	13.9	1.2	1.3	1.3	0.5	0.5
Brazil	11.3	11.3	13.9	1.1	1.1	1.3	22.4	0.5
Colombia	7.0	7.5	9.1	0.7	0.7	0.9	19.8	0.3
Trinidad & Tobago	36.4	39.0	39.3	3.5	3.8	3.8	0.5	1.3
Venezuela	31.5	32.1	31.5	3.0	3.1	3.0	-2.2	1.0
Other South/Central America	5.8	6.5	7.2	0.6	0.6	0.7	11.6	0.2
Total	151.1	155.0	158.9	14.6	15.0	15.3	2.2	5.2
Europe and Eurasia								
Azerbaijan	6.1	9.8	14.7	0.6	0.9	1.4	50.0	0.5
Denmark	10.4	9.2	10.1	1.0	0.9	1.0	9.1	0.3
Germany	15.6	14.3	13.0	1.5	1.4	1.3	-9.2	0.4

Table 14: World Natural Gas Production, cont'd.

Country	2006	2007	2008	2006	2007	2008	Change 2008 over 2007	2008 share of total
	billion cubic meters (bcm)			bcfd			%	
Italy	10.1	8.9	8.4	1.0	0.9	0.8	-6.1	0.3
Kazakhstan	23.9	26.4	30.2	2.3	2.6	2.9	13.9	1.0
Netherlands	61.6	60.5	67.5	6.0	5.9	6.5	11.2	2.2
Norway	87.6	89.7	99.2	8.5	8.7	9.6	10.4	3.2
Poland	4.3	4.3	4.1	0.4	0.4	0.4	-6.6	0.1
Romania	11.9	11.5	11.5	1.2	1.1	1.1	-0.3	0.4
Russian Federation	593.8	592.0	601.7	57.5	57.3	58.1	1.4	19.6
Turkmenistan	60.4	65.4	66.1	5.8	6.3	6.4	0.7	2.1
Ukraine	19.1	19.1	18.7	1.8	1.9	1.8	-2.3	0.6
United Kingdom	80.0	72.1	69.6	7.7	7.0	6.7	-3.7	2.3
Uzbekistan	54.5	59.1	62.2	5.3	5.7	6.0	4.9	2.0
Other Europe/Eurasia	11.5	10.8	10.3	1.1	1.0	1.0	-5.4	0.3
Total	1,050.7	1,053.3	1,087.3	101.7	101.9	104.9	2.9	35.4
Middle East								
Bahrain	11.3	11.8	13.4	1.1	1.1	1.3	13.8	0.4
Iran	108.6	111.9	116.3	10.5	10.8	11.2	3.6	3.8
Kuwait	12.5	12.1	12.8	1.2	1.2	1.2	5.5	0.4
Oman	23.7	24.1	24.1	2.3	2.3	2.3	-0.3	0.8
Qatar	50.7	63.2	76.6	4.9	6.1	7.4	20.9	2.5
Saudi Arabia	73.5	74.4	78.1	7.1	7.2	7.5	4.7	2.5
Syria	5.7	5.6	5.5	0.5	0.5	0.5	-2.8	0.2
United Arab Emirates	49.0	50.4	50.2	4.7	4.9	4.8	-0.7	1.6
Other Middle East	4.1	4.1	4.1	0.4	0.4	0.4	-1.7	0.1
Total	339.1	357.6	381.1	32.8	34.6	36.8	6.3	12.4
Africa								
Algeria	84.5	84.8	86.5	8.2	8.2	8.3	1.7	2.8
Egypt	54.7	55.7	58.9	5.3	5.4	5.7	5.4	1.9
Libya	13.2	15.3	15.9	1.3	1.5	1.5	3.6	0.5
Nigeria	28.4	35.0	35.0	2.8	3.4	3.4	-0.2	1.1
Other Africa	11.8	13.6	18.5	1.1	1.3	1.8	35.7	0.6
Total	192.6	204.4	214.8	18.6	19.8	20.7	4.8	7.0
Asia-Pacific								
Australia	38.9	40.0	38.3	3.8	3.9	3.7	-4.5	1.2
Bangladesh	15.3	16.3	17.3	1.5	1.6	1.7	6.0	0.6
Brunei	12.6	12.3	12.1	1.2	1.2	1.2	-1.2	0.4
China	58.6	69.2	76.1	5.7	6.7	7.3	9.6	2.5
India	29.3	30.1	30.6	2.8	2.9	3.0	1.4	1.0
Indonesia	70.3	67.6	69.7	6.8	6.5	6.7	2.7	2.3
Malaysia	59.7	60.8	62.5	5.8	5.9	6.0	2.5	2.0
Myanmar	12.6	13.5	12.4	1.2	1.3	1.2	-8.5	0.4
New Zealand	3.7	4.1	3.8	0.4	0.4	0.4	-6.3	0.1
Pakistan	36.1	36.5	37.5	3.5	3.5	3.6	2.6	1.2
Thailand	24.3	26.0	28.9	2.4	2.5	2.8	10.7	0.9
Vietnam	6.8	7.1	7.9	0.7	0.7	0.8	11.9	0.3
Other Asia-Pacific	10.5	13.0	14.1	1.0	1.3	1.4	8.5	0.5
Total	378.5	396.3	411.2	36.3	38.3	39.7	3.5	13.4
Total World	2,876.1	2,945.3	3,065.6	278.3	285.0	295.8	3.8	100.0

Source: Adapted from *BP Statistical Review of World Energy 2009* (June 2009). Includes data from Cedigaz.

NOTE: These data exclude gas flared or recycled.

Annual changes and shares of total are calculated using million tonnes of oil equivalent figures.

As far as possible, the data above represent standard cubic meters (measured at 15°C and 1,013 millibar).

Because they are derived from tonnes of oil equivalent using an average conversion factor, they do not necessarily equate with gas volumes expressed in specific national terms.

The difference between these world production figures and the world consumption figures in Table 13 is due to variations in stocks at storage facilities and liquefaction plants, together with unavoidable disparities in the definition, measurement, or conversion of gas supply and demand data.

Growth rates are adjusted for leap years.

Table 15a: World Natural Gas Trade Movements: Pipelines

2008 Pipeline Exports from	North America			Latin America		Europe/Former Soviet Union				
	U.S.	Canada	Mexico	Bolivia	Other LA	Belgium	Germany	Netherlands	Norway	UK
2008 Pipeline Imports to	billion cubic meters (bcm)									
North America										
United States	—	103.20	1.21	—	—	—	—	—	—	—
Canada	15.90	—	—	—	—	—	—	—	—	—
Mexico	10.28	—	—	—	—	—	—	—	—	—
South and Central America										
Argentina	—	—	—	0.89	—	—	—	—	—	—
Brazil	—	—	—	10.90	0.13	—	—	—	—	—
Chile	—	—	—	—	0.69	—	—	—	—	—
Other South/Central America	—	—	—	—	0.97	—	—	—	—	—
Europe and Former Soviet Union										
Austria	—	—	—	—	—	—	1.20	—	1.10	—
Belgium	—	—	—	—	—	—	0.70	7.90	7.95	1.70
Bulgaria	—	—	—	—	—	—	—	—	—	—
Croatia	—	—	—	—	—	—	—	—	—	—
Czech Republic	—	—	—	—	—	—	—	—	2.01	
Finland	—	—	—	—	—	—	—	—	—	—
France	—	—	—	—	—	0.80	2.40	8.00	15.90	0.60
Germany	—	—	—	—	—	—	—	20.50	26.40	1.60
Greece	—	—	—	—	—	—	—	—	—	—
Hungary	—	—	—	—	—	—	2.10	—	—	—
Ireland	—	—	—	—	—	—	—	—	—	5.00
Italy	—	—	—	—	—	—	1.10	8.70	6.00	0.70
Lithuania	—	—	—	—	—	—	—	—	—	—
Luxembourg	—	—	—	—	—	0.60	0.60	—	—	—
Netherlands	—	—	—	—	—	—	4.40	—	6.23	0.90
Poland	—	—	—	—	—	—	1.10	—	—	—
Portugal	—	—	—	—	—	—	—	—	—	—
Romania	—	—	—	—	—	—	—	—	—	—
Serbia	—	—	—	—	—	—	—	—	—	—
Slovakia	—	—	—	—	—	—	—	—	—	—
Spain	—	—	—	—	—	—	—	—	1.80	—
Switzerland	—	—	—	—	—	—	1.45	0.90	0.09	—
Turkey	—	—	—	—	—	—	—	—	—	—
United Kingdom	—	—	—	—	—	1.12	—	9.00	25.30	—
Other Europe/Former Soviet Union	—	—	—	—	—	—	0.09	—	—	—
Middle East										
Iran	—	—	—	—	—	—	—	—	—	—
Jordan	—	—	—	—	—	—	—	—	—	—
United Arab Emirates	—	—	—	—	—	—	—	—	—	—
Other Middle East	—	—	—	—	—	—	—	—	—	—
Africa										
Tunisia	—	—	—	—	—	—	—	—	—	—
Other Africa	—	—	—	—	—	—	—	—	—	—
Asia-Pacific										
Singapore	—	—	—	—	—	—	—	—	—	—
Thailand	—	—	—	—	—	—	—	—	—	—
Total	26.18	103.20	1.21	11.79	1.79	2.52	15.14	55.00	92.78	10.50

Table 15a: World Natural Gas Trade Movements: Pipelines

2008 Pipeline Imports to	2008 Pipelines Exports from												
	Europe/Former Soviet Union			Middle East		Africa				Asia-Pacific			Total
	Russian Fed.	Turkmenistan	Other Europe & Eurasia	Iran	Qatar	Algeria	Egypt	Libya	Mozambique	Indonesia	Malaysia	Myanmar	
	billion cubic meters (bcm)												
North America													
United States	—	—	—	—	—	—	—	—	—	—	—	—	104.41
Canada	—	—	—	—	—	—	—	—	—	—	—	—	15.90
Mexico	—	—	—	—	—	—	—	—	—	—	—	—	10.28
South and Central America													
Argentina	—	—	—	—	—	—	—	—	—	—	—	—	0.89
Brazil	—	—	—	—	—	—	—	—	—	—	—	—	11.03
Chile	—	—	—	—	—	—	—	—	—	—	—	—	0.69
Other S/C America	—	—	—	—	—	—	—	—	—	—	—	—	0.97
Europe and Former Soviet Union													
Austria	5.80	—	—	—	—	—	—	—	—	—	—	—	8.10
Belgium	—	—	—	—	—	—	—	—	—	—	—	—	18.25
Bulgaria	3.10	—	—	—	—	—	—	—	—	—	—	—	3.10
Croatia	1.06	—	0.23	—	—	—	—	—	—	—	—	—	1.29
Czech Republic	6.60	—	—	—	—	—	—	—	—	—	—	—	8.61
Finland	4.50	—	—	—	—	—	—	—	—	—	—	—	4.50
France	8.80	—	0.16	—	—	—	—	—	—	—	—	—	36.66
Germany	36.20	—	2.40	—	—	—	—	—	—	—	—	—	87.10
Greece	2.80	—	0.40	—	—	—	—	—	—	—	—	—	3.20
Hungary	8.90	—	0.50	—	—	—	—	—	—	—	—	—	11.50
Ireland	—	—	—	—	—	—	—	—	—	—	—	—	5.00
Italy	24.50	—	—	—	—	24.44	—	9.87	—	—	—	—	75.31
Lithuania	3.09	—	—	—	—	—	—	—	—	—	—	—	3.09
Luxembourg	—	—	—	—	—	—	—	—	—	—	—	—	1.20
Netherlands	4.33	—	2.14	—	—	—	—	—	—	—	—	—	18.00
Poland	7.20	—	1.50	—	—	—	—	—	—	—	—	—	9.80
Portugal	—	—	—	—	—	1.93	—	—	—	—	—	—	1.93
Romania	3.50	—	1.00	—	—	—	—	—	—	—	—	—	4.50
Serbia	2.15	—	—	—	—	—	—	—	—	—	—	—	2.15
Slovakia	5.60	—	—	—	—	—	—	—	—	—	—	—	5.60
Spain	—	—	0.10	—	—	8.97	—	—	—	—	—	—	10.87
Switzerland	0.35	—	0.40	—	—	—	—	—	—	—	—	—	3.19
Turkey	23.55	—	2.95	5.80	—	—	—	—	—	—	—	—	32.30
United Kingdom	—	—	—	—	—	—	—	—	—	—	—	—	35.42
Other Europe/FSU	2.38	—	0.91	—	—	0.41	—	—	—	—	—	—	3.79
Middle East													
Iran	—	6.50	0.40	—	—	—	—	—	—	—	—	—	6.90
Jordan	—	—	—	—	—	—	2.72	—	—	—	—	—	2.72
United Arab Emirates	—	—	—	—	15.40	—	—	—	—	—	—	—	15.40
Other Middle East	—	—	—	—	1.70	—	0.14	—	—	—	—	—	1.84
Africa													
Tunisia	—	—	—	—	—	1.25	—	—	—	—	—	—	1.25
Other Africa	—	—	—	—	—	0.50	—	—	3.20	—	—	—	3.70
Asia-Pacific													
Singapore	—	—	—	—	—	—	—	—	—	6.65	1.62	—	8.27
Thailand	—	—	—	—	—	—	—	—	—	—	—	8.55	8.55
Total	154.41	6.50	13.09	5.80	17.10	37.50	2.86	9.87	3.20	6.65	1.62	8.55	587.26

Source: Adapted from *BP Statistical Review of World Energy 2009* (June 2009).

NOTES: Flows are on a contractual basis and may not correspond to physical gas flows in all cases.
Data exclude trade within the Former Soviet Union, except the three Baltic States.
Data from Cedigaz (provisional).

Table 15b: World Natural Gas Trade Movements: LNG

2008 LNG Exports from	Americas		Europe		Middle East			Africa					Asia-Pacific				Total
	U.S.	Trinidad & Tobago	Belgium	Norway	Oman	Qatar	UAE	Algeria	Egypt	Equatorial Guinea	Libya	Nigeria	Australia	Brunei	Indonesia	Malaysia	
LNG Imports to	Billion cubic meters (bcm)																
Americas																	
United States	—	7.47	—	0.48	—	0.09	—	—	1.56	—	—	0.34	—	—	—	—	9.94
Dominican Republic	—	0.47	—	—	—	—	—	—	—	—	—	—	—	—	—	—	0.47
Puerto Rico	—	0.81	—	—	—	—	—	—	1.12	—	—	1.04	—	—	—	—	0.81
Argentina	—	0.33	—	—	—	—	—	—	0.08	—	—	—	—	—	—	—	0.41
Mexico	—	1.28	—	—	—	—	—	—	0.99	—	—	0.56	—	—	—	—	3.61
Europe																	
Belgium	—	0.08	-0.40	0.08	—	2.65	—	—	0.08	—	—	—	—	—	—	—	2.49
France	—	0.08	—	0.25	—	—	—	7.60	1.06	—	—	3.60	—	—	—	—	12.59
Greece	—	0.08	—	—	—	—	—	0.70	0.16	—	—	—	—	—	—	—	0.94
Italy	—	—	—	—	—	—	—	1.56	—	—	—	—	—	—	—	—	1.56
Portugal	—	—	0.05	—	—	—	—	—	—	—	—	2.58	—	—	—	—	2.63
Spain	—	4.32	0.18	0.05	0.17	5.12	—	4.90	4.91	0.08	0.53	7.47	—	—	—	—	28.73
Turkey	—	—	—	—	—	—	—	4.25	0.08	—	—	0.98	—	—	—	—	5.31
United Kingdom	—	0.47	—	—	—	0.12	—	0.37	0.08	—	—	—	—	—	—	—	1.04
Asia-Pacific																	
China	—	—	—	—	—	—	—	0.17	0.25	0.16	—	0.24	3.61	—	—	0.01	4.44
India	—	0.24	0.09	0.08	0.35	7.98	0.13	0.65	0.26	0.44	—	0.41	0.16	—	—	—	10.79
Japan	0.97	0.67	—	0.17	4.25	10.91	7.41	1.12	2.21	1.64	—	2.36	15.94	8.22	18.79	17.47	92.13
South Korea	—	0.84	0.08	—	6.04	11.62	—	0.47	2.13	1.33	—	0.16	0.53	0.98	4.06	8.31	36.55
Taiwan	—	0.22	—	—	0.09	1.10	—	0.08	0.08	1.53	—	1.36	—	—	4.55	3.92	12.07
Total	0.97	17.36	0.00	2.19	10.90	39.68	7.54	21.87	14.06	5.18	0.53	20.54	20.24	9.20	26.85	29.40	226.51

Source: Adapted from *BP Statistical Review of World Energy 2009* (June 2009).

NOTES: Data from Cedigaz (provisional). The negative entry for "from Belgium to Belgium" represents the re-export of LNG which was delivered to Zeebrugge terminal and reloaded and shipped elsewhere.

Table 16: Crude Oil Prices

	U.S. refiner acquisition cost			U.S. landed cost of imports							U.S. f.o.b. cost	
	Domestic	Imported	Composite	Total	Canada	Mexico	UK	Nigeria	Saudi Arabia	Venezuela	OPEC	Non-OPEC
Year	$/bbl											
1985	26.66	26.99	26.75	26.67	25.71	25.63	28.36	28.96	24.72	24.43	25.67	25.96
1986	14.82	14.00	14.55	13.49	13.43	12.17	14.63	15.29	12.84	11.52	12.21	12.87
1987	17.76	18.13	17.90	17.65	17.04	16.69	18.78	19.32	16.81	15.76	16.43	16.99
1988	14.74	14.56	14.67	14.08	13.50	12.58	15.82	15.88	13.37	13.66	13.43	13.05
1989	17.87	18.08	17.97	17.68	16.81	16.35	18.74	19.19	17.34	16.78	17.06	16.72
1990	22.59	21.76	22.22	21.13	20.48	19.64	22.65	23.33	21.82	20.31	20.40	20.32
1991	19.33	18.70	19.06	18.02	17.16	15.89	21.37	21.39	17.22	15.92	16.99	16.77
1992	18.63	18.20	18.43	17.75	17.04	15.60	20.63	20.78	17.48	15.13	16.87	16.66
1993	16.67	16.14	16.41	15.72	15.27	14.11	17.92	18.73	15.40	13.39	14.78	14.65
1994	15.67	15.51	15.59	15.18	14.83	14.09	16.64	17.21	15.11	13.12	14.00	14.34
1995	17.33	17.14	17.23	16.78	16.65	16.19	17.91	18.25	16.84	14.81	15.36	16.02
1996	20.77	20.64	20.71	20.31	19.94	19.64	20.88	21.95	20.49	18.59	18.94	19.65
1997	19.61	18.53	19.04	18.11	17.63	17.30	20.64	20.64	17.52	16.35	16.26	17.51
1998	13.18	12.04	12.52	11.84	11.62	11.04	13.55	14.14	11.16	10.16	10.20	11.21
1999	17.90	17.26	17.51	17.23	17.54	16.12	18.26	17.63	17.48	15.58	15.90	16.84
2000	29.11	27.70	28.26	27.53	26.69	26.03	29.26	30.04	26.58	26.05	25.56	26.77
2001	24.33	22.00	22.95	21.82	20.72	19.37	25.32	26.55	20.98	19.81	19.73	21.04
2002	24.65	23.71	24.10	23.91	22.98	22.09	26.35	26.45	24.77	21.93	22.18	22.93
2003	29.82	27.71	28.53	27.69	26.76	25.48	30.62	31.07	27.50	25.70	25.36	26.21
2004	38.97	35.90	36.98	36.07	34.51	32.25	39.28	40.95	37.11	33.79	33.95	33.58
2005	52.94	48.86	50.24	49.29	44.73	43.47	55.28	57.55	50.31	47.87	49.60	45.79
2006	62.62	59.02	60.24	59.11	53.90	53.76	67.44	68.26	59.19	57.37	59.18	55.35
2007	69.65	67.04	67.94	67.97	60.38	62.31	72.47	78.01	70.78	66.13	69.58	62.69
2008	98.47	92.77	94.74	94.63	90.57	86.45	96.95	106.07	97.06	91.06	93.15	87.15

Sources: U.S. Energy Information Administration: *Annual Energy Review*, June 26, 2009 (www.eia.doe.gov/emeu/aer/petro.html); *Monthly Energy Review*, Sept. 24, 2009 (www.eia.doe.gov/emeu/mer/contents. html); Petroleum Navigator for RAC data Aug. 26, 2009 (http: //tonto.eia.doe.gov/dnav/pet/pet_pri_rac2_dcu_nus_a.htm)

NOTES: All prices are U.S. annual average (nominal dollars).

Table 17: Petroleum Product Prices

	Retail unleaded motor gasoline			Refiner resale (wholesale)								
	Regular	Premium	All	Home Heating Oil	Motor Gasoline	Aviation Gasoline	Kerosene Jet Fuel	Kerosene	No. 2 Fuel Oil	No. 2 Diesel Fuel	Propane	Residual Fuel Oil
Year	$/gallon (nominal)			cents/gallon								
1985	1.20	1.34	1.20	77.4	83.5	113.0	79.4	87.4	77.6	77.2	39.8	57.7
1986	0.93	1.09	0.93	47.0	53.1	91.2	49.5	60.6	48.6	45.2	29.0	30.5
1987	0.95	1.09	0.96	53.1	58.9	85.9	53.8	59.2	52.7	53.4	25.2	38.5
1988	0.95	1.11	0.96	47.3	57.7	85.0	49.5	54.9	47.3	47.3	24.0	30.0
1989	1.02	1.20	1.06	56.6	65.4	95.0	58.3	66.9	56.5	56.7	24.7	36.0
1990	1.16	1.35	1.22	69.5	78.6	106.3	77.3	83.9	69.7	69.4	38.6	41.3
1991	1.14	1.32	1.20	61.8	69.9	100.1	65.0	72.2	62.2	61.5	34.9	31.4
1992	1.13	1.32	1.19	58.5	67.7	99.1	60.5	63.2	57.9	59.1	32.8	30.8
1993	1.11	1.30	1.17	55.9	62.6	96.5	57.7	60.4	54.4	57.0	35.1	29.3
1994	1.11	1.31	1.17	52.2	59.9	93.3	53.4	61.8	50.6	52.9	32.4	31.7
1995	1.15	1.34	1.21	53.0	62.6	97.5	53.9	58.0	51.1	53.8	34.4	36.3
1996	1.23	1.41	1.29	65.3	71.3	105.5	64.6	71.4	63.9	65.9	46.1	42.0
1997	1.23	1.42	1.29	60.2	70.0	106.5	61.3	65.3	59.0	60.6	41.6	38.7
1998	1.06	1.25	1.12	43.9	52.6	91.2	45.0	46.5	42.2	44.4	28.8	28.0
1999	1.17	1.36	1.22	53.6	64.5	100.7	53.3	55.0	49.3	54.6	34.2	35.4
2000	1.51	1.69	1.56	89.6	96.3	133.0	88.0	96.9	88.6	89.8	59.5	56.6
2001	1.46	1.66	1.53	77.9	88.6	125.6	76.3	82.1	75.6	77.5	54.0	47.6
2002	1.36	1.56	1.44	71.8	82.8	114.6	71.6	75.2	69.4	72.4	43.1	53.0
2003	1.59	1.78	1.64	88.2	100.2	128.8	87.1	95.5	88.1	88.3	60.7	66.1
2004	1.88	2.07	1.92	117.8	128.8	162.7	120.8	127.1	112.5	118.7	75.1	68.1
2005	2.30	2.49	2.34	172.0	167.0	207.6	172.3	175.7	162.3	173.7	93.3	97.1
2006	2.59	2.81	2.64	199.1	196.9	249.0	196.1	200.7	183.4	201.2	103.1	113.6
2007	2.80	3.03	2.85	219.0	218.2	275.8	217.1	224.9	207.2	220.3	119.4	135.0
2008	3.27	3.52	3.32	297.2	258.5	333.5	302.1	286.2	274.6	299.6	141.6	186.9

Source: U.S. Energy Information Administration, *Annual Energy Review 2008*, posted June 26, 2009. (www.eia.doe.gov/emeu/aer/petro.html).

NOTE: All prices are U.S. annual average (nominal dollars).

Table 18: Comparative Energy Prices

	Natural gas delivered to consumers, $/1000 cu ft				Retail electricity, cents/kWh			
Year	Residential	Commercial	Industrial	Electric Power	Residential	Commercial	Industrial	All
1985	6.12	5.50	3.95	3.55	7.39	7.27	4.97	6.44
1986	5.83	5.08	3.23	2.43	7.42	7.20	4.93	6.44
1987	5.54	4.77	2.94	2.32	7.45	7.08	4.77	6.37
1988	5.47	4.63	2.95	2.33	7.48	7.04	4.70	6.35
1989	5.64	4.74	2.96	2.43	7.65	7.20	4.72	6.45
1990	5.80	4.83	2.93	2.39	7.83	7.34	4.74	6.57
1991	5.82	4.81	2.69	2.18	8.04	7.53	4.83	6.75
1992	5.89	4.88	2.84	2.36	8.21	7.66	4.83	6.82
1993	6.16	5.22	3.07	2.61	8.32	7.74	4.85	6.93
1994	6.41	5.43	3.05	2.28	8.38	7.73	4.77	6.91
1995	6.06	5.05	2.71	2.02	8.40	7.69	4.66	6.89
1996	6.34	5.40	3.42	2.69	8.36	7.64	4.60	6.86
1997	6.94	5.80	3.59	2.78	8.43	7.59	4.53	6.85
1998	6.82	5.48	3.14	2.40	8.26	7.41	4.48	6.74
1999	6.69	5.33	3.10	2.62	8.16	7.26	4.43	6.64
2000	7.76	6.59	4.45	4.38	8.24	7.43	4.64	6.81
2001	9.63	8.43	5.24	4.61	8.58	7.92	5.05	7.29
2002	7.89	6.63	4.02	3.68	8.44	7.89	4.88	7.20
2003	9.63	8.40	5.89	5.57	8.72	8.03	5.11	7.44
2004	10.75	9.43	6.53	6.11	8.95	8.17	5.25	7.61
2005	12.70	11.34	8.56	8.47	9.45	8.67	5.73	8.14
2006	13.73	12.00	7.87	7.11	10.40	9.46	6.16	8.90
2007	13.06	11.32	7.68	7.31	10.65	9.65	6.39	9.13
2008	13.68	11.99	9.58	9.35	11.36	10.28	7.01	9.82

Source: U.S. Energy Information Administration: Natural Gas Navigator: http://tonto.eia.doe.gov/dnav/ng/ng_sum_lsum_dcu_nus_a.htm (Aug. 27, 2009); *Electric Power Monthly*: http://www.eia.doe.gov/cneaf/electricity/epm/table5_3.html (Sept. 11, 2009).

NOTE: All prices are U.S. annual average.

Table 19: Price History of Oil, Gas, and Gasoline

	Actual prices				Producer price index	Inflation-adjusted prices (1982=100)			
	Petroleum		Natural gas			Petroleum		Natural gas	
	Wellhead, $/bbl	Retail gasoline, $/gal	Wellhead	Residential		Wellhead, $/bbl	Retail gasoline, $/gal	Wellhead	Residential
Year			$/1000 cu ft					$/1000 cu ft	
1940	1.02	0.184	0.045	N/A	13.5	7.556	1.364	0.333	NA
1945	1.22	0.205	0.049	N/A	18.2	6.703	1.126	0.269	NA
1950	2.51	0.268	0.065	N/A	27.3	9.194	0.980	0.238	NA
1955	2.77	0.291	0.104	N/A	29.3	9.454	0.992	0.355	NA
1960	2.88	0.311	0.140	N/A	31.7	9.085	0.982	0.442	NA
1965	2.86	0.312	0.156	N/A	32.3	8.854	0.966	0.483	NA
1970	3.18	0.357	0.171	1.09	36.9	8.618	0.967	0.463	2.954
1975	7.67	0.567	0.445	1.71	58.4	12.945	0.971	0.762	2.928
1980	21.59	1.191	1.59	3.68	89.7	24.069	1.328	1.773	4.103
1981	31.77	1.311	1.98	4.29	98.0	32.418	1.338	2.020	4.378
1982	28.52	1.296	2.46	5.17	100.0	28.520	1.296	2.460	5.170
1983	26.19	1.241	2.59	6.06	101.3	25.879	1.226	2.559	5.982
1984	25.88	1.212	2.66	6.12	103.7	24.981	1.170	2.568	5.902
1985	24.09	1.202	2.51	6.12	103.2	23.366	1.166	2.435	5.930
1986	12.51	0.927	1.94	5.83	100.2	12.498	0.926	1.938	5.818
1987	15.40	0.948	1.67	5.54	102.8	14.981	0.922	1.625	5.389
1988	12.58	0.946	1.69	5.47	106.9	11.768	0.885	1.581	5.117
1989	15.86	1.021	1.69	5.64	112.2	14.135	0.910	1.506	5.027
1990	20.03	1.164	1.71	5.80	116.3	17.223	1.001	1.470	4.987
1991	16.54	1.140	1.64	5.82	116.5	14.197	0.979	1.408	4.996
1992	15.99	1.127	1.74	5.89	117.2	13.643	0.962	1.485	5.026
1993	14.25	1.108	2.04	6.16	118.9	11.985	0.932	1.716	5.181
1994	13.19	1.112	1.85	6.41	120.5	10.955	0.924	1.537	5.320
1995	14.62	1.205	1.55	6.06	124.8	11.724	0.966	1.243	4.856
1996	18.46	1.288	2.17	6.34	127.7	14.456	1.009	1.699	4.965
1997	17.23	1.291	2.32	6.94	127.6	13.503	1.012	1.818	5.439
1998	10.87	1.115	1.96	6.82	124.4	8.738	0.896	1.559	5.482
1999	15.56	1.221	2.19	6.69	125.5	12.398	0.973	1.657	5.331
2000	26.72	1.563	3.68	7.76	132.7	20.143	1.178	2.773	5.848
2001	21.84	1.531	4.00	9.63	134.2	16.274	1.141	3.070	7.183
2002	22.51	1.441	2.95	7.89	131.1	17.170	1.099	2.250	6.018
2003	27.56	1.638	4.88	9.63	138.1	19.957	1.186	3.533	6.973
2004	36.77	1.923	5.46	10.75	146.7	25.065	1.311	3.742	7.321
2005	50.28	2.338	7.33	12.70	157.4	31.944	1.485	4.657	8.069
2006	59.69	2.635	6.39	13.73	164.8	36.220	1.599	3.877	8.331
2007	66.52	2.849	6.37	13.06	172.7	38.518	1.650	3.688	7.562
2008	94.04	3.317	8.07	13.68	189.7	49.573	1.749	4.254	7.211

Sources: U.S. Energy Information Administration:
Annual Energy Review 2008 (June 26, 2009): http://www.eia.doe.gov/aer/
Monthly Energy Report (Sept. 24, 2009): http://www.eia.doe.gov/emeu/mer/
Petroleum Navigator (Aug. 26, 2009): http://tonto.eia.doe.gov/dnav/petnew/pet_pri_dfp1_k_a.htm
Capital District Regional Planning Commission, www.cdrpc.org/CPI_PPI.html

NOTES: All prices are U.S. annual averages.
Gasoline prices: 1975 and prior for leaded regular; 1976–94 for unleaded regular; 1995 forward for all grades.

Table 20: International Rig Count

Region	2003	2004	2005	2006	Dec. 2007			Dec. 2008			Dec. 2009		
					Land	Offshore	Total	Land	Offshore	Total	Land	Offshore	Total
North America													
Canada	332	440	575	456	358	2	360	359	2	361	310	2	312
United States	1,126	1,246	1,470	1,718	1,749	62	1,811	1,716	66	1,782	1,136	37	1,173
Subtotal	1,458	1,686	2,045	2,174	2,107	64	2,171	2,075	68	2,143	1,446	39	1,485
Latin America													
Argentina	60	74	81	81	87	0	87	70	0	70	55	0	55
Bolivia	5	4	5	3	2	0	2	3	0	3	4	0	4
Brazil	24	26	28	33	21	25	46	29	30	59	32	34	66
Chile	0	0	0	1	1	1	2	1	0	1	4	0	4
Colombia	11	13	21	24	40	0	40	42	0	42	31	0	31
Ecuador	7	12	14	12	9	0	9	13	0	13	10	0	10
Mexico	107	103	94	84	68	30	98	83	28	111	98	24	122
Peru	2	2	3	6	6	2	8	4	3	7	7	1	8
Trinidad	3	3	5	7	0	5	5	0	4	4	0	3	3
Venezuela	45	69	75	76	58	13	71	66	12	78	38	11	49
Other	—	—	1	2	2	0	2	1	0	1	1	0	1
Subtotal	264	306	327	329	294	76	370	312	77	389	280	73	353
Europe													
Croatia	3	3	3	1	1	0	1	0	0	0	0	1	1
Denmark	4	3	2	0	0	2	2	0	3	3	0	3	3
France	1	1	1	0	1	0	1	1	0	1	1	0	1
Germany	2	2	3	4	4	1	5	10	0	10	4	1	5
Hungary	2	2	2	3	2	0	2	3	0	3	2	0	2
Italy	7	5	3	5	4	1	5	4	0	4	3	1	4
Netherlands	5	4	5	3	0	2	2	0	5	5	3	2	5
Norway	17	14	15	15	0	19	19	0	25	25	0	22	22
Poland	11	3	2	2	2	0	2	1	0	1	3	0	3
Romania	2	2	2	2	2	1	3	16	2	18	7	0	7
Turkey	4	4	5	4	5	0	5	6	0	6	6	0	6
United Kingdom	21	16	27	22	2	20	22	4	18	22	0	14	14
Other	4	4	4	4	8	0	8	6	0	6	9	2	11
Subtotal	83	63	74	65	31	46	77	51	53	104	38	46	84
Middle East													
Iran*	39	40	38	0	0	0	0	0	0	0	0	0	0
Jordan	—	—	1	1	0	0	0	1	0	1	0	0	0
Kuwait	8	12	13	16	11	0	11	11	0	11	17	0	17
Oman	37	36	36	41	53	0	53	52	0	52	47	1	48
Qatar	8	11	14	9	3	13	16	1	9	10	1	7	8
Saudi Arabia	31	34	44	75	65	11	76	64	10	74	55	12	67
Syria	24	23	23	25	19	0	19	20	0	20	19	0	19
United Arab Emirates	16	14	15	15	10	4	14	11	3	14	7	4	11
Yemen	7	12	14	15	16	0	16	15	0	15	12	0	12
Other	1	5	2	2	0	0	0	1	0	1	2	0	2
Subtotal	171	187	200	199	177	28	205	176	22	198	160	24	184

Region	2003	2004	2005	2006	Dec. 2007			Dec. 2008			Dec. 2009		
					Land	Offshore	Total	Land	Offshore	Total	Land	Offshore	Total
Africa													
Algeria	20	20	21	25	29	0	29	27	0	27	27	0	27
Angola	7	3	3	4	2	3	5	0	6	6	2	3	5
Congo (Brazzaville)	4	2	1	1	2	1	3	1	2	3	2	1	3
Egypt	27	31	32	39	39	11	50	49	11	60	37	12	49
Gabon	3	2	3	3	3	1	4	1	0	1	2	0	2
Libya	11	10	9	13	14	1	15	14	1	15	15	1	16
Nigeria	9	11	9	11	2	7	9	3	4	7	4	4	8
South Africa	1	0	0	0	0	0	0	0	1	1	0	0	0
Sudan*	7	17	19	0	0	0	0	0	0	0	0	0	0
Tunisia	2	2	0	5	4	1	5	2	1	3	3	1	4
Other	5	3	3	5	2	2	4	3	1	4	4	1	5
Subtotal	96	101	100	106	97	27	124	100	27	127	96	23	119
Asia-Pacific													
Australia	13	14	14	22	10	11	21	15	11	26	7	10	17
Brunei	2	2	2	4	1	2	3	1	3	4	1	3	4
China: Offshore	10	12	17	17	0	22	22	0	20	20	0	29	29
India	62	76	83	85	55	29	84	55	27	82	75	29	104
Indonesia	42	51	60	53	44	22	66	46	15	61	47	11	58
Japan	4	2	3	1	1	1	2	2	0	2	3	0	3
Malaysia	14	11	12	14	0	11	11	0	19	19	0	13	13
Myanmar	10	9	9	10	7	0	7	1	1	2	3	2	5
New Zealand	2	3	6	6	4	1	5	3	1	4	5	0	5
Pakistan	12	12	11	17	19	0	19	21	0	21	18	0	18
Papua New Guinea	1	1	3	1	1	0	1	2	0	2	2	0	2
Philippines	2	3	2	1	0	1	1	3	0	3	3	0	3
Thailand	5	6	9	11	3	8	11	2	10	12	5	10	15
Vietnam	9	6	10	6	0	4	4	0	4	4	0	8	8
Other	1	4	6	4	1	2	3	0	1	1	0	0	0
Subtotal	189	212	247	252	146	114	260	151	112	263	169	115	284
Total World	**2,261**	**2,555**	**2,993**	**3,125**	**2,852**	**355**	**3,207**	**2,865**	**359**	**3,224**	**2,189**	**320**	**2,509**

Source: Baker Hughes Inc., in *Oil & Gas Journal.*

NOTES: *Iran and Sudan rig counts discontinued in Jan. 2006.
U.S. land count includes inland water rigs.
China count includes offshore rigs only.
No Iraq rig report received since the Iraq invasion of Kuwait in 1990.

Information Sources

National Oil Companies and Energy Ministries

A

Albania

Ministry of Economy, Trade and Energy
Blvd. Deshmoret e Kombit
Tirana 1001, ALBANIA
Telephone: 355 4 22 5695
Fax: 355 4 22 6717
E-mail: kabinet@mete.gov.al
Web site: www.mete.gov.al

Albpetrol Sh.A.
QPN Patos
Patos, Fieri District, ALBANIA
Telephone: 355 381 236 62
Fax: 355 381 241 68
E-mail: directory@albepetrol.net
Web site: www.albpetrol.net

National Agency of Natural Resources
Agjensia Kombetare e Burimeve Natyrore (AKBN)
Blvd. Bajram Curri
Blloku Vasil Shanto, Tirana, ALBANIA
Telephone: 355 4 225 7117
Fax: 355 4 225 7382
Web site: www.akbn.gov.al

Algeria

Ministry of Energy & Mines
Tower A, Val d'Hydra
B.P. 677 Alger Gare
80 Av. Ghermoul
Algiers, ALGERIA
Telephone: 213 21 488 526
Fax: 213 21 488 557
E-mail: info@mem-algeria.org
Web site: www.mem-algeria.org

Sonatrach
Djenane el Malik Hydra
Algiers, ALGERIA
Telephone: 213 21 54 80 11; 213 21 54 70 00
Fax: 213 21 54 77 00
E-mail: sonatrach@sonatrach.dz
Web site: www.sonatrach-dz.com

Angola

Sonangol (Sociedade Nacional de Combustiveis de Angola)
Rua 1° Congresso do MPLA, N.° 8-16
Caixa Postal 1316
Luanda, ANGOLA
Telephone: 244 226 643343
E-mail: CGI@sonangol.co.ao
Web site: www.sonangol.co.ao

Argentina

Ministry of Energy
Av. Paseo Colón 171
Capital Federal – CP (C1063ACB)
ARGENTINA
Telephone: 54 11 4349 5000
E-mail: energia@minplan.gov.ar
Web site: www.energia.gov.ar

Repsol YPF SA
Av. Pres. Roque Sáenz Peña #777
CP 1364 Buenos Aires, ARGENTINA
Telephone: 54 11 4329 2000
Fax: 54 11 4329 5717
Web site: www.ypf.com

Armenia

Ministry of Energy and Natural Resources
Government House 2
Republic Square
Yerevan 375010, ARMENIA
Telephone: 37 410 52 19 64
Fax: 37 410 52 63 65
E-mail: minenergy@minenergy.am
Web site: www.minenergy.am

Australia

Commonwealth of Australia
Department of Resources, Energy and Tourism
Industry House
10 Binara Street
Mail: GPO Box 1564
Canberra City, ACT 2601, AUSTRALIA
Telephone: 61 2 6276 1000
E-mail: ret@ret.gov.au
Web sites: www.australia.gov.au
www.ret.gov.au

Ministerial Council on Energy
Web site: mce.gov.au

Australia, New South Wales

Division of Minerals and Energy
Industry and Investment NSW
227 Elizabeth St., Level 17
Mail: GPO Box 3889
Sydney, New South Wales 2001, AUSTRALIA
Telephone: 02 8281 7777
Fax: 02 8281 7799
E-mail: information@industry.nsw.gov.au
Web site: www.industry.nsw.gov.au/energy

Australia, Northern Territory

Department of Minerals and Energy
Geological Survey: Darwin
Paspalis Centrepoint Building
Smith Street Mall
Darwin 0800, Northern Territory, AUSTRALIA
Telephone: 618 8999 5511
Web site: www.nt.gov.au/d/Minerals_Energy

Australia, Queensland

Department of Natural Resources and Water
Mineral House – 13th Floor
41 George St.
Brisbane, Queensland 4000, AUSTRALIA
GPO Box 2454
Brisbane Q 4001, AUSTRALIA
Telephone: 07 389 63356
Web site: www.nrm.qld.gov.au

Australia, South Australia

Department of Primary Industries and Resources SA
Petroleum and Geothermal Group
Level 6, 101 Grenfell St.
GPO Box 1671
Adelaide South Australia 5001, AUSTRALIA
Telephone: 61 8 8463 3204 (international)
Fax: 61 8 8463 3229 (international)
Web site: http://www.pir.sa.gov.au/petroleum

Australia, Tasmania

Department of Infrastructure, Energy and Resources
10 Murray St.
Hobart, Tasmania 7000, AUSTRALIA
Mail: GPO Box 936
Hobart, Tasmania 7001, AUSTRALIA
Telephone: 1300 135 513
E-mail: webmaster@dier.tas.gov.au
Web site: www.dier.tas.gov.au

Australia, Victoria

Department of Primary Industries (DPI)
Earth Resources Division: Melbourne Office
1 Spring St.
Mail (DPI Information Centre): GPO Box 4440
Melbourne, Victoria 3001, AUSTRALIA
Telephone: 61 3 5332 5000 (international)
Web site: www.dpi.vic.gov.au

Australia, Western Australia

Office of Energy
Level 9, Governor Stirling Tower
197 St Georges Terrace
Perth, Western Australia 6000, AUSTRALIA
Telephone: 61 08 9420 5600
Fax: 61 08 9420 5700
E-mail: enquiries@energy.wa.gov.au
Web site: www.energy.wa.gov.au

Austria

Federal Ministry of Economics and Labour
Energy and Mining Division
Stubenring 1, A-1010 Wien (Vienna), AUSTRIA
Telephone: 43 1 711 00 0
E-mail: service@bmwa.gv.at
Web site: www.bmwfj.gv.at/en

OMV AG (Aktiengesellschaft)
Trabrennstrasse 6-8
A-1020 Wien (Vienna), AUSTRIA
Telephone: 43 1 40440 0
Fax: 43 1 40440 20091
E-mail: info@omv.com
Web site: www.omv.com

Azerbaijan

AIOC (Azerbaijan International Operating Co.)
Villa Petrolea, 2 Neftchilar Prospekti (Bayil)
Baku 1003, AZERBAIJAN
Telephone: 994 12 4 97 9000
Fax: 994 12 4 97 9602
E-mail: bayatltf@bp.com
Web site: www.bp.com/caspian

Socar (State Oil Co. of Azerbaijan Republic)
73 Neftchilar Prospekti (Bayil)
Baku 1000, AZERBAIJAN
Telephone: 994 12 521 0332
Fax: 994 12 521 0383
E-mail: info@socar.az
Web site: http://socar.az/index-en.html

B

Bahamas (Commonwealth of the)

Environment, Science and Technology Commission (BEST)
P.O. Box CB
Nassau 10980, THE BAHAMAS
Telephone: 242 322 4546
Fax: 242 326 3509
E-mail: bestnbs@hotmail.com
Web site: www.best.bs

Bahrain

National Oil and Gas Authority (NOGA)
Al Mareed Street
P.O. Box 1435, Manama, BAHRAIN
Telephone: 973 1731 2644
Fax: 973 1729 3007
E-mail: info@noga.gov.bh
Web site: http://www.noga.gov.bh

Bahrain National Gas Co. (Banagas)
P.O. Box 29099, BAHRAIN
Telephone: 973 17 756222
Fax: 973 17 756991
E-mail: bng@banagas.com.bh
Web site: www.banagas.com.bh

Bahrain Petroleum Company (Bapco)
P.O. Box 25555
Awali, BAHRAIN
Telephone: 973 1770 4040
Fax: 973 1770 4070

Bangladesh

Petrobangla (Bangladesh Oil, Gas & Mineral Corp.)
Petrocentre
3 Kawranbazar
Dhaka 1215, BANGLADESH
Fax: 880 2 912 0224
Web site: www.petrobangla.bd.org

Belgium

Federal government information portal
Web site: www:belgium.be.en/

Belize

Ministry of Natural Resources and the Environment
Market Square, Belmopan, BELIZE
Telephone: 501 822 2226; 501 822 2249; 501 822 2711
Fax: 501 822 2333
E-mail: info@mnrei.gov.bz
Web site: www.mnrei.gov.bz

Bolivia

Ministry of Hydrocarbons and Energy
Communications Center Edificio La Paz, Piso 12
Av. 20 Mariscal Santa Cruz esq. c. Oruro
La Paz, BOLIVIA
Telephone: 591 2 237 4050; 4051
Fax: 591 2 214 1307
Web site: www.hidrocarburos.gov.bo

Yacimientos Petroliferos Fiscales Bolivianos (YPFB)
Central Office
Calle Well No. 185
Box 401
La Paz, BOLIVIA
Telephone: 591 2 237 0210
Web site: www.ypfb.gov.bo

Brazil

ANP (Agência Nacional do Petróleo, Gas Natural e Biocombustiveis)
Av. Rio Branco #65 – del 12 al 22 piso
Rio de Janeiro, BRAZIL 20090-004
Telephone: 55 21 2112 8100
Fax: 55 21 2112 8129
E-mail: imprensa@anp.gov.br
Web site: www.anp.gov.br

Ministry of Mines and Energy
Esplanada dos Ministerios Block "U"
Brasilia, DF, BRAZIL
Telephone: 61 3319 5555
Web site: www.mme.gov.br

Petrobras (Petroleo Brasileiro SA)
Av. República do Chile 65, 22nd Floor
Rio de Janeiro 20031-912, BRAZIL
Telephone: 55 21 2534 1510
Web Site: www.petrobras.com.br

Brunei

Office of the Prime Minister
Petroleum Unit
Bahirah Building, Jalan Menteri Besar
Darussalam BB3910, BRUNEI
Telephone: 673 2 387 102
Fax: 673 2 383 004
E-mail: brupet@brunet.bn
Web site: www.petroleum-unit.gov.bn

Bulgaria

Energy Committee
1169 Sofia, 1 Kniaz Alexander I Square
Sofia 1040, BULGARIA
Telephone: 02 939 32 01
Fax: 02 987 50 13
E-mail: kom_energy@parliament.bg
Web site: www.parliament.bg

C

Cambodia

Cambodian National Petroleum Authority (CNPA)
13-14 Confederation de la Russie
Phnom Penh 12406, CAMBODIA
Telephone: 855 023 890 330
Fax: 855 23 890 330
E-mail: duongsothy@yahoo.com
Web site: www.cnpa-cambodia.com

Cameroon

Societe Nationale des Hydrocarbures (SNH)
Box 955
Yaounde, CAMEROON
Telephone: 237 22 20 1920; 237 22 21 0430
Fax: 237 22 20 9869; 237 22 20 4651
E-mail: info@snh.cm

Canada

National Energy Board
444 Seventh Avenue SW
Calgary, Alberta
T2P 0X8 CANADA
Telephone: 403-292-4800;
1 800 899 1265
Fax: 403 292 5503
Toll-free fax: 1 877 288 8803
E-mail: info@neb-one.gc.ca
Web site: www.neb.gc.ca

Natural Resources Canada
Energy Sector
580 Booth St.
Ottawa, Ontario
K1A OE4 CANADA
Telephone: 613 995 0947
Web site: www.nrcan-rncan.gc.ca

Canada, Alberta

Alberta Department of Energy (Edmonton)
North Petroleum Plaza
10th Floor, 9945-108 Street
Edmonton, Alberta
T5K 2G6 CANADA
Telephone: 780 427 8050
Fax: 780 422 0698
Web site: www.energy.gov.ab.ca

Alberta Department of Energy (Calgary)
300, 801-6 Avenue SW
Calgary, Alberta
T2P 3W2 CANADA
Telephone: 403 297 8955
Web site: www.energy.gov.ab.ca

Alberta Energy Resources Conservation Board (ERCB)
Calgary Office
640 - 5 Avenue SW
Calgary, Alberta
T2P 3G4 CANADA
Telephone: 403 297 8311
Fax: 403 297 7336
E- mail: Inquiries@ercb.ca
Web site: www.ercb.ca

Alberta Utilities Commission
Calgary Office
Fifth Avenue Place
4th Floor, 425 First St. SW
Calgary, Alberta
T2P 3L8 CANADA
Telephone: 403 592 8845
Fax: 403 592 4406
Email: Info@auc.ab.ca
Web site: www.auc.ab.ca

Alberta Utilities Commission
Edmonton Office
10th Floor, 10055 106 Street
Edmonton, Alberta
T5J 2Y2 CANADA
Telephone: 780 427 4901
Fax: 780 427 6970
Email: Info@auc.ab.ca
Web site: www.auc.ab.ca

Canada, British Columbia

Ministry of Energy, Mines and Petroleum Resources
PO Box 9318, Stn Prov Govt
Victoria, British Columbia
V8W 9N3 CANADA
Telephone: 250 952 0241
Fax: 250 952 0627
Web site: www.gov.bc.ca/empr

Canada, Manitoba

Department of Science, Technology, Energy and Mines
Petroleum Branch
360-195 Ellice Avenue
Winnipeg, Manitoba
R3G 3P2 CANADA
Telephone: 204 945 6577
Fax: 204 945 0586
E-mail: mgi@gov.mb.ca
Web site: www.gov.mb.ca/stem/petroleum

Canada, New Brunswick

Energy Department
Brunswick Square
1 Germain Street – Suite M-100
St. John, New Brunswick
E2L 4V1 CANADA
Telephone: 506 658 3180
Fax: 506 658 3191
Web site: www.gnb.ca/0085/index-e.asp

Canada, Newfoundland & Labrador

Department of Natural Resources
7th Floor, Natural Resources Building
50 Elizabeth Avenue
P.O. Box 8700
St. John's, Newfoundland and Labrador
A1B 4J6 CANADA
Telephone: 709 729 2920
Fax: 709 729 0059
Web sites: www.gov.nf.ca
www.nr.gov.nl.ca/nr/

Canada, Northwest Territories

Department of Environment and Natural Resources
P.O. Box 1320
Yellowknife, Northwest Territories
X1A 2L9 CANADA
Telephone: 867 873 7654
Fax: 867 873 0221
Web site: www.gov.nt.ca

Canada, Nova Scotia

Department of Energy
Bank of Montreal Building, Suite 400
5151 George St.
P.O. Box 2664
Halifax, Nova Scotia
B3J 3P7 CANADA
Telephone: 902 424 4575
Fax: 902 424 0528
Web site: www.gov.ns.ca/energy

Canada, Nunavut

Department of Environment
P.O. Box 1000, Station 1300
Iqaluit, Nunavut, CANADA
X0A 0H0
Telephone: 867 975 7700
Fax: 867 975 7740
E-mail: environment@gov.nu.ca
Web site: www.gov.nu.ca/env

Canada, Ontario

Ministry of Energy and Infrastructure
900 Bay Street, 4th Floor
Hearst Block
Toronto, Ontario
M7A 2E1 CANADA
Telephone: 1 888 668 4636
Web site: www.mei.gov.on.ca

Canada, Prince Edward Island

Department of Environment, Energy and Forestry
Jones Building, 4th Floor
11 Kent Street
P.O. Box 2000
Charlottetown, Prince Edward Island
C1A 7N8 CANADA
Telephone: 902-368-5000
Fax: 902 368 5830
Web site: www.gov.pe.ca/enveng

Canada, Quebec

Department of Natural Resources
5700, 4e Avenue Ouest
B401 (Energy & Mines)
Charlesbourg, Québec
G1H 6R1 CANADA
Telephone: 418 627 6377
Fax: 418 643 0701
Web site: www.mrnf.gouv.qc.ca/english/energy

Canada, Saskatchewan

Ministry of Energy and Resources
300-2103 11th Avenue
Regina, Saskatchewan
S4P 3Z8 CANADA
Telephone: 306 787 2528
Fax: 306 787 1284
Web site: www.er.gov.sk.ca

Canada, Yukon Territory

Energy, Mines and Resources
Box 2703
Whitehorse, Yukon
Y1A 2C6 CANADA
Telephone: 867 667 3130
Fax: 867 456 3965
E-mail: emr@gov.yk.ca
Web sites: www.emr.gov.yk.ca
www.yukonoilandgas.com

Chad

Ministry of Mines and Energy
N'Djamena, CHAD
Telephone: 235 51 83 06
Fax: 235 52 75 60
E-mail: cons.mines@intnet.td
Web site: www.primature-tchad.org/

Societe de Hydrocarbures du Tchad (SHT)
BP 6179
N'Djamena, CHAD

Chile

ENAP (Empresa Nacional del Petróleo)
Av. Vitacura 2736 piso 10 Las Condes
Santiago, CHILE
Telephone: 56 2 280 3000
Fax: 56 2 280 3199
Web site: www.enap.cl

China

CNOOC Ltd. (China National Offshore Oil Corp. Ltd.)
Box 4705
No. 25 Chao Yangmen North Street
Dongcheng District
Beijing 100010, PEOPLE'S REPUBLIC OF CHINA
Telephone: 010 845 21010
Fax: 010 845 21142
E-mail: news1@cnooc.com.cn
Web site: www.cnooc.com.cn/yyww

CNPC (China National Petroleum Corp.)
9 Dongzhimen North Street, Dongcheng District
Beijing, 100007, PEOPLE'S REPUBLIC OF CHINA
Telephone: 86 10 6209 4114
Fax: 86 10 6209 4205
E-mail: admin_eng@cnpc.com.cn
Web site: www.cnpc.com.cn/en

Sinopec Corp. (China Petroleum & Chemical Corp.)
22 Chaoyangmen North Street, Chaoyang District
Beijing 100728, CHINA
Telephone: 86 10 649 90060
Fax: 86 10 649 90022
E-mail: ir@sinopec.com
Web site: http://english.sinopec.com

Colombia

Ecopetrol SA (Empresa Colombiana del Petroleos)
Edificio Principal Cr 7 No. 37-65
Bogota, COLOMBIA
Telephone: 57 1 234 4000
Fax: 57 1 234 4099
E-mail: pmdvrp@ecopetrol.com.co
Web site: www.ecopetrol.com.co/english

Ministerio de Minas y Energia de Colombia
Calle 45, No. 57-31, CAN
Bogota 80319, COLOMBIA
Telephone: 57 1 220 0300
Fax: 57 1 222 3656
Web site: www.minminas.gov.co
E-mail: menergia@minminas.gov.co

Congo (Brazzaville)

Ministry of Hydrocarbons, B.P. 2120
Brazzaville, CONGO
Telephone: 242 81 5823
Fax: 242 81 1085

SNPC (Societe Nationale des Petroles du Congo)
B.P. 188
Brazzaville, CONGO
Telephone: 242 81 0964-94 87 48
Fax: 242 81 0492-94 92 24
Web site: snpc-group.com

Congo (former Zaire)

Petrozaire
B.P. 7617
Kinshasa/Gombe, CONGO ZAIRE
Telephone: 243 12 20 344; 243 12 20 492

Croatia

INA-Industrija nafte, dd
Avenija Veceslava Holjevca 10, p.p. 555
Zagreb 10002, CROATIA
Telephone: 385 1 6450 000
E-mail: ina-besplatni.telefon@ina.hr
Web site: www.ina.hr

Cuba

Union Cuba-Petroleo (CUPET)
No. 154 e/ Bitterness and Teniente Rey
Habana Vieja
Havana 10100, CUBA
Telephone: 53 7 862 7576; 7578
E-mail: control@union.cupet.cu

Ministerio de la Industria Basica
No. 666 Ave. Salvador Allende e/ Oquendo y Soledad
Havana, CUBA
Telephone: 53 7 878 7840
Fax: 53 7 878 5345
Web site: www.cubagob.cu

Czech Republic

Urad Vlady CR
Waterfront: Edvarda Benese 4
Mala Strana, PSC 118 01
Praha 1, CZECH REPUBLIC
Telephone: 420 224 002 111
E-mail: posta@vlada.zc
Web site: www.vlada.cz

D

Denmark

Ministry of the Environment
Information Center
Strandgade 29
1401 Copenhagen K, DENMARK
Telephone: 45 70 12 02 11
E-mail: mim@mim.dk
Web site: www.mim.dk/eng/

E

Ecuador

Petroecuador
Empresa Estatal Petroleos del Ecuador
Calle Alpallana E8-86 y Av. 6 de Diciembre
Quito, ECUADOR
Telephone: 593 2 222 9043; 593 2 250 3770
Fax: 593 2 256 9738
Web site: www.petroecuador.com.ec

Ministerio of Non-Renewable Natural Resources
Av. Orellana N26-220 y Juan León Mera
Telephone: 593 2 2550 0018
Quito, ECUADOR
E-mail: info@minasypetroleos.gov.ec
Web site: www.minasypetroleos.gov.ec

OLADE (Organizacion LatinoAmericana de Energia)
Av. Mariscal Antonio José de Sucre N 58-63 y Fernández Salvador Edificio
Sector San Carlos
P.O. Box 17-11-6413 CCNU
Quito, ECUADOR
Telephone: 593 2 2598 280; 2597 995
Fax: 593 2 2531 691
E-mail: olade@olade.org.ec
Web site: www.olade.org.ec

Egypt

Arab Republic of Egypt
Ministry of Petroleum
Ahmed al Zamer Street
8th District, Nasr City
P.O. Box 7109
Cairo, EGYPT
Telephone: 20 2 2670 6401; 6402
Fax: 20 2 2670 6419; 6427
E-mail: contactus@petroleum.gov.eg
Web site: www.petroleum.gov.eg

Egyptian General Petroleum Corp. (EGPC)
Palestine Street, Sector 4, New Maadi
P.O. Box 2130
Cairo 11742, EGYPT
Telephone: 20 2 2703 1438; 20 2 2703 1439
Fax: 20 2 2703 1457
E-mail: info@egpc.com.eg
Web site: www.egpc.com.eg

Egyptian Natural Gas Holding Co. (EGAS)
85 Nasr Road, 1st District
Nasr City
P.O. Box 8064
Cairo 11371, EGYPT
Telephone: 20 2 2405 5845; 20 2 2405 5846
Fax: 20 2 2405 5876
Email: egas@egas.com.eg
Web site: www.egas.com.eg

Equatorial Guinea

Ministry of Mines, Industry and Energy
Carretera de Punta Europa
Malabo, EQUATORIAL GUINEA
Telephone: 240 09 3567
Fax: 240 09 3353
E-mail: shawd@ rpsgroup.com (David Shaw)
Web site: www.equatorialoil.com

F

Faroe Islands

Faroese Earth and Energy Directorate (Jardfeingi)
Brekkutún 1
Postsmoga 3059
FO-110 Tórshavn, FAROE ISLANDS
Telephone: 298 357000
Fax: 298 357001
Email: jardfeingi@jardfeingi.fo
Web site: www.jardfeingi.fo

Finland

Ministry of Employment and the Economy
P.O. Box 32
FI-00023 Government
Aleksanterinkatu 4, FI-00170
Helsinki, FINLAND
Telephone: 358 10 60 6000
E-mail: kirjaamo@tem.fi
Web site: www.tem.fi

France

Ministry of the Economy, Industry and Employment
139, rue de Bercy 75012
Paris, FRANCE
Web site: www.economie.gouv.fr

G

Gabon

Ministry of Mines, Energy, Oil and Hydraulic Resources
B.P. 874
Libreville, GABON
Telephone: 241 74 47 83
Fax: 241 72 49 90
E-mail: mcn@internetgabon.com
Web site: www.legabon.org

Georgia

Ministry of Economic Development
Telephone: 995 32 97 77 77
E-mail: ministry@economy.ge
Web site: www.economy.ge

Georgian Oil & Gas Corporation
21 Kakheti Highway
Tbilisi 0190, GEORGIA
Telephone: 995 32 244 040
Fax: 995 32 244 041
Web site: www.gogc.ge

Ghana

Ministry of Energy
P. O. Box T40
Stadium Post Office
Accra, GHANA
Telephone: 233 21 667 152
Fax: 233 21 668 262
Email: moen@energymin.gov.gh
Web site: www.energymin.gov.gh

Greece

Hellenic Petroleum S.A.
17th km., Athens-Corinth National Hwy.
GR-193 00 Aspropyrgos
Attikí, GREECE
Telephone: 30 210 55 39 109
Fax: 30 210 55 39 100
Web site: www.hellenic-petroleum.gr

Ministry of Development
Department of Energy and Natural Resources
80 Mihalakopoulou Street
Athens, GREECE
Telephone: 30 210 748 2770-2
E-mail: grammatia@ypan.gr
Web site: www.ypan.gr/structure_uk_c.htm

Guatemala

Ministerio de Energia y Minas
Diagonal 17 29-78, Zona 11
Las Charcas, GUATEMALA
Telephone: 502 2419 6464
E-mail: informatica@mem.gob.gt
E-mail: relapubli@mem.gob.gt
Web site: www.mem.gob.gt

H

Hungary

MOL (Hungarian Oil & Gas plc)
October 23rd Street, No. 18
H-1117 Budapest, HUNGARY
Telephone: 36 1 209 0000
E-mail: pressoffice@mol.hu
Web site: www.mol.hu

I

Iceland

Orkustofnun (National Energy Authority)
Orkugardur, Grensasvegur 9
Reykjavík 108, ICELAND
Telephone: 354 569 6000
Fax: 354 568 8896
E-mail: os@os.is
Web site: www.os.is/page/english/

India

Ministry of Petroleum & Natural Gas
Shastri Bhavan
New Delhi 110 001, INDIA
Telephone: 011 2338 3508
Fax: 011 2338 3100
E-mail: shri_prakash@nic.in
Web site: www.petroleum.nic.in

ONGC (Oil & Natural Gas Corp. Ltd.)
Tel Bhavan
Dehradun 248 003, INDIA
Telephone: 0135 275 9561 67
Web site: www.ongcindia.com

Indonesia

Pertamina
(Perusahaan Pertambangan Minyak Dan Gas Bumi Negara)
Jalan Medan Merdeka Timur No. 1 A
Jakarta-10110, INDONESIA
Telephone: 021 381 5111; 381 6111
Fax: 021 363 3585; 384 3882
E-mail: pcc@pertamina.com
Web site: www.pertamina.com

Iran

National Iranian Oil Co.
Hafez Crossing, Taleghani Avenue
Tehran, IRAN
Telephone: 98 21 615 2929
Fax: 98 21 615 3886
E-mail: managing.director@nioc.com
Web site: www.nioc.ir/english

Iraq

Ministry of Oil
Oil Complex Building
Port Saeed St.
Baghdad, IRAQ
Telephone: 946 1 727 0710
E-mail: oilministry@oil.gov.iq
Web site: www.oil.gov.iq

State Oil Marketing Organization (SOMO)
Moatasem Street, Al Zayouna Area
P.O. Box 5118
Baghdad, IRAQ
Telephone: 946 1 774 2040
Fax: 964 1 772 6574
E-Mail: newsomo@yahoo.com

Ireland

Department of Communications, Energy and Natural Resources
29-31 Adelaide Rd.
Dublin 2, IRELAND
Telephone: 353 1 678 2000
Fax: 353 1 678 2449
Web site: www.dcmnr.gov.ie

Israel

Ministry of National Infrastructures
Natural Resources Section/
Oil and Gas Exploration Division
216 Yaffo Street
POB 36148
Jerusalem 91360 ISRAEL
Web site: www.mni.gov.il/mni/en-US/default.htm
Public inquiries:
E-mail: pniot@mni.gov.il
Telephone: 972 4 864 4024
Petroleum and Mining Commissioner:
E-mail: ymimran@mni.gov.il
Telephone: 972 2 531 6070
[*See* Energy Division under Ministry of National Infrastructures on Web site]

Italy

Ente Nazionale Idrocarburi SpA (Eni SpA)
Piazzale Mattei, 1
00144 Rome, ITALY
Telephone: 39 06 59 82 1
Fax: 39 06 59 82 21 41
Web site: www.eni.it

Ministry of Economic Development
Via Molise 2
Rome 00187, ITALY
Telephone: 39 06 47 051
Web site: http://sviluppoeconomico.gov.it

Ivory Coast

Ministry of Mines and Energy
Box V 50
Abidjan, COTE D'IVOIRE
Telephone: 225 21 50 03
Fax: 225 21 53 20

Petroci Holdings
(Societe Nationale d'Operations Petrolieres de la Cote d'Ivoire)
Immeuble les Heveas, 14 Blvd. Carde
Box V 194
Abidjan, COTE D'IVOIRE
Telephone: 225 20 202 500
Fax: 225 21 68 24

J

Jamaica

Petroleum Corporation of Jamaica (PCJ)
36 Trafalgar Rd.
Kingston 10, JAMAICA, WEST INDIES
Telephone: 876 929 5380/9
Fax: 876 929 2409
E-mail: ica@pcj.com (Information & Corporate Affairs Dept.)
E-mail: gmd1@pcj.com
Web site: www.pcj.com

Japan

Japan Ministry of Economy, Trade and Industry
1-3-1 Kasumigaseki, Chiyoda-ku
Tokyo 100-8901, JAPAN
Telephone: 81 3 3501 1511
E-mail: webmail@meti.go.jp
Web site: www.meti.go.jp/english/index.html

Japan Oil, Gas and Metals National Corp. (JOGMEC)
Muza Kawasaki Central Tower
1310 Omiya-cho, Saiwai-ku, Kawasaki-shi
Kanagawa 212-8554 JAPAN
Telephone: 81 44 520 8600
Fax: 81 44 520 8710
Web site: www.jogmec.go.jp/english/index.html

Jordan

National Petroleum Company (NPC)
P.O. Box 851 634
Amman 11185, JORDAN
Telephone: 962 6 553 6912
E-mail: management@npc.com.jo
Web site: www.npc.com.jo

K

Kazakhstan

KazMunayGas National Company
22 Kabanbai Batyr Avenue
Astana 010000, KAZAKHSTAN
Telephone: 7 7172 976 305
Fax: 7 7172 976 000
E-mail: info@kmg.kz
Web site: www.kmg.kz/eng/

Ministry of Energy and Mineral Resources
37 Bejbetshilick Street
Astana, KAZAKHSTAN
Telephone: 7 7172 976 807
Fax: 7 7172 976 991
E-mail: magripa@memr.kz

Kuwait

Ministry of Oil
P.O. Box 5077
13051, KUWAIT
E-mail: alnaft@moo.gov.kw
Web site: www.moo.gov.kw

Kuwait Oil Co. (KOC)
P.O. Box 9758 Ahmadi
61008 Ahmadi, KUWAIT
Telephone: 00965 2398 9111
Fax: 00965 2398 3661
E-mail: kocinfo@kockw.com
Web site: www.kockw.com

Kuwait Petroleum Co. (KPC)
P.O. Box 26565
Postal Code 13126
Safat, KUWAIT
Telephone: 965 1 85 85 85
Fax: 965 2499 4991
E-mail: media@kpc.com.kw
Web site: www.kpc.com.kw

Kyrgyzstan

Ministry of Industry, Energy and Fuel Resources
119 Ahunbaev Street
Bishkek City 720055, KYRGYZSTAN
Telephone: 996 312 56 18 22
Web site: www.mpe.gov.kg

L

Libya

Supreme Council for Energy Affairs
Tripoli, LIBYA
(Created August 2009; no contact data available)

National Oil Corp. (NOC)
Bashir Saadawi Street
Tripoli, LIBYA
Telephone: 00218 21 444 6181
Fax: 00218 21
Web site: www.noclibya.com

Ras Lanuf Oil and Gas Processing Company (RASCO)
Subsidiary of NOC
P.O. Box 2323
Ras Lanuf Terminal
Tripoli GSPLAJ, LIBYA
Telephone: 218 21 360 5177
Fax: 218 21 360 5174
E-mail: info@raslanuf.ly
Web site: www.raslanuf.ly/en/

Lithuania

Ministry of Economy
38/2 Gedimino Ave.
Vilnius LT-2600, LITHUANIA
Telephone: 370 5 262 2416
Fax: 370 5 262 3974
E-mail: kanc@ukmin.lt
Web site: www.ukmin.lt

Lithuanian Energy Institute (LEI)
3 Breslaujos Street
LT-3035 Kaunas, LITHUANIA
Telephone: 370 37 401 801
Fax: 370 37 351 271
E-mail: skema@isag.lei.lt
Web site: www.lei.lt

M

Madagascar

Office des Mines Nationales et des Industries Strategiques (OMNIS)
21 Lalana Razanakombana
B.P. 1 BIS
Antananarivo 101, MADAGASCAR
Telephone: 261 20 22 242 83
Fax: 261 20 22 229 85
Email: omnis@simicro.mg
Web site: www.omnis-madagascar.mg

Malaysia

Petronas (Petroliam Nasional Sdn. Berhad)
Tower 1, Petronas Twin Towers
Kuala Lumpur City Centre
Kuala Lumpur 50088 MALAYSIA
Telephone: 603 2051 5000
Fax: 603 2026 5050
E-mail: webmaster@petronas.com.my
Web site: www.petronas.com.my

Malta

Ministry of Resources and Infrastructure
Oil Exploration Department
Block 3
CMR 02 Floriana, MALTA
Telephone: 356 21 237 921

Mali

Ministry of Mines, Energy and Water Resources
B.P. 238
Bamako, MALI
Telephone: 223 2 22 35 47
Fax: 223 2 22 21 60

Mexico

Pemex (Petroleos Méxicanos)
Marina Nacional 329, Col. Huasteca
C.P. 11311
México D.F., MEXICO
Telephone: 52 55 1944 2500
Web site: www.pemex.com

Mongolia

Ministry of Fuel and Energy
Baga Toiruu 6
Ulaanbaatar 46, MONGOLIA
Telephone: 976 11 129 116

Mongol Petroleum Co.
Uildverchnii Street 37
Gusamj MGT
Ulaanbaatar, MONGOLIA
Telephone: 976 1 331 176
Fax: 976 1 331 176

Petroleum Authority of Mongolia
Uildverchnii Street 37
P.O. Box 37/81
Ulaanbaatar 210337, MONGOLIA
Telephone: 976 11 631 208
Fax: 976 11 631 239
E-mail: petromon@magicnet.mn
Web site: www.mrpam.gov.mn

Morocco

Ministry of Energy, Mines, Water and the Environment
Department of Energy and Mines
Rue Abou Marouane Essaadi BP
Rabat Instituts 6208
Haut Agdal, Rabat MOROCCO
Web site: www.mem.gov.ma

National Office of Hydrocarbons and Mining (ONHYM)
Office of the General Director
5 Avenue Moulay Hassan
B.P. 99
Rabat 10050, MOROCCO
Telephone: 212 537 23 98 98
Fax: 212 537 70 94 11
Web site: www.onhym.com

Mozambique

Ministry of Energy
10 Av. 25 de Setembro
1218-3o Andar
Maputo 1381, MOZAMBIQUE
Telephone: 258 21 303 265
Fax: 258 21 313 971

Empresa Nacional des Hydrocarbonetos de Macambique (ENH)
270 Av. 25 de Setembro
Time Square, Block 1
Postal: P.O. Box 4787
Maputo, Nampula MOZAMBIQUE
Telephone: 258 21 43 0482
Fax: 258 21 32 4808

Myanmar

Ministry of Energy
Energy Planning Department
Building 6
Naypyitaw, MYANMAR
Fax: 95 1 67 411 113
E-mail: myanmoe@mptmail.net.mm
Web site: www.energy.gov.mm

N

The Netherlands

SenterNovem
Juliana van Stolberglaan 3
P.O. Box 93144
2509 AC The Hague, THE NETHERLANDS
Telephone: 31 70 373 50 00
Fax: 31 70 373 51 00
E-mail: info@senternovem.nl
Web site: http://www.senternovem.nl/english/

New Zealand

Ministry of Economic Development
New Zealand Crown Minerals
PO Box 1473
33 Bowen St., Level 5
Wellington 6140, NEW ZEALAND
Telephone: 64 3 962 6179
Fax: 64 4 471 0187
E-mail: crown.minerals@med.govt.nz
Web site: www.crownminerals.govt.nz

Nicaragua

Nicaraguan Institute of Energy (INE)
Petronic Building, 3rd Floor
Barrio Loyola
Managua 3226, NICARAGUA
Telephone: 505 222 5611
Web site: www.ine.gob.ni
E-mail: webmaster@ine.gob.ni

Nigeria

Ministry of Energy
Annex 3, Floors 3 and 4
Federal Secretariat Complex
Shehu Shagari Way, Central Area
PMB 278
Garki, Abuja, NIGERIA
Telephone: 234 09 523 9462; 9465
Fax: 234 09 523 6652
Web site: www.nigeria.gov.ng

Nigerian National Petroleum Corp. (NNPC)
NNPC Towers
Herbert Macaulay Way
Central Business District
PMB 190
Garki, Abuja, NIGERIA
Telephone: 234 09 234 8200-172008 1133
E-mail: adeola.adeqbesan@nnpcgroup.com
Web site: www.nnpcgroup.com

Norway

Ministry of Petroleum and Energy
Einar Gerhardsens plass 1
P.O. Box 8148 Dep
Oslo 0033, NORWAY
Telephone: 47 22 24 90 90
Fax: 47 22 24 95 65
E-mail: postmottak@oed.dep.no
Web site: www.regjeringen.no/en/dep/oed

Statoil ASA
Forusbeen 50
4035 Stavanger, NORWAY
Telephone: 47 51 99 00 00
Fax: 47 51 99 00 50
Web site: www.statoil.com

O

Oman

Ministry of Oil & Gas
Sultanate of Oman
P.O. Box 551
Muscat 113, OMAN
Telephone: 968 24 603 333
E-mail: omanet@omanet.om
Web site: http:www.omanet.om/english/government/ministers.asp?cat=gov

PDO (Petroleum Development Oman)
P.O. Box 81
Muscat 113 OMAN
Telephone: 968 24 675 370
Fax: 968 24 677 106
Web site: www.pdo.co.om

P

Pakistan

Ministry of Petroleum and Natural Resources
3rd Floor, A Block, Room 301, Pak Secretariat
Islamabad, PAKISTAN
Telephone: 92 51 921 0220; 92 51 920 6416
Fax: 92 51 921 3180
E-mail: info@mpnr.gov.pk
Web site: www.mpnr.gov.pk

Papua New Guinea

Department of Petroleum and Energy
P.O. Box 1993
Port Moresby, PAPUA NEW GUINEA
Telephone: 675 322 4200
Fax: 675 322 4222
E-mail: enquiry@petroleum.gov.pg
Web site: www.petroleum.gov.pg

Peru

Ministerio de Energia & Minas
Av. Las Artes Sur 260 San Borja
Lima, PERU
Telephone: 511 618 8700
Web site: www.minem.gob.pe

Petroperu (Petroleos del Peru SA)
Av. Enrique Canaval Moreyra 150
Lima 27, PERU
Telephone: 511 211 7800
Fax: 511 211 7831
Web site: www.petroperu.com.pe

Philippines

Department of Energy
Energy Center, Merritt Road
Fort Bonifacio
Taguig City
Metro Manila 1201, PHILIPPINES
Telephone: 632 840 1401
Fax: 632 840 1731
E-mail: info@doe.gov.ph
Web site: www.doe.gov.ph

Philippine National Oil Co.
PNOC Bldg. 6, Energy Center
Merritt Road
Energy Center, Fort Bonifacio
Taguig City
Metro Manila, PHILIPPINES
Telephone: 840 632 789 7662
Web site: www.pnoc.com.ph

Poland

Ministry of the Environment
52/54 Wawelska Street
Warsaw 00-922, POLAND
Telephone: 48 22 57 92 900
Fax: 48 22 57 92 505
E-mail: info@mos.gov.pl
Web site: www.mos.gov.pl

PGNiG SA (Polskie Górnictwo Naftowe i Gazownictwo)
Polish Oil & Gas Company
25 M. Kasprzaka Street
Warsaw 01-224, POLAND
Telephone.: 48 22 589 45 55
Fax: 48 22 691 82 73
E-mail: pr@pgnig.pl
Web site: www.pgnig.pl

Portugal

GALP Energia SGPS, SA
[Petroleos e Gas de Portugal]
Edifício Galp Energia
Rua Tomás da Fonseca
Lisboa 1600-209, PORTUGAL
Telephone: 351 21 724 25 00
Fax: 351 21 724 29 65
E-mail: comunicacao@galpenergia.com
Web site: www.galpenergia.com

Q

Qatar

Ministry of Foreign Affairs
Web site: http://english.mofa.gov.qa
[*see* 'Oil and Gas' under 'Economy' tab]

Qatar Petroleum
P.O. Box 3212
Doha, QATAR
Telephone: 974 440 2000
Fax: 974 483 1125
E-mail: postmaster@qp.com.qa
Web site: www.qp.com.qa

R

Romania

Ministry of Public Finance
17 Apolodor Street, Sector 5
Bucharest 010366, ROMANIA
Telephone: 40 21 319 98 19
Fax: 40 21 230 36 60
E-mail: namr@namr.ro
Web site: www.mfinante.ro.engl/index.jsp

National Agency for Mineral Resources
36-38, Mendeleev St., Sector 1
Bucharest 010366, ROMANIA
Phone: 40 21 317 00 18
Fax: 40 21 317 07 80
Web site: www.namr.ro/main_en.htm

Petrom SA
239 Calea Dorobantilor, District 1
Bucharest 010567 ROMANIA
Telephone: 40 21 406 0040
Fax: 40 21 406 0420
E-mail: press.office@petrom.com
Web site: www.petrom.ro

Russia

OAO Gazprom
16 Nametkina St., 117997
Moscow GSP-7, RUSSIA
Telephone: 7 495 719 30 01
Fax: 7 495 719 83 33
E-mail: gazprom@gazprom.ru
Web site: www.gazprom.com

OAO Rosneft Oil Co.
26/1 Sofiyskaya Embankment 1
Moscow GSP-8 117997, RUSSIA
Telephone: 7 495 777 44 22
Fax: 7 495 777 44 44
E-mail: postman@rosneft.ru
Web site: www.rosneft.com

S

Saudi Arabia

Saudi Aramco (Saudi Arabian Oil Co.)
P.O. Box 5000
Dhahran 31311, SAUDI ARABIA
Telephone: 966 3 872 0115
Fax: 966 3 873 8190
E-mail: webmaster@aramco.com
Web site: www.saudiaramco.com

Singapore

Ministry of Trade and Industry
100 High Street, #09-01
The Treasury
179434 SINGAPORE
Telephone: 65 6225 9911
Fax: 65 6332 7260
E-mail: mti_email@mti.gov.sg
Web site: www.mti.gov.sg

Ministry of the Environment and Water Resources
100 High Street, #09-01
The Treasury
179434 SINGAPORE
Telephone: 65 6225 9911
Fax: 65 6332 7260
E-mail: mti_email@mti.gov.sg
Web site: www.mti.gov.sg

Singapore Petroleum Company (SPC)
1 Maritime Square
#10-10 HarbourFront Centre, Lobby B
099253 SINGAPORE:
Telephone: 65 6276 6006
Fax: 65 6275 6006
E-mail: spccc@spc.com.sg
Web site: www.spc.com.sg

Slovakia

Slovnaft AS
Vicie Hrdlo 1
824 12 Bratislava, SLOVAKIA
Telephone: 421 2 4055 1111
Web site: www.slovnaft.sk/en

South Africa

Department of Minerals and Energy
Trevenna Campus
Corner: Meintjies and Schoeman Streets
Sunnyside, Pretoria 001, SOUTH AFRICA
Telephone: 27 012 317 8000
Fax: 27 012 322 3416
Web site: www.dme.gov.za

Petroleum Agency SA
Tygerpoort House
7 Mispel Street
Belleville, Cape Town, SOUTH AFRICA
Postal: P.O. Box 1174
Parow, Cape Town 7499, SOUTH AFRICA
Telephone: 27 21 938 3500
Fax: 27 21 938 3520
E-mail: info@petroleumagencysa.com
Web site: www.petroleumagencysa.com

South Korea

Korea Gas Corp. (Kogas)
93 Dolmaro
215 Jeongja-dong
Bundang-gu, Seongnam
Gyeonggi-do 463-754, KOREA
Telephone: 82 31 710 0014
Fax: 82 31 710 0017
E-mail: kogasmaster@kogas.or.kr
Web site: www.kogas.or.kr

Korea National Oil Corp. (KNOC)
1588-14, Gwanyang-dong, Dongan-gu, Anyang-si
Gyeonggi-do 431-711 KOREA
Telephone: 82 31 380 2114
Fax: 82 31 387 9321
E-mail: knoc_information@knoc.co.kr
Web site: www.knoc.co.kr

Spain

Comisión Nacional de Energía (CNE)
Dirección de Relaciones Documentación
Calle Alcalá, 47
28014 Madrid, SPAIN
Telephone: 91 432 96 00
Fax: 91 577 62 18
E-mail: dre@cne.es
Web site: www.cne.es

Enagás SA
Paseo de los Olmos, 19
Madrid 28005, SPAIN
Telephone: 34 902 443 700
E-mail: contacta@enagas.es
Web site: www.enagas.es

Repsol YPF SA
Paseo de la Castellana, 278-280
Madrid 28046, SPAIN
Telephone: 34 91 348 8100
Fax: 34 91 314 2821
Web site: www.repsol.com/es_en

Sudan

Ministry of Energy and Mining
P.O. Box 2649
Khartoum, SUDAN
Telephone: 249 11 80 137

Sweden

Ministry of Enterprise, Energy and Communications
Master Samuelsgatan 70
Postal: SE-103 33 Stockholm, SWEDEN
Telephone: 46 8 405 10 00
Fax: 46 8 411 36 16
Web site: www.sweden.gov.se/sb/d/2067

Syria

Ministry of Oil and Mineral Resources
Al-Adawi Khateeb Street
P.O. Box 31483
Damascus, SYRIA
Telephone: 963 11 445 1624
Fax: 963 11 446 3942
Web site: www.petroleum.gov.sy

Syrian Petroleum Co.
Dummar Province, Expansion Square
Island 19 – Building 32
P.O. Box 2849
Damascus, SYRIA
Telephone: 963 11 313 7935; 313 7913
Fax: 963 11 313 7979; 313 7977
E-mail: spccom1@scs-net.org; spccom2@scs-net.org
Web site: www.spc-sy.com

T

Taiwan

Chinese Petroleum Corp. (CPC)
No. 3 Songren Rd., Sinyi District
Taipei City 110, TAIWAN
Telephone: 02 8789 8989
Fax: 02 8789 9000
E-mail: ir@cpc.com.tw
Web site: www.cpc.com.tw

Tanzania

Ministry of Energy and Minerals
Permanent Secretary
Sokoine/Mkwepu St.
P.O. Box 2000/9152
Dar es Salaam, TANZANIA
Telephone: 255 22 211 7153
Fax: 255 22 211 6719
E-mail: madini@africaonline.co.tz
Web site: www.tanzania.go.tz/energy.htm

Thailand

Ministry of Energy (MOEN)
Energy Planning and Policy Office
12/1-2 Phetchaburi Road
Ratchathew, Bangkok 10400, THAILAND
Telephone: 66 2612 1555
Fax: 66 2612 1358
Web site: www.eppo.go.th

PTT plc (formerly Petroleum Authority of Thailand)
555 Vibhavadi Rangsit Rd., Chatuchak
Bangkok 10900, THAILAND
Telephone: 66 0 2537 2000
Fax. 66 0 2537 3498 9
E-mail: corporate@pttplc.com
Web site: www.pttplc.com/en

Timor Leste (East Timor)

General government information:
Web site: www.gov.east-timor.org

National Directorate of Petroleum and Gas
Ministry of Natural Resources, Minerals and Energy
Oil Exploration Department
Web site: www.timor-leste.gov.tl/emrd

Trinidad & Tobago

Ministry of Energy and Energy Industries
Tower C: Energy Trinidad and Tobago
International Waterfront Centre
#1 Wrightson Road
Port of Spain, TRINIDAD
Telephone: 868 623 6708
Fax: 868 625 0306
E-mail: info@energy.gov.tt
Web site: www.energy.gov.tt

National Gas Co. of Trinidad and Tobago Ltd. (NGC)
Orinoco Drive
Point Lisas Industrial Estate
P.O. Box 1127, Couva
Port of Spain, TRINIDAD
Telephone: 868 636 4662
Fax: 868 679 2384
E-mail: info@carib-link.net
Web site: www.ngc.co.tt

Petrotrin (Petroleum Co. of Trinidad & Tobago)
Administration Building
South Main Road
Pointe-a-Pierre, TRINIDAD
Telephone: 868 658 3336
Fax: 868 658 2513
E-mail: petrochair@petrotrin.com
Web site: www.petrotrin.com

Tunisia

Etap (Entreprise Tunisienne d'Activites Petrolieres)
27 bis, Ave. Khereddine Pacha
BP 83-1073 Monplaisir
1002 Tunis, TUNISIA
Telephone: 216 71 782 288
Fax: 216 71 784 092
E-mail: dexprom@etap.com.tn
Web site: www.etap.com.tn

Turkey

BOTAS Petroleum Pipeline Corp.
Bilkent Plaza A, II Blok, Bilkent
06800 Ankara, TURKEY
Telephone: 90 312 297 20 00
Fax: 90 312 266 07 33
Web site: www.botas.gov.tr

TPAO (Turkiye Petrolleri Anonim Ortakligi)
Turkish Petroleum Corporation
General Directorate
Sogutozu Mahallesi 2
Cadde No. 86
Cankaya/Ankara 06100, TURKEY
Telephone: 90 312 207 2000
Fax: 90 312 286 9000
E-mail: tpaocc@petrol.tpao.gov.tr
Web site: www.tpao.gov.tr

Turkmenistan

Ministry of Energy and Industry
6 Nurberdy Pomma St.
Ashgabat 744000, TURKMENISTAN
Telephone: 9 9312 35 49 21
E-mail: kuwwat@online.tm
Fax: 9 9312 29 06 82

U

Uganda

Ministry of Energy and Minerals Development
P.O. Box 7270
Kampala, UGANDA
Telephone: 256 41 349 010
E-mail: mubiru@energy.go.ug

Ukraine

UkrTransNafta
60 Artema St.
Kyiv 04050, UKRAINE
Telephone: 38 044 461 6971
Fax: 38 044 461 6979
E-mail: office@ukrtransnafta.com

State Committee on Geology and
Utilization of Mineral Resources
Volodymyrska 34
252601, Kyiv 34, UKRAINE
Telephone: 044 228 60 51
Fax: 044 228 60 51
E-mail: admin@brama.com
Web site: www.brama.com/ua-geology

United Arab Emirates

ADNOC (Abu Dhabi National Oil Co.)
P.O. Box 898
Abu Dhabi, UNITED ARAB EMIRATES
Telephone: 971 2 602 0000
Fax: 971 2 602 3389
Web site: www.adnoc.ae

Ministry of Energy
Petroleum & Minerals Sector
P.O. Box 59
Abu Dhabi, UNITED ARAB EMIRATES
Telephone: 971 2 626 2288
Fax: 971 2 627 2291
E-mail: mopmr@uae.gov.ae
Web sites: www.moenr.gov.ae
www.government.ae/gov/en/gov/federal/index.jsp

United Kingdom

Department of Energy and Climate Change
3 Whitehall Place
London SW1A 2AW, UNITED KINGDOM
Telephone: 0207 979 7777
Web site: www.decc.gov.uk

Department for Business Innovation and Skills (BIS)
(formerly Business, Enterprise and Regulatory Reform)
1 Victoria Street
London SW1H 0ET, UNITED KINGDOM
Web site: www.bis.gov.uk

United States

U.S. Department of Energy
1000 Independence Avenue, SW
Washington, DC 20585 USA
Telephone: 202 586 5000
Fax: 202 586 4403
E-mail: The.Secretary@hq.doe.gov
Web site: www.energy.gov

Energy Information Administration (EIA)
U.S. Department of Energy
1000 Independence Avenue, SW
Washington, DC 20585 USA
Telephone: 202 586 8800
E-mail: InfoCtr@eia.gov.us
Web site: www.eia.gov

Uruguay

Administración Nacional de Combustibles, Alcoholes
y Portland (ANCAP)
Paysandú s/n esquina Avda. del Libertador
Montevideo 1090, URUGUAY
Telephone: 598 2 902 1931
Email: webmaster@ancap.com.uy
Web site: www.ancap.com/uy

Uzbekistan

Republic of Uzbekistan
General government Web site: www.gov.uz/en

Ministry of the Economy
45A Uzbekistan Avenue
Tashkent 100003, UZBEKISTAN
Telephone: 998 71 232 6320
Web site: www.mineconomy.uz

State Committee on Geology and Mineral Resources
11 T. Shevchenko Street
Tashkent 100060, UZBEKISTAN
Telephone: 998 71 256 8653
Fax: 998 71 256 2275
E-mail: geolcom@bcc.com.uz
Web site: http://uzgeolcom.uz/en

Uzbekneftegaz State Oil and Gas Holding Company
21 Istikbol Street
Tashkent 100047, UZBEKISTAN
Telephone: 998 71 133 5757
Fax: 998 71 236 7771
Web site: www.uzneftegaz.uz

V

Venezuela

Ministerio del Poder Popular para la Energia y Petróleo
Edificio Petroleos de Venezuela, Torre Oeste
Av. Libertador con Av. Empalme Urb.
La Campiña, Parroquia El Recreo
Distrito Capital
Caracas, VENEZUELA
Telephone: 58 212 708 7001
E-mail: consultasopap@menpet.gob.ve
Web site: www.menpet.gob.ve

PdVSA (Petroleos de Venezuela)
Av. Libertador La Campina, Torre Este
Edificio Petróleos de Venezuela
Distrito Federal
Caracas, 1050-A, VENEZUELA
Telephone: 58 212 708 4111
Fax: 58 212 708 4660
E-mail: webmaster@pdvsa.com
Web site: www.pdvsa.com.ve

Vietnam

Ministry of Petroleum & Gas
193/6A Nam Ky Khoi Nghia
Ho Chi Minh City 3, VIETNAM
Telephone: 080 48137; 080 48354
Fax: 080 48299

PetroVietnam (Vietnam Oil and Gas Group)
18 Lang Ha Street
Hanoi, VIETNAM
Telephone: 84 4 3825 2526
Fax: 84 4 3826 5942
E-mail: webmaster@hn.pv.com.vn
Web site: www.pvn.vn

Y

Yemen

Ministry of Oil and Minerals
Al-Zubairy Street
P.O. Box 81
Sana'a, YEMEN
Telephone: 967 1 202 309
Fax: 967 1 202 314
Web site: www.mom.gov.ye/eng/contacte.htm

Yemen Petroleum Company
YPC Building
P.O. Box 187
Sana'a, YEMEN
Telephone: 967 1 444 046
Fax: 967 1 447 691
Web site: www.ypcye.com/english

Companies and Other Organizations Cited in Atlas Sections

Company/Organization	Location
4Gas BV	Rotterdam
A.P. Møller-Mærsk A/S (Mærsk)	Copenhagen
Abener Energia SA	Seville, Spain
Abengoa Solar	Seville, Spain
Abu Dhabi Gas Industries Ltd. (Gasco)	Abu Dhabi
Abu Dhabi Investment Authority	Abu Dhabi
Abu Dhabi National Chemicals Co.	Abu Dhabi
Abu Qir Petroleum Co.	Alexandria, Egypt
Acciona Energy SA	Navarra, Spain
Addax Petroleum Corp. (acquired by Sinopec)	Geneva, Switzerland
Adira Energy Corp.	Toronto
Adriatic LNG	Milan
Advanced Resources International, Inc.	Arlington, VA
Advanced Technology Investment Co.	Abu Dhabi
Aegean Energy SA	Athens, Greece
AES Corp.	Arlington, VA
Afar Exploration Co. LLC	Tulsa, OK
Afren PLC	London
Africa Oil Corp.	Vancouver, BC
AGA Gas AB	Sundbyberg, Sweden
Agip KCO NV (subsidiary of Eni SpA)	Kazakhstan
AGL Energy Ltd.	Sydney
Aker Exploration	Stavanger, Norway
Alabama Power Co.	Birmingham
Alaska Gas Pipeline LLC	Anchorage, AK
Alberta Geological Survey	Edmonton
Al Bukhari Group	Kuala Lumpur, Malaysia
AleAnna Resources LLC	Houston
Alstom SA	Levallois-Perret, France
Al-Waha Petrochemical Co.	Dubai
American Electric Power Co.	Columbus
American Petroleum Institute (API)	Washington
American Shale Oil LLC	Rifle, CO
American Wind Energy Association	Washington

Company/Organization	Location
America's Natural Gas Alliance	Washington
AMG Oil Ltd.	Vancouver, BC
Anadarko Petroleum Corp.	The Woodlands, TX
Angola LNG Ltd.	Luanda
Apache Canada Ltd.	Calgary
Apache Corp.	Houston
Apache Energy Ltd.	Perth, Australia
Aquamarine Power	Edinburgh
Arabian Gulf Oil Co. (AGOC)	Benghazi, Libya
Archimede Solar Energy SpA	Massa Martana, Italy
Ardent Resources Canada Ltd.	Pittsburgh, PA
Arena Exploration	The Woodlands, TX
Areva SA	Paris
Argus Media Ltd.	London
Arise Windpower AB	Laholm, Sweden
Armstrong Operating Inc.	Dickinson, ND
Arrow Energy Ltd.	Brisbane, Australia
Ascent Resources PLC	London
Aspect Energy LLC	Denver
Associated Industries of Florida	Tallahassee
Astonfield Renewable Resources Ltd.	New York City
Athabasca Oil Sands Corp.	Calgary
Atlantis Resources Corp.	London
Atomic Energy of Canada Ltd.	Mississauga, ON
AuDAX Resources Ltd.	Perth, Australia
Australian Worldwide Exploration Ltd.	Sydney
Australian Petroleum Production & Exploration Association Ltd.	Perth, Australia
Avana Petroleum	Antananarivo, Madagascar
Avner Drilling	Petah Tikva, Israel
Ayr Logistics Ltd.	Rockwall, TX
Baker Hughes Inc.	Houston
Baraka Petroleum Ltd.	Perth, Australia
Barbados National Oil Co. Ltd.	St. Philip, Barbados
Barclays Gedi Group Inc.	Palm Beach, FL
BASF SE	Ludwigshafen, Germany
Basin Electric Power Cooperative	Bismarck, ND
Batavia Energy Inc.	Toronto
Bayerngas GmbH	Munich

Company/Organization	Location
Bayfield Energy Ltd.	Guildford, UK
Beach Petroleum NL	Adelaide
Bechtel Group Inc.	San Francisco
Belize Natural Energy Ltd.	Belmopan
Bergen Oilfield Services AS	Bergen, Norway
BG Group PLC	Reading, Berkshire, UK
Bharat Petroleum Corp. Ltd.	Mumbai
BHP Billiton	Melbourne, Australia
Bill Barrett Corp.	Denver
BJ Services	Houston
Black & Veatch	Overland Park, KS
Black Marlin Energy Ltd.	Dubai
Blue Energy Ltd.	Brisbane, Australia
Blue Sky Belize Ltd.	Belize City
BNP Paribas Bank	Paris
Borders & Southern Petroleum PLC	London
Borealis AG	Vienna, Austria
Borusan Yatirim ve Pazarlama AS	Istanbul
Bow Valley Energy Ltd.	Calgary
Bowleven PLC	Edinburgh
BP PLC	London
BPC Ltd.	Stanley, Falkland Islands
BPZ Energy Inc	Houston
Brahmaputra Cracker & Polymer Ltd.	Guwahati, Assam, India
Bridge Oil Ltd.	London
Brightoil Petroleum Holdings Ltd.	Hong Kong
British Energy PLC	East Kilbride, Scotland
Brookings Institution	Washington
Brunei Darussalam Methanol Consortium	Bandar Seri Begawan, Brunei
Brunei Shell Petroleum Co.	Negara, Brunei
Bryan Research & Engineering, Inc.	Bryan, TX
Business Monitor International (BMI)	London
Cabinda Gulf Oil Co. Ltd. (CABGOC)	Luanda, Angola
Cabot Oil & Gas Corp.	Houston
Cairn Energy PLC	Edinburgh
Cairn India Ltd.	Gurgaon, India

Company/Organization	Location
Caithness Energy LLC	New York City
California Air Resources Board	Sacramento
Caligari {Petronas Caligari Sdn Bdh}	Kuala Lumpur, Malaysia
Çalik Holding AS	Istanbul
Calvalley Petroleum Inc.	Calgary
Canacol Energy Ltd.	Calgary
Canada-Newfoundland and Labrador Offshore Petroleum Board	St. John's, NL
Canada-Nova Scotia Offshore Petroleum Board	Halifax, NS
Canadian Association of Petroleum Producers	Calgary
Canadian Energy Research Institute (CERI)	Calgary
Canadian Natural Resources Ltd.	Calgary
Canadian Superior Energy Inc.	Calgary
Canamens Energy Ltd.	London
Canaport LNG Ltd.	St. John, NB
Canbriam Energy Inc.	Calgary
Candover Investments, PLC	London
Capital Oil & Gas Industries Ltd.	Lagos, Nigeria
Capital Power Corp.	Edmonton, AB
Capricorn Greenland Exploration Ltd.	Edinburgh
Carnarvon Petroleum Ltd.	Perth, Australia
CBM Asia Development Corp.	Vancouver, BC
CDS Oil & Gas Group PLC	London
CEDIGAZ	Rueil Malmaison, France
Cemex	Monterrey, Mexico
Cenovus Energy Inc.	Calgary
Center for Global Energy Studies	London
Center for Liquefied Natural Gas	Washington
Centre for Australian Weather and Climate Research	Melbourne, Australia
Centric Energy Corp.	London
Centrica PLC	Windsor, UK
Cepsa	Madrid
CEZ Group AS	Prague
CGGVeritas	Paris
Chaparral Energy Inc.	Oklahoma City
Chart Industries Inc.	Garfield Heights, OH

Company/Organization	Location
Charterhouse Capital Partners LLP	London
Chesapeake Energy Corp.	Oklahoma City
Chevron Australia Pty Ltd.	Perth, Australia
Chevron Bangladesh	Dhaka
Chevron Canada Ltd.	Vancouver, BC
Chevron Corp.	San Ramon, CA
China Datang Corp.	Beijing
China Development Bank	Beijing
China Gas Holding Ltd.	Hong Kong
China Investment Corp.	Beijing
China Ocean Shipping (Group) Co. (COSCO)	Beijing
Chiyoda Corp.	Yokohama
Cido Shipping (Japan) Co. Ltd.	Tokyo
Circle Oil PLC	Limerick, Ireland
Cirrus Energy Corp.	Calgary
CITIC Pacific	Hong Kong
Clarkson Research Services Ltd.	London
Clean Global Energy Ltd. (CGE)	Perth, Australia
Clough Australia	Perth, Australia
CLP Holdings Ltd.	Hong Kong
Coastal Energy Co.	Houston
Codelco	Santiago, Chile
Colorado School of Mines	Golden, CO
Combined Group Contracting Co.	Safat, Kuwait
Commodity Futures Trading Commission (CFTC)	Washington
Commonwealth Scientific and Industrial Research Organization (CSIRO)	Clayton South, Australia
ConocoPhillips Co.	Houston
ConocoPhillips-Karoon Gas Australia Ltd.	Mt. Martha, Australia
Contact Exploration Inc.	Calgary
Continental Resources Inc.	Enid, OK
ContourGlobal	New York City
Corridor Resources Inc.	Halifax, NS
Cosmo Oil Co. Ltd.	Tokyo
Covalent Energy Corp.	Arlington, VA
Crescent Petroleum	Sharjah, UAE
Daewoo Engineering and Construction	Seoul

Company/Organization	Location
Daewoo International Corp.	Seoul
Daewoo Shipbuilding & Marine Engineering Co. Ltd.	Okpo, South Korea
Dalia Power Energies	Israel
Dana Gas PJSC	Sharjah
Datamonitor Group	London
David Wood & Associates	Lincoln, UK
Dealogic (Holdings) PLC	London
Deep Gulf Energy LP	Houston
Delek Group Ltd.	Netanya, Israel
Deloitte LLP	New York City
Denbury Resources Inc.	Plano, TX
Dept. of Primary Industries	Melbourne, Australia
Desertec Industrial Initiative (DII) GmbH	Munich
Det Norske Veritas	Baerum, Norway
Deutsche Bank AG	Frankfurt, Germany
Development Bank of Kazakhstan	Astana
Devon Energy Corp.	Oklahoma City
Diamoco Energy	Parigny, France
Dimosia Epichirisi Paroxis Aeriou (DEPA)	Athens, Greece
Discover Petroleum	Tromsø, Norway
Dominion	Richmond, VA
Dominion Cove Point LNG	Lusby, MD
Dominion Petroleum Ltd.	Hamilton, Bermuda
DNO International ASA	Oslo
DONG Energy A/S	Fredericia, Denmark
Dor Gas Exploration (unit of Dor Alon Energy)	Yakum, Israel
Dorad Energy	Bnei Brak, Israel
Douglas-Westwood Ltd.	Canterbury, UK
Dow Chemical Co.	Midland, MI
Dragon LNG Group Ltd.	Milford Haven, Wales, UK
Drillsearch Energy Ltd.	Sydney
Dubai International Financial Centre	Dubai
Dung Quat Ship Building Industry Co.	Quang Ngai Province, Vietnam
Dushanzi Petrochemical Corp.	Dushanzi, Xinjiang, China
E.F. Durkee & Associates Inc.	Makati City, Philippines
E.O. Group Ltd.	Accra, Ghana
E.ON AG	Dusseldorf, Germany
E.ON Ruhrgas AG	Berlin
E.ON Sverige AB	Malmö, Sweden
Eaglewood Energy Inc.	Calgary
Earth Sciences Inc.	Littleton, CO
East African Exploration Ltd. (unit of Black Marlin Energy Ltd.)	Dubai
Eastern Mediterranean Gas Co.	Cairo
Eastern Star Gas Ltd.	Sydney
East-West Center	Honolulu
EcoJustice	Toronto; Vancouver, Canada
EDF Energies Nouvelles	Paris
Edison SpA	Milan
El Paso Corp.	Houston
Electricite de France (EDF)	Paris
Electricity Generating Authority of Thailand	Nonthaburi, Thailand
Elektrizitats-Gesellschaft Laufenburg (EGL)	Laufenburg, Switzerland
Emerald Energy PLC	London
Emirates National Oil Co. (ENOC)	Abu Dhabi
Enbridge Energy Partners LP	Houston
Enbridge Inc.	Calgary
EnBW Energie Baden-Wurttemberg AG	Karlsruhe, Germany
EnCana Corp.	Calgary
EnCana Oil & Gas Partnership	Bonnyville, AB
Encore Acquisition Co.	Ft. Worth, TX
Endesa SA	Madrid
Enel SpA	Rome
Enercon	Bremen, Germany
Energía de Entre Ríos SA (Enersa)	Parana, Argentina
Energy Corp. of America	Charleston, WV
Energy Partners Ltd.	New Orleans
Energy Policy Research Foundation Inc.	Washington
Energy Resources Conservation Board (ERCB)	Calgary

Company/Organization	Location
Energy Security Analysis Inc.	Wakefield, MA
Energy Transfer Partners LP	Dallas
Enhanced Oil Resources Inc.	Houston
Eni Gas Transport Deutschland SpA	Dusseldorf
Eni SpA	Rome
Ensearch Petroleum Pvt. Ltd.	Nodia, India
Ensus Group	Yarm, UK
Enterprise Products Partners LP	Houston
EnVantage Inc.	Cleveland, OH
Environmental Defense Fund	New York City
EOG Resources Inc.	Houston
Epic Energy	Melbourne, Australia
Essar Group	Mumbai
Essent NV	Arnhem, The Netherlands
Esso Australia	Melbourne, Australia
Esso Highlands Ltd.	Port Moresby, Papua New Guinea
Euro Oil Refinery	Daugavpils, Latvia
European Investment Bank (EIB)	Kirchberg, Luxembourg
European Photovoltaic Industry Association	Brussels
European Wind Energy Association	Brussels
Excelerate Energy	The Woodlands, TX
Exco Resources Inc.	Dallas
Exponent Inc.	Menlo Park, CA
Exterran	Houston
ExxonMobil Corp.	Irving, TX
FACTS Global Energy Group	Honolulu; Singapore
Falcon Oil & Gas Ltd.	Denver
Faulkner-Suits Exploration Inc.	Edmond, OK
Financial Supervisory Service	Seoul
Finnish Wind Power Association (FWPA)	Tampere, Finland
First Australian Resources Ltd.	Perth, Australia
First Calgary Petroleums Ltd.	Calgary
Fitch Ratings Ltd.	Chicago
Florida Gas Transmission Co.	Houston
Formosa Heavy Industries Corp.	Taipei
Formosa Petrochemical Corp.	Taipei
Formosa Plastics Corp.	Taipei
Fortum Oyj	Espoo, Finland
Foster Wheeler AG	Baar, Switzerland
Friedman, Billings, Ramsey & Co. Inc.	Arlington, VA
Frontline Ltd.	Oslo
Fugro Airborne Surveys	Houston
FX Energy Inc.	Salt Lake City, UT
GAIL (India) Ltd.	New Delhi
Galp Energia	Lisbon, Portugal
Gas Cities Ltd.	Sharjah, UAE
Gas Natural SDG SA	Barcelona
Gas Transmission Northwest Corp.	Portland, OR
Gasol PLC	London
Gassco AS	Kopervik, Norway
Gastar Exploration Ltd.	Houston
Gastem Inc.	Montreal
GDF Suez SA	Paris
GE Energy	Kansas City, MO
Genel Enerji AS	Ankara, Turkey
General Electric Co.	Fairfield, CT
Geoforschungszentrum GFZ	Potsdam, Germany
GeoGlobal Resources Inc.	Calgary
GeoPark Holdings Ltd.	Hamilton, Bermuda
GeoPetro Resources Co.	San Francisco
Geoplin d.o.o. Ljubljana	Ljubljana, Slovenia
GeoResources Inc.	Houston
GFZ German Research Centre for Geosciences	Potsdam
Ghana National Petroleum Corp.	Tema, Ghana
Glencore Exploration Ltd.	London
Global Carbon Capture and Storage Institute	Canberra
Global Energy Co. Ltd.	Victoria Island, Nigeria
Global Geo Services (GGS) Spectrum	Oslo
Global Witness	London
GNS Science	Lower Hutt, New Zealand
Golar LNG	Hamilton, Bermuda
Gold Oil PLC	Surrey, UK
Golden Oil Corp.	Seoul

Company/Organization	Location
Gran Tierra Energy Inc.	Calgary
Great Eastern Energy Corp. Ltd.	Gurgaon, Haryana, India
Green Dragon Gas Ltd.	Hong Kong
Greentech Energy Systems A/S	Herlev, Denmark
Groundstar Resources Ltd.	Calgary
GRTgaz Deutschland GmbH	Berlin
Grupo Alfonso Gallardo	Badajos, Spain
Grupo Guascor	Madrid
Gujarat State Petroleum Corp. (GSPC)	Gujarat, India
Gujarat State Petronet Ltd.	Gujarat, India
Gulf Finance House BSC	Manama, Bahrain
Gulf Keystone Petroleum International Ltd.	Hamilton, Bermuda
Gulf Petroleum Ltd.	Doha
GVS Netz GmbH	Stuttgart
Halliburton	Houston
Hamworthy PLC	Dorset, UK
Hanwha Chemical Corp.	Seoul
Hanwha Group	Seoul
Hardy Oil & Gas PLC	London
Harrison Lovegrove & Co. Ltd.	London
Harvest Energy Trust	Calgary
Harvest Natural Resources Inc.	Houston
Heavy Engineering Industries & Shipbuilding Co.	Safat, Kuwait
Helius Energy PLC	London
Hera SpA	Bologna, Italy
Heritage Oil Ltd. PLC	Jersey, Channel Islands, UK
Heritage Oil Corp.	Calgary
Hess Corp.	New York City
Hibernia Management and Development	St. John's, NL
Hindustan Oil Exploration Co. (HOEC)	Vadodara, India
Hindustan Petroleum Corp. Ltd.	Mumbai
Hitachi Ltd.	Tokyo
Hoang Long Joint Operating Co.	Ho Chi Minh City
Holly Corp.	Dallas
Honda Motor Co. Ltd.	Tokyo
Horizon Terminals Ltd.	Dubai

Company/Organization	Location
Hungarian Development Bank Private Ltd. Co.	Budapest
Hungarian Horizon Energy Ltd.	Budapest
Hunt Oil Co.	Dallas
Husky Energy Inc.	Calgary
Hydrogen Energy International LLC	Abu Dhabi
Hyundai Engineering and Construction Co.	Seoul
Hyundai Heavy Industries Ltd.	Ulsan, South Korea
Hyundai Oilbank Co. Ltd.	Seoul
Iberdrola SA	Bilbao, Spain
Ibeto Energy Development Co. Ltd.	Port Harcourt, Nigeria
ICAP Shipping International Ltd.	London
Idemitsu Kosan Co., Ltd.	Tokyo
Institut Francais du Petrole (IFP)	Paris
IHI Corp.	Tokyo
IHS Cambridge Energy Research Associates	Cambridge, MA
IHS Global Insight	Lexington, MA
IHS Herold Inc.	Norwalk, CT
Imperial Oil Ltd.	Calgary
Indago Petroleum Ltd.	London
Independent Energy Partners (IEP)	Denver
Indian Oil Corp. Ltd.	Mumbai
Indonesian National Shipowners' Association	Jakarta
Industrial Info Resources	Sugar Land, TX
Ineos Manufacturing Scotland Ltd.	Grangemouth, Stirlingshire, Scotland
Infield Energy Analysts	London; Houston
Infoterra	Leicester, UK
Inpex Corp.	Tokyo
Insituform Technology Inc.	Littleton, CO
INTEK Inc.	Arlington, VA
International Atomic Energy Agency (IAEA)	Vienna, Austria
International Council for Science (ICSU)	Paris
International Finance Corp.	Washington
International Maritime Associates Inc.	Washington

Company/Organization	Location
International Monetary Fund	Washington
International Petroleum Investment Council	Abu Dhabi
InterOil Corp.	Toronto
INTSOK	Oslo
IPC Petroleum Consultants Inc.	Tulsa
Iranian Offshore Oil Co.	Tehran
Ireland Electricity Supply Board	Dublin
Irish Wind Energy Association	Osberstown, Naas, Ireland
Irving Oil Ltd.	St. John, NB
Island Oil & Gas PLC	Dublin, Ireland
Isramco Inc.	Houston
ITERA Oil and Gas Co.	Moscow
Ithaca Energy Inc.	Calgary
Itochu Corp.	Osaka
Ivanhoe Energy Inc.	Calgary
Jadrova a Vyradovacia Spolocnost (JAVYS)	Bratislava, Slovakia
Japan Bank for International Cooperation (JBIC)	Tokyo
Japan Energy Corp.	Tokyo
Japan Petroleum Exploration Co. (Japex)	Tokyo
JGC Corp.	Tokyo
Jindal Petroleum Ltd.	New Delhi
John Wood Group PLC	Aberdeen, Scotland
Jordan Petroleum Refinery Co. Ltd,	Amman
JSC Transneft	Moscow
JTC Corp.	Jurong East, Singapore
Jupiter Petroleum Inc.	Calgary
Kairiki Energy Ltd.	West Perth, Australia
Karoon Gas Australia Ltd.	Melbourne, Australia
KBR Inc.	Houston
KEI Pty. (unit of Karoon Gas Australia Ltd.)	Melbourne, Australia
Kentucky Geological Survey	Lexington
Kenya Electricity Generating Co.	Nairobi
Keppel Corp. Ltd.	Singapore
Keppel Shipyard	Singapore
Kinder Morgan Energy Partners LP	Houston
Kitimat LNG Inc.	Calgary

Company/Organization	Location
Knight Capital Group Inc.	Jersey City, NJ
Kodeco Energy Co. Ltd.	Jakarta
Koninklijke BAM Groep NV	Bunnik, The Netherlands
Koninklijke Vopak NV	Rotterdam
Korea Gas Corp. Ltd.	Seongnam, South Korea
Kosmos Energy LLC	Dallas
Kuwait Energy Co. (KEC)	Salmiya, Kuwait
Kuwait Foreign Petroleum Exploration Co. (Kufpec)	Kuwait City
L&M Petroleum Ltd.	Auckland, New Zealand
Lane Energy Poland Sp.z.o.o.	Warsaw
Lario Oil & Gas Co.	Denver
Leed Petroleum PLC	London
Leni Gas & Oil PLC	London
Linde AG	Munich
Liquid Niugini Gas Ltd.	Cairns, Australia
LLOG Exploration	Metairie, LA
Lockheed Martin Corp.	Bethesda, MD
London Array Consortium	London
Los Alamos National Laboratory	Los Alamos, NM
Louis Dreyfus Commodities Brasil SA	Sao Paulo, Brazil
Louisiana State University	Baton Rouge
Lukoil Oil Co.	Moscow
Lummus Technology	Bloomfield, NJ
Lundin Petroleum AB	Stockholm
LyondellBasell Industries	Rotterdam
Madagascar Oil	Houston
Mærsk Co. Ltd.	London
Mærsk Oil UK Ltd.	London
Mærsk Olie og Gas A/S	Copenhagen
Manas Petroleum Corp.	Baar, Switzerland
MangistauMunaiGaz JSC	Aktau
Marathon Oil Corp.	Houston
Marathon Petroleum Norge	Stavanger, Norway
Marex Petroleum Corp.	Houston
Mariner Energy	Houston
Markbygden Vind AB	Tavelsjo, Sweden
Marubeni Corp.	Tokyo

Company/Organization	Location
Marubeni Oil & Gas USA	Houston
Masdar (Abu Dhabi Future Energy Co.)	Abu Dhabi
Masi Corp. Holdings Ltd.	Namaka, Nadi State, Fiji
Maurel & Prom Co. SA	Paris
McMoRan Exploration Co.	New Orleans
Medgaz	Almeria, Spain
Mediterranean Oil & Gas PLC	London
Melrose Resources PLC	Edinburgh
MEO Australia	Melbourne, Australia
Merrill Lynch Commodities, Inc.	New York City
Mesopotamia Petroleum Co. Ltd.	London
Metinvest (unit of SCM Group)	Donetsk, Ukraine
MGM Energy Corp.	Calgary
Mid-Atlantic Express LLC	Arlington, VA
Middle East Oil Refinery (MIDOR)	Cairo
MISC Berhad	Kuala Lumpur, Malaysia
Mitsubishi Chemical Holdings Corp.	Tokyo
Mitsubishi Corp.	Tokyo
Mitsubishi Heavy Industries Ltd.	Tokyo
Mitsui & Co., Ltd.	Tokyo
Mitsui Oil Exploration Co. Ltd.	Tokyo
Mittal Energy Investment Pvt Ltd.	Singapore
MOL Magyar Olaj-es Gazipari NyRt	Budapest
MOL Rt.	Budapest
Mongolsekiu LLC	Ulan Bator, Mongolia
Motiva Enterprises LLC	Houston
Mountain West Energy LLC	Orem, UT
Mouvoil SA	Zug, Switzerland
Mubadala Development Co.	Abu Dhabi
Murphy Oil Corp.	El Dorado, AR
Mustang Engineering Ltd.	Houston, Texas
Myanmar Oil & Gas Enterprise (MOGE)	Naypyitaw
Naftna Industrija Sribje (NIS)	Novi Sad, Serbia
Naftohaz	Kiev, Ukraine
Nalcor Energy	St. John's, NL
National Electricity Regulatory Commission	Kiev, Ukraine

Company/Organization	Location
National Energy Regulator of South Africa	Pretoria
National Grid PLC	London
National Iranian Oil Refining & Distribution Co.	Tehran
National Petroleum Co.	Tehran
National Thermal Power Corp. (NTPC) Ltd.	New Delhi
Nations Southeast Asia Ltd.	Calgary
NCB Capital	Brisbane, Australia
NCC Construction	Solna, Sweden
Nederlandse Gasunie NV	Groningen, The Netherlands
Neste Oil Corp.	Espoo, Finland
NetConnect Germany GmbH & Co. KG	Ratingen, Germany
Neumin Production Co.	Point Comfort, TX
New Alpha Refinery (Ghana) Ltd.	Cape Town, South Africa
New Energy Finance	London
New Guinea Energy Ltd.	Sydney
New Zealand Oil & Gas Ltd.	Wellington
New Zealand Refining Co. Ltd.	Ruakaka
Newfield Exploration Co.	Houston
Nexen Inc.	Calgary
Nexus Energy Ltd	Melbourne, Australia
NHPC Ltd.	Faridabad, Haryana, India
Nido Petroleum Ltd.	West Perth, Australia
NiGaz Energy Co. Ltd.	Abuja, Nigeria
Nigerian Liquefied Natural Gas (NLNG)	Lagos, Nigeria
Nighthawk Energy PLC	Bristol, UK
Niko Resources Ltd.	Calgary
Nippon Oil Corp.	Tokyo
Nissan Motor Co. Ltd.	Yokohama
Noble Energy Inc.	Houston
Noor Financial Investment Co.	Safat, Kuwait
Nord Stream AG	Zug, Switzerland
Nordic Oil & Gas Ltd.	Winnipeg, MB
Noreco	Stavanger, Norway
Norsk Hydro	Stavanger, Norway

Company/Organization	Location
Norske Shell AS	Tananger, Norway
Northern Oil & Gas Inc.	Wayzata, MN
Northern Petroleum PLC	London
Norwegian Energy Co. ASA	Stavanger, Norway
Norwood Resources Ltd.	Vancouver, BC
Nova Gas Transmission Ltd.	Calgary
Novaar Capital Management	George Town, Cayman Islands
Novatek	Tarko-Sale, Russia
NTPC Ltd.	New Delhi
Nunaoil AS	Nuuk, Greenland
NV Nuon	Amsterdam, The Netherlands
Oando PLC	Lagos, Nigeria
Occidental Petroleum Corp.	Los Angeles
Ocean Power Technologies	Warwick, UK
OGX Petroleo e Gas Participacoes SA	Rio de Janeiro
Oil & Gas UK	London
Oil & Gas Development Co. Ltd.	Islamabad
Oil & Natural Gas Corp. Ltd. (ONGC)	Dehradun, India
Oil India Ltd. (OIL)	Duliajan, India
Oil Search Ltd.	Sydney, Australia
Oilex Ltd.	West Perth, Australia
Oilexco Inc.	Calgary
OilMoz LDA	Maputo, Mozambique
Oiltanking Holding America	Houston
Oklahoma Geological Survey	Norman, OK
Olympic Peru Inc.	Lima
Oman Oil Co.	Muscat
OMV AG	Vienna, Austria
ONGC Campos Ltd. (subsidiary of OVL)	New Delhi
ONGC Mittal Energy Ltd.	London
ONGC Videsh Ltd. (OVL)	New Delhi
Orient Petroleum International Inc.	Islamabad, Pakistan
Oriental Energy Resources Ltd.	Abuja, Nigeria
Origin Energy Ltd.	Sydney, Australia
Oromin Explorations Ltd.	Vancouver, BC
Osceola Hydrocarbons Ltd.	Bristol, UK
Otto Energy Ltd.	Perth, Australia

Company/Organization	Location
Oxford Institute for Energy Studies	Oxford, UK
P3 Global Energy Co. Ltd.	Bangkok
PA Resources AB	Stockholm, Sweden
PACE Global Energy Services	Fairfield, VA
Pacific Gas & Electric Co.	San Francisco
Pacific Rubiales Energy Corp.	Toronto
Pakistan Petroleum Ltd.	Karachi
Palau Pacific Exploration	Brisbane, Australia
Pan Andean Resources PLC	Dublin, Ireland
Pan Orient Energy Corp.	Calgary
Pan Pacific Petroleum NL	North Sydney, Australia
Pearl Petroleum Co. Ltd. (unit of Crescent Petroleum)	Sharjah, UAE
Pebercan Inc.	Montreal
Pembina Institute & Water Matters Society of Alberta	Calgary
Perenco	London
Petral Worldwide Inc.	Houston
Petro Iran Co.	Tehran
Petro Matad Ltd.	Douglas, Isle of Man, UK
Petrobras International Braspetro BV	Santa Fe de Bogota, Colombia
Petro-Canada (merged with Suncor Energy Inc.)	Calgary
Petroceltic International PLC	Dublin, Ireland
Petrochemicals Industries Co.	Safat, Kuwait
PetroChina Co. Ltd.	Beijing
PetroChina International Investment Co. Ltd.	Beijing
Petrocomercial	Quito, Ecuador
Petrofac	London
Petrofield (M) Sdn Bhd	Kuala Lumpur, Malaysia
Petrohawk Energy Corp.	Houston
Petrojet	Cairo, Egypt
PetroKamchatka PLC	Jersey, Channel Islands, UK
Petroleum Services Association of Canada	Calgary
Petrolifera Petroleum Ltd.	Calgary
Petrolog & Associates	London

Company/Organization	Location
Petromatrix	Zug, Switzerland
Petrominerales Ltd.	Bogota
Petron Corp.	Makati City, Philippines
Petronet India Ltd.	Mumbai
Petronet LNG Ltd.	New Delhi
Petroplus Holdings AG	Zug, Switzerland
Petrovietnam Exploration and Production Corp.	Hanoi
PetroWorth Resources Inc.	Calgary
PFC Energy Inc.	Washington
Pioneer Natural Resources Co.	Dallas
PKN Orlen SA	Plock, Poland
Plains Exploration & Production Co.	Houston
Platts	New York City
Po Valley Energy Ltd.	Perth, Australia
Polska Grupa Energetyczna (PGE SA)	Warsaw
Port Dolphin Energy LLC	Tampa, FL
Potsdam Institute for Climate Impact Research	Potsdam, Germany
POSCO (Pohang Iron and Steel Co.)	Pohang, South Korea
Powerfuel PLC	Stainforth, UK
Premier Oil PLC	London
PricewaterhouseCoopers	New York City
Progress Energy Resources Corp.	Calgary
Progressive Energy Ltd.	Gloucestershire, UK
Providence Resources PLC	Dublin, Ireland
PT Ephindo	Jakarta
PT Kalila Energy Ltd	Jakarta
PT Medco Energi Internasional Tbk	Jakarta
PTT Public Co. Ltd.	Chatuchak, Thailand
PTTEPI Operating Co. Sdn Bhd	Bangkok, Thailand
Public Agenda	Washington
Pulse Tidal Ltd.	Sheffield, UK
Pure Energy Resources	Burlington, MA
Purvin & Gertz Inc.	Houston
Qatar Gas Transport Co.	Doha
Qatar Petrochemical Co.	Doha
Qatargas Operating Co. Ltd. (QatarGas)	Doha
Questerre Energy Corp.	Calgary

Company/Organization	Location
Quetzal Energy Ltd.	Toronto
Quicksilver Resources Inc.	Fort Worth
Quincy Oil & Gas Corp.	Asuncion, Paraguay
RAK Petroleum PCL	Emirate of Ras Al Khaimah
Ras Laffan LNG Co. Ltd.	Doha
Ratnagiri Gas and Power Pvt Ltd. (RGPPL)	New Delhi
RealtyTrac	Irvine, CA
Regal Petroleum Corp. Ltd.	London
Reliance Industries Ltd.	Navi Mumbai, India
REN Atlántico Terminal de GNL	Lisbon
Renewable Energy Systems Ltd.	Kings Langley, UK
RenewableUK	London
REpower Systems AG	Hamburg
Repsol-YPF SA	Madrid
Ridgewood Energy Corp.	Ridgewood, NJ
Rift Oil PLC	London
Rio Tinto Group	Melbourne, Australia
Roc Oil Co. Ltd.	Sydney
Rohm and Haas Co.	Philadelphia, PA
Rompetrol Group NV	Constanta, Romania
Rosatom (State Atomic Energy Corp.)	Moscow
Rosehill Energy Ltd.	Isle of Man, UK
Royal Dutch Shell PLC	The Hague, The Netherlands
Royal Institute of International Affairs	London
Royal Offshore LLC	Corpus Christi, TX
Royal VolkerWessels NV	Rotterdam
RSK (UK) Ltd.	Houston
RWE AG	Essen, Germany
RWE DEA	Hamburg, Germany
RWE Innogy	Essen, Germany
Ryder Scott Company Petroleum Consultants	Houston
Sabre Oil and Gas Ltd.	Aberdeen, Scotland
Sagex Petroleum ASA	Oslo
Sahara Energy Resources PLC	Lagos, Nigeria
Sahara Petrochemical Co.	Riyadh
Saipem SpA	Milan

Company/Organization	Location
Sakhalin Energy Investment Co. Ltd.	Yuzhno-Sakhalinsk, Russia
Salamander Energy	London
Samson Offshore	Tulsa
Samsung Economic Research Institute (SERI)	Seoul
Samsung Engineering Co. Ltd.	Seoul
Sandia Technologies LLC	Houston
San Leon Energy PLC	Dublin, Ireland
Santelisa Vale Bioenergia SA	Sertaozinho, Brazil
Santos Ltd.	Adelaide, Australia
Sapet Development Peru Inc.	Lima, Peru
Sapphire Energy	San Diego, CA
Saskatchewan Industry and Resources	Estevan
Sasol Ltd.	Johannesburg
Sasol Technology (Pty.) Ltd.	Johannesburg
Saudi Aramco	Dhahran, Saudi Arabia
Saudi Basic Industries Corp. (SABIC)	Riyadh
Saudi International Petrochemical Co. (Sipchem)	Khobar, Saudi Arabia
Saudi Kayan Petrochemical Co.	Al Jubail, Saudi Arabia
Saudi Trading & Research Co. (STARC) Ltd.	Khobar, Saudi Arabia
Savoy Energy Corp.	Houston
Schlumberger Ltd.	Houston
Scottish and Southern Energy PLC	Perth, Scotland
Scottish Enterprise	Glasgow
ScottishPower PLV	Glasgow
ScottishPower Renewables	Glasgow
SeaBird Exploration Norway AS	Oslo
SeaEnergy PLC	Aberdeen, Scotland
Sempra Energy	San Diego, CA
Senergy Ltd.	Aberdeen, Scotland
Serica Energy PLC	London
Setdco Group	Jakarta
Sevan Marine ASA	Arendal, Norway
SeverEnergia	Moscow
Seychelles Petroleum Co.	Victoria, Seychelles
Shaanxi Yangchang Petroleum Co. Ltd.	Yan'an City, China

Company/Organization	Location
Shannon LNG Ltd.	Dublin, Ireland
Shell Australia	Melbourne, Australia
Shell Canada Ltd.	Calgary
Shell Gas & Power Developments BV	The Hague, The Netherlands
Shell Philippines Exploration BV	Muntinlupa City, Philippines
Shell Western Exploration & Production LP	Houston
Shenhua Group Corp.	Beijing
Shenzhen Gas Corp. Ltd.	Shenzen, Guangdong Province, China
Sherritt International Corp.	Toronto
Siemens AG	Munich
Singapore Petroleum Co.	Singapore
SinoChem Corp.	Beijing
Sino Union Petroleum & Chemical International Ltd.	Kowloon, Hong Kong
Slawson Exploration Co.	Wichita, KS
Snam Rete Gas SpA	Milan
SNC-Lavalin Group Inc.	Montreal
SOCO International PLC	London
S-Oil Corp.	Seoul
Sosina Exploration Ltd.	London
South China Resources Ltd.	Makati City, Philippine
Spectra Energy Corp.	Houston
Standard Chartered Bank	London
Star Petro Energy LLC (ETA Ascon Group)	Dubai
Statoil ASA	Stavanger, Norway
Statoil Canada Ltd.	Calgary
Sterling Energy PLC	Hertfordshire, UK
St. Mary Land & Exploration Co.	Denver
Stone Energy Corp.	Lafayette, LA
Stratic Energy Corp.	Calgary
Stuart Petroleum Ltd.	Adelaide, Australia
STX Corp.	Gyeongsangnam-Do, South Korea
Subsea UK	Aberdeen, Scotland
Suez Energy International	Paris
Sui Southern Gas Co. Ltd.	Karachi

Company/Organization	Location
Sumitomo Chemical Co. Ltd.	Tokyo
Suncor Energy Inc.	Calgary
Surgutneftegaz	Surgut, Russia
Sutherland Assbill & Brennan LLP	Washington
Svevind	Tavelsjo, Sweden
Swedish Nuclear Fuel and Waste Management Co. (SKB)	Stockholm, Sweden
Swift Energy Co.	Houston
Synthetic Genomics Inc.	La Jolla, CA
Taeyoung Industry Corp.	Ulsan, South Korea
TAG Oil Ltd.	Vancouver, BC
Taiwan Generations Corp.	Taipei
Taiwan Power Co.	Taipei
Talisman Energy Inc.	Calgary
Tamoil Holdings (unit of Libya's National Oil Corp.)	Tripoli, Libya
Tap Oil Ltd.	West Perth, Australia
TAQA (Abu Dhabi National Energy Co.)	Abu Dhabi
TAQA Bratani Ltd.	Aberdeen, Scotland
TAQA North	Calgary
Technip SA	Paris
Technology International Inc.	Kingwood, TX
Teekay Corp.	Hamilton, Bermuda
Teollisuuden Voima Oyj (TVO)	Helsinki, Finland
TEPPCO Partners LP (merged with Enterprise Product Partners LP)	Houston
Terna SpA	Rome
TerraWest Energy Corp.	Vancouver, BC
Tesoro Corp.	San Antonio, TX
Tethys Petroleum Ltd.	Guernsey, Channel Islands, UK
TGS NOPEC ASA	Asker, Norway
The Carbon Trust	London
Tiroler Wasserkraft AG (TIWAG)	Innsbruck, Austria
TNK-BP Ltd.	British Virgin Islands
TNO (Netherlands Organization for Applied Scientific Research)	Delft, The Netherlands
Tokyo Electric Power Co. (TEPCO)	Tokyo
Tokyo Gas Co.	Tokyo
Tonn Energy	Dublin, Ireland
Topaz Energy and Marine Ltd.	Abu Dhabi

Company/Organization	Location
Toreador Resources Corp.	Paris
Total SA	Paris
Total Gas & Power Ltd.	London
Tower Resources PLC	London
Toyo Engineering Corp.	Chiba, Japan
Toyota Tsusho Corp.	Tokyo; Nagoya
TransAlta Corp.	Calgary
TransCanada Corp.	Calgary
Transelectrica SA	Bucharest
Transeuro Energy Corp.	Vancouver, BC
Transneft	Moscow
Transpetrol AS	Bratislava, Slovakia
Triangle Petroleum Corp.	Calgary
Tullow Oil PLC	London
Turkish Petroleum Corp.	Ankara
Turner, Mason & Co.	Dallas
Tuscarora Gas Transmission Co.	Portland, OR
UAE University	Al Ain, UAE
Ultra Petroleum Corp.	Houston
Unbridled Energy Corp.	Calgary
Union Fenosa Group	Madrid
United Coal Co.	Teays Valley, WV
University of Alberta	Edmonton
UOP	Des Plaines, IL
U.S. Geological Survey (USGS)	Reston, VA
U.S. Minerals Management Service (MMS)	Washington
U.S. Trade and Development Agency (USTDA)	Arlington, VA
UTS Energy Corp.	Calgary
VAALCO Energy Inc.	Houston
Valero Energy Corp.	San Antonio
Vattenfall AB	Stockholm, Sweden
Vega Oil SpA	Rome
Vegas Oil & Gas SA	Athens
Venezuela Iran Oil and Gas Co. (VENIROGC)	Tehran, Iran
Venoco Inc.	Denver
Venture Production PLC	Aberdeen, Scotland
Verbund-Austrian Hydro Power AG	Vienna, Austria
Verein der Kohlenimporteure e.V.	Hamburg

Company/Organization	Location
Verenex Energy Inc.	Calgary
Verenium Corp.	Cambridge, Ma
Vestas Wind Systems A/S	Randers, Denmark
Victoria Oil & Gas PLC	London
Virginia Indonesia Co. LLC (VICO)	Muara Badak, Indonesia
Vitol Group	Geneva, Switzerland
Voith AG	Heidenheim, Germany
Voyager Energy Ltd.	Calgary
Vulcan Minerals Inc.	St. John's, NL
Wavebob	Maynooth, Ireland
Western Business Roundtable	Lakewood, CO
Western Refining Inc.	El Paso, TX
White Nile Petroleum Operating Co.	Khartoum
William J. Clinton Foundation	New York City
Williams Companies Inc.	Tulsa
WINGAS GmbH	Kassel, Germany
Wintershall AG	Berlin
Wood Mackenzie Ltd.	Edinburgh
Woodside Petroleum Ltd.	Perth, Australia
World Bank	Washington
World Nuclear News	London
World Wind Energy Association	Bonn, Germany
WorleyParsons Ltd.	North Sydney, Australia
W&T Offshore	Houston
XTO Energy Inc.	Fort Worth, TX
Yam Thetis Consortium (unit of Delek Group)	Netanya, Israel
Yara Industrial Ltd.	Immingham, UK
Ziff Energy Group	Calgary
Zion Oil & Gas Inc.	Dallas

GLOSSARY

Abbreviations and Acronyms

A

ACG	Azeri-Chirag-Guneshli (Azerbaijan oil fields)
Adma Opco	Abu Dhabi Marine Operations Co.
Adnoc	Abu Dhabi National Oil Co.
AED	Alberta Economic Development
AEP	American Electric Power Co.
A.G.A.	American Gas Association
Agip KCO	Agip Kazakhstan North Caspian Operating Co., formerly OKIOC
AGPA	Alaska Gasline Port Authority
AIOC	Azerbaijan International Operating Co.
altfuels	alternative fuels including wind power and solar energy
AMBO	Albanian Macedonian Bulgarian Oil Co.
AMPCO	Atlantic Methanol Production Co. LLC
ANCAP	Administration Naciional de Combustibles, Alcohol y Portland (Uruguay)
ANP	Agencia Nacional de Petroleo, or National Petroleum Agency (Brazil)
ANWR	Arctic National Wildlife Refuge (Alaska, U.S.)
AOSP	Alberta Oil Sands Project; Alberta Oil Sands Pipeline
APG	Aboriginal Pipeline Group (Inuvik, NT)
API	American Petroleum Institute
Aramco	(Saudi) Arabian Oil Co.
ASEAN	Association of Southeast Asian Nations
ASPO	Association for the Study of Peak Oil
ATHEER	Abu Dhabi Gas Co. (United Arab Emirates)

B

BALAK	national oil and gas agency's Indonesian-language acronym
Banagas	Bahrain National Gas Co.
Banco	Bahrain National Oil Co.
Bapco	Bahrain Petroleum Co.
bbl	barrels
b/cd	barrels per calendar day
bcf	billion cubic feet
bcfd	billion cubic feet per day
bcfe	billion cubic feet equivalent
bcm	billion cubic meters
bcmd	billion cubic meters per day

bcme	billion cubic meters equivalent
b/d	barrels per day
bbl-d	barrels daily
BEDB	Brunei Economic Development Board
BGL	Bhagyanagar Gas Ltd. (India)
b/h	barrels per hour
BLM	Bureau of Land Management (U.S.)
BNA	Bundesnetzagentur (Germany)
BNDES	National Economic & Social Development Bank (Brazil)
bo/d	barrels of oil per day
boe	barrels of oil equivalent
boe/d	barrels of oil equivalent per day
BOOT	build, own, operate, and transfer
BOOTS	Bulk Oil Offshore Transfer System (U.S.)
BOTAS	Turkey government natural gas and pipeline company
bscf	billion standard cubic feet
bscfd	billion standard cubic feet per day
bscm	billion standard cubic meters
b/sd	barrels per stream day
BTC	Baku-Tbilisi-Ceyhan oil pipeline (Azerbaijan-Georgia-Turkey)
Btu	British thermal unit
bw/d	barrels of water per day

C

CAA	commercial arrangement area (Malaysia/Viet Nam)
CAC	Central Asia-Center (Former Soviet Union pipeline system)
CAGC	Canadian Association of Geophysical Contractors
CAPP	Canadian Association of Petroleum Producers
Caricom	Caribbean Community
CBM	coalbed methane
CBNG	coalbed natural gas
CBOT	Chicago Board of Trade
CD	compact disk
CEO	chief executive officer
CEPA	Canadian Energy Pipeline Association
Cepsa	Compania Espanola de Petroleos SA (Spain)
CERA	Cambridge Energy Research Associates (CERA)
CERI	Canadian Energy Research Institute
cf	cubic feet
cfd	cubic feet per day
CFE	Comision Federal de Electricidad (Federal Electricity Commission, Mexico)
cfed	cubic feet of equivalent per day
CFTC	Commodity Futures Trading Commission (U.S.)
CGES	Centre for Global Energy Studies (London)
CHDP	cricondentherm hydrocarbon dew point
Chinaoil	China National United Oil Co.
CICESE	Centro de Investigación Científica y de Educación Superior de Enseñada (Mexico)
CIS	Commonwealth of Independent States (Former Soviet Union)
CITIC	China International Trust & Investment Corporation
CLJOC	Cuu Long Joint Operating Co. (Viet Nam)
CNE	Comision Nacional de Energia (Spain)
C-NLOPB	Canada-Newfoundland Labrador Offshore Petroleum Board
CNG	compressed natural gas
CNOOC	China National Offshore Oil Corporation
CNPC	China National Petroleum Corporation
CNR	Canadian Natural Resources Ltd.
CO	carbon monoxide
CO_2	carbon dioxide
COS	China Oilfield Services Ltd.
cP	centi-Poise (viscosity measurement)
CPC	Caspian Pipeline Consortium
CPC	Chinese Petroleum Corp. (Taiwan)
CPOT	Constanta-Pancevo-Omisalj-Trieste pipeline (Romania-Croatia-Italy)
CPU	central processing unit
CPUC	California Public Utilities Commission
CRE	Energy Regulatory Commission (Mexico)
CRP	Paraguana Refining Complex (Venezuela)
CSO	Coflexip Stena Offshore
CTL	coal to liquids
cu m	cubic meters (metres)
CUCBM	China United Coalbed Methane Corp. Ltd.
CUPET	Union Cuba-Petroleo
CV	core venture (Saudi Arabia gas initiative)

D

DDCV	deep-draft caisson vessel
DEPA	Public Gas Co. (Greece)
DIGP	Dauphin Island Gathering Partners
dilbit	diluted bitumen (heavy crude oil and bitumen blended with condensate diluent)
DOE	Department of Energy (United States)
DONG	Dansk Olie ag Naturgas AS (Denmark)
DUC	Dansk Undergrunds Consortium (Denmark)
dwt	deadweight tons

E

E&D	exploration and development
E&P	exploration and production
EBRVs	Energy Bridge regasification vessels
EC	European Commission
ECO	Economic Cooperation Organization (Central Asia)
EdF	Electricité de France
EDP	Electricidade de Portugal SA
EGAS	Egyptian Natural Gas Holding Co.
EGPC	Egyptian General Petroleum Corp.
EIA	Energy Information Administration (U.S.)
EIA	environmental impact assessment
ELNG	Egyptian LNG joint venture
ENAP	Empresa Nacional de Petroleo (Chile)
Enarsa	Energía Argentina SA
Enel	Ente Nazionale Energia Elettrica SpA (Italy)
Eni	Ente Nazionale Idrocarburi SpA (Italy)
EPA	Environmental Protection Agency
EPC	engineering, procurement, and construction
EPCA	Energy Policy & Conservation Act (U.S.)
EPDC	European Pipeline Development Co. BV
EPSA	exploration and production-sharing agreement
ERC	Energy Regulatory Commission (France)
ESP	electrical submersible pump
ETAP	Enterprises Tuniseienne d'Activites Petrolieres
ETLP	extended tension leg platform
EU	European Union
EUB	Energy and Utilities Board (Alberta, Canada)
EVE	Basque Energy Authority (Spain)
EWT	extended well test
Ex-Im Bank	Export-Import Bank of the U.S.

F

FAWAG	foam-assisted water alternating gas
FCC	fluid catalytic cracking
FCCU	fluid catalytic cracking unit
FDI	foreign direct investment
FEED	front-end engineering design
FERC	Federal Energy Regulatory Commission
FNA	Frigg Norwegian Association (Norway)
fob	free on board
FPC	Formosa Petrochemical Corp.
fph	feet per hour
FPSO	floating production, storage, and off-loading
FPU	floating production units
FSRU	floating storage and regasification unit
FSU	Former Soviet Union
ft	foot
FTC	Federal Trade Commission
FTIR	Fourier transform infrared

G

GAIL	Gas Authority of India Ltd.
GALP	Petroleos e Gas du Portugal
GASCO	Abu Dhabi Gas Industries Ltd. (United Arab Emirates)
GBS	gravity-based structure
GC	Green Canyon (U.S. Gulf of Mexico)
GCC	Gulf Cooperation Council
GCDS	Gasoducto Cruz del Sur (Argentina-Uruguay)
GdF	Gaz de France SA
GDP	gross domestic product
GEA	Gas Energy Adria
GFU	Gas Negotiating Committee (Norway)
GNA	Gasoducto del Noreste Argentino (Bolivia-Argentina)
GNPOC	Greater Nile Petroleum Operating Co. (Sudan)
GOM	Gulf of Mexico
GOR	gas-oil ratio
GOSP	gas-oil separation plant
gpd	gallons per day
GSCP	Georgia Strait Crossing Project (Canada-U.S.)
GTI	Gas Technology Institute (U.S.)
GTL	gas-to-liquids
GTLB	GTL Bolivia SA
GWh	gigawatt-hours

H

H&P	Helmerich & Payne Inc.
HCU	Hydrocarbon Unit of the Bangladesh Energy and Mineral Resources Division
HDPE	high-density polyethylene
HDS	hydroskimming
HOA	heads-of-agreement
hp	horsepower
HP	Hellenic Petroleum (Greece)
HPCL	Hindustan Petroleum Corp Ltd. (India)
HPHT	high-pressure, high-temperature
HSE	health, safety, and environment

I

IADB	Inter-American Development Bank
IADC	International Association of Drilling Contractors
IAEA	International Atomic Energy Agency (Vienna)
ICO	Interconexion Centro-Occidente pipeline (Venezuela)
ID	inside diameter
IDB	Inter-American Development Bank
IEA	International Energy Agency (Paris)
IEC	Israel Electric Co.
IET	integrated exploration technology
IFP	Institut Français du Pétrole (France)
IGAT	Iran Gas Trunkline
IGCC	integrated gasification combined cycle
ILNGA	International LNG Alliance (U.S.)
IMF	International Monetary Fund
IMO	International Maritime Organization
INA	Industrija nafte dd (Croatia)
INAC	Indian and Northern Affairs Canada
INE	Instituto Nicaraguense de Energia (Nicaragua)
INGAA	Interstate Natural Gas Association of America
INOC	Iraq National Oil Co.
IOC	Indian Oil Corp.
IOC	international oil company
IPAA	Independent Petroleum Association of America
IPO	initial public offering
IRNA	Iranian news agency
ISAC	Information Sharing & Analysis Center (U.S.)
IUA	International Unitization Agreement

J

Japex	Japan Petroleum Exploration Corp.
JDA	Joint Development Area (Malaysia-Thailand, Australia-East Timor)
JIPA	joint investment production activity
JNOC	Japan National Oil Company
JOGMEC	Japan Oil, Gas and Metals National Corporation
JPSC	Joint Pipeline Security Commission (Azerbaijan, Georgia, and Turkey)
JV	joint venture

K

KGOC	Kuwait Gulf Oil Co.
KIO	Karachaganak Integrated Organization (consortium)
KKPL	Kenai Kachemak pipeline (Alaska)
km	kilometer
KNOC	Korean National Oil Corporation
KOC	Kuwait Oil Company
Kogas	Korea Gas Corporation
KPC	Kuwait Petroleum Corp.
KVA	kilo-volt-amperes
KZM	Ku-Zaap-Maloob complex (Mexico)

L

LAB	linear alkylbenzene
LION	large increase in octane number
LLDPE	linear low-density polyethylene
LNG	liquefied natural gas
LOGGS	Lincolnshire offshore gas gathering system (UK)
Logic	leading oil and gas industry competitiveness (UK)
LPG	liquefied petroleum gas
LWD	logging-while-drilling

M

m	meter
M&A	merger and acquisition
M&NP	Maritimes and Northeast Pipeline (Canada-U.S.)
Marad	Maritime Administration (U.S.)
MC	Mississippi Canyon (U.S. Gulf of Mexico)
Mcf	thousand cubic feet
MD	measured depth
MEG	monoethylene glycol
MFL	magnetic flux leakage
MHP	Market Hub Partners (U.S.)
MIBEL	Mercado Ibérico de Electricidade (Portugal, Spain)
MIDOR	Middle East Oil Refinery Ltd.
Migas	Directorate General of Oil and Gas (Indonesia)
MMBtu	million British thermal units
MMcf	million cubic feet
MMcfd	million cubic feet per day
MMcfe	million cubic feet of equivalent
MMcfed	million cubic feet of equivalent per day
MMcm	million cubic meters
MMcmd	million cubic meters per day
MMS	Minerals Management Service

MODU	mobile offshore drilling unit
MOEN	Ministry of Energy (Thailand)
MOL plc	Hungarian Oil and Gas Co.
MSC	multiple-service contract
MTBE	methyl tertiary butyl ether
mtoe	million tonnes of oil equivalent
mtpy	million tonnes per year
MW	megawatt
MWD	measurement-while-drilling
MWh	megawatt-hour

N

NAM	Nederlandse Aardolie Maatschappij BV (Netherlands)
NARUC	National Association of Regulatory Utility Commissioners (U.S.)
NCMA	North Coast Marine Area (Trinidad & Tobago)
NEGP	North European Gas Pipeline (Russia-Germany)
NELP IV	New Exploration Licensing Policy IV (India)
Nemed	North East Mediterranean deepwater concession (Egypt)
NETA	New Electricity Trading Arrangements (UK)
NGC	National Gas Company of Trinidad and Tobago Ltd.
NGO	non-governmental organization
NGSA	Natural Gas Supply Association (U.S.)
NIMBY	not in my back yard
NIOC	National Iranian Oil Company
NLNG	Nigeria Liquefied Natural Gas Co.
NNCL	Bangladesh Business Development Corp. Ltd.
NNPC	Nigerian National Petroleum Co.
NOC	National Oil Corporation (Libya)
NOGA	National Oil and Gas Authority (Bahrain)
NOIA	Newfoundland Ocean Industries Association
NO_x	nitrogen oxides
NPC	National Petroleum Company (Jordan)
NPD	Norwegian Petroleum Directorate
NPPAP	New Players Petroleum Association of the Philippines
NPR-A	National Petroleum Reserve-Alaska
NPRA	National Petrochemical & Refiners Association (U.S.)
NTN	Nova Transportadora do Nordeste SA (Brazil)
NTS	Nova Transportadora do Sudeste SA (Brazil)
NWS	North West Shelf (Australia)
NYBOT	New York Board of Trade
NYMEX	New York Mercantile Exchange
NYSE	New York Stock Exchange

O

OCA	Overlapping Claims Area (Thailand-Cambodia)
OCP	Oleoducto de Crudos Pesados Ecuador SA (consortium and pipeline)
OCTG	oil country tubular goods
OD	outer diameter
OECD	Organization for Economic Cooperation and Development
OFAC	Office of Foreign Assets Control (U.S. Department of Treasury)
OGDC	Oil & Gas Development Co. Ltd. (Pakistan)
OGIP	original gas in place
OGJ	*Oil & Gas Journal*
OKIOC	Offshore Kazakhstan International Operating Company, re-named Agip KCO
OLADE	Organizacion LatinoAmericana de Energia (Ecuador)
OLNGC	Oman Liquefied Natural Gas Co.
OML	oil mining lease (Nigeria)
OMNIS	Office of National Mines and Strategic Industries (Madagascar)
ONAREP	Office National de Recherches et d'Explorations Pétrolières (Morocco)
ONGC	Oil & Natural Gas Corporation (India)
ONHYM	National Office of Hydrocarbons and Mining (Morocco)
OPEC	Organization of Petroleum Exporting Countries
OPIC	Overseas Private Investment Corp. (U.S.)
OPL	oil prospecting license (Nigeria)
OPS	Office of Pipeline Safety (U.S.)
OSA	operating service agreement (Venezuela)

P

PADD	Petroleum Administration for Defense District (U.S.)
PALs	Plantwide Applicability Limits (U.S.)
Parco	Pak-Arab Refinery Co. Ltd.
Parsep	Proyecto de Asistencia para La Reglamentación del Sector Energético del Perú
PCO-W	Iraq Project and Contracting Office (Washington, DC), U.S. Army
PCR	production change rate
PDET	Oil Outflow and Treatment Plan (Campos basin, Brazil)
PDO	Petroleum Development Oman LLC
PDO	plan for development and operation (Norway)
PdVSA	Petróleos de Venezuela SA
PEGASO	Excellence in Environmental Management & Operational Safety Program (Brazil)

Pemex	Petroleos Mexicanos
PEP	Petroleum Exploration Permit (New Zealand)
Pertamina	Perusahaan Pertambangan Minyak Dan Gas Bumi Negara (Indonesia)
PET	polyethylene terephthalate
Petrobangla	Bangladesh Oil, Gas & Mineral Corp.
Petrobras	Petroleo Brasileiro SA
Petroecuador	Petroleos del Ecuador
Petrogal	Petroleos de Portugal EP
Petronas	Petroliam Nasional Sdn. Berhad (Malaysia)
Petroperu	Petroleos del Peru SA
PetroSA	Petroleum Oil and Gas Corp. of South Africa
Petrotrin	Petroleum Co. of Trinidad and Tobago Ltd.
PetroVietnam	Vietnam Oil & Gas Co.
PG&E	Pacific Gas & Electric Co.
PGC	Potential Gas Committee (U.S.)
PGC	Process Gas Consumers (U.S.)
PGN	PT Perusahaan Gas Negara (Persero) Tbk (Indonesia)
PGNiG	Polskie Gomictwo Naftowe I Gazownictow SA (Polish Oil & Gas Co.)
PIC	Petrochemical Industries Co. (Kuwait)
PIO	plan for installation and operation (Norway)
PIRINC	Petroleum Industry Research Foundation Inc.
PJ	petajoule (approx. = 1 trillion Btu)
PKN	Polski Koncern Naftowy (Poland)
PL	Production License (Norway)
PLEM	pipeline end manifold
PMP	Petroleum Mining Permit (New Zealand)
PNOC	Philippine National Oil Company
PNSC	Pakistan National Shipping Corporation
Poas	Petrol Ofisis (Turkey)
PoG	Perseus-over-Goodwyn (Australia)
POGC	Polish Oil and Gas Co.
PPL	petroleum prospecting license (New Zealand)
PSA	production-sharing agreement (or arrangement)
PSA	pressure swing adsorption
PSC	production-sharing contract
PSO	Pakistan State Oil
PTEN	PT Exspan Nusantara (Indonesia)
PTT	polytrimethylene terephthalate
PTT plc	Thailand government oil and gas firm (formerly Petroleum Authority of Thailand)
PTTEP	PTT Exploration and Production plc (Thailand)

R

RASCO	Ras Lanuf Oil and Gas Processing Company (Libya)
RasGas	Ras Laffan Liquefied Natural Gas Co. Ltd. (Qatar)
RFCC	residual fluid catalytic cracking
RFG	reformulated gasoline
RFR	right of first refusal
RIK	royalty-in-kind
ROSBOS	Rosetti Marino SPA-Bouygues Offshore joint venture
ROV	remotely operated vehicle
RR	railroad
RTO	regional transmission organization
RTP	rapid thermal conversion

S

Sabic	Saudi Basic Industries Corp.
SADC	South African Development Community
SAGD	steam-assisted gravity drainage
SAGE	Scottish Area Gas Evacuation
SAP	solvent-aided process
SAPREF	South African Petroleum Refineries (Pty) Ltd.
SARS	severe acute respiratory syndrome
scfd	standard cubic feet per day
SCO	synthetic crude oil
SCP	South Caucasus Pipeline
SDFI	state's direct financial interest (Norway), re-named Petoro
SEA	State Energy Administration (China)
SEEL	South-East European Line
semi	semisubmersible offshore production platform
SHT	Societe Hydrocarbures du Tchad (Chad)
Sinochem	China National Chemical Import & Export Company
Sinopec	China Petroleum & Chemical Corporation
SIR	Ivorian Refining Company (Ivory Coast)
SNG	synthetic natural gas
SNGPL	Sui Northern Gas Pipelines Ltd. (Pakistan)
SNH	National Hydrocarbon Corp. (Cameroon)
SNPC	Société Nationale des Pétroles du Congo (Brazzaville)
SO_2	sulfur dioxide
SO_x	sulfur oxides
SOCAR	State Oil Company of Azerbaijan Republic
SOEP	Sable Offshore Energy Project (Canada)
SOMO	State Oil Marketing Organization (Iraq)
Sonangol EP	Sociedade Nacionale de Combustiveis de Angola
SOTE	Trans-Ecuadorian pipeline's Spanish-language acronym
SPA	sale and purchase agreement
SPC	Singapore Petroleum Company
SPE	Society of Petroleum Engineers (U.S.)
SPLA	Sudan People's Liberation Army
SPM	single-point mooring

SPP	Slovensky plynarensky priemysel (Slovakia)
sq km	square kilometer
sq mi	square mile
SPR	Strategic Petroleum Reserve
SRS	Schmierstoff Raffinerie Salzbergen (German refinery)
SRU	sulfur-recovery unit
SSGC	Sui Southern Gas Co. Ltd. (Pakistan)
SSP	Sevan Stabilized Platform (cylindrical monohull FPSO vessel design)
Statoil	Den norske stats oljeselskap AS (Norway)
stb	standard tank barrel
STL	submerged turret loading
SVA	security vulnerability assessment (U.S.)
synbit	syncrude-bitumen blend (raw bitumen diluted with synthetic crude oil)

T

TAC	technical assistance contract
TAG	Trans Austria Gasleitung pipeline
TAP	Trans-Alpine Pipeline
TAP	Turkmenistan-Afghanistan-Pakistan pipeline
TAWN	Tariki, Ahuroa, Waihapa, and Ngaere area (New Zealand)
tcf	trillion cubic feet
tcfe	trillion cubic feet equivalent
tcm	trillion cubic meters
tcme	trillion cubic meters equivalent
TCP	Trans-Caspian Pipeline
TD	total depth
TGS	Transportadora de Gas del Sur (South American pipeline operator)
TJ	terajoules
TMD	total measured depth
TNK	Tyumen Oil Co. (Russia)
toe	tons or tonnes of oil equivalent
TPAO	Turkiye Petrolleri Anonim Ortakligi (Turkey)
tpd	tons or tonnes per day
tpy	tons or tonnes per year
TRG	Rompetrol Group NV (Netherlands)
TSB	Transportadora Sulbrasileira de Gas pipeline (Brazil)
tscm	trillion standard cubic meters
Tupras	Turkish Petroleum Refineries Corp.
TVD	true vertical depth
TVDS	true vertical depth subsea
TVK	Tiszai Vegyi Kombinat Rt. (Hungary)
TW	terawatt
TWh	terawatt-hour

U

UAE	United Arab Emirates
UDS	Ultramar Diamond Shamrock Corp.
UES	Unified Energy Systems (Russia)
UKCS	UK Continental Shelf
UKOOA	UK Offshore Operators Association Ltd.
ULCC	ultra large crude carrier
UN	United Nations
Unipec	United International Petroleum & Chemicals Company Ltd. (China)
UOG	United Arab Emirates Offsets Group
URR	ultimate resource recovery
USACE	U.S. Army Corps of Engineers
UXO	unexploded ordnance

V

Vapex	vapor extraction
VGO	vacuum gas oil
VLCC	very large crude carrier
VOCs	volatile organic compounds
VP	Vice President
VSS	vortex separation system

W

WAG	water-alternating-gas
WAGP	West Africa Gas Pipeline (Nigeria. Ghana, Togo, and Benin)
WDDM	West Delta Deep Marine (Egypt)
WLGP	Western Libyan Gas Project
WNTS	West Natuna Transportation System (Indonesia to Singapore)
Wocap	World Oil Production Capacity (model)
WOPP	White Oil Pipeline Project (Pakistan)
WTO	World Trade Organization

Y

YPFB	Yacimientos Petroliferos Fiscales Bolivianos

Z

ZIC	Zone d'Interet Commun (Angola/ Congo-Brazzaville)

North American Geographical Abbreviations

U.S. States

AL	Alabama
AK	Alaska
AZ	Arizona
AR	Arkansas
CA	California
CO	Colorado
CT	Connecticut
DC	District of Columbia
DE	Delaware
FL	Florida
GA	Georgia
HI	Hawaii
ID	Idaho
IL	Illinois
IN	Indiana
IA	Iowa
KS	Kansas
KY	Kentucky
LA	Louisiana
ME	Maine
MD	Maryland
MA	Massachusetts
MI	Michigan
MN	Minnesota
MS	Mississippi
MO	Missouri
MT	Montana
NE	Nebraska
NV	Nevada
NH	New Hampshire
NJ	New Jersey
NM	New Mexico
NY	New York
NC	North Carolina
ND	North Dakota
OH	Ohio
OK	Oklahoma
OR	Oregon
PA	Pennsylvania
RI	Rhode Island
SC	South Carolina
SD	South Dakota
TN	Tennessee
TX	Texas
UT	Utah
VT	Vermont
VA	Virginia
WA	Washington
WV	West Virginia
WI	Wisconsin
WY	Wyoming

Canada Provinces

AB	Alberta
BC	British Columbia
MB	Manitoba
NB	New Brunswick
NL	Newfoundland & Labrador
NS	Nova Scotia
NT	Northwest Territories
NU	Nunavut
ON	Ontario
PE	Prince Edward Island
QC	Quebec
SK	Saskatchewan
YT	Yukon Territory

Index

A

B

C

F

G

H

I

J

K

L

M

N

O

P

Q

R

S

T

U

V

W

X

Y

Z

Fold-out Map Legend

Use this legend for all IPE 2010 map

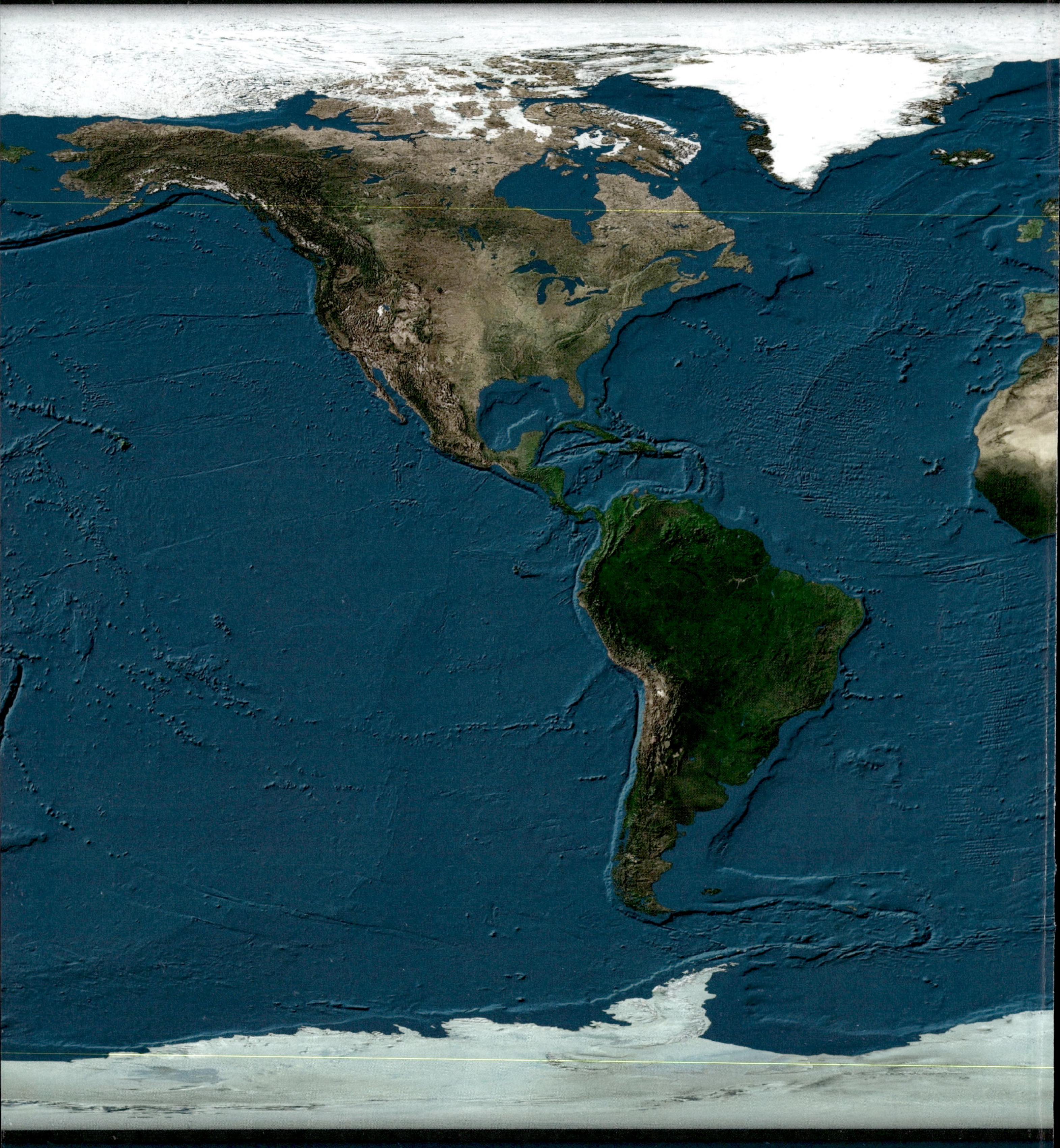